AIRCRAFT CONTROL AND SIMULATION, 2nd Edition

BRIAN L. STEVENS
Georgia Institute of Technology

FRANK L. LEWIS
The University of Texas at Arlington

JOHN WILEY & SONS, INC.

This book is printed on acid-free paper. ∞

Copyright © 2003 by John Wiley & Sons, Inc. All rights reserved.

Published by John Wiley & Sons, Inc., Hoboken, New Jersey
Published simultaneously in Canada

No part of this publication may be reproduced, stored in a retrieval system, or transmitted in any form or by any means, electronic, mechanical, photocopying, recording, scanning, or otherwise, except as permitted under Section 107 or 108 of the 1976 United States Copyright Act, without either the prior written permission of the Publisher, or authorization through payment of the appropriate per-copy fee to the Copyright Clearance Center, Inc., 222 Rosewood Drive, Danvers, MA 01923, (978) 750-8400, fax (978) 750-4470, or on the web at www.copyright.com. Requests to the Publisher for permission should be addressed to the Permissions Department, John Wiley & Sons, Inc., 111 River Street, Hoboken, NJ 07030, (201) 748-6011, fax (201) 748-6008, e-mail: permcoordinator@wiley.com.

Limit of Liability/Disclaimer of Warranty: While the publisher and author have used their best efforts in preparing this book, they make no representations or warranties with respect to the accuracy or completeness of the contents of this book and specifically disclaim any implied warranties of merchantability or fitness for a particular purpose. No warranty may be created or extended by sales representatives or written sales materials. The advice and strategies contained herein may not be suitable for your situation. You should consult with a professional where appropriate. Neither the publisher nor author shall be liable for any loss of profit or any other commercial damages, including but not limited to special, incidental, consequential, or other damages.

For general information on our other products and services or for technical support, please contact our Customer Care Department within the United States at (800) 762-2974, outside the United States at (317) 572-3993 or fax (317) 572-4002.

Wiley also publishes its books in a variety of electronic formats. Some content that appears in print may not be available in electronic books. For more information about Wiley products, visit our web site at www.wiley.com.

Library of Congress Cataloging-in-Publication Data

Stevens, Brian L., 1939–
   Aircraft control and simulation / Brian L. Stevens, Frank L. Lewis.—2nd ed.
      p. cm.
   Includes bibliographical references and index.
   ISBN 0-471-37145-9 (cloth)
   1. Airplanes—Control systems.   I. Lewis, Frank L.   II. Title.
   TL678.S74   2003
   629.135—dc21   2003043250

Printed in the United States of America

10 9 8 7 6 5 4 3 2 1

To Deane, Bill, and Richard
B.L.S.

To my sons, Chris and Roma
F.L.

CONTENTS

List of Tables	ix
List of Examples	xi
Preface	xv

1 The Kinematics and Dynamics of Aircraft Motion   1

    1.1 Introduction / 1
    1.2 Vector Kinematics / 3
    1.3 Matrix Analysis of Kinematics / 19
    1.4 Geodesy, Earth's Gravitation, Terrestrial Navigation / 34
    1.5 Rigid-Body Dynamics / 42
    1.6 Summary / 53

2 Modeling the Aircraft   59

    2.1 Introduction / 59
    2.2 Basic Aerodynamics / 60
    2.3 Aircraft Forces and Moments / 71
    2.4 Static Analysis / 99
    2.5 The Nonlinear Aircraft Model / 107
    2.6 Linear Models and the Stability Derivatives / 116
    2.7 Summary / 137

3 Modeling, Design, and Simulation Tools — 143

3.1 Introduction / 143
3.2 State-Space Models / 145
3.3 Transfer Function Models / 156
3.4 Numerical Solution of the State Equations / 172
3.5 Aircraft Models for Simulation / 181
3.6 Steady-State Flight / 187
3.7 Numerical Linearization / 201
3.8 Aircraft Dynamic Behavior / 208
3.9 Feedback Control / 216
3.10 Summary / 245

4 Aircraft Dynamics and Classical Control Design — 254

4.1 Introduction / 254
4.2 Aircraft Rigid-Body Modes / 261
4.3 The Handling-Qualities Requirements / 279
4.4 Stability Augmentation / 291
4.5 Control Augmentation Systems / 308
4.6 Autopilots / 327
4.7 Nonlinear Simulation / 349
4.8 Summary / 377

5 Modern Design Techniques — 382

5.1 Introduction / 382
5.2 Assignment of Closed-Loop Dynamics / 386
5.3 Linear Quadratic Regulator with Output Feedback / 403
5.4 Tracking a Command / 419
5.5 Modifying the Performance Index / 434
5.6 Model-Following Design / 462
5.7 Linear Quadratic Design with Full State Feedback / 478
5.8 Dynamic Inversion Design / 484
5.9 Summary / 500

6 Robustness and Multivariable Frequency-Domain Techniques — 509

6.1 Introduction / 509
6.2 Multivariable Frequency-Domain Analysis / 511
6.3 Robust Output-Feedback Design / 534
6.4 Observers and the Kalman Filter / 538

6.5　LQG/Loop-Transfer Recovery / 563
　　6.6　Summary / 587

7 Digital Control　　　　　　　　　　　　　　　　　　　　**594**

　　7.1　Introduction / 594
　　7.2　Simulation of Digital Controllers / 595
　　7.3　Discretization of Continuous Controllers / 598
　　7.4　Modified Continuous Design / 608
　　7.5　Implementation Considerations / 621
　　7.6　Summary / 629

Appendix A　F-16 Model　　　　　　　　　　　　　　　　**633**

Appendix B　Software　　　　　　　　　　　　　　　　　　**642**

Index　　　　　　　　　　　　　　　　　　　　　　　　　　**655**

LIST OF TABLES

CHAPTER 1

1.4-1 Frames and Coordinate Systems Used with Figure 1.4-2 / 37

CHAPTER 2

2.2-1 Important Wing-Planform Parameters / 68
2.3-1 Force, Moment, and Velocity Definitions / 74–75
2.3-2 Aircraft Drag Components / 80
2.5-1 The Flat-Earth, Body-Axes 6-DOF Equations / 110
2.6-1 The Force Dimensional Derivatives / 120
2.6-2 The Moment Dimensional Derivatives / 124
2.6-3 Longitudinal Dimensional versus Dimensionless Derivatives / 130
2.6-4 Lateral/Directional Dimensional versus Dimensionless Derivatives / 131
2.6-5 Importance of Longitudinal Stability Derivatives / 132
2.6-6 Importance of Lateral/Directional Derivatives / 136

CHAPTER 3

3.3-1 Network Transfer Functions and State Equations / 162
3.5-1 Aircraft Control-Surface Sign Conventions / 186
3.5-2 F-16 Model Test Case / 187
3.6-1 Trim Data for the Transport Aircraft Model / 193

LIST OF TABLES

3.6-2   Trim Data for the F-16 Model / 196
3.6-3   Trimmed Flight Conditions for the F-16 / 197
3.8-1   F-16 Model, Elevator-to-Pitch-Rate Transfer Function / 212

CHAPTER 4

4.2-1   Accuracy of Short-Period and Phugoid Formulae / 268
4.2-2   Effect of Flight-Path Angle on F-16 Modes / 277
4.2-3   Effect of Speed and Altitude on F-16 Modes / 278
4.3-1   Pilot Opinion Rating and Flying Qualities Level / 280
4.3-2   Definitions—Flying Qualities Specifications / 289
4.3-3a Short-Period Damping Ratio Limits / 289
4.3-3b Limits on $\omega_{n_{sp}}^2/(n/\alpha)$ / 290
4.3-4   Max. Roll-Mode Time Constant (sec.) / 290
4.3-5   Spiral Mode, Minimum Doubling Time / 291
4.3-6   Dutch Roll Mode Specifications / 291
4.5-1   Transfer Function Zeros versus Accelerometer Position / 315
4.5-2   Trim Conditions for Determining ARI Gain / 322

CHAPTER 5

5.2-1   Desired and Achievable Eigenvectors / 401
5.3-1   LQR with Output Feedback / 408
5.3-2   Optimal Output Feedback Solution Algorithm / 410
5.4-1   LQ Tracker with Output Feedback / 428
5.5-1   LQ Tracker with Time-Weighted PI / 440
5.7-1   LQR with State Feedback / 481

CHAPTER 6

6.4-1   The Kalman Filter / 555

CHAPTER 7

7.4-1   Padé Approximants to $e^{-s\Delta}$ for Approximation of Computation Delay / 616
7.4-2   Approximants to $(1 - e^{-sT})/sT$ for Approximation of Hold Delay / 617
7.5-1   Elements of Second-Order Modules / 629
7.5-2   Difference Equation Implementation of Second-Order Modules / 629

LIST OF EXAMPLES

CHAPTER 1

1.2-1 Coriolis Acceleration in an Earth-Fixed Frame / 12
1.2-2 Accelerometer Measurements / 13

CHAPTER 3

3.2-1 State Equations for a Mechanical System / 145
3.2-2 State Equations for an Electrical System / 147
3.4-1 Integration of the Van der Pol Equation / 178
3.6-1 Steady-State Trim for a 3-DOF Aircraft Model / 191
3.6-2 Steady-State Trim for a 6-DOF Model / 193
3.6-3 Simulated Response to an Elevator Pulse / 198
3.6-4 Simulated Response to a Throttle Pulse / 199
3.6-5 Simulation of a Coordinated Turn / 200
3.7-1 Comparison of Algebraic and Numerical Linearization / 205
3.7-2 Linearization of the F-16 Model / 206
3.8-1 F-16 Longitudinal Modes / 209
3.8-2 F-16 Lateral/Directional Modes / 210
3.8-3 F-16 Elevator-to-Pitch-Rate Transfer Function / 212
3.8-4 F-16 Elevator-to-Pitch-Rate Frequency Response / 213
3.8-5 Transport Aircraft Throttle Response / 214

xii LIST OF EXAMPLES

3.9-1 An Example of Nyquist's Stability Criterion / 226
3.9-2 Root-Locus Design Using a Lead Compensator / 232
3.9-3 Root-Locus Design of a PI Compensator / 234
3.9-4 Design of a Passive Lag Compensator / 239
3.9-5 Design of a Passive Lead Compensator / 241
3.9-6 Feedback Compensation with a Phase Lead Network / 244

CHAPTER 4

4.2-1 Lateral Modes of a Business Jet / 275
4.2-2 Mode Dependence from the Nonlinear Model / 277
4.4-1 The Effects of Pitch-Rate and Alpha Feedback / 293
4.4-2 A Pitch-SAS Design / 298
4.4-3 A Roll Damper/Yaw Damper Design / 303
4.5-1 A Pitch-Rate CAS Design / 310
4.5-2 A Normal Acceleration CAS Design / 316
4.5-3 A Lateral-Directional CAS Design / 323
4.6-1 A Simple Pitch-Attitude-Hold Autopilot / 328
4.6-2 A Pitch-Attitude Hold with Dynamic Compensation / 331
4.6-3 An Altitude-Hold Autopilot Design / 335
4.6-4 Longitudinal Control for Automatic Landing / 340
4.6-5 A Roll-Angle-Hold Autopilot / 346
4.7-1 Pitch-Rate CAS Nonlinear Simulation / 350
4.7-2 Lateral-Directional CAS Nonlinear Simulation / 353
4.7-3 Simulation of Automatic Landing / 359
4.7-4 Automatic Flare Control / 363
4.7-5 A Roll-Angle-Steering Control System / 366
4.7-6 Simulation of a Controller with Limiters / 370

CHAPTER 5

5.2-1 Selecting Eigenvectors for Decoupling / 390
5.2-2 Eigenstructure Assignment Using Dynamic Regulator / 397
5.2-3 Eigenstructure Design of Longitudinal Pitch Pointing Control / 398
5.3-1 LQR Design for F-16 Lateral Regulator / 413
5.4-1 Normal Acceleration CAS / 429
5.5-1 Constrained Feedback Control for F-16 Lateral Dynamics / 441
5.5-2 Time-Dependent Weighting Design of Normal Acceleration CAS / 443
5.5-3 Pitch-Rate Control System Using LQ Design / 444
5.5-4 Multivariable Wing Leveler / 448
5.5-5 Glide-Slope Coupler / 452

5.6-1   Automatic Flare Control by Model-Following Design / 470
5.7-1   LQR with State Feedback for Systems Obeying Newton's Laws / 482
5.8-1   Dynamic Inversion Design for Linear F-16 Longitudinal Dynamics / 489
5.8-2   Dynamic Inversion Design for Nonlinear Longitudinal Dynamics / 496

CHAPTER 6

6.2-1   MIMO Bode Magnitude Plots / 517
6.2-2   Singular Value Plots for F-16 Lateral Dynamics / 518
6.2-3   Precompensator for Balancing and Zero Steady-State Error / 524
6.2-4   Model Reduction and Stability Robustness / 527
6.3-1   Pitch-Rate Control System Robust to Wind Gusts and Unmodeled Flexible Mode / 535
6.4-1   Observer Design for Double Integrator System / 543
6.4-2   Kalman Filter Estimation of Angle of Attack in Gust Noise / 556
6.5-1   LQR/LTR Design of Aircraft Lateral Control System / 574

CHAPTER 7

7.3-1   Discrete PID Controller / 602
7.3-2   Digital Pitch-Rate Controller via BLT / 604
7.4-1   Digital Pitch-Rate Controller via Modified Continuous Design / 617
7.5-1   Antiwindup Compensation for Digital PI Controller / 624

PREFACE

This book is primarily aimed at students in aerospace engineering, at the senior and graduate level. We hope that it will also prove useful to practicing engineers, both as a reference book and as an update to their engineering education.

In addressing simulation of aerospace vehicles we have reviewed the relevant parts of classical mechanics and attempted to provide a clear, consistent notation. This has been coupled with a thorough treatment of six-degrees-of-freedom (6-DOF) motion, including a detailed discussion of attitude representation using both Euler angles and quaternions. Simulation of motion over and around the Earth requires some understanding of geodesy and the Earth's gravitation, and these topics have also been discussed in some detail within the framework of the WGS-84 datum. Familiarity with these topics is indispensable to many of the engineers working in the aerospace industry. Given this background the student can independently construct 6-DOF simulations and learn from them.

High-speed motion within the Earth's atmosphere entails aerodynamic forces and moments. We have reviewed aerodynamic modeling, and provided many graphical examples of such forces and moments for real aircraft. The small-perturbation theory of aerodynamic forces and moments is also described in detail. This study of 6-DOF motion and aerodynamic effects culminates in two realistic nonlinear aircraft models, which are then used for design and simulation examples in the rest of the book.

We have provided computer code in both MATLAB and Fortran to perform simulation and design with these models. Involvement with the models and designs will demonstrate many ideas in simulation, control theory, computer-aided design techniques, and numerical algorithms. The design examples are easily reproducible, and offer a great deal of scope to a class of students.

Before starting feedback control design we have reviewed linear systems theory, including the Laplace transform, transfer functions, and the state-space formulation. Transform theory views dynamic systems through their poles and zeros and leads to many convenient graphical and back-of-the-envelope design techniques, while state-space techniques are ideally suited to computer-aided design. We have attempted to pass "seamlessly" between the two formulations.

Classical control design is illustrated through many examples performed on the aircraft models using transform domain techniques supported by an underlying state-space model. Modern design in the later chapters simply uses the state-space models.

Finally, we note that the choice of topics herein is influenced by our experience in the broader area of guidance, navigation, and control (GNC). Very few engineers entering the aerospace industry will find themselves designing flight control systems, and those few will take part in the design of only two or three such systems in their careers. Instead, they will find themselves involved in a broad spectrum of projects, where a good grasp of classical mechanics, dynamics, coordinate transformations, geodesy, and navigation will be invaluable. The importance of modeling and simulation cannot be overstated. Large sums of money are spent on mathematical modeling and digital simulation before any hardware is built.

The first author wishes to acknowledge the help of colleagues in Aerospace Engineering at Georgia Tech, and the support of Provost Charles L. Liotta. Profs. J. V. R. Prasad and E. N. Johnson read the manuscript and made many helpful, constructive suggestions. Prof. C. V. Smith provided invaluable help with Chapter 1 during many hours of interesting discussion. The computer support of B. H. Hudson at the Georgia Tech Research Institute is also gratefully acknowledged. Both authors wish to thank the staff of John Wiley & Sons for their painstaking preparation of the manuscript.

BRIAN L. STEVENS
Georgia Institute of Technology

FRANK L. LEWIS
University of Texas at Arlington

CHAPTER 1

THE KINEMATICS AND DYNAMICS OF AIRCRAFT MOTION

1.1 INTRODUCTION

In this chapter the end point will be the equations of motion of a rigid aircraft moving over the oblate, rotating Earth. The flat-Earth equations, describing motion over a small area of a nonrotating Earth, with constant gravity, will be derived as a special case. To reach this end point we will use the vector analysis of classical mechanics to set up the equations of motion, matrix algebra to describe operations with coordinate systems, and concepts from geodesy, gravitation, and navigation to introduce the effects of the Earth's shape and mass attraction.

The moments and forces acting on the vehicle, other than the mass attraction of the Earth, will be abstract until Chapter 2 is reached. At this stage the equations can be used to describe the motion of any type of aerospace vehicle, including an Earth satellite, provided that suitable force and moment models are available. The term *rigid* means that structural flexibility is not allowed for, and all points in the vehicle are assumed to maintain the same relative position at all times. This assumption is good enough for flight simulation in most cases, and good enough for flight control system design provided that we are not trying to design a system to control structural modes or to alleviate aerodynamic loads on the aircraft structure.

The vector analysis needed for the treatment of the equations of motion often causes difficulties for the student, particularly the concept of the angular velocity vector. Therefore, a review of the relevant topics is provided. In some cases we have gone beyond the traditional approach to flight mechanics. For example, quaternions have been introduced because of their "all-attitude" capability and numerical advantages in simulation and control. They are now widely used in simulation, robotics, guidance and navigation calculations, attitude control, and graphics animation. Topics from

2 THE KINEMATICS AND DYNAMICS OF AIRCRAFT MOTION

geodesy (a branch of mathematics dealing with the shape of the Earth), gravitation (the mass attraction effect of the Earth), and navigation have also been introduced. This is because aircraft can now fly autonomously at very high altitudes and over long distances and there is a need to simulate navigation of such vehicles.

The equations of motion will be organized as a set of simultaneous first-order differential equations, explicitly solved for the derivatives. For n dependent variables, X_i, and m control inputs, U_i, the general form will be:

$$\dot{X}_1 = f_1(X_1, X_2, \ldots, X_n, U_1, \ldots, U_m)$$
$$\vdots \tag{1.1-1}$$
$$\dot{X}_n = f_n(X_1, X_2, \ldots, X_n, U_1, \ldots, U_m),$$

where the functions f_i are the nonlinear functions that can arise from modeling real systems. If the variables X_i constitute the smallest set of variables that, together with given inputs U_i, completely describe the behavior of the system, then the X_i are a set of *state variables* for the system. Equations (1.1-1) become a *state-space* description of the system. The functions f_i are required to be single-valued continuous functions. Equations (1.1-1) are often written symbolically as:

$$\dot{X} = f(X, U), \tag{1.1-2}$$

where the *state vector* X is an $(n \times 1)$ column array of the n state variables, the *control vector* U is an $(m \times 1)$ column array of the control variables, and f is an array of nonlinear functions. The nonlinear state equations (1.1-1), or a subset of them, usually have one or more *equilibrium points* in the multidimensional state and control space, where the derivatives vanish. The equations are often approximately linear for small perturbations from equilibrium, and can be written in matrix form as the *linear state equation*:

$$\dot{x} = Ax + Bu \tag{1.1-3}$$

Here, the lowercase notation for the state and control vectors indicates that they are perturbations from equilibrium, although the derivative vector contains the actual values (i.e., perturbations from zero). The "A matrix" is square and the "B matrix" has dimensions determined by the number of states and controls.

The state-space formulation will be described in more detail in Chapters 2 and 3. At this point we will simply note that a major advantage of this formulation is that the nonlinear state equations can be solved numerically. The simplest numerical solution method is *Euler integration*, given by:

$$X = X + f(X, U)\delta t, \tag{1.1-4}$$

where "=" indicates replacement of X in computer memory by the value on the right-hand side of the equation. The *integration time-step*, δt, must be made small enough that, for every δt interval, U can be approximated by a constant, and $\dot{X}\delta t$ provides a good approximation to the increment in the state vector. This numerical integration allows the state vector to be stepped forward in time, in time-increments of δt, to obtain a *time-history* simulation (Problem 1.5-2).

1.2 VECTOR KINEMATICS

Definitions and Notation

Kinematics can be defined as the study of the motion of objects without regard to the mechanisms that cause the motion. The motion of physical objects can be described by means of vectors in three dimensions, and in performing kinematic analysis with vectors we will make use of the following definitions:

> *Frame of Reference*: a rigid body or set of rigidly related points that can be used to establish distances and directions (denoted by F_i, F_e, etc.). In general, a subscript used to indicate a frame will be lowercase, while a subscript used to indicate a point will be uppercase.
>
> *Inertial Frame*: a frame of reference in which Newton's laws apply. Our best inertial approximation is probably a "helio-astronomic" frame in which the center of mass (cm) of the sun is a fixed point, and fixed directions are established by the normal to the plane of the ecliptic and the projection on that plane of certain stars that appear to be fixed in position.
>
> *Vector*: a vector is an abstract geometrical object that has both magnitude and direction. It exists independently of any coordinate system. The vectors used here are Euclidean vectors that exist only in three-dimensional space.
>
> *Coordinate System*: a measurement system for locating points in space, set up within a frame of reference. We may have multiple coordinate systems (with no relative motion) within one frame of reference, and we sometimes loosely refer to them as "frames."

In choosing a notation the following facts must be taken into account. For position vectors, the notation should specify the two points whose relative position the vector describes. Velocity and acceleration vectors are relative to a frame of reference, and the notation should specify the frame of reference as well as the moving point. The derivative of a vector depends on the observer's frame of reference, and this frame must be specified in the notation. A derivative may be taken in a different frame from that in which a vector is defined, so the notation may require two frame designators with one vector. We will use the following notation:

> Vectors will be in boldface typefonts.
> A right subscript will be used to designate two points for a position vector, and a point and a frame for a velocity or acceleration vector. A "/" in a subscript will mean "with respect to."

4 THE KINEMATICS AND DYNAMICS OF AIRCRAFT MOTION

A left superscript will specify the frame in which a derivative is taken, and the dot notation will indicate a derivative.

A right superscript on a vector will specify a coordinate system. It will therefore denote an array of the components of that vector in the specified system.

Vector length will be denoted by single bars, for example, $|\mathbf{p}|$.

Examples of the notation are:

$\mathbf{p}_{A/B} \equiv$ position vector of point A with respect to point B
$\mathbf{v}_{A/i} \equiv$ velocity of point A in frame i (F_i)
$^b\dot{\mathbf{v}}_{A/i} \equiv$ derivative of $\mathbf{v}_{A/i}$ taken in F_b
$\mathbf{v}^c_{A/i} \equiv (\mathbf{v}_{A/i})^c \equiv$ components of $\mathbf{v}_{A/i}$ in coordinate system c
$^b\dot{\mathbf{v}}^c_{A/i} \equiv$ components in system c of the derivative in F_b

The components of a vector will be denoted by subscripts that indicate the coordinate system, or by the vector symbol with subscripts x, y, and z. All component arrays will be column arrays unless otherwise indicated by the transpose symbol, a right superscript T. For example,

$$\mathbf{p}^b_{A/B} = \begin{bmatrix} x_b \\ y_b \\ z_b \end{bmatrix} \text{ or } \mathbf{v}^b = \begin{bmatrix} v_x \\ v_y \\ v_z \end{bmatrix} = \begin{bmatrix} v_x & v_y & v_z \end{bmatrix}^T$$

are arrays of components in a coordinate system b.

The Derivative Vector

The derivative of a vector can be defined in the same way as the derivative of a scalar:

$$\frac{d\mathbf{p}_{A/B}}{dt} = \lim_{\delta t \to 0} \left[\frac{\mathbf{p}_{A/B}(t + \delta t) - \mathbf{p}_{A/B}(t)}{\delta t} \right]$$

This is a new vector created by the changes in length and direction of $\mathbf{p}_{A/B}$. Different answers will be obtained for the derivative depending on how the observer's frame is rotating. As another example of the notation above, consider

$$^i\dot{\mathbf{p}}_{A/B} = \text{derivative of the vector } \mathbf{p}_{A/B}, \text{ taken in frame } i$$

Note that if $\mathbf{p}_{A/B}$ is a position vector, the derivative is a velocity vector only if it is taken in the frame in which B is a fixed point. Similarly, the derivative of a velocity vector is an acceleration vector only if it is taken in the frame in which the velocity vector is defined.

If the derivative of a general vector $\mathbf{v}$ is taken in frame a, the components of the derivative vector in a coordinate system fixed in frame a are given by the rates of change of the components of $\mathbf{v}$ in that coordinate system. For example, if

$$\mathbf{v}^c = \begin{bmatrix} v_x & v_y & v_z \end{bmatrix}^T,$$

where system c is fixed in frame a, then

$$^a\dot{\mathbf{v}}^c = \begin{bmatrix} \dot{v}_x & \dot{v}_y & \dot{v}_z \end{bmatrix}^T$$

The vector derivative deserves special attention, and is discussed further in connection with angular velocity.

Vector Properties

Vectors are independent of any coordinate system, but some vector operations yield *pseudo-vectors* that are not independent of a "handedness" convention. For example, the result of the *vector cross-product* operation is a vector whose direction depends on whether a right-handed or left-handed convention is being used. We will always use the right-hand rule in connection with vector direction. Similarly, we will always use Cartesian coordinate systems that are right-handed. Figure 1.2-1 shows a vector $\mathbf{p}$ and a reference coordinate system (fixed in some frame) used to describe the direction of $\mathbf{p}$. The axes of the coordinate system are aligned with the unit vectors $\mathbf{i}, \mathbf{j}, \mathbf{k}$, which (in that order) form a right-handed set (i.e., $\mathbf{i} \times \mathbf{j} = \mathbf{k}$). The direction of $\mathbf{p}$, relative to this coordinate system, is described by the three *direction angles* α, β, γ. The *direction cosines* of $\mathbf{p}$—$\cos\alpha$, $\cos\beta$, and $\cos\gamma$—give the projections of $\mathbf{p}$ on the coordinate axes, and two applications of the theorem of Pythagoras yield

$$|\mathbf{p}|^2 \cos^2\alpha + |\mathbf{p}|^2 \cos^2\beta + |\mathbf{p}|^2 \cos^2\gamma = |\mathbf{p}|^2$$

Therefore, the direction cosines satisfy

$$\cos^2\alpha + \cos^2\beta + \cos^2\gamma = 1 \tag{1.2-1}$$

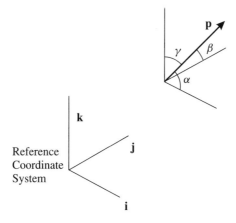

Figure 1.2-1 The direction angles of a vector.

Addition and subtraction of vectors can be defined independently of coordinate systems by means of geometrical constructions (the "parallelogram law"). The *dot product* of two vectors is a scalar defined by

$$\mathbf{u} \cdot \mathbf{v} = |\mathbf{u}||\mathbf{v}| \cos \theta, \qquad (1.2\text{-}2)$$

where θ is the included angle between the vectors (it may be necessary to translate the vectors so that they intersect). The dot product is commutative and distributive; thus,

$$\mathbf{u} \cdot \mathbf{v} = \mathbf{v} \cdot \mathbf{u}$$

and

$$(\mathbf{u} + \mathbf{v}) \cdot \mathbf{w} = \mathbf{u} \cdot \mathbf{w} + \mathbf{v} \cdot \mathbf{w}$$

The principal uses of the dot product are to find the projection of a vector, to establish orthogonality, and to find length. For example, if (1.2-2) is divided by $|\mathbf{v}|$, we have the projection of $\mathbf{u}$ on $\mathbf{v}$,

$$(\mathbf{u} \cdot \mathbf{v})/|\mathbf{v}| = |\mathbf{u}| \cos \theta$$

If $\cos \theta = 0$, $\mathbf{u} \cdot \mathbf{v} = 0$, and the vectors are said to be *orthogonal*. If a vector is dotted with itself, then $\cos \theta = 1$ and we obtain the square of its length.

Orthogonal unit vectors satisfy the dot product relationships

$$\mathbf{i} \cdot \mathbf{i} = \mathbf{j} \cdot \mathbf{j} = \mathbf{k} \cdot \mathbf{k} = 1$$
$$\mathbf{i} \cdot \mathbf{j} = \mathbf{j} \cdot \mathbf{k} = \mathbf{k} \cdot \mathbf{i} = 0$$

Using these relationships, the dot product of two vectors can be expressed in terms of components,

$$\mathbf{u} \cdot \mathbf{v} = u_x v_x + u_y v_y + u_z v_z, \qquad (1.2\text{-}3)$$

where the vector are taken in any orthogonal Cartesian coordinate system.

The *cross-product* of $\mathbf{u}$ and $\mathbf{v}$, denoted by $\mathbf{u} \times \mathbf{v}$, is a vector $\mathbf{w}$ that is normal to the plane of $\mathbf{u}$ and $\mathbf{v}$ and is in a direction such that $\mathbf{u}, \mathbf{v}, \mathbf{w}$ (in that order) form a right-handed system (again, it may be necessary to translate the vectors so that they intersect). The length of $\mathbf{w}$ is defined to be $|\mathbf{u} \times \mathbf{v}| = |\mathbf{u}||\mathbf{v}| \sin \theta$, where θ is the angle between $\mathbf{u}$ and $\mathbf{v}$.

It has the following properties:

$$\mathbf{u} \times \mathbf{v} = -(\mathbf{v} \times \mathbf{u}) \qquad \text{(anticommutative)}$$
$$a(\mathbf{u} \times \mathbf{v}) = (a\mathbf{u}) \times \mathbf{v} = \mathbf{u} \times (a\mathbf{v}) \qquad \text{(associative)}$$

$$\mathbf{u} \times (\mathbf{v} + \mathbf{w}) = (\mathbf{u} \times \mathbf{v}) + (\mathbf{u} \times \mathbf{w}) \quad \text{(distributive)}$$

$$\mathbf{u} \cdot (\mathbf{v} \times \mathbf{w}) = \mathbf{v} \cdot (\mathbf{w} \times \mathbf{u}) = \mathbf{w} \cdot (\mathbf{u} \times \mathbf{v}) \quad \text{(scalar triple product)}$$

$$\mathbf{u} \times (\mathbf{v} \times \mathbf{w}) = \mathbf{v}(\mathbf{w} \cdot \mathbf{u}) - \mathbf{w}(\mathbf{u} \cdot \mathbf{v}) \quad \text{(vector triple product)} \quad (1.2\text{-}4)$$

As an aid for remembering the form of the triple products, note the cyclic permutation of the vectors involved. Alternatively, the vector triple product can be remembered phonetically using "ABC = BAC − CAB."

The cross-products of the unit vectors describing a right-handed orthogonal coordinate system satisfy the equations

$$\mathbf{i} \times \mathbf{i} = \mathbf{j} \times \mathbf{j} = \mathbf{k} \times \mathbf{k} = 0$$

$$\mathbf{i} \times \mathbf{j} = \mathbf{k}$$

$$\mathbf{j} \times \mathbf{k} = \mathbf{i}$$

$$\mathbf{k} \times \mathbf{i} = \mathbf{j}$$

Also remember that $\mathbf{j} \times \mathbf{i} = -\mathbf{i} \times \mathbf{j} = -\mathbf{k}$, and so on. From these properties we can derive a formula for the cross-product of two vectors; a convenient way of remembering the formula is to write it so that it resembles the expansion of a determinant.

The mnemonic is

$$\mathbf{u} \times \mathbf{v} = \begin{vmatrix} \mathbf{i} & \mathbf{j} & \mathbf{k} \\ u_x & u_y & u_z \\ v_x & v_y & v_z \end{vmatrix} = \mathbf{i}\begin{vmatrix} u_y & u_z \\ v_y & v_z \end{vmatrix} - \mathbf{j}\begin{vmatrix} u_x & u_z \\ v_x & v_z \end{vmatrix} + \mathbf{k}\begin{vmatrix} u_x & u_y \\ v_x & v_y \end{vmatrix}, \quad (1.2\text{-}5)$$

where subscripts x, y, z indicate components in a coordinate system whose axes are aligned respectively with the unit vectors $\mathbf{i}, \mathbf{j}, \mathbf{k}$.

An example of the use of the cross-product is to find the moment $\mathbf{M}$ of a force $\mathbf{F}$, acting at a point whose position vector is $\mathbf{r}$; the vector moment about the origin of $\mathbf{r}$ is given by

$$\mathbf{M} = \mathbf{r} \times \mathbf{F}$$

Other examples are given in the following subsections.

Rotation of a Vector

It is intuitively obvious that a vector can be made to point in an arbitrary direction by means of a single rotation around an appropriate axis. Here we follow Goldstein (Goldstein, 1980) to derive a formula for vector rotation.

Consider Figure 1.2-2, in which a free vector $\mathbf{u}$ has been rotated to form a new vector $\mathbf{v}$ by defining a rotation axis along a unit vector $\mathbf{n}$ and performing a left-handed rotation through μ around $\mathbf{n}$. NV and NU have been constructed to find the projections of $\mathbf{v}$ and $\mathbf{u}$ on the rotation axis and hence identify μ. A vector expression for $\mathbf{v}$ is

8 THE KINEMATICS AND DYNAMICS OF AIRCRAFT MOTION

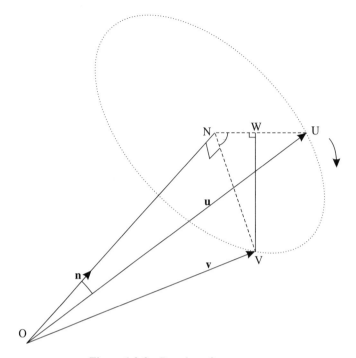

Figure 1.2-2 Rotation of a vector.

$$\mathbf{v} = \overrightarrow{ON} + \overrightarrow{NW} + \overrightarrow{WV}$$

$$= (\mathbf{u} \cdot \mathbf{n})\,\mathbf{n} + \frac{(\mathbf{u} - (\mathbf{u} \cdot \mathbf{n})\,\mathbf{n})}{|\mathbf{u} - (\mathbf{u} \cdot \mathbf{n})\,\mathbf{n}|}\, NV \cos \mu + \frac{(\mathbf{u} \times \mathbf{n})}{|\mathbf{u}| \sin \phi}\, NV \sin \mu$$

Now,

$$NV = NU = |\mathbf{u} - (\mathbf{u} \cdot \mathbf{n})\,\mathbf{n}| = |\mathbf{u}| \sin \phi$$

Therefore,

$$\mathbf{v} = \mathbf{n}(\mathbf{n} \cdot \mathbf{u}) + \cos \mu \,(\mathbf{u} - \mathbf{n}(\mathbf{n} \cdot \mathbf{u})) - \sin \mu \,(\mathbf{n} \times \mathbf{u})$$

or,

$$\mathbf{v} = (1 - \cos \mu)\,\mathbf{n}\,(\mathbf{n} \cdot \mathbf{u}) + \cos \mu \,\mathbf{u} - \sin \mu \,(\mathbf{n} \times \mathbf{u}) \qquad (1.2\text{-}6)$$

Equation (1.2-6) is sometimes called the *rotation formula*; it shows that, after choosing $\mathbf{n}$ and μ, we can operate on $\mathbf{u}$ with dot and cross-product operations to get the desired rotation.

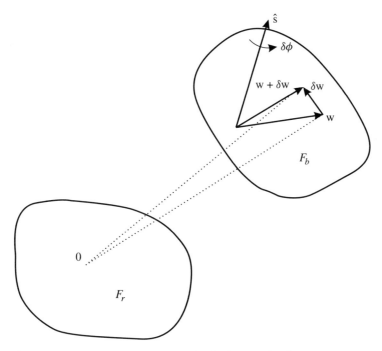

Figure 1.2-3 A vector derivative in a rotating frame.

Vector Derivatives and the Angular Velocity Vector

Figure 1.2-3 shows a vector **w** that is fixed in a frame F_b, and F_b is rotating with respect to a reference frame F_r. The derivative of **w** taken in F_r is nonzero if **w** is changing direction and/or changing length when observed from F_r, and is independent of translational motion between the frames. The change in direction with respect to F_r can be found by using the rotation theorem. In the figure, let $\hat{\mathbf{s}}$ be a unit vector parallel to the instantaneous axis of rotation at time t. To an observer in F_r, **w** becomes a new vector $\mathbf{w} + \delta\mathbf{w}$ at time $t + \delta t$, due to the small rotation $\delta\phi$. The rotation formula can be used to find $\delta\mathbf{w}$. Then, by taking the limit of $\delta\mathbf{w}/\delta t$ as δt becomes infinitesimal, the derivative of **w** in F_r can be found. The rotation formula, with small-angle approximations and positive $\delta\phi$ right-handed around $\hat{\mathbf{s}}$, gives

$$\frac{\delta\mathbf{w}}{\delta t} \approx \left(\hat{\mathbf{s}}\frac{\delta\phi}{\delta t}\right) \times \mathbf{w}$$

Taking the limit as $\delta t \to 0$,

$$^r\dot{\mathbf{w}} = (\hat{\mathbf{s}}\,\dot{\phi}) \times \mathbf{w}$$

The quantity in parentheses has the properties of a vector, with direction along the axis of rotation and magnitude equal to the angular rotation rate. It is defined to be

10 THE KINEMATICS AND DYNAMICS OF AIRCRAFT MOTION

the instantaneous *angular velocity vector*, $\boldsymbol{\omega}_{b/r}$, of F_b with respect to F_r. A right-handed rotation around $\hat{\mathbf{s}}$ corresponds to a positive angular velocity vector. If $\mathbf{w}$ is also changing in length in F_b, we must add this effect to the right-hand side of the above equation, so that

$$^r\dot{\mathbf{w}} = {}^b\dot{\mathbf{w}} + \boldsymbol{\omega}_{b/r} \times \mathbf{w} \qquad (1.2\text{-}7)$$

Equation (1.2-7) is sometimes called the *equation of Coriolis* (Blakelock, 1965) and will be an essential tool in developing equations of motion from Newton's laws. It is much more general than is indicated above, and applies to any physical quantity that has a vector representation. The derivatives need not even be taken with respect to time. Angular velocity can be defined as the vector that relates the derivatives of any arbitrary vector in two different frames, according to (1.2-7). In our context we have a physical interpretation of this vector as a right-handed angular rate around a directed axis with, in general, both rate and direction changing with time. An alternative derivation of the angular velocity vector can be found in many texts (McGill and King, 1995; Kane, 1983).

Some formal properties of the angular velocity vector are:

(i) It is a unique vector that relates the derivatives of a vector taken in two different frames.
(ii) It satisfies the relative motion condition $\boldsymbol{\omega}_{b/a} = -\boldsymbol{\omega}_{a/b}$.
(iii) It is additive over multiple frames, for example, $\boldsymbol{\omega}_{c/a} = \boldsymbol{\omega}_{c/b} + \boldsymbol{\omega}_{b/a}$ (this is not true of angular acceleration).
(iv) Its derivative is the same in either frame, ${}^a\dot{\boldsymbol{\omega}}_{b/a} = {}^b\dot{\boldsymbol{\omega}}_{b/a}$. This is made evident by using (1.2-7) to find the derivative of $\boldsymbol{\omega}$.

A common problem is the determination of an angular velocity vector after the frames have been defined in a practical application. This can be achieved by finding one or more intermediate frames in which an axis of rotation and an angular rate are physically evident. Then the additive property can be invoked to combine the intermediate angular velocities. An example of this is given later, with the "rotating-Earth" equations of motion of an aerospace vehicle.

Velocity and Acceleration in Moving Frames

Figure 1.2-4 shows a point P moving with respect to two frames F_a and F_b, with fixed points O and Q, respectively. Suppose that we wish to relate the velocities in the two frames and also the accelerations. First, we must relate the position vectors shown in the figure, and then take derivatives in F_a to introduce velocity:

$$\mathbf{r}_{P/O} = \mathbf{r}_{Q/O} + \mathbf{r}_{P/Q} \qquad (1.2\text{-}8)$$

$$^a\dot{\mathbf{r}}_{P/O} = {}^a\dot{\mathbf{r}}_{Q/O} + {}^a\dot{\mathbf{r}}_{P/Q} \qquad (1.2\text{-}9)$$

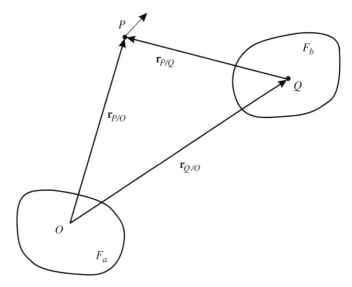

Figure 1.2-4 Velocity and acceleration in moving frames.

Starting from the left-hand side of Equation (1.2-9), the first two terms are velocities in F_a but the last term involves the position of P relative to a fixed point in F_b, with the derivative taken in F_a. Let **v** with an appropriate subscript represent a velocity vector. Then, by applying the equation of Coriolis, Equation (1.2-9) gives

$$\mathbf{v}_{P/a} = \mathbf{v}_{Q/a} + \mathbf{v}_{P/b} + \boldsymbol{\omega}_{b/a} \times \mathbf{r}_{P/Q} \tag{1.2-10}$$

As an application of Equation (1.2-10), let F_a be an inertial reference frame and F_b a body moving with respect to the reference frame. Assume that a navigator on the moving body determines, from an onboard inertial navigation system, his velocity in the inertial reference frame ($\mathbf{v}_{Q/a}$) and his inertial angular velocity vector ($\boldsymbol{\omega}_{b/a}$). Also, using a radar set, he measures the velocity of P in F_b ($\mathbf{v}_{P/b}$) and the position of P with respect to Q ($\mathbf{r}_{P/Q}$). He can then use Equation (1.2-10) to calculate the velocity of the object in the inertial reference frame and, knowing the equation of motion in the inertial frame, predict its trajectory. The word *measure* should always evoke the thought "coordinate system?" and Equation (1.2-10) cannot be evaluated without choosing coordinate systems for this example. In Section 1.3 it will become clear how the coordinate systems calculations can be performed.

We next find the acceleration of P by taking derivatives of (1.2-10) in F_a. Starting from the left, the first two terms are velocities in F_a and these become accelerations in F_a. The third term is a velocity in F_b and must be differentiated by the equation of Coriolis. The last term involving a cross-product can be differentiated by the "product rule," and the derivative of angular velocity is an angular acceleration vector, denoted by $\boldsymbol{\alpha}$. Therefore, denoting translational acceleration vectors by **a**, (1.2-10) yields,

12 THE KINEMATICS AND DYNAMICS OF AIRCRAFT MOTION

$$\mathbf{a}_{P/a} = \mathbf{a}_{Q/a} + (\mathbf{a}_{P/b} + \boldsymbol{\omega}_{b/a} \times \mathbf{v}_{P/b}) + \boldsymbol{\alpha}_{b/a} \times \mathbf{r}_{P/Q} + \boldsymbol{\omega}_{b/a} \times (\mathbf{v}_{P/b} + \boldsymbol{\omega}_{b/a} \times \mathbf{r}_{P/Q})$$

Regrouping terms, we get,

$$\mathbf{a}_{P/a} = \underbrace{\mathbf{a}_{P/b}}_{\substack{\text{total} \\ \text{accl.}}} + \underbrace{\mathbf{a}_{Q/a} + \boldsymbol{\alpha}_{b/a} \times \mathbf{r}_{P/Q} + \boldsymbol{\omega}_{b/a} \times (\boldsymbol{\omega}_{b/a} \times \mathbf{r}_{P/Q})}_{\text{Centripetal accl.}} + \underbrace{2\boldsymbol{\omega}_{b/a} \times \mathbf{v}_{P/b}}_{\substack{\text{Coriolis} \\ \text{accl.}}}$$

$$\text{Transport accln. of } P \text{ in } F_a \qquad (1.2\text{-}11)$$

The term labeled "transport acceleration" is the acceleration in F_a of a fixed point in F_b that is instantaneously coincident with P. This is evident because the two remaining right-hand-side terms vanish when P is fixed in F_b. Note that (1.2-10) can be written as

$$\mathbf{v}_{P/a} = \mathbf{v}_{P/b} + (\mathbf{v}_{Q/a} + \boldsymbol{\omega}_{b/a} \times \mathbf{r}_{P/Q}),$$

where the term in parentheses is the velocity in F_a of a fixed point in F_b that is instantaneously coincident with P. Therefore, the acceleration equation does not have the same form as this velocity equation because of the "Coriolis acceleration" term.

Example 1.2-1: Coriolis Acceleration in an Earth-Fixed Frame. As an example of the application of (1.2-11), let F_b be fixed in the Earth, and let F_a also translate with the Earth but be nonrotating (i.e., chosen to be an approximation to an inertial frame). Let P be a point moving over the surface of the Earth, and let the points Q and O coincide, at the Earth's cm, so that the acceleration $\mathbf{a}_{Q/a}$ vanishes and $\mathbf{r}_{P/Q}$ is a geocentric position vector. The Earth's angular velocity is quite closely constant and so the derivative of $\boldsymbol{\omega}_{b/a}$ vanishes. This leaves only the relative acceleration, centripetal acceleration, and Coriolis acceleration terms. Solving for the relative acceleration gives:

$$\mathbf{a}_{P/b} = \mathbf{a}_{P/a} - \boldsymbol{\omega}_{b/a} \times (\boldsymbol{\omega}_{b/a} \times \mathbf{r}_{P/Q}) - 2\boldsymbol{\omega}_{b/a} \times \mathbf{v}_{P/b} \qquad (1.2\text{-}12)$$

For a particle of mass m at P, the relative acceleration corresponds to an "apparent force" on the particle and produces the trajectory observed by a stationary observer on the Earth. The true acceleration ($\mathbf{a}_{P/a}$) corresponds to "true" forces (e.g., mass attraction, drag), therefore,

$$\text{apparent force} = \text{true force} - m\boldsymbol{\omega}_{b/a} \times (\boldsymbol{\omega}_{b/a} \times \mathbf{r}_{P/Q}) + 2m\mathbf{v}_{P/b} \times \boldsymbol{\omega}_{b/a}$$

The second term on the right is the "centrifugal" force, directed normal to the angular velocity vector. The third term is usually referred to as the Coriolis force and will cause a ballistic trajectory over the Earth to curve to the left or right. A stationary observer on the Earth might realize that the Earth is not an inertial frame by seeing this curvature, which is really just the kinematic effect of the Earth's rotation.

An often quoted example of the Coriolis force is the circulation of winds around a low-pressure area (a cyclone) on the Earth. The true force is radially inward along the pressure gradient. In the Northern Hemisphere, for example, the Earth's angular velocity vector points outward from the Earth's surface and, whichever way the velocity vector $\mathbf{v}_{P/b}$ is directed, the Coriolis force is directed to the right of $\mathbf{v}_{P/b}$. Therefore, in the Northern Hemisphere the winds spiral inward in a counterclockwise direction around a cyclone.

Example 1.2-2: Accelerometer Measurements. This example will illustrate the principle of an accelerometer and the contribution of angular motion to the linear acceleration at a point away from the cm of a rigid body. Figure 1.2-5 shows a very simple accelerometer mounted on a rigid body, and aligned so as to measure z-axis components in the body-fixed coordinate system shown. The accelerometer consists of a "proof mass," m, a suspension spring, a viscous damper for the motion of the mass, and a means of measuring its displacement. The proof mass is constrained to move in one dimension only, in this case, in the body-z direction. Point P is the deflected position of the cm of the proof mass, $R(x_R, 0, 0)$ is the rest-position, and d is the deflection. Applying Equation (1.2-11) to find the acceleration of P in the inertial reference frame F_i yields

$$\mathbf{a}_{P/i} = \mathbf{a}_{P/b} + \mathbf{a}_{CM/i} + \boldsymbol{\alpha}_{b/i} \times \mathbf{r}_{P/Q} + \boldsymbol{\omega}_{b/i} \times (\boldsymbol{\omega}_{b/i} \times \mathbf{r}_{P/Q}) + 2\boldsymbol{\omega}_{b/i} \times \mathbf{v}_{P/b}$$

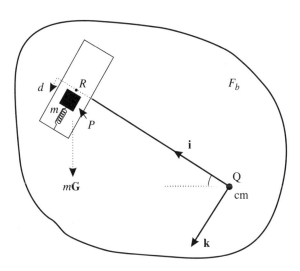

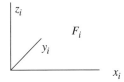

Figure 1.2-5 An accelerometer on a rigid body.

Now write this equation in terms of orthogonal unit vectors, **i**, **j**, **k**, fixed in F_b, with

$$\mathbf{a}_{P/i} = ap_x\mathbf{i} + ap_y\mathbf{j} + ap_z\mathbf{k} \quad \mathbf{a}_{P/b} = \ddot{d}\mathbf{k} \quad \mathbf{r}_{P/Q} = x_R\mathbf{i} + d\mathbf{k} \quad \mathbf{v}_{P/b} = \dot{d}\mathbf{k}$$

$$\mathbf{a}_{CM/i} \equiv a_x\mathbf{i} + a_y\mathbf{j} + a_z\mathbf{k} \quad \boldsymbol{\alpha}_{b/i} \equiv \alpha_x\mathbf{i} + \alpha_y\mathbf{j} + \alpha_z\mathbf{k} \quad \boldsymbol{\omega}_{b/i} \equiv \omega_x\mathbf{i} + \omega_y\mathbf{j} + \omega_z\mathbf{k}$$

and consider only the components along **k**:

$$ap_z = \ddot{d} + a_z - \alpha_y x_R - d\left(\omega_x^2 + \omega_y^2\right) + x_R\omega_x\omega_z$$

The z-component of force required to produce this acceleration is given by $m(ap_z)$, and is provided by the mass attraction force ($m\mathbf{G}$) toward the Earth's cm (Section 1.4), spring force, and viscous-damping force, that is,

$$m(ap_z) = mG_z - k_sd - b\dot{d},$$

where k_s is the accelerometer spring constant and b is the accelerometer viscous damping constant. Equating these two force expressions and rearranging terms gives

$$\ddot{d} + \frac{b}{m}\dot{d} + \frac{k_s}{m}d - d\left(\omega_x^2 + \omega_y^2\right) = G_z - (a_z - \alpha_y x_R + x_R\omega_x\omega_z)$$

Note that the last term on the right is the transport acceleration of point R in F_i.

The reading of the accelerometer is represented by d. The derivatives of d come into play when the accelerometer has to respond to a changing acceleration; here we will focus on the steady-state behavior with a constant acceleration input and neglect the derivatives. The variable position of the proof mass will cause a measurement error through the term $d(\omega_x^2 + \omega_y^2)$; this is eliminated in high-sensitivity accelerometers by using a "force rebalancing" technique, which measures the force required to maintain d and its derivatives very close to zero.

Acceleration in the z direction corresponds to negative d, so the steady-state accelerometer reading is

$$\text{reading} \propto \left(a_z' - G_z\right), \quad (1.2\text{-}13a)$$

where

$$a_z' = (a_z - \alpha_y x_R + x_R\omega_x\omega_z) \quad (1.2\text{-}13b)$$

Equation (1.2-13a) shows that, in general, this type of accelerometer responds to the pertinent component of $(\mathbf{a} - \mathbf{G})$ at its location. If the mass of the rigid body plus accelerometer is M, and the applied "contact force" (i.e., not counting the gravitational field force) is $\mathbf{F}$, the accelerometer responds to $(\mathbf{F} + M\mathbf{G})/M - \mathbf{G}$, or simply $\mathbf{F}/M$. This quantity is a *specific force*, denoted by $\mathbf{f}$, and so an accelerometer measures a component of the specific contact force given by

$$\mathbf{f} = \mathbf{a} - \mathbf{G} \qquad (1.2\text{-}14)$$

in whatever acceleration units are chosen.

As an example of Equation (1.2-14) consider a stationary accelerometer on the surface of the Earth, with its sensitive axis aligned with a plumb-bob measurement of the vertical. Neglecting the inertial acceleration of the Earth's cm, the term $\mathbf{a}$ will be the small centripetal acceleration due to the Earth's rotation. The $\mathbf{G}$ term depends on distance from Earth's cm, but is close to 9.8 m/s^2 in magnitude and directed toward the Earth's cm (see Section 1.4). The measurement $\mathbf{f}$ is the specific force due to the upward reaction of the Earth on the accelerometer, and will be exactly equal to the negative of the weight $m\mathbf{g}$ divided by the mass m (i.e., it will be $-\mathbf{g}$), where $\mathbf{g}$ is the local *gravity* vector. Alternatively, if the accelerometer is in free fall above the Earth, $\mathbf{a} = \mathbf{G}, \mathbf{f} = 0$, and the accelerometer reading is zero.

The accelerometer reading can be made dimensionless by dividing by $|\mathbf{g}|$, and accelerometers are commonly calibrated to read 1.0 g-units when stationary on the Earth's surface and having their sensitive axis parallel to the plumb-bob vertical. Therefore, the accelerometer measurement of specific force, at any location, can be obtained by multiplying the scale reading by the gravity value used for calibration, in whatever units are desired. It is evident that $\mathbf{G}$ must be known accurately to get an accurate value of acceleration from an accelerometer measurement of specific force (Section 1.4).

Quaternions and Vectors

Here we will show that the vector rotation formula can be expressed much more compactly in terms of quaternions. W. R. Hamilton (1805–1865) introduced the quaternion form:

$$x_0 + x_1 i + x_2 j + x_3 k$$

with,

$$i^2 = j^2 = k^2 = ijk = -1, \quad ij = k, jk = i, ki = j = -ik$$

in an attempt to generalize complex numbers in a plane to three dimensions. Quaternions obey the normal laws of algebra, except that multiplication is not commutative. Multiplication is defined by the associative law, for example, if,

$$r = (p_0 + p_1 i + p_2 j + p_3 k) \times (q_0 + q_1 i + q_2 j + q_3 k)$$

then,

$$r = p_0 q_0 + p_0 q_1 i + p_0 q_2 j + p_0 q_3 k + p_1 q_0 i + p_1 q_1 i^2 + \ldots$$

By using the rules for i, j, k products, and collecting terms, the answer can be written in various forms, for example,

$$\begin{bmatrix} r_0 \\ r_1 \\ r_2 \\ r_3 \end{bmatrix} = \begin{bmatrix} p_0 & -p_1 & -p_2 & -p_3 \\ p_1 & p_0 & -p_3 & p_2 \\ p_2 & p_3 & p_0 & -p_1 \\ p_3 & -p_2 & p_1 & p_0 \end{bmatrix} \begin{bmatrix} q_0 \\ q_1 \\ q_2 \\ q_3 \end{bmatrix}$$

Alternatively, by interpreting i, j, k as unit vectors, the quaternion can be treated as $(q_0 + \mathbf{q})$, where $\mathbf{q}$ is the quaternion vector part, with components q_1, q_2, q_3, along $\mathbf{i}, \mathbf{j}, \mathbf{k}$. We will write the quaternion as an array, formed from q_0 and the vector components, thus

$$p = \begin{bmatrix} p_0 \\ \mathbf{p}^r \end{bmatrix} \quad q = \begin{bmatrix} q_0 \\ \mathbf{q}^r \end{bmatrix}, \qquad (1.2\text{-}15)$$

where components of the vector are taken in a reference system r, to be chosen when the quaternion is applied. The above multiplication can be written as

$$p * q = \begin{bmatrix} p_0 q_0 - \mathbf{p} \cdot \mathbf{q} \\ (p_0 \mathbf{q} + q_0 \mathbf{p} + \mathbf{p} \times \mathbf{q})^r \end{bmatrix}, \qquad (1.2\text{-}16)$$

where "$*$" indicates quaternion multiplication. We will use (1.2-15) and (1.2-16) as the definitions of quaternions and quaternion multiplication. Quaternion properties can now be derived using ordinary vector operations.

Quaternion Properties

(i) Quaternion Noncommutativity
Consider the following identity:

$$p * q - q * p = \begin{bmatrix} 0 \\ (\mathbf{p} \times \mathbf{q} - \mathbf{q} \times \mathbf{p})^r \end{bmatrix} = \begin{bmatrix} 0 \\ 2(\mathbf{p} \times \mathbf{q})^r \end{bmatrix}$$

It is apparent that, in general,

$$p * q \neq q * p$$

(ii) The Quaternion Norm
The norm of a quaternion is defined to be the sum of the squares of its elements:

$$\text{norm}(q) = \sum_{i=0}^{i=3} q_i^2$$

(iii) Norm of a Product
Using the definition of the norm, and vector operations, it is straightforward to show (Problem 1.2-9) that the norm of a product is equal to the product of the individual norms:

$$\text{norm}(p * q) = \text{norm}(p) \times \text{norm}(q)$$

(iv) Associative Property over Multiplication The associative property:

$$(p * q) * r = p * (q * r)$$

is proven in a straightforward manner.

(v) The Quaternion Inverse Consider the following product,

$$\begin{bmatrix} q_0 \\ \mathbf{q}^r \end{bmatrix} * \begin{bmatrix} q_0 \\ -\mathbf{q}^r \end{bmatrix} = \begin{bmatrix} q_0^2 + \mathbf{q} \cdot \mathbf{q} \\ (q_0\mathbf{q} - q_0\mathbf{q} - \mathbf{q} \times \mathbf{q})^r \end{bmatrix} = \begin{bmatrix} \sum q_i^2 \\ 0 \\ 0 \\ 0 \end{bmatrix}$$

We see that multiplying a quaternion by another quaternion, which differs only by a change in sign of the vector part, produces a quaternion with a scalar part only. A quaternion of the latter form will have very simple properties in multiplication (i.e., multiplication by a constant) and, when divided by the quaternion norm, will serve as the "identity quaternion." Therefore, the inverse of a quaternion is defined by

$$q^{-1} = \begin{bmatrix} q_0 \\ \mathbf{q}^r \end{bmatrix}^{-1} = \frac{1}{\text{norm}(q)} \begin{bmatrix} q_0 \\ -\mathbf{q}^r \end{bmatrix} \qquad (1.2\text{-}17)$$

However, we will work entirely with unit-norm quaternions, thus simplifying many expressions.

(vi) Inverse of a Product The inverse of a quaternion product is given by the product of the individual inverses in the reverse order. This can be seen as follows:

$$(p * q)^{-1} = \frac{1}{\text{norm}(p * q)} \begin{bmatrix} p_0 q_0 - \mathbf{p} \cdot \mathbf{q} \\ -(p_0\mathbf{q} + q_0\mathbf{p} + \mathbf{p} \times \mathbf{q})^r \end{bmatrix}$$

$$= \frac{1}{\text{norm}(q)} \begin{bmatrix} q_0 \\ -\mathbf{q}^r \end{bmatrix} * \begin{bmatrix} p_0 \\ -\mathbf{p}^r \end{bmatrix} \frac{1}{\text{norm}(p)}$$

Therefore,

$$(p * q)^{-1} = q^{-1} * p^{-1}$$

Vector Rotation by Quaternions

A quaternion can be used to rotate a Euclidean vector in the same manner as the rotation formula, and the quaternion rotation is much simpler in form. The vector part of the quaternion is used to define the rotation axis, and the scalar part to define the angle of rotation. The rotation axis is specified by its direction cosines in the

18 THE KINEMATICS AND DYNAMICS OF AIRCRAFT MOTION

reference coordinate system, and it is convenient to impose a unity norm constraint on the quaternion. Therefore, if the direction angles of the axis are α, β, and γ, and a measure of the rotation angle is δ, the rotation quaternion is written as

$$q = \begin{bmatrix} \cos\delta \\ \cos\alpha\,\sin\delta \\ \cos\beta\,\sin\delta \\ \cos\gamma\,\sin\delta \end{bmatrix} = \begin{bmatrix} \cos\delta \\ \sin\delta\,\mathbf{n}^r \end{bmatrix}, \qquad (1.2\text{-}18)$$

where $\mathbf{n}$ is a unit vector along the rotation axis,

$$\mathbf{n}^r = \begin{bmatrix} \cos\alpha & \cos\beta & \cos\gamma \end{bmatrix}^T$$

and,

$$\mathrm{norm}(q) = \cos^2\delta + \sin^2\delta\,(\cos^2\alpha + \cos^2\beta + \cos^2\gamma) = 1$$

This formulation also guarantees that there is a unique quaternion for every value of δ in the range ± 180 degrees, thus encompassing all possible rotations.

Now consider the form of the transformation, which must involve multiplication. For compatibility of multiplication between vectors and quaternions, a Euclidean vector is written as a quaternion with a scalar part of zero, thus

$$u = \begin{bmatrix} 0 \\ \mathbf{u}^r \end{bmatrix}$$

The result of the rotation must also be a quaternion with a scalar part of zero, the transformation must be reversible by means of the quaternion inverse, and Euclidean length must be preserved. The transformation $v = q*u$ obviously does not satisfy the first of these requirements. Therefore, we consider the transformations:

$$v = q*u*q^{-1} \quad \text{or} \quad v = q^{-1}*u*q,$$

which are reversible by performing the inverse operations on v. The second of these transformations leads to the convention most commonly used:

$$v = q^{-1}*u*q = \begin{bmatrix} q_0(\mathbf{q}\cdot\mathbf{u}) - (q_0\mathbf{u} - \mathbf{q}\times\mathbf{u})\cdot\mathbf{q} \\ ((\mathbf{q}\cdot\mathbf{u})\mathbf{q} + q_0(q_0\mathbf{u} - \mathbf{q}\times\mathbf{u}) + (q_0\mathbf{u} - \mathbf{q}\times\mathbf{u})\times\mathbf{q})^r \end{bmatrix},$$

which reduces to

$$v = q^{-1}*u*q = \begin{bmatrix} 0 \\ (2\mathbf{q}(\mathbf{q}\cdot\mathbf{u}) + (q_0^2 - \mathbf{q}\cdot\mathbf{q})\mathbf{u} - 2q_0(\mathbf{q}\times\mathbf{u}))^r \end{bmatrix} \qquad (1.2\text{-}19)$$

Therefore, this transformation meets the requirement of zero scalar part. Also, because of the properties of quaternion norms, the Euclidean length is preserved. For a match with the rotation formula, we require:

rotation formula	*quaternion rotation*
$(1 - \cos \mu) \, \mathbf{n}(\mathbf{n} \cdot \mathbf{u})$	$2 \sin^2 \delta \, \mathbf{n}(\mathbf{n} \cdot \mathbf{u})$
$\cos \mu \, \mathbf{u}$	$(\cos^2 \delta - \sin^2 \delta) \, \mathbf{u}$
$-\sin \mu \, (\mathbf{n} \times \mathbf{u})$	$-2 \cos \delta \, \sin \delta \, (\mathbf{n} \times \mathbf{u})$

The corresponding terms agree if $\delta = \mu/2$ and half-angle trigonometric identities are applied. Therefore, the quaternion

$$q = \begin{bmatrix} \cos(\mu/2) \\ \sin(\mu/2)\mathbf{n}^r \end{bmatrix} \tag{1.2-20a}$$

and transformation

$$q^{-1} * u * q \tag{1.2-20b}$$

give a left-handed rotation of a vector $\mathbf{u}$ through an angle μ, around $\mathbf{n}$, when μ is positive.

1.3 MATRIX ANALYSIS OF KINEMATICS

Properties of Linear Transformations

Before studying matrix representation of kinematic relationships, we will review some pertinent matrix theory. Consider the matrix equation

$$v = Au, \tag{1.3-1}$$

where v and u are $(n \times 1)$ matrices (e.g., vector component arrays) and A is an $(n \times n)$ constant matrix, not necessarily nonsingular. Each element of v is a linear combination of the elements of u, and so this equation is a *linear transformation* of the matrix u. Next, suppose that in an analysis we change to a new set of variables through a reversible linear transformation. If L is the matrix of this transformation, then L^{-1} must exist (i.e., L is nonsingular) for the transformation to be reversible, and the new variables corresponding to u and v are

$$u_1 = Lu, \qquad v_1 = Lv$$

Therefore, the relationship between the new variables must be

$$v_1 = LAu = LAL^{-1}u_1 \tag{1.3-2a}$$

The transformation LAL^{-1} is a *similarity transformation* of the original coefficient matrix A. A special case of this transformation occurs when the inverse of the matrix L is given by its transpose (i.e., L is an orthogonal matrix) and the similarity transformation becomes a *congruence transformation*, LAL^T.

As an important example of a linear transformation, consider the linear state equation (1.1-3) with a nonsingular change of variables $z = Lx$. The state equation in terms of the z-variables is

$$\dot{z} = (LAL^{-1})z + (LB)u \qquad (1.3\text{-}2b)$$

and L can be chosen so that the state equations have a much simpler form, as shown below.

Eigenvalues and Eigenvectors

A square-matrix linear transformation has the property that vectors exist whose components are only scaled by the transformation. If **v** is such an "invariant" vector, its components must satisfy the equation

$$Av = \lambda v, \qquad v(n \times 1), \qquad (1.3\text{-}3)$$

where A is the transformation matrix and λ is a (scalar) constant of proportionality. A rearrangement of (1.3-3) gives the set of homogeneous linear equations

$$(A - \lambda I)v = 0, \qquad (1.3\text{-}4)$$

which has a non-null solution for v if, and only if, the determinant of the coefficient matrix is zero (Strang, 1980); that is,

$$|A - \lambda I| = 0 \qquad (1.3\text{-}5)$$

This determinant is an n-th order polynomial in λ, called the *characteristic polynomial* of A, so there may be up to n distinct solutions for λ. Each solution, λ_i, is known as an *eigenvalue* or *characteristic value* of the matrix A. The associated invariant vector defined by (1.3-3) is known as a *right eigenvector* of A (the left eigenvectors of A are the right eigenvectors of its transpose A^T).

In the mathematical model of a physical system, a reversible change of internal variables does not usually change the behavior of the system if observed at the same outputs. An example of this is the invariance of the eigenvalues of a linear system, described by Equation (1.1-3), under the similarity transformation (1.3-2b). After the similarity transformation, the eigenvalues are given by

$$\left|(\lambda I - LAL^{-1})\right| = 0,$$

which can be rewritten as

$$\left|(\lambda LL^{-1} - LAL^{-1})\right| = 0$$

The determinant of a product of square matrices is equal to the product of the individual determinants; therefore,

$$|L| \times |(\lambda I - A)| \times |L^{-1}| = 0 \tag{1.3-6}$$

This equation is satisfied by the eigenvalues of the matrix A, so the eigenvalues are unchanged by the transformation.

Now consider a special similarity transformation that will reduce the linear equations to a canonical (standard) form. First, consider the case when all of the n eigenvalues of the coefficient matrix A are distinct. Then the n eigenvectors $\mathbf{v}_i$ can be shown to form a linearly independent set; therefore, their components can be used to form the columns of a nonsingular transformation matrix. This matrix is called the *modal matrix*, M, and

$$M = [v_1 \; v_2 \cdots v_n]$$

According to the eigenvector/eigenvalue defining equation (1.3-3), if M is a modal matrix, we find that

$$AM = MJ \quad \text{and} \quad J = \text{diag}(\lambda_1 \cdots \lambda_n),$$

or

$$M^{-1}AM = J \tag{1.3-7a}$$

When some of the eigenvalues of A are repeated (i.e., multiple), it may not be possible to find a set of n linearly independent eigenvectors. Also, in the case of repeated eigenvalues, the result of the similarity transformation (1.3-7a) is in general a *Jordan-form matrix* (Wilkinson and Golub, 1976). In this case the matrix J may have some unit entries on the superdiagonal. These entries are associated with blocks of repeated eigenvalues on the main diagonal.

As an example, the linear state equation (1.3-2b), with $L^{-1} = M$, becomes

$$\dot{z} = Jz + M^{-1}Bu \tag{1.3-7b}$$

This corresponds to a set of state equations with minimal coupling between them. For example, if the eigenvalue λ_i is of multiplicity 2, and the associated Jordan block has a superdiagonal 1, we can write the corresponding equations as

$$\dot{z}_i = \lambda_i z_i + z_{i+1} + b'_i u$$
$$\dot{z}_{i+1} = \lambda_i z_i + b'_{i+1} u \tag{1.3-7c}$$

The variables z_i are called the *modal coordinates*. When the eigenvalues are all distinct, the modal coordinates yield a set of uncoupled first-order differential equations.

The Scalar Product

If $\mathbf{u}^a$ and $\mathbf{v}^a$ are column arrays of the same dimension, their scalar product is $(\mathbf{u}^a)^T \mathbf{v}^a$, for example,

$$(\mathbf{u}^a)^T \mathbf{v}^a = \begin{bmatrix} u_x & u_y & u_z \end{bmatrix} \begin{bmatrix} v_x \\ v_y \\ v_z \end{bmatrix} = u_x v_x + u_y v_y + u_z v_z \qquad (1.3\text{-}8a)$$

and this result is identical to (1.2-3) obtained from the vector dot product. The scalar product allows us to find the norm of a column matrix:

$$|\mathbf{v}^a| = \left((\mathbf{v}^a)^T \mathbf{v}^a\right)^{1/2} \qquad (1.3\text{-}8b)$$

In Euclidean space this is the length of a vector.

The Cross-Product Matrix

Suppose that the cross-product $\boldsymbol{\omega} \times \mathbf{v}$ is to be evaluated in system a, where $\boldsymbol{\omega}$ and $\mathbf{v}$ have components given by

$$\boldsymbol{\omega}^a = \begin{bmatrix} P \\ Q \\ R \end{bmatrix} \qquad \mathbf{v}^a = \begin{bmatrix} x \\ y \\ z \end{bmatrix}$$

Then it is easy to show (Problem 1.3-1), using the determinant formula for the cross-product, that

$$(\boldsymbol{\omega} \times \mathbf{v})^a = \begin{bmatrix} 0 & -R & Q \\ R & 0 & -P \\ -Q & P & 0 \end{bmatrix} \begin{bmatrix} x \\ y \\ z \end{bmatrix} \equiv \Omega^a \mathbf{v}^a \qquad (1.3\text{-}9)$$

The same idea can be applied to the vector triple product. For example,

$$(\boldsymbol{\omega} \times (\boldsymbol{\omega} \times \mathbf{v}))^a = \begin{bmatrix} 0 & -R & Q \\ R & 0 & -P \\ -Q & P & 0 \end{bmatrix}^2 \begin{bmatrix} x \\ y \\ z \end{bmatrix} \equiv (\Omega^a)^2 \mathbf{v}^a \qquad (1.3\text{-}10)$$

The symbol Ω will be used throughout to denote the *cross-product matrix* corresponding to the operation $(\boldsymbol{\omega} \times \,)$ when $\boldsymbol{\omega}$ is an angular velocity vector. For other vectors, a tilde symbol over the vector will be used to denote the cross-product matrix. A cross-product matrix is skew-symmetric, that is,

$$\Omega^T = -\Omega \equiv -\tilde{\omega} \qquad (1.3\text{-}11)$$

and therefore the square of the cross-product matrix is symmetric. Note that in the general case the matrix operations must be written in the same order as the vector operations, but may be performed in any order (the associative property for matrix multiplication).

Coordinate Rotation

When the rotation formula (1.2-6) is resolved in a coordinate system a, the result is

$$\mathbf{v}^a = \left[(1 - \cos \mu) \, \mathbf{n}^a (\mathbf{n}^a)^T + \cos(\mu) I - \sin(\mu) \, \tilde{\mathbf{n}}^a \right] \mathbf{u}^a, \qquad (1.3\text{-}12)$$

where $\mathbf{n}^a (\mathbf{n}^a)^T$ is a square matrix, I is the identity matrix, and $\tilde{\mathbf{n}}^a$ is a cross-product matrix. This formula was developed as an "active" vector operation in that a vector was being rotated to a new position by means of a left-handed rotation about the specified unit vector. In component form, the new array can be interpreted as the components of a new vector in the same coordinate system, or as the components of the original vector in a new coordinate system, obtained by a right-handed coordinate rotation around the specified axis. This can be visualized in Figure 1.3-1, which shows a b coordinate system obtained by a right-handed rotation around the z-axis. If the vector is next given a left-handed rotation through μ, then (x_b, y_b) will become the components in the original system. Taking the coordinate-system rotation viewpoint, and combining the matrices in (1.3-12) into a single coefficient matrix, this linear transformation can be written as

$$\mathbf{u}^b = C_{b/a} \, \mathbf{u}^a \qquad (1.3\text{-}13)$$

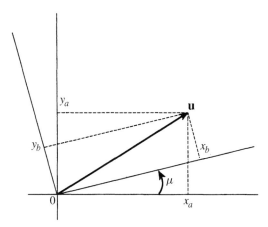

Figure 1.3-1 A plane rotation of coordinates.

Here $C_{b/a}$ is a matrix that transforms the components of the vector **u** from system a to system b, and is called a *direction cosine matrix*, or simply a *rotation matrix*.

We will look briefly at some of the properties of the rotation matrix, and then at how it may be determined in applications. A coordinate rotation must leave the length of a vector unchanged. The change of length under the rotation above is

$$\left(\mathbf{u}^b\right)^T \mathbf{u}^b = \left(C_{b/a}\mathbf{u}^a\right)^T C_{b/a}\mathbf{u}^a = \left(\mathbf{u}^a\right)^T C_{b/a}^T C_{b/a} \mathbf{u}^a$$

and the length is preserved if

$$C_{b/a}^T C_{b/a} = I = C_{b/a} C_{b/a}^T \qquad (1.3\text{-}14)$$

This is the definition of an orthogonal matrix, and it makes the inverse matrix particularly easy to determine ($C^{-1} = C^T$). It also implies that the columns (and also the rows) of the rotation matrix form an orthonormal set:

$$C = [c_1 \ c_2 \ c_3] \qquad c_i^T c_j = \begin{cases} 0, & i \neq j \\ 1, & i = j \end{cases}$$

Also, since

$$c_1 = C \begin{bmatrix} 1 \\ 0 \\ 0 \end{bmatrix}$$

columns of the rotation matrix give us the components, in the new system, of a unit vector in the old system.

If a vector is expressed in a new coordinate system by a sequence of rotations as

$$\mathbf{u}^d = C_{d/c} C_{c/b} C_{b/a} \mathbf{u}^a, \qquad (1.3\text{-}15)$$

then the inverse operation is given by

$$\mathbf{u}^a = \left(C_{d/c}C_{c/b}C_{b/a}\right)^{-1} \mathbf{u}^d = C_{b/a}^{-1}C_{c/b}^{-1}C_{d/c}^{-1}\mathbf{u}^d = C_{b/a}^T C_{c/b}^T C_{d/c}^T \mathbf{u}^d$$

or

$$\mathbf{u}^a = \left(C_{d/c}C_{c/b}C_{b/a}\right)^T \mathbf{u}^d = \left(C_{d/a}\right)^T \mathbf{u}^d \qquad (1.3\text{-}16)$$

A better understanding of coordinate rotations can be obtained by examining the eigenvalues of the (3 × 3) rotation matrix. Goldstein (1980) shows that any nontrivial rotation matrix has one, and only one, eigenvalue equal to +1, and that this corresponds to a theorem proved by Leonhard Euler (1707–1783) for a rigid body. The other two eigenvalues are a complex conjugate pair with unit magnitude, and can be written as ($\cos \phi \pm j \sin \phi$). Therefore, using a similarity transformation, and writing separate equations for the real and imaginary parts, it is possible to transform any rotation matrix C to the form of a plane rotation matrix P, for example,

$$P = \begin{bmatrix} \cos\phi & \sin\phi & 0 \\ -\sin\phi & \cos\phi & 0 \\ 0 & 0 & 1 \end{bmatrix} \qquad (1.3\text{-}17)$$

The matrix (1.3-17) corresponds to a single rotation through an angle ϕ about the z-axis. It shows that the orientation of one coordinate frame with respect to another is uniquely determined by a single rotation about a unique axis (the Euler axis), and this is the essence of Euler's theorem. This principle is used as the basis of the quaternion representation of rotation.

Summary of Rotation Matrix Properties

(i) Successive rotations can be described by the product of the individual rotation matrices; cf. (1.3-15).
(ii) Rotation matrices are not commutative, for example, $C_{c/b} C_{b/a} \neq C_{b/a} C_{c/b}$.
(iii) Rotation matrices are orthogonal matrices, for example, (1.3-14).
(iv) The determinant of a rotation matrix is unity.
(v) A nontrivial rotation matrix has one, and only one, eigenvalue equal to unity.

Euler Rotations

The direction cosine matrix is so-called because its elements can be determined from dot products that involve the direction cosines between corresponding axes of the new and old coordinate systems. Here we will determine the rotation matrix in a way that is better suited to visualizing aircraft orientation.

The orientation of one Cartesian coordinate system with respect to another can always be described by three successive rotations, and the angles of rotation are called the *Euler angles* (or Eulerian angles). These angles are specified in various ways in different fields of science, and the reader should be aware that there are small differences in many of the formulae in the literature as a result of this. In the aerospace field the rotations are performed, in a specified order, about each of the three Cartesian axes in succession. That is, they are performed in each of the three coordinate planes, and are therefore called *plane rotations*.

Figure 1.3-1 shows a plane rotation, in which coordinate system b has been rotated relative to system a. The systems are right-handed, with the z-axis coming out of the page, and the rotation is a right-handed rotation about the z-axis, through the angle μ. Assume that the components of the vector **u** are known in system a, and that we need to know its components in b. Equation (1.3-12) readily gives the rotation matrix, or simple trigonometry can be applied to the figure (Problem 1.3-2); the result is

$$\begin{bmatrix} x_b \\ y_b \\ z_b \end{bmatrix} = \begin{bmatrix} \cos\mu & \sin\mu & 0 \\ -\sin\mu & \cos\mu & 0 \\ 0 & 0 & 1 \end{bmatrix} \begin{bmatrix} x_a \\ y_a \\ z_a \end{bmatrix} \qquad (1.3\text{-}18)$$

Henceforth the plane rotation matrix will be written immediately by inspection. The unit and zero elements correspond to the coordinate that does not change, and the remaining elements are always cosines on the main diagonal and sines off the diagonal (so that zero rotation produces the identity matix). The negative sine element always occurs on the row above the one containing the unit element when the second system is reached by a right-handed rotation (note that the third row is considered as being above the first row). Note that changing the sign of the rotation angle yields the matrix transpose.

Three-dimensional coordinate rotations can now be built up as a sequence of plane rotations. The fact that the individual rotations are not commutative can be checked by performing sequences of rotations with any convenient solid object. Therefore, although the order of the sequence can be chosen arbitrarily, the same order must be maintained ever after. For example, standard aircraft practice is to describe the aircraft orientation by the z, y, x (also called 3, 2, 1) right-handed rotation sequence that is required to get from a reference system on the surface of the Earth into alignment with an aircraft body-fixed coordinate system. Therefore, starting from the reference system, the sequence of rotations is:

1. Right-handed rotation about the z-axis (positive ψ)
2. Right-handed rotation about the new y-axis (positive θ)
3. Right-handed rotation about the new x-axis (positive ϕ)

The reference system, on the Earth, normally has its z-axis pointing down and the aircraft axes are normally aligned forward, starboard, and down. Starting with the aircraft axes aligned with the corresponding reference axes, we see that this sequence corresponds first to a right-handed rotation around the aircraft z-axis, which is a positive "yaw." This is followed by a right-handed rotation around the aircraft y-axis, which is a positive "pitch," and a right-handed rotation around the aircraft x-axis, which is a positive "roll." Therefore, the rotations are often described as a yaw-pitch-roll sequence, starting from the reference system.

The plane rotation matrices can be written down immediately with the help of the rules established above. Thus, abbreviating cosine and sine to c and s, and using r and b to denote reference and body systems, we get

$$\mathbf{u}^b = \begin{bmatrix} 1 & 0 & 0 \\ 0 & c\phi & s\phi \\ 0 & -s\phi & c\phi \end{bmatrix} \begin{bmatrix} c\theta & 0 & -s\theta \\ 0 & 1 & 0 \\ s\theta & 0 & c\theta \end{bmatrix} \begin{bmatrix} c\psi & s\psi & 0 \\ -s\psi & c\psi & 0 \\ 0 & 0 & 1 \end{bmatrix} \mathbf{u}^r \quad (1.3\text{-}19)$$

Let $C_{b/r}$ denote the complete transformation from the reference system to the body system. Then, multiplying out these transformations, we get:

$$C_{b/r} = \begin{bmatrix} c\theta\, c\psi & c\theta\, s\psi & -s\theta \\ (-c\phi\, s\psi + s\phi\, s\theta\, c\psi) & (c\phi\, c\psi + s\phi\, s\theta\, s\psi) & s\phi\, c\theta \\ (s\phi\, s\psi + c\phi\, s\theta\, c\psi) & (-s\phi\, c\psi + c\phi\, s\theta\, s\psi) & c\phi\, c\theta \end{bmatrix} \quad (1.3\text{-}20)$$

This matrix represents a standard transformation, and will be used throughout the text.

The Euler angles are not unique for a given orientation. For example, imagine an aircraft performing a vertical loop with the pilot's head inside the loop. This could be represented by a pitch angle that is continuous in the range $-\pi < \theta \leq \pi$, and zero roll and yaw angles. Alternatively, we can restrict the pitch-attitude angle to $\pm\pi/2$ and, when the pitch attitude reaches $\pi/2$, we can allow the roll and yaw angles to change abruptly by π radians (inverted and heading in the opposite direction). The pitch attitude will then begin to decrease, passing through zero when the plane is at the top of the loop and reaching $-\pi/2$ when it is nose down, at which point the roll and yaw angles change back to zero. This is a more mathematically convenient choice, and so pitch is normally restricted to $\pm\pi/2$. The Euler angles are then unique, apart from the case when the pitch is exactly $\pm\pi/2$ and the roll and yaw are undefined during their abrupt transition.

Matrix Kinematic Relationships for Rotation

Given a set of time-varying Euler angles describing a rotating frame, it is not difficult to determine the components of the angular velocity vector. For example, let the orientation of a coordinate system in frame F_b, relative to a system in reference frame F_r, be described by the aircraft standard yaw (ψ), pitch (θ), roll (ϕ) sequence of Euler rotations. Also, let the Euler angles have derivatives $\dot{\psi}, \dot{\theta}, \dot{\phi}$. Starting from F_r, using two intermediate frames whose relative angular velocities are given by the Euler angle rates, and the additive property of angular velocity, we obtain

$$\boldsymbol{\omega}_{b/r}^b = \begin{bmatrix} \dot{\phi} \\ 0 \\ 0 \end{bmatrix} + C_\phi \left(\begin{bmatrix} 0 \\ \dot{\theta} \\ 0 \end{bmatrix} + C_\theta \begin{bmatrix} 0 \\ 0 \\ \dot{\psi} \end{bmatrix} \right),$$

where C_ϕ and C_θ are the right-handed plane rotations through the particular Euler angles, as given in Equation (1.3-19). After multiplying out the matrices, the final result is

$$\boldsymbol{\omega}_{b/r}^b \equiv \begin{bmatrix} P \\ Q \\ R \end{bmatrix} = \begin{bmatrix} 1 & 0 & -s\theta \\ 0 & c\phi & s\phi\, c\theta \\ 0 & -s\phi & c\phi\, c\theta \end{bmatrix} \begin{bmatrix} \dot{\phi} \\ \dot{\theta} \\ \dot{\psi} \end{bmatrix}, \qquad (1.3\text{-}21)$$

where P, Q, R are standard symbols for, respectively, the roll, pitch, and yaw rate components of the aircraft angular-velocity vector. The inverse transformation is

$$\begin{bmatrix} \dot{\phi} \\ \dot{\theta} \\ \dot{\psi} \end{bmatrix} = \begin{bmatrix} 1 & t\theta\, s\phi & t\theta\, c\phi \\ 0 & c\phi & -s\phi \\ 0 & s\phi/c\theta & c\phi/c\theta \end{bmatrix} \begin{bmatrix} P \\ Q \\ R \end{bmatrix} \qquad (1.3\text{-}22a)$$

We will use the following matrix notation for these equations:

$$\dot{\Phi} = H(\Phi)\omega^b_{b/r} \qquad (1.3\text{-}22b)$$

Equations (1.3-21) and (1.3-22) will be referred to as the *Euler kinematical equations*. Note that the Euler-angle derivatives are each in a different coordinate system, and so the array of derivatives does not represent the components of a vector. Therefore, the equations do not represent coordinate transformations, and the coefficient matrices are not orthogonal matrices. Note also that Equations (1.3-22) have a singularity when $\theta = \pm\pi/2$. In addition, if these equations are used in a simulation, the Euler-angle rates may integrate up to values outside the Euler-angle range. Therefore, logic to deal with this problem must be included in the computer code. Despite these disadvantages the Euler kinematical equations are commonly used in aircraft simulation.

An alternative set of kinematic equations can be derived as follows. The reference system to body-fixed coordinate system transformation was

$$\mathbf{u}^b = C_{b/r}\,\mathbf{u}^r$$

Performing the operations of matrix multiplication in terms of the columns of C shows us that a vector $\mathbf{c}_i$, whose (time-varying) components in F_b are given by the i-th column of $C_{b/r}$, represents a (fixed) unit vector in F_r. Now, applying the equation of Coriolis to the derivative of this vector in the two frames, we have

$$0 = {}^r\dot{\mathbf{c}}_i = {}^b\dot{\mathbf{c}}_i + \omega_{b/r} \times \mathbf{c}_i \qquad i = 1, 2, 3$$

Resolving in F_b,

$$0 = {}^b\dot{\mathbf{c}}_i^b + \Omega^b_{b/r}\mathbf{c}_i^b \qquad i = 1, 2, 3$$

The term ${}^b\dot{\mathbf{c}}_i^b$ is the derivative of the i-th column of $C_{b/r}$. If we combine the three equations into one matrix equation, the result is

$$\dot{C}_{b/r} = -\Omega^b_{b/r}\,C_{b/r} \qquad (1.3\text{-}23)$$

These equations are known as *Poisson's kinematical equations* or, in inertial navigation, as the *strapdown equation*. Whereas Equations (1.3-22) deal with the Euler angles, this equation deals directly with the elements of the rotation matrix. The components P, Q, R, of the angular velocity vector are, of course, contained in Ω. Compared to the Euler kinematical equations, the strapdown equation has the advantage of being singularity-free and the disadvantage of a large amount of redundancy (nine scalar equations).

When the strapdown equation is used in a simulation, the Euler angles are not directly available and must be calculated from the direction cosine matrix as follows. Let the elements of the rotation matrix (1.3-20) be denoted by c_{ij}. Then for this definition of Euler angles and rotation order, we see that

$$\theta = -\sin^{-1}(c_{13})$$
$$\phi = \operatorname{atan2}(c_{23}, c_{33}) \quad (1.3\text{-}24)$$
$$\psi = \operatorname{atan2}(c_{12}, c_{11}),$$

where atan2( ) is the four-quadrant inverse tangent function, available in most programming languages. These equations automatically put the Euler angles into the ranges discussed earlier.

Derivative of an Array

It is interesting to consider formulae for the derivative of an array, and look for a parallel to the equation of Coriolis. Starting from a time-varying coordinate transformation of the components of a general vector,

$$\mathbf{u}^b = C_{b/a}\,\mathbf{u}^a$$

with coordinate systems a and b fixed in F_a and F_b, differentiate the arrays on both sides of the equation. Differentiating the $\mathbf{u}^b$ array is equivalent to taking the derivative in F_b with components taken in system b, therefore,

$$^b\dot{\mathbf{u}}^b = C_{b/a}\,^a\dot{\mathbf{u}}^a + \dot{C}_{b/a}\,\mathbf{u}^a$$

or,

$$^b\dot{\mathbf{u}}^b = {}^a\dot{\mathbf{u}}^b + \dot{C}_{b/a}\,\mathbf{u}^a$$

Now use the Poisson equations to replace $\dot{C}_{b/a}$ (note that we used the equation of Coriolis to derive the Poisson equations, but they could have been derived in other ways),

$$^b\dot{\mathbf{u}}^b = {}^a\dot{\mathbf{u}}^b - \Omega^b_{b/a} C_{b/a} \mathbf{u}^a$$

or,

$$^b\dot{\mathbf{u}}^b = {}^a\dot{\mathbf{u}}^b + \Omega^b_{a/b}\mathbf{u}^b \quad (1.3\text{-}25)$$

Equation (1.3-25) is Equation (1.2-7) (the equation of Coriolis) resolved in coordinate system b.

Quaternion Coordinate Rotation

Referring to the quaternion rotation formulae (1.2-20) and the discussion of Equation (1.3-13), we again take the viewpoint that positive μ is a right-handed coordinate rotation rather than a left-handed rotation of a vector. We will define the quaternion that performs the coordinate rotation from system a to system b to be $q_{b/a}$, therefore,

$$q_{b/a} \equiv \begin{bmatrix} \cos(\mu/2) \\ \sin(\mu/2)\mathbf{n}^r \end{bmatrix} \quad (1.3\text{-}26a)$$

and the coordinate transformation is

$$\mathbf{u}^b = q_{b/a}^{-1} * \mathbf{u}^a * q_{b/a} \quad (1.3\text{-}26b)$$

Equation (1.3-26b) can take the place of the direction cosine matrix transformation (1.3-13), and the coordinate transformation is thus achieved by a single rotation around an axis aligned with the quaternion vector $\mathbf{n}\sin(\mu/2)$. Euler's theorem shows that the same coordinate rotation can be achieved by a plane rotation around the unique axis corresponding to an eigenvector of the rotation matrix. Therefore, the vector $\mathbf{n}$ must be parallel to this eigenvector, and so

$$\mathbf{n}^b = C_{b/a}\,\mathbf{n}^a = \mathbf{n}^a,$$

which shows that the quaternion vector part has the same components in system a or system b. In (1.3-26a) the reference coordinate system r may be either a or b. We will postpone, for the moment, the problem of finding the rotation quaternion without finding the direction cosine matrix and its eigenstructure, and instead examine the properties of the quaternion transformation.

Performing the inverse transformation to (1.3-26b) shows that

$$(q_{b/a})^{-1} = q_{a/b} \quad (1.3\text{-}27)$$

Also, for multiple transformations,

$$\mathbf{u}^c = q_{c/b}^{-1} * q_{b/a}^{-1} * \mathbf{u}^a * q_{b/a} * q_{c/b}, \quad (1.3\text{-}28)$$

which, because of the associative property, means that we can also perform this transformation with the single quaternion given by

$$q_{c/a}^{-1} = q_{c/b}^{-1} * q_{b/a}^{-1}$$

or,

$$q_{c/a} = q_{b/a} * q_{c/b} \quad (1.3\text{-}29)$$

The quaternion coordinate transformation (1.3-26b) actually involves more arithmetic operations than premultiplication of $\mathbf{u}^a$ by the direction cosine matrix. However, when the coordinate transformation is evolving with time, the time-update of the quaternion involves differential equations (following shortly) that are numerically preferable to the Euler kinematical equations and more efficient than the Poisson kinematical equations. In addition, the quaternion formulation avoids the singularity of the Euler equations.

In simulation and control, we often choose to keep track of orientation with a quaternion and construct the direction cosine matrix from the quaternion as needed.

It is easy to construct the quaternion for a simple plane rotation, but for a compound rotation (e.g., yaw, pitch, and roll combined) the quaternion rotation axis is not evident. Therefore, we initialize the quaternion from Euler angles or the direction cosine matrix. We now derive the relationships between the quaternion and the Euler angles and direction cosine matrix.

Direction Cosine Matrix from Quaternion

If we write the quaternion rotation formula (1.2-19) in terms of array operations, using the vector part of the quaternion, we get

$$\mathbf{u}^b = \left[2\mathbf{q}^a (\mathbf{q}^a)^T + \left(q_0^2 - (\mathbf{q}^a)^T \mathbf{q}^a \right) I - 2q_0 \tilde{\mathbf{q}}^a \right] \mathbf{u}^a \qquad (1.3\text{-}30)$$

The cross-product matrix $\tilde{\mathbf{q}}^a$ is given by

$$\tilde{\mathbf{q}}^a = \begin{bmatrix} 0 & -q_3 & q_2 \\ q_3 & 0 & -q_1 \\ -q_2 & q_1 & 0 \end{bmatrix} \qquad (1.3\text{-}31)$$

Now, evaluating the complete transformation matrix in (1.3-30), we find that

$$C_{b/a} = \begin{bmatrix} \left(q_0^2 + q_1^2 - q_2^2 - q_3^2 \right) & 2(q_1 q_2 + q_0 q_3) & 2(q_1 q_3 - q_0 q_2) \\ 2(q_1 q_2 - q_0 q_3) & \left(q_0^2 - q_1^2 + q_2^2 - q_3^2 \right) & 2(q_2 q_3 + q_0 q_1) \\ 2(q_1 q_3 + q_0 q_2) & 2(q_2 q_3 - q_0 q_1) & \left(q_0^2 - q_1^2 - q_2^2 + q_3^2 \right) \end{bmatrix}$$

$$(1.3\text{-}32)$$

This expression for the rotation matrix, in terms of quaternion parameters, corresponds to Equations (1.3-26) and the single right-handed rotation around $\mathbf{n}$, through the angle μ. Equation (1.3-32) is independent of any choice of Euler angles. We now show how a quaternion may be determined for any given sequence of Euler rotations.

Quaternion from Euler Angles

For the yaw, pitch, roll sequence described by (1.3-19) the quaternion formulation is

$$\mathbf{v}^b = q_{roll}^{-1} q_{pitch}^{-1} q_{yaw}^{-1} \mathbf{v}^r q_{yaw} q_{pitch} q_{roll}$$

If we think of this equation as three successive transformations, with pairs of quaternions, the rotation axes for the quaternions are immediately evident:

$$q_{yaw} = \begin{bmatrix} \cos(\psi/2) \\ 0 \\ 0 \\ \sin(\psi/2) \end{bmatrix} \quad q_{pitch} = \begin{bmatrix} \cos(\theta/2) \\ 0 \\ \sin(\theta/2) \\ 0 \end{bmatrix} \quad q_{roll} = \begin{bmatrix} \cos(\phi/2) \\ \sin(\phi/2) \\ 0 \\ 0 \end{bmatrix}$$

These transformations can be multiplied out, using quaternion multiplication, with only a minor amount of pain. The result is

$$q_0 = \pm(\cos\phi/2\ \cos\theta/2\ \cos\psi/2 + \sin\phi/2\ \sin\theta/2\ \sin\psi/2)$$
$$q_1 = \pm(\sin\phi/2\ \cos\theta/2\ \cos\psi/2 - \cos\phi/2\ \sin\theta/2\ \sin\psi/2)$$
$$q_2 = \pm(\cos\phi/2\ \sin\theta/2\ \cos\psi/2 + \sin\phi/2\ \cos\theta/2\ \sin\psi/2)$$
$$q_3 = \pm(\cos\phi/2\ \cos\theta/2\ \sin\psi/2 - \sin\phi/2\ \sin\theta/2\ \cos\psi/2)$$

(1.3-33)

and these are the elements of $q_{b/r}$. A plus or minus sign has been added to these equations because neither (1.3-26b) nor (1.3-32) is affected by the choice of sign. The same choice of sign must be used in all of Equations (1.3-33).

Quaternion from Direction Cosine Matrix

The quaternion parameters can also be calculated from the elements $\{c_{ij}\}$ of the general direction cosine matrix. If terms on the main diagonal of (1.3-32) are combined, the following relationships are obtained:

$$4q_0^2 = 1 + c_{11} + c_{22} + c_{33}$$
$$4q_1^2 = 1 + c_{11} - c_{22} - c_{33}$$
$$4q_2^2 = 1 - c_{11} + c_{22} - c_{33}$$
$$4q_3^2 = 1 - c_{11} - c_{22} + c_{33}$$

(1.3-34a)

These relationships give the magnitudes of the quaternion elements but not the signs. The off-diagonal terms in (1.3-32) yield the additional relationships

$$4q_0 q_1 = c_{23} - c_{32} \qquad 4q_1 q_2 = c_{12} + c_{21}$$
$$4q_0 q_2 = c_{31} - c_{13} \qquad 4q_2 q_3 = c_{23} + c_{32}$$
$$4q_0 q_3 = c_{12} - c_{21} \qquad 4q_1 q_3 = c_{13} + c_{31}$$

(1.3-34b)

From the first set of equations, (1.3-34a), the quaternion element with the largest magnitude (at least one of the four must be nonzero) can be selected. The sign associated with the square root can be chosen arbitrarily, and then this variable can be used as a divisor with (1.3-34b) to find the remaining quaternion elements. An interesting quirk of this algorithm is that the quaternion may change sign if the algorithm is restarted with a new set of initial conditions. This will have no effect on the rotation matrix given in (1.3-32). Algorithms like this are discussed in Shoemake (1985) and Shepperd (1978).

The Quaternion Kinematical Equations

When two frames are in relative angular motion, and we wish to keep track of the relative orientation by means of a quaternion, a method is required for continuously updating the quaternion. This takes the form of a differential equation for the quaternion, with the coefficients determined from the relative angular rates. The equation is analagous to the Euler and Poisson kinematical equations.

Let the orientation of a rotating frame F_b, relative to a reference frame F_r, be given, at time t, by the quaternion $q_{b/r}(t)$. Also, as in Figure 1.2-3, let the instantaneous angular velocity of F_b be in the direction of a unit vector $\hat{s}$, with magnitude ω. Then, in a small time interval δt, the quaternion $\delta q_{b/r}$, which describes the incremental coordinate rotation around $\hat{s}$, can be found by using small angle approximations in (1.3-26a):

$$\delta q_{b/r}(\delta t) \approx \begin{bmatrix} 1 \\ \hat{s}^b \, \omega \delta t / 2 \end{bmatrix}$$

At time $t + \delta t$ the rotation is given by the quaternion $q_{b/r}(t + \delta t)$, where

$$q_{b/r}(t + \delta t) = q_{b/r}(t) * \delta q_{b/r}(\delta t)$$

(Note that the order of the multiplication matches (1.3-29)). By definition the derivative of $q_{b/r}(t)$ is, temporarily omitting the subscripts,

$$\frac{dq}{dt} = \lim_{\delta t \to 0} \frac{q(t) * [\delta q - I_q]}{\delta t},$$

where I_q is the identity quaternion. Substituting for δq gives

$$\frac{dq}{dt} = \tfrac{1}{2} q(t) * \begin{bmatrix} 0 \\ \hat{s}^b \omega \end{bmatrix} = \tfrac{1}{2} q(t) * \boldsymbol{\omega}^b$$

This result can be written formally as:

$$\dot{q}_{b/r} = \tfrac{1}{2} q_{b/r} * \boldsymbol{\omega}^b_{b/r} \tag{1.3-35}$$

Replacing the quaternion multiplication by matrix multiplication, Equation (1.3-35) can be put into the form

$$\dot{q} = \tfrac{1}{2} \begin{bmatrix} 0 & -\boldsymbol{\omega}^T \\ \boldsymbol{\omega} & -\Omega \end{bmatrix} \begin{bmatrix} q_0 \\ \mathbf{q} \end{bmatrix}$$

Writing this out in full, using the body-system components of $\boldsymbol{\omega}_{b/r}$, gives

$$\begin{bmatrix} \dot{q}_0 \\ \dot{q}_1 \\ \dot{q}_2 \\ \dot{q}_3 \end{bmatrix} = \frac{1}{2} \begin{bmatrix} 0 & -P & -Q & -R \\ P & 0 & R & -Q \\ Q & -R & 0 & P \\ R & Q & -P & 0 \end{bmatrix} \begin{bmatrix} q_0 \\ q_1 \\ q_2 \\ q_3 \end{bmatrix} \quad (1.3\text{-}36)$$

This equation is widely used in simulation of rigid-body motion, and in discrete form it is used in digital attitude control systems (e.g., for satellites) and for inertial navigation digital processing.

1.4 GEODESY, EARTH'S GRAVITATION, TERRESTRIAL NAVIGATION

Geodesy is a branch of mathematics that deals with the shape and area of the Earth. Some ideas and facts from geodesy are needed to simulate the motion of an aerospace vehicle around the Earth. In addition, some knowledge of the Earth's gravitation is required. Useful references are *Encyclopaedia Britannica* (1987), Heiskanen and Moritz (1967), Kuebler and Sommers (1981), NIMA (1997), and Vanicek and Krakiwsky (1982).

The Shape and Gravitation of the Earth, WGS-84

Simulation of high-speed flight over large areas of the Earth's surface, with accurate equations of motion and precise calculation of position, requires an accurate model of the Earth's shape, rotation, and gravity. The shape of the Earth can be well modeled by an ellipsoid of revolution (i.e., a spheroid). The polar radius of the Earth is approximately 21 km less than the equatorial radius, so the generating ellipse must be rotated about its minor axis, to produce an oblate spheroidal model. Organizations from many countries participate in making accurate measurements of the parameters of such models. In the United States the current model is the Department of Defense World Geodetic System 1984, or WGS-84, and the agency responsible for supporting this model is the National Imagery and Mapping Agency (NIMA) (NIMA, 1997). The Global Positioning System (GPS) relies on WGS-84 for the ephemerides of its satellites.

The equipotential surface of the Earth's gravity field that coincides with the undisturbed mean sea level, extended continuously underneath the continents, is called the *geoid*. Earth's irregular mass distribution causes the geoid to be an undulating surface, and this is illustrated in Figure 1.4-1. Note that the *local vertical* is defined by the direction in which a plumb-bob hangs and is accurately normal to the geoid. The angle that it makes with the spheroid normal is called the *deflection of the vertical*, and is usually less than 10 arc-sec (the largest deflections over the entire Earth are about 1 arc-min).

The WGS-84 spheroid has its center at the Earth's cm, and was originally (1976–1979 data) a least-squares best fit to the geoid. More recent estimates have slightly changed the "best fit" parameters, but the current WGS-84 spheroid now uses the

GEODESY, EARTH'S GRAVITATION, TERRESTRIAL NAVIGATION   35

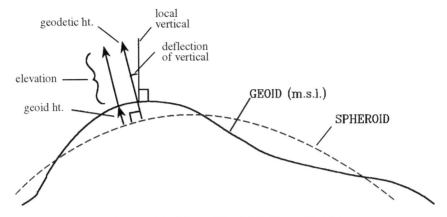

Figure 1.4-1 The geoid and definitions of height.

original parameters as its defining values. Based on a 1° by 1° (latitude, longitude) worldwide grid, the rms deviation of the geoid from the spheroid is only about 30 m! Figure 1.4-2 shows the oblate spheroidal model of the Earth, with the oblateness greatly exaggerated. In the figure, a and b are, respectively, the semimajor and semiminor axes of the generating ellipse. Two other parameters of the ellipse (not

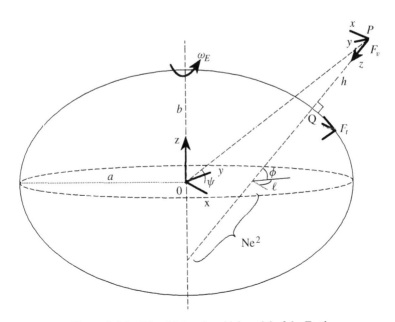

Figure 1.4-2 The oblate spheroidal model of the Earth.

shown) are its flattening, f, and its eccentricity, e. The WGS-84 defined and derived values are:

$$a \equiv 6{,}378{,}137.0 \text{ m} \qquad \text{(defined)} \qquad (1.4\text{-}1a)$$

$$f = \frac{a-b}{a} \equiv 1/298.257223563 \qquad \text{(defined)} \qquad (1.4\text{-}1b)$$

$$b = 6{,}356{,}752 \text{ m} \qquad \text{(derived)} \qquad (1.4\text{-}1c)$$

$$e = \frac{(a^2 - b^2)^{1/2}}{a} \approx .08181919 \qquad \text{(derived)} \qquad (1.4\text{-}1d)$$

Two additional parameters are used to define the complete WGS-84 reference frame; these are a fixed Earth rotation rate, ω_E, and the Earth's gravitational constant (GM) with the mass of the atmosphere included. In WGS-84 they are defined to be:

$$\omega_E \equiv 7.2921150 \times 10^{-5} \text{ rad/s} \qquad (1.4\text{-}1e)$$

$$GM \equiv 3986004.418 \times 10^8 \text{ m}^3/\text{s}^2 \qquad (1.4\text{-}1f)$$

The ω_E value is the sidereal rate of rotation, that is, the "inertial" rate relative to the "fixed" stars (Kaplan, 1981).

Frames and Coordinate Systems

Table 1.4-1 shows the frames and coordinate systems that will be used with Figure 1.4-2. Note that a "geographic" coordinate system has its axes aligned east, north, and up (ENU), or north, east, and down (NED), where "up" or "down" means along the spheroid normal at the system location. It is also called a local navigational system, and the symbol n is used to denote the components in this system. The stability and wind axes systems will be defined in Chapter 2.

Geocentric Coordinates

Geocentric coordinates are referenced to the common origin of the ECI and ECEF systems. Thus, in Figure 1.4-2, the dashed line *OP* represents the *geocentric radius* of *P*, and the angle ψ (measured positive north) is the *geocentric latitude* of *P*. Longitude is measured in the equatorial plane, from one axis of the ECI or ECEF system, to the projection of *P* on the equatorial plane. We will assume that the ECEF *x*-axis points to the zero-longitude meridian, and the *terrestrial longitude*, ℓ, (positive east) is shown in the figure. *Celestial longitude*, λ, is measured from the ECI *x*-axis, which is aligned with some celestial reference direction such as a line from the Sun's cm to the Earth's position in orbit at vernal equinox. In a given time interval, an increment in celestial longitude is equal to the increment in terrestrial longitude plus the increment in Earth's rotation angle. This can be written as

$$\lambda - \lambda_0 = \ell - \ell_0 + \omega_E t, \qquad (1.4\text{-}2)$$

TABLE 1.4-1 Frames and Coordinate Systems Used with Figure 1.4-2

Frame of Reference	Coordinate Systems
F_i, an "inertial" frame, nonrotating but translating with Earth's cm	ECI (Earth-centered inertial), origin at Earth's cm, axes in the equatorial plane and along the spin axis
F_e, a frame defined by the "rigid" Earth	ECEF (Earth-centered, Earth-fixed), axes in the equatorial plane and along the spin axis; Tangent-plane system, a geographic system with its origin on the Earth's surface
F_v, a frame translating with the vehicle cm, in which north, east, and down, represent fixed directions	Vehicle-carried system, a geographic system with its origin at the vehicle cm
F_b, a "body" frame defined by the "rigid" vehicle	Vehicle body-fixed system, origin at vehicle cm, axes aligned with vehicle reference directions; Vehicle stability-axes system; Vehicle wind-axes system

where λ_0 and ℓ_0 are the values at $t = 0$. Absolute celestial longitude is often unimportant, and $\lambda_0 \equiv 0$ can be used.

Geodetic Coordinates

Referring to Figure 1.4-2, *geodetic position* over the surface of the Earth, as used for maps and navigation, is determined by using a normal to the spheroid. *Geodetic latitude,* ϕ, is the angle that the normal makes with the geodetic equatorial plane, and is positive in the Northern Hemisphere. *Geodetic height, h,* is the height above the spheroid, along the normal, as shown in Figure 1.4-1. It can be determined from a database of tabulated geoid height versus latitude and longitude, plus the elevation above mean sea level (msl). The elevation above msl is in turn obtained from a barometric altimeter, or from the land elevation (in a *hypsographic database*) plus the altitude above land (e.g., radar altimeter).

Navigation Calculations

Two important parameters of the spheroid are required for navigation calculations, namely, the radii of curvature. The *meridian radius of curvature, M*, is the radius of curvature in a meridian plane, that is, the radius of curvature of the generating ellipse. Calculations of the radius of curvature for an ellipse can be found in calculus texts and, in terms of geodetic latitude, it is easy to show that M is given by

$$M = \frac{a(1 - e^2)}{(1 - e^2 \sin^2 \phi)^{3/2}} \tag{1.4-3}$$

A radius of curvature, integrated with respect to angle, gives arc length. In this case the integral cannot be found in closed form, and it is much easier to compute distance over the Earth approximately using spherical triangles. The usefulness of the radius of curvature lies in calculating components of velocity. Thus, at geodetic height, h, the geographic-system north component of velocity over the Earth is related to latitude rate by

$$V_N = (M + h)\dot{\phi} \tag{1.4-4}$$

The *prime vertical radius of curvature, N*, is the radius of curvature in a plane containing the spheroid normal and a normal to the meridian plane. It is equal to the distance along the normal, from the spheroid surface to the semiminor axis, and is given by

$$N = \frac{a}{(1 - e^2 \sin^2 \phi)^{1/2}} \tag{1.4-5}$$

Again, N is useful for calculating velocity components. If we take the component of N parallel to the equatorial plane, we obtain the radius of a constant-latitude circle. Therefore, the geographic-system east component of velocity over the Earth is related to longitude rate by

$$V_E = (N + h) \cos(\phi) \dot{\ell} \tag{1.4-6}$$

Cartesian position coordinates (ECI or ECEF) can be readily calculated from the prime vertical radius of curvature. The projection of N on the x-y plane gives the x and y components. The z-component can be found by dividing N into its parts (Problem 1.4-3) above and below the x-y plane:

$$N = \underset{\text{(below } x\text{-}y)}{Ne^2} + \underset{\text{(above } x\text{-}y)}{N(1 - e^2)} \tag{1.4-7}$$

Therefore, ECEF position can be calculated from geodetic coordinates by

$$\mathbf{p}^e = \begin{bmatrix} (N + h) \cos(\phi) \cos(\ell) \\ (N + h) \cos(\phi) \sin(\ell) \\ \left[N(1 - e^2) + h\right] \sin(\phi) \end{bmatrix}, \tag{1.4-8}$$

where superscript e indicates ECEF coordinates. Position in ECI coordinates is of the same form as (1.4-8), but with celestial longitude λ replacing terrestrial longitude ℓ.

The reverse of the above transformation is to find the geodetic coordinates from Cartesian coordinates. An exact formula exists but requires the solution of a quartic equation in $\tan(\phi)$ (Vanicek and Krakiwsky, 1982). Therefore, an iterative algorithm is often used. Referring to Figure 1.4-2, we see that

$$\sin \phi = \frac{z}{N(1 - e^2) + h} \tag{1.4-9}$$

Using the large triangle whose hypotenuse is $(N+h)$, and sides $\sqrt{(x^2+y^2)}$, $[z + Ne^2 \sin(\phi)]$, we can write

$$\tan \phi = \frac{[z + Ne^2 \sin \phi]}{\sqrt{(x^2 + y^2)}} \quad (1.4\text{-}10)$$

If (1.4-9) is substituted for $\sin(\phi)$ in (1.4-10) and simplified, we obtain

$$\tan \phi = \frac{z}{\sqrt{(x^2 + y^2)}[1 - Ne^2/(N+h)]}$$

Because N is a function of ϕ, this formula is implicit in ϕ, but it can be used in the following iterative algorithm for the geodetic coordinates:

$$\ell = \operatorname{atan2}(y, x)$$

$$h = 0, \quad N = a$$

$$\begin{aligned}
\rightarrow \phi &= \tan^{-1}\left[\frac{z}{(x^2+y^2)^{1/2}\,[1 - Ne^2/(N+h)]}\right] \\
N &= \frac{a}{(1 - e^2 \sin^2 \phi)^{1/2}} \quad (1.4\text{-}11) \\
(N+h) &= \frac{\sqrt{(x^2+y^2)}}{\cos \phi}
\end{aligned}$$

$$h = (N + h) - N$$

Latitudes of $\pm 90°$ must be dealt with as a special case, but elsewhere the iterations converge very rapidly.

Earth-Related Coordinate Transformations

The aircraft equations of motion will require the coordinate rotation matrices between the three systems defined above. The rotation between ECEF and ECI is a plane rotation around the z-axis. Equation (1.4-2) gives the rotation angle as

$$\mu = \lambda_0 - \ell_0 + \omega_E t$$

Therefore, the rotation from ECI to ECEF can be written as

$$C_{e/i} = \begin{bmatrix} \cos \mu & \sin \mu & 0 \\ -\sin \mu & \cos \mu & 0 \\ 0 & 0 & 1 \end{bmatrix}, \quad (1.4\text{-}12)$$

where subscripts i and e will be used, respectively, to indicate ECI and ECEF systems.

When going from the ECEF to a geographic system, the convention is to perform the longitude rotation first. Now imagine the ECEF system moved to the equator at the correct longitude; a left-handed rotation through 90 degrees, around the y-axis, is needed to get the x-axis pointing north and the z-axis down. It is then only necessary to move to the correct latitude and fall into alignment with the NED system by means of an additional left-handed rotation around the y-axis, through the latitude angle. Therefore, the transformation is

$$C_{n/e} = \begin{bmatrix} c\phi & 0 & s\phi \\ 0 & 1 & 0 \\ -s\phi & 0 & c\phi \end{bmatrix} \begin{bmatrix} 0 & 0 & 1 \\ 0 & 1 & 0 \\ -1 & 0 & 0 \end{bmatrix} \begin{bmatrix} c\ell & s\ell & 0 \\ -s\ell & c\ell & 0 \\ 0 & 0 & 1 \end{bmatrix}$$

or

$$C_{n/e} = \begin{bmatrix} -s\phi\, c\ell & -s\phi\, s\ell & c\phi \\ -s\ell & c\ell & 0 \\ -c\phi\, c\ell & -c\phi\, s\ell & -s\phi \end{bmatrix}, \qquad (1.4\text{-}13)$$

where the n subscript indicates a geographic (local navigational) system.

Gravitation

The term *gravitation* denotes a mass attraction effect, as distinct from *gravity*, meaning the combination of mass attraction and centrifugal force experienced by a body constrained to move with the Earth's surface. Our most accurate equations of motion will contain the centripetal acceleration as a separate term.

The WGS-84 datum includes an amazingly detailed model of the Earth's gravitation. This model is in the form of a (scalar) potential function V, such that a component of specific mass-attraction force along each of three axes can be found from the gradients of the potential function. The current potential function, for use with WGS-84, is Earth Gravitational Model 1996 (EGM96). This has 130,676 coefficients and is intended for very precise satellite and missile calculations. The largest coefficient is two orders of magnitude bigger than the next coefficient and, if we retain only the largest coefficient, the result is still a very accurate model. Neglecting these coefficients removes the dependence on terrestrial longitude, leaving the following potential function:

$$V = \frac{GM}{r}\left[1 - 0.5 J_2 (a/r)^2 (3\sin^2\psi - 1)\right] \qquad (1.4\text{-}14)$$

in which r is radial distance from the Earth's cm, and ψ is the geocentric latitude. The Earth's gravitational constant GM is the product of the Earth's mass and the universal gravitational constant of the inverse square law. Its EGM96 value, with the mass of the atmosphere included, was given in Equation (1.4-1f). The constant J_2 is given by

$$J_2 = -\sqrt{5}\bar{C}_{2,0} = 1.0826267 \times 10^{-3}, \tag{1.4-15}$$

where $\bar{C}_{2,0}$ is the actual EGM96 coefficient.

The gradients of the potential function are easily evaluated in geocentric coordinates (the same as NED coordinates but with the z-axis pointing to the cm). When this is done and the results are transformed into the ECEF system, we obtain the following gravitation model:

$$\mathbf{G}^e = \frac{-GM}{r^2} \begin{bmatrix} \left[1 + 1.5 J_2 (a/r)^2 (1 - 5\sin^2\psi)\right] p_x/r \\ \left[1 + 1.5 J_2 (a/r)^2 (1 - 5\sin^2\psi)\right] p_y/r \\ \left[1 + 1.5 J_2 (a/r)^2 (3 - 5\sin^2\psi)\right] p_z/r \end{bmatrix}, \tag{1.4-16}$$

where r is equal to the length of the geocentric position vector $\mathbf{p}_e$, whose ECEF components are p_x, p_y, p_z. This model is accurate to about 30–35 $\times\ 10^{-3}$ cm/s^2 rms, but local deviations can be quite large. Note that the x and y components are identical because there is no longitude dependence. The geocentric latitude is given by

$$\sin\psi = p_z/|\mathbf{p}_e|$$

The model can also be converted to geodetic coordinates. A useful relationship is

$$\tan\psi = (1-n)\tan\phi, \quad \text{where} \quad n = Ne^2/(N+h) \tag{1.4-17}$$

The weight of an object on Earth is determined by the gravitational attraction ($m\mathbf{G}$) minus the force needed to produce the centripetal acceleration at the Earth's surface ($m\boldsymbol{\omega}_{e/i} \times (\boldsymbol{\omega}_{e/i} \times \mathbf{p}_e)$). Dividing the weight of the object by its mass gives the *gravity* vector $\mathbf{g}$. Therefore, the vector equation for $\mathbf{g}$ is

$$\mathbf{g} = \mathbf{G} - \boldsymbol{\omega}_{e/i} \times (\boldsymbol{\omega}_{e/i} \times \mathbf{p}_e) \tag{1.4-18}$$

As noted earlier, at the Earth's surface $\mathbf{g}$ is accurately normal to the geoid and defines the local vertical. When Equation (1.4-16) is substituted for $\mathbf{G}$ in (1.4-18), and the equation is resolved in the NED system, we find that $\mathbf{g}$ is almost entirely along the down axis with a variable north component of only a few micro-g's. This is a modeling error, since deflection of the vertical is not explicitly included in the model. The down component of $\mathbf{g}$ given by the model, at the Earth's surface, varies sinusoidally from 9.780 m/s^2 at the equator to 9.806 m/s^2 at 45° geodetic latitude, and 9.832 m/s^2 at the poles. When a constant value of gravity is to be used (e.g., in a simulation), the value at 45° latitude is taken as the standard value of gravity (actually defined to be 9.80665 m/s^2). Our simplified "flat-Earth" equations of motion will use a $\mathbf{g}$ vector that has only a down component, and is measured at the Earth's surface.

1.5 RIGID-BODY DYNAMICS

Angular Motion

By using the vehicle cm as a reference point, the rotational dynamics of the aircraft can be separated from the translational dynamics (Wells, 1967). Here, we develop the equations for the rotational dynamics, which will be the same for both the flat-Earth and oblate-rotating-Earth equations of motion. The following definitions will be needed:

F_i = an inertial reference frame
F_b = a body-fixed frame in the rigid vehicle
$\mathbf{v}_{CM/i}$ = velocity of vehicle cm in F_i
$\boldsymbol{\omega}_{b/i}$ = angular velocity of F_b with respect to F_i
$\mathbf{M}_{A,T}$ = sum of aerodynamic and thrust moments at the cm

The moment is generated by aerodynamic effects, by any reaction-control thrusters, and by any components of the engine thrust not acting through the cm (e.g., due to thrust-vectoring control).

Let the angular momentum vector of a rigid body, in the inertial frame and taken about the cm, be denoted by $\mathbf{h}$. It is shown in textbooks on classical mechanics (Goldstein, 1980) that the derivative of $\mathbf{h}$, taken in the inertial frame, is equal to the vector torque or moment $\mathbf{M}_{A,T}$, applied about the cm. Therefore, analogously to Newton's law for translational momentum, we write

$$\mathbf{M}_{A,T} = {}^i\dot{\mathbf{h}} \qquad (1.5\text{-}1)$$

In order to determine the angular momentum vector, consider an element of mass δm with position vector $\mathbf{r}$ relative to the cm. Its inertial velocity is given by

$$\mathbf{v} = \mathbf{v}_{CM/i} + \boldsymbol{\omega}_{b/i} \times \mathbf{r}$$

The angular momentum of this particle, about the cm, is the moment of the translational momentum about the cm, or

$$\delta \mathbf{h} = \mathbf{r} \times \mathbf{v}\delta m = \mathbf{r} \times \mathbf{v}_{CM/i}\delta m + \mathbf{r} \times (\boldsymbol{\omega}_{b/i} \times \mathbf{r})\,\delta m$$

When this equation is integrated over all the elements of mass, the first term will disappear. This is because $\mathbf{v}_{CM/i}$ is constant for the purposes of the integration and can be taken outside the integral, and the integral of $\mathbf{r}dm$ is zero by the definition of the cm. If the vector triple product formula is applied to the remaining term, and the integration is over all of the elements of mass, the result is:

$$\mathbf{h} = \boldsymbol{\omega}_{b/i} \int (\mathbf{r} \cdot \mathbf{r})\,dm - \int \mathbf{r}(\mathbf{r} \cdot \boldsymbol{\omega}_{b/i})\,dm$$

We must now choose a coordinate system in which to perform the integration. The easiest choice is a body-fixed system in which **r** has constant components. Therefore, let

$$\boldsymbol{\omega}_{b/i}^{b} = \begin{bmatrix} P \\ Q \\ R \end{bmatrix}, \quad \text{and} \quad \mathbf{r}^{b} = \begin{bmatrix} x \\ y \\ z \end{bmatrix}$$

so that

$$d\mathbf{h}^{b} = \begin{bmatrix} P \\ Q \\ R \end{bmatrix} (x^{2} + y^{2} + z^{2}) \, dm - \begin{bmatrix} x \\ y \\ z \end{bmatrix} (Px + Qy + Rz) \, dm$$

and the integral over the whole vehicle can be written as

$$\mathbf{h}^{b} = \begin{bmatrix} P \int (y^{2} + z^{2}) \, dm - Q \int xy \, dm - R \int xz \, dm \\ Q \int (x^{2} + z^{2}) \, dm - R \int yz \, dm - P \int yx \, dm \\ R \int (x^{2} + y^{2}) \, dm - P \int zx \, dm - Q \int zy \, dm \end{bmatrix}$$

The various integrals in the angular momentum components are defined to be the moments and cross-products of inertia, for example,

$$\text{moment of inertia about } x\text{-axis} = J_{xx} = \int (y^{2} + z^{2}) \, dm$$

$$\text{cross-product of inertia } J_{xy} \equiv J_{yx} = \int xy \, dm$$

On substituting these definitions into the angular momentum, we obtain expressions for the components of $\mathbf{h}^{b}$ that are bilinear in P, Q, R, and the inertias. This allows $\mathbf{h}^{b}$ to be written as the matrix product:

$$\mathbf{h}^{b} = \begin{bmatrix} J_{xx} & -J_{xy} & -J_{xz} \\ -J_{xy} & J_{yy} & -J_{yz} \\ -J_{xz} & -J_{yz} & J_{zz} \end{bmatrix} \begin{bmatrix} P \\ Q \\ R \end{bmatrix} \equiv J^{b} \boldsymbol{\omega}_{b/i}^{b} \qquad (1.5\text{-}2)$$

The matrix J will be referred to as the *inertia matrix* of the rigid body. It can be calculated or experimentally determined, and is a constant matrix when calculated in body-fixed coordinates for a body with a fixed distribution of mass. It was necessary to choose a coordinate system to obtain this matrix and, consequently, it is not possible to obtain a vector equation of motion that is completely coordinate-free. In more advanced treatments this paradox is avoided by the use of tensors. Note also that J is a real symmetric matrix, and therefore has special properties that are discussed below.

With the angular momentum expressed in terms of the inertia matrix and angular velocity vector of the complete rigid body, Equation (1.5-1) can be evaluated. Since the inertia matrix is known, and constant in the body frame, it will be convenient to replace the derivative in (1.5-1) by a derivative taken in the body frame:

$$\mathbf{M}_{A,T} = {}^i\dot{\mathbf{h}} = {}^b\dot{\mathbf{h}} + \boldsymbol{\omega}_{b/i} \times \mathbf{h}$$

Now, differentiating (1.5-2) in F_b, with J constant, and taking body-fixed components, we obtain,

$$\mathbf{M}^b_{A,T} = J^b \, {}^b\dot{\boldsymbol{\omega}}^b_{b/i} + \Omega^b_{b/i} J^b \boldsymbol{\omega}^b_{b/i},$$

where $\Omega^b_{b/i}$ is a cross-product matrix for $\boldsymbol{\omega}^b_{b/i}$. A rearrangement of this equation gives the *state equation for angular velocity*:

$$ {}^b\dot{\boldsymbol{\omega}}^b_{b/i} = \left(J^b\right)^{-1} \left[\mathbf{M}^b_{A,T} - \Omega^b_{b/i} J^b \boldsymbol{\omega}^b_{b/i}\right] \tag{1.5-3}$$

This state equation is widely used in simulation and analysis of rigid-body motion, from satellites to ships. It can be solved numerically for $\boldsymbol{\omega}^b_{b/i}$ given the inertia matrix and the torque vector, and its features will now be described.

The assumption that the inertia matrix is constant is not always completely true. For example, with aircraft the inertias may change slowly as fuel is transferred and burned. Also, the inertias may change abruptly if an aircraft is engaged in dropping stores. These effects can usually be adequately accounted for in a simulation by simply changing the inertias in (1.5-3) without accounting for their rates of change. As far as aircraft control system design is concerned, point designs are done for particular flight conditions, and interpolation between point designs can be used when the aircraft mass properties change. This is more likely to be done to deal with movement of the vehicle cm and the resultant effect on static stability.

The inverse of the inertia matrix occurs in (1.5-3), and because of symmetry this has a relatively simple form:

$$J^{-1} = \frac{1}{\Delta} \begin{bmatrix} k_1 & k_2 & k_3 \\ k_2 & k_4 & k_5 \\ k_3 & k_5 & k_6 \end{bmatrix}, \tag{1.5-4}$$

where

$$k_1 = \left(J_{yy}J_{zz} - J_{yz}^2\right)/\Delta \qquad k_2 = \left(J_{yz}J_{zx} + J_{xy}J_{zz}\right)/\Delta$$
$$k_3 = \left(J_{xy}J_{yz} + J_{zx}J_{yy}\right)/\Delta \qquad k_4 = \left(J_{zz}J_{xx} - J_{zx}^2\right)/\Delta$$
$$k_5 = \left(J_{xy}J_{zx} + J_{yz}J_{xx}\right)/\Delta \qquad k_6 = \left(J_{xx}J_{yy} - J_{xy}^2\right)/\Delta$$

and

$$\Delta = J_{xx}J_{yy}J_{zz} - 2J_{xy}J_{yz}J_{zx} - J_{xx}J_{yz}^2 - J_{yy}J_{zx}^2 - J_{zz}J_{xy}^2$$

A real symmetric matrix has real eigenvalues and, furthermore, a repeated eigenvalue of order p still has associated with it p linearly independent eigenvectors. Therefore, a similarity transformation can be found that reduces the matrix to a real diagonal form. In the case of the inertia matrix this means that we can find a set of coordinate axes in which the inertia matrix is diagonal. These axes are called the *principal axes*.

In principal axes the inverse of the inertia matrix is also diagonal and the angular velocity state equation takes its simplest form, known as *Euler's equations of motion*. If the torque vector has body-axes components given by

$$\mathbf{M}^b_{A,T} = \begin{bmatrix} \ell \\ m \\ n \end{bmatrix}, \qquad (1.5\text{-}5)$$

then Euler's equations are

$$\dot{P} = \frac{(J_y - J_z)\, QR}{J_x} + \frac{\ell}{J_x}$$

$$\dot{Q} = \frac{(J_z - J_x)\, RP}{J_y} + \frac{m}{J_y} \qquad (1.5\text{-}6)$$

$$\dot{R} = \frac{(J_x - J_y)\, PQ}{J_z} + \frac{n}{J_z}$$

The equations involve cyclic permutation of the rate and inertia components; they are inherently coupled because angular rates about any two axes produce an angular acceleration about the third. This *inertia coupling* has important consequences for aircraft maneuvering rapidly at high angles of attack; we examine its effects in Chapter 4. The stability properties of the Euler equations are interesting and will be studied in Problem (1.5-1).

The angular velocity state equation is again simplified when applied to aircraft since for most aircraft the x-z plane is a plane of symmetry. Under this condition, for every product $y_i z_j$ or $y_i x_j$ in an inertia computation there is a product that is identical in magnitude but opposite in sign. Therefore, only the J_{xz} cross-product of inertia is nonzero. A notable exception is an oblique-wing aircraft (Travassos et al., 1980), which does not have a plane of symmetry. Under the plane-of-symmetry assumption the inertia matrix and its inverse reduce to

$$J = \begin{bmatrix} J_x & 0 & -J_{xz} \\ 0 & J_y & 0 \\ -J_{xz} & 0 & J_z \end{bmatrix}, \quad J^{-1} = \frac{1}{\Gamma} \begin{bmatrix} J_z & 0 & J_{xz} \\ 0 & \Gamma/J_y & 0 \\ J_{xz} & 0 & J_x \end{bmatrix}, \qquad (1.5\text{-}7)$$

where $\Gamma = (J_x J_z - J_{xz}^2)$ and the double-subscript notation on the moments of inertia has been dropped.

46   THE KINEMATICS AND DYNAMICS OF AIRCRAFT MOTION

If the angular velocity state equation (1.5-3) is expanded using the torque vector (1.5-5) and the simple inertia matrix given by (1.5-7), the result is:

$$\Gamma \dot{P} = J_{xz}\left[J_x - J_y + J_z\right] PQ - \left[J_z(J_z - J_y) + J_{xz}^2\right] QR + J_z \ell + J_{xz} n$$

$$J_y \dot{Q} = (J_z - J_x) PR - J_{xz}(P^2 - R^2) + m \qquad (1.5\text{-}8)$$

$$\Gamma \dot{R} = \left[(J_x - J_y)J_x + J_{xz}^2\right] PQ - J_{xz}\left[J_x - J_y + J_z\right] QR + J_{xz} \ell + J_x n$$

In the analysis of angular motion we have so far neglected the angular momentum of any spinning rotors. Technically this violates the rigid-body assumption, but the resulting equations are valid. Note that, strictly, we require axial symmetry of the spinning rotors, otherwise the position of the vehicle cm will vary. This is not a restrictive requirement because it is also a requirement for dynamically balancing the rotors. The effects of the additional angular momentum may be quite significant. For example, a number of World War I aircraft had a single "rotary" engine that had a fixed crankshaft and rotating cylinders. The gyroscopic effects caused by the large angular momentum of the engine gave these aircraft tricky handling characteristics. In the case of a small jet with a single turbofan engine on the longitudinal axis, the effects are smaller. To represent the effect, a constant vector can be added to the angular momentum vector in (1.5-2). Thus,

$$\mathbf{h}^b = J^b \omega_{b/i}^b + \begin{bmatrix} h_x \\ h_y \\ h_z \end{bmatrix} \qquad (1.5\text{-}9a)$$

If this analysis is carried through, the effect is to add the following terms, respectively, to the right-hand sides of the three equations (1.5-8):

$$J_z(Rh_y - Qh_z) + J_{xz}(Qh_x - Ph_y)$$
$$-Rh_x + Ph_z \qquad (1.5\text{-}9b)$$
$$J_{xz}(Rh_y - Qh_z) + J_x(Qh_x - Ph_y)$$

To complete the set of equations for angular motion, a kinematic equation is required that describes the rigid-body orientation. The changing orientation is a result of the nonzero (in general) angular rates that satisfy the state equation (1.5-3). The kinematics may be described by:

(a) Euler's kinematical equations (1.3-21/22)
(b) Poisson's kinematical equations (1.3-23)
(c) the quaternion kinematical equations (1.3-35/36)

For example, the quaternion state equation

$$\dot{q}_{b/i} = \tfrac{1}{2} q_{b/i} * \omega_{b/i}^{b} \tag{1.5-10}$$

can be solved simultaneously with the angular velocity state equation, for a total of seven state variables. The direction cosine matrix $C_{b/i}$ and the Euler angles will be required in a complete simulation, and these can be derived from the quaternion using (1.3-32) and (1.3-24). The quaternion can be initialized from the initial Euler angles using (1.3-33).

This completes the discussion of the angular motion dynamics. We now turn our attention to the translational motion of the cm.

Translational Motion of the Center of Mass

The first step will be to find the inertial acceleration of the vehicle cm so that Newton's second law may be applied. The frames and coordinate systems related to Figure 1.4-2 are required. These include the ECI system fixed in the inertial frame F_i, the ECEF system fixed in F_e, the NED geographic system in F_v, and the vehicle body-fixed system in frame F_b. In addition to the definitions in the angular motion subsection, we must define the following vectors:

$\mathbf{p}_{CM/O}$ = vehicle cm position relative to ECI origin
$\mathbf{v}_{CM/i} = {}^{i}\dot{\mathbf{p}}_{CM/O}$ = velocity of the cm in F_i
$\mathbf{v}_{CM/e} = {}^{e}\dot{\mathbf{p}}_{CM/O}$ = velocity of the cm in F_e
$\omega_{x/y}$ = angular velocity of frame x with respect to frame y
$\mathbf{F}_{A,T}$ = vector sum of aerodynamic and thrust forces at cm

Note that because $\mathbf{p}_{CM/O}$ is a position vector in both F_i and F_e, $\mathbf{v}_{CM/i}$ and $\mathbf{v}_{CM/e}$ are both velocity vectors.

By using the equation of Coriolis to relate the derivatives of $\mathbf{p}_{CM/O}$ in F_i and F_e, we find that

$$\mathbf{v}_{CM/i} = {}^{i}\dot{\mathbf{p}}_{CM/O} = \mathbf{v}_{CM/e} + \omega_{e/i} \times \mathbf{p}_{CM/O} \tag{1.5-11}$$

Newton's second law applied to the motion of the cm, and neglecting the rate of change of mass of the vehicle, gives

$$(1/m)\,\mathbf{F}_{A,T} + \mathbf{G} = {}^{i}\dot{\mathbf{v}}_{CM/i}$$

where m is the mass of the vehicle.

Now differentiate (1.5-11) and substitute for the right-hand side of this equation. The Earth's angular velocity vector is constant for the differentiation and, introducing the derivative of $\mathbf{v}_{CM/e}$ taken in the body frame, gives

$$(1/m)\,\mathbf{F}_{A,T} + \mathbf{G} = {}^{b}\dot{\mathbf{v}}_{CM/e} + \omega_{b/i} \times \mathbf{v}_{CM/e} + \omega_{e/i} \times {}^{i}\dot{\mathbf{p}}_{CM/O}$$

Substitute (1.5-11) for the inertial position derivative:

48 THE KINEMATICS AND DYNAMICS OF AIRCRAFT MOTION

$$(1/m)\,\mathbf{F}_{A,T} + \mathbf{G} = {}^b\dot{\mathbf{v}}_{CM/e} + (\boldsymbol{\omega}_{b/i} + \boldsymbol{\omega}_{e/i}) \times \mathbf{v}_{CM/e} + \boldsymbol{\omega}_{e/i} \times (\boldsymbol{\omega}_{e/i} \times \mathbf{p}_{CM/O})$$

or

$${}^b\dot{\mathbf{v}}_{CM/e} = (1/m)\,\mathbf{F}_{A,T} + \mathbf{g} - (\boldsymbol{\omega}_{b/i} + \boldsymbol{\omega}_{e/i}) \times \mathbf{v}_{CM/e}, \quad (1.5\text{-}12)$$

where

$$\mathbf{g} = \mathbf{G} - \boldsymbol{\omega}_{e/i} \times (\boldsymbol{\omega}_{e/i} \times \mathbf{p}_{CM/O})$$

The kinematic equation (1.5-11) and the dynamic equation (1.5-12) will provide two matrix state equations. When (1.5-11) is resolved in the ECI system, it relates the derivatives of the ECI components of the inertial position vector to themselves and the velocity in F_e. The velocity in F_e must be obtained from the simultaneous solution of (1.5-12). Similarly, Equation (1.5-12) must be resolved in the vehicle body-fixed coordinate system so that it relates derivatives of body-fixed components to themselves and the aerodynamic and thrust forces. The body-fixed components are needed to determine the aerodynamic and thrust forces. Equation (1.5-12) also requires the inertial position vector and the angular velocity vector, and is therefore coupled to (1.5-11) and the angular velocity equation (1.5-3). Because (1.5-11) and (1.5-12) are resolved in different coordinate systems, the direction cosine matrix $C_{b/i}$, obtained from the quaternion (1.5-10), is needed to transform components between these equations. This means that the translational motion equations must be solved simultaneously with the angular motion equations. The complete set of equations of motion, in component form, will be assembled after we have examined the importance of the Earth-rotation terms in (1.5-12).

Significance of the Earth-Rotation Terms

The third term on the right-hand side of (1.5-12) is where these equations will differ from the flat-Earth equations that we will derive. Using the additive property of angular velocities, Equation (1.5-12) can also be written as

$${}^b\dot{\mathbf{v}}_{CM/e} = (1/m)\,\mathbf{F}_{A,T} + \mathbf{g} - (\boldsymbol{\omega}_{b/v} + \boldsymbol{\omega}_{v/e} + 2\boldsymbol{\omega}_{e/i}) \times \mathbf{v}_{CM/e} \quad (1.5\text{-}13)$$

The angular velocity $\boldsymbol{\omega}_{b/v}$ is null when the vehicle is not maneuvering; the angular velocity $\boldsymbol{\omega}_{v/e}$ can be obtained from geodetic latitude and longitude rates, which can in turn be determined from the NED components of $\mathbf{v}_{CM/e}$ using Equations (1.4-4) and (1.4-6). The term $2\boldsymbol{\omega}_{e/i} \times \mathbf{v}_{CM/e}$ is the Coriolis acceleration, introduced in Equation (1.2-11).

We will use $\boldsymbol{\omega}_{v/e}$ to illustrate how an angular velocity vector can be determined in general. The additive property of angular velocity can be used in conjunction with intermediate frames whose angular velocities are more easily determined. Imagine an intermediate vehicle-carried frame, F_{vi}, at the same longitude as the vehicle in question, but at zero latitude. Summation of angular velocities gives

$$\boldsymbol{\omega}_{v/e} = \boldsymbol{\omega}_{v/vi} + \boldsymbol{\omega}_{vi/e}$$

The geographic components of these angular velocities are easily found from the latitude and longitude rates. The transformation from the geographic system in F_{vi} to that in F_v is a left-handed rotation around the east axis through the latitude angle ϕ. Therefore we obtain

$$\boldsymbol{\omega}_{v/e}^n = \begin{bmatrix} 0 \\ -\dot{\phi} \\ 0 \end{bmatrix} + \begin{bmatrix} c\phi & 0 & s\phi \\ 0 & 1 & 0 \\ -s\phi & 0 & c\phi \end{bmatrix} \begin{bmatrix} \dot{\ell} \\ 0 \\ 0 \end{bmatrix}$$

The latitude and longitude rates are given by (1.4-4) and (1.4-6) and are often adequately approximated by letting $(M + h) = (N + h) = R$, with $R = 21 \times 10^6$ ft. Then

$$\boldsymbol{\omega}_{v/e}^n = 1/R \begin{bmatrix} V_E \\ -V_N \\ -V_E \tan\phi \end{bmatrix} \qquad (1.5\text{-}14a)$$

In ECEF or ECI coordinates the angular velocity $\boldsymbol{\omega}_{e/i}$ has only a z component, equal to the Earth's spin rate, ω_E. The NED components can be found by using the coordinate rotation (1.4-13):

$$\boldsymbol{\omega}_{e/i}^n = \begin{bmatrix} \omega_E \cos\phi \\ \phi \\ -\omega_E \sin\phi \end{bmatrix} \qquad (1.5\text{-}14b)$$

Now, adding (1.5-14a) to twice (1.5-14b), we can form a cross-product matrix to premultiply the NED components of $\mathbf{v}_{CM/e}$, and hence evaluate the cross-product terms in the equation of motion (1.5-13) when the vehicle is not maneuvering. Letting the NED components of $\mathbf{v}_{CM/e}$ be $[V_N \, V_E \, V_D]^T$, the result is very cumbersome, so we will take the special case where the vehicle is flying due east at constant altitude; then $V_N = V_D = 0$. The result is

$$\left(\Omega_{v/e}^n + 2\Omega_{e/i}^n\right)\mathbf{v}_{CM/e}^n = \begin{bmatrix} V_E^2 \tan(\phi)/R + 2V_E\omega_E \sin\phi \\ 0 \\ V_E^2/R + 2V_E\omega_E \cos\phi \end{bmatrix} \qquad (1.5\text{-}15)$$

The centripetal term V_E^2/R is equal to the Coriolis term $2V_E\omega_E$ at about 3000 ft/s. At zero latitude, the down component in (1.5-15) is about 0.9 ft/s^2, which is to be compared with $g_D = 32.2$ ft/s^2. In the "flat-Earth" equations of motion, the Coriolis terms are omitted and the equations have significant errors for velocities over the Earth with magnitude greater than about 2000 ft/s.

The Oblate Rotating-Earth Equations of Motion

We now return to the task of assembling a set of state equations. Position derivatives are obtained from the kinematic equation (1.5-11), resolved in ECI coordinates.

THE KINEMATICS AND DYNAMICS OF AIRCRAFT MOTION

Derivatives of translational velocity components are found from the Newton's law equation (1.5-12) in body-fixed coordinates, and derivatives of angular velocity components are found from the angular velocity equation (1.5-3) as it stands. A time-varying coordinate rotation from ECI to body-fixed coordinates will be needed, and so we will apply the quaternion differential equations (1.5-10), and construct the direction cosine matrix from the quaternion using Equation (1.3-32). This leads to the following set of state equations:

$$C_{b/i} = fn(q_{b/i}) \quad [\text{auxiliary eqn., see 1.3-32}] \quad (1.5\text{-}16a)$$

$$\dot{q}_{b/i} = \tfrac{1}{2} q_{b/i} * \omega^b_{b/i} \quad (1.5\text{-}16b)$$

$$^i\dot{\mathbf{p}}^i_{CM/O} = C_{i/b} \mathbf{v}^b_{CM/e} + \Omega^i_{e/i} \mathbf{p}^i_{CM/O} \quad (1.5\text{-}16c)$$

$$^b\dot{\mathbf{v}}^b_{CM/e} = (1/m)\mathbf{F}^b_{A,T} - \left(\Omega^b_{b/i} + \Omega^b_{e/i}\right) \mathbf{v}^b_{CM/e} + C_{b/i}\mathbf{g}^i \quad (1.5\text{-}16d)$$

$$^b\dot{\omega}^b_{b/i} = (J^b)^{-1}\left[\mathbf{M}^b_{A,T} - \Omega^b_{b/i} J^b \omega^b_{b/i}\right] \quad (1.5\text{-}16e)$$

The state vector for this set of simultaneous differential equations is given by

$$X^T = \left[\left(\mathbf{p}^i_{CM/O}\right)^T (q_{b/i})^T \left(\mathbf{v}^b_{CM/e}\right)^T \left(\omega^b_{b/i}\right)^T\right]$$

The auxiliary equation for the direction cosine matrix must first be calculated from the state vector, before the state equations can be evaluated. Given the mass properties of the vehicle (m and J^b), and the forces and moments, all of the terms on the right-hand side of Equations (1.5–16) should be determined by the state vector.

On the right-hand side of (1.5-16c) $\Omega^i_{e/i}$ is the cross-product matrix for the ECI components of Earth's angular velocity vector $[0, 0, \omega_E]^T$. In Equation (1.5-16d) a new cross-product matrix for the body-fixed components of this angular velocity is needed. This is given by a similarity transformation:

$$\Omega^b_{e/i} = C_{b/i} \Omega^i_{e/i} C_{i/b} \quad (1.5\text{-}17)$$

since each cross-product matrix is a linear transformation of components in one coordinate system. A simpler calculation is to form a cross-product matrix for

$$\left(\omega^b_{b/i} + C_{b/i}\omega^i_{e/i}\right)$$

and postmultiply it by $\mathbf{v}^b_{CM/e}$.

For convenience the gravity vector has been left in terms of ECI components. Because our gravitation model has no longitude dependence, these can be used instead of ECEF components. If the gravity term must be computed accurately at high altitude, or over a wide range of latitude, we must use

$$\mathbf{g}^b = C_{b/i}\left(\mathbf{G}^i - (\Omega^i_{e/i})^2 \mathbf{p}^i_{CM/O}\right) \quad (1.5\text{-}18)$$

with the gravitation model (1.4-16).

In Equations (1.5-16d) and (1.5-16e) models of the aerodynamic and thrust forces and moments are needed, as derived in Chapter 2. This is where the control inputs enter the model, as throttle settings, aerodynamic control surface deflections, and so on. These forces and moments also depend on the velocity of the vehicle relative to the surrounding air. Therefore, we define a relative velocity vector, $\mathbf{v}_{rel}$, by

$$\mathbf{v}_{rel} = \mathbf{v}_{CM/e} - \mathbf{v}_{W/e}, \qquad (1.5\text{-}19a)$$

where $\mathbf{v}_{W/e}$ is the wind velocity taken in F_e. Since the wind is normally specified in NED components, and body-fixed components are required for aerodynamic calculations, we will calculate $\mathbf{v}_{rel}^b$ from

$$\mathbf{v}_{rel}^b = \mathbf{v}_{CM/e}^b - C_{b/n}\,\mathbf{v}_{W/e}^n \qquad (1.5\text{-}19b)$$

Equation (1.5-19b) requires $C_{b/n}$, from which we can also find yaw, pitch, and roll Euler angles that describe the attitude of the vehicle relative to the vehicle-carried geographic system. These are usually needed for control purposes, and are calculated as follows.

The algorithm (1.4-11) can be used to find geodetic altitude and latitude, and celestial longitude is found from the inertial position vector. Geodetic altitude can be used to determine the atmospheric properties required for the aerodynamic calculations. The direction cosine matrix $C_{b/n}$ can be computed using

$$C_{b/n} = C_{b/i}\,C_{i/n}, \qquad (1.5\text{-}20)$$

where $C_{i/n}$ is found from (1.4-13) with λ replacing ℓ. Equations (1.3-24) can be used to obtain the attitude Euler angles from $C_{b/n}$:

$$\begin{aligned}
\phi &= \operatorname{atan2}(c_{23}, c_{33}) \\
\theta &= -\sin^{-1}(c_{13}) \\
\psi &= \operatorname{atan2}(c_{12}, c_{11})
\end{aligned} \qquad (1.5\text{-}21)$$

Finally, the terrestrial longitude can be calculated from celestial longitude using Equation (1.4-2). Hence position over the Earth is specified.

Equations (1.5-16) and the auxiliary equations (1.5-17) through (1.5-21) constitute the equations needed to simulate the motion of a vehicle around the oblate rotating Earth. They should be used when an accurate simulation is required for a vehicle flying faster than about 2000 ft/s over the Earth, or when accurate long-distance navigation is being simulated. They apply to any type of rigid aerospace vehicle; differences between vehicles begin to appear when the various forces and torques, acting on the vehicle, are modeled. For example simulation of a satellite might require models for gravity-gradient torque, radiation pressure, and residual atmospheric drag. In the next subsection we will derive the so-called flat-Earth equations of motion as a subset of these equations.

The Flat-Earth Equations of Motion

For low-speed flight simulation of aircraft flying over a small region of the Earth, and with no requirement for precise simulation of position, it is usual to neglect the centripetal and Coriolis terms in Equation (1.5-13), as described earlier. Neglecting the centripetal term is equivalent to pretending that the Earth is flat ($R \to \infty$ in (1.5-14a)), and neglecting the Coriolis term is equivalent to assuming that the Earth is an inertial frame. The vehicle-carried frame F_v now has zero angular velocity relative to F_e, and so $\omega_{b/v} \equiv \omega_{b/e}$, and the geographic coordinate system in F_v remains aligned with a tangent-plane system in the vicinity of the vehicle. Vehicle attitude can be described, relative to the tangent plane, by yaw, pitch, and roll angles and a direction-cosine matrix $C_{b/n}$ (vehicle body with respect to the local navigational system). Position can conveniently be measured from the origin of the tangent-plane system, T. Therefore, Equations (1.5-16) become:

$$C_{b/n} = fn(\Phi) \quad \text{(from 1.3-20)} \tag{1.5-22a}$$

$$^e\dot{\mathbf{p}}^n_{CM/T} = C_{n/b}\mathbf{v}^b_{CM/e} \tag{1.5-22b}$$

$$\dot{\Phi} = H(\Phi)\omega^b_{b/e} \quad \text{(from 1.3-22)} \tag{1.5-22c}$$

$$^b\dot{\mathbf{v}}^b_{CM/e} = (1/m)\,\mathbf{F}^b_{A,T} + C_{b/n}\mathbf{g}^n - \Omega^b_{b/e}\mathbf{v}^b_{CM/e} \tag{1.5-22d}$$

$$^b\dot{\omega}^b_{b/e} = (J^b)^{-1}\left[\mathbf{M}^b_{A,T} - \Omega^b_{b/e}J^b\omega^b_{b/e}\right], \tag{1.5-22e}$$

where

$\mathbf{p}_{CM/T}$ = vehicle cm position relative to tangent-system origin
$\mathbf{v}_{CM/e} = {}^e\dot{\mathbf{p}}_{CM/T}$ = cm velocity vector in F_e
$\omega_{b/e}$ = angular velocity of F_b with respect to F_e
Φ = Euler angles of body-fixed system relative to NED system

The equation for the velocity vector relative to the surrounding air becomes

$$\mathbf{v}^b_{rel} = \mathbf{v}^b_{CM/e} - C_{b/n}\,\mathbf{v}^n_{W/e} \tag{1.5-23}$$

Gravity appears in the velocity state equation, and in tangent system components this is

$$\mathbf{g}^t = [0 \quad 0 \quad g_D] \tag{1.5-24}$$

with g_D equal to the standard gravity (9.80665 m/s^2), or the local value. The state vector can be seen to be

$$X^T = \left[\left(\mathbf{p}^n_{CM/T}\right)^T \quad \Phi^T \quad \left(\mathbf{v}^b_{CM/e}\right)^T \quad \left(\omega^b_{b/e}\right)^T\right] \tag{1.5-25}$$

The vector form of the relative velocity equation is (1.5-19a). If this equation is differentiated in the body-fixed frame, and used to eliminate $\mathbf{v}_{CM/e}$ and its derivative from the vector equation for the translational acceleration, we obtain

$$^b\dot{\mathbf{v}}_{rel} = (1/m)\,\mathbf{F}_{A,T} + \mathbf{g} - \boldsymbol{\omega}_{b/e} \times \mathbf{v}_{rel} - {}^e\dot{\mathbf{v}}_{W/e}, \qquad (1.5\text{-}26a)$$

where a term $\boldsymbol{\omega}_{b/e} \times \mathbf{v}_{W/e}$ has been canceled from each side of the equation. The last term on the right can be used as a way of introducing gust inputs into the model, or can be set to zero for steady winds. Taking the latter course, and introducing components in the body-fixed system, gives

$$^b\dot{\mathbf{v}}^b_{rel} = (1/m)\,\mathbf{F}^b_{A,T} + C_{b/n}\mathbf{g}^n - \Omega^b_{b/e}\mathbf{v}^b_{rel} \qquad (1.5\text{-}26b)$$

This equation is an alternative to (1.5-22d), and Equation (1.5-22b) must then be modified to use the sum of the relative and wind velocities.

The dynamic behavior of the rigid vehicle is determined by the force and moment equations and the attitude kinematic equation. There is only a weak dependence on altitude, and hence only a weak coupling to Equation (1.5-22b). Furthermore, it is shown in Chapter 2 that the aerodynamic forces and moments depend on the velocity relative to the air mass, with only a weak dependence on altitude. Therefore, the dynamic behavior is essentially determined by $\mathbf{v}_{rel}$, or its negative, the *relative wind*, and is independent of the steady wind velocity. In Chapter 2 we will use (1.5-26) to make a model that is suitable for studying the dynamic behavior.

Equations (1.5-22b–d) are twelve coupled, nonlinear, first-order differential equations. Chapter 2 shows how they can be "solved" analytically. Chapter 3 shows how they can be solved simultaneously by numerical integration for the purposes of flight simulation. Coupling exists because angular acceleration integrates to angular velocity, which determines the Euler angle rates, which in turn determine the direction cosine matrix. The direction cosine matrix is involved in the state equations for position and velocity, and position (the altitude component) and velocity determine aerodynamic effects which determine angular acceleration. Coupling is also present through the translational velocity. These interrelationships will become more apparent in Chapter 2.

1.6 SUMMARY

This chapter provides sufficient background material to enable the reader to deal with many of the dynamical problems that occur in the modern aerospace industry. Classical mechanics is the key to the analysis and solution of many of these problems, and we have reviewed and used many of the vector operations from classical mechanics. Coordinate rotations are everywhere in the analysis and in the software used for computer control of many systems. Therefore, we have attempted to provide an easy approach to setting up the rotations. The use of quaternions for coordinate rotation is very popular in satellite and missile problems, and in computer graphics. We have provided background material in this area because the quaternion avoids any numerical singularity problems with "all-attitude" simulation of aircraft.

The gravity model presented here is more detailed than usual for aircraft simulations and gives the reader an introduction to the more detailed modeling that would be needed to simulate accurate navigation, or simulate spacecraft and launch and reentry vehicles. The rotating-Earth equations of motion may be applied to vehicles intended to reach hypersonic speeds and perhaps go into orbit, or to slowly moving aircraft near the surface of the Earth. The design of these vehicles requires large computer simulations involving the equations of motion, and controlling them may require programming onboard computers with algorithms that employ many of the concepts described here.

REFERENCES

Blakelock, J. H. *Automatic Control of Aircraft and Missiles.* New York: Wiley, 1965.

Encyclopaedia Britannica. 15th ed., vol. 17, *Macropaedia.* Chicago: Encyclopaedia Britannica, 1987, pp. 530–539.

Goldstein, H. *Classical Mechanics.* 2d ed. Reading, Mass.: Addison-Wesley, 1980.

Heiskanen, W. A., and H. Moritz. *Physical Geodesy.* San Francisco and London: W. H. Freeman, 1967.

Kane, T. R., P. W. Likins, and D. A. Levinson. *Spacecraft Dynamics.* New York: McGraw-Hill, 1983.

Kaplan, G. H. The IAU Resolutions on Astronomical Constants, Time Scales, and the Fundamental Reference Frame. Circular no. 163. United States Naval Observatory, Washington, D.C., 10 December 1981.

Kuebler, W., and S. Sommers. "The Role of the Earth's Shape in Navigation: An Example." *Journal of the Institute of Navigation.* 28, no. 1 (spring 1981).

McGill, D. J., and W. W. King. *An Introduction to Dynamics.* 3d ed. Boston: PWS Engineering, 1995.

NIMA. "Department of Defense World Geodetic System 1984, Its Definition and Relationships with Local Geodetic Systems." National Imagery and Mapping Agency Technical Report 8350.2 3d ed., 4 July 1997.

Shepperd, S. W. "Quaternion from Rotation Matrix." *AIAA Journal of Guidance and Control* 1, no. 3 (May–June 1978): 223–224.

Shoemake, K. "Animating Rotation with Quaternion Curves." *Computer Graphics* 19, no. 3 (1985): 245–254.

Strang, G. *Linear Algebra and Its Applications.* New York: Academic, 1980.

Travassos, R. H., N. K. Gupta, K. W. Iliffe, and R. Maine. "Determination of an Oblique Wing Aircraft's Aerodynamic Characteristics." Paper no. 80-1630. *AIAA Atmospheric Flight Mechanics Conference*, 1980, p. 608.

Vanicek, P., and E. J. Krakiwsky. *Geodesy: The Concepts.* Amsterdam: North-Holland, 1982.

Wells, D., *Theory and Problems of Lagrangian Dynamics.* Schaum's Outline Series. New York: McGraw-Hill, 1967.

Wilkinson, J. H., and G. Golub. "Ill Conditioned Eigensystems and the Computation of the Jordan Canonical Form." *SIAM Review.* 18, (October 1976): 578–619.

PROBLEMS

Section 1.2

1.2-1 Prove the scalar triple product formula.

1.2-2 Prove the vector triple product formula.

1.2-3 If $\mathbf{u}, \mathbf{v}, \mathbf{w}$ are bound vectors (i.e., they have a common origin), show that $\mathbf{u} \cdot (\mathbf{v} \times \mathbf{w})$ represents the signed volume of the parallepiped that has $\mathbf{u}, \mathbf{v}, \mathbf{w}$ as adjacent edges.

1.2-4 Show that $\mathbf{u} \times (\mathbf{v} \times \mathbf{w}) + \mathbf{v} \times (\mathbf{w} \times \mathbf{u}) + \mathbf{w} \times (\mathbf{u} \times \mathbf{v}) = 0$.

1.2-5 If two particles moving with constant velocity are described by the position vectors

$$\mathbf{p} = \mathbf{p}_0 + \mathbf{v}t, \quad \text{and} \quad \mathbf{s} = \mathbf{s}_0 + \mathbf{w}t$$

(a) show that the shortest distance between their trajectories is given by:

$$d = |(\mathbf{s}_0 - \mathbf{p}_0) \cdot (\mathbf{w} \times \mathbf{v})| / (|\mathbf{w} \times \mathbf{v}|)$$

(b) find the shortest distance between the particles themselves.

1.2-6 If the vectors $\mathbf{u}, \mathbf{v}$ in the rotation formula (1.2-6) are known, what can be determined mathematically about the unit vector $\mathbf{n}$?

1.2-7 (a) Write a vector equation for the specific force at the cm of a moving vehicle, in terms of the gravitational, centripetal, and Coriolis acceleration vectors, and the derivative in F_e of vehicle velocity $\mathbf{v}_{CM/e}$. (b) Rewrite the equation in terms of the derivative taken in F_v, $\boldsymbol{\omega}_{v/i}$ and $\boldsymbol{\omega}_{e/i}$ (see Table 1.4-1). (c) Write the matrix equation for the NED components of the velocity in F_v. (d) Explain how this equation could be used to perform inertial navigation using Equations (1.4-3–6) and measurements from three accelerometers on a servo-driven, NED-aligned platform.

1.2-8 Compare the Coriolis deflections of a mass reaching the ground for the following two cases:

(a) thrown vertically upward with initial velocity u

(b) dropped, with zero initial velocity, from the maximum height reached in (a).

1.2-9 Show that, for a quaternion product, the norm of the product is equal to the product of the individual norms.

1.2-10 Compare the operation count $(+, -, \times, \div)$ of the vector rotation formula (1.2-6) with that of the quaternion formula (1.2-20b).

1.2-11 If a coordinate system b is rotating at a constant rate with respect to a system a, and only the components of the angular velocity vector in system b are given, find an expression for the quaternion that transforms coordinates from b to a.

Section 1.3

1.3-1 Derive the cross-product matrix used in Equation (1.3-9).

1.3-2 Derive the plane rotation matrix given in Equation (1.3-18)
  (a) by using Equation (1.3-12);
  (b) by trigonometry from Figure 1.3-1.

1.3-3 A "compound" rotation can be represented by a sequence of plane rotations, but the plane rotations do not commute. Start with an airplane heading north in level flight and draw two sequences of pictures to illustrate the difference between a yaw, pitch, roll sequence, and a roll, yaw, pitch sequence. Let the rotations (Euler angles) be yaw $\psi = -90°$, pitch $\theta = -45°$, and roll $\phi = 45°$. State the final orientation.

1.3-4 Find the rotation matrix corresponding to (1.3-20) if the reference system has its z-axis pointing up, not down.

1.3-5 Show that the rotation matrix between two coordinate systems can be calculated from a knowledge of the position vectors of two different objects if the position vectors are known in each system. Specify the rotation matrix in terms of the solution of a matrix equation. Show how this technique could be used to determine vehicle attitude by taking telescope bearings on two known stars, given a star catalog.

1.3-6 Find the eigenvalues of the rotation matrix (1.3-17).

Section 1.4

1.4-1 The ECI position coordinates of a celestial object are (x, y, z). Determine the ENU (east, north, up) position coordinates of the object with respect to a tracking station on the surface of the Earth at celestial longitude λ, geodetic latitude ϕ, and sea-level altitude. Assume a spherical Earth (of radius R), align the ECI system with its z-axis pointing up toward the North Pole, and assume that the ENU frame is obtained by rotating first through the longitude angle and then through the latitude angle. Assume also that longitude is measured east from the ECI x-axis. Show that the Earth's radius appears in only one of the required coordinates.

1.4-2 Starting from a calculus-textbook definition of radius of curvature, and the equation of an ellipse, derive the formula (1.4-3) for the meridian radius of curvature.

1.4-3 Use the fact that the prime vertical radius of curvature is equal to the distance along the normal, from the spheroid surface to the semiminor axis, together with the equation of the generating ellipse, and an expression for the gradient of a normal, to derive the formula (1.4-5) for N. Also confirm Equation (1.4-7) for the two parts of N.

1.4-4 Program the iterative calculation of geodetic coordinates, (1.4-11) and use some test cases to demonstrate that it converges very quickly to many decimal digits.

PROBLEMS

1.4-5 Derive the formula (1.4-16), for **G**, starting from the potential function, V, in Equation (1.4-14). Use a geocentric coordinate system as mentioned in the text.

1.4-6 Derive the formula (1.4-17) for geocentric latitude in terms of geodetic latitude by using the geometry of the generating ellipse.

1.4-7 Starting from (1.4-16), write and test a program to evaluate $|\mathbf{g}|$ and $|\mathbf{G}|$ as functions of geodetic latitude and altitude. Plot them both on the same axes, against latitude ($0 \to 90°$). Do this for $h = 0$ and 30,000 m.

1.4-8 Derive the conditions for a body to remain in a geostationary orbit of the Earth. Use the gravity model and geodetic data to determine the geostationary altitude. What are the constraints on the latitude and inclination of the orbit?

Section 1.5

1.5-1 Derive a set of linear state equations from Equations (1.5-6) by considering perturbations from a steady-state condition with angular rates P_e, Q_e, and R_e. Find expressions for the eigenvalues of the coefficient matrix when only one angular rate is nonzero, and show that there is an unstable eigenvalue if the moment of inertia about this axis is either the largest or the smallest of the three inertias. Deduce any practical consequences of this result.

1.5-2 Use Euler's equations of motion (1.5-6) and the Poisson kinematical equations (1.3-23) to simulate the angular motion of a brick tossed in the air and spinning. Write a MATLAB program using Euler integration (1.1-4) to integrate these equations over a 300 s interval, using an integration step of 10 ms. Let the brick have dimensions $8 \times 5 \times 2$ units, corresponding to x, y, z axes at the center of mass. The moments ℓ, m, n are all zero, and the initial conditions are:

(a) $\phi = \theta = \psi = 0$, $P = 0.1$, $Q = 0$, $R = 0.001$ rad/s
(b) $\phi = \theta = \psi = 0$, $P = 0.001$, $Q = 0$, $R = 0.1$ rad/s
(c) $\phi = \theta = \psi = 0$, $P = 0.0$, $Q = 0.1$, $R = 0.001$ rad/s

Plot the three angular rates (deg/s) on one graph, and the three Euler angles (in deg) on another. Which motion is stable and why?

1.5-3 Repeat 1.5-2, but use the Euler kinematical equations (1.3-22a) to represent attitude. Add logic to the program to restrict the Euler angles to the ranges described in Section 1.3.

1.5-4 An aircraft is to be mounted on a platform with a torsional suspension so that its moment of inertia, I_{zz}, can be determined. Treat the wings as one piece, equal to one-third of the aircraft weight, and placed on the fuselage one-third back from the nose.

(a) Find the distance of the aircraft cm from the nose, as a fraction of the fuselage length.

(b) The aircraft weight is 80,000 lbs, the wing planform is a rectangle 40 ft by 16 ft, and the planview of the fuselage is a rectangle 50 ft by 12 ft.

Assuming uniform density, calculate the aircraft moment of inertia (in slug-ft^2).

(c) Calculate the period of oscillation (in s) of the platform if the torsional spring constant is 10,000 lb-ft/rad.

1.5-5 The propeller and crankshaft of a single-engine aircraft have a combined moment of inertia of 45 slug-ft^2 about the axis of rotation, and are rotating at 1500 rpm clockwise when viewed from in front. The moments of inertia of the aircraft are roll: 3000 slug-ft^2, pitch: 6700 slug-ft^2, yaw: 9000 slug-ft^2. If the aircraft rolls at 100 deg/s, while pitching at 20 deg/s, determine the angular acceleration in yaw. All inertias and angular rates are body-axes components.

CHAPTER 2

MODELING THE AIRCRAFT

2.1 INTRODUCTION

Model building is a fundamental process. An aircraft designer has a mental model of the type of aircraft that is needed, uses physical models to gather wind tunnel data, and designs with mathematical models that incorporate the experimental data. The modeling process is often iterative; a mathematical model based on the laws of physics will suggest what experimental data should be taken, and the model may then undergo considerable refinement in order to fit the data. In building the mathematical model we recognize the onset of the law of diminishing returns and build a model that is good enough for our purposes, but has known limitations. Some of these limitations involve uncertainty in the values of parameters. Later we attempt to characterize this uncertainty mathematically and allow for it in control system design.

Actually, because of the high cost of building and flight testing a real aircraft, the importance of the mathematical models goes far beyond design. The mathematical model is used, in conjunction with computer simulation, to evaluate the performance of the prototype aircraft and hence improve the design. It can also be used to drive training simulators, to reconstruct the flight conditions involved in accidents, and to study the effects of modifications to the design. Furthermore, mathematical models are used in all aspects of the aircraft design (e.g., structural models for studying stress distribution and predicting fatigue life).

All of the chapters following this one will make use of the mathematical models presented in this chapter in some form, and thus demonstrate the importance of modeling in the design of aircraft control systems. The rigid-body equations of motion that were derived in Chapter 1 form the skeleton of the aircraft model. In this chapter we add some muscles to the skeleton by modeling the aerodynamic forces and moments

60 MODELING THE AIRCRAFT

that drive the equations. By the end of the chapter we will have the capability, given the basic aerodynamic data, to build mathematical models that can be used for computer simulation or for control systems design. We start by considering some basic elements of aerodynamics.

2.2 BASIC AERODYNAMICS

In the aerospace industry it is necessary for a wide range of specialists to work together; thus flight control engineers must be able to work with the aerodynamicists as well as with structural and propulsion engineers. Each must have some understanding of the terms and mathematical models used by the other. This is becoming increasingly important as designers seek to widen aircraft performance envelopes by integrating the many parts of the whole design process. Furthermore, at the prototype stage the controls designer must work closely with the test pilots to make the final adjustments to the control systems. This may take many hours of simulator time and flight-testing.

Airfoil Section Aerodynamics

The mathematical model used by the control engineer will usually contain aerodynamic data for the aircraft as a whole. However, to gain the necessary insight, we start by examining the aerodynamic forces on an airfoil.

Figure 2.2-1 shows the cross section of an airfoil (a theoretical body of infinite length shaped to produce lift when placed in an airflow) and defines some of the terms used. The flowfield around the airfoil is represented by the *streamlines* shown in the figure (for a steady flow, the flow direction at any point is tangential to the streamline passing through that point). The figure illustrates attached flow, that is, the streamlines follow the surface of the airfoil and do not reverse direction anywhere over the surface. This is a two-dimensional situation; the cross section is constant and the length

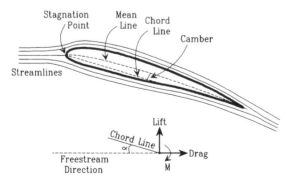

Figure 2.2-1 Definitions associated with an airfoil.

of this airfoil is infinite, so that the flowfield does not change in the direction perpendicular to the plane of the diagram. The initial direction of the flowfield is defined by the freestream velocity vector. This is the velocity measured ahead of the airfoil at a sufficient distance that the flow there is unaffected by the presence of the airfoil.

The shape of the airfoil determines its aerodynamic properties, and some of the important geometrical parameters are shown in the figure. The *chord line* is a straight line drawn from the leading edge to the trailing edge, and is the reference line for describing the shape. An airfoil may be symmetric or, more usually, asymmetric with respect to the chord line. The *mean line* (or *camber line*) is a line joining the leading edge to the trailing edge and having a desired shape. The airfoil is constructed on this camber line by drawing perpendiculars and placing the upper and lower surfaces equal distances above and below the camber line according to a chosen distribution of airfoil thickness. The shape of the camber line, the thickness distribution, and the leading-edge radius combine to determine the aerodynamic properties and the useful speed range.

Two different physical mechanisms contribute to producing an aerodynamic force. First, each element of surface area, multiplied by the pressure at that position, leads to an elemental force normal to the airfoil surface. When this calculation is integrated over the whole surface, the resultant force is, in general, nonzero, except, for example, in the idealized case of laminar flow around a symmetrical airfoil pointed directly into the flow. Second, for each element of surface area there is a layer of the fluid (air) in contact with the surface, and not moving relative to the surface. When the flow is laminar we can visualize layers of fluid farther from the surface moving progressively faster, and the molecular forces between layers, per unit area, constitute the *shear stress*. Shear stress multiplied by the element of area leads to an elemental force tangential to the surface. When the shear forces are integrated over the whole surface, the resultant force is defined to be the *skin friction*. The skin friction force will be proportional to the *wetted area* (area in contact with the fluid) of the airfoil. When the flow is *turbulent* (i.e., the motion at any point is irregular and eddies are formed) over some or all of the airfoil surface, the physical mechanism is harder to visualize but we still define a skin friction force, although the mathematical model is different. The combination of the pressure force and the skin friction force is the resultant aerodynamic force on the airfoil.

Now imagine that the airfoil is pivoted about an axis perpendicular to the cross section, passing through the chord line at an arbitrary distance back from the leading edge. The angle that the chord line makes with the freestream velocity vector is the airfoil *angle of attack,* usually denoted by α (hereinafter referred to as "alpha") and shown as a positive quantity in the figure. In our hypothetical experiment, let the freestream velocity vector be constant in magnitude and direction, and the ambient temperature and pressure be constant. In this situation, the only remaining variable that influences the aerodynamic forces is alpha. Also, elementary mechanics tells us that in this situation the aerodynamic effects can be represented by a force acting at the axis, and normal to it (because of symmetry), and a couple acting around the axis.

The aerodynamic force is conventionally resolved into two perpendicular components, the *lift* and *drag* components, shown in the figure. Lift is defined to be perpendicular to the freestream velocity vector, and drag is parallel to it. Lift and drag normally increase as alpha is increased. An aerodynamic moment is also indicated in the figure, and the positive reference direction is shown there. By definition, the moment is zero when the axis is chosen to pass through the center of pressure (cp) of the airfoil (i.e., the cp is the point through which the total force can be thought to be acting). This is not a particularly convenient location for the axis since experiments show that the location of the cp changes significantly with alpha. There is another special location for the axis: the *aerodynamic center* (ac) of the airfoil. The ac is a point at which the aerodynamic moment tends to be invariant with respect to alpha (within some range of alpha). It is normally close to the chord line, about one quarter-chord back from the leading edge, and moves back to the half-chord position at supersonic speeds. As alpha is varied through positive and negative values, the cp moves in such a way that the moment about the ac remains constant. For the cambered airfoil shown in Figure 2.2-1, the moment about the ac will be a nose-down (negative) moment, as shown in Figure 2.2-2, curve 1.

The aerodynamic center is important when we come to consider the stability of the airfoil in an airflow. It is obvious (by reductio ad absurdum) that if we move the pivot axis forward of the ac we will measure a negative pitching moment that becomes more negative as alpha is increased. This is shown in curve 2 of Figure 2.2-2; point B on this curve is the angle of attack where the pitching moment becomes zero. If we attempt to increase alpha away from point B, a negative pitching moment is generated; conversely, decreasing alpha generates a positive moment. These are restoring moments that tend to hold alpha at the value B. Therefore, neglecting any moment due to its weight, the airfoil will settle into a *stable equilibrium* condition at point B when allowed to pivot freely about a point forward of the aerodynamic center.

When the axis is at the aerodynamic center, as in curve 1 of the figure, there is a stable equilibrium at point A. This point is at a large negative value of alpha outside

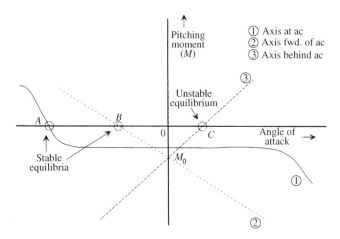

Figure 2.2-2 Airfoil moment about different axes.

the normal range of operation. When we place the pivot axis behind the ac, as in curve 3, the pitching moment increases with alpha. There is an equilibrium condition at point C, but this is an *unstable equilibrium* because any small perturbation in alpha creates a moment that drives the angle of attack out of this region. With the sign conventions we have chosen, we see that a stable equilibrium is associated with a negative slope to the pitching moment curve and unstable equilibrium with a positive slope. If the airfoil had to support the weight of an aircraft, a stable equilibrium point would normally have to occur at a positive angle of attack. This would require curve 2 to be shifted upward (i.e., M_0 positive); in practice, the horizontal tail of the aircraft provides the additional positive pitching moment required.

The stability of this hypothetical experiment has been analyzed by considering the *static* balance of the pitching moment and the effect of small perturbations. The condition of a steady-state moment tending to restore equilibrium is known as *positive stiffness* and, in this case, positive pitch stiffness is associated with a *negative slope* of the pitching-moment versus alpha curve. In this experiment only a single degree of freedom was involved: rotation around a fixed axis. The static analysis was sufficient to determine whether the equilibrium was stable or unstable (and to determine the stability boundary), but not sufficient to determine the dynamics of the motion when the equilibrium was disturbed.

The motion of an airplane in the vertical plane involves three degrees of freedom, one rotational (the pitching motion described above) and two translational (horizontal and vertical velocity components). An analysis of the stability of the motion requires that a steady-state trajectory be defined and an analysis of small perturbations in the motion be performed. From this analysis the *dynamic modes* (i.e., the time-dependent behavior of the system in response to an impulsive input) can be determined. A pilot's ability to control an airplane is linked to the stability of the modes, so *dynamic stability* is of critical importance. Dynamic stability analyses will be performed in later chapters. Here we simply note that positive stiffness is not sufficient to ensure dynamic stability, but the aircraft dynamic stability conditions will later be seen to be dominated by the static stability condition.

We must now describe the mathematical models for the forces and moments on an airfoil, and include the situation where the parameters of the flowfield may vary. It is shown in textbooks on aerodynamics (Kuethe and Chow, 1984) that, for a body of given shape with a given orientation to the freestream flow, the forces and moments are proportional to the product of freestream mass density, ρ, the square of the freestream airspeed, V_T, and a characteristic area for the body. The product of the first two quantities has the dimensions of pressure and it is convenient to define the *dynamic pressure*, $\bar{q}$, by

$$\bar{q} = \tfrac{1}{2} \rho V_T^2 \quad \text{(pressure units)} \tag{2.2-1}$$

and note that this is also the kinetic energy per unit volume. In the *standard atmosphere* model (U.S. Standard Atmosphere, 1976) the mass density ρ is 2.3769×10^{-3} slugs/ft³ at sea level (1.2250 kg/m³) and, as an example, the dynamic pressure at 300 mph (440 ft/s) at sea level is

$$\bar{q} = 0.5 \times 0.002377 \times 440^2 = 230 \text{ lb/ft}^2 \text{ (psf)}$$

This dynamic pressure of about 1.6 lb/in^2 (psi) is to be compared with the static pressure of approximately 14.7 psi at sea level. By dividing a measured (or calculated) aerodynamic force by the product of dynamic pressure and an arbitrarily chosen reference area, we determine dimensionless coefficients that represent the ability of the airfoil to produce lift, or drag. In the case of an aerodynamic moment we must also divide by an arbitrarily chosen reference length. The dimensionless coefficients are called *aerodynamic coefficients* and depend on the shape of the airfoil and its angle of attack.

An aerodynamic coefficient is also a function of the freestream viscosity, which is a measure of a fluid's resistance to rate of change of shape. In addition, the aerodynamic coefficient depends on how much the fluid is compressed in the flow around the airfoil. If this dependence is expressed in terms of two appropriate parameters, then *geometrically similar* airfoils (i.e., same shape, same definition of reference area, but not necessarily the same size) will have the same aerodynamic coefficient when they are at the same angle of attack in two different flowfields, providing that the two *similarity parameters* are the same for each. This assumes that the effect of surface roughness is negligible, and that there is no effect from turbulence in the freestream airflow. Matching of the two sets of similarity parameters is required for wind tunnel results to carry over to full-sized aircraft. The two conventional similarity parameters will now be described.

The nature of the boundary-layer viscous flow is determined by a single freestream dimensionless parameter, the *Reynolds number*, R_e, given by

$$R_e = (\rho \, \ell \, V_T)/\mu, \tag{2.2-2}$$

where ℓ is some characteristic length and μ is the viscosity of the fluid. Note that the viscosity varies greatly with the temperature of the fluid but is practically independent of the pressure. The characteristic length is usually the airfoil chord or, for an aircraft, the mean chord of the wing. Reynolds numbers obtained in practice vary from a few hundred thousand to several million. The flow in the boundary layer is laminar at low Reynolds numbers and, at some critical Reynolds number of the order of a few hundred thousand, it transitions to turbulent flow with a corresponding increase in the skin friction drag.

For most airplanes in flight, the boundary layer flow is turbulent over most of the wing airfoil, except for close to the leading edge. The NACA 6-series airfoils, designed in the 1930s and 1940s to promote laminar flow, showed a significant drag reduction in wind tunnel tests but this usually could not be maintained in the face of the surface contamination and production roughness of practical wings.

The dynamic pressure is an increment of pressure on top of the static pressure. The fractional change in volume, which is a measure of how much the fluid is compressed, is given by dividing the dynamic pressure by the bulk modulus of elasticity (which has the units of pressure). Physics texts show that the speed of sound in a fluid is given by the square root of the quotient of the modulus of elasticity over the mass density. Therefore, when the dynamic pressure is divided by the modulus of elasticity, we obtain a dimensionless quantity equal to one-half of the square of the *freestream Mach number, M*, defined by

$$M = V_T/a, \qquad (2.2\text{-}3)$$

where a is the speed of sound at the ambient conditions. At sea level in the standard atmosphere, a is equal to 1117 ft/s (340 m/s, 762 mph). Freestream Mach number is the second similarity parameter, and the aerodynamic coefficients are written as functions of alpha, Reynolds number, and Mach number. The Mach number ranges of interest in aerodynamics are

$$\begin{aligned}
\text{subsonic speeds:} &\quad M < 1.0 \\
\text{transonic speeds:} &\quad 0.8 \leq M \leq 1.2 \\
\text{supersonic speeds:} &\quad 1.0 < M < 5.0 \\
\text{hypersonic speeds:} &\quad 5.0 \leq M
\end{aligned} \qquad (2.2\text{-}4)$$

The *compressibility effects*, described above, may begin to have a noticeable influence on an aerodynamic coefficient at a freestream Mach number as low as 0.3. By definition pressure disturbances propagate through a fluid at the speed of sound, and an approaching low-speed aircraft can be heard when it is still some distance from the observer. When the Mach number reaches unity at some point in the flow, pressure disturbances at that point can no longer propagate ahead. The wavefront remains fixed to the aerodynamic body at that point, and is called a *shock wave*. At still higher Mach numbers the wavefront is inclined backward in the flow and forms a *Mach cone* with its apex at the source of the pressure disturbance. The Mach number will in general reach unity at some point on the airfoil surface when the freestream velocity is still subsonic. This freestream Mach number, called the *critical Mach number*, defines the beginning of *transonic flow* for an airfoil or wing. Because of the formation of shock waves and their interaction with the boundary layer, the aerodynamic coefficients can vary with Mach number in a complex manner in the transonic regime. For example, at a freestream Mach number slightly greater than the critical Mach number, a sharp increase in drag coefficient occurs. This is called the *drag divergence Mach number*. In the supersonic regime the aerodynamic coefficients tend to change less erratically with Mach number, and in the hypersonic regime the aerodynamic effects eventually become invariant with Mach number.

We are now in a position to write down the mathematical models for the magnitudes of the forces and moments shown in Figure 2.2-1. The measurements are typically made at some point in the airfoil close to the ac (usually at the quarter-chord point). The force components and the moment of the couple are modeled by the following equations, involving lift, drag, and moment *section coefficients* C_ℓ, C_d, and C_m, respectively:

66 MODELING THE AIRCRAFT

$$\text{lift per unit span} = \bar{q}\, c\, C_\ell\, (\alpha, M, R_e)$$
$$\text{drag per unit span} = \bar{q}\, c\, C_d\, (\alpha, M, R_e) \qquad (2.2\text{-}5)$$
$$\text{pitching moment per unit span} = \bar{q}\, c^2\, C_m\, (\alpha, M, R_e)$$

The reference length for this infinitely long airfoil section is the chord length, c, and the product $\bar{q}c$ has the dimensions of force per unit length.

Consider first the variation of section aerodynamic coefficients with alpha. The dimensionless *lift coefficient*, C_ℓ, measures the effectiveness of the airfoil at producing lift. This coefficient is linear in alpha at low values of alpha, and positive at zero angle of attack for cambered airfoils. The *lift-curve slope* has a theoretical value of 2π per radian for thin airfoils at low subsonic Mach numbers. The drag equation has the same form as the lift equation, and the *drag coefficient*, C_d, is usually parabolic in alpha, in the region where the lift coefficient is linear in alpha. The drag coefficient is commonly presented as a function of lift coefficient. Typical plots of lift and drag coefficients, with representative values, are shown, respectively, in Figures 2.2-3a and b. The moment equation is different from the lift and drag equations in that it requires an additional length variable to make it dimensionally correct. The airfoil chord, c, is used once again for this purpose. A typical plot of the *pitching moment coefficient*, C_m, is also shown in Figure 2.2-3a.

Now consider the variation of these coefficients at higher values of alpha. Wind tunnel flow-visualization studies show that, at high values of alpha, the flow can no longer follow the upper surface of the airfoil and becomes detached. There is a region above the upper surface, near the trailing edge, where the velocity is low and the flow reverses direction in places in a turbulent motion. As the angle of attack is increased farther, the beginning of the region of separated flow moves toward the leading edge of the airfoil. The pressure distribution over the airfoil is changed in such a way that the lift component of the aerodynamic force falls off rapidly and the drag component

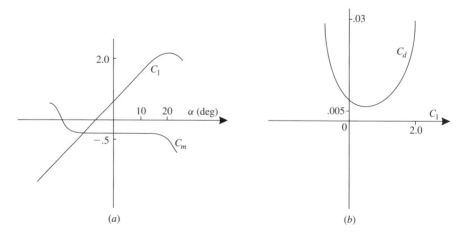

Figure 2.2-3 Typical plots of lift, drag, and moment coefficients.

increases rapidly. The airfoil is said to be *stalled*, and this condition is normally avoided in flight. The pitching moment (about the axis through the aerodynamic center) also changes rapidly, typically becoming more negative.

Next consider the effect of Reynolds number. The lift-curve slope is essentially independent of R_e when $R_e \approx 10^6$ to 10^7 (where normal, manned aircraft fly), but is significantly reduced when $R_e \approx 10^5$ (which may be reached by miniature and unmanned vehicles). The maximum (stall) lift coefficient tends to increase with R_e, even at high values. The drag curve is affected by R_e in that its minimum value is larger at lower R_e; also, near stall the drag coefficient is increased by lower R_e. The pitching moment is similar to the lift coefficient in that it is independent of Reynolds number in the linear region, at high R_e, but not independent in the stall region.

Finally, consider the effect of Mach number on the aerodynamic coefficients. In the case of the lift coefficient, both the lift-curve slope and the maximum lift are changed when compressibility effects begin to occur. The theoretical values for lift-curve slope are modified by the Prandtl-Glauert correction (Anderson, 1991; Kuethe and Chow, 1984):

$$\frac{dC_E}{d\alpha} = \frac{2\pi}{\sqrt{(1 - M^2)}}, \quad M < 1 \tag{2.2-6a}$$

$$\frac{dC_L}{d\alpha} = \frac{4}{\sqrt{(M^2 - 1)}}, \quad M > 1 \tag{2.2-6b}$$

In the transonic region, the lift-curve slope of a thin airfoil will generally pass through a smooth peak, while that of a thick airfoil will show a more complicated variation. The maximum lift coefficient falls with increasing Mach number in the supersonic regime. For the drag coefficient, the effect of increasing subsonic Mach number is to bodily raise the drag curve shown in Figure 2.2-3b; the drag coefficient then falls off somewhat with increasing supersonic Mach number.

The effect of Mach number on the pitching moment coefficient is due to a rearward shift of the airfoil cp with Mach number. This causes a shift in position of the airfoil aerodynamic center. At low subsonic Mach numbers it is usually at a distance back from the leading edge equal to about 25 percent of the chord. In the transonic region its position may change erratically, and at higher speeds it tends to shift aft to the 50 percent chord position. Therefore, if the pitching moment is measured at the quarter-chord position, the slope, with alpha, changes from zero to a negative value as the Mach number is increased from subsonic to supersonic values.

Finite Wings

Real wings are finite in length and involve "three-dimensional" aerodynamics. When a wing is producing lift, the air tends to flow around the tip, from the high-pressure region under the wing to the low-pressure region above the upper surface. This circulation of the air creates a vortex motion at the tips so that, behind the wing, the sheet of air that is deflected downward by airfoil action curls up at the edges to form a *vortex sheet*. The energy that goes into creating the vortex motion leads to an increase in the force needed to push the wing through the air, that is, an increase in drag. In addition, the leakage around the tips creates a spanwise component of flow and reduces the lift-curve slope compared to that of a "two-dimensional" airfoil. Thus, there is a decrease in the *lift-over-drag* ratio compared to the airfoil. Many aircraft use wing-tip devices and aerodynamic "fences" on the wing to reduce these detrimental effects.

A complete wing may have straight or curved leading and trailing edges, or it may consist of two identical halves that are swept back toward the tips. The chord may be constant or reduced toward the wing tip (wing taper), and different airfoil sections may be used over different parts of the span. The "planform" of a wing has a large impact on its aerodynamic properties. Among the most important parameters of the planform are the aspect ratio and the leading-edge sweep angle. These and other parameters are defined in Table 2.2-1.

An explanation of the calculation of the mean aerodynamic chord can be found in various aerodynamics texts (e.g., Dommasch et al., 1967). The aspect ratio is equivalent to a measure of span relative to chord; for complete aircraft, values range from about 30 (some sailplanes), through 14 (Lockheed U-2) and 7 (Boeing 747), down to about 3 (fighter aircraft), and even lower for delta wings. High-aspect-ratio wings act more like the "two-dimensional" airfoil, while low-aspect wings have greatly reduced lift-curve slope and lift-over-drag. High lift-over-drag is needed for efficient cruise performance (passenger jets), long-duration flight (military reconnaissance), and shallow glide angle (sailplanes). A low aspect ratio simplifies structural design problems for high-g aircraft, permits very high roll rates, and reduces *supersonic wave-drag* (described later).

Prandtl's lifting line theory (Anderson, 1991) provides a simple expression for the lift-curve slope of a straight high-aspect-ratio finite wing in incompressible flow, in terms of aspect ratio and the lift-curve slope of the corresponding airfoil section. This formula can be combined with the Prandtl-Glauert corrections of Equations (2.2-6) to give a formula that applies to subsonic compressible flow. The transonic lift-curve slope is hard to predict but, in the supersonic regime, the lift-curve slope can be approximated as a constant (4.0) divided by the Prandtl-Glauert correction factor, as in Equation (2.2-6b). For low-aspect-ratio wings ($AR < 4$), slightly more complicated formulae are available (Anderson, 1999). Wing sweep further complicates the picture. A lift-curve slope formula can be derived for subsonic swept wings by introducing the cosine of the sweep angle into the above-described formulae. For supersonic swept wings the behavior of the lift-curve slope depends on whether the sweep of the wing

TABLE 2.2-1: Important Wing-Planform Parameters

b = wing span (i.e., tip to tip)	λ = taper ratio (tip chord/root chord)
c = wing chord (varies along span)	
$\bar{c}$ = mean aerodynamic chord (mac)	Λ = leading-edge sweep angle
S = wing area (total)	$AR = b^2/S$ = aspect ratio

puts its leading edge inside or outside of the shock wave from the apex of the swept wing, and no convenient formulae are available. The above facts are clearly described in much more detail by Anderson (1999).

Delta-shaped wing planforms behave in a fundamentally different way than conventional wings. When producing lift, a delta wing has a strong vortex rolling over the full length of each leading edge. The vortices are stable, in the sense that they remain in place over a wide range of alpha, and contribute to lower pressure over the upper surface. The lift-curve of the delta wing is slightly nonlinear, with the slope increasing at first as alpha increases. The average lift-curve slope is only about half that of a conventional wing, but the stall angle of attack is about twice as big. A delta wing has been used on various fighter aircraft because it can provide the large sweep angle needed for supersonic flight, and can also attain a normal peak lift coefficient through the *vortex lift*. A degree of vortex lift similar to that of a delta wing can be obtained from a conventional swept wing if, near the wing root, the leading edge is carried forward with a sharp-edged extension having a sweep angle near 90 degrees. This leading-edge extension generates a vortex that trails back over the inboard wing panels and keeps the flow attached to the wing at high alpha.

Aircraft Configurations

A conventional aircraft uses airfoil sections for the wings, horizontal tail, vertical tail, and possibly for additional surfaces such as horizontal canards (notable exceptions to this configuration are the flying wing aircraft, such as the Northrop YB series [Anderson, 1976] and the more modern B-2 bomber). The close proximity of the wings and fuselage, and of the wing and tail surfaces, creates interference effects, so that the total aerodynamic force is not given by the sum of the forces that would be obtained from the individual surfaces acting alone. In addition, the fuselage of the airplane provides some lift and a considerable amount of drag. Therefore, the aerodynamic coefficients of a complete aircraft must be found from wind tunnel measurements and computational fluid dynamics (CFD). Anderson (1999) cites a study that measured the subsonic lift-curve slope of a wing-fuselage combination as a function of the ratio of fuselage diameter to wingspan (d/b). The study showed that for a range of d/b from zero to 0.6, the lift-curve slope was within 5 percent of that of the wing alone. This was because of the lift of the fuselage, and because of favorable crossflows induced on the wing by the fuselage. A further conclusion was that the lift of the wing-body combination could be approximated by using the lift coefficient of the wing alone, with a reference area given by the planform area of the wing projected through the fuselage. This is the usual definition of the wing planform reference area.

Figure 2.2-4 shows a number of distinctive planforms. Low-speed aircraft, ranging from light general-aviation types to military heavy-lift transport aircraft, have stiff moderate-aspect-ratio wings with no sweepback (cf. Figure 2.2-4*d*). Aircraft designed to reach transonic speeds and beyond have highly swept wings. The effect of the sweep is to postpone the *transonic drag rise*, since the component of the airflow perpendicular to the leading edge has its speed reduced by the cosine of the sweep angle. Large jet airliners designed to cruise efficiently at high subsonic Mach numbers

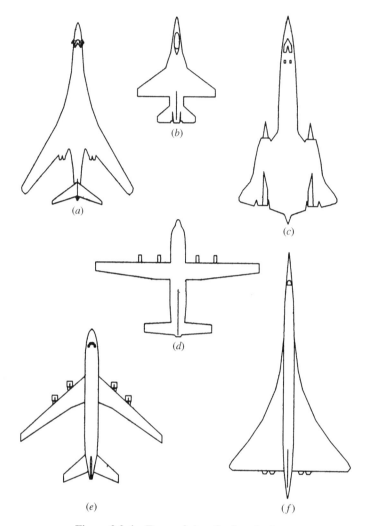

Figure 2.2-4 Types of aircraft wing planform.

have swept wings with a high aspect ratio (Figure 2.2-4e). This produces the highest ratio of lift to induced drag (the increase in drag that occurs when lift is produced). In the case of high-speed fighter aircraft, the requirement for low supersonic wave drag and high maneuverability causes a dramatic change to very-low-aspect-ratio wings (Figure 2.2-4b). The stubby wings allow the aircraft structure to be designed to withstand very high lift forces during maneuvers. They also reduce the moment of inertia about the longitudinal axis and the aerodynamic damping moments during rolling, thus promoting a high roll acceleration and a high maximum roll rate.

Wing sweep has the disadvantage of reducing the lift-curve slope of the wing (i.e., less lift at a given alpha) and producing suboptimal performance at low speeds. A way to overcome this when a high lift-to-drag ratio is required over a wide envelope is to use a variable-sweep wing, as exemplified by the F-14 and B-1B aircraft (Figure 2.2-4a). This is a heavy and costly solution. For commercial aircraft that are usually optimized for one cruise condition, the most common method of achieving adequate lift at low speeds is to increase the camber and area of the wing by means of leading- and trailing-edge devices (slats and flaps). These may then be deployed manually for landing. More specialized solutions are to use an automatic maneuvering flap, as in the case of the F-16 leading-edge flap, which is deployed automatically as a function of angle of attack when the Mach number is low. More recently the concept has been taken to its logical conclusion in the *mission adaptive wing* (DeCamp, 1987), tested on an F-111 aircraft.

Wing planforms that create vortex lift are shown in Figures 2.2-4b, c, and f, representing the F-16, SR-71, and Concorde aircraft, respectively.

The F-16 has sharp-edged, highly swept forebody *strakes* to generate the vortices. The design goal was to achieve maximum maneuverability through the use of vortex lift. The Concorde has an *ogee* wing with very large initial sweep angle, with the design aim of increasing the lift at low speed and reducing the movement of the aerodynamic center between low-speed and supersonic cruise conditions. The high angle of attack needed to get the low-speed vortex lift would obscure the pilot's view of the runway, and this problem was solved by using the droop nose. Some description of the design of these wings can be found in the AIAA case studies (Droste and Walker, no date; Rech and Leyman, no date). The SR-71 Mach 3-plus, high-altitude, strategic reconnaisance aircraft (Drendel, 1982) has a blended wing-body with chines. This blending reduces wing-body wave interference drag at cruise speed, while vortex lift effects may be useful during takeoff and landing.

Vortices are also shed from a conventional forebody at high alpha, and a long forebody overhang (as in the case of the shark nose on the F-5) presents difficult design problems. This is because any slight asymmetry in the shed vortices causes pressure differentials at the nose and leads to a relatively large (and unpredictable) yawing moment because of the long lever arm from the aircraft center of mass.

2.3 AIRCRAFT FORCES AND MOMENTS

The equations of motion derived in Chapter 1 are driven by the aerodynamic forces and moments acting at the cm of the complete rigid aircraft. In Section 2.2 we have covered enough basic aerodynamics to understand how these forces and moments come about. We now begin to examine how they can be measured and expressed.

Definition of Axes and Angles

The aerodynamic forces and moments on an aircraft are produced by the relative motion with respect to the air and depend on the orientation of the aircraft with respect to the airflow. In a uniform airflow these forces and moments are unchanged after a rotation around the freestream velocity vector. Therefore, only two orientation angles

72   MODELING THE AIRCRAFT

(with respect to the relative wind) are needed to specify the aerodynamic forces and moments. The angles that are used are the angle of attack (alpha) and the *sideslip angle* (beta). They are known as the *aerodynamic angles* and will now be formally defined for an aircraft. Note that the aerodynamic forces and moments are also dependent on angular rates, but for the moment we are concerned only with orientation.

Figure 2.3-1 shows an aircraft with the relative wind on its right side (i.e., sideslipping), with three right-handed (forward, starboard, and down) coordinate systems with a common origin at the aircraft cm, and with aerodynamic angles α and β. The *body-fixed coordinate system* has its x-axis parallel to the fuselage reference line (used in the blueprints), and its z-axis in the (conventional-) aircraft plane of symmetry. The angle of attack is denoted by α_{frl} when measured to the fuselage reference line from the projection of the relative wind on the body x-z plane. It is positive when the relative wind is on the underside of the aircraft. The sideslip angle is measured to the relative wind vector from the same projection. It is positive when the relative wind is on the right side of the airplane.

The angle of attack is also given the symbol α_0 when measured to the aircraft zero-lift line (where aircraft lift is zero, with neutral controls and no sideslip). We will simply write "α" throughout, and mean α_{frl} unless otherwise stated. For an aircraft in steady-state flight (Section 2.6) the "equilibrium" angle of attack will be denoted by α_e, and the equilibrium sideslip angle is normally zero.

In Figure 2.3-1, α_e defines the orientation of the *stability-axes* coordinate system, which is used for analyzing the effect of perturbations from steady-state flight. As can be seen from the figure, it is obtained from the body-fixed system by a left-handed

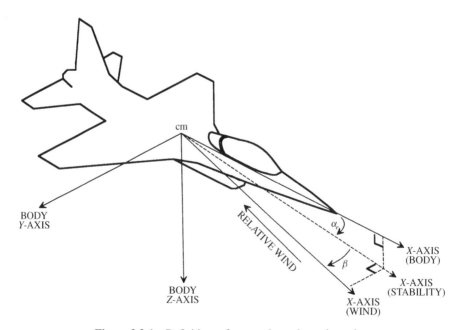

Figure 2.3-1 Definitions of axes and aerodynamic angles.

rotation, through α_e, around the body y-axis. The *wind-axes system* is obtained from the stability-axes system by a rotation around the z-axis that aligns the wind x-axis directly into the relative wind. A left-handed wind-axes system, aligned backward, left, and "up" relative to the aircraft, has been used in the past for wind tunnel data (Pope, 1954). *Lift, L, drag, D*, and *cross-wind force, C*, were defined naturally in these axes as the aerodynamic force components along the respective positive axes. The notation for our right-handed coordinate systems is given in Table 2.3-1, in the next subsection.

Following the rules for finding rotation matrices, the rotation matrices from body to stability, and stability to wind axes are

$$C_{s/b} = \begin{bmatrix} \cos\alpha_e & 0 & \sin\alpha_e \\ 0 & 1 & 0 \\ -\sin\alpha_e & 0 & \cos\alpha_e \end{bmatrix} \qquad (2.3\text{-}1a)$$

$$C_{w/s} = \begin{bmatrix} \cos\beta & \sin\beta & 0 \\ -\sin\beta & \cos\beta & 0 \\ 0 & 0 & 1 \end{bmatrix} \qquad (2.3\text{-}1b)$$

and the combined rotation from body to wind is

$$C_{w/b} = \begin{bmatrix} \cos\alpha_e \cos\beta & \sin\beta & \sin\alpha_e \cos\beta \\ -\cos\alpha_e \sin\beta & \cos\beta & -\sin\alpha_e \sin\beta \\ -\sin\alpha_e & 0 & \cos\alpha_e \end{bmatrix} \qquad (2.3\text{-}2)$$

This transformation will also be used without the subscript e when converting instantaneous wind-axes components into body axes, and vice-versa.

Definition of Forces and Moments

Table 2.3-1 defines the symbols that will be needed for aircraft force, moment, and velocity components. The subscripts A or T on the force and moment vectors indicate, respectively, aerodynamic or thrust effects. In the case of the aerodynamic forces, there are no specific symbols for stability-axes components but, as indicated in (2.3-3b), the stability axes have two components that are unchanged from the other axes. These dimensionless coefficients are defined in the next section. Note that C_N and C_X are, respectively, the *normal force* and *axial force* coefficients; C_N is the negative of the body-axes force coefficient C_Z.

A useful notation scheme is to use lowercase symbols to indicate small perturbations on the "uppercase" variables. Unfortunately, aircraft moments are almost universally denoted by lowercase symbols, as shown in (2.3-4) and (2.3-5). Also, the same symbols are commonly used for the dimensionless moment coefficients regardless of coordinate system, and the coordinate system must be explicitly stated.

Thrust components are shown in (2.3-5); note that a sideforce component can be

TABLE 2.3-1: Force, Moment, and Velocity Definitions

Aerodynamic Forces:

$$\mathbf{F}_A^w \equiv \begin{bmatrix} -D \\ -C \\ -L \end{bmatrix} = C_{w/b} \begin{bmatrix} X_A \\ Y_A \\ Z_A \end{bmatrix} \equiv C_{w/b} \mathbf{F}_A^b \qquad (2.3\text{-}3a)$$

Dimensionless Force Coefficients:

$$\begin{array}{cccc} & x & y & z \\ \text{Wind:} & C_D & C_C & C_L \\ \text{Stability:} & * & C_Y & C_L \\ \text{Body:} & C_X & C_Y & C_Z \quad (-C_N) \end{array} \qquad (2.3\text{-}3b)$$

Aerodynamic Moments:

$$\mathbf{M}_A^b = \begin{bmatrix} \ell \\ m \\ n \end{bmatrix} \quad \mathbf{M}_A^s = \begin{bmatrix} \ell_s \\ m \\ n_s \end{bmatrix} \quad \mathbf{M}_A^w = \begin{bmatrix} \ell_w \\ m_w \\ n_w \end{bmatrix} \quad (n_w = n_s) \qquad (2.3\text{-}4a)$$

Dimensionless Moment Coefficients:

$$C_\ell, C_m, C_n \qquad (2.3\text{-}4b)$$
$$\text{(same notation in all systems)}$$

Thrust Force and Moment:

$$\mathbf{F}_T^b = \begin{bmatrix} X_T \\ Y_T \\ Z_T \end{bmatrix} \quad \mathbf{M}_T^b = \begin{bmatrix} m_{x,T} \\ m_{y,T} \\ m_{z,T} \end{bmatrix} \qquad (2.3\text{-}5)$$

Relative Velocity Components:

$$\mathbf{v}_{rel}^b \equiv \begin{bmatrix} U' \\ V' \\ W' \end{bmatrix} = C_{b/w} \mathbf{v}_{rel}^w = C_{b/w} \begin{bmatrix} V_T \\ 0 \\ 0 \end{bmatrix} = \begin{bmatrix} V_T \cos\alpha \cos\beta \\ V_T \sin\beta \\ V_T \sin\alpha \cos\beta \end{bmatrix} \qquad (2.3\text{-}6a)$$

Aerodynamic Angles:

$$\tan(\alpha) = \frac{W'}{U'}; \quad \sin(\beta) = \frac{V'}{V_T}; \quad V_T = |\mathbf{v}_{rel}| \qquad (2.3\text{-}6b)$$

Absolute Velocity Components:

$$\mathbf{v}_{CM/e}^b = \begin{bmatrix} U \\ V \\ W \end{bmatrix} \qquad (2.3\text{-}6c)$$

TABLE 2.3-1: (*continued*)

Angular Velocity Components (r denotes any ref. frame):

$$\boldsymbol{\omega}_{b/r}^b = \begin{bmatrix} P \\ Q \\ R \end{bmatrix} \quad \boldsymbol{\omega}_{b/r}^s = \begin{bmatrix} P_s \\ Q \\ R_s \end{bmatrix} \quad \boldsymbol{\omega}_{b/r}^w = \begin{bmatrix} P_w \\ Q_w \\ R_w \end{bmatrix} \quad (R_w = R_s) \quad (2.3\text{-}7)$$

Control Surface Deflections:

  Elevator: δ_e Aileron: δ_a Rudder: δ_r Flap: δ_F

  Throttle Position: δ_t

  Control Vector: $\bar{U} = [\delta_t \ \delta_e \ \delta_a \ \delta_r \ \delta_F \ \ldots]^T$

---

produced by unbalanced engine power because in a multiengine aircraft, the engines may be toed-in to align them with the airflow from the forebody. Also, the thrust axis is often slightly tilted with respect to the body x-axis, and so a z-component of thrust can result. In the case of VTOL or V/STOL aircraft, the z component of thrust will be particularly important. Models of propeller-driven aircraft must include several important force and moment effects.

In Equation (2.3-6) primes are used to denote velocity components relative to the atmosphere, as opposed to "inertial" components. In the wind system the relative velocity vector $\mathbf{v}_{rel}$ has only an x component V_T, and so $V_T = |\mathbf{v}_{rel}|$. In (2.3-6b) the aerodynamic angles have been found from the interrelationships of the components in (2.3-6a). The control vector of the nonlinear state-space model has been denoted by $\bar{U}$, in this chapter only, to distinguish it from a velocity component.

Force and Moment Coefficients

The forces and moments acting on the complete aircraft are defined in terms of dimensionless aerodynamic coefficients in the same manner as for the airfoil section. The situation is now three-dimensional, and the coefficients are functions of the two aerodynamic angles, as well as Mach and Reynolds numbers. Furthermore, an aircraft is a flexible structure and its shape is deformed by the influence of high dynamic pressure, with consequent changes in the aerodynamic coefficients. If Mach and altitude are specified, together with a temperature and density model of the atmosphere, then Reynolds number and dynamic pressure can be determined. Therefore, the aircraft aerodynamic coefficients are, in practice, specified as functions of the aerodynamic angles, Mach, and altitude (in the standard atmosphere). In addition, control surface deflections and propulsion system effects cause changes in the coefficients. A control surface deflection, δ_s, effectively changes the camber of a wing, which changes the lift, drag, and moment. Consequently, we write the dependence of an aerodynamic coefficient as

$$C_{(\ )} = C_{(\ )}(\alpha, \beta, M, h, \delta_s, T_c), \quad (2.3\text{-}8a)$$

where T_c is a *thrust coefficient* (defined later). Other factors that change the coefficients are configuration changes (e.g., landing gear, external tanks, etc.) and ground-proximity effects. In terms of wind-axes components, we have the following coefficients:

$$\text{drag, } D = \bar{q}\, S\, C_D$$
$$\text{lift, } L = \bar{q}\, S\, C_L$$
$$\text{crosswind force, } C = \bar{q}\, S\, C_C$$
$$\text{rolling moment, } \ell_w = \bar{q}\, S\, b\, C_\ell \quad\quad (2.3\text{-}8b)$$
$$\text{pitching moment, } m_w = \bar{q}\, S\, \bar{c}\, C_m$$
$$\text{yawing moment, } n_w = \bar{q}\, S\, b\, C_n$$

Exactly equivalent definitions are used for body or stability-axes components, with the symbols given in Table 2.3-1. In Equation (2.3-8a), as a rough generality, *longitudinal coefficients* (lift, drag, pitching moment) are primarily dependent on alpha, and in the *lateral-directional coefficients* (roll, yaw, and sideforce) beta is equally as important as alpha.

Equation (2.3-8a) implies a complicated functional dependence that would have to be modeled as a "lookup-table" in a computer. The vast majority of aircraft have flight envelopes restricted to small angles of attack and/or low Mach numbers. For these aircraft, the functional dependence will be simpler and any given coefficient might be broken down into a sum of simpler terms, with linearity assumed in some terms.

The coefficients considered so far are *static coefficients*, that is, they would be obtained from measurements on a stationary model in a wind tunnel (other methods are considered later). It is also necessary to model the aerodynamic effects when an airplane maneuvers. In general terms this requires a differential equation model of the aerodynamic force or moment. To determine if this level of complexity is warranted, we examine maneuvering flight more closely, in two categories. First, consider maneuvers that are slow enough that the flowfield around the aircraft is able to adjust in step with the maneuver, and so that the maneuver-induced translational velocities of points on the aircraft cause changes in the local aerodynamic angles that are still in the linear regime. The aerodynamic forces or moments can then be modeled as linearly proportional to the angular rate that produced them. Linearization is usually associated with taking a partial derivative, and in this case the coefficient of proportionality is called an *aerodynamic derivative*. The aerodynamic derivatives will be described in the next subsection.

In the second category are maneuvers in which an airplane can significantly change its orientation in a time interval that is comparable with the time required for the flowfield around the aircraft to readjust. These *unsteady aerodynamic effects* lead to time-dependence in the aerodynamic coefficients and much more complicated mathematical models. For example, when a very maneuverable aircraft is pitched up rapidly, and the angle of attack reaches a value near to stall, the lift generated by the wing may briefly exceed that predicted by the static lift curve. This *dynamic lift* occurs because flow separation takes a finite time to progress from the trailing edge of the wing to the leading edge. The effect can be modeled by making the lift coefficient satisfy a first-order differential equation involving angle-of-attack rate, "alpha-dot" (Goman and Khrabrov, 1994). Another example of possible unsteady aerodynamic behavior is *wing-rock* (McCormick, 1995).

The Aerodynamic Derivatives

The aerodynamic derivatives can be subdivided into two categories. First, when the body-fixed frame has a constant angular velocity vector, every point on the aircraft has a different translational velocity in the geographic frame and, taking body-axes components, the aerodynamic angles could be computed at any point, using the equivalent of Equation (2.3-6b). For example, a roll-rate P would create translational velocity components $\pm Pb/2$ at the wing tips. When $P > 0$ this would cause the angle of attack to be reduced by approximately $Pb/(2V_T)$ at the left wing tip, and increased by the same amount at the right wing tip. This would in turn create a skew-symmetric variation in lift across the full span of the wings and, assuming that the wing is not stalled across most of the span, produce a negative rolling moment. Because the moment opposes the roll rate P, the coefficient relating the rolling moment to the roll rate is called a *damping derivative*.

The quantity $Pb/(2V_T)$ is given the symbol $\hat{p}$ and is thought of as a dimensionless roll rate. In a continuous roll, with the aircraft cm moving in a straight line, the wing tips move along a helical path and $Pb/(2V_T)$ is the *helix angle*. The helix angle is a useful figure of merit for roll control power and has been evaluated and compared for a variety of aircraft (Perkins and Hage, 1949; Stinton, 1996). The mathematical model for the dimensionless damping force, or moment ΔC, is of the form:

$$\Delta C_{(\ )} = C_{(\ )}(\alpha, \beta, M, h, \delta_s, T_c) \times \frac{k}{2V_T} \times \text{rate} \qquad (2.3\text{-}9a)$$

The constant k in the dimensionless rate, in Equation (2.3-9a), is either the wingspan (for roll and yaw rates) or the wing mean aerodynamic chord (for pitch rate). The coefficient $C_{(\ )}$ is one of the following p, q, or r derivatives,

$$C_{\ell_p} \quad C_{m_q} \quad C_{n_r} \qquad (2.3\text{-}9b)$$

$$C_{\ell_r} \quad C_{n_p} \qquad (2.3\text{-}9c)$$

$$C_{L_q} \quad C_{Y_p} \quad C_{Y_r}, \qquad (2.3\text{-}9d)$$

which relate the increments in the moments or forces to the yawing, pitching, and rolling rates. Names are given to the derivatives later. The dimensionless forces and moments are converted to actual forces and moments as in Equations (2.3-8b). Some possible derivatives have been omitted, for example, the effect of pitch rate on drag is usually insignificant. The moment derivatives are the source of the important damping effects on the natural modes of the aircraft.

The second category of aerodynamic derivatives is the *acceleration derivatives*. When the aircraft has translational acceleration, the aerodynamic angles have nonzero first derivatives that can be found by differentiating Equations (2.3-6b). Thus

$$\dot{\alpha} = \frac{U' \dot{W}' - W' \dot{U}'}{(U')^2 + (W')^2} \qquad (2.3\text{-}10a)$$

and

$$\dot{\beta} = \frac{\dot{V}' V_T - V' \dot{V}_T}{V_T \left[(U')^2 + (W')^2\right]^{1/2}} \qquad (2.3\text{-}10b)$$

where

$$\dot{V}_T = \frac{U' \dot{U}' + V' \dot{V}' + W' \dot{W}'}{V_T} \qquad (2.3\text{-}10c)$$

The main steady-aerodynamic effect of the changing aerodynamic angles is that, as the flowfield around the wings and fuselage changes, there is a small airspeed-dependent delay before the changes in downwash and sidewash are felt at the tail. A first-order approximation in modeling these effects is to make the resulting force and moment increments directly proportional to the aerodynamic-angle rates. Therefore, the following acceleration derivatives are commonly used,

$$\text{alpha-dot derivatives:} \quad C_{L_{\dot{\alpha}}} \; C_{m_{\dot{\alpha}}} \qquad (2.3\text{-}11)$$

These derivatives are used in an equation of exactly the same form as Equation (2.3-9a). The beta-dot derivatives, used to model the delay in the change in sidewash at the vertical tail, are less commonly used.

Aerodynamic Coefficient Measurement and Estimation

The static aerodynamic coefficients can be measured in a wind tunnel using an aircraft scale-model mounted on a rigid "sting," to which strain gages have been attached. An older wind tunnel may use a "balance" rather than strain gages. Rigid mounting in a wind tunnel allows *untrimmed coefficients* to be measured, that is, nonzero aerodynamic moments can be measured as the aerodynamic angles are changed or control surfaces are moved.

Specially equipped wind tunnels allow the model to be subjected to an oscillatory motion (Queijo, 1971) so that damping and acceleration derivatives can be measured. Unfortunately, as might be expected, the results are dependent on the frequency of the oscillation. Empirical criteria have been formulated to determine frequency limits below which a quasi-steady assumption (i.e., instantaneous flowfield readjustment) can be made about the flow (Duncan, 1952).

The second important method of measuring aerodynamic coefficients is through *flight-test*. In this case *trimmed coefficients* are measured by using the control surfaces to make perturbations from the trimmed steady-state flight condition (Maine and Iliffe, 1980). The typical results are curves of a coefficient plotted against Mach, with altitude as a parameter, for a specified aircraft weight and cm position. The dependence on altitude comes about through the variation of alpha with altitude for a given Mach number, through aeroelastic effects changing with dynamic pressure, and, possibly, through Reynolds number effects. To convert to untrimmed coefficients, which are functions of the aerodynamic angles, Mach, and altitude, the trimmed angle of attack must also be recorded in the same form. The flight-test results can then be cross-plotted to obtain untrimmed coefficients. The untrimmed coefficients are required when building an aircraft model that is intended to function over a wide range of flight conditions; the trimmed coefficients are used to build small-perturbation models for control systems design or *handling qualities* studies.

Other ways of determining aerodynamic coefficients include the use of *CFD* computer codes or a combination of empirical data and theory built into a computer program such as the Stability and Control Datcom (Hoak et al., 1970). The input data must include a geometrical description of the aircraft. There are also simple formulae based on assumptions of linearity that can be used to estimate the aerodynamic derivatives. Some of these will be described in subsequent sections.

Component Buildup

The aerodynamic coefficients have a complex dependence on a large number of variables, and this creates both modeling problems and measurement problems. For example, a computer model might be created in the form of a data lookup-table in five dimensions (five independent variables). It would be difficult to design an interpolation algorithm for this table or to set up a data measurement system (e.g., wind tunnel measurements), and very little physical insight would be available to help. It is advantageous to build up an aerodynamic coefficient from a sum of components that provide physical insight, require just a single type of test and wind tunnel model, and are convenient to handle mathematically (e.g., fewer dimensions, linearizable, etc.). We will now take each of the aerodynamic coefficients in turn and examine their functional dependence and how this can be modeled.

Drag Coefficient, C_D

The drag coefficient of the complete aircraft is of paramount importance to the aircraft designer. Low drag provides better performance in terms of range, fuel economy, and maximum speed, and designers take pains to estimate the total drag accurately. By the same token we should understand how to make a good mathematical model of the drag. In general, the drag force is a combination of friction drag and drag caused when the integral of pressure over the whole surface area of the body is nonzero. Table 2.3-2 shows the total drag of an aircraft, composed of friction drag and various constituent parts of the pressure drag.

This is not a linear superposition of independent effects; the proportions of the three components will change with flight conditions, and they cannot necessarily be separated and measured individually. The *parasite drag* is called *profile drag* when

TABLE 2.3-2: Aircraft Drag Components

Parasite Drag = Friction Drag + Form Drag (flow separation)
+ Induced Drag (effect of wing-tip vortices, finite wing)
+ Wave Drag (effect of shock waves on pressure distribution)
Total Drag

applied to an airfoil section; it is the sum of skin friction and *form drag*. Form drag is simply the pressure drag caused by flow separation at high alpha. *Induced drag* (also called *vortex drag*) is the pressure drag caused by the tip vortices of a finite wing when it is producing lift. *Wave drag* is the pressure drag when shock waves are present over the surface of the aircraft. The total drag may be broken down into other different components according to the experimental situation. The resulting components will only be meaningful when used in the correct context. For example, *interference drag* is the difference between the summed drag of separate parts of the aircraft and the total drag when these parts are combined. It is a result of mutual interference between the flows over the different parts of the aircraft. Other terms include *drag due to lift* and *zero-lift drag* used for the complete aircraft.

Now consider, one by one, the drag terms from Table 2.3-2. The aircraft parasite drag is virtually all skin-friction drag when the aircraft wing is not stalled. The amount of skin friction drag will depend on the wetted area of the aircraft. The wetted area can range from several times the wing planform area, down to approximately twice the planform area in the case of a flying-wing aircraft. However, as we have already seen, the value of the airplane drag coefficient is calculated based on the wing planform area. The flow in the boundary layer will ordinarily be mostly turbulent in normal flight, but this will depend to a small extent on the lift coefficient. In laminar flow the drag coefficient for skin friction is inversely proportional to the square root of the Reynolds number; in turbulent flow it decreases more slowly as the Reynolds number increases. The Reynolds number increases in proportion to airspeed, but dynamic pressure increases with the square of airspeed. Therefore, we expect to see an increase in skin friction drag with airspeed, although it will become a smaller fraction of the total drag at higher speeds. For example, the skin friction of a supersonic fighter may be about 50 percent of the total drag at subsonic speed, and about 25 percent at supersonic speed (Whitford, 1987). The skin friction drag coefficient is found to vary parabolically with lift coefficient (Perkins and Hage, 1949).

Turning now to induced drag, the drag coefficient for the induced drag of a high-aspect unswept wing, in subsonic flow, can be modeled as (Perkins and Hage, 1949; Anderson, 1999):

$$C_{D_i} = C_L^2/(\pi \ e \ AR) \qquad (2.3\text{-}12)$$

The *efficiency factor*, e, is close to unity, and aspect ratio is the important design parameter. This equation provides a guide to minimizing the induced drag of a complete aircraft, but the difficulties of constructing a light, high-aspect wing tend to limit the aspect ratio to values of 10 or lower, with the exceptions mentioned earlier.

Finally, consider wave drag. As in the case of an airfoil, an airplane will have a critical Mach number when the flow reaches supersonic speed at some point on the surface, and the airplane drag coefficient begins to rise. The drag-divergence Mach number is the corner point or "knee" of the increasing drag coefficient curve, and is reached next. A shock wave pattern is now established over the airplane and the total drag now includes wave drag. The drag coefficient continues to rise, peaks at about the end of the transonic regime, and falls off in the manner of the Prandtl-Glauert formula. Figure 2.3-2a shows the transonic drag rise for a particular fighter aircraft. The peak drag can be minimized by using a combination of three techniques. First, wing sweep (up to about 70 degrees) is used to reduce the component of the relative wind that is normal to the leading edge of the wing. This has the effect of shifting the drag rise curve to the right and merging it into the supersonic part.

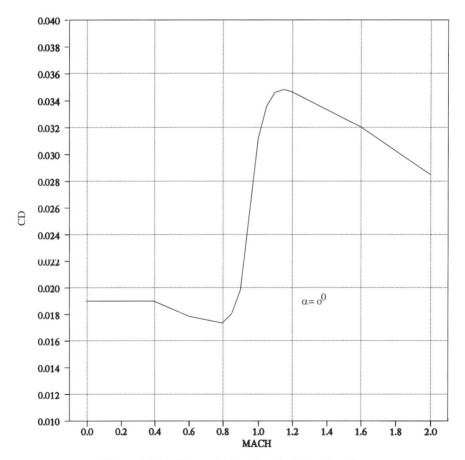

Figure 2.3-2a Transonic drag rise of a fighter aircraft.

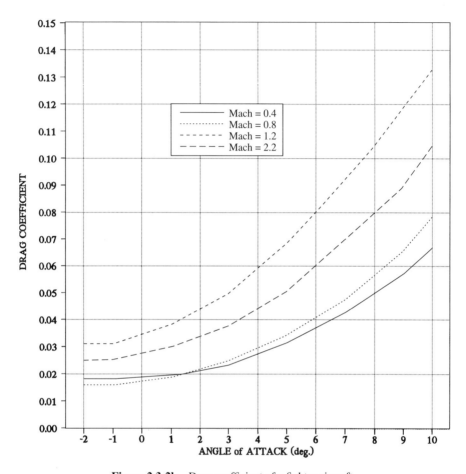

Figure 2.3-2b Drag coefficient of a fighter aircraft.

The drag rise becomes less steep, the peak of the curve becomes less sharp, and its height is reduced. Second, supersonic aircraft use thin airfoils, with thickness-over-chord ratios down to about 5 percent; these airfoils have lower wave drag and higher critical Mach numbers than thick airfoils. Finally, if the cross-sectional area of the complete airplane is made to vary smoothly with the distance from nose or tail, then the drag peak can be significantly reduced. This is R. T. Whitcomb's famous *area rule* (Anderson, 1999), and it leads to a fighter fuselage with a pinched waist at the point where the wings begin.

At constant Mach and below stall, the three types of drag described above each have a component that varies with the square of lift coefficient and a component that is independent of lift coefficient. Therefore, below stall, the complete airplane drag can be written as

$$C_D(C_L, M) = k(M)\left(C_L - C_{L_{DM}}\right)^2 + C_{DM}(M), \qquad (2.3\text{-}13)$$

where $k(M)$ is a proportionality constant that changes with Mach. This parabolic equation matches the actual drag variation quite accurately; it is known as the *drag polar*. Note that the minimum drag C_{DM} can occur at a nonzero value $C_{L_{DM}}$ of the lift coefficient.

If we consider lift beyond the stall, up to an angle of attack of 90 degrees, two or possibly three values of alpha can correspond to a given lift coefficient, and the drag is in general different for each of these values of alpha. Therefore, for high-alpha simulation, we model drag as a function of alpha. Because lift is quite linear as a function of alpha below stall, the plot of drag coefficient is still parabolic in this region. Figure 2.3-2b shows the untrimmed baseline drag coefficient of the same supersonic fighter aircraft as used for Figure 2.3-2a, plotted against alpha with Mach as a parameter. In the figure, the drag varies parabolically with alpha, and varies with Mach number in the same way as in Figure 2.3-2a.

In addition to the above effects, we can expect the drag coefficient to be dependent on altitude, sideslip, control surface and flap deflections, landing gear extension, and possibly ground effect. Altitude dependence (with Mach) allows for the effect of Reynolds number on the skin friction drag.

With the above facts and Equation (2.3-8a) in mind, we might expect a drag coefficient model consisting of a "baseline" component plus drag increments for control surfaces and gear, of the form:

$$C_D = C_D(\alpha, \beta, M, h) + \Delta C_D(M, \delta_e) + \Delta C_D(M, \delta_r) + \Delta C_D(\delta_F)$$
$$+ \Delta C_D(\text{gear}) + \cdots \qquad (2.3\text{-}14)$$

With aircraft that operate with little sideslip, the sideslip dependence can be treated as a separate increment.

Lift Coefficient, C_L

The lift coefficient of the complete aircraft is determined by the wings, fuselage, and horizontal tail, and their mutual interference effects. Nevertheless it varies with alpha and Mach in a way similar to that described earlier for the finite wing. The variation of lift coefficient with alpha is usually quite linear until near the stall, when it drops sharply and then may rise again, before falling to zero when alpha is near 90 degrees. The peak value of the lift coefficient may be as great as 3 for a highly cambered wing, but the increased drag of a highly cambered wing is not acceptable for high-speed aircraft. These aircraft use thin wings with not much camber and get their lift from the higher dynamic pressure, or from effectively increasing the camber with leading and trailing-edge flaps to get lift at low speed. Ground effect produces greater lift for a given drag; it is usually negligible beyond one wingspan above the ground.

The slope of the lift curve increases with aspect ratio and with reduction in the wing leading-edge sweep-angle. Light aviation aircraft and large passenger jets can have wing aspect ratios greater than 7, compared to 3 to 4 for a fighter aircraft. Increasing the wing sweep-angle has the desirable (for high-speed aircraft) effect of delaying the transonic drag rise, and the sweep angle may lie between roughly 25 and 60 degrees. Since lift-curve slope is an important factor in determining the response to turbulence, some military aircraft with a requirement for very low-altitude high-speed flight tolerate the expense and weight of variable-sweep wings. Compressibility effects also change the slope of the lift curve. Airfoil section theory predicts that at subsonic Mach numbers the slope should vary as $(1 - M^2)^{-1/2}$, and at supersonic Mach numbers as $(M^2 - 1)^{-1/2}$, and this kind of behavior is observed in practical wings.

Dependence of lift on sideslip is usually small until the magnitude of the sideslip reaches several degrees, and since large values of sideslip only occur at low speed, this effect will typically be modeled as a separate Mach-independent correction to the baseline lift. The dependence on altitude is small and will be neglected here, and the dependence on control surface deflection is specific to the particular surface. Therefore, we will focus on the remaining three variables.

The *thrust coefficient*, T_C, normally applies to propeller aircraft and is used to account for propeller wash over the wings, fuselage, and vertical fin. It is defined by normalizing engine thrust in the same way as the nondimensional coefficients; thus

$$T_C = \text{thrust}/\bar{q} S_D, \qquad (2.3\text{-}15)$$

where S_D is the area of the disc swept out by a propeller blade. The propeller slipstream increases the airspeed over the wings, changes the angle of downwash behind the wing (which affects the angle of attack of the horizontal tail), and changes the dynamic pressure at the tail. The effect on the airplane lift curve can be very significant; Figure 2.3-3a shows the lift curve of a turboprop transport aircraft with four engines mounted directly on the wing. At high thrust coefficient, the figure shows a major increase in the peak lift coefficient and a shift of the peak to higher alpha. More information on power effects can be found in Perkins and Hage (1949), and Stinton (1983).

Figure 2.3-3b shows the effect of Mach number on the lift curve of a fighter aircraft. Note that the slope of the lift curve at first increases with Mach number and then decreases. An additional effect (not shown) is that the peak lift coefficient decreases with increasing supersonic Mach number.

The normal force coefficient is often a more convenient quantity than lift coefficient. The normal force coefficient will usually rise with alpha, nearly monotonically, all the way to 90 degrees angle of attack, whereas lift coefficient shows the complicated stall behavior. Unfortunately, its partner, the axial force coefficient, displays very complicated behavior in the same range of alpha, and may change sign a few times over the range of alpha. The rotation matrix (2.3-2) and the definitions in Table 2.3-1 give the following expressions for the lift and drag coefficients in terms of the body-axes coefficients:

$$C_D = -\cos\alpha \, \cos\beta \, C_X - \sin\beta \, C_Y + \sin\alpha \, \cos\beta \, C_N$$
$$C_L = \sin\alpha \, C_X + \cos\alpha \, C_N$$

and at low alpha $C_L \approx C_N$. Figures 2.3-3c and d show the low-Mach, high-alpha, normal and axial force coefficients for the F-4E aircraft.

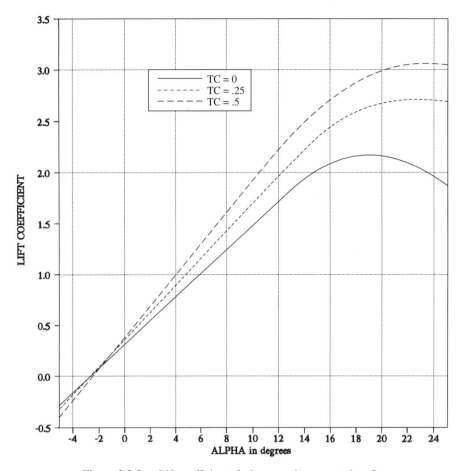

Figure 2.3-3a Lift coefficient of a low-speed transport aircraft.

A general model for lift coefficient may be of the form:

$$C_L = C_L(\alpha, \beta, M, T_c) + \Delta C_L(\delta_F) + \Delta C_{L_{ge}}(h), \qquad (2.3\text{-}16)$$

where $\Delta C_{L_{ge}}(h)$ is the increment of lift in ground effect.

Sideforce Coefficient, C_Y

In the case of a symmetrical aircraft, sideforce is created mainly by sideslipping motion (i.e., $\beta \neq 0$) and by rudder deflection. Figure 2.3-4 shows the sideforce coefficient for the F-4B and -C aircraft (Chambers and Anglin, 1969) for alpha equal to zero and 40 degrees, and with linear interpolation for other values of alpha. Note that positive sideslip leads to negative sideforce because positive sideslip corresponds to the relative wind on the right-hand side of the nose. The high-alpha curve does not

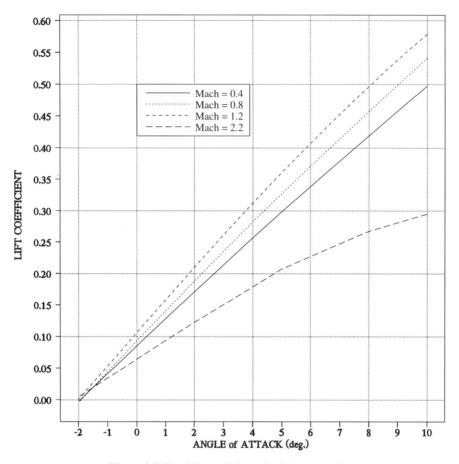

Figure 2.3-3b Lift coefficient of a fighter aircraft.

pass through the origin possibly because of asymmetry in the wind tunnel model, or anomalies in the measurements. Note that at high subsonic speeds very little sideslip is possible without exceeding the hinge-moment limit of the rudder, or the structural limit of the vertical fin.

The sideforce model for a high-performance aircraft is typically of the form:

$$C_Y = C_Y(\alpha, \beta, M) + \Delta C_{Y_{\delta_r}}(\alpha, \beta, M, \delta_r) + \Delta C_{Y_{\delta_a}}(\alpha, \beta, M, \delta_a)$$
$$+ \frac{b}{2V_T}\left[C_{Y_p}(\alpha, M)P + C_{Y_r}(\alpha, M)R\right] \quad (2.3\text{-}17a)$$

Additional corrections are added for flaps, gear, and the like. The last two terms are linear in the angular rates, and the other terms are linearized whenever acceptable accuracy is achieved; thus,

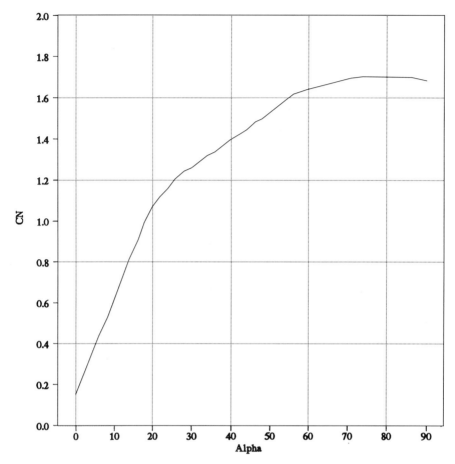

Figure 2.3-3c Normal force coefficient of the F-4E aircraft.

$$C_Y(\alpha, \beta, M) \approx C_{Y_\beta}(\alpha, M) \times \beta$$

$$\Delta C_{Y_{\delta_r}}(\alpha, \beta, M, \delta_r) \approx C_{Y_{\delta_r}}(\alpha, \beta, M) \times \delta_r \quad (2.3\text{-}17b)$$

$$\Delta C_{Y_{\delta_a}}(\alpha, \beta, M, \delta_a) \approx C_{Y_{\delta_a}}(\alpha, \beta, M) \times \delta_a$$

These terms have been linearized "around the origin," that is, for a symmetrical aircraft, the sideforce can be expected to go to zero when the sideslip is zero and the rudder and aileron are in their neutral positions.

Rolling Moment

Rolling moments are created by sideslip alone, by the control action of the ailerons and the rudder, and as damping moments resisting rolling and yawing motion.

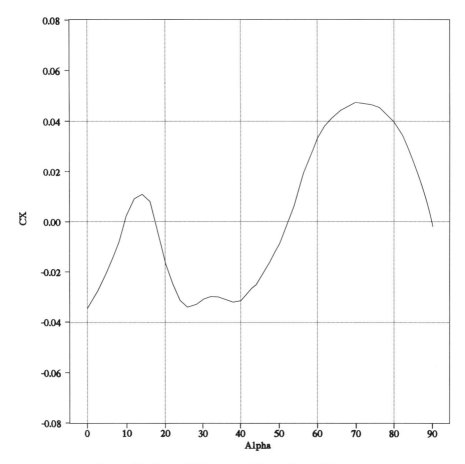

Figure 2.3-3d Axial force coefficient of the F-4E aircraft.

Consider first the effect of sideslip; if a right-wing-down roll disturbance occurs, and is not corrected (stick fixed), then the effect of gravity will be to start a positive sideslip. If the aircraft aerodynamics are such that positive sideslip causes a positive rolling moment, then the roll angle will increase farther. This is an unstable situation. We see that, for positive stiffness in roll, the slope of the rolling moment versus sideslip curve should be negative. Therefore, it is useful to understand the aerodynamic effects that determine the behavior of the rolling moment coefficient with sideslip; this will be our baseline term in the rolling moment coefficient buildup.

The baseline rolling-moment coefficient is primarily a function of sideslip, alpha, and Mach, and can be written as $C_\ell(\beta, \alpha, M)$. Figure 2.3-5 is a plot of the rolling moment coefficient for the F-4B at low Mach number; it shows that, for small sideslip, the coefficient is approximately linear with beta, but changes in alpha can cause a significant change in slope. Also, at low alpha, sideslip greater than 20 degrees can cause a loss of stability in roll. In general, the effect of sideslip is to create a lateral

AIRCRAFT FORCES AND MOMENTS   **89**

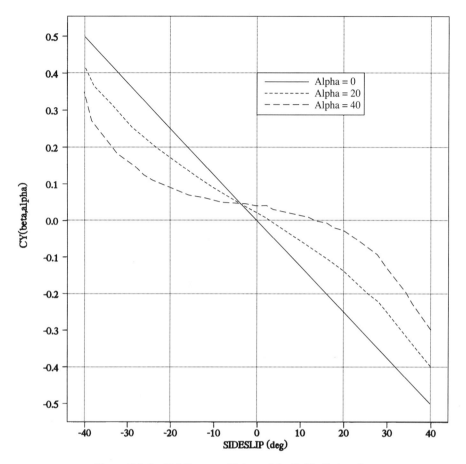

Figure 2.3-4 Sideforce coefficient of the F-4B, C aircraft.

component of the relative wind, and there are three separate effects of this lateral component on the horizontal aerodynamic surfaces. These will now be described.

First, note that the angle by which the wings of an aircraft are canted up above the body-axes x-y plane is called the *dihedral angle*, and a negative dihedral angle is called an *anhedral angle*. Dihedral is often very noticeable on small low-wing (wing root attached at the bottom of the fuselage) aircraft, while a well-known example of anhedral is the Harrier (AV-8B) aircraft. Dihedral (or anhedral) angles give one wing a positive angle of attack (in a spanwise direction) to the lateral component of the relative wind, and the other wing receives a similar negative angle of attack. Referring to Figure 2.3-1, it is easy to see that positive beta creates a negative rolling moment when the wings have positive dihedral. This same effect applies to the horizontal tail.

The second effect of sideslip on the horizontal surfaces occurs when they are swept back. In this case the relative wind is more nearly perpendicular to the leading edge of the windward wing than is the case for the leeward wing. Therefore, the windward

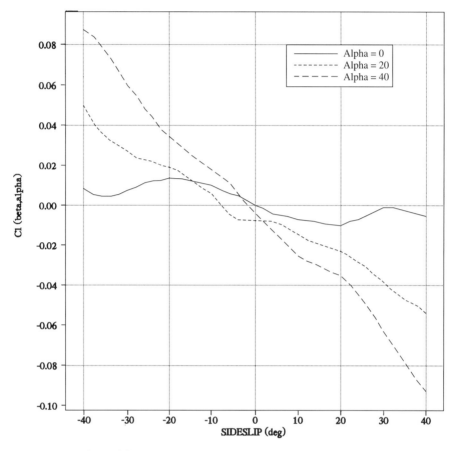

Figure 2.3-5 Rolling moment coefficient of the F-4B aircraft.

wing develops more lift, and the outcome is again a negative rolling moment for positive beta.

The third effect of sideslip on the horizontal surfaces is that on the windward side of the fuselage some of the lateral airflow is diverted up and over the fuselage, and some is diverted under the fuselage. This flow will modify the angle of attack of the wings, depending on their position on the fuselage. Above the centerline of the fuselage, the upward component of the relative wind is increased. Therefore, for a high-wing aircraft, the angle of attack of that wing is increased (assuming that it was operating at a positive alpha). For a low-wing aircraft the upward component of the relative wind would be reduced by the effect of the air flowing down and under the fuselage, and the angle of attack of that wing would be reduced. Thus for low-wing aircraft, positive sideslip creates a positive contribution to rolling moment, and for high-wing aircraft it creates a negative contribution.

Finally, the lateral component of the relative wind acting on the vertical tail will generate a rolling moment about the cm. Depending on the aircraft angle of attack, and the location of the center of pressure of the vertical tail, this rolling moment could be positive or negative. Usually positive beta will produce a negative rolling moment component.

Of all the above effects, only the fuselage effect on a low-wing airplane led to a positive increment in rolling moment in response to a positive increment in beta. This can be a strong effect, and is responsible for a loss of stability in roll. Low-wing airplanes usually have noticeable positive dihedral in order to provide positive roll stiffness. The airplanes will then have an inherent tendency to fly with wings level.

For a high-performance aircraft the rolling moment model will typically be of the form:

$$C_\ell = C_\ell(\alpha, \beta, M) + \Delta C_{\ell_{\delta a}}(\alpha, \beta, M, \delta_a) + \Delta C_{\ell_{\delta r}}(\alpha, \beta, M, \delta_r)$$
$$+ \frac{b}{2V_T}\left[C_{\ell_p}(\alpha, M)P + C_{\ell_r}(\alpha, M)R\right], \qquad (2.3\text{-}18a)$$

where C_{ℓ_p} is the *roll-damping derivative*. The rolling moment dependence on β, and the aileron and rudder, can often be linearized around the origin:

$$C_\ell(\alpha, \beta, M) \approx C_{\ell_\beta}(\alpha, M) \times \beta$$
$$\Delta C_{\ell_{\delta a}}(\alpha, \beta, M, \delta_a) \approx C_{\ell_{\delta a}}(\alpha, \beta, M) \times \delta_a \qquad (2.3\text{-}18b)$$
$$\Delta C_{\ell_{\delta r}}(\alpha, \beta, M, \delta_r) \approx C_{\ell_{\delta r}}(\alpha, \beta, M) \times \delta_r,$$

where C_{ℓ_β} is the *dihedral derivative* that determines static stability in roll, and $C_{\ell_{\delta a}}$ and $C_{\ell_{\delta r}}$ are roll *control derivatives*.

Figure 2.3-6 shows the stability-axes, trimmed, roll-damping derivative for the F-4C. The data are Mach-dependent because of compressibility, and altitude-dependent because the trimmed angle of attack changes with altitude, and because of aeroelastic changes with dynamic pressure.

Control Effects on Rolling Moment

We now briefly examine the control moment terms in Equations (2.3-18), with respect to their dependence on alpha and Mach. The rudder is intended to provide directional control (yaw), so the "cross-control" effect on rolling moment is an unwanted effect. This effect comes about because the center of pressure of the rudder is normally above the longitudinal axis.

Conventional ailerons mounted outboard on the trailing edge of the wings become ineffective and can reverse their net effect as high subsonic speeds are approached. This is because the aileron lift component produced by a downward deflection twists the wing in the direction that reduces its angle of attack and hence reduces the wing-lift component. Spoilers, which are uncambered surfaces deflected upward above the after surface of the wing, "spoil" the lift on that portion of the wing and thus provide roll control. The twisting effect on the wing is reduced and control reversal

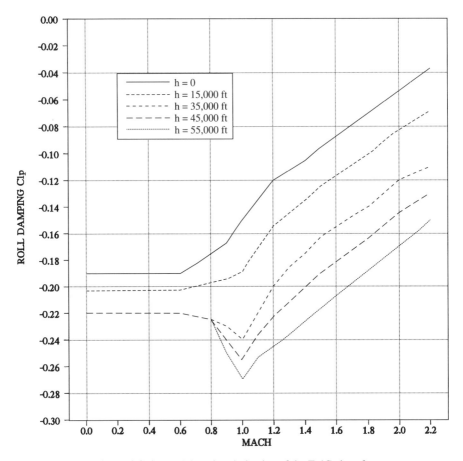

Figure 2.3-6 Roll damping derivative of the F-4C aircraft.

can be avoided. Spoilers are commonly combined with ailerons in such a way that one aileron and the opposite spoiler operate simultaneously, and the ailerons deflect downward only. Mounting the ailerons farther inboard reduces the effect of wing twist but also reduces their moment arm. However, the X-29 forward-swept-wing aircraft is an example of combined inboard and outboard "flaperons" being made to work very effectively up to high alpha (Kandebo, 1988).

The effectiveness of both ailerons and spoilers is reduced by cross-flows on the wing and hence by wing sweep. Therefore, for swept-wing aircraft, an additional rolling moment is obtained by using differential control of the horizontal-tail control surfaces (e.g., most modern fighter aircraft).

Pitching Moment

The baseline pitching moment coefficient may typically be written as $C_m(\alpha, M, T_c)$ for a low-speed aircraft, or $C_m(\alpha, M, h)$ for a high-speed jet aircraft where aeroelastic

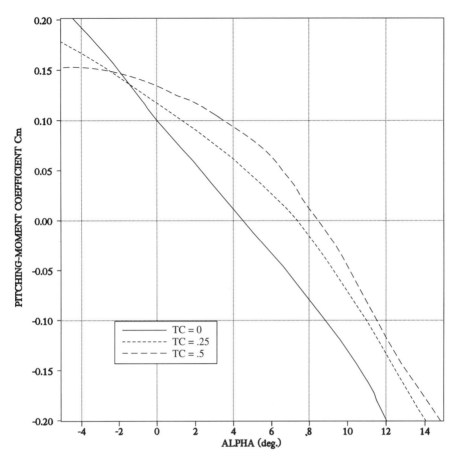

Figure 2.3-7a Pitching moment coefficient of a low-speed transport aircraft.

effects are included. Figure 2.3-7a illustrates the dependence of this coefficient on α and T_c for the low-speed turbo-prop transport aircraft. The figure shows that, as alpha increases, the nose-down (restoring) moment becomes steadily stronger. At low freestream angles of attack and high thrust coefficient, the propeller wash tends to make the effective angle of attack independent of the freestream direction, and the moment curve has only a small negative slope (reduced pitch stiffness).

Figure 2.3-7b shows a baseline moment coefficient that is representative of a supersonic jet-trainer. In this case the parameter is Mach number, and the slope of the moment curve gets steeper with increasing Mach because of the rearward shift of the wing-body aerodynamic center. This increasing pitch stiffness is detrimental to maneuverability and to the lift-over-drag ratio; it is discussed further in the pitch static-stability section.

For a high-performance aircraft the pitching moment coefficient will be built up in the form:

94 MODELING THE AIRCRAFT

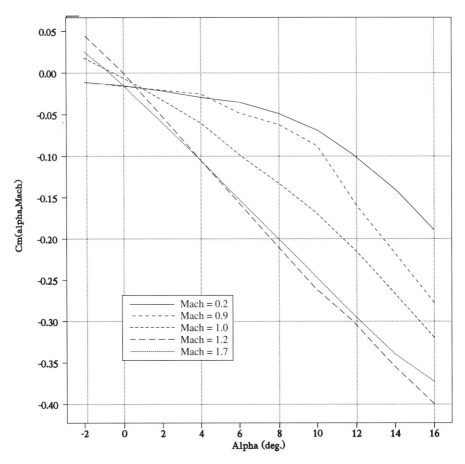

Figure 2.3-7b Pitching moment coefficient of a jet trainer aircraft.

$$C_m = C_m(\alpha, M, h, \delta_F, T_c) + \Delta C_{m_{\delta e}}(\alpha, M, h, \delta_e) + \frac{\bar{c}}{2V_T}\left[C_{m_q}Q + C_{m_{\dot\alpha}}\dot\alpha\right]$$

$$+ \frac{x_R}{\bar{c}}C_L + \Delta C_{m_{\text{thrust}}}(\delta_t, M, h) + \Delta C_{m_{\text{gear}}}(h) \quad (2.3\text{-}19a)$$

In the baseline term, all five variables are unlikely to be present simultaneously. The M and h variables imply a high-speed aircraft, while T_c implies a low-speed propeller aircraft. Also, the effect of wing-flap deflection, δ_F, may be treated as a separate increment. The elevator increment term may often be linearized around the origin:

$$\Delta C_{m_{\delta e}}(\alpha, M, h, \delta_e) \approx C_{m_{\delta e}}(\alpha, M, h) \times \delta_e, \quad (2.3\text{-}19b)$$

where $C_{m_{\delta e}}$ is the *elevator control power*. The pitch damping derivative C_{m_q} and the alpha-dot acceleration derivative will also be functions of alpha, Mach, and altitude, and are discussed in Section 2.6. The purpose of the term $(x_R C_L)/\bar{c}$ is to correct for any x-displacement (x_R) of the aircraft cm from the aerodynamic data reference position. If x_R is not zero, the lift force will provide a contribution to the pitching moment. This is considered in more detail in the next section. The last two terms represent, respectively, the effect of the engine thrust vector not passing through the aircraft cm, and the moment due to landing-gear doors and landing gear. This last term is dependent on height above ground because of ground effect.

Control Effects on Pitching Moment

A conventional elevator, for a subsonic aircraft, consists of a movable surface at the trailing edge of the horizontal tail. In addition, the horizontal tail may move as a whole, or a "tab" on the elevator may move, so that the elevator deflection can be trimmed to zero in various flight conditions. In transonic and supersonic flight a shock wave attached to the horizontal tail would render this type of elevator ineffective. Therefore, on supersonic aircraft the complete horizontal stabilizer surface moves (i.e., a "stabilator") in response to control stick or trim button signals. As indicated above, elevator (or stabilator) control power is dependent on Mach and altitude because compressibility and aeroelastic effects cause the elevator effectiveness to decrease with increasing Mach number and dynamic pressure.

An aft tail experiences a downwash effect from the wing and a reduction in dynamic pressure. These are alpha-dependent effects and can be included in the control power term as implied above. However, for a propeller aircraft, the dynamic pressure at the tail is strongly dependent on thrust coefficient, and may be greatly increased. This can be modeled by multiplying the elevator control power by a *tail efficiency factor*, η, which is a function of alpha, thrust coefficient, flap deflection, and ground effect.

$$\eta(\alpha, T_c, \delta_F, h) \equiv \bar{q}_{TAIL}/\bar{q} \qquad (2.3\text{-}20)$$

The tail efficiency factor of a propeller aircraft may exceed 2.0 at high values of thrust coefficient.

Yawing Moment

Yawing moments are created by sideslip, by the action of the rudder, by propeller effects, by unbalanced thrust in a two-engine aircraft, and, to a lesser extent, by differences in drag between the ailerons, and by asymmetric aerodynamic effects at high alpha (e.g., "vortex shedding").

The sideslip dependence has three components. A small component is created by wing sweep: positive sideslip creates a positive yawing moment because the right wing becomes more nearly perpendicular to the freestream direction and develops more lift and drag. Second, the fuselage produces a strong negative yawing moment when in positive sideslip (see, for example, Perkins and Hage, 1949). Third, directional stability demands that the aircraft should tend to weathercock into the relative wind; therefore, it is the job of the vertical tail to provide a strong yawing moment of the same sign as beta. This moment is computed from the moment arm of the tail about the cm, and the "lift" generated by the vertical tail when in sideslip. The overall result of these effects is that the yawing moment is quite linear in beta, for low values of sideslip.

When the aircraft is at a high angle of attack, the fuselage yawing moment can become more adverse, and at the same time the dynamic pressure at the tail may be reduced, eventually resulting in a loss of directional stability. Figure 2.3-8a shows low-speed, high-alpha, yawing moment data for the F-4B,C aircraft, and clearly shows the loss of directional stability at high alpha.

A rotating propeller produces several different "power effects," which are best included in the propulsion model (Perkins and Hage, 1949; Ribner, 1943). A conventional tractor propeller has a destabilizing effect in yaw, while a pusher propeller has

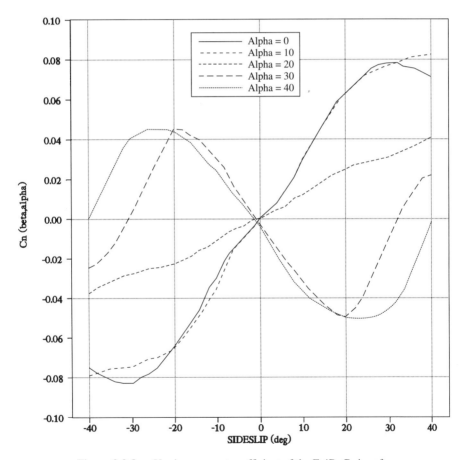

Figure 2.3-8a Yawing moment coefficient of the F-4B, C aircraft.

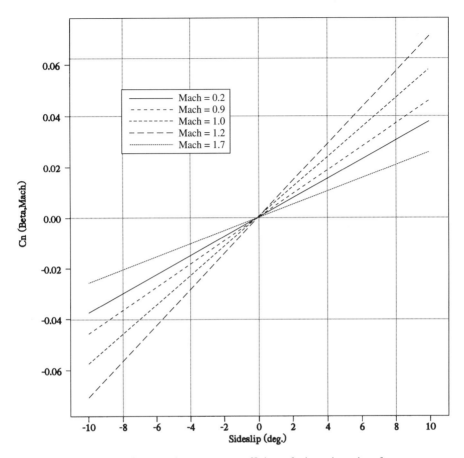

Figure 2.3-8b Yawing moment coefficient of a jet trainer aircraft.

a stabilizing effect. The slipstream of a tractor propeller strongly affects the dynamic pressure at the tail of the airplane, and the swirl of the slipstream modifies the flow over the fuselage and tail. Thus the baseline yawing moment can have a strong dependence on thrust coefficient.

Finally, with a high-speed aircraft, compressibility effects can cause the slope of the yawing moment versus beta curve to be a function of Mach number. Figure 2.3-8b shows the effect of Mach on the yawing moment of the jet-trainer aircraft at low alpha.

The yawing moment coefficient for a high-performance aircraft will be of the form,

$$C_n = C_n(\alpha, \beta, M, T_c) + \Delta C_{n_{\delta_r}}(\alpha, \beta, M, \delta_r) + \Delta C_{n_{\delta_a}}(\alpha, \beta, M, \delta_a)$$

$$+ \frac{b}{2V_T}\left[C_{n_p}(\alpha, M)P + C_{n_r}(\alpha, M)R\right], \tag{2.3-21a}$$

where C_{n_r} is the *yaw-damping derivative*. The thrust coefficient in the baseline term is appropriate for a propeller aircraft. The yawing moment dependence on β, and the rudder and aileron, can often be linearized around the origin:

$$C_n(\alpha, \beta, M, T_c) \approx C_{n_\beta}(\alpha, M, T_c) \times \beta$$
$$\Delta C_{n_{\delta_r}}(\alpha, \beta, M, \delta_r) \approx C_{n_{\delta_r}}(\alpha, \beta, M) \times \delta_r \quad (2.3\text{-}21b)$$
$$\Delta C_{n_{\delta_a}}(\alpha, \beta, M, \delta_a) \approx C_{n_{\delta_a}}(\alpha, \beta, M) \times \delta_a,$$

where C_{n_β} is the *yaw stiffness derivative* that determines the directional stability, and $C_{n_{\delta_r}}$, $C_{n_{\delta_a}}$ are yaw control derivatives.

Control Effects on Yawing Moment

The rudder usually forms a part of the trailing edge of the vertical tail; when deflected, it provides a strong yawing moment and some rolling moment. Its purpose is to create sideslip (e.g., for crosswind landing), or remove sideslip (e.g., to coordinate a turn). The vertical tail is no longer a symmetric airfoil section when the rudder is deflected, and then begins to produce "lift." The resulting sideforce is such that deflection of the rudder trailing edge to the right produces a positive yawing moment. All-moving vertical fins are sometimes used for rudder control, as, for example, on the SR-71, where large yawing moments can occur as a result of an engine "unstart."

Like the horizontal tail, the vertical tail and rudder can be affected by wing downwash and blanketed at high angles of attack. A tail efficiency factor can be used to model the effect, as in the pitching moment equation. Wing flap deflection can also significantly change the downwash at the rudder.

Differential deflection of the ailerons and spoilers also produces a yawing moment because of the difference in drag between the two sets. As described earlier, roll control can be obtained in a number of different ways, and the cross-control effects on yawing moment can vary.

Data Handling

It should be clear from the foregoing description of aerodynamic forces and moments that the aerodynamic database for a given aircraft can become rather large. It may range from roughly fifty data tables for a relatively simple piloted simulation model, to several thousand tables for an aircraft like the Lockheed-Martin F-22. Many of the tables will have four independent variables, and could contain over 10,000 data points; the whole database could contain a few million points.

A large aerodynamic database must be handled efficiently within an organization; it represents thousands of hours of planning, model testing, flight testing, and computer simulation. It must be kept current, with all changes fully documented, and be accessible to different users. The control engineer will have access to the database through a computer workstation, and will be able to call up the appropriate force and moment routines for the equations of motion. An example of a small database has been given by Nguyen (Nguyen et al., 1979) for low-speed F-16 model data, taken

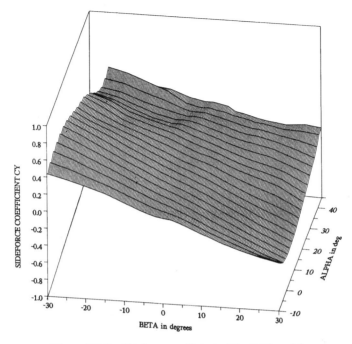

Figure 2.3-9 Sideforce coefficient of the F-16 model.

at the NASA Dryden and Langley Research Centers. A three-dimensional plot made from one of the two-dimensional (two independent variables) tables of this F-16 data is shown in Figure 2.3-9. A reduced data set derived from this report is listed in the appendices and is used for the F-16 model given in Chapter 3.

Aerodynamic lookup-table data are discrete, whereas aircraft models require data at arbitrary values of the independent variables. This problem is solved by using an interpolation algorithm with the data. In the appendices we have provided a simple interpolation algorithm for use with the F-16 data.

Basic aerodynamic data are often rough (scattered data points). This is because of the inaccuracies associated with measuring aerodynamic data, the sensitivity to small changes in the independent variables, and fusion of data from different sources or test runs. The data can be smoothed and regenerated at new uniform increments of the independent variables as required, for example, by means of a "spline" algorithm (IMSL, 1980; Press et al, 1989).

2.4 STATIC ANALYSIS

In steady-state flight the forces and moments acting on an aircraft are constant (i.e., static) when taken in the body-fixed frame. Static analysis provides the basic information for sizing and configuring the aircraft and evaluating its performance, and lays the groundwork for dynamic analysis. A "static stability" analysis is used to determine if the aircraft will return to a steady-state flight condition after being subjected to a small atmospheric disturbance. For example, an incremental increase in Mach number can produce a net increase in thrust-minus-drag and cause a relatively slow departure, in speed, from the equilibrium condition. In contrast, an unstable departure in pitch could be too fast for the pilot to control, and could lead to structural failure. The static stability analysis is so-called because rate-dependent effects are not considered, and it is usually performed for the special case of wings-level, nonturning flight. Dynamic stability in all of the motion variables can be determined from the eigenvalues of the linearized equations of motion, and is considered in Chapters 3 and 4. It can easily be performed in other steady-state flight conditions, for example, a steady turn.

Static Equilibrium

Here we consider only wings-level, zero-sideslip flight. Suitable coordinate systems for this analysis are the body-fixed axes and the stability axes (now coincident with the wind axes). We must bear in mind that the origin of these systems, the aircraft cm, is not a fixed point. The cm will move as fuel is drawn from different tanks, or because of cargo movement or stores being dropped. Aerodynamic data are referred to a fixed point, typically the point inside the fuselage where a line joining the quarter-chord points, in the wing roots, intersects the plane of symmetry.

Figure 2.4-1 shows the forces and moments on the aircraft. In the figure, $R(x_R, 0, z_R)$ is the reference point for the aerodynamic moment data, C is the aircraft cm, and T is the quarter-chord point (in the plane of symmetry) of the horizontal tail. $\mathbf{F}_R$ is the resultant aerodynamic force on the aircraft, L and D are its lift and drag components, and $\mathbf{M}_R$ is the total aerodynamic moment at R. With respect to the aircraft cm, the position vectors of R and the quarter-chord point in the horizontal

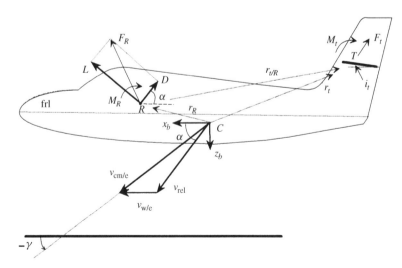

Figure 2.4-1 Diagram for calculating pitching moment.

tail are, respectively, $\mathbf{r}_R$ and $\mathbf{r}_t$. The chord line of the horizontal tail has an incidence angle i_t to the fuselage reference line. The thrust vector $\mathbf{F}_T$ (not shown) is assumed to lie in the plane of symmetry, tilted up at an angle α_T to the fuselage reference line, and does not necessarily pass through the cm.

To determine the equilibrium conditions, the direction of the gravity vector must be known relative to the aircraft. The *flight-path angle* γ shown in the figure is the angle that the velocity vector $\mathbf{v}_{CM/e}$ makes with the NE plane, and is positive when the aircraft is climbing. For simplicity the wind velocity will be taken to be zero so that $\mathbf{v}_{rel} = \mathbf{v}_{CM/e}$, and alpha and the flight-path angle will determine orientation relative to the gravity vector. Then summing force components along the x and z stability axes, yields

$$F_T \cos(\alpha_{frl} + \alpha_T) - D - mg_D \sin \gamma = 0 \qquad (2.4\text{-}1a)$$

$$F_T \sin(\alpha_{frl} + \alpha_T) + L - mg_D \cos \gamma = 0 \qquad (2.4\text{-}1b)$$

The moment at the cm is given by:

$$\mathbf{M}_{CM} = \mathbf{M}_R + \mathbf{r}_R \times \mathbf{F}_R + \mathbf{M}_p, \qquad (2.4\text{-}2)$$

where $\mathbf{M}_p$ is the pitching moment created directly by the engines. Using body-axes components in the cross-product yields the equilibrium equation:

$$0 = M_{CM} = M_R + x_R F_N + z_R F_X + M_p, \qquad (2.4\text{-}3)$$

where the normal force, F_N, and axial force, F_X, are given by

$$F_N = L \cos \alpha_{frl} + D \sin \alpha_{frl}$$
$$F_X = L \sin \alpha_{frl} - D \cos \alpha_{frl}$$

Now divide (2.4-3) by $(\bar{q}S\bar{c})$ to obtain dimensionless moment coefficients,

$$C_{m_{CM}} = C_{m_R} + \frac{x_R}{\bar{c}} C_N + \frac{z_R}{\bar{c}} C_X + C_{m_p}, \qquad (2.4\text{-}4)$$

where C_{m_p} is the thrust moment made dimensionless by dividing by $(\bar{q}S\bar{c})$, and

$$C_N = \left[C_L \cos \alpha_{frl} + C_D \sin \alpha_{frl}\right] \approx C_L \text{ in cruise,}$$
$$C_X = \left[C_L \sin \alpha_{frl} - C_D \cos \alpha_{frl}\right] \qquad (2.4\text{-}5)$$

In equilibrium, the left-hand side of Equation (2.4-4) is zero, and in cruise conditions, $\cos \alpha_{frl} \approx 1$, $\sin \alpha_{frl} \approx \alpha_{frl}$, $C_L \gg C_D$, and $C_N \gg C_X$. Normally the coordinates x_R and z_R are both small, and either one could be zero.

In *performance analysis* we solve the nonlinear equilibrium equations (2.4-1) and (2.4-4), for a given flight condition (true airspeed and altitude), using an iterative computer algorithm (Problem 2.4-1). The data required are $F_T(M, h, \delta_t)$, aerodynamic data—$C_L(\alpha, M, \delta e)$, $C_D(\alpha, M, \delta e)$, $C_{m_R}(\alpha, M, \delta e)$—and an atmosphere model to determine mass density and the speed of sound at any altitude. Effects such as flaps and gear can be included if required. If the effect of elevator deflection on lift and drag is ignored, then the force equations are independent of the moment equation.

Effect of the Horizontal Tail

Now suppose that $\mathbf{F}_{wb}$ and $\mathbf{M}_{R,wb}$ are the aerodynamic force and moment vectors at R when the horizontal tail is removed from the aircraft. The flow over the wing-body combination creates a downwash effect at the horizontal tail position and a change in dynamic pressure, both of which are dependent on flight conditions. These effects will be modeled later. Let $\mathbf{F}_t$ and $\mathbf{M}_{c/4,t}$ be the force and moment vectors measured at the quarter-chord point of the isolated horizontal tail when it is placed in the same flowfield that exists at the tail position of the wing-body combination. Also, assume that putting the horizontal tail back on the wing-body combination does not significantly modify the wing-body flowfield. With these assumptions, we can write

$$\mathbf{M}_{CM} = \mathbf{M}_{R,wb} + \mathbf{M}_{c/4,t} + \mathbf{r}_R \times \mathbf{F}_{wb} + \mathbf{r}_t \times \mathbf{F}_t + \mathbf{M}_p \qquad (2.4\text{-}6a)$$

$$\mathbf{F}_R = \mathbf{F}_{wb} + \mathbf{F}_t \qquad (2.4\text{-}6b)$$

In order to have the tail position specified in terms of a fixed vector, let

$$\mathbf{r}_t = \mathbf{r}_R + \mathbf{r}_{t/R},$$

where $\mathbf{r}_{t/R}$ is shown in the figure. Then from Equations (2.4-6a) and (2.4-6b),

$$\mathbf{M}_{CM} = \mathbf{M}_{R,wb} + \mathbf{M}_{c/4,t} + \mathbf{r}_R \times \mathbf{F}_R + \mathbf{r}_{t/R} \times \mathbf{F}_t + \mathbf{M}_p \qquad (2.4\text{-}7)$$

The aerodynamic moment vector at the cm has the same y component M in either body or stability axes; other components are zero because of the symmetrical flight condition. Equation (2.4-7) will be evaluated using the body-fixed components of the reference point and the horizontal tail. It will be assumed that the aerodynamic-data reference point, tail quarter-chord point, and center of mass are at the same height in the fuselage, so that the z components disappear from the equation. This allows conclusions to be drawn about the effect of the longitudinal position of the cm on static stability, with much less cumbersome equations; it is also usually a good approximation in practice. Equation (2.4-7) then yields the scalar equation:

$$M_{CM} = M_{R,wb} + M_{c/4,t} + x_R \left[L \cos(\alpha_{frl}) + D \sin(\alpha_{frl}) \right]$$
$$+ (x_t - x_R) \left[L_t \cos(\alpha_{frl} - \epsilon) + D_t \sin(\alpha_{frl} - \epsilon) \right] + M_p \qquad (2.4\text{-}8)$$

In this equation ϵ is the *downwash angle* at the horizontal tail, and is usually positive with a magnitude of a few degrees. It represents the effect of the wings and fuselage on the direction of flow at the tail, and is a function of the aircraft angle of attack and thrust effects (Roskam, 1979). The tail lift, L_t, and drag, D_t, are defined relative to the direction $(\alpha_{frl} - \epsilon)$. In addition to the downwash effect at the tail, the airflow over the tail is modified in speed by the effect of the wings and body. This effect is modeled by the tail efficiency (η, Equation 2.3-20).

The tail lift and drag are computed from $\bar{q}_t$ and tail reference area S_t. Therefore, dividing Equation (2.4-8) by $(\bar{q}\bar{c}S)$ yields

$$C_{m_{CM}} = C'm_R + \frac{x_R}{\bar{c}} \left[C_L \cos(\alpha_{frl}) + C_D \sin(\alpha_{frl}) \right]$$
$$- \eta \bar{V}_H \left[C_{L_t} \cos(\alpha_{frl} - \epsilon) + C_{D_t} \sin(\alpha_{frl} - \epsilon) \right] + C_{m_p}, \quad (2.4\text{-}9)$$

where

$$C'_{m_R} = C_{m_{R,wb}} + \eta \frac{\bar{c}_t S_t}{\bar{c}S} C_{m_{c/4,t}} \quad (2.4\text{-}10)$$

and $\bar{V}_H$ is a modification of the *horizontal tail volume ratio* (Equation 2.6-35), given by:

$$\bar{V}_H = -\frac{(x_t - x_R)S_t}{\bar{c}S} \quad (2.4\text{-}11)$$

$\bar{V}_H$ is constant, and positive for an aft tail, because its numerator contains the distance of the reference point ahead of the tail. Note that, in the normal range of alpha, the drag terms in (2.4-9) can be discarded. A study of the moment equation will tell us how much elevator deflection is required to trim the aircraft, and the effect of movement of the aircraft cm on trimmed elevator deflection.

Static Stability Analysis in Pitch

We focus here on the pitching moment equation and the requirements on the aircraft configuration for static stability in pitch. Static directional and rolling stability are considered in Section 2.6, in conjunction with the stability derivatives.

The moment balance around the pitch axis of the aircraft is critical to both performance and stability. If the lift force generated by the wings and body creates a large moment about the cm, the horizontal tail must carry a significant load. If this is a download, the overall effect is additional drag with a net reduction in lift and reduced load-carrying efficiency. In Section 2.2 we saw that the moment about the airfoil aerodynamic center was constant and relatively small, and that for positive pitch stiffness, the axis about which the airfoil pivoted needed to be ahead of the ac. Therefore, the cm of the aircraft should be ahead of the ac of the wing-body combination, and the pitch stiffness of the complete aircraft must be analyzed. We also saw in Section 2.2 that the zero-alpha moment M_0 needed to be positive in order to obtain equilibrium with a positive angle of attack (and therefore provide the design lift). For a conventional aircraft this is achieved by giving the horizontal tail a negative incidence, so that it provides a positive contribution to the total pitching moment. Efficiency can be improved by reducing the pitch stiffness (this is done in relaxed static stability [RSS] designs), but then the movement of the cm must be more carefully controlled and the flight control system may need to be designed to provide artificial stability.

As in Section 2.2, to determine the static stability in pitch we need to find the slope of the pitching-moment versus alpha curve. Therefore, we must differentiate, with respect to α_{frl}, the total pitching moment at the center of mass as given by (2.4-9). In this equation each trigonometric function is multiplied by an aerodynamic coefficient that is also a function of alpha. In addition, tail efficiency and downwash angle are functions of alpha, and differentiation produces a very cumbersome expression. Nevertheless, the expression can be simplified by making use of the relationships:

$$\alpha_t = \alpha_{frl} + i_t - \epsilon$$

$$\partial \alpha_t / \partial \alpha_{frl} = 1 - \partial \epsilon / \partial \alpha_{frl}$$

(2.4-12a)

and, for the wing and body, the approximations:

$$\left.\begin{array}{r} C_L \sin(\alpha_{frl}) \\ C_D \cos(\alpha_{frl}) \\ C_{D_\alpha} \sin(\alpha_{frl}) \end{array}\right\} \ll C_{L_\alpha} \cos(\alpha_{frl}) \qquad (2.4\text{-}12b)$$

and for the tail:

$$\left.\begin{array}{r} (\partial \epsilon/\partial \alpha_{frl}) C_{L_t} \sin(\alpha_{frl} - \epsilon) \\ (\partial \epsilon/\partial \alpha_{frl}) C_{D_t} \cos(\alpha_{frl} - \epsilon) \\ (1 - \partial \epsilon/\partial \alpha_{frl}) C_{D_{\alpha,t}} \sin(\alpha_{frl} - \epsilon) \end{array}\right\} \ll (1 - \partial \epsilon/\partial \alpha_{frl}) C_{L_{\alpha,t}} \cos(\alpha_{frl} - \epsilon)$$

(2.4-12c)

These are normally very good approximations. Equation (2.4-9) can now be differentiated with respect to α_{frl}, and the above approximations applied, leading to

$$C_{m_\alpha} = \frac{\partial C'_{m_R}}{\partial \alpha} + \frac{x_R}{\bar{c}} C_{L_\alpha} \cos(\alpha_{frl})$$

$$- \bar{V}_H \left[\eta \left(1 - \frac{\partial \epsilon}{\partial \alpha}\right) C_{L_{\alpha,t}} + \frac{\partial \eta}{\partial \alpha} C_{L_t} \right] \cos(\alpha_{frl} - \epsilon) + \frac{\partial C_{m_p}}{\partial \alpha}, \quad (2.4\text{-}13)$$

where the first term in this equation is given by

$$\frac{\partial C'_{m_R}}{\partial \alpha} = \frac{\partial C_{m_{R,wb}}}{\partial \alpha} + \frac{\bar{c}_t S_t}{\bar{c} S}\left[\frac{\partial \eta}{\partial \alpha} C_{m_{c/4,t}} + \eta \frac{\partial C_{m_{c/4,t}}}{\partial \alpha} \right] \qquad (2.4\text{-}14)$$

We will also make use of (2.4-9) with the drag terms neglected:

$$C_{m_{CM}} \approx C'_{m_R} + \frac{x_R}{\bar{c}} C_L \cos(\alpha_{frl}) - \eta \bar{V}_H C_{L_t} \cos(\alpha_{frl} - \epsilon) + C_{m_p} \qquad (2.4\text{-}15)$$

For positive stiffness in pitch, of a conventional aircraft, Equation (2.4-13) must yield a negative value for C_{m_α}, and this must occur in equilibrium ($C_{m_{CM}} = 0$ in Equation (2.4-15)) at a positive angle of attack. An examination of the terms in these equations will show how this is possible.

On the right-hand side of Equation (2.4-15), the first term (C'_{m_R}) will be small and negative for a normally cambered airfoil, the second term will be negative when the cm is forward of the reference point (which is a requirement for $C_{m_\alpha} < 0$), and so the third term must be chosen to overcome these negative contributions. The volume coefficient $\bar{V}_H$ is positive, and so C_{L_t} is made negative by giving the tail a negative incidence and/or using an upside-down cambered airfoil. The incidence is chosen so that the total pitching moment is positive at small angles of attack, and becomes zero at the desired positive value of α_{frl}. Trim adjustments are made with an elevator "tab" or by using an "all-flying" tail, and control adjustments by using the elevator to effectively change the camber of the horizontal tail airfoil. The remaining term C_{m_p} will be variable with flight conditions, but the thrust line must be kept close enough to the cm to keep it small.

In the C_{m_α} equation (2.4-13) the first term on the right-hand side is small because the reference points are close to aerodynamic centers. The second term is of major importance; if the cm is aft of the aerodynamic reference point, x_R is positive and this term provides a positive (destabilizing) contribution to C_{m_α}.

The third term contains tail efficiency (always positive) and the complement of the "downwash slope." A good deal of information is available about the derivative of the downwash angle with respect to alpha (Roskam, 1979). Its value depends on the distance of the tail from the wing, and on Mach number, and is typically about 0.5 at low subsonic Mach numbers. Since lift-curve slope is always positive below the stall, the term will provide a negative contribution to C_{m_α}.

The fourth term contains the derivative of tail efficiency with respect to α_{frl}. Tail efficiency can be strongly dependent on alpha, thrust coefficient, and flap setting. For example, the four-engine, turbo-prop heavy transport aircraft, whose pitching moment is given in Figure 2.3-7a, has the tail at the same height on the fuselage as the wing, and relatively close to the wing. For this aircraft the tail efficiency rises rapidly to a peak at several degrees alpha, and then falls rapidly. The height of the peak increases strongly with thrust coefficient, and can be higher than $\eta = 2.0$, but is reduced by increasing amounts of flap. The slope $\partial \eta / \partial \alpha$ can be greater than 10/radian, with no flaps, high T_c, and at a few degrees alpha. At zero thrust coefficient, the tail efficiency is slightly less than unity, and the slope $\partial \eta / \partial \alpha$ is approximately constant and slightly negative. Since C_{L_t} had to be made negative, this fourth term is destabilizing for normal thrust coefficients and alpha. This could have been observed from physical reasoning, since we know that the tail efficiency behavior makes the tail more effective at producing a download as alpha increases.

The fifth term in (2.4-13) is the derivative of the propulsion system moment coefficient with angle of attack. Power effects are very complex, especially for propeller aircraft. The existing mathematical models do not necessarily give very good results, and experimental data from powered models are needed. The reader is referred to the literature for more details (Perkins and Hage, 1949; Ribner, 1943; Stinton, 1983).

In summary, an aft-tail aircraft will become statically unstable in pitch if the cm

106 MODELING THE AIRCRAFT

is moved too far aft by incorrect loading. Conversely, if we regard the aerodynamic-data reference point as movable, and at the aircraft aerodynamic center, then as the aerodynamic center moves aft with increasing subsonic Mach number, an aircraft will become more stable in pitch. The lift-to-drag penalty becomes worse as the aerodynamic center moves aft, and for high-performance military aircraft there is a strong incentive to use "relaxed static stability."

Neutral Point

The *neutral point* is the cm position for which $C_{m_\alpha} = 0$. It is therefore an "aerodynamic center" for the whole aircraft. To find a relationship involving the neutral point, we return to the tail-on moment equation, Equation (2.4-4). If we now differentiate this equation with respect to α_{frl}, we obtain

$$C_{m_\alpha} = \frac{\partial C_{m_R}}{\partial \alpha} + \frac{x_R}{\bar{c}} C'_N(\alpha) + \frac{z_R}{\bar{c}} C'_X(\alpha) + \frac{\partial C_{m_P}}{\partial \alpha}, \qquad (2.4\text{-}16)$$

where the primes on C_N and C_X denote their derivatives with respect to alpha, which can be calculated by differentiating Equations (2.4-5).

Let $x_{R/np}$ and $z_{R/np}$ be the coordinates of the reference point when the body-fixed coordinate system has its origin at the neutral point. Then by definition of the neutral point, C_{m_α} becomes zero when we insert these coordinates into Equation (2.4-16). If we solve the resulting equation for the derivative of C_{m_R}, and substitute it into Equation (2.4-16), we obtain

$$C_{m_\alpha} = \frac{(x_R - x_{R/np})}{\bar{c}} C'_N(\alpha) + \frac{(z_R - z_{R/np})}{\bar{c}} C'_X(\alpha) \qquad (2.4\text{-}17)$$

This equation for C_{m_α} holds over the complete range of alpha and Mach. An additional independent equation is needed if we require a solution for the coordinates of the neutral point. A universally used approximation is obtained by neglecting the z component in this equation, and by using the approximations (2.4-12) in C'_N. The result is

$$C_{m_\alpha} = \frac{(x_R - x_{R/np})}{\bar{c}} C_{L_\alpha} \cos(\alpha_{frl}) \qquad (2.4\text{-}18)$$

The x-difference in this equation is the position of the aerodynamic-data reference point with respect to the neutral point. It can be written in terms of distances measured in the aft direction, from the leading edge of the wing mac to the cm and to the neutral point. Thus, let distance measured aft from the leading edge be divided by $\bar{c}$ and denoted by h. Then, Equation (2.4-18) becomes

$$C_{m_\alpha} = -(h_{np} - h_{cm}) C_{L_\alpha} \cos(\alpha_{frl}) \qquad (2.4\text{-}19)$$

In this equation the h-difference in parentheses is called the *static margin*; h_{cm} might typically be 0.25 (chords), and h_{np} might typically be 0.30 (chords), and then the static margin would be 0.05.

A conventionally balanced aircraft is usually designed to have a minimum (worst-case loading) positive static margin of between 3 percent and 5 percent (0.03 to 0.05). This is for safety reasons and to allow some margin for cm variations with load conditions. Aircraft that operate into the transonic and supersonic regions pay a price for this low-speed static stability. The aerodynamic center of an airfoil tends to shift aft from $0.25\bar{c}$ toward $0.5\bar{c}$ in going from high subsonic speeds to supersonic speeds (see Section 2.2). This causes a corresponding movement in the aircraft neutral point and a large increase in the static margin. The undesirable consequences are increased trim drag (and therefore reduced range or fuel economy) and reduced maneuverability. Some modern military aircraft (notably the F-16) have minimized these penalties by using a reduced, or negative, static margin at subsonic speeds. Since negative pitch stiffness normally leads to dynamic instability in pitch, these aircraft use an automatic control system to restore pitch stability. This is described in later chapters.

2.5 THE NONLINEAR AIRCRAFT MODEL

In this section the aerodynamic force and moment models will be combined with the vector equations of motion to obtain aircraft models for simulation and for analytical purposes. For simplicity, the treatment will be limited to the flat-Earth equations of motion. First, the vector equations will be expanded with the translational-velocity state equation expressed in terms of velocity components in the aircraft body-fixed system. The resulting equations are well conditioned when all of the these components become zero (e.g., hovering motion, or sitting on the runway), and body-axes equations are the best choice for general flight simulation.

On the other hand, for the purposes of linearizing the equations of motion and studying the dynamic behavior, it is better to have the velocity equation in terms of stability or wind-axes variables: airspeed and aerodynamic angles. A convenient way to introduce these variables is to treat the stability and wind axes as being fixed to frames that are rotating relative to the vehicle-body frame. The angular velocity vector then involves alpha-dot or beta-dot, and these become state derivatives in the state-space model. In addition, the drag, lift, and crosswind force each appear in separate state equations and, under certain conditions, the equations decouple into two sets describing, separately, the *longitudinal motion* (pitching and translation in the geographic vertical plane) and *lateral-directional motion* (rolling, sideslipping, and yawing). The "stability" or "wind-axes" equations are therefore useful for deriving simpler, small perturbation models that can be used for linear analysis and design.

Model Equations

(i) Body-Axes Equations For convenience the flat-Earth equations of motion (1.5-22) are repeated here:

$$C_{b/n} = fn(\Phi) \qquad (2.5\text{-}1a)$$

108 MODELING THE AIRCRAFT

$$^e\dot{\mathbf{p}}^n_{CM/T} = C_{n/b} \mathbf{v}^b_{CM/e} \tag{2.5-1b}$$

$$\dot{\Phi} = H(\Phi)\boldsymbol{\omega}^b_{b/e} \tag{2.5-1c}$$

$$^b\dot{\mathbf{v}}^b_{CM/e} = (1/m)\,\mathbf{F}^b_{A,T} + C_{b/n}\mathbf{g}^n - \Omega^b_{b/e}\mathbf{v}^b_{CM/e} \tag{2.5-1d}$$

$$^b\dot{\boldsymbol{\omega}}^b_{b/e} = (J^b)^{-1}\left[\mathbf{M}^b_{A,T} - \Omega^b_{b/e}J^b\boldsymbol{\omega}^b_{b/e}\right] \tag{2.5-1e}$$

with the auxiliary equation,

$$\mathbf{v}^b_{rel} = \mathbf{v}^b_{CM/e} - C_{b/n}\mathbf{v}^n_{W/e} \tag{2.5-2}$$

Let the NED components of the position vector, and the body-axes components of the velocity vector be given by, respectively,

$$\mathbf{p}^n_{CM/T} = [p_N \ \ p_E \ \ p_D]^T, \qquad \mathbf{v}^b_{CM/e} = [U \ \ V \ \ W]^T$$

The body-axes components of the angular velocity vector and the Euler angles are

$$\boldsymbol{\omega}^b_{b/e} = [P \ \ Q \ \ R]^T \qquad \Phi = [\phi \ \ \theta \ \ \psi]^T$$

Therefore, the state vector for the body-axes equations is

$$X = [p_N \ \ p_E \ \ p_D \ \ \phi \ \ \theta \ \ \psi \ \ U \ \ V \ \ W \ \ P \ \ Q \ \ R]^T \tag{2.5-3}$$

Given a value for the state vector, the direction cosine matrix can be found from (2.5-1a), and the state equations (2.5-1b) and (2.5-1c) can be evaluated immediately. They are shown in expanded form in Table 2.5-1, with the substitution $\dot{h} = -\dot{p}_D$ for vertical velocity.

The remaining state equations (2.5-1d) and (2.5-1e) require the aerodynamic forces and moments, and therefore a calculation of the relative wind. Let the wind have NED components

$$\mathbf{v}^n_{W/e} = [W_N \ \ W_E \ \ W_D]^T$$

Then (2.5-2) can be used to find the vehicle velocity relative to the surrounding air:

$$\mathbf{v}^b_{rel} \equiv [U' \ \ V' \ \ W']^T$$

For lack of a convenient alternative, let the wind enter the model through the control vector, so a typical control vector will be

$$\bar{U} = [\delta_t, \ \delta_e, \ \delta_a, \ \delta_r, \ W_N, \ W_E, \ W_D]^T \tag{2.5-4}$$

Additional inputs can be created as needed for flaps, gear, speed-brake, and so on, and for derivatives of wind velocity components.

The dynamic pressure, Mach number, and aerodynamic angles must now be calculated from the true airspeed (Table 2.3-1):

$$V_T = \left|\mathbf{v}_{rel}^b\right| \qquad (2.5\text{-}5a)$$

$$\bar{q} = \tfrac{1}{2}\,\rho\,(h)\,V_T^2 \qquad (2.5\text{-}5b)$$

$$M = M(V_T, h) \qquad (2.5\text{-}5c)$$

$$\alpha = \tan^{-1}(W'/U') \qquad (2.5\text{-}5d)$$

$$\beta = \sin^{-1}(V'/V_T), \qquad (2.5\text{-}5e)$$

where a model of the standard atmosphere is used to calculate $\rho(h)$ and $M(V_T, h)$. Next compute installed thrust and the body-axes aerodynamic coefficients, transforming from stability or wind axes, as necessary, and using the component-buildup equations from Section 2.3. The state-derivatives alpha-dot and beta-dot cannot yet be calculated, so we must either neglect $C_{L_{\dot\alpha}}$ and $C_{Y_{\dot\beta}}$ or use approximate values of alpha-dot and beta-dot (e.g., from the last simulation time-step). Control-surface deflections must come either directly from the pilot via the control vector, or from additional state-variable models representing actuator dynamics. The aerodynamic and thrust forces can now be calculated:

$$C_{(\ )} = C_{(\ )}(\alpha, \beta, M, h, \delta_s, P, Q, R) \qquad (2.5\text{-}6a)$$

$$\mathbf{F}_{A,T}^b = \mathbf{F}_T^b + \bar{q}S\,[C_X, C_Y, C_Z]^T \qquad (2.5\text{-}6b)$$

and the translational-velocity state equation (2.5-1d) can be evaluated. This "force equation" is expanded in Table 2.5-1.

The aerodynamic-angle rates, alpha-dot and beta-dot (Equations 2.3-10) can now be found. This can be done by differentiating Equation (2.5-2) and substituting the Poisson kinematical equations for the derivative of the direction-cosine matrix. Derivatives of the wind components can be included if the wind is not steady. The aerodynamic and thrust moments can be calculated in the form:

$$C_{(\ )} = C_{(\ )}(\alpha, \beta, M, h, \delta_s, \dot\alpha, \dot\beta, P, Q, R) \qquad (2.5\text{-}7a)$$

$$\mathbf{M}_{A,T}^b = \mathbf{M}_T^b + \bar{q}S\,[bC_\ell, \bar{c}C_m, bC_n]^T \qquad (2.5\text{-}7b)$$

and the angular-velocity state equation (2.5-1e) can be evaluated. The expansion of this "moment equation" is repeated in Table 2.5-1. This completes the body-axes 6-DOF equations.

(ii) Wind or Stability-Axes Equations

A nonlinear model in terms of the state variables

$$X^T = [V_T\ \beta\ \alpha\ \phi\ \theta\ \psi\ P_s\ Q\ R_s] \qquad (2.5\text{-}8)$$

TABLE 2.5-1: The Flat-Earth, Body-Axes 6-DOF Equations

FORCE EQUATIONS

$$\dot{U} = RV - QW - g_D \sin\theta + (X_A + X_T)/m$$
$$\dot{V} = -RU + PW + g_D \sin\phi \cos\theta + (Y_A + Y_T)/m$$
$$\dot{W} = QU - PV + g_D \cos\phi \cos\theta + (Z_A + Z_T)/m$$

KINEMATIC EQUATIONS

$$\dot{\phi} = P + \tan\theta \, (Q \sin\phi + R \cos\phi)$$
$$\dot{\theta} = Q \cos\phi - R \sin\phi$$
$$\dot{\psi} = (Q \sin\phi + R \cos\phi)/\cos\theta$$

MOMENT EQUATIONS

$$\Gamma\dot{P} = J_{xz}\left[J_x - J_y + J_z\right]PQ - \left[J_z(J_z - J_y) + J_{xz}^2\right]QR + J_z\ell + J_{xz}n$$
$$J_y\dot{Q} = (J_z - J_x)PR - J_{xz}(P^2 - R^2) + m$$
$$\Gamma\dot{R} = \left[(J_x - J_y)J_x + J_{xz}^2\right]PQ - J_{xz}\left[J_x - J_y + J_z\right]QR + J_{xz}\ell + J_x n$$
$$\cdot \, \Gamma = J_x J_z - J_{xz}^2$$

NAVIGATION EQUATIONS

$$\dot{p}_N = Uc\theta c\psi + V(-c\phi s\psi + s\phi s\theta c\psi) + W(s\phi s\psi + c\phi s\theta c\psi)$$
$$\dot{p}_E = Uc\theta s\psi + V(c\phi c\psi + s\phi s\theta s\psi) + W(-s\phi c\psi + c\phi s\theta s\psi)$$
$$\dot{h} = Us\theta - Vs\phi c\theta - Wc\phi c\theta$$

---

will be constructed here and, as described in the Section 2.5 introduction, will be found to have advantages for linearization and decoupling. The variables V_T, α, β describe the magnitude and direction of the relative wind, and P_s, Q, R_s are stability-axes components of the aircraft angular-velocity vector. Equations (1.5-26) show that the flat-Earth dynamic equations are dependent only on the relative velocity, $\mathbf{v}_{rel}$, and they are an appropriate starting point here.

(a) Force Equations As an expedient way of deriving these equations, imagine the stability axes as fixed in a new "stability frame" with angular velocity vector $-\dot{\alpha}\mathbf{j}$ with respect to the body frame. Similarly, imagine the wind axes as fixed in a new "wind frame" with angular velocity $\dot{\beta}\mathbf{k}$ with respect to the stability frame. The flat-Earth force equation (1.5-26a), with steady wind, reduces to

THE NONLINEAR AIRCRAFT MODEL

$$^b\dot{\mathbf{v}}_{rel} = (1/m)\,\mathbf{F}_{A,T} + \mathbf{g} - \boldsymbol{\omega}_{b/e} \times \mathbf{v}_{rel} \qquad (2.5\text{-}9)$$

Let the derivative taken in the body frame be replaced with a derivative taken in the wind frame:

$$^w\dot{\mathbf{v}}_{rel} + \boldsymbol{\omega}_{w/b} \times \mathbf{v}_{rel} = (1/m)\,\mathbf{F}_{A,T} + \mathbf{g} - \boldsymbol{\omega}_{b/e} \times \mathbf{v}_{rel} \qquad (2.5\text{-}10)$$

Resolving these vectors in wind-axes gives the matrix equation:

$$^w\dot{\mathbf{v}}_{rel}^w + \Omega_{w/b}^w \mathbf{v}_{rel}^w = (1/m)\,\mathbf{F}_{A,T}^w + C_{w/b} C_{b/n} \mathbf{g}^n - \Omega_{b/e}^w \mathbf{v}_{rel}^w \qquad (2.5\text{-}11)$$

The cross-product matrix on the left-hand side can be determined as follows. If alpha-dot is greater than zero, the angle of attack is increasing and the stability frame is undergoing a left-handed rotation about the body y-axis, relative to the body frame. Also, if beta-dot is greater than zero, the wind frame is undergoing a right-handed rotation around the stability z-axis, relative to the stability frame. Therefore,

$$\boldsymbol{\omega}_{s/b}^s = \boldsymbol{\omega}_{s/b}^b = [0 \quad -\dot{\alpha} \quad 0]^T \qquad (2.5\text{-}12a)$$

$$\boldsymbol{\omega}_{w/s}^w = \boldsymbol{\omega}_{w/s}^s = [0 \quad 0 \quad \dot{\beta}]^T \qquad (2.5\text{-}12b)$$

Now

$$\boldsymbol{\omega}_{w/b} = \boldsymbol{\omega}_{w/s} + \boldsymbol{\omega}_{s/b} \qquad (2.5\text{-}13)$$

and so

$$\boldsymbol{\omega}_{w/b}^w = \begin{bmatrix} 0 \\ 0 \\ \dot{\beta} \end{bmatrix} + \begin{bmatrix} c\beta & s\beta & 0 \\ -s\beta & c\beta & 0 \\ 0 & 0 & 1 \end{bmatrix} \begin{bmatrix} 0 \\ -\dot{\alpha} \\ 0 \end{bmatrix} = \begin{bmatrix} -\dot{\alpha}s\beta \\ -\dot{\alpha}c\beta \\ \dot{\beta} \end{bmatrix} \qquad (2.5\text{-}14)$$

Then the left-hand side of the force equation becomes

$$^w\dot{\mathbf{v}}_{rel}^w + \Omega_{w/b}^w \mathbf{v}_{rel}^w = \begin{bmatrix} \dot{V}_T \\ \dot{\beta} V_T \\ \dot{\alpha} V_T \cos\beta \end{bmatrix} \qquad (2.5\text{-}15)$$

This array contains the derivatives of the first three state variables in (2.5-8).

The aerodynamic and thrust force term can now be calculated as follows. Again for simplicity, assume that the thrust vector lies in the x_b-z_b plane, but is inclined at an angle α_T to the fuselage reference line (so that positive α_T corresponds to a component of thrust in the negative z_b direction). Then it is easy to write the stability axes components of thrust and transform to wind axes:

$$\mathbf{F}_{A,T}^{w} = \begin{bmatrix} c\beta & s\beta & 0 \\ -s\beta & c\beta & 0 \\ 0 & 0 & 1 \end{bmatrix} \begin{bmatrix} F_T \cos(\alpha + \alpha_T) \\ 0 \\ F_T \sin(\alpha + \alpha_T) \end{bmatrix} - \begin{bmatrix} D \\ C \\ L \end{bmatrix}, \qquad (2.5\text{-}16)$$

where

$$F_T = |\mathbf{F}_T|$$

In wind axes the gravity term is given by

$$\mathbf{g}^w = C_{w/b} C_{b/n} \begin{bmatrix} 0 \\ 0 \\ g_D \end{bmatrix} \equiv \begin{bmatrix} g_1 \\ g_2 \\ g_3 \end{bmatrix},$$

where the components are

$$g_1 = g_D(-c\alpha\,c\beta\,s\theta + s\beta\,s\phi\,c\theta + s\alpha\,c\beta\,c\phi\,c\theta) = -g_D \sin(\gamma)$$
$$g_2 = g_D(c\alpha\,s\beta\,s\theta + c\beta\,s\phi\,c\theta - s\alpha\,s\beta\,c\phi\,c\theta)$$
$$g_3 = g_D(s\alpha\,s\theta + c\alpha\,c\phi\,c\theta) \qquad (2.5\text{-}17)$$

(See (3.6-2) to introduce γ into the first equation.) The remaining cross-product is given by

$$\Omega_{b/e}^{w} \mathbf{v}_{rel}^{w} = \begin{bmatrix} 0 & V_T R_w & -V_T Q_w \end{bmatrix}^T, \qquad (2.5\text{-}18)$$

where $R_w = R_s$ and $Q_w = (-P_s \sin\beta + Q \cos\beta)$.

When all of these terms are assembled, the force equations are

$$m\dot{V}_T = F_T \cos(\alpha + \alpha_T) \cos\beta - D + mg_1$$
$$m\dot{\beta} V_T = -F_T \cos(\alpha + \alpha_T) \sin\beta - C + mg_2 - mV_T R_s \qquad (2.5\text{-}19)$$
$$m\dot{\alpha} V_T \cos\beta = -F_T \sin(\alpha + \alpha_T) - L + mg_3 + mV_T(Q\cos\beta - P_s \sin\beta)$$

It is evident that, if lift and cross-wind force include a linear dependence on the state-derivatives alpha-dot and beta-dot, respectively, the equations can be solved for these state derivatives. Unfortunately this requires a nonzero airspeed V_T. However, we will show that these equations are useful for constructing a small-perturbation model of aircraft dynamics.

(b) Moment Equations In the moment equation (2.5-1e) the derivative taken in the body frame can be replaced with a derivative taken in the stability or wind frame; the form of the equation is the same in either case. The stability frame will be shown later to be a more convenient choice, thus,

$$^s\dot{\boldsymbol{\omega}}^b_{b/e} + (\boldsymbol{\omega}_{s/b} \times \boldsymbol{\omega}_{b/e})^b = (J^b)^{-1}\left[\mathbf{M}^b_{A,T} - \Omega^b_{b/e} J^b \boldsymbol{\omega}^b_{b/e}\right] \qquad (2.5\text{-}20)$$

Now change from body-axes to stability-axes components, insert a cross-product matrix for $(\omega \times)^s$, and solve for the derivatives:

$$^s\dot{\boldsymbol{\omega}}^s_{b/e} = -\Omega^s_{s/b}\boldsymbol{\omega}^s_{b/e} + (J^s)^{-1}\left[\mathbf{M}^s_{A,T} - \Omega^s_{b/e} J^s \boldsymbol{\omega}^s_{b/e}\right], \qquad (2.5\text{-}21)$$

where the stability-axes inertia matrix is

$$J^s = C_{s/b}\, J^b\, C_{b/s} \qquad (2.5\text{-}22)$$

Equation (2.5-22) has an extra term compared to the body-axes moment equation and offers no advantages for simulation. It does have advantages for deriving a small-perturbation model because it introduces alpha-dot into the small-perturbation moment equations in a formal manner, as will be shown in the next section. If we had wished to introduced beta-dot it would have been necessary to convert to wind axes (Stevens and Lewis, 1992).

Consider the terms in the stability-axes moment equation; starting with the inertia matrix. We will restrict ourselves to aircraft having a plane of symmetry, so that the body-axes inertia matrix is given by Equation (1.5-7). When the transformation (2.5-22) is performed, the matrix is found to have the same structure in stability axes:

$$J^s = \begin{bmatrix} J'_x & 0 & -J'_{xz} \\ 0 & J'_y & 0 \\ -J'_{xz} & 0 & J'_z \end{bmatrix}, \qquad (2.5\text{-}23)$$

where

$$J'_x = J_x \cos^2 \alpha + J_z \sin^2 \alpha - J_{xz} \sin 2\alpha$$

$$J'_y = J_y$$

$$J'_z = J_x \sin^2 \alpha + J_z \cos^2 \alpha + J_{xz} \sin 2\alpha$$

$$J'_{xz} = \tfrac{1}{2}(J_x - J_z)\sin 2\alpha + J_{xz} \cos 2\alpha$$

Furthermore, the inverse of this matrix is easily found, and is again of the same form:

$$(J^s)^{-1} = \frac{1}{\Gamma}\begin{bmatrix} J'_z & 0 & J'_{xz} \\ 0 & \Gamma/J'_y & 0 \\ J'_{xz} & 0 & J'_x \end{bmatrix} \qquad (2.5\text{-}24)$$

with

$$\Gamma = J'_x J'_z - J'^2_{xz} \quad (= J_x J_z - J^2_{xz}, \quad \text{Problem 2.5-3})$$

Note that in wind axes J^w is a full matrix, so that working in stability axes is considerably more convenient, provided that we can neglect beta-dot derivatives in the moment equations. Other terms in the moment equation are:

$$^s\dot{\omega}^s_{b/e} = [\dot{P}_s \quad \dot{Q} \quad \dot{R}_s]^T$$

$$\omega^s_{s/b} = [0 \quad -\dot{\alpha} \quad 0]^T \tag{2.5-25}$$

$$\Omega^s_{s/b}\omega^s_{b/e} = \dot{\alpha}[-R_s \quad 0 \quad P_s]^T$$

The stability-axes moment equations (2.5-21) can now be written in component form as

$$\begin{bmatrix} \dot{P}_s \\ \dot{Q} \\ \dot{R}_s \end{bmatrix} = -\dot{\alpha} \begin{bmatrix} -R_s \\ 0 \\ P_s \end{bmatrix} + \frac{1}{\Gamma} \begin{bmatrix} J'_z & 0 & J'_{xz} \\ 0 & \Gamma/J'_y & 0 \\ J'_{xz} & 0 & J'_x \end{bmatrix} \left(\begin{bmatrix} \ell_s \\ m \\ n_s \end{bmatrix} - \Omega^s_{b/e} J^s \omega^s_{b/e} \right) \tag{2.5-26}$$

The last term is of the same form as the corresponding term in the body-axes moment equations; it will not need to be expanded.

Decoupling of the Nonlinear Equations/3-DOF Longitudinal Model

Most aircraft spend most of their flying time in a wings-level steady-state flight condition and, since the model of the 3-DOF motion in the NED vertical plane is much simpler than the 6-DOF model, it is worthwhile investigating the equations of motion under the wings-level flight condition. Referring to the force equations (2.5-19), if the roll angle ϕ is zero, the gravity terms are greatly simplified:

$$g_1 = -g_D \sin \gamma = -g_D \cos \beta \sin(\theta - \alpha)$$

$$g_2 = g_D [\sin \theta \cos \alpha - \cos \theta \sin \alpha] \sin \beta = g_D \sin \beta \sin(\theta - \alpha)$$

$$g_3 = g_D [\sin \alpha \sin \theta - \cos \alpha \cos \theta] = g_D \cos(\theta - \alpha) \tag{2.5-27}$$

When the sideslip is small, the flight-path angle is given by the difference between pitch attitude and angle of attack, and so the gravity terms become

$$g_1 = -g_D \sin(\gamma)$$

$$g_2 = \beta g_D \sin(\gamma) \tag{2.5-28}$$

$$g_3 = g_D \cos(\gamma)$$

and the force equations (2.5-19) reduce to

$$m\dot{V}_T = F_T \cos(\alpha + \alpha_T) - D - mg_D \sin\gamma$$
$$m\dot{\beta}V_T = -\beta F_T \cos(\alpha + \alpha_T) - C + \beta mg_D \sin\gamma - mV_T R_s \quad (2.5\text{-}29)$$
$$m\dot{\alpha}V_T = -F_T \sin(\alpha + \alpha_T) - L + mg_D \cos\gamma + mV_T(Q - \beta P_s)$$

The first and third equations describe longitudinal motion and, when beta is negligible, they are independent of the second (sideslip) equation.

Decoupling of the longitudinal motion also occurs in the attitude equations and the moment equations. It can be seen from the kinematic equations in Table 2.5-1 that when the roll angle is zero,

$$\dot{\theta} = Q \quad (2.5\text{-}30)$$

The body-axes moment equations in Table 2.5-1 show that, if the roll and yaw rates (P and R) are small, the pitching moment equation is not coupled to the rolling and yawing moment equations, and

$$J_Y \dot{Q} = m \quad (2.5\text{-}31)$$

Therefore, we can obtain a model for pure longitudinal motion by adding Equations (2.5-30) and (2.5-31) to the decoupled longitudinal force equations:

$$m\dot{V}_T = F_T \cos(\alpha + \alpha_T) - D - mg_D \sin(\theta - \alpha)$$
$$m\dot{\alpha}V_T = -F_T \sin(\alpha + \alpha_T) - L + mg_D \cos(\theta - \alpha) + mV_T Q$$
$$\dot{\theta} = Q \quad (2.5\text{-}32)$$
$$\dot{Q} = m/J_Y$$

The state vector for these equations is

$$X = [V_T \ \alpha \ \theta \ Q]^T \quad (2.5\text{-}33)$$

A common alternative model uses flight-path angle as a state variable in place of pitch attitude:

$$m\dot{V}_T = F_T \cos(\alpha + \alpha_T) - D - mg_D \sin(\gamma)$$
$$m\dot{\gamma}V_T = F_T \sin(\alpha + \alpha_T) + L - mg_D \cos(\gamma) \quad (2.5\text{-}34)$$
$$\dot{\alpha} = Q - \dot{\gamma}$$
$$\dot{Q} = m/J_y$$

These longitudinal models are used for a variety of purposes, from performance analysis to automatic control system design. If the lift and drag forces are linearized for small perturbations from a specified flight condition, we obtain linear longitudinal equations that are the same as those derived in the next section by a formal linearization of the complete 6-DOF equations, followed by decoupling.

2.6 LINEAR MODELS AND THE STABILITY DERIVATIVES

When we perform a computer simulation to evaluate the performance of an aircraft with its control systems, we almost invariably use a nonlinear model. Also, the linear equations needed for control systems design will mostly be derived by numerical methods from the nonlinear computer model. Because the nonlinear state models are difficult to handle without the use of a digital computer, most of the early progress in understanding the dynamics of aircraft and the stability of the motion came from studying linear algebraic small-perturbation equations. G. H. Bryan (1911) introduced the idea of perturbed forces and moments with respect to a "steady-state" flight condition, and this approach is still in use. The small-perturbation equations are linear equations derived algebraically from nonlinear equations like those of Section 2.5. In these equations the nonlinear aerodynamic coefficients are replaced by terms involving the aerodynamic derivatives described briefly in Section 2.3.

There are two good reasons, apart from their historical importance, for algebraically deriving the small-perturbation equations. First, the aerodynamic derivatives needed for the linear equations can be estimated relatively quickly (Hoak et al., 1970) before nonlinear aerodynamic data become available. Second, the algebraic small-perturbation equations provide a great deal of insight into the relative importance of the various aerodynamic derivatives under different flight conditions and their effect on the stability of the aircraft motion. In preparation for deriving the linear equations we now examine the concept of a steady-state flight condition.

Singular Points and Steady-State Flight

In the preceding section, when the body-axes force equations were used, alpha-dot or beta-dot force dependence created a difficulty in that the state equations became implicit in the derivatives of the states alpha and beta. This problem was solved in an ad hoc manner by using the wind-axes equations and collecting linear alpha-dot or beta-dot terms on one side of the equations. In this section where the goal is to derive linear equations algebraically, we take a more general approach, starting with *implicit state equations* in the general form

$$f(\dot{X}, X, \bar{U}) = 0, \qquad (2.6\text{-}1)$$

where f is an array of n scalar nonlinear functions f_i, as in (1.1-1).

In the theory of nonlinear systems (Vidyasagar, 1978) the concept of a *singular point*, or *equilibrium point*, of an autonomous (no external control inputs) time-invariant system is introduced. The coordinates of a singular point of the implicit nonlinear state equations are given by a solution, $X = X_e$, which satisfies

$$f(\dot{X}, X, \bar{U}) = 0, \quad \text{with } \dot{X} \equiv 0; \; \bar{U} \equiv 0 \text{ or constant} \qquad (2.6\text{-}2)$$

This idea has strong intuitive appeal; the system is "at rest" when all of the derivatives are identically zero, and then one may examine the behavior of the system near the singular point by slightly perturbing some of the variables. If, in the case of an aircraft model, the state trajectory departs rapidly from the singular point in response to a small perturbation in, say, pitch attitude, the human pilot is unlikely to be able to control this aircraft.

Steady-state aircraft flight can be defined as a condition in which all of the force and moment components in the body-fixed coordinate system are constant or zero. It follows that the aerodynamic angles add the angular rate components must be constant, and their derivatives must be zero. It must be assumed that the aircraft mass remains constant. In the case of the round-Earth equations, minor circles (and the major circle around the equator) are the only trajectories along which gravity remains constant in magnitude.

Assuming that the flat-Earth equations are satisfactory for all of our control system design purposes, the definition allows steady wings-level flight and steady turning flight. Furthermore, if the change in atmospheric density with altitude is neglected, a wings-level climb and a climbing turn are permitted as steady-state flight conditions. In this case the NED position equations do not couple back into the equations of motion and need not be used in finding a steady-state condition. Therefore, the steady-state conditions that are important to us for control system design can be defined in terms of the remaining nine state variables of the flat-Earth equations as follows:

Steady-State Flight

$$\dot{P}, \dot{Q}, \dot{R}, \text{ and } \dot{U}, \dot{V}, \dot{W} \; (\text{or } \dot{V}_T, \dot{\beta}, \dot{\alpha}) \equiv 0, \text{ controls fixed} \qquad (2.6\text{-}3a)$$

with the following additional constraints according to the flight condition:

STEADY WINGS-LEVEL FLIGHT: $\phi, \dot{\phi}, \dot{\theta}, \dot{\psi} \equiv 0 \quad (\therefore P, Q, R \equiv 0)$

STEADY TURNING FLIGHT: $\dot{\phi}, \dot{\theta} \equiv 0, \quad \dot{\psi} \equiv$ TURN RATE

STEADY PULL-UP: $\phi, \dot{\phi}, \dot{\psi} \equiv 0, \quad \dot{\theta} \equiv$ PULL-UP RATE

STEADY ROLL: $\dot{\theta}, \dot{\psi} \equiv 0, \quad \dot{\phi} \equiv$ ROLL RATE $\qquad (2.6\text{-}3b)$

The steady-state conditions $\dot{P}, \dot{Q}, \dot{R} \equiv 0$ require the angular rates to be zero or constant (as in steady turns), and therefore the aerodynamic and thrust moments must be zero or constant. The conditions $\dot{U}, \dot{V}, \dot{W} \equiv 0$ require the airspeed, angle of attack, and sideslip angle to be constant, and hence the aerodynamic forces must be zero or constant. Therefore, the steady-state pull-up (or push-over) and steady-state roll conditions can only exist instantaneously. However, it is useful to be able to linearize the aircraft dynamics in these flight conditions since the control systems must operate there.

While a pilot may not find it very difficult to put an aircraft into a steady-state flight condition, the mathematical model requires the solution of the simultaneous nonlinear equations (2.6-2). In general, because of the nonlinearity, a steady-state solution can only be found by using a numerical method on a digital computer. Multiple solutions can exist, and a feasible solution will emerge only when practical constraints are placed on the variables. We consider this problem in Chapter 3, and assume here that a solution X_e, $\bar{U}_e$ is known for the desired flight condition.

Linearization

The implicit nonlinear equations will be written as

$$f_1(\dot{X}, X, \bar{U}) = 0$$
$$f_2(\dot{X}, X, \bar{U}) = 0 \quad (2.6\text{-}4)$$
$$\vdots \quad \vdots$$
$$f_9(\dot{X}, X, \bar{U}) = 0$$

and will be obtained from the wind-axes force equations, kinematic equations, and stability-axes moment equations, by moving all nonzero terms to the right-hand side of the equations. The reduced state vector is

$$X = [V_T \ \beta \ \alpha \ \phi \ \theta \ \psi \ P_s \ Q_s \ R_s]^T \quad (2.6\text{-}5a)$$

The control vector, given by (2.5-4), is reduced here to

$$\bar{U} = [\delta_t \ \delta_e \ \delta_a \ \delta_r]^T \quad (2.6\text{-}5b)$$

We now consider small perturbations from the steady-state condition X_e, $\bar{U}_e$ and derive a set of linear constant-coefficient state equations. If we expand the nonlinear state equations (2.6-4) in a Taylor series about the equilibrium point $(X_e, \bar{U}_e)$, and keep only the first-order terms, we find that the perturbations in the state, state derivative, and control vectors must satisfy

$$\nabla_{\dot{X}} f_1 \ \delta\dot{X} + \nabla_X f_1 \ \delta X + \nabla_U f_1 \ \delta\bar{U} = 0$$
$$\vdots \quad \vdots \quad \vdots \quad \vdots \quad \vdots \quad \vdots$$
$$\nabla_{\dot{X}} f_9 \ \delta\dot{X} + \nabla_X f_9 \ \delta X + \nabla_U f_9 \ \delta\bar{U} = 0 \quad (2.6\text{-}6)$$

In this equation ∇ (del, or nabla) represents a row array of first partial derivative operators, for example,

$$\nabla_X f_i \equiv \begin{bmatrix} \dfrac{\partial f_i}{\partial X_1} & \dfrac{\partial f_i}{\partial X_2} & \cdots & \dfrac{\partial f_i}{\partial X_n} \end{bmatrix}$$

Each term in (2.6-6) is a scalar product; thus, $\nabla_X f_1 \, \delta X$ is the total differential of f_1 due to simultaneous perturbations in all the elements of the state vector.

Equations (2.6-6) can now be written in implicit linear state-variable form as

$$E\dot{x} = Ax + Bu \qquad (2.6\text{-}7)$$

Lowercase notation has been used to indicate that x and u are perturbations from the equilibrium values of the state and control vectors. The coefficient matrices

$$E = -\begin{bmatrix} \nabla_{\dot{X}} f_1 \\ \vdots \\ \nabla_{\dot{X}} f_9 \end{bmatrix}_{\substack{\bar{U}=\bar{U}_e \\ X=X_e}} \quad A = \begin{bmatrix} \nabla_x f_1 \\ \vdots \\ \nabla_x f_9 \end{bmatrix}_{\substack{\bar{U}=\bar{U}_e \\ X=X_e}} \quad B = \begin{bmatrix} \nabla_U f_1 \\ \vdots \\ \nabla_U f_9 \end{bmatrix}_{\substack{\bar{U}=\bar{U}_e \\ X=X_e}} \qquad (2.6\text{-}8)$$

are called *Jacobian matrices* and must be calculated at the equilibrium point. If E is nonsingular, (2.6-7) can be rewritten as an explicit set of linear state equations, but we will see later that this is not necessarily the most convenient way to use the implicit state equations.

The Jacobian matrices E, A, B will be evaluated three rows at a time, corresponding, respectively, to the wind-axes force equations (f_1 to f_3), kinematic equations (f_4 to f_6), and moment equations (f_7 to f_9). The evaluation will be for the steady, level flight condition, with the additional constraint of no sideslip ($\beta = 0$). The latter condition greatly simplifies the algebra involved in the linearization and leads to "lat-long" decoupling. Therefore, the equilibrium (steady-state) conditions are

STEADY-STATE CONDITIONS:

$$\beta, \phi, P, Q, R \equiv 0 \qquad (2.6\text{-}9)$$

all derivatives $\equiv 0$

$$V_T = V_{Te}, \alpha = \alpha_e, \theta = \theta_e, \psi = 0$$

The algebra can be further reduced by taking advantage of some features of the equations. Thus, when differentiating products containing $\cos\beta$ or $\cos\phi$, all of the resulting $\sin\beta$ or $\sin\phi$ terms will disappear when we apply the $\beta = 0$ and $\phi = 0$ equilibrium conditions. Therefore, the $\cos\beta$ or $\cos\phi$ terms can be set to unity before differentiation. Similarly, a $\cos\beta$ or $\cos\phi$ in the denominator of a quotient term can be set to unity. Also, if two or more terms with equilibrium values of zero (e.g., $\sin\beta$, $\sin\phi$) occur in a product term, this product can be discarded before differentiation.

The Linearized Force Equations The first three rows of the linear equations (2.6-7) will now be obtained by performing the gradient operations, shown in (2.6-8),

120 MODELING THE AIRCRAFT

on the nonlinear force equations (2.5-19). All of the terms in (2.5-19) will be moved to the right-hand side of the equations. First, we find the partial derivatives with respect to $\dot{X}$ and use the steady-state condition (2.6-9). The thrust is assumed to be independent of the state derivatives; this gives

$$-\begin{bmatrix} \nabla_{\dot{X}} f_1 \\ \nabla_{\dot{X}} f_2 \\ \nabla_{\dot{X}} f_3 \end{bmatrix}_{X=X_e} = \begin{bmatrix} m\nabla_{\dot{X}}\dot{V}_T + \nabla_{\dot{X}} D \\ mV_T \nabla_{\dot{X}}\dot{\beta} + \nabla_{\dot{X}} C \\ mV_T \nabla_{\dot{X}}\dot{\alpha} + \nabla_{\dot{X}} L \end{bmatrix} \qquad (2.6\text{-}10)$$

A term such as $\nabla_{\dot{X}} \dot{V}_T$ is simply a row array with unity in the position corresponding to the $\dot{V}_T$ state derivative, and zeros elsewhere. The other terms, such as $\nabla_{\dot{X}} L$, are row arrays containing all of the partial derivatives of the forces with respect to the state derivatives.

The partial derivatives of the aerodynamic forces and moments with respect to other variables are the *aerodynamic derivatives*, first introduced in Section 2.3. Table 2.6-1 defines the derivatives that are normally significant in the force equations. These derivatives are called the *dimensional derivatives*, and later we will introduce a related set of derivatives that have been made dimensionless in the same way that the aerodynamic coefficients are made dimensionless. The dimensional derivatives are given the symbols X, Y, and Z to indicate which force component is involved (the symbols D, C, and L are also used). Their subscripts indicate the quantity with respect to which the derivative is taken (subscripts for the controls were defined in 2.6-5*b*).

TABLE 2.6-1: The Force Dimensional Derivatives

X-AXIS	Y-AXIS	Z-AXIS
$X_V = \dfrac{-1}{m} \dfrac{\partial D}{\partial V_T}$	$Y_\beta = \dfrac{1}{m}\left(\dfrac{\partial C}{\partial \beta} - D\right)$	$Z_V = \dfrac{-1}{m}\dfrac{\partial L}{\partial V_T}$
$X_{T_V} = \dfrac{1}{m}\dfrac{\partial F_T}{\partial V_T}$	$Y_p = \dfrac{1}{m}\dfrac{\partial C}{\partial P}$	$Z_\alpha = \dfrac{-1}{m}\left(D + \dfrac{\partial L}{\partial \alpha}\right)$
$X_\alpha = \dfrac{1}{m}\left(L - \dfrac{\partial D}{\partial \alpha}\right)$	$Y_r = \dfrac{1}{m}\dfrac{\partial C}{\partial R}$	$Z_{\dot{\alpha}} = \dfrac{-1}{m}\dfrac{\partial L}{\partial \dot{\alpha}}$
$X_{\delta e} = \dfrac{-1}{m}\dfrac{\partial D}{\partial \delta_e}$	$Y_{\delta r} = \dfrac{1}{m}\dfrac{\partial C}{\partial \delta_r}$	$Z_q = \dfrac{-1}{m}\dfrac{\partial L}{\partial Q}$
$X_{\delta t} = \dfrac{1}{m}\dfrac{\partial F_T}{\partial \delta_t}$	$Y_{\delta a} = \dfrac{1}{m}\dfrac{\partial C}{\partial \delta_a}$	$Z_{\delta e} = \dfrac{-1}{m}\dfrac{\partial L}{\partial \delta_e}$

LINEAR MODELS AND THE STABILITY DERIVATIVES

For the purpose of deriving the linear equations, only the derivatives shown in the table will be assumed to be nonzero. Therefore, the terms $\nabla_{\dot{X}} D$ and $\nabla_{\dot{X}} C$ in (2.6-10) will now be dropped (additional terms will be dropped later). Note that the components involved in the partial derivatives are wind-axes components, except for the engine thrust F_T. This force belongs naturally to the aircraft-body axes, and it only appears in the wind-axes equations in conjunction with trigonometric functions of the aerodynamic angles.

We will interpret (2.6-10) in terms of the dimensional derivatives. The array $\nabla_{\dot{X}} L$ contains only the derivative $Z_{\dot{\alpha}}$ (multiplied by m) in the $\dot{\alpha}$ position, so (2.6-10) can now be rewritten as

$$-\begin{bmatrix} \nabla_{\dot{X}} f_1 \\ \nabla_{\dot{X}} f_2 \\ \nabla_{\dot{X}} f_3 \end{bmatrix}_{X = X_e} = m \begin{bmatrix} 1 & 0 & 0 & 0 & 0 & 0 & 0 & 0 \\ 0 & V_{T_e} & 0 & 0 & 0 & 0 & 0 & 0 \\ 0 & 0 & V_{T_e} - Z_{\dot{\alpha}} & 0 & 0 & 0 & 0 & 0 \end{bmatrix} \quad (2.6\text{-}11)$$

Next, using (2.5-19), form the partial derivatives with respect to X and apply the steady-state conditions (2.6-9). The result is

$$\begin{bmatrix} \nabla_X f_1 \\ \nabla_X f_2 \\ \nabla_X f_3 \end{bmatrix} =$$

$$\begin{bmatrix} -F_T \sin(\alpha_e + \alpha_T) \nabla_X \alpha + \cos(\alpha_e + \alpha_T) \nabla_X F_T - m g_D \cos \gamma_e \nabla_X (\theta - \alpha) - \nabla_X D \\ -F_T \cos(\alpha_e + \alpha_T) \nabla_X \beta + m g_D (\sin \gamma_e \nabla_X \beta + \cos \theta_e \nabla_X \phi) - \nabla_X C - m V_{T_e} \nabla_X R_s \\ -F_T \cos(\alpha_e + \alpha_T) \nabla_X \alpha - \sin(\alpha_e + \alpha_T) \nabla_X F_T + m g_D \sin \gamma_e \nabla_X (\alpha - \theta) - \nabla_X L + m V_{T_e} \nabla_X Q \end{bmatrix}$$
(2.6-12)

This result can be further reduced by using the steady-state conditions, obtained by setting the left-hand side of (2.5-19) to zero, to replace some groups of terms by the steady-state lift and drag forces. Thus, the partial derivatives evaluated at the equilibrium point are

$$\begin{bmatrix} \nabla_X f_1 \\ \nabla_X f_2 \\ \nabla_X f_3 \end{bmatrix}_{\substack{\bar{U} = \bar{U}_e \\ X = X_e}} = \begin{bmatrix} \cos(\alpha_e + \alpha_T) \nabla_X F_T - \nabla_X D + L_e \nabla_X \alpha - m g_D \cos \gamma_e \nabla_X \theta \\ -\nabla_X C - D_e \nabla_X \beta + m g_D \cos \theta_e \nabla_X \phi - m V_{T_e} \nabla_X R_s \\ -\sin(\alpha_e + \alpha_T) \nabla_X F_T - \nabla_X L - D_e \nabla_X \alpha - m g_D \sin \gamma_e \nabla_X \theta + m V_{T_e} \nabla_X Q \end{bmatrix},$$
(2.6-13)

where $\alpha_e, \theta_e, \gamma_e, L_e$, and D_e are the steady-state values. Note that there is no steady-state sideforce. If this expression is interpreted in terms of the derivatives from Table 2.6-1, we obtain for the right-hand side:

122 MODELING THE AIRCRAFT

$$m \begin{bmatrix} X_V + X_{T_V}\cos(\alpha_e + \alpha_T) & 0 & X_\alpha & 0 & -g_D\cos\gamma_e & 0 & 0 & 0 & 0 \\ 0 & Y_\beta & 0 & g_D\cos\theta_e & 0 & 0 & Y_p & 0 & Y_r - V_{T_e} \\ Z_V - X_{T_V}\sin(\alpha_e + \alpha_T) & 0 & Z_\alpha & 0 & -g_D\sin\gamma_e & 0 & 0 & V_{T_e} + Z_q & 0 \end{bmatrix} \quad (2.6\text{-}14)$$

This matrix constitutes the top three rows of A in (2.6-7).

It only remains to obtain the partial derivatives of the force equations with respect to the control vector $\bar{U}$. The partial derivatives are

$$\begin{bmatrix} \nabla_U f_1 \\ \nabla_U f_2 \\ \nabla_U f_3 \end{bmatrix} = \begin{bmatrix} \cos(\alpha + \alpha_T)\nabla_U F_T - \nabla_U D \\ \nabla_U Y \\ -\sin(\alpha + \alpha_T)\nabla_U F_T - \nabla_U L \end{bmatrix} \quad (2.6\text{-}15)$$

Now, inserting the relevant dimensional derivatives and the equilibrium values of the angles, we obtain

$$\begin{bmatrix} \nabla_U f_1 \\ \nabla_U f_2 \\ \nabla_U f_3 \end{bmatrix}_{\substack{X=X_e \\ \bar{U}=\bar{U}_e}} = m \begin{bmatrix} X_{\delta t}\cos(\alpha_e + \alpha_T) & X_{\delta e} & 0 & 0 \\ 0 & 0 & Y_{\delta a} & Y_{\delta r} \\ -X_{\delta t}\sin(\alpha_e + \alpha_T) & Z_{\delta e} & 0 & 0 \end{bmatrix} \quad (2.6\text{-}16)$$

and these are the top three rows of B in (2.6-7).

This completes the linearization of the force equations. Note that the positions of the zero elements correspond to the beginnings of the anticipated decoupling in (2.6-7). One of the assumptions contributing to this decoupling is that the partial derivatives of drag with respect to the lateral/directional controls (ailerons and rudder) can be neglected. In practice aileron and rudder deflections do cause non-negligible changes in drag, but this assumption does not have any significant consequences on the linearized dynamics.

The Linearized Kinematic Equations We will now determine the second block of three rows in (2.6-7). The nonlinear kinematic relationship between the Euler-angle rates and the stability-axes rates P_s, Q, and R_s is obtained from Table 2.5-1 and the transformation matrices $C_{b/s}$. Thus,

$$\dot{\Phi} = H(\Phi)C_{b/s}\boldsymbol{\omega}_{b/e}^s \quad (2.6\text{-}17)$$

There are no aerodynamic forces or moments involved in these equations, and it is easy to see that the contribution to the E matrix is given by

$$-\begin{bmatrix} \nabla_{\dot{X}} f_4 \\ \nabla_{\dot{X}} f_5 \\ \nabla_{\dot{X}} f_6 \end{bmatrix} = \begin{bmatrix} 0 & 0 & 0 & 1 & 0 & 0 & 0 & 0 & 0 \\ 0 & 0 & 0 & 0 & 1 & 0 & 0 & 0 & 0 \\ 0 & 0 & 0 & 0 & 0 & 1 & 0 & 0 & 0 \end{bmatrix} \quad (2.6\text{-}18)$$

Next we determine the contributions of the kinematic equations to the A matrix. Equations (2.6-17) are linear in P_s, Q, and R_s, so all partial derivatives of the coefficient matrix elements will be eliminated when we set $P_s = Q = R_s = 0$. It only remains to evaluate the coefficient matrices under the steady-state conditions. The result is

$$H(\Phi)C_{b/s} = \begin{bmatrix} c\alpha + t\theta c\phi s\alpha & t\theta s\phi & -s\alpha + t\theta c\phi c\alpha \\ -s\phi s\alpha & c\phi & -s\phi c\alpha \\ c\phi s\alpha/c\theta & s\phi/c\theta & c\phi c\alpha/c\theta \end{bmatrix} \quad (2.6\text{-}19)$$

Inserting the steady-state conditions in this matrix and applying some trigonometric identities, we see that

$$\begin{bmatrix} \nabla_X f_4 \\ \nabla_X f_5 \\ \nabla_X f_6 \end{bmatrix}_{\substack{\bar{U}=\bar{U}_e \\ X=X_e}} = \begin{bmatrix} 0 & 0 & 0 & 0 & 0 & 0 & c\gamma_e/c\theta_e & 0 & s\gamma_e/c\theta_e \\ 0 & 0 & 0 & 0 & 0 & 0 & 0 & 1 & 0 \\ 0 & 0 & 0 & 0 & 0 & 0 & s\alpha_e/c\theta_e & 0 & c\alpha_e/c\theta_e \end{bmatrix} \quad (2.6\text{-}20)$$

The partial derivatives of the kinematic variables with respect to the control vector are all zero, so this completes the linearization of the kinematic equations. Note that the force and moment equations are independent of the heading angle ψ in the NED geographic frame, so the third kinematic equation is not really needed in the linear model.

The Linearized Moment Equations Here we determine the last three rows of the linear state equations (2.6-7). The starting point for this linearization is the stability-axes moment equations (2.5-26), with all terms moved to the right-hand side. The moment partial derivatives that are normally considered important are contained in Table 2.6-2; the table defines the moment dimensional derivatives. These dimensional derivatives are given the symbols L, M, and N to denote, respectively, rolling, pitching, and yawing moments, and their subscripts indicate the quantity with respect to which the derivative is taken. These include all six of the state variables that determine the translational and rotational rates, the four control variables, and the angular rate alpha-dot. The derivatives with respect to beta-dot have been omitted from the table because they are usually unimportant and are difficult to measure. The effect of beta-dot on yawing moment may sometimes be important, and the derivative can be estimated with methods given in the USAF DATCOM (Hoak et al., 1970). It is convenient to include the moment of inertia for the corresponding axis in the definition of the dimensionless coefficient. Therefore, each derivative has the dimensions of angular acceleration divided by the independent variable dimensions (s^{-1}, s^{-2}, $ft^{-1}s^{-1}$, or none).

We will assume, as in Section 2.5, that the engine thrust vector lies in the x_b-z_b plane and therefore contributes only a pitching moment $m_{y,T}$ to the stability-axes moment equations. This is not an accurate assumption for a propeller aircraft, and

TABLE 2.6-2: The Moment Dimensional Derivatives

ROLL	PITCH	YAW
$L_\beta = \dfrac{1}{J'_X} \dfrac{\partial \ell}{\partial \beta}$	$M_V = \dfrac{1}{J'_Y} \dfrac{\partial m_A}{\partial V_T}$	$N_\beta = \dfrac{1}{J'_Z} \dfrac{\partial n_A}{\partial \beta}$
$L_p = \dfrac{1}{J'_X} \dfrac{\partial \ell}{\partial P}$	$M_\alpha = \dfrac{1}{J'_Y} \dfrac{\partial m_A}{\partial \alpha}$	$N_p = \dfrac{1}{J'_Z} \dfrac{\partial n_A}{\partial P}$
$L_r = \dfrac{1}{J'_X} \dfrac{\partial \ell}{\partial R}$	$M_{\dot\alpha} = \dfrac{1}{J'_Y} \dfrac{\partial m_A}{\partial \dot\alpha}$	$N_r = \dfrac{1}{J'_Z} \dfrac{\partial n_A}{\partial R}$
$L_{\delta a} = \dfrac{1}{J'_X} \dfrac{\partial \ell}{\partial \delta_a}$	$M_q = \dfrac{1}{J'_Y} \dfrac{\partial m_A}{\partial Q}$	$N_{\delta a} = \dfrac{1}{J'_Z} \dfrac{\partial n_A}{\partial \delta_a}$
$L_{\delta r} = \dfrac{1}{J'_X} \dfrac{\partial \ell}{\partial \delta_r}$	$M_{\delta e} = \dfrac{1}{J'_Y} \dfrac{\partial m_A}{\partial \delta_e}$	$N_{\delta r} = \dfrac{1}{J'_Z} \dfrac{\partial n_A}{\partial \delta_r}$
	$M_{T_V} = \dfrac{1}{J'_Y} \dfrac{\partial m_T}{\partial V_T}$	$N_{T_\beta} = \dfrac{1}{J'_Z} \dfrac{\partial n_T}{\partial \beta}$
	$M_{T_\alpha} = \dfrac{1}{J'_Y} \dfrac{\partial m_T}{\partial \alpha}$	
	$M_{\delta t} = \dfrac{1}{J'_Y} \dfrac{\partial m_T}{\partial \delta_t}$	

there will be a number of power effects (Stinton, 1983; Ribner, 1943). These include a rolling moment due to propeller torque reaction, which is a function of throttle setting, and moments and forces that depend on the total angle of attack of the propeller, which is a function of alpha and beta. The table shows derivatives for thrust moment varying with speed, alpha, throttle position, and sideslip. For simplicity, the derivatives with respect to alpha and beta will be omitted from our equations.

The stability-axes moment equations (2.5-26) are repeated here, with all nonzero terms moved to the right-hand side:

$$0 = \begin{bmatrix} f_7 \\ f_8 \\ f_9 \end{bmatrix} = \begin{bmatrix} -\dot{P}_s + \dot\alpha R_s \\ -\dot{Q} \\ -\dot{R}_s - \dot\alpha P_s \end{bmatrix} + \frac{1}{\Gamma} \begin{bmatrix} J'_z & 0 & J'_{xz} \\ 0 & \Gamma/J'_y & 0 \\ J'_{xz} & 0 & J'_x \end{bmatrix} \left(\begin{bmatrix} \ell_s \\ m \\ n_s \end{bmatrix} - \Omega^s_{b/e} J^s \omega^s_{b/e} \right)$$

(2.6-21)

To find the block of E-matrix terms, all the moment-equation terms that involve state derivatives must be examined. These are

$$\dot{P}_s, \ \dot{Q}, \ \dot{R}_s, \ \dot\alpha R_s, \ \dot\alpha P_s, \ m$$

The two alpha-dot terms are of degree 2 in the variables that are set to zero in the steady state. Therefore, the corresponding partial derivatives vanish from the E matrix, leaving only four terms:

$$-\begin{bmatrix} \nabla_{\dot{X}} f_7 \\ \nabla_{\dot{X}} f_8 \\ \nabla_{\dot{X}} f_9 \end{bmatrix}_{\substack{\bar{U}=\bar{U}_e \\ X=X_e}} = \begin{matrix} \dot{v}_T & \dot{\beta} & \dot{\alpha} & \dot{\phi} & \dot{\theta} & \dot{\psi} & \dot{P}_s & \dot{Q} & \dot{R}_s \\ \begin{bmatrix} 0 & 0 & 0 & 0 & 0 & 0 & 1 & 0 & 0 \\ 0 & 0 & -M_{\dot{\alpha}} & 0 & 0 & 0 & 0 & 1 & 0 \\ 0 & 0 & 0 & 0 & 0 & 0 & 0 & 0 & 1 \end{bmatrix} \end{matrix} \quad (2.6\text{-}22)$$

The contributions of the moment equations to the A and B matrices must now be found. In Equation (2.6-21), on the right, the derivatives of the angular rates are constants for the purposes of partial differentiation; the alpha-dot terms are of degree 2 in the variables of the steady-state condition, as is the last term on the right. This leaves only the term comprising the product of the inertia matrix and the moment vector. The stability-axes inertia matrix is a function of alpha, but its derivative will be multiplied by a moment vector that is null in steady-state nonturning flight. Therefore, the only partial derivatives that are of interest are given by the inertia matrix terms multiplied by the partial derivatives of the moments:

$$\begin{bmatrix} \nabla_X f_7 \\ \nabla_X f_8 \\ \nabla_X f_9 \end{bmatrix}_{\substack{\bar{U}=\bar{U}_e \\ X=X_e}} = \begin{bmatrix} (J'_Z \nabla_X \ell_s + J'_{XZ} \nabla_X n_s)/\Gamma \\ (\nabla_X m)/J'_Y \\ (J'_{XZ} \nabla_X \ell_s + J'_X \nabla_X n_s)/\Gamma \end{bmatrix} \quad (2.6\text{-}23)$$

and

$$\begin{bmatrix} \nabla_U f_7 \\ \nabla_U f_8 \\ \nabla_U f_9 \end{bmatrix}_{\substack{\bar{U}=\bar{U}_e \\ X=X_e}} = \begin{bmatrix} (J'_Z \nabla_U \ell_s + J'_{XZ} \nabla_U n_s)/\Gamma \\ (\nabla_U m)/J'_Y \\ (J'_{XZ} \nabla_U \ell_s + J'_X \nabla_U n_s)/\Gamma \end{bmatrix} \quad (2.6\text{-}24)$$

When the partial derivatives in (2.6-23) and (2.6-24) are interpreted in terms of the dimensional derivatives in Table 2.6-2, we obtain the last three rows of the A matrix,

$$\begin{bmatrix} \nabla_X f_7 \\ \nabla_X f_8 \\ \nabla_X f_9 \end{bmatrix}_{\substack{\bar{U}=\bar{U}_e \\ X=X_e}} = \quad (2.6\text{-}25)$$

$$\begin{bmatrix} 0 & \mu L_\beta + \sigma_1 N_\beta & 0 & 0 & 0 & 0 & \mu L_p + \sigma_1 N_p & 0 & \mu L_r + \sigma_1 N_r \\ M_V + M_{T_V} & 0 & M_\alpha + M_{T\alpha} & 0 & 0 & 0 & 0 & M_q & 0 \\ 0 & \mu N_\beta + \sigma_2 L_\beta & 0 & 0 & 0 & 0 & \mu N_p + \sigma_2 L_p & 0 & \mu N_r + \sigma_2 L_r \end{bmatrix}$$

and the last three rows of the B matrix,

$$\begin{bmatrix} \nabla_U f_7 \\ \nabla_u f_8 \\ \nabla_u f_9 \end{bmatrix}_{\substack{\bar{U}=\bar{U}_e \\ X=X_e}} = \begin{bmatrix} 0 & 0 & \mu L_{\delta a} + \sigma_1 N_{\delta a} & \mu L_{\delta r} + \sigma_1 N_{\delta r} \\ M_{\delta t} & M_{\delta e} & 0 & 0 \\ 0 & 0 & \mu N_{\delta a} + \sigma_2 L_{\delta a} & \mu N_{\delta r} + \sigma_2 L_{\delta r} \end{bmatrix} \quad (2.6\text{-}26)$$

In these equations the constants μ and σ_i are given by

$$\mu = (J'_Z J'_X)/\Gamma \quad \sigma_1 = (J'_Z J'_{XZ})/\Gamma \quad \sigma_2 = (J'_X J'_{XZ})/\Gamma \quad (2.6\text{-}27)$$

The cross-product of inertia is normally small in magnitude compared to the moments of inertia, so the parameter μ is quite close to unity, and the σ_i are much smaller than unity.

The Decoupled Linear State Equations

All of the information for the coefficient matrices of the linear state equations (2.6-7) has now been obtained. An inspection of the coefficient blocks shows that the longitudinal- and lateral-directional equations are decoupled (although the lateral-directional equations do depend on steady-state longitudinal quantities such as γ_e and θ_e). Therefore, rather than attempt to assemble the complete equations, we will collect the longitudinal- and lateral-directional equations separately.

If the longitudinal state and control variables are ordered as follows, additional potential decoupling will become apparent. Thus we choose the longitudinal state and input vectors as

$$x = [\alpha \ q \ v_T \ \theta]^T \qquad u = [\delta_e \ \delta_t]^T \quad (2.6\text{-}28)$$

The longitudinal equations are obtained from the first and last rows of (2.6-11), (2.6-14), and (2.6-16) (divided through by m), the middle rows of (2.6-18) and (2.6-20), and the middle rows of (2.6-22), (2.6-25), and (2.6-26). The longitudinal coefficient matrices are now given by

$$E = \begin{bmatrix} V_{T_e} - Z_{\dot{\alpha}} & 0 & 0 & 0 \\ -M_{\dot{\alpha}} & 1 & 0 & 0 \\ 0 & 0 & 1 & 0 \\ 0 & 0 & 0 & 1 \end{bmatrix} \quad B = \begin{bmatrix} Z_{\delta e} & -X_{\delta t} \sin(\alpha_e + \alpha_T) \\ M_{\delta e} & M_{\delta t} \\ X_{\delta e} & X_{\delta t} \cos(\alpha_e + \alpha_T) \\ 0 & 0 \end{bmatrix}$$

$$A = \begin{bmatrix} Z_\alpha & V_{T_e} + Z_q & Z_V - X_{T_V} \sin(\alpha_e + \alpha_T) & -g_D \sin \gamma_e \\ M_\alpha + M_{T_\alpha} & M_q & M_V + M_{T_V} & 0 \\ X_\alpha & 0 & X_V + X_{T_V} \cos(\alpha_e + \alpha_T) & -g_D \cos \gamma_e \\ 0 & 1 & 0 & 0 \end{bmatrix}$$

$$(2.6\text{-}29)$$

We see that E is block-diagonal and does not contribute to any coupling between the α, q and v_T, θ pairs of variables. Furthermore, E is nonsingular for nonhovering flight because, although $Z_{\dot{\alpha}}$ can be positive, it is normally much smaller in magnitude than V_T.

The A matrix has several null elements and, in level flight, the (1, 4) element is zero. In trimmed flight, at low Mach numbers, the moment derivatives in element (2, 3) are zero (see next section). Finally, the (1, 3) element is small compared to the other elements of the first row and can often be neglected. Under the above conditions, the angle-of-attack and pitch-rate differential equations have no dependence on the speed, pitch-attitude perturbations (but not vice-versa). The solution of these equations, with the elevator and throttle inputs fixed, is a "stick-fixed" mode of oscillation known as the short-period mode (Chapter 3).

In the same vein, the (3, 1) element of the B matrix (drag due to elevator deflection) is usually negligible, and the pitching moment due to throttle inputs, element (2, 2), is zero if the x-z plane component of the engine thrust vector passes through the aircraft center of mass (this is not true for aircraft such as the B-747 and B-767 [Roskam, 1979]). Also, the (1, 2) element of B may often be neglected because of the small sine component. Under these conditions, the elevator input controls only the alpha-pitch-rate dynamics, the throttle input controls only the speed-pitch-attitude dynamics, and transfer-function analysis is simplified (Chapter 4).

The lateral/directional states and controls are

$$x = [\beta \quad \phi \quad p_s \quad r_s]^T \qquad u = [\delta_a \quad \delta_r]^T, \tag{2.6-30}$$

where the state ψ has been dropped. The state equations are obtained from the second rows of (2.6-11), (2.6-14), and (2.6-16), the first rows of (2.6-18) and (2.6-20), and the first and third rows of (2.6-22), (2.6-25), and (2.6-26). The resulting coefficient matrices are

$$E = \begin{bmatrix} V_{T_e} & 0 & 0 & 0 \\ 0 & 1 & 0 & 0 \\ 0 & 0 & 1 & 0 \\ 0 & 0 & 0 & 1 \end{bmatrix} \qquad B = \begin{bmatrix} Y_{\delta a} & Y_{\delta r} \\ 0 & 0 \\ L'_{\delta a} & L'_{\delta r} \\ N'_{\delta a} & N'_{\delta r} \end{bmatrix}$$

$$A = \begin{bmatrix} Y_\beta & g_D \cos\theta_e & Y_p & Y_r - V_{T_e} \\ 0 & 0 & c\gamma_e/c\theta_e & s\gamma_e/c\theta_e \\ L'_\beta & 0 & L'_p & L'_r \\ N'_\beta & 0 & N'_p & N'_r \end{bmatrix}, \tag{2.6-31}$$

where primed moment derivatives are defined (McRuer et al., 1973) by:

$$L'_\beta = \mu L_\beta + \sigma_1 N_\beta \qquad L'_p = \mu L_p + \sigma_1 N_p \qquad L'_r = \mu L_r + \sigma_1 N_r$$

$$N'_\beta = \mu N_\beta + \sigma_2 L_\beta \qquad N'_p = \mu N_p + \sigma_2 L_p \qquad N'_r = \mu N_r + \sigma_2 L_r$$

$$L'_{\delta a} = \mu L_{\delta a} + \sigma_1 N_{\delta a} \qquad L'_{\delta r} = \mu L_{\delta r} + \sigma_1 N_{\delta r}$$

$$N'_{\delta a} = \mu N_{\delta a} + \sigma_2 L_{\delta a} \qquad N'_{\delta r} = \mu N_{\delta r} + \sigma_2 L_{\delta r}$$

(2.6-32)

The inverse of the E matrix is diagonal and exists for nonzero airspeed. Its efffect is simply to divide the right-hand side of the beta-dot equation by airspeed. Therefore, although the original nonlinear equations were assumed implicit, the linear equations can now be made explicit in the derivatives. The coefficient matrices depend on the steady-state angle of attack and pitch attitude in both cases. Although they nominally apply to small perturbations about a wings-level, steady-state flight condition, the equations can be used satisfactorily for perturbed roll angles of several degrees.

In this chapter we will be content with simply deriving the coefficient matrices for the linear state equations; the equations will not be used until Chapter 3. The remainder of the chapter will be devoted to expressing the dimensional stability derivatives, used in the coefficient matrices, in terms of derivatives of the dimensionless aerodynamic coefficients defined in (2.3-8b). The resulting "dimensionless derivatives" have the advantage that they are less dependent on the specific aircraft and flight condition, and more dependent on the geometrical configuration of an aircraft. Methods have been developed to estimate the dimensionless derivatives, and they can be used to compare and assess different design configurations.

The Dimensionless Stability and Control Derivatives

The dimensional aerodynamic derivatives are simply a convenient set of coefficients for the linear equations. We must now relate them to the *dimensionless stability derivatives* used by stability and control engineers and found in aerodynamic data. The way in which the stability derivatives are made dimensionless depends on whether the independent variable for the differentiation is angle, angular rate, or velocity. This will be illustrated by example before we tabulate the derivatives.

Consider the derivative X_V in Table 2.6-1; this derivative is taken with respect to airspeed. The drag force depends on airspeed both through dynamic pressure and through the variation of the aerodynamic drag coefficient with airspeed. Therefore, using the definition of X_V and the drag equation from (2.3-8b), we have

$$X_V = \frac{-1}{m}\left[\frac{\partial \bar{q}}{\partial V_T}SC_D + \bar{q}S\frac{\partial C_D}{\partial V_T}\right] = \frac{-\bar{q}S}{mV_{T_e}}(2C_D + C_{D_V}),$$

where $C_{D_V} \equiv V_{T_e}(\partial C_D/\partial V_T)$ is the dimensionless speed damping derivative.

Next consider a derivative that is taken with respect to angular rate, C_{m_q}. The dimensionless rate-damping derivatives were defined in Section 2.3 and can now be related to the dimensional derivatives. Making use of the definition of pitching moment coefficient in (2.3-8b), we have

$$M_q = \frac{\bar{q}S\bar{c}}{J'_Y}\frac{\partial C_m}{\partial Q} = \frac{\bar{q}S\bar{c}}{J'_Y}\frac{\bar{c}}{2V_{T_e}}C_{m_q}, \qquad \text{where} \quad C_{m_q} \equiv \frac{2V_{T_e}}{\bar{c}}\frac{\partial C_m}{\partial Q}$$

The "dimensionless" stability derivatives taken with respect to angle actually have dimensions of deg^{-1} when expressed in degrees rather than radians.

Tables 2.6-1 and 2.6-2 include six thrust derivatives (X_{T_V}, X_{δ_t}, M_{T_V}, M_{T_α}, M_{δ_t}, N_{T_β}). The corresponding dimensionless derivatives can be defined by expressing the thrust force and moment components in terms of dimensionless coefficients. For example, a pitching moment component due to thrust can be written as $M_T = \bar{q}S\bar{c}C_{m_T}$. Values for the thrust derivatives would be found by referring to the "installed thrust" data for the specific engine and determining the change in thrust due to a perturbation in the variable of interest. In the case of the derivatives with respect to V_T and throttle setting, it is probably most convenient to work directly with the dimensional derivatives. Determination of the thrust derivatives with respect to α and β is more complicated; a readable explanation is given in Roskam (1979).

Following the lines of the examples above, the longitudinal dimensionless stability and control derivatives and the lateral/directional dimensionless stability and control derivatives corresponding to Tables 2.6-1 and 2.6-2 are given in Tables 2.6-3 and 2.6-4. Some of the thrust derivatives have been omitted because of lack of space, and because of their limited utility.

The dimensionless stability derivatives are in general very important to both the aircraft designer and the stability and control engineer. They provide information about the natural stability of an aircraft, about the effectiveness of the control surfaces, and about the maneuverability. They correlate with the geometrical features of the aircraft and thereby facilitate the preliminary design process. The typical variation of many of the stability derivatives with flight conditions (e.g., speed, angle of attack, sideslip angle) is known to the designer, and he or she can therefore anticipate the design problems in different parts of the flight envelope. Information on the importance of the stability derivatives, the accuracy with which they can be estimated, and their variation with flight conditions can be found in stability and control textbooks (Roskam, 1979; Etkin, 1972; Perkins and Hage, 1949) and in the USAF DATCOM (Hoak et al., 1970). Stability derivatives at certain flight conditions, for a number of different aircraft, are also given in these books, in Blakelock (1965), in McRuer et al. (1973), and in various other texts.

In the next subsection we briefly describe the significance of various dimensionless derivatives and their variation with flight conditions. This information will be utilized in Chapter 4 when aircraft dynamic modes are analyzed.

Description of the Longitudinal Dimensionless Derivatives

The names and relative importance of the longitudinal stability derivatives are shown in Table 2.6-5, starting with the most important.

The stability derivatives are estimated from geometrical properties, from the slopes of the aerodynamic coefficients, or from perturbed motion of an aircraft in flight-test or a model in a wind tunnel. The aerodynamic coefficients are, in general, nonlinear

TABLE 2.6-3: Longitudinal Dimensional versus Dimensionless Derivatives

$$X_V = \frac{-\bar{q}S}{mV_{T_e}}(2C_{D_e} + C_{D_V}) \qquad C_{D_V} \equiv V_{T_e}\frac{\partial C_D}{\partial V_T}$$

$$X_{T_V} = \frac{\bar{q}S}{mV_{T_e}}(2C_{T_e} + C_{T_V}) \qquad C_{T_V} \equiv V_{T_e}\frac{\partial C_T}{\partial V_T}$$

$$X_\alpha = \frac{\bar{q}S}{m}(C_{L_e} - C_{D_\alpha}) \qquad C_{D_\alpha} \equiv \frac{\partial C_D}{\partial \alpha}$$

$$X_{\delta e} = \frac{-\bar{q}S}{m}C_{D_{\delta e}} \qquad C_{D_{\delta e}} \equiv \frac{\partial C_D}{\partial \delta_e}$$

$$Z_V = \frac{-\bar{q}S}{mV_{T_e}}(2C_{L_e} + C_{L_V}) \qquad C_{L_V} \equiv V_{T_e}\frac{\partial C_L}{\partial V_T}$$

$$Z_\alpha = \frac{-\bar{q}S}{m}(C_{D_e} + C_{L_\alpha}) \qquad C_{L_\alpha} \equiv \frac{\partial C_L}{\partial \alpha}$$

$$Z_{\dot\alpha} = \frac{-\bar{q}S\bar{c}}{2mV_{T_e}}C_{L_{\dot\alpha}} \qquad C_{L_{\dot\alpha}} \equiv \frac{2V_{T_e}}{\bar{c}}\frac{\partial C_L}{\partial \dot\alpha}$$

$$Z_q = \frac{-\bar{q}S\bar{c}}{2mV_{T_e}}C_{L_q} \qquad C_{L_q} \equiv \frac{2V_{T_e}}{\bar{c}}\frac{\partial C_L}{\partial Q}$$

$$Z_{\delta e} = \frac{-\bar{q}S}{m}C_{L_{\delta e}} \qquad C_{L_{\delta e}} \equiv \frac{\partial C_L}{\partial \delta_e}$$

$$M_V = \frac{\bar{q}S\bar{c}}{J_Y V_{T_e}}(2C_{m_e} + C_{m_V}) \qquad C_{m_V} \equiv V_{T_e}\frac{\partial C_m}{\partial V_T}$$

$$M_{T_V} = \frac{\bar{q}S\bar{c}}{J_Y V_{T_e}}(2C_{m_T} + C_{m_{T_V}}) \qquad C_{m_{T_V}} \equiv V_{T_e}\frac{\partial C_{m_T}}{\partial V_T}$$

$$M_\alpha = \frac{\bar{q}S\bar{c}}{J_Y}C_{m_\alpha} \qquad C_{m_\alpha} \equiv \frac{\partial C_m}{\partial \alpha}$$

$$M_{\dot\alpha} = \frac{\bar{q}S\bar{c}}{J_Y}\frac{\bar{c}}{2V_{T_e}}C_{m_{\dot\alpha}} \qquad C_{m_{\dot\alpha}} \equiv \frac{2V_{T_e}}{\bar{c}}\frac{\partial C_m}{\partial \dot\alpha}$$

$$M_q = \frac{\bar{q}S\bar{c}}{J_Y}\frac{\bar{c}}{2V_{T_e}}C_{m_q} \qquad C_{m_q} \equiv \frac{2V_{T_e}}{\bar{c}}\frac{\partial C_m}{\partial Q}$$

$$M_{\delta e} = \frac{\bar{q}S\bar{c}}{J_Y}C_{m_{\delta e}} \qquad C_{m_{\delta e}} \equiv \frac{\partial C_m}{\partial \delta_e}$$

TABLE 2.6-4: Lateral/Directional Dimensional versus Dimensionless Derivatives

$$Y_\beta = \frac{\bar{q}S}{m}C_{Y_\beta} \qquad C_{Y_\beta} \equiv \frac{\partial C_Y}{\partial \beta}$$

$$Y_p = \frac{\bar{q}Sb}{2mV_{T_e}}C_{Y_p} \qquad C_{Y_p} \equiv \frac{2V_{T_e}}{b}\frac{\partial C_Y}{\partial P}$$

$$Y_r = \frac{\bar{q}Sb}{2mV_{T_e}}C_{Y_r} \qquad C_{Y_r} \equiv \frac{2V_{T_e}}{b}\frac{\partial C_Y}{\partial R}$$

$$Y_{\delta r} = \frac{\bar{q}S}{m}C_{Y_{\delta r}} \qquad C_{Y_{\delta r}} \equiv \frac{\partial C_Y}{\partial \delta_r}$$

$$Y_{\delta a} = \frac{\bar{q}S}{m}C_{Y_{\delta a}} \qquad C_{Y_{\delta a}} \equiv \frac{\partial C_Y}{\partial \delta_a}$$

$$L_\beta = \frac{\bar{q}Sb}{J'_X}C_{\ell_\beta} \qquad C_{\ell_\beta} \equiv \frac{\partial C_\ell}{\partial \beta}$$

$$L_p = \frac{\bar{q}Sb}{J'_X}\frac{b}{2V_{T_e}}C_{\ell_p} \qquad C_{\ell_p} \equiv \frac{2V_{T_e}}{b}\frac{\partial C_\ell}{\partial P}$$

$$L_r = \frac{\bar{q}Sb}{J'_X}\frac{b}{2V_{T_e}}C_{\ell_r} \qquad C_{\ell_r} \equiv \frac{2V_{T_e}}{b}\frac{\partial C_\ell}{\partial R}$$

$$L_{\delta a} = \frac{\bar{q}Sb}{J'_X}C_{\ell_{\delta a}} \qquad C_{\ell_{\delta a}} \equiv \frac{\partial C_\ell}{\partial \delta_a}$$

$$L_{\delta r} = \frac{\bar{q}Sb}{J'_X}C_{\ell_{\delta r}} \qquad C_{\ell_{\delta r}} \equiv \frac{\partial C_\ell}{\partial \delta_r}$$

$$N_\beta = \frac{\bar{q}Sb}{J'_Z}C_{n_\beta} \qquad C_{n_\beta} \equiv \frac{\partial C_n}{\partial \beta}$$

$$N_p = \frac{\bar{q}Sb}{J'_Z}\frac{b}{2V_{T_e}}C_{n_p} \qquad C_{n_p} \equiv \frac{2V_{T_e}}{b}\frac{\partial C_n}{\partial P}$$

$$N_r = \frac{\bar{q}Sb}{J'_Z}\frac{b}{2V_{T_e}}C_{n_r} \qquad C_{n_r} \equiv \frac{2V_{T_e}}{b}\frac{\partial C_n}{\partial R}$$

$$N_{\delta a} = \frac{\bar{q}Sb}{J'_Z}C_{n_{\delta a}} \qquad C_{n_{\delta a}} \equiv \frac{\partial C_n}{\partial \delta_a}$$

$$N_{\delta r} = \frac{\bar{q}Sb}{J'_Z}C_{n_{\delta r}} \qquad C_{n_{\delta r}} \equiv \frac{\partial C_n}{\partial \delta_r}$$

TABLE 2.6-5: Importance of Longitudinal Stability Derivatives

C_{L_α}	Lift-curve slope	(determines ride quality)
C_{m_α}	Pitch stiffness	(< 0 for static stability)
C_{m_q}	Pitch damping	(< 0 for short-period damping)
C_{m_V}	Tuck derivative	(< 0 gives unstable tuck)
$C_{m_{\dot\alpha}}$	Alpha-dot derivative	(less important than C_{m_q})
C_{D_V}	Speed damping	(if > 0 can mitigate unstable C_{m_V})
C_{D_α}	Drag versus alpha slope	
C_{L_v}	Lift versus speed slope	
$C_{L_{\dot\alpha}}$	Acceleration derivative for lift	
C_{L_q}	Pitch-rate dependent lift	

functions, and so for a given aircraft the stability derivatives vary with the aerodynamic angles (α, β), Mach number (compressibility effect), thrust (power effect), and dynamic pressure (aeroelastic effects). Descriptions of these variations, and methods of estimating the derivatives can be found in the literature (Roskam, 1979; Perkins and Hage, 1949; Queijo, 1971). Stability derivatives obtained from flight test are usually presented in graphs that apply to trimmed-flight conditions at, for example, a given altitude with varying Mach number. Therefore, a sequence of points along a particular curve would correspond to different combinations of thrust, angle-of-attack, and elevator setting. This is acceptable to the flying-qualities engineer, but presents a difficulty to the simulation engineer seeking to build a lookup-table for that derivative.

Plots of aerodynamic coefficients, particularly those of high-speed aircraft, can exhibit both small-scale fluctuations and regimes of widely different behavior. Differentiation exaggerates such effects, and so it is easier to generalize about the behavior of aerodynamic coefficients than about the stability derivatives. Furthermore, the stability derivatives do not provide an adequate model of aircraft behavior for large amplitude maneuvers and very nonlinear regimes such as stall. Stability derivative information is more readily available than aerodynamic coefficient data, and is appropriate for linear models for stability analysis and flight control system design, but is limited in its applicability to flight simulation. We now summarize the typical behavior of the most important derivatives in the normal flight regimes.

Lift-Curve Slope The derivative C_{L_α} is called the lift-curve slope; it is important because it determines how turbulent changes in alpha translate into changes in lift, and hence determines the level of comfort for the pilot. In the same manner, it affects the maneuverability of the aircraft. It also affects the damping of the pitching motion of the aircraft when subjected to sudden disturbances, as will be shown in Chapter 4. This influences the pilot's opinion of the *handling qualities* of the airplane.

The lift-curve slope is approximately independent of alpha, and typically in the range 1 to 8 (per radian) for the linear region of the lift curve below stall. When the wing is producing a large amount of lift, wing twist will reduce the local angle of attack of the wing panels according to distance out from the wing root. This will tend to reduce the lift-curve slope as alpha increases.

As explained in Section 2.3, compressibility effects also change C_{L_α} significantly; below the critical Mach number it increases with Mach, and at supersonic speeds it decreases with Mach. In the transonic range it may pass smoothly through a maximum (e.g., fighter-type wings), or may show a dip (thick, higher-aspect wings with no sweep). Wing sweep-back has the effect of reducing the lift-curve slope and making the curve of C_{L_α} versus Mach less peaked. The propulsion system can also have a strong effect on the C_{L_α}, as can be visualized from Figure 2.3-3a.

Pitch Stiffness Derivative The derivative C_{m_α} is the slope of the curve of the static pitching-moment coefficient, around the cm, versus alpha (see Section 2.4, also Figure 2.3-7), with controls neutral. Its importance was demonstrated in Section 2.4. It also plays an important role in the dynamic behavior of pitching motion, as shown in Chapter 4.

Section 2.4 shows that the pitch stiffness will increase as the aerodynamic center moves aft with increasing Mach number and, depending on cm position, will also be affected by changes in wing C_{L_α} with Mach. The second important component of C_{m_α} contains the lift-curve slope of the horizontal tail. Again, the lift curve slope varies with Mach, but this may not have a very great effect on C_{m_α}, particularly in the case of a thin, swept, low-aspect "all-flying" tail. The lift-curve slope is multiplied by the tail efficiency factor, and this will tend to decrease with increasing alpha, to an extent depending on the degree of coupling between wing and tail. The third component of C_{m_α} is the derivative with respect to alpha of the pitching moment at the "wing-body aerodynamic center." A true wing-body aerodynamic center may not exist, and so this term is nonzero and difficult to determine. Roskam (1979) states that C_{m_α} will normally lie in the range -3 to $+1$ rad^{-1}.

Pitch Damping Derivative, C_{m_q} The pitch damping derivative, C_{m_q}, was introduced in Section 2.3. This derivative is normally negative, and determines the moment that opposes any pitch rate. It provides the most important contribution to the damping of the dynamic behavior in pitch (see Chapter 4), and hence is intimately involved in aircraft handling qualities.

The pitch damping is not given by the slope of an aerodynamic coefficient; it must be estimated from oscillatory motion of the aircraft or aircraft model, or calculated. The main physical mechanism involved is that pitch rate determines translational rate of the horizontal tail perpendicular to the relative wind. This changes the tail angle of attack, tail "lift," and hence the tail moment about the cm. When the induced translational rate is small compared to true airspeed, the change in tail angle of attack will be linearly related to pitch rate. Therefore, the pitch damping moment is invariably modeled as linearly proportional to pitch rate through C_{m_q}.

134   MODELING THE AIRCRAFT

A very simple expression for the pitch damping derivative C_{m_q} can be obtained by calculating the horizontal tail increment in lift due to a pitch-rate-induced translational velocity at the tail. Equation (2.3-9a) gives this derivative as

$$C_{m_q} = \frac{2V_{T_e}}{\bar{c}} \frac{\Delta C_m}{Q} \tag{2.6-33}$$

Let the moment arm of the tail ac about the aircraft cm be ℓ_t. The increment in lift of the tail is, in dimensionless form,

$$\Delta C_{L_t} = C_{L_{\alpha,t}} \tan^{-1}\left(Q\ell_t/V_{T_e}\right) \approx C_{L_{\alpha,t}}\left(Q\ell_t/V_{T_e}\right) \tag{2.6-4}$$

Now remember that the dimensional pitching moment coefficient is obtained by multiplying the dimensionless moment coefficient by $(S\bar{c})$. Therefore this lift must be converted to a nondimensional pitching moment by multiplying by the *horizontal tail volume ratio*,

$$V_H = (S_t \ell_t)/(S\bar{c}) \tag{2.6-35}$$

From the above three equations, noting that a positive pitch rate gives the tail a downward motion, a positive lift component, and therefore a negative contribution to aircraft pitching moment, we obtain

$$C_{m_q} = -2V_H C_{L_{\alpha,t}}\,(\ell_t/\bar{c}) \tag{2.6-36}$$

This equation neglects any pitch damping effect from the wings and fuselage, applies only for small alpha, and does not model any compressibility, aeroelastic, or thrust-dependent effects. It can be made to include some thrust and wing-downwash effects by including the tail efficiency factor (Equation (2.3-20)).

Figure 2.6-1 shows the variation of the pitch damping and acceleration derivatives for the jet trainer. These derivatives change quite dramatically with Mach in the transonic region, they are independent of alpha until stall is approached, and the pitch damping is somewhat dependent on altitude through aeroelastic effects. They can also be expected to be dependent on elevator deflection and movement of the tail aerodynamic center.

Tuck Derivative   The effect of speed variations on pitching moment is contained in the "tuck derivative," C_{m_V}. This derivative can also be written in terms of Mach:

$$C_{m_V} = \frac{V_{T_e}}{a} \frac{\partial C_m}{\partial (V_T/a)} = M \frac{\partial C_m}{\partial M} \tag{2.6-37}$$

The derivative will be negligible at low subsonic speeds when compressibility effects are absent, and at supersonic speeds when the aerodynamic center has ceased to move. In the transonic region we would expect to find a negative value as the ac moves

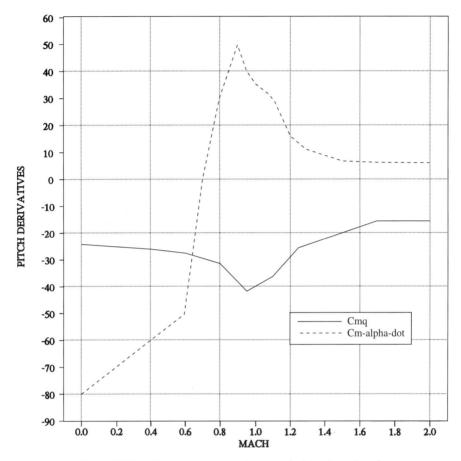

Figure 2.6-1 Pitching moment derivatives of a jet trainer aircraft.

aft but, in fact, C_{m_V} can be positive for some aircraft. The derivative changes quite abruptly as the transonic regime is reached and, if it is negative, the aircraft will tend to pitch-down as speed increases. Gravity will then tend to further increase the speed, leading to an unstable pitch down or "tuck under" effect. The tuck may be particularly troublesome if the elevator control effectiveness is simultaneously decreasing with Mach. The transonic drag rise helps to mitigate an unstable tuck characteristic. Values of the tuck derivative range between about -0.4 and $+0.6$ (Roskam, 1979).

Speed Damping Derivative The speed damping derivative, C_{D_V}, can also be written in terms of Mach:

$$C_{D_V} = \frac{V_T}{a} \frac{\partial C_D}{\partial (V_T/a)} = M \frac{\partial C_D}{\partial M} \qquad (2.6\text{-}38)$$

Like the tuck derivative, the speed damping derivative is a compressibility effect. It is negligible at low Mach numbers, rises to a peak with the transonic drag rise, then changes rapidly to negative values as the drag coefficient peaks and falls off with Mach, after the transonic regime. Values typically lie in the range -0.01 to $+0.30$ (Roskam, 1979).

Description of the Lateral-Directional Dimensionless Derivatives

The lateral/directional stability derivatives are shown in Table 2.6-6, starting with the most important. The more important derivatives are discussed below.

Dihedral Derivative The *dihedral derivative*, C_{l_β}, is the slope of the rolling moment versus sideslip curve. Section 2.3 showed that this slope should be negative to achieve positive stiffness in roll, and that wings with positive dihedral could provide this. However, too much positive stiffness in roll tends to reduce the damping of the aircraft dynamic behavior in a yawing-rolling motion (the dutch roll mode, see Chapter 4), and the designer must find a compromise in the value of C_{l_β}. In some aircraft, wing sweep produces a C_{l_β} that is too negative and the aircraft may have negative dihedral (anhedral) of the wings or horizontal tail to offset this effect (e.g., F-4 and AV8-B aircraft). The value of C_{l_β} is typically in the range -0.4 to $+0.1$ per radian (Roskam, 1979) and may change significantly with Mach number in the transonic range.

Yaw Stiffness Derivative The *yaw-stiffness derivative*, C_{n_β}, is the slope of the curve of yawing moment due to sideslip (Section 2.3), and it is associated with

TABLE 2.6-6: Importance of Lateral/Directional Derivatives

C_{ℓ_β}	Dihedral derivative	(< 0 for positive stiffness)
C_{n_β}	Yaw stiffness	(> 0 for positive stiffness)
C_{ℓ_p}	Roll damping	(< 0 for roll damping)
C_{n_r}	Yaw damping	(< 0 for yaw damping)
C_{n_p}	Yawing moment due to roll rate	
C_{ℓ_r}	Rolling moment due to yaw rate	
C_{Y_β}	Sideforce due to sideslip	
C_{Y_r}	Sideforce due to yaw rate	
C_{Y_p}	Sideforce due to roll rate	
$C_{n_{\dot\beta}}$	Yawing moment due to sideslip rate	
$C_{Y_{\dot\beta}}$	Sideforce due to sideslip rate	
$C_{\ell_{\dot\beta}}$	Rolling moment due to sideslip rate	

weathercock stability (tendency to head into the relative wind). It must be positive for positive stiffness in yaw, and it is principally determined by the size of the vertical tail. Weathercock stability can be lost at high dynamic pressure, due to structural deformation of the vertical tail, and aircraft have been known to "swap ends" in flight.

C_{n_β} plays a major role in the aircraft dutch roll mode (Chapter 4). Its value is typically in the range 0 to 0.4 rad^{-1} (Roskam, 1979), tending to fall off and possibly even become negative at transonic to supersonic Mach numbers. It may also become negative at high angles of attack when the vertical tail becomes immersed in the wake from the wings and body. Achieving a suitable value of C_{n_β} is a consideration in the initial sizing of the vertical tail of an aircraft.

Roll Damping Derivative The *roll damping derivative*, C_{ℓ_p}, was introduced in Section 2.3, and is chiefly due to the variation of angle of attack along the wing span when the aircraft is rolling. The rolling moment produced by the differential lift between the two wings will be linearly proportional to roll rate until stall begins on the outer wing panels. This derivative is positive, except possibly in a spin, and usually lies in the range -0.8 to -0.1 per radian. It thus provides a moment that damps rolling motion, plays the major roll in the response of the aircraft to aileron inputs (see roll time-constant, Chapter 4), and determines the associated handling qualities. It is determined from small amplitude rolling motion measurements. When considering maximum roll rate, the helix angle (Section 2.3) is the more important parameter. C_{ℓ_p} is, in general, a function of Mach number, altitude (because of aeroelastic effects), and alpha.

The Yaw Damping Derivative The *yaw damping derivative*, C_{n_r}, was introduced in Section 2.3, and assumes a linear relationship between yaw rate and the yawing moment it produces. It is mainly determined by the vertical tail, and is always negative except possibly in a spin. A simple calculation, analogous to the calculation of pitch damping, gives

$$C_{n_r} = -2V_v C_{L_{\alpha,vt}}(\ell_t/b), \qquad (2.6\text{-}39)$$

where V_v is a volume ratio for the vertical tail. It is the most important parameter in the airplane dutch roll mode (Chapter 4), and many aircraft must use an automatic control system to augment C_{n_r} (Chapter 4) because of inadequate dutch roll damping.

2.7 SUMMARY

In this chapter we have described how the aerodynamic forces and moments acting on an aircraft are created, how they are modeled mathematically, and how the data for the models are gathered. We have related these forces and moments to the equations of motion of a rigid aircraft that were derived in Chapter 1. The transformation of the equations of motion into a different set of coordinates has been demonstrated, and also the derivation of a nonlinear model for longitudinal motion only. Steady-state flight conditions have been defined. It has been shown that the equations of motion can be linearized around a steady-state condition and that they can then be separated into two decoupled sets. One of these sets describes the longitudinal motion of an aircraft, and the other describes the lateral/directional motion. The linear equations have been expressed in terms of the aerodynamic derivatives, and the significance of these derivatives has been explained. In Chapter 3 we develop a number of powerful analytical and computational tools and use them in conjunction with the aircraft models developed here.

REFERENCES

Anderson, F. *Northrop: An Aeronautical History*. Los Angeles: Northrop Corporation, 1976.

Anderson, J. D., Jr. *Fundamentals of Aerodynamics*. 2d ed. New York: McGraw-Hill, 1991.

———. *Aircraft Performance and Design*. New York: McGraw-Hill, 1999.

Babister, A. W. *Aircraft Stability and Control*. Oxford: Pergamon, 1961.

Blakelock, J. H. *Automatic Control of Aircraft and Missiles*. New York: Wiley, 1965.

Bryan, G. H. *Stability in Aviation*. London: Macmillan, 1911.

Chambers, J. R., and E. L. Anglin. "Analysis of Lateral-Directional Stability Characteristics of a Twin-Jet Fighter Airplane at High Angles of Attack." *NASA Tech. Note D-5361*. Washington, D.C.: NASA, 1969.

DeCamp, R. W., R. Hardy, and D. K. Gould. "Mission Adaptive Wing." *Society of Automotive Engineers Paper 872419*, 1987.

Dommasch, D. O., S. S. Sherby, and T. F. Connolly. *Airplane Aerodynamics*. 4th ed. New York: Pitman, 1967.

Drendel, L. *SR-71 Blackbird in Action*. Carrollton, Tex.: Squadron/Signal Publications, 1982.

Droste, C. S., and J. E. Walker. "The General Dynamics Case Study on the F-16 Fly-by-Wire Flight Control System." *AIAA Professional Case Studies*, no date.

Duncan, W. J. *The Principles of the Control and Stability of Aircraft*. Cambridge: Cambridge University Press, 1952.

Etkin, B. *Dynamics of Atmospheric Flight*. New York: Wiley, 1972.

Goman, M., and A. Khrabrov. "State-Space Representation of Aerodynamic Characteristics of an Aircraft at High Angles of Attack." *Journal of Aircraft* 31, no. 5 (September–October 1994): 1109–1115.

Hoak, D. E., et al. *USAF Stability and Control DATCOM*. Flight Control Division, Air Force Flight Dynamics Laboratory, Wright Patterson Air Force Base, Ohio, September 1970.

IMSL. *Library Contents Document*. 8th ed. International Mathematical and Statistical Libraries, Inc., 7500 Bellaire Blvd., Houston, TX 77036, 1980.

Kandebo, S. W. "Second X-29 Will Execute High-Angle-of-Attack Flights." *Aviation Week and Space Technology* (October 31, 1988): 36–38.

Kuethe, A. M., and C. Y. Chow. *Foundations of Aerodynamics*. 4th ed. New York: Wiley, 1984.

Maine, R. E., and K. W. Iliffe. "Formulation and Implementation of a Practical Algorithm for Parameter Estimation with Process and Measurement Noise." Paper 80-1603. *AIAA Atmospheric Flight Mechanics Conference*, August 11–13, 1980, pp. 397–411.

McCormick, B. W. *Aerodynamics, Aeronautics, and Flight Mechanics*. New York: Wiley, 1995.

McFarland, R. E. "A Standard Kinematic Model for Flight Simulation at NASA Ames." *NASA CR-2497*. Washington, D.C.: NASA, January 1975.

McRuer, D., I. Ashkenas, and D. Graham. *Aircraft Dynamics and Automatic Control*. Princeton, N.J.: Princeton University Press, 1973.

Nguyen, L. T., et al. "Simulator Study of Stall/Post-Stall Characteristics of a Fighter Airplane with Relaxed Longitudinal Static Stability." *NASA Tech. Paper 1538*. Washington D.C.: NASA, December 1979.

Perkins, C. D., and R. E. Hage. *Airplane Performance Stability and Control*. New York: Wiley, 1949.

Pope, A. *Wind Tunnel Testing*. New York: Wiley, 1954.

Press, W. H., B. P. Flannery, S. A. Teukolsky, and W. T. Vetterling. *Numerical Recipes: The Art of Scientific Computing*. New York: Cambridge University Press, 1989.

Queijo, M. J. "Methods of Obtaining Stability Derivatives." *Performance and Dynamics of Aerospace Vehicles*. Washington, D.C.: NASA-Langley Research Center, NASA, 1971.

Rech, J., and C. S. Leyman. "A Case Study by Aerospatiale and British Aerospace on the Concorde." *AIAA Professional Case Studies*, no date.

Ribner, H. S. *NACA Reports 819, 820*. Langley, Va., 1943.

Roskam, J. *Airplane Flight Dynamics and Automatic Flight Controls*. Lawrence, Kans.: Roskam Aviation and Engineering Corp., 1979.

Scott, W. B. "X-28 Proves Viability of Forward-swept Wing." *Aviation Week and Space Technology* (October 31, 1988): 36–42.

Stevens, B. L., and F. L. Lewis. *Aircraft Control and Simulation*. 1st ed. New York: John Wiley & Sons, Inc., 1992.

Stinton, D. *Flying Qualities and Flight Testing of the Airplane*. AIAA Educational Series, 1996.

———. *Design of the Airplane*. Distributed by AIAA, 1983.

U.S. Standard Atmosphere. U.S. Extensions to the ICAO Standard Atmosphere. Washington, D.C.: U.S. Government Printing Office, 1976.

Vidyasagar, M. *Nonlinear Systems Analysis*. Englewood Cliffs, N.J.: Prentice Hall, 1978.

Whitford, R. *Design for Air Combat*. London: Jane's, 1987.

Yuan, S. W. *Foundations of Fluid Mechanics*. Englewood Cliffs, N.J.: Prentice Hall, 1967.

PROBLEMS

Section 2.2

2.2-1 An airfoil is tested in a subsonic wind tunnel. The lift is found to be zero at a geometric angle of attack $\alpha = -1.5°$. At $\alpha = 5°$, the lift coefficient is measured as 0.52. Also, at $\alpha = 1°$ and $7.88°$, the moment coefficients about the center of gravity are measured as -0.01 and 0.05, respectively. The center of gravity is located at $0.35c$. Calculate the location of the aerodynamic center and the value of Cm_{ac}.

Section 2.3

2.3-1 An aircraft is flying wings-level at constant altitude, at a speed of 500 ft/s, with an angle of attack of 8 degrees and a sideslip angle of −5 degrees, when it runs into gusty wind conditions. Determine the new "instantaneous" angles of attack and sideslip angle for the following cases:

(i) a horizontal gust of 20 ft/s from *left to right* along the body y-axis

(ii) a horizontal of 50 ft/s, from dead astern

(iii) a gust of 30 ft/s, *from the right*, with velocity vector in the y-z plane, and at an angle of 70 degrees below the x-y plane

2.3-2 Derive expressions for the derivatives of V_T, alpha, and beta, in terms of U', V', and W' and their derivatives. Check the results against Equations (2.3-10).

2.3-3 Consult the literature to find information on the significance and numerical values of the helix angle achieved by different types of fighter aircraft. Find some graphs of roll rate versus equivalent airspeed and calculate some values of helix angle. Explain the shape of the graph.

2.3-4 Program the functions for the body-axes force coefficients CX and CZ, as given in Appendix A, for the F-16 model. Write another program to use these data and plot a set of curves of lift coefficient as a function of alpha (for $-10° \leq \alpha \leq 50°$), with elevator deflection as a parameter (for $\delta e = -25°, 0°, 25°$). Determine the angle of attack at which maximum lift occurs.

2.3-5 Program the body-axes moment coefficient CM, as given in Appendix A, for the F-16 model. Write another program to plot a set of curves of pitching moment as a function of alpha (for $-10° \leq \alpha \leq 50°$), with elevator deflection as a parameter (for $\delta e = -25°, 0°, 25°$). Comment on the pitch stiffness and on the elevator control power.

2.3-6 Program the F-16 engine thrust model, function THRUST, in Appendix A. Write a program to plot the thrust as a function of power setting (0–100 percent), with altitude as a parameter (for $h = 0, 25\text{kft}, 50\text{kft}$), at Mach 0.6. Also, plot thrust against Mach number, at 100 percent power, with altitude as a parameter (for $h = 0, 25\text{kft}, 50\text{kft}$). Comment on these characteristics of the jet engine.

Section 2.4

2.4-1 Solve numerically the nonlinear longitudinal-equilibrium equations, to determine the angle of attack and elevator deflection (both in deg) of the following small airplane, for level ($\gamma = 0$) steady-state flight at 90 ft/s. Assume $g = 32.2$ ft/s.

Atmospheric density $= 2.377 * 10^{-3}$ slugs/ft^3 (assumed constant)

Weight = 2300 lbs, inertia (slug-ft^2): $I_{yy} = 2094$

Wing reference area, $S = 175$ ft^2. Mean aerodynamic chord, $\bar{c} = 4.89$ ft

Thrust angle, $\alpha_T = 0$.

Lift: $C_L = 0.25 + 4.58 * \alpha$ (alpha in rads)

Drag: $C_D = 0.038 + 0.053 * C_L * C_L$

Pitch: $C_m = 0.015 - 0.75 * \alpha - 0.9 * \delta_e$ (alpha, δ_e, in rads)

Pitch damping coefficient, $C_{m_q} = -12.0$ (per rad/s)

2.4-2 Derive Equation (2.4-13), including all of the missing steps.

2.4-3 An aircraft is flying at 30,000 ft ($\rho = 8.9068 * 10^{-4}$) and has a wing lift coefficient of 1.0 and a tail lift coefficient of 1.2. The wing surface area and tail surface area are 600 ft^2 and 150 ft^2, respectively. The mean aerodynamic chord of the wing is 10 ft. The mean aerodynamic center of the wing is 10 ft ahead of the cm. The pitching moment coefficient about the aerodynamic center of the wing is -0.05. The tail is made up of a symmetric airfoil cross section; take the tail efficiency as $\eta = 1.0$. Determine the distance of the tail aerodynamic center from the cm for trimmed flight. If the aircraft weighs 50,000 lbs, calculate the air speed for trimmed level flight.

Section 2.5

2.5-1 Make a block diagram of the flat-Earth vector equations of motion (2.5-1), including wind inputs, pilot control inputs, and terrestrial position calculations. Blocks included should be vector integration, moment generation, force generation, atmosphere model with Mach and dynamic pressure calculation, addition, subtraction, and cross-product. The diagram should show all of the variables that would be needed in a high-fidelity simulation.

2.5-2 Repeat Problem 2.5-1 for the oblate rotating-Earth equations of motion.

2.5-3 Show that the quantity Γ in Equation (2.5-24) can be calculated from either body-axes or wind-axes quantities, with the same formula.

2.5-4 Expand the flat-Earth vector-form equations of motion, Equations (2.5-1), into scalar equations. Check the results against Table 2.5-1.

Section 2.6

2.6-1 Work through the derivation of the coefficient matrices for the linearized force equations [Equations (2.6-11), (2.6-14), and (2.6-16)], filling in all of the steps.

2.6-2 Fill in all of the steps in the derivation of the coefficient matrix, Equation (2.6-20), for the linearized kinematic equations.

2.6-3 Fill in all of the steps in the derivation of the coefficient matrices for the linearized moment equations [Equations (2.6-22), (2.6-25), and (2.6-26)].

2.6-4 Write a program to calculate (approximately) the derivative of a function of a single variable (assumed to be continuous), given discrete values of the function. Use the program with the lookup-table from Problem 2.3-5, to estimate the derivative C_{m_α} at the values of $\alpha = 0°$, $10°$, $20°$, and $30°$ (when $\delta e = 0^0$). Determine whether the aircraft has positive pitch stiffness at these angles of attack.

CHAPTER 3

MODELING, DESIGN, AND SIMULATION TOOLS

3.1 INTRODUCTION

In this chapter we will look more closely at continuous-time state-space models, their properties, and how they are derived from physical systems. This will lead to numerical methods and algorithms for computer software that can be applied to the many tasks associated with the simulation of an aerospace vehicle and design of its control systems. The software tools will provide the capability to trim aircraft models for steady-state flight, perform digital flight simulation, extract linear state-space and transfer function descriptions of aircraft models, and perform linear control system design. These operations are illustrated in Figure 3.1-1.

In the figure the nonlinear state and output equations are, respectively,

$$\dot{X} = f(X, U), \quad X(n \times 1), U(m \times 1) \qquad (3.1\text{-}1a)$$

$$Y = g(X, U), \quad Y(p \times 1), \qquad (3.1\text{-}1b)$$

where f and g represent arrays of continuous, single-valued functions. The linear versions of these equations are

$$\dot{x} = Ax + Bu \qquad (3.1\text{-}2a)$$

$$y = Cx + Du \qquad (3.1\text{-}2b)$$

An output equation is required because the state variables may not all be physical variables or directly accessible. Hence there is a need to represent physically measurable quantities by the output variables Y_i or y_i.

144 MODELING, DESIGN, AND SIMULATION TOOLS

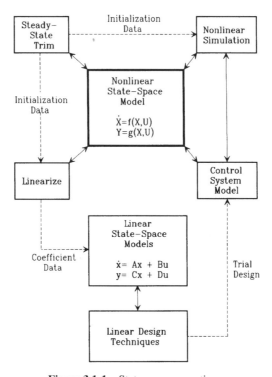

Figure 3.1-1 State-space operations.

Referring to Figure 3.1-1, the behavior of a real system can be simulated by solving the nonlinear model equations. The mathematical theory for numerical solution of ordinary differential equations (ODEs) is mature, and is cast in the state-space canonical form. Therefore, a state-space model is a prerequisite for simulation. Second, real systems operate around design points or equilibrium conditions, and are nonlinear to varying degrees as they deviate from equilibrium. By linearizing the nonlinear model around these design points the powerful tools of linear systems theory can be used to analyze the system and perform control systems design. The state-space formulation lends itself to linearization around equilibrium points. Third, the matrix formulation of the linear state equations readily handles multiple-input multiple-output (MIMO) systems, with single-input, single-output (SISO) systems as a special case. We will see that the MIMO description is essential for some aspects of aircraft dynamics. Lastly, if additional equations are to be coupled to the model, for example, to simulate an automatic control system, this is easily done when the controller equations are also in state-space form.

Also in this chapter, we have provided source code for state-space models of two different aircraft. These models will be used to illustrate aircraft dynamic behavior, and for control system design examples in Chapter 4. The final topic of the chapter is a review of the classical control theory and design techniques that will be used in Chapter 4.

3.2 STATE-SPACE MODELS

ODEs are our most powerful method for modeling continuous-time lumped-parameter dynamic systems. Continuous-time implies that the variables are uniquely defined at all moments in time within a specified interval, except possibly at a countable set of points. Lumped-parameter implies that each of the interconnected elements that make up the model responds immediately to its excitation. This is in contrast to distributed-parameter systems, in which disturbances propagate through the system as waves. Distributed-parameter systems are described by partial differential equations. Real dynamic systems can behave as lumped or distributed parameter systems, depending on the frequency spectrum of their excitation. For example, an aircraft will respond partly as a distributed-parameter system to a sudden wind gust that excites the flexible bending-modes of the wings and fuselage. For wind disturbances that are less abrupt, it will respond according to our rigid-body equations of motion. For large, flexible aircraft such as passenger jets, the flexible modes can be low enough in frequency to approach or overlap with the rigid body modes. We will restrict ourselves to lumped-parameter models described by ODEs, and these ODEs allow us to describe a wide range of nonlinear and time-varying systems.

The continuous-time state-space equations are a "canonical form" of the ODE description of continuous-time lumped-parameter dynamic systems. That is, they represent a form to which all the members of the class can be reduced by appropriate transformations. Reduction to this canonical form allows a wide range of modeling, simulation, and design problems to be handled within a common framework of algorithms and software tools as shown in Figure 3.1-1.

Models of Mechanical and Electrical Systems

There are many ways of deriving state equations for these systems. Here we give two examples in which we choose the state variables according to the energy storage elements, and then use an appropriate technique for finding the state equations.

Example 3.2-1: State Equations for a Mechanical System. Figure 3.2-1 shows a simple mechanical system to illustrate the derivation of state equations. The input to the system is the spring displacement $u(t)$, and the output is the displacement $y(t)$ of the mass m. The mass slides with negligible friction, and the springs are linear with stiffness k_i and zero mass. The dampers also have no mass, and only dissipate energy. They produce a reaction force equal to the viscous constant d_i times the rate

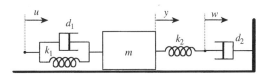

Figure 3.2-1 A mechanical system.

at which they are extending or contracting. The auxiliary variable $w(t)$ is needed so that equations can be written for the associated spring and damper. There are three independent energy storage elements, and state variables can be assigned accordingly:

$$x_1 = (u - y), \quad \text{compression of spring } k_1$$
$$x_2 = (y - w), \quad \text{compression of spring } k_2$$
$$x_3 = \dot{y}, \quad \text{translational rate of } m$$

Equations involving the state variables can now be written by inspection:

$$\dot{x}_1 = \dot{u} - x_3$$
$$d_2(x_3 - \dot{x}_2) = k_2 x_2$$
$$m\dot{x}_3 = k_1 x_1 + d_1(\dot{u} - x_3) - k_2 x_2$$

These equations can be put into the form:

$$\dot{x} = Ax + B_1 \dot{u}$$

To reduce this to state-space form, let

$$z = x - B_1 u$$

Then

$$\dot{z} = Ax = Az + (AB_1) u$$

and these are state equations in standard form. This approach will work for most simple mechanical systems. The technique used to remove $\dot{u}$ from the first set of equations will also work for the equation:

$$\dot{x} = Ax + B_1 \dot{u} + B_0 u \qquad \blacksquare$$

In any physical system, the stored energy cannot be changed instantaneously because this would require infinite power. This provides a way of directly finding the coefficient matrices of the linear state equations, which is most easily illustrated with an electrical circuit.

In an electrical circuit, energy is stored in the magnetic fields of inductors and in the electric fields of capacitors. Inductor currents and capacitor voltages can be chosen as state variables because these quantities determine the stored energy. Therefore, the current through an inductor, or voltage across a capacitor, cannot change instantaneously.

For any given unexcited circuit, if we could place an ideal unit-step voltage generator in series with each capacitor in turn, inside its terminals, its state variable would jump to 1.0 volts at $t = 0$. Instantaneously, all of the capacitors would act like short circuits, and all of the inductors like open circuits. The capacitor currents and inductor voltages are proportional to the derivatives of their state variables ($i = C(dv/dt)$, $v = L(di/dt)$). Therefore, all of the state derivatives can be found, when all but one of the state variables are zero, by analyzing a much simpler circuit in which capacitors have been replaced by short circuits and inductors by open circuits. This gives one column of the A matrix. Similarly, placing a unit-step current generator in parallel with each inductor in turn gives the columns of the A matrix corresponding to the inductor-current state variables. This procedure is illustrated in our second example.

Example 3.2-2: State Equations for an Electrical System. Here we will find the A and B matrix elements, a_{ij}, b_i, for the "bridged-T" circuit shown in Figure 3.2-2. In the figure the state variables x_1, x_2, and x_3 have been assigned to the capacitor voltages and inductor current. Imagine a unit-step voltage generator placed in series with C_1, as indicated by the dashed circle, and let $t = 0^+$ indicate that the generator has switched from zero to 1 volt. Then at $t = 0^+$,

$$x_1(0^+) = 1.0, \quad x_2(0^+) = 0.0, \quad x_3(0^+) = 0.0$$

The defining equation for a capacitor and the linear state equation then give

$$i_1(0^+) \equiv C_1 \dot{x}_1(0^+) = C_1 a_{11} x_1(0^+) \quad \text{or} \quad a_{11} = \frac{1}{C_1} \frac{i_1(0^+)}{x_1(0^+)}$$

The conductance $i_1(0^+)/x_1(0^+)$ is easily found from the resistive circuit obtained when C_2 is replaced by a short circuit, and L by an open circuit. The result is

$$a_{11} = \frac{-1}{C_1 \left(R_1 + \dfrac{R_2 R_4}{R_2 + R_4} \right)}$$

The same voltage generator also gives

$$v_2(0^+) \equiv L\dot{x}_2(0^+) = La_{21} x_1(0^+) \quad \text{or} \quad a_{21} = \frac{1}{L} \frac{v_2(0^+)}{x_1(0^+)} = \frac{1}{L}$$

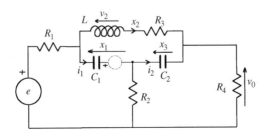

Figure 3.2-2 State variables for an electrical circuit.

and

$$a_{31} = \frac{i_2(0^+)}{C_2 x_1(0^+)} = \frac{-1}{C_2(R_1 + R_4 + (R_1 R_4)/R_2)}$$

The second and third columns of the A matrix can be found, respectively, by putting a unit-step current generator in parallel with L and a unit-step voltage generator in series with C_2. In the same manner, the B matrix can be found by letting the input voltage $e(t)$ come from a unit-step voltage generator. The C and D matrices can be found in an ad hoc way; in this case an expression for the output voltage is

$$v_o = (x_2 + C_2 \dot{x}_3) R_4$$

Substitution of the state equation for $\dot{x}_3$ into this expression yields C and D. ■

This technique will always work for linear time-invariant electric circuits, though any mutual-induction coupling causes complications. Alternative techniques include setting up the Kirchoff loop or nodal equations and reducing these to state equations (Nise, 1995).

Reduction of Differential Equations to State-Space Form

Consider the following nonlinear, scalar ODE:

$$\ddot{y} + f(\dot{y}) + g(y) = h(u) \qquad (3.2\text{-}1)$$

Here $u(t)$ is a known input, and $y(t)$ is the response of the system described by this ODE. The functions f, g, h are arbitrary, known nonlinear functions, and may be the result of manipulating a preceding nonlinear equation to reduce the coefficient of the highest derivative to unity. Suppose that we convert this differential equation to two simultaneous integral equations by writing:

$$y = \int \dot{y}\, dt$$

$$\dot{y} = \int [h(u(t)) - f(\dot{y}) - g(y)]\, dt \qquad (3.2\text{-}2)$$

Now consider how these equations might be solved. The variable y may be, for example, a position variable, but the functional form of the solution is the same if we use any other quantity as an *analog* of position. In an *analog computer* a device is available to perform integration of voltages with respect to time, and voltage is the analog of whatever physical variable we wish to simulate. We can make a diagram of the analog computer connections if we draw a box representing the integration operation, symbols representing summation and multiplication of variables, and lines showing which variables are subjected to each operation. Thus Figure 3.2-3 shows a

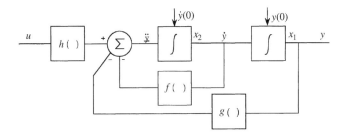

Figure 3.2-3 A SISO simulation diagram.

simulation diagram (or simply a *block diagram*) of Equations (3.2-2). If an analog computer is connected in this way, and switched on with the correct initial conditions on the integrators, it will effectively solve Equations (3.2-1) or (3.2-2).

Referring to Figure 3.2-3, if we simultaneously break the input connections of all of the integrators (in an analog computer the inputs would be connected to "signal ground"), the outputs of the integrators will remain constant at the values they had at the instant of breaking the connections. If the continuation of the input signal $u(t)$ is available, the analog computer simulation can be restarted at any time and all of the signals will assume the values they had when the connections were broken, and the simulation can continue with no information lost. Therefore, the integrator output variables satisfy our earlier definition of state variables. A knowledge of their values at any time instant, together with the input signal, completely defines the state of the system.

Now starting from the right-hand side of Figure 3.2-3, let state variables x_1 and x_2 be defined as the integrator outputs. Therefore, by inspection of the diagram, or Equations (3.2-2), the state equations for the second-order ODE (3.2-1) are

$$\dot{x}_1 = x_2$$
$$\dot{x}_2 = h(u) - f(x_2) - g(x_1) \tag{3.2-3}$$

This technique can easily be extended to find n state variables for an n-th order differential equation. It is not necessary to draw the block diagram, because the technique simply amounts to defining $y(t)$ and its derivatives up to the $(n-1)$-th as state variables, and then writing the equation for the n-th derivative by inspection. State variables chosen in this way are often called *phase variables*.

Next consider a more difficult example where the block diagram will be helpful. This time the ODE model will involve derivatives of the input variable, but will be linear with constant coefficients, as in the following second-order example:

$$\ddot{y} + a_1 \dot{y} + a_0 y = b_2 \ddot{u} + b_1 \dot{u} + b_0 u \tag{3.2-4}$$

The coefficient of the highest derivative of y can always be made equal to unity by dividing all of the other coefficients. Let the operator p^n indicate differentiation n-times with respect to time, and write this equation as

$$p^2(y - b_2 u) + p(a_1 y - b_1 u) + (a_0 y - b_0 u) = 0$$

Now turn this into an integral equation for y,

$$y = b_2 u + \int (b_1 u - a_1 y) dt + \iint (b_0 u - a_0 y) d\tau dt$$

A simulation diagram can be drawn by inspection of this equation, and is shown in Figure 3.2-4. A set of state equations can again be found by assigning state variables to the outputs of the integrators.

This example can be extended to the general case of an n-th order differential equation, with all derivatives of the input present. The differential equation for the general case is

$$p^n y + \sum_{i=0}^{n-1} a_i p^i y = \sum_{i=0}^{n} b_i p^i u \tag{3.2-5}$$

and the state equations are

$$\frac{d}{dt}\begin{bmatrix} x_1 \\ x_2 \\ \cdots \\ \cdots \\ x_n \end{bmatrix} = \begin{bmatrix} -a_{n-1} & 1 & 0 & \ldots & 0 \\ -a_{n-2} & 0 & 1 & \ldots & 0 \\ \cdots & \cdots & \cdots & \cdots & \cdots \\ -a_1 & 0 & 0 & \ldots & 1 \\ -a_0 & 0 & 0 & \ldots & 0 \end{bmatrix} \begin{bmatrix} x_1 \\ x_2 \\ \cdots \\ \cdots \\ x_n \end{bmatrix} + \begin{bmatrix} (b_{n-1} - a_{n-1} b_n) \\ (b_{n-2} - a_{n-2} b_n) \\ \cdots \\ \cdots \\ (b_0 - a_0 b_n) \end{bmatrix} u(t) \tag{3.2-6}$$

If the highest derivative on the right-hand side of (3.2-5) is the m-th, then for real systems $m \leq n$. In the cases where $m = n$ this is a practical approximation in situations when very-high-frequency effects can be neglected. Therefore $b_i \equiv 0$ for $i > m$, and when $m < n$ a group of n terms disappears from the right-hand side of (3.2-6).

This form of the linear state equations is known as the *observer canonical form*. The A matrix has a structure called a *companion form*, which is known to be "ill

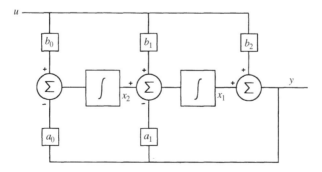

Figure 3.2-4 A general SISO simulation diagram.

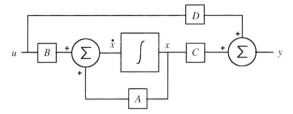

Figure 3.2-5 State equations simulation diagram.

conditioned" for numerical computation, but is useful for theoretical results. The technique used here to obtain the simulation and state equations can be extended to include time-varying ODE coefficients (Laning and Battin, 1956), but there is no general method for deriving the simulation diagram when the ODE is nonlinear.

Finally, let us take a look at a simulation diagram representation of the linear state equations. Figure 3.2-5 shows a diagram that represents the linear state and output equations, as given by Equations (3.1-2). In this diagram the lines, or signal paths, carry several variables simultaneously, the coefficient boxes represent matrix operations, and the integrator box represents multiple integrators simultaneously processing all of the input signals individually. This also represents a valid way to wire up an analog computer; the advantages of this form of system model will become apparent in this chapter.

Time-Domain Solution of LTI State Equations

When the state equations are linear and time-invariant (LTI) they can be solved analytically. Elementary differential equations texts show that a linear first-order ODE can be solved by using an exponential "integrating factor" to yield an exact derivative. When the equation is also time-invariant, the integrating factor reduces to a simple exponential function. An analogous method can be used to solve the set of n first-order LTI state equations.

The matrix exponential e^{At}, where A is a square constant matrix, is defined by the matrix series:

$$e^{At} \equiv I + At + \frac{A^2 t^2}{2!} + \frac{A^3 t^3}{3!} + \dots \qquad (3.2\text{-}7)$$

The series is uniformly convergent and can be differentiated or integrated term by term, resulting in properties analogous to the scalar exponential function:

$$\frac{d}{dt}\left(e^{At}\right) = A e^{At}$$

and

$$\int e^{At} dt = A^{-1} e^{At}, \quad \text{if } A^{-1} \text{ exists}$$

It is also evident from (3.2-7) that $exp(At)$ is commutative with A, that is,

$$Ae^{At} = e^{At}A$$

Now using e^{At} as an integrating factor, the state equation (3.1-2a) can be written as

$$\frac{d}{dt}\left(e^{-At}x(t)\right) = e^{-At}Bu(t)$$

If this equation is integrated, with the constant of integration determined from initial conditions at $t = t_0$, we obtain

$$x(t) = e^{A(t-t_0)}x(t_0) + \int_{t_0}^{t} e^{A(t-\tau)}Bu(\tau)\,d\tau \tag{3.2-8}$$

The first component of this solution is the *homogeneous component*, which is the response to the initial conditions $x(t_0)$, and $e^{A(t-t_0)}$ takes the state from time t_0 to t and is called the *transition matrix*. The second component is the *forced component*, which is the response to the input $u(t)$. The integral on the right-hand side of (3.2-8) is a convolution integral, the time-domain equivalent of transform multiplication in the frequency domain.

The solution (3.2-8) is of little computational value to us because of the difficulties of finding analytical expressions for the transition matrix for systems of all but the lowest order. Also the convolution integral is inconvenient to evaluate for any but the simplest input functions. However, this solution does lead to a discrete-time recursion formula that is useful. A recursion formula becomes practical when we consider a short time interval T, over which the input can be approximated by a simple function. Therefore, we look for a discrete time formula by considering a time interval from $t = kT$ to $t = (k+1)T$, where k is a positive integer. In Equation (3.2-8), let $t_0 = kT$ and $t = (k+1)T$ and make a change of variable $\lambda = (\tau - kT)$, then

$$x(k+1) = e^{AT}x(k) + e^{AT}\int_{0}^{T} e^{-A\lambda}Bu(\lambda + kT)d\lambda, \tag{3.2-9}$$

where T is implied in the argument of x.

The integral in (3.2-9) can be evaluated by a variety of methods, for example, the trapezoidal rule, or Simpson's rule. We will take a simple stepped approximation to $u(t)$, such that $u(\lambda + kT) \approx u(kT)$ for $0 \leq \lambda < T$ (this is called a zero-order hold [ZOH] approximation); $u(kT)$ can then be taken out of the integrand. The remaining integral can be evaluated by considering term-by-term integration of the matrix exponential, and the result is

$$x(k+1) = e^{AT}x(k) + Q(T)Bu(k), \tag{3.2-10}$$

where $Q(T)$ is given by

$$Q(T) = T\left[I + \frac{AT}{2!} + \frac{A^2T^2}{3!} + \frac{A^3T^3}{4!} + \cdots\right] \quad (3.2\text{-}11a)$$

or, when A^{-1} exists,

$$Q(T) = A^{-1}\left(e^{AT} - I\right) \quad (3.2\text{-}11b)$$

Equation (3.2-10) is a discrete-time recursion formula. It can be used as an alternative to numerical integration of the state equations when the equations are linear, and e^{AT} has been found analytically or numerically. The matrix exponential e^{AT} is called the *discrete-time transition matrix*. Methods of computing the transition matrix are described in the literature (Healey, 1973; Moler and Van Loan, 1978; Zakian, 1970), and methods of computing integrals involving the matrix exponential (e.g., $Q(T)$) are described by Van Loan (1978). Commercial software is available to compute the transition matrix (e.g., MATLAB "expm").

Modal Decomposition

In Section 1.3 the modal coordinates were introduced to show the connection between eigenvalues and the natural modes of a dynamic system. It is possible to use the additional information contained in the eigenvectors to determine which variables are involved in a given mode and what inputs will excite the mode. The time-domain solution (3.2-8) of the LTI state equation can be used for this purpose.

The continuous-time transition matrix can be expressed in terms of eigenvalues and eigenvectors in the following way. The similarity between the A matrix and (in general) a Jordan form matrix can be used to express an arbitrary power of A as

$$A^k = \left(MJM^{-1}\right)\left(MJM^{-1}\right)\cdots = MJ^kM^{-1}$$

When this is done for every term in the matrix exponential series, and when J is assumed to be diagonal (distinct eigenvalues), the result is

$$e^{At} = Me^{Jt}M^{-1} = [v_1\ v_2 \cdots v_n] \begin{bmatrix} e^{\lambda_1 t} & & & \\ & e^{\lambda_2 t} & & \\ & & \ddots & \\ & & & e^{\lambda_n t} \end{bmatrix} \begin{bmatrix} w_1^T \\ w_2^T \\ \vdots \\ w_n^T \end{bmatrix}, \quad (3.2\text{-}12)$$

where v_i is the i-th column of M (the i-th eigenvector), and w_i^T is the i-th row of M^{-1}. Then, by definition, "vectors" w_i are orthonormal with the eigenvectors, that is,

$$w_i^T v_j = \begin{cases} 0, & i \neq j \\ 1, & i = j \end{cases} \quad (3.2\text{-}13)$$

It is also easy to show that the vectors w_i are actually the *left eigenvectors* of A, that is, the right eigenvectors of A^T. If (3.2-12) is postmultiplied by the initial-condition vector x_0, the homogeneous part of the solution of the continuous-time state equation is obtained:

$$e^{At}x_0 = [v_1 \; v_2 \cdots v_n] \begin{bmatrix} e^{\lambda_1 t}\left(w_1^T x_0\right) \\ e^{\lambda_2 t}\left(w_2^T x_0\right) \\ \cdots \\ e^{\lambda_n t}\left(w_n^T x_0\right) \end{bmatrix}, \qquad (3.2\text{-}14)$$

where the terms $(w_i^T x_0)$ are scalar products. Equation (3.2-14) can be rewritten as

$$e^{At}x_0 = \sum_{i=1}^{n} \left(w_i^T x_0\right) e^{\lambda_i t} v_i$$

If this same procedure is followed, but with (3.2-12) postmultiplied by $Bu(\tau)$, the forced component of the response is obtained. The complete response is therefore given by

$$x(t) = \sum_{i=1}^{n} v_i e^{\lambda_i t} \left(w_i^T x_0\right) + \sum_{i=1}^{n} v_i \int_0^t e^{\lambda_i(t-\tau)} \left(w_i^T Bu(\tau)\right) d\tau \qquad (3.2\text{-}15)$$

In effect, Equation (3.2-15) uses the n linearly independent eigenvectors as a basis for the n-dimensional space, associates a characteristic mode with each basis vector, and shows the fixed component of $x(t)$ in each direction. If, for example, the initial condition vector lies in a direction such that a scalar product $(w_i^T x_0)$ is zero, the mode $e^{\lambda_i t}$ will not appear in the homogeneous response. According to (3.2-13), this will occur if the initial-condition vector lies along any eigenvector other then the i-th. Similarly, if the scalar product $(w_i^T Bu(\tau))$ is zero, the mode $e^{\lambda_i t}$ will not be excited by the input. In Chapters 5 and 6 we discuss the related idea of "controllability" of the modes and show how it is determined by the A and B matrices. If we form the output vector by premultiplying (3.2-15) by the C matrix, we see that whether or not a mode appears in the output depends on the C and A matrices. This is the concept of "observability," also described in Chapters 5 and 6.

Equation (3.2-15) also shows that if we examine the i-th eigenvector, its nonzero elements will indicate to what extent each state variable participates in the i-th mode. The relative involvement of the different variables is complicated by the fact that the eigenvector elements can, in general, each have different units.

Laplace Transform Solution of LTI State Equations

The Laplace transform (LT) maps real functions of time into functions of the complex variable s, which is written in terms of its real and imaginary parts as $s = \sigma + j\omega$, and has the dimensions of a complex frequency variable. In the complex frequency domain (or s domain) the functions of s can be manipulated algebraically into recognizable, known transforms and then mapped back into the time domain. Laplace transform theory is thoroughly covered in many undergraduate texts (Ogata, 1998), and here we will only review two important points concerning applicability.

Two different ways of applying the LT will now be described. First, in general, analysis of a dynamic system will produce a set of simultaneous integro-differential equations. These equations should be transformed immediately, so that the initial condition terms that appear from applying the LT differentiation and integration theorems represent the initial stored energy. If the integral terms are removed by differentiation, derivatives of the system input may appear. These give rise to extra initial condition terms when the differential equations are transformed, and can make it difficult to solve for all of the required initial conditions on the dependent variable. Therefore, we avoid transforming the general ODE in Equation (3.2-5), if there are nonzero initial conditions and derivatives on the right-hand side.

The second method of using the LT applies to initially unexcited systems; no initial condition terms appear after transforming. If the equations have been differentiated so that derivatives of the input appear, the input initial condition terms must cancel with the initial conditions on the output. Therefore, if the system is described by the differential equation (3.2-15), this equation can be transformed with zero initial conditions. For circuits there is actually no need to write the differential equations because, with no initial stored energy, the system elements can be represented by transform impedances (or admittances). Circuit analysis rules will then yield s-domain equations that can be solved for the output transform. This method will be addressed in the next section.

We will denote Laplace transforms by uppercase symbols, thus

$$X(s) = \mathscr{L}[x(t)], \quad \text{and} \quad U(s) = \mathscr{L}[u(t)]$$

The LTI state equations have no derivatives of $u(t)$, and can be solved by Laplace transforming (3.1-2a):

$$sX(s) - x(0) = AX(s) + BU(s)$$

$$\therefore \quad X(s) = (sI - A)^{-1}[x(0) + BU(s)] \quad (3.2\text{-}16)$$

and so

$$Y(s) = C(sI - A)^{-1}[x(0) + BU(s)] + DU(s) \quad (3.2\text{-}17)$$

Because there are no input derivatives this solution requires n initial conditions on $x(t)$ only, and these would specify the initial stored energy in our earlier examples.

If we compare the transform solution for $X(s)$ with the time domain solution for $x(t)$, Equation (3.2-8), we see that the transition matrix is given by

$$e^{At} = \mathscr{L}^{-1}\left[(sI - A)^{-1}\right] \quad (3.2\text{-}18)$$

The Laplace transform solutions, (3.2-16) and (3.2-17), are not well suited to machine computation, and hand computation involves a prohibitive amount of labor for other than low-order dynamic systems. Therefore, the Laplace transform solutions are mainly of interest as a complex number description of system properties, as we will now see.

3.3 TRANSFER FUNCTION MODELS

Derivation of Transfer Functions; Poles and Zeros

Consider the system described by the n-th order ODE (3.2-5), and transform with zero initial conditions. Solving algebraically for $Y(s)$:

$$Y(s) = \frac{b_m s^m + b_{m-1} s^{m-1} + \ldots + b_1 s + b_0}{s^n + a_{n-1} s^{n-1} + \ldots + a_1 s + a_0} U(s), \quad m \leq n \quad (3.3\text{-}1)$$

The polynomial rational function relating $Y(s)$ to $U(s)$ is the *transfer function* of this SISO system. If we have obtained the transform $U(s)$ then, using the partial fractions technique (see below), the right-hand side of (3.3-1) can be broken down into a sum of transforms corresponding to known time functions, and hence $y(t)$ can be found as a sum of time functions.

From (3.2-17) we see that the transfer function obtained from the LTI state equations is a matrix expression and, for a MIMO system with p outputs and m inputs, the $(p \times m)$ transfer function matrix is given by

$$G(s) = C(sI - A)^{-1} B + D \quad (3.3\text{-}2)$$

It is easy to show that a transfer function matrix is unchanged by a nonsingular transformation of the state variables. Equation (1.3-2b) represents such a transformation, and if the coefficient matrices from that equation are substituted into (3.3-2), the result is

$$G(s) = CL^{-1} \left(sI - LAL^{-1}\right)^{-1} LB + D$$
$$= C \left[L^{-1} \left(sI - LAL^{-1}\right) L\right]^{-1} B + D$$
$$= C(sI - A)^{-1} B + D,$$

which is (3.3-2) again. Therefore, we can choose a new set of state variables for a system, and the transfer function will be unchanged.

We will now review some other important properties of transfer functions. A matrix inverse can be expressed in terms of the adjoint matrix with its elements divided by its determinant, and so (3.3-2) can be written as

$$G(s) = \frac{C \, adj(sI - A) B + D|sI - A|}{|sI - A|} \quad (3.3\text{-}3)$$

The transfer function from the j-th input to the i-th output is the i, j-th element of $G(s)$, and this is the SISO transfer function $G_{ij}(s)$. A SISO transfer function can therefore be written as

$$G_{ij}(s) = \frac{c_i\, adj(sI - A)b_j + d_{ij}\,|sI - A|}{|sI - A|}, \qquad (3.3\text{-}4)$$

where c_i and b_j are, respectively, the i-th row of C and the j-th column of B. This transfer function is a rational function of two polynomials. The elements of the adjoint $adj(sI - A)$ are, by definition, cofactors of $|sI - A|$, and are therefore polynomials in s of degree $(n - 1)$ or lower. The determinant $|sI - A|$ is a polynomial of degree n. When (3.3-4) is written out as the ratio of two polynomials, it will correspond exactly to the SISO transfer function in (3.3-1) that we obtained from the n-th order ODE.

In (3.3-4), when $d_{ij} = 0$, the *relative degree* (denominator degree minus numerator degree) of this transfer function is unity or higher. When $d_{ij} \neq 0$, the relative degree is zero and, referring to the simulation diagram in Figure 3.2-5, we see that d_{ij} forms a "direct-feed" path from input to output. This means that the system output immediately begins to follow an input, and then the modes of the system respond and begin to modify the output.

If the polynomials in the transfer function (3.3-4) are factored, we obtain

$$G_{ij}(s) = \frac{k(s + z_1)(s + z_2)\ldots(s + z_m)}{(s + p_1)(s + p_2)\ldots(s + p_n)} \qquad (3.3\text{-}5a)$$

or equivalently,

$$G_{ij}(s) = \frac{a_1}{s + p_1} + \frac{a_2}{s + p_2} + \ldots + \frac{a_n}{s + p_n} \qquad (3.3\text{-}5b)$$

The denominator factors are the factors of $|sI - A|$, and it is evident from (3.3-3) and (3.3-4) that all of the individual SISO transfer functions have the same denominator factors, given by the roots of the n-th degree polynomial equation

$$|sI - A| = 0 \qquad (3.3\text{-}6)$$

The roots $\{-p_i\}$ are called the *poles* of the transfer function and, at these values of s, the transfer function becomes infinite in magnitude.

Equation (3.3-6) is also the defining equation for the eigenvalues of the A matrix. Therefore, the system poles are given by the eigenvalues of A. We know from Chapter 1 that the eigenvalues of a real system are real or occur in complex conjugate pairs and, according to Equation (3.2-15), determine the natural modes of a system. The position of the poles in the complex s-plane will determine the time constant of a real mode, or the frequency of oscillation and exponential damping factor of a complex mode. Also, poles in the right-half s-plane will correspond to exponentially growing functions (unstable behavior). For this reason graphical operations in the s-plane are important to us.

Equation (3.3-5b) is the *partial fraction expansion* of the transfer function, and a coefficient a_i is the *residue* in the pole at $-p_i$ (Ogata, 1998). In the case of complex poles the partial fractions combine as conjugate pairs. Poles of multiplicity k require a numerator of degree $(k-1)$, and can be further broken down into a finite expansion in inverse powers of $(s+p_i)$.

The *zeros* of the individual SISO transfer functions are the positions in the s-plane where their magnitudes become zero, that is, the roots $\{-z_i\}$ of the numerator polynomial of (3.3-4). The number of zeros of each SISO transfer function will range from none to n, depending on the relative degree of the transfer function. Equations (3.3-3) and (3.3-4) show that the transfer function zeros depend on the B, C, and D matrices, and Equation (3.2-15) shows how the B and C matrices, respectively, play a role in the excitation of a mode and its appearance in the system output. When a response transform is expanded in partial fractions, we see that the partial fraction coefficients depend on the numerator polynomial and hence on the zeros. The partial fraction terms correspond to the modes and the zeros determine how strongly the modes are represented in the response. If all of the poles of the transform of the system input coincide with zeros of the SISO transfer function, there will be no forced response at the output of the system.

It is known that the values of polynomial roots are very sensitive to changes in the polynomial coefficients, and Equations (3.3-3) and (3.3-4) are the starting points of algorithms used to change the computation of zeros into a much more numerically stable eigenvalue problem (Emami-Naeini and Van Dooren, 1982). Transfer function-related analysis and design tools are based on poles and zeros, and we will have little use for the polynomial form of the transfer function.

When $G_{ij}(s)$ is in the factored form (3.3-5a), with all coefficients of s equal to unity, or expressed as the ratio of two monic polynomials, the coefficient k is known as the *static loop sensitivity*. Note that if there are no poles or zeros at the s-plane origin then, when $s=0$, the magnitude of the transfer function (the dc gain) is finite and is determined by k and the zero and pole positions. If the relative degree is zero, k is the value of the transfer function at large values of s (the high-frequency gain).

From this point on we will drop the subscripts on $G(s)$, and it will be obvious from the context whether G represents a matrix or a scalar transfer function.

Interpretation of the SISO Transfer Function

The complex exponential function e^{st}, $s = \sigma + j\omega$, possesses time derivatives of all orders, all of the same form, and if we could apply it as an input to the SISO system described by the ODE (3.2-5), the particular solution of the ODE would be a time function of the same form. Furthermore, the solution would be given by an expression exactly like the transfer function (3.3-1), and we could use the response to e^{st} as a definition of a transfer function. The effect of the system on a specific exponential function $e^{s_1 t}$ could be found by evaluating $G(s_1)$, given by

$$G(s_1) = \frac{k(s_1+z_1)(s_1+z_2)\ldots(s_1+z_m)}{(s_1+p_1)(s_1+p_2)\ldots(s_1+p_n)} \quad (3.3\text{-}7)$$

The numerator and denominator factors in this transfer function can be represented in magnitude and phase by vectors in the s-plane, drawn from the zeros and poles,

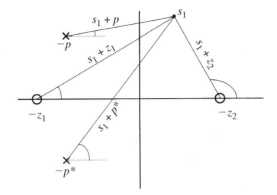

Figure 3.3-1 s-plane vectors representing pole-zero factors.

respectively, to the point s_1. This is illustrated by the example shown in Figure 3.3-1 and, in general,

$$|G(s_1)| = k \frac{\text{product of lengths of vectors from zeros to } s_1}{\text{product of lengths of vectors from poles to } s_1} \quad (3.3\text{-}8a)$$

and

$$\begin{aligned} \angle G(s_1) &= \text{sum of angles of vectors from zeros to } s_1 \\ &\quad - \text{sum of angles of vectors from poles to } s_1 \end{aligned} \quad (3.3\text{-}8b)$$

Because complex poles and zeros occur in conjugate pairs,

$$G\left(s_1^*\right) = G^*(s_1), \quad (3.3\text{-}9)$$

where "*" denotes the conjugate, and this is clearly illustrated by drawing the appropriate vectors. This interpretation of the transfer function is particularly useful when s_1 is a point on the s-plane $j\omega$ axis. A real sinusoid can be represented as

$$\cos(\omega_1 t) = \tfrac{1}{2}\left(e^{j\omega_1 t} + e^{-j\omega_1 t}\right) \quad (3.3\text{-}10)$$

The particular solution of the ODE, with this input, is given by

$$y(t) = \tfrac{1}{2}\left[G(j\omega_1)e^{j\omega_1 t} + G(-j\omega_1)e^{-j\omega_1 t}\right]$$
$$\therefore y(t) = \operatorname{Re}\{G(j\omega_1)e^{j\omega_1 t}\} = |G(j\omega_1)|\cos\left(\omega_1 t + \angle G(j\omega_1)\right), \quad (3.3\text{-}11)$$

where Re is the real-part operator, and we have made use of (3.3-9). The sinusoidal input (3.3-10) was not switched on at some particular time; mathematically it has existed for all time, and the solution (3.3-11) represents the *steady-state response* to a sinusoidal input, of the system whose transfer function is $G(s)$.

The plots of $|G(j\omega)|$ and $\angle G(j\omega)$, as ω is varied from low to high frequency, are called the magnitude and phase of the system *frequency response*. The vector interpretation (3.3-8) shows that if there is a pair of complex poles near the imaginary axis, the vectors drawn from these poles will become very short in length over some range of frequencies, and there will be a peak in the magnitude of $G(j\omega)$ and rapid changes in its phase. This is the phenomenon of *resonance*, in which a natural mode of a system is excited by the input to the system. Conversely, if there is a pair of complex zeros close to the $j\omega$ axis, the magnitude will pass through a minimum and the phase will again change rapidly. These effects are discussed more thoroughly in the section on frequency response.

A transfer function carries some very basic information about the way in which an aircraft (or any other system) will respond that is usually not obvious to the student. Two theorems that are fundamental in interpreting the transfer function are the Laplace transform initial and final value theorems:

$$\text{initial value:} \quad f(0^+) \equiv \lim_{t \to 0} f(t) = \lim_{s \to \infty} sF(s) \quad (3.3\text{-}12a)$$

$$\text{final value:} \quad f(\infty) \equiv \lim_{t \to \infty} f(t) = \lim_{s \to 0} sF(s) \quad (3.3\text{-}12b)$$

As an example of these theorems, consider the response of a system to a unit-step function. A useful notation (DeRusso, 1965) is

$$\text{unit-step function:} \quad U_{-1}(t) = \begin{cases} 0, & t < 0 \\ 1, & t > 0 \end{cases}$$

$U_{-1}(t)$ is undefined at $t = 0$, and has a Laplace transform $1/s$ (Ogata, 1998). The symbols U_{-2}, U_{-3} denote, respectively, a unit-ramp and unit-parabola, and U_0 denotes a unit impulse function. Now let the unit step, occurring at $t = 0$, be the input to the transfer function:

$$G(s) = \frac{-(s-\alpha)}{s^2 + s + 1}, \quad \alpha > 0 \quad (3.3\text{-}13)$$

The transfer function has a relative degree $r = 1$, and this makes the initial value of the step response zero:

$$f(0^+) = \lim_{s \to \infty} \left[s \frac{-(s-\alpha)}{s^2+s+1} \frac{1}{s} \right] = 0$$

The Laplace transform differentiation theorem tells us that the derivative of the step response will involve a transfer function of relative degree zero; therefore, we do not expect it to have an initial value of zero:

$$\mathcal{L}[\dot{f}(t)] = sF(s) - f(0^+) = sF(s)$$

and the initial value of the derivative is

$$\dot{f}(0^+) = \lim_{s \to \infty} [s^2 F(s)] = -1$$

The final value of the step response is

$$f(\infty) = \lim_{s \to 0} [sF(s)] = \alpha$$

The transfer function (3.3-13) has a sign difference between its behavior at small and large s; $G(0) = \alpha$ (positive dc gain), and $G(\infty) = -1/s$. From the above analysis, the consequences are that the step response starts out in the negative direction but finishes with a positive value. The transfer-function numerator factor $(s - \alpha)$, $\alpha > 0$, corresponds to a zero in the right-half s-plane, and this is the cause of the above behavior.

If a transfer function contains right-half-plane zeros, it is called *non-minimum-phase* (NMP), and the initial response to a step input may have the opposite sign to the final response (depending on the number of NMP zeros). This is an undesirable type of response from the point of view of a human operator. NMP zeros are also undesirable in feedback controller design since, as we will see later from "root-locus" plots, a right-half plane zero tends to attract the closed-loop poles to the right-half s-plane. These types of zeros occur when there are two or more different paths to the system output, or two or more different physical mechanisms, producing competing output components.

When there are left-half plane zeros near to the origin, these tend to promote an overshoot in the response to a step input, which is again undesirable. Problem 3.3-2 illustrates the effects of both NMP zeros and left-half plane zeros close to the origin.

By writing the simple differential equations for an ideal integrator or differentiator, and transforming them, we can derive their transfer functions. Thus the transfer function of an integrator consists of a single pole at the origin, and a differentiator corresponds to a single zero at the origin. In a block diagram using transform-domain quantities, we will represent integrators and differentiators by boxes containing, respectively, $1/s$ and s.

Finally, when transfer function poles and zeros are close together in the s-plane, the residue in the poles tends to be small (i.e., the coefficients of the corresponding partial fraction terms are small). A pole can be effectively canceled out of the transfer function by a nearby zero in this way.

Transfer Function Examples and Standard Forms

Table 3.3-1 shows a number of electrical networks that are used as either models or compensating networks in control systems design (see Chapter 4). Their voltage transfer functions (assuming no source and output loading effects) have been derived by representing the network elements by their Laplace transform impedances (i.e., $1/sC$ for a capacitor, sL for an inductor). They can then be analyzed in the same way as dc circuits. The transfer functions have been written in *standard form*. This requires all numerator and denominator factors to be written as either $(s\tau)$, $(s\tau + 1)$, where τ is a time constant, or the second-order standard form given below. State equations are shown for the networks and can be derived from the transfer functions by the methods given earlier.

TABLE 3.3-1: Network Transfer Functions and State Equations

NETWORK	TRANSFER FUNCTION	STATE EQUATIONS
SIMPLE LAG	$\dfrac{1}{1+s\tau}, \quad \tau = CR$	$\dot{x} = \dfrac{u-x}{\tau}$ $y = x$
QUADRATIC LAG	$\dfrac{\omega_n^2}{s^2 + 2\zeta\omega_n s + \omega_n^2}$ $\omega_n^2 = \dfrac{1}{LC} \quad \zeta = \dfrac{R}{2}\sqrt{\dfrac{C}{L}}$	$\dot{x}_1 = x_2$ $\dot{x}_2 = -\omega_n^2 x_1 - 2\zeta\omega_n x_2 + \omega_n^2 u$ $y = x_1$
SIMPLE LEAD	$\dfrac{s\tau}{1+s\tau}, \quad \tau = CR$	$\dot{x} = \dfrac{u-x}{\tau}$ $y = u - x$
LEAD COMPENSATOR	$\dfrac{s+z}{s+p}$ $z = \dfrac{1}{CR_1}$ $\dfrac{z}{p} = \dfrac{R_2}{R_1+R_2}$	$\dot{x} = u - px$ $y = u + (z-p)x$
LAG COMPENSATOR	$\dfrac{p}{z}\dfrac{s+z}{s+p}$ $z = \dfrac{1}{CR_2}$ $\dfrac{p}{z} = \dfrac{R_2}{R_1+R_2}$	$\dot{x} = u - px$ $y = \dfrac{p}{z}[u + (z-p)x]$

TRANSFER FUNCTION MODELS

Four of the networks in Table 3.3-1 have only a single energy storage element, are modeled with a single state variable, and hence have only a single real pole in their transfer functions. The standard form for a transfer function factor corresponding to a single real pole or zero is $(\tau s + 1)$. The pole or zero is at $s = -1/\tau$, and this factor is dimensionless. As an example, we will derive the transfer function of the network identified as a "simple lead."

With the restriction that any load connected to the output of the network must draw negligible current, the same current flows in the series (connecting input and output) branch as in the shunt (across the output terminals) branch. The voltage transfer function $G(s)$ is then simply the impedance of the shunt branch divided by the sum of the shunt and series impedances:

$$G(s) = \frac{Y(s)}{U(s)} = \frac{R}{R + 1/sC} = \frac{s\tau}{s\tau + 1}, \quad \text{where } \tau = CR \quad (3.3\text{-}14)$$

This transfer function has a zero at the s-plane origin and a pole at $s = -1/\tau$. We could immediately write down the differential equation relating input and output voltages, and recognize that the derivative of the input is present (the transfer function relative degree is zero). We will therefore find the state equations by a method that is similar to that used for Equation (3.2-6).

Rewrite (3.3-14) with an auxiliary variable $Z(s)$, as

$$\frac{Y(s)}{s\tau} = \frac{U(s)}{s\tau + 1} \equiv Z(s) \quad (3.3\text{-}15)$$

Now draw a simulation diagram with, in general, a chain of integrators whose outputs, starting from the last one, are Z, sZ, s^2Z, and so on. Here we need only a single integrator. The $U(s)$ equation in (3.3-15) gives

$$s\tau \, Z(s) = U(s) - Z(s),$$

which allows the input connections of the integrator to be established, as shown in Figure 3.3-2. Similarly the $Y(s)$ equation in (3.3-15) allows the simulation diagram output connections to be established. The final step is to assign state variables to the outputs of the integrators, and write the state equations by inspection of the simulation diagram. In this case, Figure 3.3-2 gives the result shown in Table 3.3-1,

$$\dot{x} = (u - x)/\tau, \qquad y = (u - x)$$

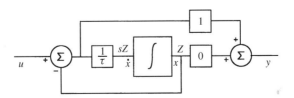

Figure 3.3-2 Simulation diagram for a simple lead.

This method of finding state equations from transfer functions or ODEs extends readily to higher-order systems; it leads to an A matrix in companion form. Therefore, for practical purposes we restrict it to low-order systems.

Next consider the *quadratic-lag* circuit in Table 3.3-1. This has two energy storage elements and requires two state variables. Because there is again only a single loop, the voltage transfer function can again be found from the branch impedances:

$$G(s) = \frac{1/sC}{sL + R + 1/sC} = \frac{1/(LC)}{s^2 + s(R/L) + 1/(LC)}$$

When a transfer function has the possibility of a pair of complex poles or zeros, it is usually convenient to represent these by a real second-order factor rather than a pair of complex first-order factors. A second-order transfer function factor is written as $(s^2 + 2\zeta\omega_n s + \omega_n^2)$, where ω_n is called the *natural frequency* and ζ is called the *damping ratio*. Using this form, the above transfer function becomes

$$G(s) = \frac{\omega_n^2}{s^2 + 2\zeta\omega_n s + \omega_n^2}, \qquad (3.3\text{-}16)$$

where

$$\omega_n = 1/\sqrt{(LC)}, \qquad \zeta = \tfrac{1}{2}R\sqrt{(C/L)}$$

Equation (3.3-16) is the standard second-order form for a complex pole pair; for complex zeros this form is inverted. Note that it has a dc gain of unity. Transfer functions can always be written in terms of the standard forms, and in the next sections we explore the properties of some standard forms rather than specific systems.

Frequency Response

Frequency response was defined in connection with Equations (3.3-10) and (3.3-11). Here we look at the frequency response of some standard-form transfer functions. An example of a first-order transfer function is

$$G(s) = \frac{s\tau_1}{s\tau_2 + 1} \quad \text{or} \quad G(s) = \frac{s\tau_1/\tau_2}{s + 1/\tau_2} \qquad (3.3\text{-}17)$$

The first transfer function is in standard form for plotting frequency response, and the second matches the vector representation described earlier. Visualizing the vectors, we can see immediately that the frequency response starts from zero magnitude and 90 degrees leading phase and, at high frequencies, it becomes constant in magnitude with zero phase angle. The transfer function itself shows that the high-frequency value of the magnitude is equal to τ_1/τ_2.

Using the first form of the transfer function, with $s = j\omega$, the magnitude and phase are given by

$$|G(j\omega)| = \frac{\omega\tau_1}{\sqrt{(1+\omega^2\tau_2^2)}}, \quad \angle G(j\omega) = \pi/2 - \tan^{-1}(\omega\tau_2) \quad (3.3\text{-}18)$$

An octave is a two-to-one frequency interval, and a decade is a ten-to-one interval; experience shows that the extent of the frequency range of interest for practical systems is usually a few decades. If the frequency response plots are made with a logarithmically spaced frequency scale, each decade occupies the same width, and features that would be lost on a linear scale are visible.

In the case of the magnitude plot, it is found that plotting the logarithm of the magnitude is very convenient for engineering purposes. This is because overall gain can be found by adding log-magnitudes, but also because very often mechanical, electrical, or physiological effects are more nearly linearly related to the logarithm of a power ratio than to the direct power ratio. An example is the Weber-Fechner law of psychology, which states that the human ear responds logarithmically. The logarithmic units most commonly used are the bel and the decibel (1 bel = 10 dB); the decibel is given by 10 times the common logarithm of the relevant power ratio, or 20 times the corresponding amplitude ratio. In engineering measurements, 0.1 dB represents good resolution, and a 60 to 80 dB range is roughly the limit of linear operation for many systems. Plots of decibel magnitude and linear phase, plotted against logarithmically spaced frequency, are known as *Bode plots*.

Taking the log-magnitude in Equation (3.3-18) gives

$$20\log_{10}(|G(j\omega)|) = 20\log_{10}(\omega\tau_1) - 10\log_{10}(1+\omega^2\tau_2^2)$$

The first term on the right increases by 20 dB for every ten-fold increase in ω; k-fold increases in frequency all occupy the same width on a logarithmic-spaced frequency scale. Therefore, this term has a straight-line Bode plot with a slope of 20 dB per decade (6 dB/octave). The second term on the right can be approximated as follows:

$$10\log_{10}(1+\omega^2\tau_2^2) \begin{cases} \approx 20\log_{10}(\omega\tau_2), & \omega^2\tau_2^2 \gg 1 \\ = 3.01 \text{ dB}, & \omega\tau_2 = 1 \\ \approx 0 \text{ dB}, & \omega^2\tau_2^2 \ll 1 \end{cases}$$

These results show that this term has asymptotes given by a 0 dB line at low frequency, and a line with a slope of 20 dB/decade at high frequency. At the "corner frequency" (or break frequency), $\omega = 1/\tau_2$, the term is 3 dB from the 0 dB asymptote, and at an octave above and below the corner frequency, it is 1 dB from its asymptotes. The phase plot asymptotically approaches 90 degrees at low frequencies and 0 degrees at high frequencies, and passes through 45 degrees at the corner frequency. It is much more spread out, being about 6 degrees from its asymptotic values at a decade above and below the corner frequency. These dB values and phase values are to be subtracted because this term came from the denominator of the transfer function. Exact Bode plots of the transfer function (3.3-17), with $\tau_1 = 10$ and $\tau_2 = 2$, are shown in Figure 3.3-3.

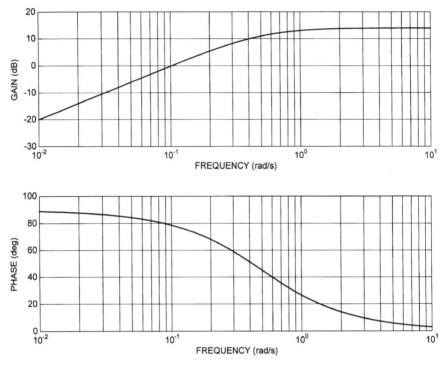

Figure 3.3-3 Bode plots for a simple-lead.

Consider next a quadratic transfer-function factor:

$$s^2 + 2\zeta\omega_n s + \omega_n^2 \qquad (3.3\text{-}19a)$$

$$= (s + \zeta\omega_n)^2 + \omega_n^2\left(1 - \zeta^2\right) \qquad (3.3\text{-}19b)$$

The quadratic formula shows that this factor represents complex conjugate roots when $\zeta^2 < 1$, and (3.3-19b) shows that the roots are given by

$$s = -\zeta\omega_n \pm j\omega_n\sqrt{\left(1 - \zeta^2\right)} \equiv -\zeta\omega_n \pm j\omega_d, \qquad (3.3\text{-}20)$$

where $\omega_d \equiv \omega_n(1 - \zeta^2)^{1/2}$ is the *damped frequency*. Figure 3.3-4 shows the s-plane vectors that could be used to evaluate the frequency response of a quadratic factor with complex roots. Complex poles are shown in the figure, but the following results also apply to complex zeros. The *resonant frequency*, ω_r, is the frequency at which the product of the lengths of the vectors is at a minimum, and is given by the imaginary axis intersection of the semicircle whose diameter is the line joining the poles. This is because the vectors drawn to an imaginary-axis point are the sides of a constant area triangle (constant base, constant height). At the point $j\omega_r$ the angle at the apex

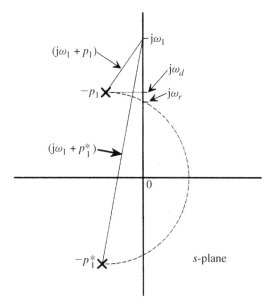

Figure 3.3-4 Geometric properties of a quadratic lag.

of the triangle reaches a maximum of 90 degrees, so the product of its two sides is at a minimum (area = product of two sides and sine of included angle). By constructing another right triangle, whose hypotenuse is a line from $-\zeta\omega_n$ on the real axis to $j\omega_r$ on the imaginary axis, we find that ω_r is given by

$$\omega_r = \omega_n \sqrt{\left(1 - 2\zeta^2\right)} \qquad (3.3\text{-}21)$$

and so the resonant frequency approaches the natural frequency as the damping ratio becomes small. There is no peak or dip in the frequency response of a complex pair of poles or zeros when $\zeta > 1/\sqrt{2}$.

We can apply these results to the quadratic-lag standard form, (3.3-16). Its magnitude and phase are given by

$$|G(j\omega)| = \frac{\omega_n^2}{\sqrt{\left[\left(\omega_n^2 - \omega^2\right)^2 + 4\zeta^2\omega^2\omega_n^2\right]}} \qquad (3.3\text{-}22a)$$

$$\angle G(j\omega) = -\text{atan2}\left(2\zeta\omega/\omega_n,\ 1 - \omega^2/\omega_n^2\right) \qquad (3.3\text{-}22b)$$

The four-quadrant inverse-tangent function is necessary because the phase angle, by which the output lags behind the input, lies between zero and 180 degrees. At resonance, the magnitude of the quadratic-lag standard form is found by substituting (3.3-21) in (3.3-22a):

$$|G(j\omega_r)| = \frac{1}{2\zeta\sqrt{(1-\zeta^2)}}, \quad \zeta < 1/\sqrt{2} \qquad (3.3\text{-}23)$$

Figure 3.3-5 shows the Bode magnitude and phase plots for a quadratic lag, with $\omega_n = 1.0$. The asymptotes of the magnitude plot are now found to be a zero dB line and a line with a slope of -40 dB/decade, intersecting the zero dB line at ω_n. When the damping ratio is small there is a large deviation from the asymptotes near ω_n.

Finally, consider a transfer function $(s+z)/(s+p)$. This has corner frequencies at $\omega = z$ and $\omega = p$, and the s-plane vectors show that its gain varies from z/p at zero frequency to unity at infinite frequency. If $z < p$, the gain will rise to unity and, if $z > p$, the gain will fall to unity. Rising gain is accompanied by a leading phase angle, and vice versa. On a logarithmic frequency scale, the maximum or minimum of the phase shift occurs midway between the pole and zero frequencies, and this is the geometric mean $\sqrt{(pz)}$. Other properties of this transfer function are derived in Section 3.9, where it is used for control system "compensation." Figure 3.3-6 shows Bode plots for the leading-phase case, $z < p$.

Various systems, including control systems, audio amplifiers, and sensors and measurement devices, can have their performance specified in terms of frequency response. The usual criteria are the *bandwidth*, *peak magnification*, and amount of

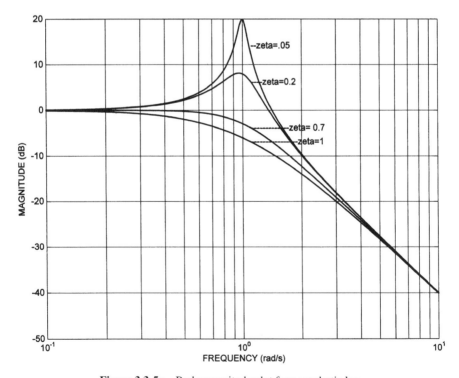

Figure 3.3-5a Bode magnitude plot for a quadratic lag.

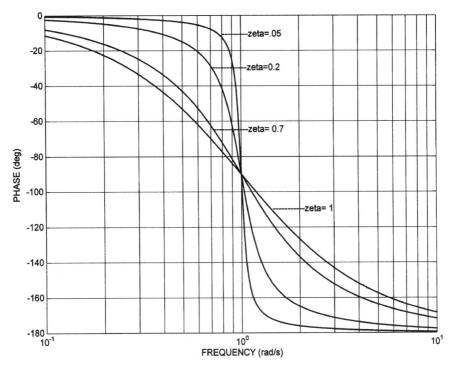

Figure 3.3-5b Bode phase plot for a quadratic lag.

phase shift at some frequency. A system whose frequency response extends down to zero frequency, and "rolls off" with increasing frequency (e.g., an integrator, or a simple-lag), is called a *low-pass* system. Most control systems behave this way. Similar definitions apply to *high-pass* and *band-pass* systems. If a low-pass system has a level frequency response at low frequency, we define the bandwidth to be the frequency at which the gain has fallen by 3 dB from its low-frequency value. As an example, the quadratic lag is a low-pass transfer function, it may have a resonant peak before it rolls off, and it can be shown to be "3 dB down" at the frequency:

$$\omega_B = \omega_n \sqrt{\left[(1 - 2\zeta^2) + \sqrt{(4\zeta^4 - 4\zeta^2 + 2)}\right]} \quad (3.3\text{-}24)$$

Time Response

Here we will look briefly at the step response of the simple-lag and quadratic-lag transfer functions. The transfer function of the simple lag is given in Table 3.3-1; the transform of a unit step input, occurring at $t = 0$, is $U(s) = 1/s$, and so the output transform is

$$Y(s) = \frac{1/\tau}{s + 1/\tau} \frac{1}{s} \equiv \frac{1}{s} + \frac{-1}{s + 1/\tau}$$

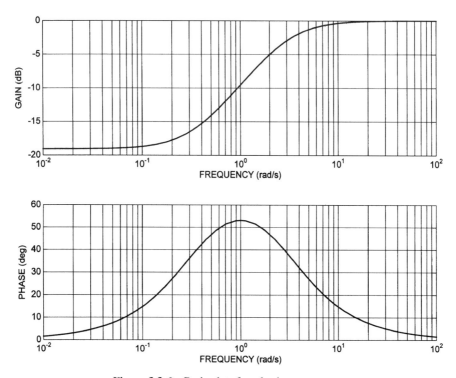

Figure 3.3-6 Bode plots for a lead compensator.

The partial fraction terms on the right correspond to known time functions, and so $y(t)$ can be written down directly:

$$y(t) = \left(1 - e^{-t/\tau}\right) U_{-1}(t), \tag{3.3-25}$$

where the unit-step $U_{-1}(t)$ serves to define the answer to be zero for $t < 0$. Equation (3.3-25) shows that the response of a simple real-pole transfer function, to a step input, is an exponential growth from zero to a final value given by the dc gain times the magnitude of the step.

The unit-step response of the quadratic lag (3.3-16) is given by

$$Y(s) = \frac{\omega_n^2}{(s + \zeta\omega_n)^2 + \omega_n^2(1 - \zeta^2)} \frac{1}{s} \equiv \frac{1}{s} - \frac{s + 2\zeta\omega_n}{(s + \zeta\omega_n)^2 + \omega_n^2(1 - \zeta^2)},$$

where the partial fraction coefficients were determined by the method of "comparing coefficients." The solution can now be written down from a knowledge of the Laplace transforms of sine and cosine functions of time, and the complex-domain shifting theorem. Using a trigonometric identity to combine the sine and cosine terms gives

$$y(t) = \left[1 - \frac{e^{-\zeta \omega_n t}}{\sqrt{(1-\zeta^2)}} \sin(\omega_d t + \phi)\right] U_{-1}(t), \qquad (3.3\text{-}26)$$

where

$$\phi = \cos^{-1}(\zeta)$$

Plots of this answer are shown in Figure 3.3-7 for several values of ζ. The graphs were plotted with $\omega_n = 1$; they apply to any natural frequency if the horizontal scale is treated as "normalized time" $\omega_n t$. The case $\zeta = 1$ is defined to be *critically damped*; when the damping ratio is less than 1, the step response has an overshoot and the poles are complex.

These results are useful because very often a dynamic system has a *dominant pair* of complex poles (Nise, 1995), which essentially determine its behavior. The damping of a system can be specified by the *maximum overshoot*, or *settling time*, of its step response, while system speed of response can be specified by the *rise time*, or the *peak time*, of its step response. These performance figures can be related to the damping ratio and natural frequency of a dominant pair (Dorf and Bishop, 2001).

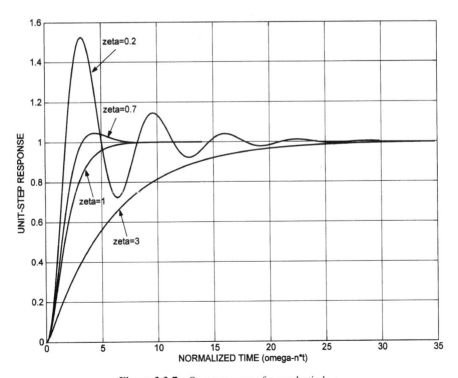

Figure 3.3-7 Step response of a quadratic lag.

3.4 NUMERICAL SOLUTION OF THE STATE EQUATIONS

Introduction

The aircraft state equations are nonlinear, depend on experimentally determined data, and are subjected to arbitrary input signals. An analytical solution is out of the question, and numerical methods must be used to compute an aircraft trajectory. In general the state vector of a physical system will move in a smooth, continuous manner in the n-dimensional state-space because the state variables describe the energy stored in a physical system, and an instantaneous change in energy would require infinite power. Therefore, derivatives of the state variables will exist, and a Taylor series expansion can be used to predict the motion.

Numerical evaluation of the continuous trajectory implies that, given the initial condition $X(t_0)$ and control input $U(t)$, we must calculate discrete sequential values of the state:

$$X(t_0 + kT), \qquad k = 1, 2 \ldots \qquad (3.4\text{-}1a)$$

that satisfy the state equations

$$\dot{X}(t) = f(X(t), U(t)) \qquad (3.4\text{-}1b)$$

This is called the *initial value problem*, and the *time-step* T is usually chosen to be a fixed size. The state equations are not autonomous since the control input is an external input, and the time-step must be made small enough that the control input can be approximated by a constant value during any interval kT to $(k+1)T$. There are two classes of numerical solution methods for the initial-value problem, Runge-Kutta (RK) methods and linear multistep methods (LMMs), and these will now be described.

Runge-Kutta (RK) Methods

Consider the simplest ODE initial-value problem: a single first-order autonomous differential equation with a specified boundary condition,

$$\frac{dx}{dt} = f(x, t), \qquad x(t_0) = x_0 \qquad (3.4\text{-}2)$$

The problem of finding the discrete solution values for (3.4-2) has an obvious connection to the Taylor series:

$$x(t_0 + T) = x(t_0) + T\dot{x}(t_0) + \frac{T^2}{2!}\ddot{x}(t_0) + \ldots \qquad (3.4\text{-}3)$$

The simplest RK method is *Euler integration*, which merely truncates the Taylor series after the first derivative. The Euler formula applied to (3.4-2) is

$$x_E(t_0 + T) \approx x(t_0) + Tf(x(t_0), t_0) \qquad (3.4\text{-}4)$$

This formula is not very accurate unless very small time-steps are used, and furthermore it can easily be improved upon, as follows.

In *trapezoidal integration* an estimate of the function derivative at the end of the time-step is obtained from the Euler formula; then the average of the derivatives at the beginning and end of the time-step is used to make a more accurate Euler step. The equations for a step forward from time t to $(t + T)$ are

$$x_E(t + T) = x(t) + Tf(x(t), t)$$
$$\dot{x}_E(t + T) = f(x_E(t + T), t + T)$$
$$x_T(t + T) = x(t) + \frac{T}{2}[\dot{x}(t) + \dot{x}_E(t + T)], \tag{3.4-5}$$

where subscripts E and T indicate, respectively, Euler and trapezoidal steps. For reasons that will soon become clear, these equations are commonly written as

$$k_1 = Tf(x, t)$$
$$k_2 = Tf(x + k_1, t + T)$$
$$x_T(t + T) = x(t) + \tfrac{1}{2}(k_1 + k_2) \tag{3.4-6}$$

This algorithm can be shown to agree with the first three Taylor series terms, that is, up to and including the second derivative term. Therefore, this trapezoidal integration formula is said to be of order two, and it gives an improvement in accuracy over the Euler first-order method. RK algorithms are an extension of (3.4-6) to higher orders, and the general form is

$$k_1 = Tf(x, t)$$
$$k_2 = Tf(x + \beta_1 k_1, t + \alpha_1 T)$$
$$k_3 = Tf(x + \beta_2 k_1 + \beta_3 k_2, t + \alpha_2 T)$$
$$k_4 = Tf(x + \beta_4 k_1 + \beta_5 k_2 + \beta_6 k_3, t + \alpha_3 T)$$
$$\vdots$$
$$x_{RK}(t + T) = x(t) + \gamma_1 k_1 + \gamma_2 k_2 + \gamma_3 k_3 + \ldots \tag{3.4-7}$$

Implicit RK algorithms also exist, wherein a coefficient k_i occurs on both sides of one of the equations above. The constants α_i, β_i, and γ_i are chosen so that a particular RK scheme agrees with the Taylor series to as high an order as possible. A great deal of algebraic effort is needed to derive higher-order (greater than four) RK algorithms, and the constants are not unique for a given order. An algorithm that dates from the end of the nineteenth century, and is still popular, is Runge's fourth-order rule, which uses the constants

$$\alpha_1 = \alpha_2 = \beta_1 = \beta_3 = \tfrac{1}{2}$$

$$\alpha_3 = \beta_6 = 1$$

$$\beta_2 = \beta_4 = \beta_5 = 0$$

$$\gamma_1 = \gamma_4 = 1/6, \qquad \gamma_2 = \gamma_3 = 1/3 \tag{3.4-8}$$

In this case only one previous k value appears in each of the k equations in (3.4-7), thus making a simpler algorithm. This algorithm has been used for most of our examples, and computer code for the general case of n simultaneous nonlinear state equations is given in the example below.

An important feature of the RK methods is that the only value of the state vector that is needed is the value at the beginning of the time-step; this makes them well suited to the ODE initial value problem. The amount of computation involved is governed by the number of derivative evaluations using the state equations, performed during each time-step. The number of derivative evaluations depends on the order chosen. For example, a fourth-order RK algorithm cannot be achieved with fewer than four derivative evaluations. For a given overall accuracy in a time response calculation, there is a trade-off between many small steps with a low-order method, and fewer steps but more derivative evaluations with a higher-order method. This led mathematicians to consider the problem of estimating the error in the computed solution function at each time step. Such an error estimate can be used to control the step size automatically in order to meet a specified accuracy. Algorithms that combine RK integration with error estimation include Runge-Kutta-Merson (RKM), Runge-Kutta-England, and Runge-Kutta-Gill; computer codes are commonly available. In terms of (3.4-7) the coefficients for the RKM scheme, for example, are:

$$\alpha_1 = \beta_1 = 1/3$$

$$\alpha_2 = 1/3, \quad \beta_2 = \beta_3 = 1/6 \tag{3.4-9}$$

$$\alpha_3 = 1/2, \quad \beta_4 = 1/8, \quad \beta_5 = 0, \quad \beta_6 = 3/8$$

$$\alpha_4 = 1, \quad \beta_7 = 1/2, \quad \beta_8 = 0, \quad \beta_9 = -3/2, \quad \beta_{10} = 2$$

$$\gamma_1 = 1/6, \quad \gamma_2 = \gamma_3 = 0, \quad \gamma_4 = 2/3, \quad \gamma_5 = 1/6$$

and the estimated error is

$$E \approx \frac{1}{30}[2k_1 - 9k_3 + 8k_4 - k_5]$$

Linear Multistep Methods (LMMs)

In the LMMs the solution function is a linear combination of past values of the function and its derivatives, as described by the linear difference equation,

$$x(n+1) = \sum_{r=0}^{n} \alpha_r x(n-r) + T \sum_{r=-1}^{n} \beta_r \dot{x}(n-r), \qquad (3.4\text{-}10)$$

where $x(i)$ indicates the value of x at time iT, with i an integer. If β_{-1} is nonzero, the algorithm is an implicit algorithm because the solution $x(n+1)$ is needed to evaluate $\dot{x}(n+1)$ on the right-hand side. Otherwise the algorithm is explicit. The implicit equation must be solved at each time step. LMMs can be designed to require less computation than RK methods because a number of past values can be kept in storage as the computation proceeds. Because of the requirements for past values the LMMs are not self-starting, and an RK method, for example, could be used to generate the starting values.

The LMM algorithms can be created in a number of different ways. For instance if the scalar state equation (3.4-2) is written as an integral equation over the time interval nT to $(n+k)T$, the result is

$$x(n+k) = x(n) + \int_{nT}^{(n+k)T} f(x,t)\, dt \qquad (3.4\text{-}11)$$

There are many finite-difference formulae for evaluating a definite integral, and this approach leads to the *Newton-Coates integration formulae* (Isaacson and Keller, 1966; Ralston, 1965). Two examples are

$$x(n+1) = x(n-1) + 2T\dot{x}(n) \qquad (3.4\text{-}12a)$$

$$x(n+1) = x(n-1) + \frac{T}{3}[\dot{x}(n+1) + 4\dot{x}(n) + \dot{x}(n-1)] \qquad (3.4\text{-}12b)$$

The first formula uses the *midpoint rule* for the area represented by the integral and is explicit, while the second uses *Simpson's rule* and is implicit. Implicit and explicit formulae can be used together in a *predictor-corrector algorithm* (Hamming, 1962). The explicit formula is the predictor, used to obtain an approximate value of the solution; and the implicit formula is the corrector equation, which is solved (by iteration) to obtain a more accurate solution.

LMMs of any order can be derived directly from (3.4-10). When $\alpha_r \equiv 0$ for $r > 0$, the *Adams-Bashforth-Moulton* (ABM) *formulae* are obtained.

We now give two examples. Assume that equation (3.4-10) has the terms

$$x(n+1) = \alpha_0 x(n) + T[\beta_0 \dot{x}(n) + \beta_1 \dot{x}(n-1)] \qquad (3.4\text{-}13)$$

Now write Taylor series expansions for the terms that are not taken at time nT:

$$x(n+1) = x(n) + T\dot{x}(n) + \frac{T^2}{2!}\ddot{x}(n) + \dots$$

$$\dot{x}(n-1) = \dot{x}(n) - T\ddot{x}(n) + \frac{T^2}{2!}\dddot{x}(n)\dots$$

Substitute these expressions in (3.4-13) and equate powers of T on both sides of the resulting equation; this gives

$$T^0: \quad 1 = \alpha_0$$
$$T^1: \quad 1 = \beta_0 + \beta_1$$
$$T^2: \quad \tfrac{1}{2} = -\beta_1$$

Therefore, (3.4-13) yields the second-order ABM formula

$$x(n+1) = x(n) + \frac{T}{2}[3\dot{x}(n) - \dot{x}(n-1)] \qquad (3.4\text{-}14)$$

This requires only one state equations evaluation per time-step, and it has often been used for simulation. The higher-order methods also require only one derivative evaluation per time-step, and the third-order ABM is

$$x(n+1) = x(n) + \frac{T}{12}[23\,\dot{x}(n) - 16\,\dot{x}(n-1) + 5\,\dot{x}(n-2)] \qquad (3.4\text{-}15)$$

The implicit formulae may be derived in the same way; they give improved accuracy and can also provide an error estimate. They are commonly used in the predictor-corrector form, and this requires two derivative evaluations per step.

Stability, Accuracy, and Stiff Systems

In developing numerical algorithms it is always necessary to consider how computational errors are magnified. If, in pursuit of greater accuracy, one blindly attempts to create higher-order LMM formulae, it is quite possible that the algorithm will be unstable and errors will grow with time. Stability can be determined by analyzing a finite-difference equation associated with the integration algorithm. This analysis (Shampine and Gordon, 1975) is beyond the scope of this chapter and we simply note that the specific algorithms described above are stable.

The RK stability properties are different from those of the LMMs. In the case of the RK algorithms, a reduction in time-step size will eventually eliminate an instability, although the required step size may be unreasonably small. Example 3.6-5 is an example of a reduction in step size eliminating an instability. When a set of state equations is being integrated, the required step size will be determined by the smallest time constant (i.e., the fastest component) of the solution function. A system with a very wide spread of time constants is known as a *stiff system*, and a very large number of RK steps may be necessary to yield only a small part of the complete solution. Other techniques are required for stiff systems (see below).

Choice of Integration Algorithm

The most important feature of the RK methods is that they directly solve the initial value problem. That is, no past values are needed to start the integration. This, of course, exactly matches the philosophy of the state-space formulation in which all of the information describing the "state" of the system is contained in the state vector at any given time instant. The full significance of these facts can only be appreciated when a simulation containing discrete events is considered. This is a common practical engineering situation. For instance, at a given time a new subsystem may be activated, or at a certain value of some variable, the equations of motion may change because limiting or saturation behavior occurs. This means that previous states are less relevant; the information they carry may now apply to only a part of the complete system. This fact favors the RK methods over the multistep methods, and we will return to these points later. The disadvantages of the RK methods are that the error expressions are complex, they are inefficient when dealing with stiff systems, and more derivative evaluations are required for a given order than is the case with LMMs. The tremendous increases in computing power in recent years have made these disadvantages much less significant for small to medium-sized simulations. Such simulations are commonly run with a fixed time step that has been found (by trial and error) to be adequate for the required accuracy and is also determined by other discrete event considerations.

The important features of LMMs are that higher-order methods are obtained for a given number of derivative evaluations, and an accurate expression for the integration error can usually be obtained. These methods come into their own on very large systems of equations, large stiff systems, and when there is no hard-limiting behavior or topological changes due to switching. The Software package ODE-PACK (Hindmarsh, 1982) is available for large and stiff problems, and it handles equations in standard explicit form or in linearly implicit form. For nonstiff problems it uses the implicit ABM methods, and for stiff problems it uses a backward difference formula and improves on the Gear algorithms (Gear, 1971) that have long been used for stiff systems. These algorithms have been used on atmosphere models with more than 10,000 simultaneous ODEs; the spread of time constants in the problem ranged from milliseconds to years, thus making the equations extremely stiff.

Time-History Simulation

Here we will show how the integration techniques can be used to determine a *state trajectory*, that is, the motion of the tip of the state vector as a function of time in the n-dimensional space. This is usually called *time-history simulation*. Our state-space dynamic equations are already in the best form for simulation, either non-real-time simulation or real-time simulation (e.g., in a flight simulator); it is only necessary to couple them with the integration algorithm.

In general a simulation will also need to process discrete-time calculations, that is, calculations in which the signals are only defined at the "sample instants." Such signals may arise from simulating a digital computer or sampling external signals. The numerical integration algorithms are based on the assumption that external inputs to the state equations will remain constant during an integration step. Therefore, the integration routines effectively impose a "zero-order data-hold" (ZOH) on the sampled signals. The ZOH is described in more detail in Chapter 7.

178 MODELING, DESIGN, AND SIMULATION TOOLS

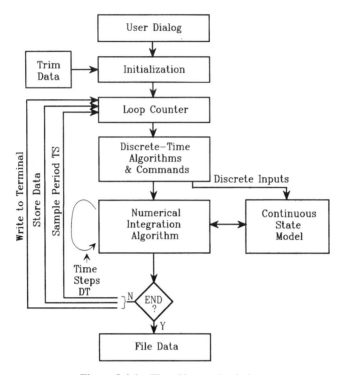

Figure 3.4-1 Time-history simulation.

Figure 3.4-1 shows how a non-real-time simulation program may be organized. Two separate functions or subroutines are needed for the dynamic models. One function contains the continuous-time state equations, and another function contains preprogrammed discrete-time commands and any discrete-time algorithms used for digital control. Simulation time is controlled by a for-loop, and the basic increment of time is the integration time-step DT. The *sample period* for the discrete dynamics TS can conveniently be chosen to be an integer multiple of DT. Alternatively, TS may be the basic time increment, and the integration algorithm may integrate over TS while adaptively adjusting its step-size to attain a specified integration accuracy. Periodic sampling is not essential, and the adaptive integration may continue until a discrete input occurs. The following example shows that a simple time-history simulation is very easy to perform.

Example 3.4-1: Integration of the Van der Pol Equation. A simple time-history program "NLSIM" written in MATLAB code is shown below; it prompts the user for the name of the m-file containing the state equations and the name of an initial condition file. The convention used is that the state equations function will always have the arguments "time," X, and U, in that order. The initial condition file will be a text file with a ".dat" extension, which can also be read by other programming languages; it should have a different name from the state equations file. The ".dat" extension must be entered at the MATLAB prompt because otherwise MATLAB assumes an ".m" extension.

```
% NLSIM.M Nonlinear Simulation
   clear all
   % global % add variables as needed
   name= input('Enter Name of State Equations m-file : ','s');
   icfile= input('Enter Name of i.c. File : ','s');
   tmp= dlmread(icfile,',');
   n=tmp(1); m=tmp(2);
   x=tmp(3:n+2); u=tmp(n+3:n+m+2);
   stat=fclose('all');
   runtime= input('Enter Run-Time : ');
   dt = input('Enter Integration Time-step : ');
   N=runtime/dt; k=0; NP= fix ( max(1,N/500) ); time=0.;
   xd= feval(name,time,x,u); % Set variables in state equations
   %save=u(2); % For Example 3.6-3 only
   for i=0:N
      time=i*dt;
      if rem(i,NP)==0
        k=k+1;
        y(k,1)= x(1);          % record data as needed
        y(k,2)= x(2);
        %y(k,3)=
      end
      %if time>=2             % For Example 3.6-3
      %   u(2)=save;
      %elseif time>=1.5
      %   u(2)=save-2;
      %elseif time>=1.0
      %   u(2)=save+2;
      %else
      %   u(2)=save;
      %end
      [x]= RK4(name,time,dt,x,u);
   end
   t= NP*dt*[0:k-1];
   figure(1)
   plot(y(:,1), y(:,2))       % For Van der Pol
   grid on
   axis([-3,3,-4,5])
   xlabel('X(1)')
   ylabel('X(2)')
   text(-1.8,3.2,'(-2,3)')
```

The fourth-order Runge-Kutta algorithm, with the constants given in (3.4-8), is

```
   function [xnew]= RK4(f,time,dt,xx,u)
   xd=feval(f,time,xx,u);
   xa=xd*dt;
   x =xx + 0.5*xa;
   t =time + 0.5*dt;
   xd=feval(f,t,x,u);
```

180 MODELING, DESIGN, AND SIMULATION TOOLS

```
q = xd*dt;
x = xx + 0.5*q;
xa= xa + 2.0*q;
xd= feval(f,t,x,u);
q = xd*dt;
x = xx + q;
xa= xa + 2.0*q;
time= time + dt;
xd= feval(f,time,x,u);
xnew= xx + (xa + xd*dt)/6.0;
```

The state equations used as an example are those of the Van der Pol oscillator, which exhibits interesting nonlinear behavior,

```
% File VDPOL.m
function [xd]= vdpol(time,x,u)
xd= [x(2)    -u(1)*(x(1)^2-1)*x(2)-x(1)];
```

The control input u = 0.8 was used as the parameter that controls the dynamic behavior of the Van der Pol oscillator. For this example the initial condition file

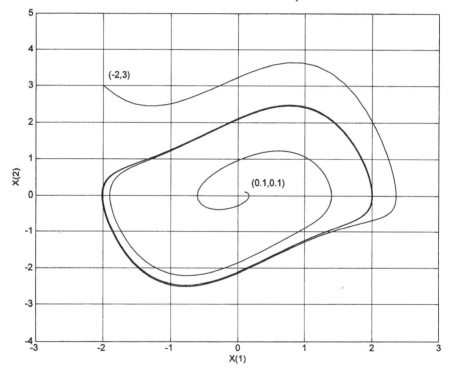

Figure 3.4-2 A Van der Pol limit cycle.

This function calculates the state derivative vector xd from the state vector x and the control vector u, the formal argument "time" is unused. It is compatible with the time-history program given in Example 3.4-1. The control inputs $u(3)$ and $u(4)$ are

VDP.dat contained the number of states and controls, the initial state, and the control input, as follows:

2, 1, .1, .1, .8

Figure 3.4-2 shows state x_2 plotted against state x_1 and is called a *phase portrait*. Two different sets of initial conditions are shown in the figure, and in both cases the state trajectories approach the same closed contour. The resulting constant amplitude oscillation is called a *limit cycle*. This example is studied further in Problem 3.4-1, and NLSIM.m is used for aircraft simulation in Section 3.6. ∎

3.5 AIRCRAFT MODELS FOR SIMULATION

Simulation Issues

In Section 3.4 we used MATLAB to illustrate simple nonlinear simulation, but this *interpreted* code executes one to two orders of magnitude more slowly than *compiled* code. In time-history simulation we wish to use a fixed sample period of 5 to 50 ms for the purposes of adequately sampling external inputs, generating random inputs, and interfacing with discrete-time controllers. The integration step size must be less than or equal to the discrete-time sample period, and depends on the accuracy required and the stiffness of the dynamics. With this constraint, any improvement in speed of execution must come from linking MATLAB with compiled code. Rather than use this approach we have chosen to present a simple aircraft model in MATLAB code, and a more complicated model in Fortran code. The choice of Fortran produces readable code. The reader has the option of converting the Fortran code to MATLAB (which is relatively easy to do), using MATLAB with compiled code (free compilers can be found), or running the Fortran code. A Fortran simulation program, TRESP, was written in the same form as the MATLAB program in Example 3.4-1 and using the RK4 integrator, but using subroutines for the continuous dynamics and the discrete dynamics, and with more comprehensive interactive capabilities. This program was used for the F-16 flight simulation examples in Section 3.6 and Chapter 4.

A Simple Longitudinal Model

This model has only three degrees of freedom (i.e., translation and pitching motion in the vertical plane): it has fixed aerodynamic coefficients and is representative of a medium-size transport aircraft at a low-speed flight condition. Data are also provided for the effects of extending landing gear and flaps. The aircraft weighs 162,000 lb (one-half fuel, partial cargo), and it has two turboprop engines, each developing 30,000 lb of static thrust at sea level. The wing area is 2170 ft^2, wing span 140 ft, length 90 ft, and pitch-axis inertia 4.1×10^6 slug-ft^2. The model is illustrated in Figure 3.5-1, programmed as a MATLAB function.

```
function [xd]= transp(time,x,u)
% Medium-sized transport aircraft, longitudinal dynamics.
%
  S=2170.0; CBAR=17.5; MASS=5.0E3; IYY= 4.1E6;
  TSTAT=6.0E4; DTDV  =-38.0; ZE    = 2.0; CDCLS= .042;
  CLA  = .085; CMA   =-.022; CMDE =-.016;  % per degree
  CMQ  =-16.0; CMADOT= -6.0; CLADOT= 0.0;   % per radian
  RTOD = 57.29578; GD=32.17;
  THTL =u(1);
  ELEV =u(2);
  XCG  = u(3);
  LAND = u(4);
  VT   = x(1);                            % TAS in fps
  ALPHA= RTOD*x(2);                       % A.O.A.
  THETA= x(3);                            % PITCH ATTITUDE
  Q    = x(4);                            % PITCH RATE
  H    = x(5);                            % ALTITUDE
%
  [MACH,QBAR]= ADC(VT,H);
  QS   = QBAR*S;
  SALP= sin (x(2));    CALP= cos(x(2));
  GAM = THETA - x(2); SGAM= sin (GAM); CGAM= cos(GAM);
  if LAND == 0                            % CLEAN
    CL0= .20; CD0= .016;
    CM0= .05; DCDG= 0.0; DCMG= 0.0;
  elseif LAND == 1                        % LANDING FLAPS & GEAR
    CL0= 1.0; CD0= .08;
    CM0= -.20; DCDG= .02; DCMG= -.05;
  else
    disp('Landing Gear & Flaps ?')
  end
  THR= (TSTAT+VT*DTDV) * max(THTL,0);     % THRUST
  CL=CL0+CLA*ALPHA;                       % NONDIM. LIFT
  CM=DCMG+CM0+CMA*ALPHA+CMDE*ELEV+CL*(XCG-.25); % MOMENT
  CD=DCDG+CD0+CDCLS*CL*CL;                % DRAG POLAR
%
% STATE EQUATIONS NEXT
  xd(1) = (THR*CALP-QS*CD)/MASS - GD*SGAM;
  xd(2) =(-THR*SALP-QS*CL+MASS*(VT*Q+GD*CGAM))/(MASS*VT+QS*CLADOT);
  xd(3) = Q;
  D     = .5*CBAR*(CMQ*Q+CMADOT*xd(2))/VT; % PITCH DAMPING
  xd(4) = (QS*CBAR*(CM + D) + THR*ZE)/IYY; % Q-DOT
  xd(5) = VT*SGAM;                         % VERTICAL SPEED
  xd(6) = VT*CGAM;                         % HORIZNTL. SPEED
```

Figure 3.5-1 Transport aircraft model.

used, respectively, to set the *x*-axis position of the cg and the landing configuration switch. For the aircraft models we will use the customary term *center of gravity* (cg) synonymously with cm, although technically a cg does not exist if a body does not have a spherically symmetrical distribution of mass.

Miscellaneous model outputs can be made available by setting up global variables. The aerodynamic derivatives are in stability axes and have "per degree" units except for the pitch damping coefficients (C_{m_q}, $C_{m_{\dot{\alpha}}}$), which are per radian per second. There is provision for a $C_{L_{\dot{\alpha}}}$ derivative, but it is zero in this case. Lift is calculated from a linear lift curve and the stall is not modeled, while drag is calculated from the nonlinear drag polar. The elevator deflection is in degrees, and the throttle input is in the zero to 1 range. Atmospheric density (and hence dynamic pressure) is calculated in the function ADC (air data computer, see Appendix A) from the temperature variation of the standard atmosphere (Yuan, 1967). The engine thrust is modeled as decreasing linearly with airspeed, to approximate the characteristics of a propeller-driven aircraft. The thrust vector does not pass through the cg (the perpendicular distance from the vector to the cg is Z_E), and therefore throttle changes will tend to cause pitching motion of the aircraft. Other parts of the model are either self-evident or can be understood by referring to the descriptions of aerodynamic effects in Chapter 2. This model will be used later for illustrative examples.

A Six-Degrees-of-Freedom Nonlinear Aircraft Model

The mathematical model given here uses the wind tunnel data from NASA-Langley wind tunnel tests on a subscale model of an F-16 airplane (Nguyen et al,. 1979). The data apply to the speed range up to about $M = 0.6$, and were used in a NASA-piloted simulation to study the maneuvering and stall/poststall characteristics of a relaxed static-stability airplane. Because of the application and the ease of automated data collection, the data cover a very wide range of angle of attack (-20 to 90 degrees), and of sideslip angle (-30 to 30 degrees). However, the present state of the art does not allow accurate dynamic modeling in the poststall region, and in addition the aircraft has insufficient pitching moment control for maneuvering at angles of attack beyond about 25 degrees. Therefore, for use here, we have reduced the range of the data to $-10° \leq \alpha \leq 45°$, and approximated the beta dependence in some cases.

The F-16 has a leading-edge flap that is automatically controlled as a function of alpha and Mach and responds rapidly as alpha changes during maneuvering. In the speed range for which the data are valid, the Mach-dependent variation of the flap is small, and so we have eliminated this dependence. Then, neglecting the dynamics of the flap actuator and assuming that the flap is dependent on alpha only, we have merged all of the independent flap data tables into the rest of the tabular aerodynamic data. The effect of the flap deflection limits (but not the rate limits) is still present in the reduced data. These steps have greatly reduced the size of the database and made it feasible to present the data here (Appendix A). The approximate model constructed from these data exhibits steady-state flight trim conditions, and corresponding dynamic modes, that are close to those of the full (50 lookup-table) model.

The F-16 model has been programmed as a Fortran subroutine in a form similar to the MATLAB model. The code is shown in Figure 3.5-2; all subroutines and functions called by the model are included in Appendix A. Note that English units have been used here, rather than SI units. State variables V_T, α, and β have been used instead of the velocity components U, V, and W, for ease of comparison with the linear small-perturbation equations. For serious simulation purposes it would be preferable to change to states U, V, and W. The quantities XCGR and HX are, respectively, x-coordinate of the reference cg position, and engine angular momentum (assumed constant at 160 slug-ft^2).

```
      SUBROUTINE F(TIME,X,XD)
      REAL X(*), XD(*), D(9), MASS
      COMMON/PARAM/XCG
      COMMON/CONTROLS/THTL,EL,AIL,RDR
      COMMON/OUTPUT/AN,ALAT,AX,QBAR,AMACH,Q,ALPHA
      PARAMETER (AXX=9496.0, AYY= 55814.0, AZZ=63100.0, AXZ= 982.0)
      PARAMETER (AXZS=AXZ**2, XPQ=AXZ*(AXX-AYY+AZZ),
     &                                         GAM=AXX*AZZ-AXZ**2)
      PARAMETER (XQR= AZZ*(AZZ-AYY)+AXZS, ZPQ=(AXX-AYY)*AXX+AXZS)
      PARAMETER ( YPR= AZZ - AXX )
      PARAMETER (WEIGHT= 25000.0, GD= 32.17, MASS= weight/gd)

      DATA S,B,CBAR,XCGR,HX/300,30,11.32,0.35,160.0/
      DATA RTOD / 57.29578/
C
C Assign state & control variables
C
      VT= X(1); ALPHA= X(2)*RTOD; BETA= X(3)*RTOD
      PHI=X(4); THETA= X(5); PSI= X(6)
      P= X(7);  Q= X(8); R= X(9); ALT= X(12); POW= X(13)
C
C Air data computer and engine model
C
      CALL ADC(VT,ALT,AMACH,QBAR); CPOW= TGEAR(THTL)
      XD(13) = PDOT(POW,CPOW);      T= THRUST(POW,ALT,AMACH)
C
C  Look-up tables and component buildup
C
      CXT = CX (ALPHA,EL)
      CYT = CY (BETA,AIL,RDR)
      CZT = CZ (ALPHA,BETA,EL)
      DAIL= AIL/20.0; DRDR= RDR/30.0
      CLT = CL(ALPHA,BETA) + DLDA(ALPHA,BETA)*DAIL
     &                     + DLDR(ALPHA,BETA)*DRDR
      CMT = CM(ALPHA,EL)
      CNT = CN(ALPHA,BETA) + DNDA(ALPHA,BETA)*DAIL
     &                     + DNDR(ALPHA,BETA)*DRDR
C
C  Add damping derivatives :
C
      TVT= 0.5/VT;  B2V= B*TVT;  CQ= CBAR*Q*TVT
      CALL DAMP(ALPHA,D)
      CXT= CXT + CQ   * D(1)
      CYT= CYT + B2V *  ( D(2)*R + D(3)*P )
      CZT= CZT + CQ   * D(4)
      CLT= CLT + B2V *  ( D(5)*R + D(6)*P )
      CMT= CMT + CQ   * D(7) + CZT * (XCGR-XCG)
      CNT= CNT + B2V*(D(8)*R + D(9)*P) - CYT*(XCGR-XCG) * CBAR/B
C
C Get ready for state equations
C
      CBTA  = COS(X(3));  U=VT*COS(X(2))*CBTA
      V= VT * SIN(X(3));  W=VT*SIN(X(2))*CBTA
```

AIRCRAFT MODELS FOR SIMULATION

```
      STH    = SIN(THETA); CTH= COS(THETA);   SPH= SIN(PHI)
      CPH    = COS(PHI)  ; SPSI= SIN(PSI);    CPSI= COS(PSI)
      QS     = QBAR * S  ; QSB= QS * B;       RMQS= QS/MASS
      GCTH   = GD * CTH  ; QSPH= Q * SPH
      AY     = RMQS*CYT  ; AZ= RMQS * CZT
C
C  Force equations
C
      UDOT   = R*V - Q*W - GD*STH + (QS * CXT + T)/MASS
      VDOT   = P*W - R*U + GCTH * SPH + AY
      WDOT   = Q*U - P*V + GCTH * CPH + AZ
      DUM    = (U*U + W*W)
      XD(1)  = (U*UDOT + V*VDOT + W*WDOT)/VT
      XD(2)  = (U*WDOT - W*UDOT) / DUM
      XD(3)  = (VT*VDOT- V*XD(1)) * CBTA / DUM
C
C  Kinematics
C
      XD(4)  = P + (STH/CTH)*(QSPH + R*CPH)
      XD(5)  = Q*CPH - R*SPH
      XD(6)  = (QSPH + R*CPH)/CTH
C
C  Moments
C
      ROLL   = QSB*CLT
      PITCH  = QS *CBAR*CMT
      YAW    = QSB*CNT
      PQ     = P*Q
      QR     = Q*R
      QHX    = Q*HX
      XD(7)  = ( XPQ*PQ - XQR*QR + AZZ*ROLL + AXZ*(YAW + QHX) )/GAM
      XD(8)  = ( YPR*P*R - AXZ*(P**2 - R**2) + PITCH - R*HX )/AYY
      XD(9)  = ( ZPQ*PQ - XPQ*QR + AXZ*ROLL + AXX*(YAW + QHX) )/GAM
C
C  Navigation
C
      T1= SPH * CPSI; T2= CPH * STH; T3= SPH * SPSI
      S1= CTH * CPSI; S2= CTH * SPSI; S3- T1 * STH - CPH * SPSI
      S4= T3  * STH + CPH * CPSI; S5= SPH * CTH; S6= T2*CPSI + T3
      S7= T2  * SPSI - T1; S8= CPH * CTH
C
      XD(10)   = U * S1 + V * S3 + W * S6    ! North speed
      XD(11)   = U * S2 + V * S4 + W * S7    ! East speed
      XD(12)   = U * STH -V * S5 - W * S8    ! Vertical speed
C
C  Outputs
C
      AN= -AZ/GD; ALAT= AY/GD
      RETURN
      END
```

Figure 3.5-2 Model of the F-16 aircraft.

The aerodynamic force and moment component buildup follows the outline presented in Section 2.3 except that body axes are used. For example, CX(alpha, EL) is a function subprogram that computes the nondimensional force coefficient for the body x-axis and is a function of angle of attack and elevator deflection. The total force coefficients for the three axes are CXT, CYT, and CZT. As shown in the appendix, the component functions typically contain a two-dimensional data lookup-table and a linear interpolation routine. We have used as much commonality as possible in the data tables and interpolation routines and have provided an interpolator that will also extrapolate beyond the limits of the tables. Therefore, a simulation may recover without loss of all data despite temporarily exceeding the limits of a lookup-table.

Engine Model The NASA data include a model of the F-16 afterburning turbofan engine, in which the thrust response is modeled with a first-order lag, and the lag time constant is a function of the actual engine power level (POW) and the commanded power (CPOW). This time constant is calculated in the function PDOT, whose value is the rate of change of power, while the state variable X_{13} represents the actual power level. The function TGEAR (throttle gearing) relates the commanded power level to the throttle position (0 to 1.0) and is a linear relationship apart from a change of slope when the military power level is reached at 0.77 throttle setting. The variation of engine thrust with power level, altitude, and Mach number is contained in the function THRUST.

Sign Convention for Control Surfaces The sign conventions used in the model follow a common industry convention and are given in Table 3.5-1.

Testing the Model When constructing this model a simple program should be written to exercise each of the aerodynamic lookup-tables individually, and plot the data, before the tables are used with the model. The range of the independent variables should be chosen to ensure that both extrapolation and interpolation are performed correctly. A simple check on the complete model can be obtained by writing another program to set the parameter, input, and state vectors to the arbitrarily chosen values given in Table 3.5-2. The resulting values of the derivative vector should then agree with those given in the table.

TABLE 3.5-1: Aircraft Control-Surface Sign Conventions

	Deflection	Sense	Primary Effect
Elevator	trailing-edge down	positive	negative pitching moment
Rudder	trailing-edge left	positive	negative yawing moment
Ailerons	right-wing trailing-edge down	positive	negative rolling moment

TABLE 3.5-2: F-16 Model Test Case

Index (i)	PARAM	$U(i)$	$X(i)$	$\dot{X}(i)$
1	.4	0.9	500	−75.23724
2		20	0.5	−0.8813491
3		−15	−0.2	−0.4759990
4		−20	−1	2.505734
5			1	0.3250820
6			−1	2.145926
7			0.7	12.62679
8			−0.8	0.9649671
9			0.9	0.5809759
10			1000	342.4439
11			900	−266.7707
12			10000	248.1241
13			90	−58.68999

Next we must bring this model under control by finding a combination of values of the state and control variables that correspond to a steady-state flight condition. Unlike a real pilot who is constantly receiving visual and other cues, this is quite difficult for us and will be the subject of the next section. In the next section steady-state trim data will be given for both wings-level, non-sideslipping flight and turning flight. Therefore, the longitudinal equations can be tested alone before all the equations are brought into play.

3.6 STEADY-STATE FLIGHT

Steady-state flight is important because it provides an initial condition for flight simulation, and a flight condition in which we can linearize the aircraft dynamics (see Section 3.7). Figure 3.11 shows how a steady-state "trim" program fits into the state-space context. A generic trim program links to any nonlinear model and produces a file containing the steady-state values of the control and state vectors for use by the time-history and linearization programs. Steady-state flight was defined in Section 2.6 and was shown to require the solution of a set of nonlinear simultaneous equations derived from the state model. Now we are faced with the problem of actually calculating the values of the state and control vectors that satisfy these equations. This cannot be done analytically because of the very complex functional dependence of the aerodynamic data. Instead, it must be done with a numerical algorithm that iteratively adjusts the independent variables until some solution criterion is met. The solution will be approximate but can be made arbitrarily close to the exact solution by tightening up the criterion. Also, the solution may not be unique—for example, steady-state level flight at a given engine power level can in general correspond to two different airspeeds and angles of attack. Our knowledge of aircraft behavior will allow us to specify the required steady-state condition so that the trim algorithm converges on an appropriate, if not unique, solution.

One of the first things that must be decided is how to specify the steady-state condition, how many of the state and control variables can be chosen independently, and what constraints exist on the remaining variables. A computer program can then be written so that the specification variables are entered from the keyboard, and the independent variables are adjusted by the numerical algorithm that solves the nonlinear equations, while the remaining variables are determined from the constraint equations.

For steady-state flight we expect to be able to specify the altitude and the velocity vector (i.e., speed and flight-path angle) within the limits imposed by engine power. Then, assuming that the aircraft configuration (i.e., flap settings, landing gear up or down, speed brake deployed, etc.) is prespecified, for a conventional aircraft we expect that a unique combination of the control inputs and the remaining state variables will exist. All of the control variables (throttle, elevator, aileron, and rudder) enter the model only through tabular aerodynamic data, and we cannot, in general, determine any analytical constraints on these control inputs. Therefore, these four control inputs must be adjusted by our numerical algorithm. This is not the case for the state variables.

Since only the NED altitude component of the tangent-frame position vector is relevant and can be prespecified, we can temporarily eliminate the three position states from consideration. Consider first steady translational flight. The state variables ϕ, P, Q, R are all identically zero, and the orientation ψ can be specified freely; this only leaves V_T, α, β (or U, V, W) and θ to be considered. The sideslip angle cannot be specified freely; it must be adjusted by our trim algorithm to zero out any sideforce. This leaves the variables V_T, α, and θ; the first two are interrelated through the amount of lift needed to support the weight of the aircraft; therefore, only two may be specified independently (θ, and either V_T or α). We usually wish to impose a flight-path angle (γ) constraint on the steady-state condition, so we will finally choose to specify V_T and γ.

Because the atmospheric density changes with altitude, a steady-state flight condition does not strictly include a nonzero flight-path angle. Nevertheless, it is useful to be able to determine a trimmed condition for a nonzero flight-path angle at any given altitude, since rate of climb (ROC) can then be determined and linearized dynamic models can be obtained for nonzero flight-path angles. We will therefore derive a general rate-of-climb constraint; this constraint will allow a nonzero roll angle so that it can also be applied to steady-state turning flight.

Steady-state turning flight must now be considered; the variables ϕ, P, Q, and R will no longer be set to zero. The turn can be specified by the Euler angle rate $\dot{\psi}$; this is the rate at which the aircraft's heading changes (the initial heading can still be freely specified). Then, given values of the attitude angles ϕ and θ, the state variables P, Q, and R can be determined from the kinematic equation (1.3-21). The required value of θ can be obtained from the ROC constraint if the value of ϕ is known, and we next consider the determination of ϕ.

The roll angle (ϕ) for the steady-state turn can be freely specified, but then, in general, there will be a significant sideslip angle and the turn will be a "skidding" turn. The pilot will feel a force pushing him or her against the side of the cockpit, the passengers' drinks will spill, and the radius of the turn will be unnecessarily large. In a "coordinated" turn the aircraft is rolled at an angle such that there is no component of aerodynamic force along the body y-axis. This condition is used as the basis of the turn coordination constraint derived below. The turn coordination constraint will be found to involve both θ and ϕ; therefore, it must be solved simultaneously with the ROC constraint.

Chapters 1 and 2 have shown, via the flat-Earth equations, that the dynamic behavior of the aircraft is determined by the relative wind ($-\mathbf{v}_{rel}$) and height in the atmosphere, and hence by the variables V_T, α, β, and h. The behavior is essentially independent of the wind velocity $\mathbf{v}_{W/e}$. Therefore, when we wish to determine a steady-state flight condition for studying the dynamics, we will set the wind velocity to zero.

The Rate-of-Climb Constraint

Equation (1.5-23), with no wind, gives,

$$\mathbf{v}^n_{CM/e} = C_{n/b}\, C_{b/w}\, \mathbf{v}^w_{rel}$$

In the flat-Earth equations the rate of climb is simply $V_T \sin \gamma$, and this is the *NED*, negative-z, component of the velocity in the tangent frame. Therefore, the above equation yields:

$$V_T \begin{bmatrix} * \\ * \\ -\sin\gamma \end{bmatrix} = C_{n/b} C_{b/w} \begin{bmatrix} V_T \\ 0 \\ 0 \end{bmatrix} \qquad (3.6\text{-}1)$$

The asterisks indicate "don't care" components, and if this equation is expanded and then arranged to solve for θ (Problem 3.6-3), the results are

$$\sin\gamma = a\sin\theta - b\cos\theta, \qquad (3.6\text{-}2)$$

where

$$a = \cos\alpha\cos\beta, \qquad b = \sin\phi\sin\beta + \cos\phi\sin\alpha\cos\beta$$

Now, solving for θ, we find

$$\tan\theta = \frac{ab + \sin\gamma \sqrt{[a^2 - \sin^2\gamma + b^2]}}{a^2 - \sin^2\gamma}, \qquad \theta \neq \pm\pi/2 \qquad (3.6\text{-}3)$$

As a check, in wings-level, non-sideslipping flight this equation reduces to $\theta = \alpha + \gamma$.

The Turn-Coordination Constraint

In a perfectly coordinated turn the components of force along the aircraft body-fixed y-axis sum to zero, and in addition we have the steady-state condition $\dot{V} = 0$. Then, from Table 2.5-1,

$$0 = -RU + PW + g_D \sin\phi \cos\theta$$

Now use Equation (1.3-21), with $\dot{\phi} = \dot{\theta} = 0$, to introduce the "turn rate" $\dot{\psi}$ in place of P and R. Also use Equation (2.3-6a) to introduce V_T in place of U and W; the result is

$$-V_T \dot{\psi} \cos\beta [\cos\alpha \cos\theta \cos\phi + \sin\alpha \sin\theta] = g_D \sin\phi \cos\theta$$

If we define

$$\mathcal{G} \equiv \dot{\psi} V_T / g_D,$$

which is the centripetal acceleration (in g's), then the constraint can be written as

$$\sin\phi = \mathcal{G} \cos\beta (\sin\alpha \tan\theta + \cos\alpha \cos\phi) \qquad (3.6\text{-}4)$$

This is the required coordination constraint; it can be used in conjunction with (3.6-3) to trim the aircraft for turning flight with a specified rate of climb. If we can now solve (3.6-3) and (3.6-4) simultaneously for the state variables ϕ and θ, our numerical trim algorithm need only vary α and β (in addition to the four controls). The simultaneous solution is quite cumbersome but can be shown to be

$$\tan\phi = \mathcal{G} \frac{\cos\beta}{\cos\alpha} \frac{(a - b^2) + b \tan\alpha \sqrt{\left[c(1 - b^2) + \mathcal{G}^2 \sin^2\beta\right]}}{a^2 - b^2(1 + c \tan^2\alpha)}, \qquad (3.6\text{-}5)$$

where

$$a = 1 - \mathcal{G} \tan\alpha \sin\beta, \quad b = \sin\gamma / \cos\beta, \quad \text{and} \quad c = 1 + \mathcal{G}^2 \cos^2\beta$$

The value of ϕ given by (3.6-5) can now be used to solve (3.6-3) for θ. Note that when the flight path angle γ is zero, (3.6-5) reduces to

$$\tan\phi = \frac{\mathcal{G} \cos\beta}{\cos\alpha - \mathcal{G} \sin\alpha \sin\beta} \qquad (3.6\text{-}6)$$

and when β is small, this reduces to

$$\tan\phi = \frac{\mathcal{G}}{\cos\alpha} = \frac{\dot{\psi} V_T}{g_D \cos\alpha} = \frac{\dot{\psi} V_T}{g_D \cos\theta} \qquad (3.6\text{-}7)$$

Equation (3.6-7) applies to a level, non-sideslipping turn, and can be found from a simplified analysis given in standard texts. This completes the description of the flight-path constraints; we next show how a trim program may be constructed and provide examples of trimming the aircraft models.

The Steady-State Trim Algorithm

The steady-state flight conditions are determined by solving the nonlinear state equations for the state and control vectors that make the state derivatives $\dot{U}, \dot{V}, \dot{W}$ (or $\dot{V}_T, \dot{\alpha}, \dot{\beta}$), and $\dot{P}, \dot{Q}, \dot{R}$, identically zero. A convenient way to do this, with a readily available numerical algorithm, is to form a scalar *cost function* from the sum of the squares of the derivatives above. A function minimization algorithm can then be used to adjust the control variables and the appropriate state variables, to minimize this scalar cost. Examples of suitable algorithms are the IMSL routine "ZXMWD" (IMSL, 1980), and the SIMPLEX algorithm (Press et al., 1986; Nelder and Mead, 1964). Figure 3.6-1 illustrates how the complete trim algorithm may be organized. Only the cost function is tailored to a specific aircraft or set of state equations. We now give a simple trim example using the transport aircraft model.

Example 3.6-1: Steady-State Trim for a 3-DOF Aircraft Model. In this example we will construct a simple 3-DOF trim program and use it on the transport aircraft model in Figure 3.5-1. It is only necessary to choose the speed and altitude,

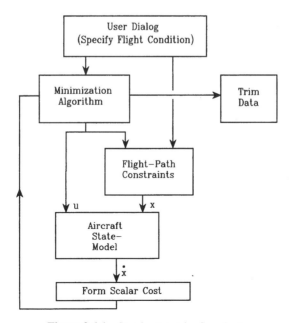

Figure 3.6-1 Steady-state trim flowchart.

set the pitch rate state to zero, and adjust the throttle and elevator controls and the angle of attack state. Instead of the ROC constraint, we can specify the flight-path angle and constrain the pitch attitude state to be equal to the angle of attack plus the flight-path angle. A simple MATLAB program is as follows:

```
% TRIM.m
clear all
global x u gamma
x(1)=input('Enter Vt : ');
x(5)= input('Enter h : ');
gamma=input('Enter Gamma (deg.) : ')/57.29578;
name= input('Name of Cost function file ? : ',' s');
cg= 0.25; land=1;   % 0=clean 1=gear+flaps
u=[0.1 -10 cg land];
x(2)= .1;              % Alpha, initial guess
x(3)=x(2) +gamma;      % Theta
x(4)=0;                % Pitch rate
x(6)=0;
s0=[u(1) u(2) x(2)];
% Now initialize any other states and get initial cost
    disp(['Initial cost = ',num2str( feval(name,s0) ) ])
    [s,fval]=fminsearch(name,s0) ;
    x(2)=s(3); x(3)=s(3)+gamma;
    u(1)=s(1); u(2)=s(2) ;
    disp(['minimum cost = ',num2str(fval)])
    disp(['minimizing vector= ',num2str(s)])
    temp=[length(x),length(u),x,u];
    name= input('Name of output file ? : ',' s') ;
    dlmwrite(name,temp);
```

and a cost function for the transport aircraft model is:

```
% Cost Function for 3-DOF Aircraft
function [f]=cost(s);
global x u gamma
u(1)= s(1);
u(2)= s(2);
x(2)= s(3);
x(3)= x(2)+ gamma;
time= 0.0;
[xd]=transp(time,x,u);
f= xd(1)^2 + 100*xd(2)^2 + 10*xd(4)^2;
```

The Matlab function "fminsearch" performs the minimization, and is actually a Nelder and Mead Simplex algorithm. The results obtained for level flight with cg $= 0.25\bar{c}$ (cg is the variable representing c.g. position), and flaps and landing gear retracted, are shown in Table 3.6-1. The cost function can be reduced to less than 1E-30, but anything below about 1E-12 causes negligible changes in the states and controls. The weighting on the derivatives in the cost function was experimental, and makes little difference to the results.

STEADY-STATE FLIGHT

TABLE 3.6-1: Trim Data for the Transport Aircraft Model

alt (ft)	speed (ft/s)	initial cost	final cost	throttle	elevator (deg.)	alpha (deg.)
0	170	28.9	< 1E-20	0.297	−25.7	22.1
0	500	3.54	< 1E-20	0.293	2.46	0.580
30k	500	10.8	< 1E-20	0.204	−4.10	5.43

∎

The trim program for Example 3.6-1 can easily be modified for other experiments, such as trimming for a specific alpha by varying the airspeed (Problem 3.6-4). We next consider the slightly more difficult problem of trimming a 6-DOF model, with additional dynamics such as an engine model that must also be put into a steady-state condition. This will be illustrated with the F-16 model using the code in Appendix B.

Example 3.6-2: Steady-State Trim for a 6-DOF Model. The following cost function subprogram has been specifically tailored to the F-16 model, but is representative of the 6-DOF case in general.

```
function cost(s)
parameter (nn=20)
real s(*)
common/state/x(nn),xd(nn)
common/controls/thtl,el,ail,rdr
thtl  = s(1)
el    = s(2)
x(2)  = s(3)
ail   = s(4)
rdr   = s(5)
x(3)  = s(6)
x(13) = tgear(thtl)
call constr(x)
call f(time,x,xd)
cost = xd(1)**2 + 100*(xd(2)**2 + xd(3)**2) + 10*(xd(7)**2
&    + xd(8)**2 + xd(9)**2)
return
end
```

This cost function is specific to the F-16 model because of the assignment statement for X_{13}. An examination of the F-16 model will show that this statement sets the derivative $\dot{X}_{13}$ to zero and hence puts the engine dynamics into the steady state. Any other dynamics in the aircraft model besides the rigid-body dynamics must be put into the steady-state condition in this way. In our original large F-16 model, this was done for the leading-edge flap actuator and its phase-lead network.

In this cost function, unlike the previous case, the state variables X_4 through X_9 (excluding X_6) are continually assigned new values in the constraint routine

CONSTR. This routine implements the rate-of-climb and turn coordination constraints that were derived earlier. In the cost the aerodynamic-angle rates $\dot{\alpha}$ and $\dot{\beta}$ have been weighted the most heavily, the angular rate derivatives $\dot{P}, \dot{Q}, \dot{R}$ have medium weights, and the derivative $\dot{V}_T$ has the least weight. Again, the weights are uncritical.

We will now use this cost function to determine the steady-state conditions in a coordinated turn, performed by the F-16 model. The cg location of the model is at $0.35\bar{c}$, and the aircraft dynamics are unstable in pitch in the chosen flight condition. The turn would stress a pilot since it involves a sustained normal acceleration of 4.5 g's.

The trim program dialog and keyboard inputs are shown in Figure 3.6-2 as they would appear on a terminal display. Note that entering a "/" in response to a Fortran read statement causes the program to use the last values assigned to the variable. This allows the minimization to be picked up from where it was stopped if the final cost function was not low enough. In the run shown, the cost function was reduced by almost 10 orders of magnitude after 1000 function calls. Execution is very fast, and this is a reasonable number of calls.

The cost function can always be reduced to 1×10^{-10} or less; lower values are useful simply for checking consistency of results. The most effective way to use the simplex algorithm is to perform 500 to 1000 iterations, and if the cost is not

```
? Altitude (ft)                                         : 0
? Air Speed (ft/s) and Climb Angle (deg)                : 502,0
? Roll, Pull-Up, and Turn rates (rad/s)                 : 0,0,.3

Turn Radius (ft)= 1.6733E+03    Approx Roll Angle (deg)= 77.94
? Coordinated Turn (def. = Y) : /
? Required No. of Trim Iterations (def. =1000) : /

? Guess    :  Throttle,  Elevator,  Ailerons,  Rudder   : /
Computed   :  8.35E-01   -1.48E+00  9.54E-02   -4.11E-01

        Angle of Attack   1.37E+01      Sideslip Angle   2.92E-02
            Pitch Angle   2.87E+00         Roll Angle   7.83E+01
    Normal Acceleration   4.65E+00      Lateral Accln.  -5.02E-06
       Dynamic Pressure   3.00E+02        Mach Number   4.50E-01

Initial Cost Function 1.85E+01,     Final Cost Fn.   3.98E-09
? More Iterations (def = Y) : N

? Enter "M" to modify this trim
        "R" to restart
        "/" to file data/quit : /

? Name of Output File (def= None) : /

? Enter "M" for Menu
        "/" to quit    : /
```

Figure 3.6-2 Terminal display for trim.

acceptable, to reinitialize the step size of the minimization algorithm before each new set of iterations. More trim iterations were later performed on this example and the cost function reached a lower limit of 5.52E-13 (the trim program and model use only single-precision arithmetic); no significant changes occurred in the numerical values given above. The final state and control vectors placed in the output file were as follows:

$$X_1 = 5.020000\ E+02,\quad X_2 = 2.392628\ E-01,\quad X_3 = 5.061803\ E-04,$$
$$X_4 = 1.366289\ E+00,\quad X_5 = 5.000808\ E-02,\quad X_6 = 2.340769\ E-01,$$
$$X_7 = -1.499617\ E-02,\quad X_8 = 2.933811\ E-01,\quad X_9 = 6.084932\ E-02,$$
$$X_{10} = 0.000000\ E+00,\quad X_{11} = 0.000000\ E+00,\ X_{12} = 0.000000\ E+00,$$
$$X_{13} = 6.412363\ E+01,$$

$$U_1 = 8.349601 E - 01,\quad U_2 = -1.481766 E + 00,\quad U_3 = 9.553108 E - 02,$$
$$U_4 = -4.118124 E - 01$$

This trim will be used for a flight simulation example in a following subsection, and in Section 3.7 to illustrate coupling effects in the aircraft dynamics. ■

Trimmed Conditions for Studying Aircraft Dynamics

The steady-state performance of an airplane can be investigated very thoroughly from a set of trimmed flight conditions. The specific fuel consumption, rate of climb, various critical speeds for takeoff and landing, radius of turn, and so on, can all be determined for a number of different flight conditions. We have not provided enough modeling detail for all of these investigations, but the model and the trim program could be further developed if required.

Here we will examine the trimmed level flight conditions over a range of speed. The F-16 is balanced to minimize trim drag, and for straight and level flight across the speed range of our model, the change in the trimmed elevator deflection is very small and varies erratically. At very low speeds, and therefore low dynamic pressure, a high value of the lift coefficient is needed to support the aircraft weight. This causes high induced drag, and because of the large angle of attack, the engine thrust must support a large component of the aircraft weight. Therefore, the throttle setting must increase at low speeds. The throttle setting also increases as transonic speeds are approached because of the increasing drag, and thus the throttle setting versus speed curve must pass through a minimum.

Data for trimmed level flight at sea level, with the nominal cg position, are given in Table 3.6-2. As the speed is lowered, the angle of attack increases, the leading-edge flap reaches its limit (at about $\alpha = 18$ deg, although no longer visible in the data), and the trimmed throttle setting begins to increase from its very low value. The model can be trimmed until alpha reaches about 45 degrees, when a rapid increase in trimmed elevator deflection occurs, quickly reaching the deflection limit.

TABLE 3.6-2: Trim Data for the F-16 Model

speed	130	140	150	170	200	260	300	350	400
thrtl.	0.816	0.736	0.619	0.464	0.287	0.148	0.122	0.107	0.108
AOA	45.6	40.3	34.6	27.2	19.7	11.6	8.49	5.87	4.16
elev.	20.1	−1.36	0.173	0.621	0.723	−0.090	−0.591	−0.539	−0.591

speed	440	500	540	600	640	700	800	ft/s
thrtl.	0.113	0.137	0.160	0.200	0.230	0.282	0.378	per unit
AOA	3.19	2.14	1.63	1.04	0.742	0.382	−0.045	degrees
elev.	−0.671	−0.756	−0.798	−0.846	−0.871	−0.900	−0.943	degrees

Figure 3.6-3 shows throttle setting plotted against airspeed. This curve is not the same as the airplane "power-required" curve because the engine characteristics are also included in it. Nevertheless, we shall loosely refer to it as the *power curve*. It shows clearly the minimum throttle setting. For a propeller-driven plane this is the condition for *best endurance* (but not best range) at the given altitude. For a jet plane

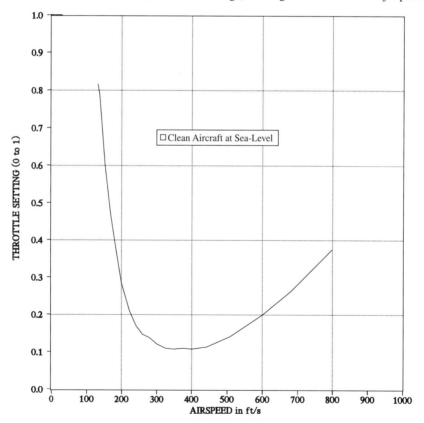

Figure 3.6-3 F-16 model trimmed power curve.

TABLE 3.6-3: Trimmed Flight Conditions for the F-16

Nom. Condition: $h = 0$ ft, $\bar{q} = 300$ psf, $Xcg = .35\bar{c}$, $\dot{\phi} = \dot{\theta} = \dot{\psi} = \gamma = 0$

		CONDITION			
Variable	Nominal	$Xcg = 0.3\bar{c}$	$Xcg = +0.38\bar{c}$	$Xcg = +0.3\bar{c}$ $\dot{\psi} = 0.3$ r/s	$Xcg = -0.3\bar{c}$ $\dot{\theta} = 0.3$ r/s
V_T (ft/s)	502.0	502.0	502.0	502.0	502.0
α (rad)	0.03691	0.03936	0.03544	0.2485	0.3006
β (rad)	−4.0E-9	4.1E-9	3.1E-8	4.8E-4	4.1E-5
ϕ (rad)	0	0	0	1.367	0
θ (rad)	0.03691	0.03936	0.03544	0.05185	0.3006
P (r/s)	0	0	0	−0.01555	0
Q (r/s)	0	0	0	0.2934	0.3000
R (r/s)	0	0	0	0.06071	0
THTL(0-1)	0.1385	0.1485	0.1325	0.8499	1.023
EL (deg)	−0.7588	−1.931	−0.05590	−6.256	−7.082
AIL(deg)	−1.2E-7	−7.0E-8	−5.1E-7	0.09891	−6.2E-4
RDR(deg)	6.2E-7	8.3E-7	4.3E-6	−0.4218	0.01655

the fuel consumption is more strongly related to thrust than power, so this is no longer true. For more details on the static performance information that can be derived from a power-available curve, see Dommasch et al. (1967).

The region to the left of the minimum of the power-required curve is known as the *back side of the power curve*. If the aircraft is operating on the back side of the power curve, opening the throttle produces an increase in altitude, *not* an increase in speed. The speed is then controlled by the elevator. This region of operation may be encountered in the landing phase of flight (e.g., carrier landings).

Table 3.6-3 presents another set of trimmed conditions for the F-16 model; these will be used for the simulation examples in this chapter and for controller design in subsequent chapters. The F-16 model aerodynamic data were referenced to the $0.35\bar{c}$ x-position, and this is the "nominal" position for the cg. The nominal speed and altitude were chosen to give a representative flight condition suitable for later examples and designs. The table contains data for the nominal condition, a forward-cg condition, an aft-cg condition, and steady-state turn and pull-up conditions with a forward cg. The forward and aft-cg cases have been included for a later demonstration of the effect of cg position on stability. A forward-cg location has been used for the two maneuvering cases so that the effects of the maneuver can be illustrated without the additional complication of unstable dynamics.

Flight Simulation Examples

Here we give two flight simulation examples using the MATLAB simulation program from Section 3.4 with the transport aircraft model, and one example using the F-16 model with a Fortran version of the simulation program.

198 MODELING, DESIGN, AND SIMULATION TOOLS

Example 3.6-3: Simulated Response to an Elevator Pulse. The transport aircraft model was trimmed for level flight in the "clean" condition at sea level, with $xcg = 0.25$ and a true airspeed of 250 ft/s, using the trim program given in this section. The state and control vectors were:

$$U^T = [0.1845 \quad -9.2184]; \qquad X^T = [250 \quad 0.16192 \quad 0.16192 \quad 0 \quad 0]$$

A time-history simulation was performed using the program NLSIM.m, as given in Example 3.4-1, and with the above initial conditions. RK4 integration with a step-size of 20 ms was used. An elevator *doublet* pulse of 2 degrees from 1 to 1.5 s and -2 degrees from 1.5 to 2 s was superimposed on the trimmed elevator deflection, using the code that was shown disabled in Example 3.4-1. A doublet is bidirectional with a mean value of zero, and is intended to restore the original flight conditions when it ends.

Figure 3.6-4 shows the pitch attitude and angle-of-attack responses to the elevator doublet. The initial pitch responses do not match (in shape or duration) the elevator disturbance that caused them. Instead, the responses are characteristic of the aircraft and represent a *natural mode* of the aircraft dynamics, in which alpha and theta vary

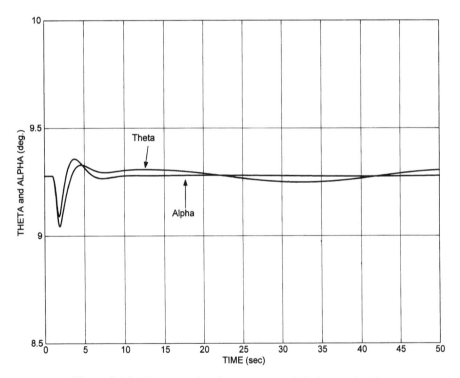

Figure 3.6-4 Transport aircraft response to a $\pm 2°$ elevator doublet.

together, thus causing very little change in the flight-path angle. This mode is known as the *short-period mode*. If we inspect the other longitudinal variables, we will find that airspeed and altitude are almost constant, and only alpha, theta, and pitch-rate vary. When the short-period response dies out, at about 10 s, alpha becomes constant and pitch-rate becomes zero. There remains a small amplitude, very lightly damped oscillation in which the aircraft gains altitude, with increasing pitch-attitude and a positive flight-path angle, and decreasing speed, and then reverses this motion. This is the *phugoid mode* of an aircraft. The short-duration elevator doublet may cause very little excitation of the phugoid mode if that mode is better damped than is the case here. ∎

Example 3.6-4: Simulated Response to a Throttle Pulse. In this example we will use the transport aircraft with the same trim conditions as Example 3.6-3, and superimpose a doublet pulse on the steady-state throttle setting. The doublet will have the value 0.1 from 1 to 4 s, and -0.1 from 4 to 7 s. Figure 3.6-5 shows the response. The angle of attack is barely affected, but the pitch attitude exhibits the phugoid oscillation that was observed in Example 3.6-3. An examination of the speed, altitude, and flight-path angle variables shows that they vary in unison with

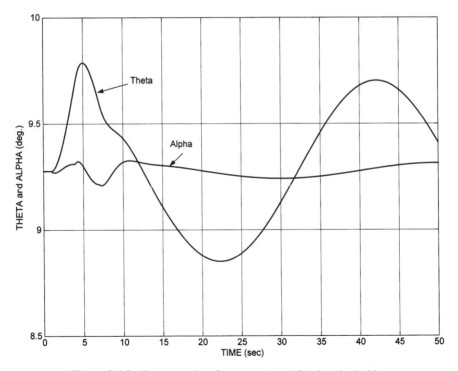

Figure 3.6-5 Transport aircraft response to a ± 0.1 throttle doublet.

theta. Therefore, we conclude that the thrust disturbance has excited the phugoid mode, with very little effect on the short-period mode. ∎

Example 3.6-5: Simulation of a Coordinated Turn. This example is a time-history simulation of a steady-state coordinated turn, using the F-16 model with the trim data from Example 3.6-2. The simulation data from the TRESP program are presented in Figure 3.6-6. The aircraft is turning at 0.3 rad/s, and therefore turns through 54 rad or about 8.6 revolutions in the 180s simulation.

Figure 3.6-7 shows the ground track of the aircraft and shows that the eight circles fall exactly over each other. In Section 3.8 we will see that the aircraft dynamics have quite a wide spread of time constants, and in this flight condition, there is an unstable mode with a time constant of about 1.7 s. Unless the integration time-step is reduced below about 0.02 s, the fourth-order Runge-Kutta routine eventually diverges when integrating this example.

```
Initial condition filename ?   (def.= none): EX362

Number of states and outputs to be recorded ?
# States (def.= 0) : 2
Which ones ? : 10   11
# Outputs (def.= 0) : /
Variable (V) or fixed (F) step integration ? (def= F) : /
Length of run (sec) ? : 180
Print time-interval (on screen) ? : 10
Plotting time-interval ? : .5
Sample period (integration step) ? : .01

    TIME        X-10        X-11
 0.00E+00    0.00E+00    0.00E+00
 1.00E+01    2.36E+02    3.33E+03
 2.00E+01   -4.68E+02    6.65E+01
 3.00E+01    6.90E+02    3.20E+03
 4.00E+01   -8.97E+02    2.61E+02
 5.00E+01    1.09E+03    2.94E+03
 6.00E+01   -1.26E+03    5.68E+02
 7.00E+01    1.40E+03    2.59E+03
 8.00E+01   -1.51E+03    9.62E+02
 9.00E+01    1.60E+03    2.16E+03
 1.00E+02   -1.65E+03    1.41E+03
 1.10E+02    1.67E+03    1.70E+03
 1.20E+02   -1.66E+03    1.89E+03
 1.30E+02    1.61E+03    1.22E+03
 1.40E+02   -1.53E+03    2.35E+03
 1.50E+02    1.42E+03    7.87E+02
 1.60E+02   -1.28E+03    2.76E+03
 1.70E+02    1.11E+03    4.22E+02
 1.80E+02   -9.21E+02    3.07E+03

 Enter   0 to file data              1 to quit
         2 to restart                3 to pick other states
         4 to change integration     5 to change run time   :
```

Figure 3.6-6   Simulation results for F-16 model.

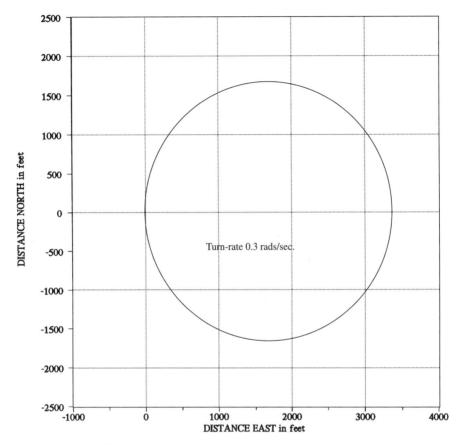

Figure 3.6-7 The ground track of a coordinated turn. ■

The foregoing examples have illustrated digital simulation using nonlinear continuous time dynamic equations, with control inputs applied in discrete time (i.e., changing only at the sampling instants). In the next section we will derive linear dynamic equations; these offer no advantages for simulation, but do allow a variety of analytical tools to be applied to the dynamics.

3.7 NUMERICAL LINEARIZATION

Theory of Linearization

In Section 2.6 we linearized the aircraft implicit nonlinear state equations algebraically, and obtained LTI state equations corresponding to a given flight condition. This linearization was specific to aircraft equations, and was only tractable under the restrictions of wings-level, non-sideslipping steady-state flight. Now we will introduce a numerical linearization algorithm that can be applied to any nonlinear model in the same explicit state-space form that was used with numerical integration.

A multivariate Taylor-series expansion of the explicit state equations (3.4-1b), around a point (X_e, U_e), gives

$$\dot{X} + \delta\dot{X} = f(X_e, U_e) + \frac{\partial f}{\partial X}\delta X + \frac{\partial f}{\partial U}\delta U + \text{h.o.t.},$$

where the partial derivative terms denote Jacobian matrices (as in Section 2.6), and the perturbations

$$\delta X \equiv (X - X_e), \qquad \delta U \equiv (U - U_e)$$

are "small." In the series h.o.t. denotes higher-order terms, which will be neglected. If X_e and U_e are equilibrium solutions, obtained from the trim program, then

$$0 = \dot{X} = f(X_e, U_e)$$

and so

$$\delta\dot{X} = \frac{\partial f}{\partial X}\delta X + \frac{\partial f}{\partial U}\delta U \qquad (3.7\text{-}1)$$

This equation is in the form of the LTI state equation,

$$\dot{x} = Ax + Bu, \qquad (3.7\text{-}2)$$

where the lowercase symbols denote perturbations from the equilibrium, but $\dot{x}$ is the actual value of the derivative vector.

The method of estimating the first partial derivatives, which make up the Jacobian matrices, will be illustrated with a function of a single variable $z = g(v)$. Using Taylor series expansions of g, around $v = v_e$, we obtain

$$z_1 \equiv g(v_e + h) = g(v_e) + h\frac{\partial g}{\partial v}(v_e) + \frac{h^2}{2!}\frac{\partial^2 g}{\partial v^2}(v_e) + \text{h.o.t.}$$

$$z_{-1} \equiv g(v_e - h) = g(v_e) - h\frac{\partial g}{\partial v}(v_e) + \ldots$$

Then it is easy to see that

$$\left.\frac{\partial g}{\partial v}\right|_{v=v_e} = \frac{z_1 - z_{-1}}{2h} - \frac{h^2}{3!}\frac{\partial^3 g}{\partial v^3}(v_e) - \text{h.o.t.} \qquad (3.7\text{-}3)$$

and neglecting terms of order h^2 and higher leaves a very simple approximation for the first partial derivative.

A higher-order approximation can be found by writing the Taylor series for

$$z_2 = g(v_e + 2h)$$

and

$$z_{-2} = g(v_e - 2h)$$

It can then be shown that

$$\left. \frac{\partial g}{\partial v} \right|_{v=v_e} = \frac{8(z_1 - z_{-1}) - (z_2 - z_{-2})}{12h} + O\left(h^4\right) \qquad (3.7\text{-}4)$$

Therefore, by using four values of the function g, we can obtain an estimate of the first partial derivative that includes Taylor-series terms through h^3.

Algorithm and Examples

When turning the formulae for the partial derivatives into a numerical algorithm one must determine what size of perturbation can be considered "small" in Equation (3.7-1). The perturbations may often be around an equilibrium value of zero, so it is not always possible to choose some fraction of the equilibrium value. Instead, one can start with a fairly arbitrary initial perturbation and progressively reduce it until the algorithm obtained from (3.7-3) or (3.7-4) converges on some value for the derivative. Figure 3.7-1 shows a flowchart for numerical linearization, and a simple MATLAB program is given below.

```
% File LINZE.m
clear all
name = input('Enter Name of State Eqns. File : ','s');
tfile= input('Enter Name of Trim File : ','s');
tmp= dlmread(tfile,',');
n=tmp(1); m=tmp(2); x=tmp(3:n+2);
u=tmp(n+3:m+n+2); tol=1e-6; time=0.;
mm= input('Number of control inputs to be used ? : ');
dx=0.1*x;
for i=1:n              % Set Perturbations
    if dx(i)==0.0;
        dx(i)=0.1;
    end
end
last=zeros(n,1); a=zeros(n,n);
for j=1:n
    xt=x;
    for i=1:10
        xt(j)=x(j)+dx(j);
        xd1= feval (name,time,xt,u);
        xt(j)=x(j)-dx(j);
        xd2= feval (name,time,xt,u);
```

```
      a(:,j)= (xd1-xd2)'/(2*dx(j));
      if max( abs(a(:,j)-last)./abs( a(:,j) + 1e-12 ) )<tol;
         break
      end
      dx(j)= 0.5*dx(j);
      last = a(:,j);
   end
   %column=j
   iteration=i;
   if iteration==10
      disp('not converged on A, column',num2str(j))
   end
end
dlmwrite('A.dat',a);
```

This program computes one column of A at a time; a fractional error is computed for each element of the column array, and the largest fractional error must satisfy a convergence tolerance. If the algorithm has not converged after dividing the initial perturbation by 2^9, the user is informed and must deal with the problem by increasing the tolerance, or linearizing at a slightly different flight condition. To save space the

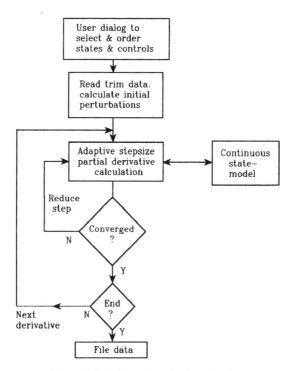

Figure 3.7-1 Flowchart for linearization.

calculation of the B matrix is not shown; note that not all of the control variables should be perturbed (e.g. the landing- gear switch) so the user must enter a value for the variable *mm*.

For a differentiable function, perturbations that are too large produce a (Taylor-series) truncation error, while perturbations that are too small cause a round-off error due to the finite precision computer arithmetic. Many simulations are written with single-precision arithmetic, and round-off error will then be much more significant than with MATLAB double-precision arithmetic. Difficulties are also caused by discontinuities in the function or its derivative. For example, if a simulation variable has reached a "hard" limit, this will cause a discontinuity in its derivative. A more sophisticated linearization algorithm may be designed to find an optimal-size perturbation, and to detect discontinuities (see, for example, Taylor and Antoniotti, 1993). The student should add the calculations for the $B, C,$ and D matrices to the above program (see Problem 3.7-2). A linearization program "Jacob," for the Fortran model, is given in Appendix B.

Example 3.7-1: Comparison of Algebraic and Numerical Linearization.

This example uses the transport-aircraft longitudinal model of Section 3.5. The model contains an alpha-dot contribution to the pitching moment, the thrust vector is offset from the cg by the amount ZE, and the engine thrust varies with speed. It therefore provides a good check on the results of the algebraic linearization in Section 2.6.

A short program was written to evaluate the longitudinal state-equations coefficient matrices (2.6-29), using the formulae in Tables 2.6-1 to 2.6-3 (see Problem 3.7-3). The program contains the dimensionless stability derivatives given in the transport aircraft model and reads the steady-state trim data from a data file. It calculates the A and B matrices in (2.6-30) and then premultiplies them by the inverse of the E matrix. The new A and B matrices were printed out for comparison with numerical linearization results.

The model was trimmed in the clean condition at a large angle of attack, and in climbing flight, so that $\sin \alpha$ and $\sin \gamma$ terms contributed significantly to the results. The trim condition was $cg = 0.25\bar{c}, h = 0$ ft, $V_T = 200$ ft/s, and $\gamma = 15°$. This condition required an angle of attack of 13.9 degrees and a throttle setting of 1.01 (i.e., slightly beyond maximum power!). The algebraic linearization program gave

$$E^{-1}A = \begin{bmatrix} v_T & \alpha & \theta & q \\ -2.7337E-02 & 1.6853E+01 & -3.1074E+01 & 0.0000E+00 \\ -1.4167E-03 & -5.1234E-01 & -4.1631E-02 & 1.0000E+00 \\ 0.0000E+00 & 0.0000E+00 & 0.0000E+00 & 1.0000E+00 \\ -1.1415E-04 & -4.9581E-01 & 4.8119E-03 & -4.2381E-01 \end{bmatrix}$$

$$E^{-1}B = \begin{bmatrix} \delta_t & \delta_e \\ 1.0173E+01 & 0.0000E+00 \\ -1.2596E-02 & 0.0000E+00 \\ 0.0000E+00 & 0.0000E+00 \\ 2.7017E-02 & -7.0452E-03 \end{bmatrix}$$

The algebraic linearization did not account for an altitude state and the consequent coupling of the equations through the atmosphere model. Therefore, only the first four states were selected when the numerical linearization was performed. The numerical linearization produced the following results:

$$A = \begin{bmatrix} -2.7337E-02 & 1.6852E+01 & -3.1073E+01 & 0.0000E+00 \\ -1.4168E-03 & -5.1232E-01 & -4.1630E-02 & 1.0000E+00 \\ 0.0000E+00 & 0.0000E+00 & 0.0000E+00 & 1.0000E+00 \\ -1.1415E-04 & -4.9583E-01 & 4.8118E-03 & -4.2381E-01 \end{bmatrix}$$

$$B = \begin{bmatrix} 1.0173E+01 & 0.0000E+00 \\ -1.2596E-02 & 0.0000E+00 \\ 0.0000E+00 & 0.0000E+00 \\ 2.7017E-02 & -7.0452E-03 \end{bmatrix}$$

These results are in very close agreement with the algebraic linearization results; the largest discrepancy is a difference of 2 in the fifth digit. ∎

The stability derivatives and numerical linearization play complementary roles. The stability derivatives are useful for preliminary design; they can be estimated from a geometrical description of an aircraft, and can be used to calculate the modes and stability of the aircraft. When an aircraft configuration has been chosen, and enough data have been gathered to build a mathematical model, numerical linearization can be used to perform control systems design and to obtain a dynamic description of the aircraft in other than wings-level non-sideslipping flight.

Example 3.7-2: Linearization of the F-16 Model. In Chapter 2 it was shown that under the conditions of small perturbations from steady-state, wings-level, non-sideslipping flight, the rigid-aircraft equations of motion could be split into two uncoupled sets. These were the longitudinal equations that involve the variables speed, alpha, pitch attitude, and pitch rate, and the lateral-directional equations that involve beta, roll angle, and roll and yaw rates. The Jacob program makes it easy to demonstrate this decoupling and to show that coupling occurs when the sideslip and roll angles are nonzero. A good example is provided by two steady-state flight conditions that differ only in terms of roll angle. In Table 3.6-3, the wings-level pull-up with $\dot{\theta} = 0.3$ rad/s and the coordinated turn at 0.3 rad/s are both at 300 psf dynamic pressure with zero sideslip and similar angles of attack (and almost identical normal acceleration). The Jacobian matrices for these two steady-state conditions will now be compared.

The Fortran linearization program makes provision for reordering states and for choosing subsets of states. The north and east geographic position states, and the geographic heading state ψ have no effect on the dynamic behavior; their derivatives are a function of the other states, but these states themselves are not coupled back into the state equations. Also, the altitude state only enters the aircraft equations through the atmosphere model and dynamic pressure, and in this case it has negligible coupling to the other states. Therefore, we will not select these state variables for the LTI model.

To illustrate decoupling, Jacob was used to reorder the remaining nine states into longitudinal states v_T, α, θ, q, and the engine power state (POW), followed by lateral directional states β, ϕ, p, r. The inputs were ordered as $\delta_t, \delta_e, \delta_a, \delta_r$, and the outputs as a_n, q, and α. The results for the steady-state pull-up are shown in Figure 3.7-2.

$$A = \left[\begin{array}{ccccc:cccc}
 & v_T & \alpha & \theta & q & \text{pow} & \beta & \phi & p & r \\
-.127 & -235 & -32.2 & -9.51 & .314 & -.0028 & .00126 & 5\text{E}-5 & 2\text{E}-4 \\
-7\text{E}-4 & -.969 & 0 & .908 & -2\text{E}-4 & 1.5\text{E}-5 & 0 & -4\text{E}-5 & -1\text{E}-5 \\
0 & 0 & 0 & 1 & 0 & 0 & 0 & 0 & 0 \\
9\text{E}-4 & -4.56 & 0 & -1.58 & 0 & 9.2\text{E}-5 & 0 & 0 & -.00287 \\
0 & 0 & 0 & 0 & -5.00 & 0 & 0 & 0 & 0 \\
\hdashline
1\text{E}-8 & 2\text{E}-5 & 3\text{E}-6 & 8\text{E}-7 & -3\text{E}-8 & -.322 & .0612 & .298 & -.948 \\
0 & 0 & 0 & 0 & 0 & 0 & .0930 & 1.00 & .310 \\
-3\text{E}-7 & -.00248 & 0 & 3\text{E}-4 & 0 & -62.5 & 0 & -3.00 & 1.99 \\
-3\text{E}-6 & -.00188 & 0 & .00254 & 0 & 7.67 & 0 & -.262 & -.629
\end{array}\right]$$

$$B = \left[\begin{array}{cc:cc}
\delta_t & \delta_e & \delta_a & \delta_r \\
0 & -.244 & 6\text{E}-6 & 2\text{E}-5 \\
0 & -.00209 & 0 & 0 \\
0 & 0 & 0 & 0 \\
0 & -.199 & 0 & 0 \\
1087 & 0 & 0 & 0 \\
\hdashline
0 & 2\text{E}-8 & 3\text{E}-4 & 8\text{E}-4 \\
0 & 0 & 0 & 0 \\
0 & 0 & -.645 & .126 \\
0 & 0 & -.0180 & -.0657
\end{array}\right] \qquad D = \begin{bmatrix} 0 & .0333 & 0 & 0 \\ 0 & 0 & 0 & 0 \\ 0 & 0 & 0 & 0 \end{bmatrix}$$

$$C = \begin{bmatrix} .0208 & 15.2 & 0 & 1.45 & 0 & -4.5\text{E}-4 & 0 & 0 & 0 \\ 0 & 0 & 0 & 1.00 & 0 & 0 & 0 & 0 & 0 \\ 0 & 57.3 & 0 & 0 & 0 & 0 & 0 & 0 & 0 \end{bmatrix} \begin{array}{c} a_n \\ q \\ \alpha \end{array}$$

Figure 3.7-2 Jacobian matrices for a pull-up.

$$A = \begin{bmatrix} -.090 & -169 & -31.2 & -7.75 & .318 & : & 31.4 & -7.73 & 5\text{E}-4 & 2\text{E}-3 \\ -5\text{E}-4 & -1.05 & .0151 & .903 & -2\text{E}-4 & : & 3\text{E}-4 & -.0607 & -5\text{E}-4 & -1\text{E}-4 \\ 0 & 0 & 0 & .203 & 0 & : & 0 & -.300 & 0 & -.979 \\ 1\text{E}-3 & 1.26 & 0 & -1.66 & 0 & : & 1\text{E}-3 & 0 & .0589 & -.0157 \\ 0 & 0 & 0 & 0 & -5.00 & : & 0 & 0 & 0 & 0 \\ \hdashline -1\text{E}-4 & 1.4\text{E}-4 & -.0032 & 7\text{E}-6 & -3\text{E}-7 & : & -.322 & .0130 & .248 & -.961 \\ 0 & 0 & .300 & .0508 & 0 & : & 0 & 0 & 1.00 & .0105 \\ -3\text{E}-4 & .0578 & 0 & -.0469 & 0 & : & -59.4 & 0 & -3.19 & 1.64 \\ 5\text{E}-5 & -.0617 & 0 & .0123 & 0 & : & 8.88 & 0 & -.299 & -.564 \end{bmatrix}$$

Figure 3.7-3 Jacobian matrix for $4.5g$ turn.

These results are rounded to three significant digits, except for numbers less than 0.001, which are rounded to only one significant digit. The A and B matrices have been partitioned to separate the longitudinal and lateral states and controls, and it is evident that the expected decoupling does indeed exist. In the A matrix, the exact relationship $\dot{\theta} = q$, when $\phi = 0$, is evident on the third row. There is a small amount of coupling of $\dot{V}_T$ to β and ϕ, and of $\dot{q}$ to r, but the other terms are one to two orders of magnitude smaller. The B matrix shows that the aileron and rudder have essentially no effect on the longitudinal states, and the throttle and elevator have no effect on the lateral/directional states.

The D matrix has a nonzero entry, corresponding to the elevator to normal acceleration transfer function, because accelerations are directly coupled to forces and therefore to control surface deflection. Other expected features are the reciprocal of the engine time constant (at full power) as the only nonzero entry on the fifth row of the A matrix, and the value of g appearing in the (1, 3) position. The reader should compare these results with the state equations derived in Chapter 2 and determine the significance of the numerical values.

If we next use the Jacobian program to determine the A matrix for the $4.5g$ coordinated turn, we obtain the matrix shown in Figure 3.7-3.

A comparison of this matrix with the previous A matrix shows that several strong coupling terms have now appeared in the upper right block. Less pronounced coupling has appeared in the lower left block. These couplings can be understood by referring to the nonlinear equations (2.5-19), the nonlinear moment equations, and the Euler angle equations (1.3-22a), with ϕ nonzero. In the next section we develop the tools to determine how the dynamic behavior is changed by this coupling, and in Chapter 4 we consider the implications of these changes in the dynamics. ■

3.8 AIRCRAFT DYNAMIC BEHAVIOR

In Examples 3.6-3 and 3.6-4 two different characteristic modes of the aircraft dynamics were excited separately, by applying different inputs. One of the variables (θ) observed in the simulation was found to be involved in both modes, the other observed variable (α) was essentially involved in only one. In this section we will study the dynamic behavior analytically through modal decomposition and through analysis of the transfer functions.

Modal Decomposition Applied to Aircraft Dynamics

In Section 3.6 the classical phugoid and short-period aircraft modes were illustrated by nonlinear simulation. The complete set of modes of a conventional aircraft will now be illustrated by modal decomposition using a linear F-16 model. The second set of trim conditions in Table 3.6-3 will be used, that is, straight and level flight with stable dynamics. A Jacobian A matrix must first be found for this flight condition, and not all of the thirteen states in the full A matrix will be needed.

Once again we will drop the north and east geographic position states and the geographic heading state, which do not affect the dynamics. These states correspond to the integrals of linear combinations of other states, and if retained in the A matrix, will produce zero eigenvalues (poles at the origin). The engine power state couples into the dynamics (through V_T) but is not influenced by any other states. Left in the A matrix, it will produce an eigenvalue of -1.0, corresponding to the reciprocal of the 1 s engine time constant, and again does not affect the aircraft modes.

There is also clear decoupling of the lateral and longitudinal dynamics in this flight condition. Therefore, the modal decomposition will be demonstrated using two separate reduced Jacobian matrices. Note that the method of deriving the A matrix by perturbing the state variables assumes that the control inputs are constant. Therefore, the modes derived in the analysis are "stick-fixed" modes, that is, the control surfaces are implicitly assumed to be locked in position. This assumption will hold most accurately for fully powered (as opposed to power-boosted or unpowered) control surfaces; these control systems are called *irreversible*.

Example 3.8-1: F-16 Longitudinal Modes. The IMSL eigenvalue subroutine EIGRF (IMSL, 1980), with double precision, was used to produce the following results. Other sources of eigenvalue/eigenvector routines are readily available (Press et al., 1986; MATLAB, 1990). A simple driver program was written, and each pair of eigenvectors was normalized by dividing all elements by the complex number corresponding to the element of greatest magnitude. The longitudinal-dynamics Jacobian matrix for the F-16 model in straight and level flight at 502.0 ft/s with a cg position of $0.3\bar{c}$ is given by

$$A = \begin{bmatrix} v_T & \alpha & \theta & q \\ -2.0244E-02 & 7.8763E+00 & -3.2170E+01 & -6.5020E-01 \\ -2.5372E-04 & -1.0190E+00 & 0.0000E+00 & 9.0484E-01 \\ 0.0000E+00 & 0.0000E+00 & 0.0000E+00 & 1.0000E+00 \\ 7.9472E-11 & -2.4982E+00 & 0.0000E+00 & -1.3861E+00 \end{bmatrix}$$

The four states give rise to two complex-conjugate pairs of eigenvalues, which correspond to two stable oscillatory modes. The eigenvalues are:

$-1.2039 \pm j1.4922$ (short-period mode, $T = 4.21s$, $\zeta = 0.628$)

$-0.0087297 \pm j0.073966$ (phugoid mode, $T = 84.9s$, $\zeta = 0.117$)

The periods of these modes are separated by more than an order of magnitude, so they are easily identifiable as the short-period and phugoid modes. The phugoid mode is very lightly damped ($\zeta = 0.117$), but its period is so long that a pilot would have no difficulty in damping out a phugoid oscillation. The short-period mode is reasonably well damped in this particular flight condition, and the aircraft response to elevator commands would be acceptable to the pilot.

The corresponding eigenvectors are two complex conjugate pairs given by:

$$\begin{array}{cc} \text{short-period} & \text{phugoid} \end{array}$$

$$\begin{bmatrix} 1.0 \\ 0.090 \\ 0.059 \\ 0.0092 \end{bmatrix} \pm j \begin{bmatrix} 0 \\ 0.017 \\ 0.054 \\ 0.15 \end{bmatrix} \quad \begin{bmatrix} 1.0 \\ -9.6E-5 \\ -3.8E-4 \\ 1.7E-4 \end{bmatrix} \pm j \begin{bmatrix} 0 \\ 5.0E-7 \\ 2.3E-3 \\ 8.4E-6 \end{bmatrix} \begin{array}{c} v_T \\ \alpha \\ \theta \\ q \end{array}$$

Both pairs of eigenvectors are dominated by the element corresponding to airspeed and the relative involvement of the other variables is difficult to assess. Nevertheless, the results show that the variables α and q are involved relatively weakly in the phugoid mode as compared to the short period, and the phugoid mode involves mostly V_T and θ. This agrees with the conclusions drawn from the nonlinear simulation examples.

The relative involvement of different variables in the dynamic modes can be determined more precisely if the dynamic equations are made dimensionless, so that the eigenvectors are also dimensionless. This requires the introduction of time scaling (Etkin, 1972). Additional information can be extracted from the eigenvectors if they are plotted in the complex plane so that their phase relationships can be observed (Etkin, 1972). ■

Example 3.8-2: F-16 Lateral/Directional Modes. The Jacobian matrix for the lateral/directional dynamics of the F-16 model, in straight and level flight at 502.0 ft/s, with a cg position of $0.3\bar{c}$, is given by

$$\begin{array}{cccc} \beta & \phi & p & r \end{array}$$

$$A = \begin{bmatrix} -3.2200E-01 & 6.4032E-02 & 3.8904E-02 & -9.9156E-01 \\ 0.0000E+00 & 0.0000E+00 & 1.0000E+00 & 3.9385E-02 \\ -3.0919E+01 & 0.0000E+00 & -3.6730E+00 & 6.7425E-01 \\ 9.4724E+00 & 0.0000E+00 & -2.6358E-02 & -4.9849E-01 \end{bmatrix}$$

This time there are two real eigenvalues and a complex-conjugate pair. They are:

$-0.4399 \pm j3.220$ (dutch roll mode, $T = 1.95s$, $\zeta = 0.135$)
-3.601 (roll subsidence mode, $\tau = 0.28$ s)
-0.0128 (spiral mode, $\tau = 77.9$ s)

The eigenvectors are:

$$\underset{Dutch\ Roll\ Mode}{\begin{bmatrix} -0.11 \\ -0.037 \\ 1.0 \\ -0.29 \end{bmatrix} \pm j \begin{bmatrix} -0.097 \\ -0.30 \\ 0 \\ 0.33 \end{bmatrix}} \quad \underset{Roll\ Mode}{\begin{bmatrix} -0.0020 \\ -0.28 \\ 1 \\ 0.015 \end{bmatrix}} \quad \underset{Spiral\ Mode}{\begin{bmatrix} 0.0032 \\ 1 \\ -0.015 \\ 0.063 \end{bmatrix}} \quad \begin{array}{c} \beta \\ \phi \\ p \\ r \end{array}$$

The oscillatory mode involves all of the variables, and is a rolling and yawing motion with some sideslipping. This motion has been likened to the motion of a drunken skater, and is called the *dutch roll mode*. The aircraft rudder produces both rolling and yawing moments, and a rudder pulse will excite this mode. The eigenvalues show that the dutch-roll period is quite short ($T = 1.95$ s) and the oscillation is very lightly damped ($\zeta = 0.135$). This would make landing in gusty wind conditions difficult for the pilot, and in a passenger aircraft, passengers sitting near the tail would be very uncomfortable in turbulent conditions.

The second mode is simply a stable exponential mode and clearly involves mostly roll rate and a corresponding roll angle; it is known as the *roll subsidence mode*. The aircraft roll angle response to lateral control inputs is an important part of the handling qualities requirements. This mode, derived from the linear model, will not allow the aircraft maximum roll rate to be calculated but does give a good idea of how quickly the aircraft will start to roll. In this case the time constant of 0.28 s indicates a fast roll response.

The third mode is also a stable exponential mode but is distinguished by a much longer time constant (78 s). It involves more roll angle and yaw rate than the roll mode and is known as the *spiral mode*. The small amount of sideslip shows that the spiral mode can be a coordinated motion. In some aircraft the spiral mode may be unstable, and stability can be built into a design by using wing dihedral (see Chapter 2). An unstable spiral mode can cause an aircraft to get into an ever-steeper, but coordinated, spiral dive. ∎

Interpretation of Aircraft Transfer Functions

We now look at the aircraft dynamic behavior through various transfer functions. More specifically, we will look for pole zero cancellations to determine what modes remain involved in a particular response. We will also look at the frequency response of a transfer function, in order to improve our understanding of the correlation between the frequency and time domains.

212   MODELING, DESIGN, AND SIMULATION TOOLS

Example 3.8-3: F-16 Elevator-to-Pitch-Rate Transfer Function. For this example a full thirteen-state Jacobian A matrix was obtained for the straight and level flight conditions used in Examples 3.8-1 and 3.8-2. The B and C Jacobian matrices were also obtained, and from these matrices the elevator-to-pitch-rate transfer function was selected. For this transfer function the corresponding D-matrix element is zero.

Table 3.8-1 shows the static loop sensitivity and poles and zeros resulting from double-precision computations, rounded to seven digits. The poles and zeros have been ordered to suit the purposes of this example. It is evident from the table that all of the poles, except for two complex pairs and one real pole, have been canceled (to at least six digits). These remaining poles correspond to the phugoid and short-period modes and the altitude pole, as identified in the table. The lateral/directional dynamics are completely unaffected by the elevator input because of the symmetry of the wings-level flight condition. The first three sets of poles are at the origin and represent the integration of velocity components that lead to the north, east, and directional (ψ) states. These are canceled by zeros at the origin. The engine pole is canceled exactly because the engine-lag model is driven only by the throttle input.

The elevator-to-pitch-rate transfer function is

$$\frac{q}{\delta_e} = \frac{-10.45s(s+0.9871)(s+0.02179)}{(s+1.204 \pm j1.492)(s+0.007654 \pm j0.07812)} \frac{\text{deg/s}}{\text{deg}}$$

The phugoid mode has a natural frequency of 0.079 rad/s and a damping ratio of 0.10; the corresponding figures for the short-period mode are 1.9 rad/s and 0.63. This transfer function has a dc gain of zero (because of the zero at the origin), indicating

TABLE 3.8-1: F-16 Model, Elevator-to-Pitch-Rate Transfer Function

Static Loop Sensitivity = -10.453 (deg. units)				
ZEROS		POLES		
Real Part	Imag. Part	Real Part	Imag. Part	
0.0000E+00	0.0000E+00	0.0000E+00	0.0000E+00	N
0.0000E+00	0.0000E+00	0.0000E+00	0.0000E+00	E
0.0000E+00	0.0000E+00	0.0000E+00	0.0000E+00	ψ
4.3987E-01	3.2200E+00	−4.3987E-01	3.2200E+00	dutch
−4.3987E-01	−3.2200E+00	−4.3987E-01	−3.2200E+00	dutch
−3.6009E+00	0.0000E+00	−3.6009E+00	0.0000E+00	roll
−1.2835E-02	0.0000E+00	−1.2835E-02	0.0000E+00	spiral
−8.8010E-04	0.0000E+00	−2.0874E-03	0.0000E+00	Alt.
−1.0000E+00	0.0000E+00	−1.0000E+00	0.0000E+00	Engine
0.0000E+00	0.0000E+00	−1.2040E+00	1.4923E+00	SP
−2.1785E-02	0.0000E+00	−1.2040E+00	−1.4923E+00	SP
−9.8713E-01	0.0000E+00	−7.6538E-03	7.8119E-02	PHUG
		−7.6538E-03	−7.8119E-02	PHUG

that a constant elevator deflection will not sustain a steady pitch rate. If the phugoid poles are canceled with the zero at the origin and the zero at s = -0.02, a *short-period approximation* transfer function is obtained:

$$\frac{q}{\delta_e} = \frac{-10.45(s + 0.9871)}{(s + 1.204 \pm j1.492)} \frac{\deg/s}{\deg}$$

This transfer function has a finite dc gain and shows that constant elevator deflection tends to produce constant pitch rate over an interval of time that is short compared to the phugoid period. The short-period approximation will be used in controller designs in Chapter 4, and its validity will be demonstrated. In the next example the short-period approximation will be examined in the frequency domain. ∎

Example 3.8-4: F-16 Elevator-to-Pitch-Rate Frequency Response. The poles and zeros of the elevator-to-pitch-rate transfer function, given in Example 3.8-3, will now be used to generate the corresponding frequency-response plots. Figure 3.8-1a shows a Bode plot of the magnitude response of both the complete transfer

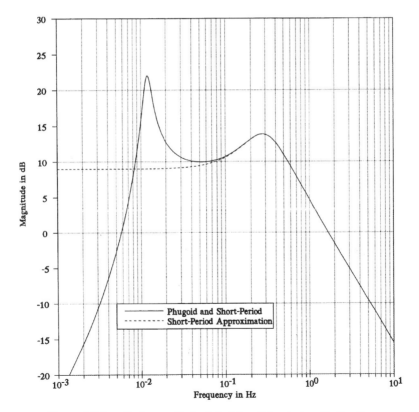

Figure 3.8-1a Bode magnitude plot of the pitch-rate transfer function.

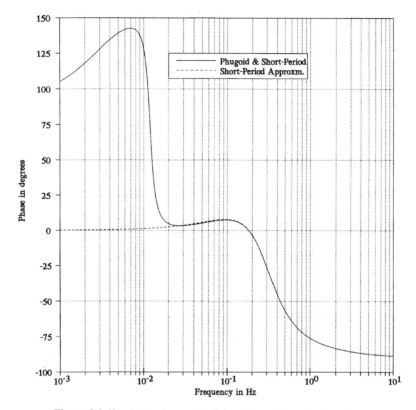

Figure 3.8-1b Bode phase plot of the pitch-rate transfer function.

function and the short-period approximation. The phase plots are shown in Figure 3.8-1b. The magnitude plot shows a large peak in the response at a frequency close to the natural frequency of the lightly damped phugoid mode, and a smaller peak due to the more heavily damped short-period mode.

Both the magnitude and phase plots show that the short-period approximation is a good approximation to the pitch-rate transfer function at frequencies above about 0.03 Hz. The upper cutoff or corner frequency of the short-period transfer function is about 0.8 Hz, and this gives some feel for the speed of response in pitch when different aircraft are compared. Note that the exact phase plot starts at +90 degrees due to the zero at the origin, rises toward 180 degrees because of the additional phase lead of the zero at $s = -0.02$, and then falls back rapidly because of the 180 degree lag effect of the phugoid poles. The zero at $s = -0.99$ causes another small lead effect before the lag of the short-period poles takes over; the high-frequency asymptotic phase shift is -90 degrees because the relative degree of the transfer function is unity. ∎

Example 3.8-5: Transport Aircraft Throttle Response. In this example we examine the throttle-to-speed transfer function for the transport aircraft model in Section 3.5, using the same flight condition as Examples 3.6-3 and 3.6-4. The model was trimmed for level flight at sea level in the clean configuration, with $x_{cg} = 0.25\bar{c}$ and a true airspeed of 250 ft/s, and the following Jacobian matrices were determined:

$$A = \begin{bmatrix} \overset{v_T}{-1.6096E-02} & \overset{\alpha}{1.8832E+01} & \overset{\theta}{-3.2170E+01} & \overset{q}{0.0000E+00} & \overset{h}{5.4000E-05} \\ -1.0189E-03 & -6.3537E-01 & 0.0000E+00 & 1.0000E+00 & 3.7000E-06 \\ 0.0000E+00 & 0.0000E+00 & 0.0000E+00 & 1.0000E+00 & 0.0000E+00 \\ 1.0744E-04 & -7.7544E-01 & 0.0000E+00 & -5.2977E-01 & -4.1000E-07 \\ 0.0000E+00 & -2.5000E+02 & 2.5000E+02 & 0.0000E+00 & 0.0000E+00 \end{bmatrix}$$

$$B = \begin{bmatrix} \overset{\delta_t}{9.9679E+00} \\ -6.5130E-03 \\ 0.0000E+00 \\ 2.5575E-02 \\ 0.0000E+00 \end{bmatrix} \qquad C = [1 \ 0 \ 0 \ 0 \ 0] \quad (v_T)$$

The altitude state h has very small coupling to the other states and was initially neglected. The throttle-to-speed transfer function (with the elevator fixed), as determined from the states v_T, α, θ, q, was found to be

$$\frac{V_T}{\delta_t} = \frac{9.968(s - 0.0601)(s + 0.6065 \pm j0.8811)}{(s + 2.277E - 4 \pm j0.1567)(s + 0.5904 \pm j0.8811)} \tag{1}$$

As expected, this transfer function essentially involves only the phugoid mode, and when the short-period poles are canceled with the nearby complex zeros, we are left with the approximation

$$\frac{V_T}{\delta_t} = \frac{9.968(s - 0.0601)}{(s \mid 2.277E - 4 \perp j0.1567)} \tag{2}$$

The poles and zeros of (2) are quite close to the origin and the relative degree is unity, so throttle inputs are initially integrated. However, the phugoid mode will soon take over and hide this effect under a very lightly damped oscillation in speed. In addition, the NMP zero indicates that there are competing physical mechanisms at work. It may be remembered that the engine thrust line is offset below the cg, and this will cause the aircraft to tend to pitch up and consequently slow down in response to a sudden increase in throttle. Furthermore, at this relatively low speed the aircraft is trimmed with a large amount of "up-elevator," so that any initial increase in speed tends to create an increase in the nose-up pitching moment and again counteract the increase in speed. These facts can be confirmed by changing the engine offset and by trimming the model at higher speeds where less elevator is required. The NMP zero can thus be made to move to the origin and into the left-half plane.

In general, when the throttle is opened, the extra thrust may produce an increase in speed and/or a gain in altitude, and the phugoid mode is associated with the subsequent interchange of potential and kinetic energy. In this case we see that the positive static loop sensitivity, and single NMP zero, correspond to a negative dc gain. Therefore, when the throttle is opened a very lightly damped phugoid oscillation will be initiated, starting with an increase in speed but with a mean value corresponding to a lower speed. The increased thrust will therefore be converted to an increase in altitude. This can be confirmed with a time-history simulation, by applying a step throttle input to the linear model from which transfer function (1) was obtained.

Now we consider a more accurate transfer function model of the aircraft. If the aircraft altitude state is included in the A matrix, it is found that because of the atmosphere model, there are small coupling terms from altitude to several other states. The transfer function corresponding to (1) then becomes

$$\frac{V_T}{\delta_t} = \frac{9.968(s - 0.01506)(s - 0.04528)(s + 0.6066 \pm j0.8814)}{(s + 3.305E - 5)(s + 0.5905 \pm j0.8813)(s + 6.788E - 5 \pm j0.1588)} \quad (3)$$

The very slow altitude pole (at $s = -3E - 5$) has now appeared in the transfer function. An additional NMP zero is also present, and the dc gain of the transfer function is now positive. The physical explanation is that since the decrease in atmospheric density with altitude is now modeled, the tendency to gain altitude is reduced and the speed will now increase in response to a throttle increase.

Simulation results (see Problem 3.8-2) show that for the linear model without the altitude state, the average airspeed (averaged over the phugoid period) decreases in response to a throttle step. When the altitude state is included, the average airspeed decreases at first, and then increases. The altitude increases in either case. The response of the nonlinear model with a relatively small throttle step increment (10 percent increase) agrees closely with the linear model with altitude included. ∎

3.9 FEEDBACK CONTROL

Introduction

A major portion of this book is concerned with performing feedback control design on aircraft dynamics, and in this section we review the relevant classical control theory and design techniques.

In the context of controlling dynamic systems, feedback is defined as returning to the input of a system a signal obtained from its output, as shown in Figure 3.9-1. If the signal is fed back with the intention of canceling the effect of an input that produced it, we have negative feedback as indicated by the minus sign at the *summing junction* in the figure. The system to be controlled is called the *plant*, denoted by G_p, and the feedback connection forms a *closed loop* around the plant. Components that are added to make the feedback control work effectively are represented by the *compensator* G_c. The block labeled H may represent additional compensation and/or a measurement transducer, thus the output of H may be the electrical analog of the physical *output*

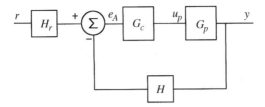

Figure 3.9-1 Feedback control: single-loop configuration.

variable y. The *command input* r may be a different physical variable (e.g., r may be the pilot's control stick force or deflection and y may be g's) and the *command prefilter* H_r will have a conversion factor such that the same kinds of physical quantity are compared at the summing junction.

When r and y are the same physical quantities they can still have different scale factors, for example, a system may be designed to make $y = 10r$. A general definition of *control error*, e, is

$$e = r - y \qquad (3.9\text{-}1)$$

Very often H_r and H will have different dynamics, but the same low-frequency gain. Then the control error is independent of input amplitude and is a more practical measure of performance. In addition, when the dynamic effects in H and H_r lie outside the frequency range of interest for the plant and compensator, they can be replaced by identity transformations. If H is replaced by the identity operation, the system is said to have *unity feedback*. The output of the summing junction is the "actuating" signal e_A, and in a well-designed control system it will normally be very small compared to the summed inputs. When H and H_r can be replaced by identity operations, e_A is identical to the control error.

One of the reasons for using negative feedback is to "regulate" the output of the plant, that is, to hold the output constant at a "set point," as in an aircraft "altitude-hold" autopilot. Another reason is to make the output "track" (i.e., follow) a changing command, as in making an aircraft track the pilot's pitch-rate command. In these *reg ulator* and *tracker* applications, negative feedback can change the nature of the plant behavior, for example, the plant may exhibit integrating behavior (e.g., as in steering a vehicle) whereas the closed-loop system simply follows its commands. Also, negative feedback can stabilize an unstable plant (e.g., the X-29 and F-117 aircraft), improve an unsatisfactory system-response, allow us to trade gain for increased bandwidth and better linearity, and reduce the effects of extraneous inputs (i.e., "disturbances"). The closed-loop transient response can be made to have dominant modes that are known to provide acceptable *handling qualities* for a human operator (from back-hoe operator to aircraft pilot), or even make one aircraft simulate another (e.g., Gulfstream trainer for space shuttle pilots). To achieve these benefits it is usually necessary to include some additional compensator dynamics within the feedback loop. In the rest of this section we will see how to choose a suitable type of compensator, how to perform a design with the compensator in place, and how to evaluate the control system design.

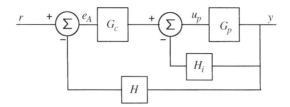

Figure 3.9-2 Feedback control with an inner loop.

Feedback Configurations and Closed-Loop Equations

We will analyze feedback configurations of the form shown in Figure 3.9-1, and with additional loops as in Figure 3.9-2. In Figure 3.9-2 the dynamics G_c, in the *forward path*, represent a *cascade compensator* (i.e., in "series" with the plant). The "inner-loop" feedback H_i represents *feedback compensation* and may correspond to *rate feedback* from a tachometer or a rate gyro, when y is an angular position.

In Figure 3.9-2, the individual dynamics may be described by transfer functions or state equations, and from these we need to be able to construct a dynamic description of the complete system, with input r and output y. To achieve this we must first reduce the inner loop to a single transfer function or state-space description, cascade the result with the dynamics G_c, and then reduce the outer loop by the same method as the inner loop.

First, consider the basic problem of cascading two dynamic systems. We already know that systems described by transfer functions may be cascaded by multiplying together the transfer functions and, in matrix form, we must ensure compatible dimensions in the matrix multiplication. We will now derive a formula for cascading state-space dynamic equations. In Figure 3.9-3, state-space system "1" is followed by system "2" in cascade. If we include both sets of state variables in one "augmented" state vector, an inspection of the block diagram gives the following equations:

$$\begin{bmatrix} \dot{x}_1 \\ \dot{x}_2 \end{bmatrix} = \begin{bmatrix} A_1 & 0 \\ B_2 C_1 & A_2 \end{bmatrix} \begin{bmatrix} x_1 \\ x_2 \end{bmatrix} + \begin{bmatrix} B_1 \\ B_2 D_1 \end{bmatrix} u_1 \qquad (3.9\text{-}2a)$$

$$y_2 = \begin{bmatrix} D_2 C_1 & C_2 \end{bmatrix} \begin{bmatrix} x_1 \\ x_2 \end{bmatrix} + [D_2 D_1] u_1 \qquad (3.9\text{-}2b)$$

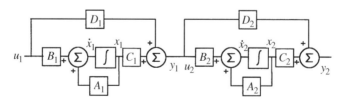

Figure 3.9-3 Cascaded state-space systems.

These operations are numerically stable and easy to perform or incorporate in software design tools (e.g., MATLAB "Series" command).

Next, consider a single feedback loop with forward path dynamics G and feedback-path dynamics H (Figure 3.9-1 with $H_r = I$ and $G_c G_p$ combined as G). Let G represent a $(p \times m)$ transfer function matrix, and H a $(m \times p)$ transfer function matrix. $H(s)$ may have some null rows or columns, depending on the choice of feedback connections. Simple matrix algebra gives the relationships

$$E_A(s) = R - HGE_A \qquad (3.9\text{-}3a)$$

$$Y(s) = GE_A = G(I + HG)^{-1} R(s) \qquad (3.9\text{-}3b)$$

so the closed-loop transfer function matrix is

$$G_{CL}(s) = G(I + GH)^{-1} \qquad (3.9\text{-}4a)$$

and in the SISO case,

$$G_{CL}(s) = \frac{G}{1 + GH}, \qquad (3.9\text{-}4b)$$

where $G(s)H(s)$ is called the *loop transfer function*. For frequency ranges over which the SISO loop transfer function has large magnitude, the closed-loop transfer function is given by

$$G_{CL}(s) \approx \frac{1}{H} \qquad (3.9\text{-}4c)$$

and this property is used to provide a precisely defined transfer function when G is large in magnitude but not well defined (e.g., operational amplifiers).

In the state-space case, if we draw diagrams of the form used in Figure 3.9-3 for both G and H, it is easy to derive the following closed-loop state equations:

$$\begin{bmatrix} \dot{x}_G \\ \dot{x}_H \end{bmatrix} = \begin{bmatrix} A_G - B_G D_H C_G & -B_G C_H \\ B_H C_G & A_H \end{bmatrix} \begin{bmatrix} x_G \\ x_H \end{bmatrix} + \begin{bmatrix} B_G \\ 0 \end{bmatrix} r \qquad (3.9\text{-}5a)$$

$$y = [C_G \ 0] \begin{bmatrix} x_G \\ x_H \end{bmatrix} + [0]\, u \qquad (3.9\text{-}5b)$$

Here the subscripts G and H indicate, respectively, the forward path and feedback path dynamics. We have also taken the special case of $D_G \equiv 0$ since the plant dynamics are normally low-pass. Again, these operations are very simple numerically and are preferable to working with polynomials (the MATLAB command "feedback" will perform these operations).

The forward-path dynamics may include a cascade compensator and a plant and, in the special case where the feedback path contains only a gain matrix K, the above equations reduce to

$$\dot{x} = (A_G - B_G K C_G)x + B_G r \qquad (3.9\text{-}6a)$$

$$y = C_G x \qquad (3.9\text{-}6b)$$

The closed-loop A-matrix A_{CL} is given by

$$A_{CL} = (A_G - B_G K C_G) \qquad (3.9\text{-}6c)$$

and the other matrices are unchanged.

Design software is usually arranged to produce a value for the feedback gain K. If we wish to obtain a unity-feedback design, we must scale the command r by the same gain as the feedback (i.e., K). The closed-loop system is then equivalent to a unity feedback system with a gain K in the error path; this is illustrated in Figure 3.9-4. The closed-loop equations, with $D \equiv 0$, are easily seen to be

$$\dot{x} = (A - BKC)x + BKr \qquad (3.9\text{-}7a)$$

$$y = Cx \qquad (3.9\text{-}7b)$$

Therefore, we can design unity-feedback systems with the same software, as long as we finally postmultiply the original B matrix by K.

In Chapter 4 we will perform a classical control design on a MIMO plant, namely, the aircraft lateral/directional dynamics. These have two inputs, aileron and rudder deflections, and two or more outputs. The classical design approach is iterative, closing one loop at a time and then repeating the process until the design is satisfactory. Here we will note how the loops can interact. The closed-loop MIMO transfer function is given in Equation (3.9-4a) and, closing any individual SISO loop through GH, in general, changes both the poles and zeros of any other SISO transfer function in G_{CL}. Various types of zeros are defined for a transfer function matrix (MacFarlane and Karcanias, 1976; Desoer and Schulman, 1974), and different subsets of these zeros will appear in the various SISO transfer functions. The zeros of a particular element of the transfer function matrix will not all appear in the corresponding SISO transfer function because of pole-zero cancellations. The actual SISO zeros can be found by the method of "coupling numerators" (McRuer et al., 1973).

Now let us return to the SISO case, with the closed-loop transfer function (3.9-4b). Also, let G and H be represented by monic polynomials as

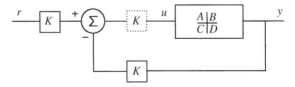

Figure 3.9-4 Transforming to unity feedback.

$$G(s) = k_G \frac{N_G(s)}{D_G(s)}; \qquad H(s) = k_H \frac{N_H(s)}{D_H(s)}$$

Then

$$\frac{Y(s)}{R(s)} = \frac{G}{1+GH} = \frac{k_G N_G D_H}{D_G D_H + k_G k_H N_G N_H} \qquad (3.9\text{-}8)$$

and $K = k_G k_H$ is the static loop sensitivity of the loop transfer function $G(s)H(s)$. We will often simply refer to K as the loop gain. Equation (3.9-8) shows that the closed-loop zeros are the combined zeros of G and poles of H. The closed-loop poles are given by the zeros of the *characteristic equation*,

$$1 + G(s)H(s) = 0 \qquad (3.9\text{-}9a)$$

that is, the roots of the *characteristic polynomial*,

$$D_G D_H + K N_G N_H \qquad (3.9\text{-}9b)$$

Therefore, the positions of the closed-loop poles vary with K. The frequency response of the loop transfer function GH, and the behavior, with K, of the roots of the characteristic polynomial are the basis of the two common design techniques of classical control theory. These will be described shortly.

This subsection has examined the closed-loop equations in various forms. Our last development will be the closed-loop equations for signals, other than the command, injected into the loop. An example is an aircraft model with control inputs applied to the control surfaces, and wind gusts entering into the aerodynamic calculations as disturbances. In general, a MIMO plant model will be required with inputs for both disturbances and controls. Figure 3.9-5 shows an example where the plant G_P is treated as a two-input, two-output system. One input-output pair is used by the feedback loop (e.g., for feedback of angle of attack to the elevator), and the other pair has been created to determine the effect of the disturbance on some other output variable (e.g., effect of wind gusts on normal acceleration).

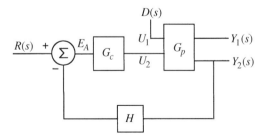

Figure 3.9-5 Disturbance input, transfer function description.

Referring to Figure 3.9-5, the transfer-function equations are

$$Y(s) = G_p \begin{bmatrix} 1 \\ 0 \end{bmatrix} D(s) + G_p \begin{bmatrix} 0 \\ 1 \end{bmatrix} (G_c R(s) - [0 \quad G_c H] Y(s))$$

or, solving for $Y(s)$,

$$Y(s) = \left(I + G_p \begin{bmatrix} 0 & 0 \\ 0 & G_c H \end{bmatrix} \right)^{-1} G_p \left(\begin{bmatrix} 1 \\ 0 \end{bmatrix} D + \begin{bmatrix} 0 \\ G_c \end{bmatrix} R \right) \quad (3.9\text{-}10)$$

Control inputs are multiplied by G_c, and higher gain in G_c will give better rejection of disturbance inputs. Equation (3.9-10) requires an inversion of a polynomial matrix and is inconvenient for numerical computations.

Figure 3.9-6 shows a state-space description of a compensator cascaded with a plant, and a disturbance input to the plant. An analysis of the diagram is greatly simplified by the absence of a direct feedthrough path in the compensator, and this matches the aircraft situation if the control surface actuators are included with the compensator model. By inspection, the state equations are

$$\begin{bmatrix} \dot{x}_p \\ \dot{x}_c \end{bmatrix} = \begin{bmatrix} A_p & B_p C_c \\ -B_c C_p & A_c - B_c D_p C_c \end{bmatrix} \begin{bmatrix} x_p \\ x_c \end{bmatrix} + \begin{bmatrix} B_p \\ -B_c D_p \end{bmatrix} d + \begin{bmatrix} 0 \\ B_c \end{bmatrix} r \quad (3.9\text{-}11a)$$

$$y = \begin{bmatrix} C_p & D_p C_c \end{bmatrix} \begin{bmatrix} x_p \\ x_c \end{bmatrix} + \begin{bmatrix} D_p \end{bmatrix} d \quad (3.9\text{-}11b)$$

These equations can be used to determine the transfer function (or poles and zeros, or frequency response) from the disturbance input to the response variable of interest. Aircraft in flight are subjected to random disturbances from discrete wind gusts and continuous turbulence, and these types of disturbances are modeled with a *power spectral density function* (PSDF) (see *MIL-F-8785C*). A measure of a stationary (statistics independent of time) random signal is its *root-mean-square* (rms) value, and this can be calculated from the area under the PSDF (Brown and Hwang,

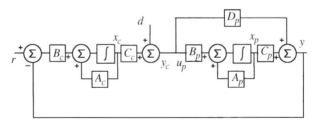

Figure 3.9-6 Disturbance input, state-space description.

1997). The aircraft response PSDF must be calculated from the squared-magnitude of the appropriate transfer function, multiplied by the PSDF of the disturbance. Alternatively, a model of the disturbance can be included in the nonlinear state equations, and the response to the disturbance can be found by simulation.

Steady-State Error and System Type

An analysis of Figure 3.9-1 with SISO transfer functions G and H in the forward and feedback paths (with $H_r = I$ and $G = G_c G_p$), using the error definition (3.9-1), gives the error transfer function:

$$\frac{E(s)}{R(s)} = \frac{1 + G(H-1)}{1 + GH} \qquad (3.9\text{-}12a)$$

With unity feedback, this reduces to

$$\frac{E(s)}{R(s)} = \frac{1}{1 + G} \qquad (3.9\text{-}12b)$$

First consider the unity feedback case, and think of a regulator, where "steady state" implies a constant output. If the forward path transfer function exhibits integrating action, then a steady-state error cannot exist. Electronic operational amplifier circuits can provide almost perfect integration and so, if necessary, we may decide to include an integrator in the error path. In the case of a tracker, the steady-state error will depend on the type of command input and so we must analyze this situation.

In (3.9-12b) let there be q pure integrations in G, after any cancellations with zeros at the origin, and let $G = s^{-q} G'$. Then,

$$\frac{E(s)}{R(s)} = \frac{1}{1 + G} = \frac{s^q}{s^q + G'} \qquad (3.9\text{-}13)$$

where $G'(0)$ is finite. Let the system be stable, and have a polynomial input $(t^n/n!)$ $U_{-1}(t)$ with Laplace transform $1/s^{n+1}$. Then the final value theorem gives the following expression for the steady-state control error:

$$e_{SS}(t) = \operatorname*{Lim}_{s \to 0} s E(s) = \operatorname*{Lim}_{s \to 0} \frac{s^{q-n}}{(s^q + G')} \qquad (3.9\text{-}14)$$

Therefore, in order to track a polynomial input of degree n, with finite steady-state error, the control system must have n pure integrations in the forward path. Such a system is called a *type-n* control system. An additional integration will reduce the steady-state error to zero, while one less integration will cause the steady-state error to grow without bound. Each integration adds 90 degrees of phase lag and makes the control system progressively more difficult to stabilize, so that systems are restricted to type-2 or lower in practice. The value of the finite steady-state error is when $q = n = 0$ (step input), $\qquad e_{ss} = \dfrac{1}{(1 + G(0))}$ (3.9-15)

when $q = n \geq 1$ (ramp, parabola, etc.), $\quad e_{ss} = 1/G'(0)$

The step, ramp, and parabolic *error coefficients* are defined as follows,

$$K_p \equiv \lim_{s \to 0} G(s), \quad \text{and error to a unit-step} = \frac{1}{(1 + K_p)}$$

$$K_v \equiv \lim_{s \to 0} sG(s), \quad \text{and error to a unit ramp} = \frac{1}{K_v} \qquad (3.9\text{-}16)$$

$$K_a \equiv \lim_{s \to 0} s^2 G(s), \quad \text{and error to a unit parabola} = \frac{1}{K_a}$$

and are used to specify performance requirements. Now consider the non-unity-feedback case and, as discussed in connection with (3.9-1), let $H(0) = 1$. The formula corresponding to (3.9-14) is

$$e_{ss}(t) = \lim_{s \to 0} \frac{s^{q-n} + G'(H-1)/s^n}{s^q + G'H} \qquad (3.9\text{-}17)$$

and for comparison with the unity-feedback case, let $q = n$. Then,

if $q = n = 0$, $\quad e_{ss} = \left[\dfrac{1 + G(H-1)}{(1 + GH)} \right]_{s=0} = \dfrac{1}{1 + G(0)} \qquad (3.9\text{-}18a)$

and if $q = n > 0$,

$$e_{ss} = \lim_{s \to 0} \left[\frac{1 + G'(H-1)/s^n}{G'H} \right] \qquad (3.9\text{-}18b)$$

Because of the condition $H(0) = 1$, the steady-state error with a constant input, Equation (3.9-18a), is the same as the unity-feedback case. With polynomial inputs, Equation (3.9-18b) shows that the steady-state error depends on the limit of $(H-1)/s^n$ as s becomes zero. If $H(s)$ is written as a ratio of polynomials, then $H(0) = 1$ guarantees that $(H-1)$ has at least one free s to cancel with s^n. Therefore, unlike the unity-feedback case, the error can become infinite with a parabolic input ($n = 2$). This is illustrated in Example 3.9-6.

Practical command inputs may contain derivatives of all orders for short periods of time, so that the tracking error may grow and then decrease again. System type requirements and error coefficients are preliminary design considerations, and these ideas are used later in the design examples, and in Chapter 4.

System type can be misleading in some situations. If, for example, the plant includes an electric motor, the integration of motor speed to angular position is a "kinematic" integration. If there are no other integrations in the forward path, the

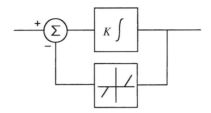

Figure 3.9-7 Integrator windup protection.

error signal must become large enough to overcome the static frictional torques of the motor and load before the motor will begin to turn. Therefore, this system will not behave like a type-1 system.

Another problem encountered with integral control of real systems is *integrator windup*. An electronic integrator *saturates* when its output gets close to the circuit positive or negative supply voltages. If the plant becomes temporarily nonlinear (e.g., "rate saturation") before saturation occurs in the integrator, then depending on the command signal, the integrator may begin to integrate a large error signal that takes its output farther beyond the plant saturation level. When the plant comes out of saturation, or the command reverses, it may take some significant time before this excessive output is removed and linear control is regained. Figure 3.9-7 shows an anti-windup arrangement. When the output of the integrator reaches the plant saturation value, it exceeds the threshold of the dead-zone device. The resulting feedback turns the integrator into a fast lag transfer-function. Anti-windup arrangements are used in both analog and digital aircraft flight-control systems. A related problem can occur when switching between different control system modes. All energy-storage elements must be initialized so that unwanted sudden movements of the control surfaces do not occur.

Stability

A familiar example of feedback causing instability is provided by the public address (PA) system of an auditorium. When an acoustical signal from the loudspeakers is received at the microphone, and the gain and phase around the acoustical path are such that the signal reinforces itself, the loudspeakers produce a loud whistle. This is probably the most natural intuitive way to understand feedback stability. Thus we might examine the frequency response of the loop transfer function GH to determine if the gain is greater than unity when the phase-lag has reached 180 degrees; that is, we must look for the condition $GH(j\omega) = -1$. This corresponds to finding a root of $(1 + GH) = 0$ on the s-plane $j\omega$ axis, which is the stability boundary.

In 1932 H. Nyquist used *the principle of the argument* from complex variable theory (Phillips, 1961), applied to $F(s) = 1 + GH(s)$, to develop a test for stability. A semicircular "test" contour, of "infinite" radius, is used to enclose the right-half s-plane. According to the principle of argument, as s traverses the closed test contour in a clockwise direction, the increment in the argument of $F(s)$ is $N \times (2\pi)$, where

$N = (P - Z)$, and P and Z are, respectively, the number of poles and zeros of $F(s)$ inside the test contour. We see that poles and zeros of $F(s) = 1 + GH$ are, respectively, the open-loop and closed-loop poles, and N is the number of counterclockwise encirclements of the s-plane origin.

Rather than count the encirclements of the origin by $1 + GH$ we can, more conveniently, count the encirclements of the *critical point* $(-1 + j0)$ by $GH(s)$. P and Z are both, in general, greater than or equal to zero, and so N may be a positive or a negative integer. Since the test contour encloses the whole right-half s-plane, we have a closed-loop stability test by finding Z, given by

$$Z = P - N \tag{3.9-19}$$

or,

unstable *CL* poles = # unstable *OL* poles − # *CCW* encirclements

The test contour, known as the Nyquist D-contour, can be indented with infinitesimal semicircles to exclude open loop poles on the $j\omega$ axis. Note that some authors define N to be the number of clockwise encirclements, and they reverse the sign of N in (3.9-19).

Example 3.9-1: An Example of Nyquist's Stability Criterion. Let the open-loop transfer function be given by

$$G(s)H(s) = \frac{K(s+2)(s+4)}{s(s^2 - 4s + 13)}, \quad K > 0$$

Figure 3.9-8a shows the Nyquist D-contour, indented with a semicircle to avoid the pole at the origin. Figure 3.9-8b shows the Nyquist plot; letters have been used to

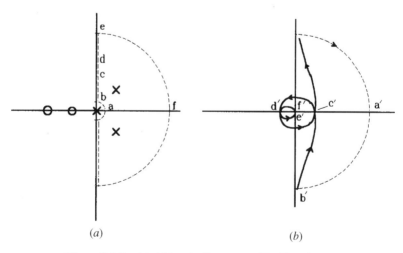

Figure 3.9-8 (*a*) A Nyquist D-contour; (*b*) a Nyquist plot.

mark corresponding points on the two plots. The indentation can be represented by the equation $s = re^{j\theta}$, with $r \to 0$ and $-\pi/2 \le \theta \le \pi/2$, as an aid to establishing the corresponding points. Imagine $G(s)H(s)$ represented by vectors drawn from the poles and zeros to a starting point at a on the D-contour. When $s = a$, the net angle of the vectors is zero and the magnitude of $GH(a)$ approaches infinity as r becomes zero; this gives the corresponding point a' on the GH plot. When $s = b$, the angle of the vector from the pole at the origin has become 90 degrees, but the net angle of the other vectors is close to zero; this gives the point b'. The part of the D-contour from b to e corresponds to real frequencies, and the frequency response $GH(j\omega)$ could be measured with test equipment if the system were not unstable. The relative degree is unity so, as ω increases, $GH(j\omega)$ approaches zero magnitude with a phase angle of -90 degrees (or $+270$ degrees). Let $s = c$ and d be the points where the phase of GH passes through zero and 180 degrees, respectively. From $s = e$ to $s = f$, the phase of $GH(s)$ returns to zero, while the magnitude remains infinitesimal. The remainder of the D-contour uses conjugate values of s, and the remaining half of the Nyquist plot is the conjugate of the part already drawn.

The D-contour shows that $P = 2$, and the Nyquist plot shows that the number of counterclockwise encirclements of the critical point is $N = 0$ or $N = 2$, depending on the magnitude of $GH(j\omega)$ at d'. This in turn depends on the loop gain K and so,

$$\text{small } K \to Z = 2 - 0 = 2 \quad \text{(closed-loop unstable)}$$
$$\text{large } K \to Z = 2 - 2 = 0 \quad \text{(closed-loop stable)}$$

This is the opposite of the common behavior, in which a system becomes unstable when the gain is increased too much, and this behavior is known as *conditional stability*. At $s = c$ and d, $\text{Im}[GH(j\omega)] = 0$; then solving $\text{Re}[GH(j\omega)] = -1$, evaluated at the higher value of ω, gives the value of K at the stability boundary. A Nyquist plot can be obtained with the following MATLAB code:

```
num=[1 6 8]; den=[1 -4 13 0];
w= logspace(-1,1,400);      % 0.1 Hz to 10 Hz, 400 points
k= 6;                        % Stable k
nyquist(k*num,den,2*pi*w)
```

■

Most practical control systems are open-loop stable, so that $Z = -N$, and therefore we require $N = 0$ for stability. Also, we need only consider the positive $j\omega$ axis of the D-contour, since the negative $j\omega$ axis gives a conjugate locus in the GH-plane, and the infinite semicircle maps to the origin of the GH-plane (because the relative degree of the transfer function of a real compensator and plant is greater than zero). A few rough sketches will show that, under these conditions, *if the locus of GH is plotted as the frequency is varied from $\omega = 0$ to $\omega = \infty$, the closed-loop system is unstable if the critical point lies to the right of the locus*. An example of the Nyquist plot of a stable type-1, relative-degree three, system is shown in Figure 3.9-13. These restricted conditions for stability agree with the intuitive criterion that the magnitude of the loop transfer function should be less than unity when its phase-lag is 180 degrees.

Stability criteria, other than Nyquist's test, mostly involve testing the characteristic equation (CE) directly for roots in the right-half s-plane. A necessary but not sufficient condition for stability is that all of the coefficients of the CE should have the same sign and be nonzero (see also Descarte's rule of signs [D'Azzo and Houpis, 1988]). Routh's test (Dorf and Bishop, 2001) uses the coefficients of the CE and provides more information, in that the number of right-half plane roots and the stability boundary can be determined. In the state-space context the roots of the CE are the eigenvalues of the A matrix, but then we must go to the trouble of solving the characteristic equation.

Types of Compensation

The discussion of stability, and some Nyquist sketches for simple systems that are open-loop stable, should lead to some ideas about the frequency-domain properties required of a compensator. Alternatively, we might look at a compensator as a means of adding extra terms to the system characteristic equation so that the roots can be moved to desirable locations in the left-half s-plane. In frequency-domain terms, a compensator should produce phase-lead in a frequency range where the lag of the plant is approaching 180 degrees and the gain is near unity, or it should cut the gain when the phase-lag is approaching 180 degrees. For a minimum-phase transfer function phase-lead is associated with rising gain, and this approximates the characteristics of a differentiator. Differentiation accentuates the noise on a signal, and so practical compensators should be designed to produce phase lead and rising gain only over a limited frequency range.

Let us now examine a compensator with a single differentiation, a proportional-plus-derivative, PD, compensator, which can be approximated in real systems. The transfer function is

$$G_c(s) = K_P + K_D s \quad (3.9\text{-}20)$$

Equation (3.9-4b) gives the unity-feedback closed-loop transfer function as

$$\frac{Y}{R} = \frac{(K_P + K_D s)G_p}{1 + (K_P + K_D s)G_p} \quad (3.9\text{-}21)$$

The characteristic equation now contains the proportional and derivative terms K_P and K_D, and it may be possible to achieve satisfactory closed-loop poles. However, in addition to the noise problem, there is now a closed-loop zero at $s = -K_P/K_D$, and this zero can cause a large overshoot in the step response unless the plant poles are heavily damped.

As an alternative to PD compensation, consider Figure 3.9-2 with unity feedback, simple proportional control, and inner-loop rate-feedback:

$$G_c = K_P; \ H_i = K_r s$$

Then, using (3.9-4b) to close the inner loop first, the overall closed-loop transfer function is found to be

$$\frac{Y}{R} = \frac{K_P G_p}{1 + (K_P + K_r s) G_p} \qquad (3.9\text{-}22)$$

Therefore, with rate feedback, we can achieve the same closed-loop poles as PD control, but without accentuating noise in the error channel, and without the troublesome closed-loop zero.

A practical cascade compensator that only approximates PD control, and satisfies the practical requirement of relative degree greater than, or equal to zero, is the simple "phase-lead" compensator shown in Table 3.3-1. The numerator $(s + z)$, on its own, represents a derivative term plus a proportional term, which is equivalent to a zero at $s = -z$. The pole is at $s = -p$, with $p > z$, and if we were to compare the Bode plots of $(s/z + 1)$ and $(s/z + 1)/(s/p + 1)$ we would see that the derivative action begins to disappear as the second corner frequency $\omega = p$ is approached. The practical limit $(p/z) < 10$ is usually observed to avoid greatly accentuating noise.

Phase-lead compensation is effective and inexpensive; inner-loop rate-feedback incurs the cost of a rate sensor and may not be physically appropriate for a particular plant. A practical rate sensor also has limited bandwidth, and its transfer function pole(s) will appear as closed-loop zeros. However, these zeros are likely to be much farther from the s-plane origin then the lead-compensator zero, and therefore less troublesome in terms of causing overshoot.

The subsection on steady-state error and system type explained the need for "integral control." Unfortunately, "pure" integral control has some detrimental effects on closed-loop transient response. First, a pole at the s-plane origin is destabilizing because it adds a constant 90 degree phase lag to the loop transfer function. Second, an open-loop pole at the origin may become a slow closed-loop pole (see the root-locus section). To overcome the phase-lag problem we use "proportional plus integral," PI, control in the cascade compensator. The compensator transfer function is $(k_p + k_i/s)$ or, equivalently, $k_p(s + k_i/k_p)/s$. The Bode plot of this transfer function shows that the phase-lag disappears at high frequency.

If we use Figure 3.9-1 with unity feedback, $H_r = 1$, and a cascaded PI compensator the closed-loop transfer function is

$$\frac{Y}{R} = \frac{(sK_P + K_i)G_p}{s + (sK_P + K_i)G_p} \qquad (3.9\text{-}23)$$

The PI control has introduced a closed-loop zero at $s = -K_i/K_P$, and again this may cause an excessive overshoot in the step-response. To see what inner-loop feedback can do for us, in Figure 3.9-2 let

$$G_c = K_i/s; \quad H_i = K_f; \quad H = 1$$

so that the closed-loop transfer function becomes

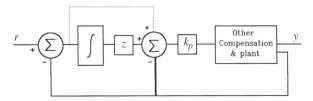

Figure 3.9-9 PI compensation with no closed-loop zero.

$$\frac{Y}{R} = \frac{K_i G_p}{s + (sK_f + K_i)G_p} \qquad (3.9\text{-}24)$$

The inner-loop proportional feedback, combined with pure integral control, has the same characteristic equation as PI control but has eliminated the closed-loop PI zero. Another way of looking at this is shown in Figure 3.9-9. The signal fed back to the plant input is unchanged if the PI proportional path (lightly dotted line) is removed and the feedback path shown with the heavy line is added. The overall closed-loop transfer function has changed because the input signal, r, no longer sees a proportional path. We can also see that the inner-loop feedback would remove the effect of an integration in the plant, and so this modification may reduce the system type. Proportional-plus-derivative control can be interpreted in a similar manner.

A lag compensator is also shown in Table 3.3-1; it has a pole and a zero, with the pole closer to the origin. If the pole is placed very close to the origin it can be thought of as an approximation to PI compensation, although we usually choose the zero position in a different way from PI compensation. By using the s-plane vector interpretation of the lag compensator, or drawing its Bode plots, we see that it provides a reduction in gain at high frequency (hf) without the 90 degree asymptotic phase lag. It can be thought of as a way of alleviating stability problems caused by phase lag at hf, or as a way of boosting low-frequency (lf) gain relative to hf gain in order to improve the position-error coefficient in a type-0 system.

The compensators described above may be used in combination and, for example, two stages of phase-lead compensation can provide more lead than a single stage, for the same increase in gain.

SISO Root-Locus Design

In this subsection we introduce our first classical design technique: *root-locus* design, devised by W. R. Evans in 1948. The root-locus technique provides a graphical method of plotting the locii of the roots of a polynomial when a coefficient in the polynomial is varied. It can be applied directly to the characteristic equation of a closed-loop control system, to determine when any poles become lightly damped or unstable and to determine the effects of adding compensator poles and zeros.

Consider the following polynomial equation in the complex variable s,

$$s^n + a_{n-1}s^{n-1} + \cdots + a_j s^j + \cdots + a_1 s + a_0 = 0$$

Suppose that we wish to examine the movement of the roots when the coefficient a_j is varied. The root-locus rules of construction can be applied by writing the equation as

$$1 + \frac{a_j s}{(s^n + \cdots + a_0) - a_j s} = 0 \qquad (3.9\text{-}25)$$

The characteristic equation (3.9-9a) can be written in this form, as,

$$1 + \frac{K\,N(s)}{D(s)} = 0, \qquad (3.9\text{-}26)$$

where the monic polynomials $N(s)$ and $D(s)$ contain, respectively, the known open-loop zeros and poles (n poles and m zeros), and the static loop sensitivity K is to be varied.

Equation (3.9-26) is the equation that is satisfied on the locii of the closed-loop poles, that is, on the "branches" of the root-locus plot. It can be rewritten as

$$\frac{K\,N(s)}{D(s)} = -1 \qquad (3.9\text{-}27)$$

from which we get the "angle condition"

$$\angle N(s) - \angle D(s) = \begin{cases} (2r+1)\pi, & K > 0 \\ r(2\pi), & K < 0 \end{cases} \quad r = 0, \pm 1, \pm 2, \ldots \qquad (3.9\text{-}28)$$

and the "magnitude condition"

$$|K| = \frac{|D(s)|}{|N(s)|} = \frac{\prod (\text{lengths of vectors from poles})}{\prod (\text{lengths of vectors from zeros})} \qquad (3.9\text{-}29)$$

When there are no zeros, the denominator of (3.9-29) is unity. These two conditions are the basis of most of the root-locus rules, which are now enumerated:

1. Number of branches = number of open-loop poles (n).
2. The root-locus plot is symmetrical about the s-plane real axis.
3. For $K > 0$, sections of the real axis to the left of an odd number of poles and zeros are part of the locus. When K is negative we have the so-called *zero-angle root locus*, which is on the axis to the left of an even number of poles and zeros.
4. The n branches start (when $K = 0$) at the open-loop poles, and end (when $K = \infty$) on the m open-loop zeros, or at infinity (if $n > m$).
5. Branches that go to infinity approach asymptotes given by

$$\angle \text{asymptotes} = \pm \frac{(2r+1)\pi}{(n-m)}, \qquad r = 0, \pm 1, \pm 2, \ldots$$

$$\text{real-axis intersection of asymptotes} = \frac{\Sigma(\text{finite poles}) - \Sigma(\text{finite zeros})}{(n-m)}$$

6. If two real-axis branches meet as K is increased, they will break away to form a complex pair of poles. Similarly, two complex branches may arrive at the same real-axis point and become a real pair. Break-away and arrival points can be found by solving (3.9-27) for K, and then finding the values of s that satisfy $\partial K / \partial s = 0$ with s treated as a real variable.
7. Root-locus branches meet or leave the real axis at 90 degrees.
8. If a "test point" is very close to a complex pole or zero, all of the vectors from the other poles and zeros can be approximated by drawing them to that pole or zero. The angle of the remaining vector, found from the angle condition (3.9-28), gives the angle of departure or arrival of the root-locus branch for the pole or zero in question.
9. Imaginary-axis crossing points can be found by replacing s by $j\omega$ in the characteristic equation, and solving the separate real and imaginary conditions that result. Alternatively, the root-locus angle condition can be applied, or a standard test for stability (e.g., Routh-Hurwitz) can be used.
10. Constant net damping: When the relative degree $(n-m)$ of the loop transfer function is greater than unity, then, if some branches are moving left, others must be moving right.

Software is available to construct root-locus plots (e.g., MATLAB "rlocus" and "rltool"), but the above rules allow us to anticipate the effects of proposed compensators. We will now illustrate root-locus design by means of some examples.

Example 3.9-2: Root-Locus Design Using a Lead Compensator. In this example we will show how a phase-lead compensator can stabilize an unstable system, but the compensator will be chosen to illustrate the root-locus rules rather than to produce the "best" control system design. This example can be done more easily using transfer functions, but we wish to develop familiarity with the state-space approach, for later applications.

Let the plant be type-2, with transfer function:

$$G(s) = \frac{100}{s^2(s+10)}$$

To provide yet another technique for obtaining state equations, the transfer function was expanded as a sum of partial fraction terms, and state variables were chosen to be the integrator outputs in the simulation diagram representation of each partial-fraction term, as in Section 3.2. The plant A, B, C, D matrices are

$$ap = \begin{bmatrix} 0 & 1 & 0 \\ 0 & 0 & 0 \\ 0 & 0 & -10 \end{bmatrix} \quad bp = \begin{bmatrix} 0 \\ 1 \\ 1 \end{bmatrix} \quad cp = [10 \quad -1 \quad 1] \quad dp = [0]$$

The compensator state-space description is given in Table 3.3-1. Equations (3.9-2) can easily be used to cascade the compensator with the plant, but here we will illustrate the use of the "series" command in a MATLAB program:

```
ap= [ 0 1 0; 0 0 0; 0 0 -10];
bp= [0; 1; 1]; cp= [10 -1 1]; dp= [0];      % Plant
z= .6; p= 9;                                 % Compr. Zero & pole
ac= [-p]; bc= [1]; cc= [z-p]; dc= [1];       % Lead comp.
[a b c d] = series(ac,bc,cc,dc,ap,bp,cp,0);  % Comp. + plant
k= linspace(0,10,2000);
r= rlocus(a,b,c,d,k); plot(r)
grid on
```

The root-locus plot is shown in Figure 3.9-10, with the compensator pole at $s = -9$ and the zero at $s = -0.6$. Without the compensator the two branches from

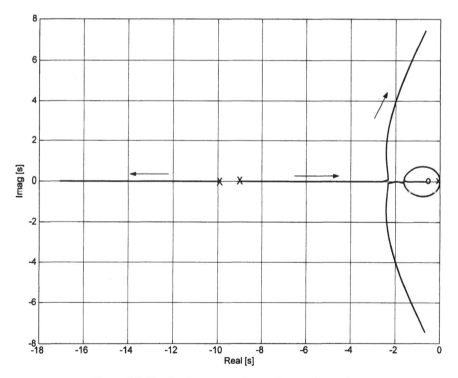

Figure 3.9-10 Lead compensation on the root-locus plot.

the double pole at the origin would immediately move into the right-half s-plane, while the real pole at $s = -10$ moves left (i.e., constant net damping). The effect of placing the compensator zero near to the origin, with its pole well to the left, is strong enough to pull the two branches from the origin into the real axis. The branch that approaches the compensator zero represents a closed-loop pole close to the origin, and hence a slow closed-loop mode. The "strength" of this mode (i.e., residue in the pole) will depend on how close the pole gets to the zero, but in a practical design the compensator zero would be placed farther to the left. The other branch from the origin moves left and meets the compensator pole. They break away from the real axis and move toward the right-half plane (i.e., constant net damping again), and approach 60 degree asymptotes. It is worthwhile to check the root-locus rules, one by one, against this example. All of the rules are illustrated except the "angle-of-departure" rule. ■

This phase-lead example will be repeated as Example 3.9-5, done in the frequency domain, and with more emphasis on practical design considerations. In general, possible root-locus design techniques include placing the compensator zero on, or to the left of, the second real plant pole from the origin, or placing it at the real part of a desired complex pair. The compensator pole position may then be adjusted to give a closed-loop dominant pair a desired frequency or damping. The closed-loop step response should be checked and the design may be modified by moving the pole position, or by moving both the pole and zero keeping the ratio p/z constant.

PI compensator design will be illustrated next, by the following root-locus example.

Example 3.9-3: Root-Locus Design of a PI Compensator. Let the plant and PI-compensator transfer functions be

$$G_p = \frac{1}{(s+3)(s+6)} \qquad G_c = \frac{K(s+z)}{s}$$

The design goals will be to obtain a dominant complex pole-pair with damping ratio of $1/\sqrt{2}$, together with the highest possible ramp error coefficient. The root-locus plot will show the trade-offs in the design, and a simulation will be used to check that the closed-loop step response is like that of a quadratic lag with $\zeta = 1/\sqrt{2}$. A MATLAB program is:

```
z= 2; num= [1 z]; den= [1 9 18 0];      % Choose z
[a,b,c,d]= tf2ss(num,den);              % Compr. + Plant
k= linspace(0,50,2000);
r= rlocus(a,b,c,d,k); plot(r), grid on  % Root locus

sgrid(.707,0)
axis=([-8,1,-8,8])
rlocfind(a,b,c,d)                       % Find K for zeta=.707

sys1= ss(a,18*b,c,d);                   % K=18
sys2= feedback(sys1,1,-1);              % Close loop
step(sys2,3)                            % Step response
grid on
```

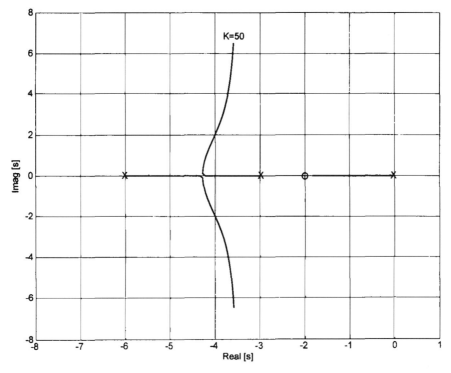

Figure 3.9-11 PI compensation on the root-locus plot.

Figure 3.9-11 is the root-locus plot with $z = 2$. The relative degree of the loop transfer function is 2, and so the asymptotes are at 90 degrees to the real axis. The damping of the complex poles can become very small, but the system can never become unstable. The ramp error coefficient is

$$K_v = \lim_{s \to 0} sG(s) = \frac{Kz}{(3)(6)}$$

If we make z small, the error coefficient will be small. In addition, the root-locus plot shows that there will be a slow closed-loop pole trapped near the origin. If we place the PI zero to the left of the plant pole at $s = -3$, the complex poles will break away from the axis between $s = 0$ and $s = -3$. This could produce a dominant pair of poles, but they may be too close to the origin for a fast, well-damped response. Therefore, we might try $1 < z < 4$ while adjusting K to give a damping ratio $\zeta = 0.707$ and checking K_v. When this is done, K_v is found to peak when $z = 3$ and $K = 18$. The zero then cancels the slowest plant pole and the closed-loop dynamics are second-order with the desired damping ratio. The step response is shown in Figure 3.9-12. In general, the best position for the zero should be determined on a case-by-case basis, using considerations similar to those above.

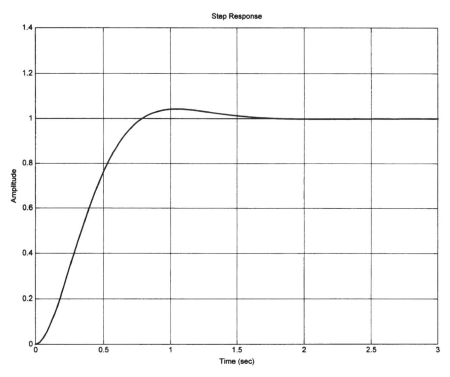

Figure 3.9-12 Step response with PI compensation.

A lag compensator (see Table 3.3-1) can be used to increase the value of a control-system error coefficient (K_p, K_v, K_a, ...), without appreciably affecting stability. The lag-compensator pole can be placed close to the origin, and the lag-compensator zero placed not far to the left of the pole. At low frequencies, the compensator gain is given by the length of the zero-vector divided by that of the pole-vector. Time constants up to about 100 s are practicable, so the pole could be placed at $s = -0.01$. Then, placing the zero at $s = -0.1$ will give a low-frequency gain of 10.0. At high frequency these two pole and zero vectors are close together and have little effect on the dynamics. This technique traps a slow pole near to the origin, as was noted in Example 3.9-2. Note that to get greater than unity low-frequency compensator gain, an amplifier is required. An alternative approach to lag compensation is to increase the loop gain as much as possible, and solve problems of high-frequency instability by using the lag compensator to cut the high-frequency gain without adding much phase lag. The term *lag compensator* is unfortunate in that, unlike a simple lag transfer function, its ultimate phase lag is zero. This technique is useful in situations where "unmodeled" high-frequency dynamics are causing instability, and trial and error compensation is used. Lag compensation is better illustrated in the frequency domain than on the root-locus plot.

In summary, the root-locus technique works well with low-order dynamics, and is especially useful as a "back-of-the-envelope" analysis or design technique. With a large number of poles and zeros it becomes necessary to switch to the frequency-domain techniques illustrated in the next subsection.

Frequency-Domain Design

In frequency-domain design we plot the frequency response of the loop transfer function, and use ideas related to Nyquist, Bode, and Nichols plots (Franklin et al., 2002) to arrive at appropriate parameters for one of the standard compensator transfer functions.

The Nyquist stability test leads to useful analysis and design ideas. Some control loops contain pure delay effects, for example, signal-propagation delays in a transmission medium or "transport delays" due to piping, belt-feed devices, and so on. In the aircraft case, we have computational delays in a flight-control computer, and decision and reaction time delays in the human pilot. A pure delay has the transcendental transfer function ke^{-sT}. This function can easily be plotted in a graphical frequency-response design format but, for root-locus design, it can only be approximated as a rational polynomial function (Franklin et al., 2000). In the case of nonlinear plants, the *describing function* technique allows us to analyze stability and *limit-cycle* oscillations by using a movable critical point on the Nyquist plot (West, 1960). Other important Nyquist-related design tools are the *gain* and *phase margins*; these will be illustrated here, and applied in Chapter 4.

If the open-loop frequency-response locus passes close to the point $(-1 + j0)$, the stability boundary is being approached and the system transient response is likely to be underdamped. The gain margin of a feedback loop is the increase in gain that can be allowed before the loop becomes unstable. It can be calculated by finding the gain at the phase-crossover frequency, as illustrated in Figure 3.9-13. The phase margin is the number of degrees by which the phase angle of GH exceeds -180 degrees, when $|GH| = 1.0$. It can be calculated from the gain-crossover shown in Figure 3.9-13. As a rule of thumb a phase margin of 30 to 60 degrees will be required to obtain a good closed-loop transient response, and this should be accompanied by a gain margin of 6 to 15 dB. When closing a feedback loop produces only an underdamped complex pair of poles, the closed-loop damping ratio is related to the phase margin by $\zeta \approx PM^0/100$, for phase margins up to about 70 degrees (Franklin et al., 2002). This relationship also holds approximately, if the closed dynamics are dominated by a complex pair.

In classical frequency-domain design, lead, lag, and PI cascade compensators are used, in conjunction with gain and phase margins, to achieve satisfactory closed-loop designs. We will first review the frequency domain properties of these compensators. Table 3.3-1 shows passive networks that implement lead and lag compensation (see also Section 3.3), and the lead and lag transfer functions can both be written (apart from a gain constant) as

$$G_C(s) = \frac{s+z}{s+p} \qquad \begin{array}{l} p > z \equiv \text{lead} \\ p < z \equiv \text{lag} \end{array} \qquad (3.9\text{-}30)$$

Inspection of this transfer function shows that the high-frequency (hf) gain is 1.0, and the low-frequency (lf) gain is z/p, with a phase angle of zero in both cases. The

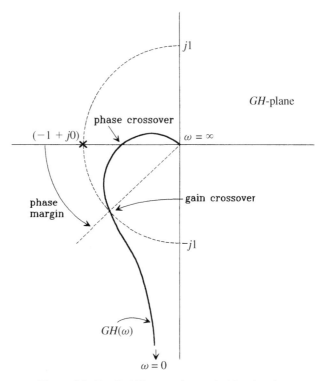

Figure 3.9-13 Stability margins on the Nyquist plot.

polar plot of $G_c(j\omega)$ is a semicircle above (lead) or below (lag) the positive real axis, with its diameter determined by the lf and hf gains (Problem 3.9-9). This is shown in Figure 3.9-14 for the lead transfer function. The figure shows that the maximum lead angle ϕ_M and the corresponding gain are given by

$$\sin \phi_{MAX} = \frac{1 - z/p}{1 + z/p} \qquad (3.9\text{-}31a)$$

$$|G(\phi = \phi_M)| = \sqrt{(z/p)} \qquad (3.9\text{-}31b)$$

For the passive lag-compensator this gain must be multiplied by p/z, giving,

$$|G(\phi = \phi_M)| = \sqrt{(p/z)} \quad \text{(lag comp.)} \qquad (3.9\text{-}31c)$$

The Bode plot shows that the frequency of maximum lead or lag is the geometric mean of the corner frequencies:

$$\omega_{\phi_M} = \sqrt{(pz)} \qquad (3.9\text{-}31d)$$

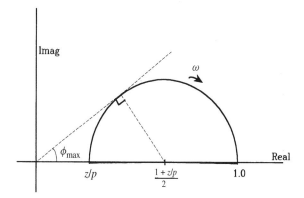

Figure 3.9-14 Lead compensator polar plot.

The design techniques with these compensators are illustrated in the following examples.

Example 3.9-4: Design of a Passive Lag Compensator. This system has unity feedback, with a loop transfer function

$$G_c(s)G_p(s) = \frac{p_c}{z_c}\frac{(s+z_c)}{(s+p_c)}\frac{K}{s(s+1)(s+15)(s+20)}$$

and the closed-loop requirements will be a velocity-error coefficient of $K_v \geq 13$ and a phase margin of 25 degrees.

A loop gain of $K = 4000$ meets the K_v requirement. Also, the Bode plot of G_p shows that the phase angle is -150 degrees at $\omega \equiv \omega_1 = 1.25$ rad/s. If the compensated loop transfer function has unit magnitude at this frequency, and the compensator produces only about 5 degrees lag, then the phase margin requirement will be satisfied. A passive lag compensator has a gain close to (p_c/z_c), and about 5 degree lag, at one decade above the upper corner frequency. Therefore, we now choose the compensator zero to be $Z_c = 0.1\omega_1 = 0.125$ rad/s. At ω_1 the magnitude of the plant transfer function is 6.5, and so we require $p_c/z_c = 1/6.5$. This gives the compensator pole frequency as $p_c = 0.125/6.5$, or about .02.

Parts of the following MATLAB code were used to produce the Bode and Nyquist plots shown in Figures 3.9-15a and b, and also a step response.

```
den= conv([1 1 0],[1 35 300]); num=[4000];    % Plant
nc= conv([1 .125], num);
dc=conv(6.5*[1 .02],den);                      % Plant + Compr.
margin(num,den); hold on margin(nc,dc)         % Margins

w= 2*pi*logspace(-.5,1,400);                   % Code for Nyquist Plots
[re,im]=nyquist(num,den,w);                    % Uncompensated
plot(re,im)
```

```
grid on
axis([-4,.5,-1,.4])
hold on
w=2*pi*logspace(-.8,1,400);
[re,im]=nyquist(nc,dc,w);        % Compensated
plot(re,im)

sys=tf(nc,dc);                   % Code for closed loop step
sys2=feedback(sys,1,-1);         % close loop
step(sys2)
```

The Bode plots in Figure 3.9-15a show that, above about 0.1 rad/s, the lag compensator has cut the gain by a constant amount and, at the gain-crossover frequency (1.27 rad/s), it adds negligible phase lag. This stabilizes the system, and the compensated phase margin is almost exactly equal to the design value of 25 degrees. The phase lag of the compensator can be seen to be concentrated in the range 0.01 to 0.5 rad/s. The Nyquist plots in Figure 3.9-15b show the unstable uncompensated system and the stable compensated system. Because of the small phase margin the closed-loop step response is lightly damped (overshoot > 50 percent), and the design could easily be repeated to increase the phase margin.

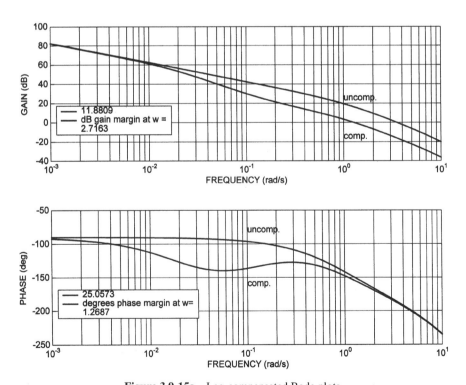

Figure 3.9-15a   Lag-compensated Bode plots.

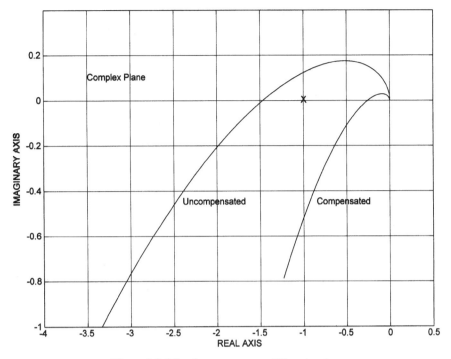

Figure 3.9-15b Lag-compensated Nyquist plot.

Example 3.9-5: Design of a Passive Lead Compensator. In this example we will use a passive lead compensator to stabilize an unstable type-2 system. The design specifications will be to achieve a phase margin of 45 degrees with the highest possible acceleration error coefficient K_a and a compensator pole/zero ratio not greater than 10. Using the passive lead transfer-function from Table 3.3-1, the loop transfer function will be

$$G_c(s)G_p(s) = \frac{(s+z_c)}{(s+p_c)} \frac{K}{s^2(s+5)}$$

The usual starting point for lead-compensator design is to choose the frequency of maximum lead to be equal to the phase-margin frequency of the plant. A Bode plot of the plant only, with $K=1$, shows that the phase-margin frequency and the phase margin are, respectively,

$$\omega_\phi = 0.45 \text{ rad/s}$$

$$\phi_M = -5.1° \text{ (unstable)}$$

The required compensator phase lead is obtained from the design specification, with an allowance of an extra 5 degrees:

242 MODELING, DESIGN, AND SIMULATION TOOLS

$$\phi_{MAX} = (45° - (-5.1°)) + 5° \approx 55°$$

Therefore, from (3.9-31a), the compensator zero/pole ratio is

$$\frac{z_c}{p_c} = \frac{1 - \sin(\phi_{MAX})}{1 + \sin(\phi_{MAX})} = 10.05$$

By setting

$$\omega_{\phi_M} = \omega_\phi$$

as noted above, the compensator equation (3.9-31d) gives

$$z_c = \omega_\phi \sqrt{(z_c/p_c)} = 0.142 \text{ rad/s}$$

$$p_c = 10 z_c = 1.42 \text{ rad/s}$$

The compensated phase-margin can now be checked, and Figure 3.9-16 shows the compensated and uncompensated Bode plots, as well as the gain and phase margins. The phase margin is only 42.3 degrees but, if we adjust the compensator to move the peak of the phase curve to the left, it will coincide with the gain crossover and the phase margin will be adequate. It is also evident that the gain margin is bigger than required and, if we raise the loop gain, the phase margin will improve without

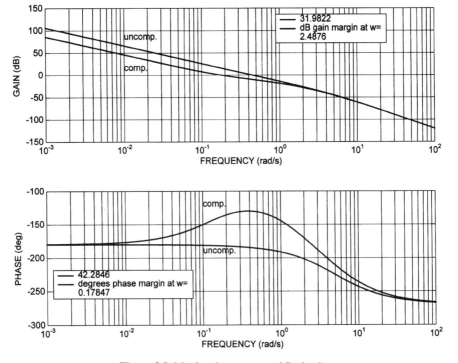

Figure 3.9-16 Lead-compensated Bode plots.

FEEDBACK CONTROL   **243**

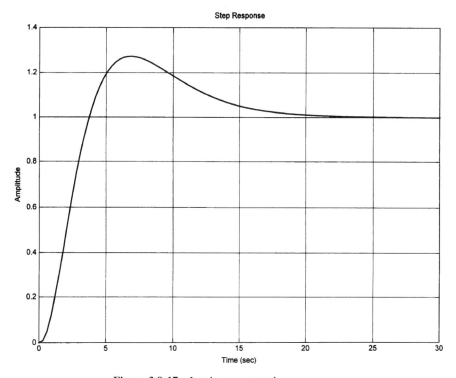

Figure 3.9-17 Lead-compensated step response.

changing the compensator. For the next design iteration $K = 3$ was used, and the peak of the phase curve occurred exactly at the gain crossover, with a phase margin of 50 degrees and gain margin of 22.4 dB.

The phase margin of 50 degrees is greater than required by our specification, and we could either retain $K = 3$ and retune the compensator using a reduced p/z ratio, or increase K and retune the compensator using the same p/z ratio. For example, we can achieve $\phi_M = 45°$, with $K = 13$ if we keep $(p_c/z_c) = 10$ and use $z_c = 0.31$. Another consideration is that the acceleration error coefficient is given by

$$K_a = \frac{z_c}{p_c} \frac{K}{5}$$

and should be checked as we trade gain K with compensator pole/zero ratio. However, even with the generous margins of 50 degrees in phase and 22.4 in gain, the step response, plotted in Figure 3.9-17, has a large overshoot. The next example illustrates a way to overcome this problem. ■

In summary, a phase-lead cascade compensator has the effect of increasing closed-loop bandwidth, thereby producing a faster system. It usually provides a moderate increase in error coefficient, and an overshoot in the closed-loop step response.

244 MODELING, DESIGN, AND SIMULATION TOOLS

The frequency-domain design techniques that have been illustrated above do not distinguish between transfer functions in the forward path or in the feedback path. If the cascade compensator is in the forward path, its zeros appear as zeros of the closed-loop transfer function. If it is in the feedback path, its poles will appear as closed-loop zeros. In the above example, the lead compensator zero, close to the origin, caused a large overshoot in the step response. If the same compensator is moved to the feedback path, the resulting closed-loop zero will be much farther to the left and the overshoot will be reduced. This technique has been used in aircraft and missile control systems.

Another technique that can be used to overcome the effect of the closed-loop zero is to cancel it with a pole of the prefilter H_r. Alternatively, instead of forward path lead-compensation, another compensation technique that similarly speeds up the system response can be used, for example, inner-loop rate-feedback.

Example 3.9-6: Feedback Compensation with a Phase Lead Network.
Here we will use the results of Example 3.9-5 with the lead compensator in the feedback path. To demonstrate the effectiveness of this technique the loop-gain has been increased to $K = 106$, when the phase margin is only 30 degrees (the optimum compensator is now $z_c = 0.93$, $p_c = 9.3$). The compensator dc gain has been increased to unity by multiplying the B and D matrices by p/z and, to maintain $K = 106$, the plant gain has been reduced by z/p. The following MATLAB code will generate step, ramp, and parabolic responses:

```
ap=[0 1 0; 0 0 1; 0 0 -5];      k=106;              % Plant
bp=[0; 0; k*z/p];  cp=[1 0 0];  dp=[0];
p=9.3;  z=.93;                                      % Compensator
ac=[-p];  bc=[p/z];  cc=[z-p];  dc=[p/z];
[a,b,c,d]= feedback(ap,bp,cp,dp,ac,bc,cc,dc,-1); % Close loop
t=[0:.005:6];
u=ones(length(t),1);            % Step Input
%u=t';                           % Ramp Input
%u=[0.5*t.^2]';                  % Parabolic Input
[y,x]=lsim(a,b,c,d,u,t);        % Time history
plot(t,y,t,u)
```

Figure 3.9-18 shows the unit-step response. The overshoot is about 5 percent, compared to 55 percent when the compensator is in the forward path, and the speed of response is about the same in each case.

When Equation (3.9-17) is applied, with $q = 2$, we find:

$$\text{step input, } n = 0: \quad e_{ss} = 0$$
$$\text{ramp input, } n = 1: \quad e_{ss} = (p/z - 1)/p = 0.968$$
$$\text{parabolic input, } n = 2: \quad e_{ss} = \infty$$

These results can be confirmed by simulation using the code given above. Therefore the system has effectively been reduced to a type-1 system. Depending on the design specifications, this may be perfectly acceptable. ∎

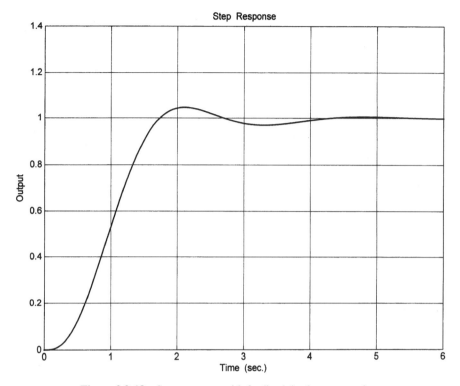

Figure 3.9-18 Step response with feedback lead-compensation.

3.10 SUMMARY

In this chapter we have developed all of the components shown in Figure 3.1-1. Two nonlinear state-space aircraft models have been provided in the form of source code. Programs for trimming, linearization, and time-response simulation have been described, and some source code is given in Appendix B. All of the development has been illustrated with applications to aircraft, so that the reader should be well prepared for aircraft control system design in Chapter 4. Our review of linear systems and feedback control has been limited to theory and techniques that we use in the text. For additional background material, the reader should consult some of the current control theory texts (Kailath, 1980; Kuo, 1987; D'Azzo and Houpis, 1988; Brogan, 1991; Nise, 1995; Dorf and Bishop, 2001; Ogata, 2002; Franklin et al., 2002).

REFERENCES

Brogan, W. L. *Modern Control Theory*. 3d ed. Englewood Cliffs, N.J.: Prentice Hall, 1991.

Brown, R. G., and P.Y.C. Hwang. *Introduction to Random Signals and Applied Kalman Filtering*. 3d ed. New York: Wiley, 1997.

D'Azzo, J. J., and C. H. Houpis. *Linear Control System Analysis and Design*. 3d ed. New York: McGraw-Hill, 1988.

Desoer, C. A., and J. D. Schulman. "Zeros and Poles of Matrix Transfer Functions and Their Dynamical Interpretation." *IEEE Transactions on Circuits and Systems*, CAS-21, pp. 3–8, 1974.

DeRusso, P. M., R. J. Roy, and C. M. Close. *State Variables for Engineers*. New York: Wiley, 1965, p. 397.

Dommasch, D. O., S. S. Sherby, and T. F. Connolly. *Airplane Aerodynamics*. 4th ed. Belmont, Calif.: Pitman, 1967.

Dorf, R. C., and R. H. Bishop, *Modern Control Systems*. 9th ed. Upper Saddle River, N.J.: Prentice Hall, 2001.

Emami-Naeini, A., and P. Van Dooren. "Computation of Zeros of Linear Multivariable Systems." *Automatica* 18, no. 4 (1982): 415–30.

Etkin, B. *Dynamics of Atmospheric Flight*. New York: Wiley, 1972.

Franklin, G. F., J. D. Powell, and A. Emami Naeini. *Feedback Control of Dynamic Systems*. 4th ed. Upper Saddle River, N.J.: Prentice Hall, 2002.

Gear, C. W. *Numerical Initial Value Problems in Ordinary Differential Equations*. Englewood Cliffs, N.J.: Prentice Hall, 1971.

Hamming, R. W. *Numerical Methods for Scientists and Engineers*. New York: McGraw-Hill, 1962.

Healey, M. "Study of Methods of Computing Transition Matrices." *Proceedings of the IEE* 120, no. 8 (August 1973): 905–912.

Hindmarsh, A. C. "Large Ordinary Differential Equation Systems and Software." *IEEE Control Systems Magazine* (December 1982): 24–30.

IMSL. *Library Contents Document*. 8th ed. International Mathematical and Statistical Libraries, Inc., 7500 Bellaire Blvd., Houston, TX, 77036, 1980.

Isaacson, E., and H. B. Keller. *Analysis of Numerical Methods*. New York: Wiley, 1966.

Kailath, T. *Linear Systems*. Englewood Cliffs, N.J.: Prentice Hall, 1980.

Kuo, B. C. *Automatic Control Systems*. Englewood Cliffs, N.J.: Prentice Hall, 1987.

Laning, J. H., and R. H. Battin, *Random Processes in Automatic Control*. New York: McGraw-Hill, 1956, Appendix C.

Laub, A. J., and B. C. Moore. "Calculation of Transmission Zeros Using QZ Techniques." *Automatica* 14 (1978): 557–66.

MacFarlane, A.G.J., and N. Karcanias. "Poles and Zeros of Linear Multivariable Systems, A Survey of the Algebraic, Geometric, and Complex-variable Theory." *International Journal of Control* 24 (1976): 33–74.

MATLAB *User's Guide*. Natick, Mass.: MathWorks, Inc., 1990.

McRuer, D., I. Ashkenas, and D. Graham. *Aircraft Dynamics and Automatic Control*. Princeton, N.J.: Princeton University Press, 1973.

MIL-F-8785C. "U.S. Dept. of Defense Military Specification: Flying Qualities of Piloted Airplanes," November 5, 1980.

Moler, C., and C. Van Loan. "Nineteen Dubious Ways to Compute the Exponential of a Matrix." *SIAM Review* 20, no. 4 (October 1978): 801–36.

Nelder, J. A., and R. Mead. "A Simplex Method for Function Minimization." *Computer Journal* 7 (1964): 308–13.

Nguyen, L. T., et al. "Simulator Study of Stall/Post-Stall Characteristics of a Fighter Airplane with Relaxed Longitudinal Static Stability." *NASA Technical Paper 1538.* Washington, D.C.: NASA, December 1979.

Nise, N. S. *Control Systems Engineering.* 2d ed. Menlo Park, Ca.: Addison-Wesley, 1995.

Ogata, K. *System Dynamics.* 3d ed. Upper Saddle River, N.J.: Prentice Hall, 1998.

———. *Modern Control Engineering.* 4th ed. Upper Saddle River, N.J.: Prentice Hall, 2002.

Phillips, E. G. *Functions of a Complex Variable.* 8th ed. Edinburgh: Oliver and Boyd, 1961.

Press, W. H., B. P. Flannery, S. A. Teukolsky, and W. T. Vetterling. *Numerical Recipes: The Art of Scientific Computing.* New York: Cambridge University Press, 1986.

Ralston, A. *A First Course in Numerical Analysis.* New York: McGraw-Hill, 1965.

Shampine, L. F., and M. K. Gordon. *Solution of Ordinary Differential Equations: The Initial Value Problem.* San Francisco: W. H. Freeman, 1975.

Taylor, J. H., and A. J. Antoniotti. "Linearization Algorithm for Computer-Aided Control Engineering." *IEEE Control Systems Magazine* 13, no. 2 (April 1993).

Van Loan, C. F. "Computing Integrals Involving the Matrix Exponential." *IEEE Transactions on Automatic Control* AC-23, no. 3 (June 1978).

West, J. C. *Analytical Techniques for Nonlinear Control Systems.* London: The English Universities Press, 1960.

Yuan, S. W. *Foundations of Fluid Mechanics.* Englewood Cliffs, N.J.: Prentice Hall, 1967.

Zakian, V. "Rational Approximants to the Matrix Exponential." *Electronics Letters* 6, no. 5 (December 10, 1970): 814–15.

PROBLEMS

Section 3.2

3.2-1 Given the mechanical system in Figure 3.2-1, add another mass, m_2, at the junction of k_2 and d_2. Let the mass have negligible friction to ground (other than d_2). Find a set of state equations for this system and write out the A, B, C, D coefficient matrices. The input is u, and there are two outputs: y and w.

3.2-2 Repeat Problem 3.2-1 with an additional spring, k_3, connected from the input, u, to the mass, m_2.

3.2-3 For the bridged-T circuit shown in Figure 3.2-2, follow the method given in Example 3.2-2, and find expressions for the rest of the elements of the A, B, C, and D matrices.

3.2-4 Use the technique from Example 3.2-2 to find a set of state and output equations for the quadratic-lag circuit in Table 3.3-1.

3.2-5 Given the differential equation

$$2\ddot{y} + 3\dot{y} + 4y = 4\ddot{u} + 6\dot{u} + u$$

turn it into an integral equation for y, draw a simulation diagram, assign state variables to the outputs of the integrators, and find a set of coefficient matrices A, B, C, D for the state equations. Show that your A and B matrices agree with Equation 3.2-6.

3.2-6 Apply Equation (3.2-9) to two sample periods and use Simpson's rule to obtain an approximation to the integral. Then obtain a recursion formula for $x(k)$.

3.2-7 Given the A matrix

$$A = \begin{bmatrix} 0 & 1 & 0 \\ 0 & 0 & 1 \\ -2 & -4 & -3 \end{bmatrix}$$

find, by hand, the eigenvalues, eigenvectors, and a modal matrix. Use (3.2-12) to find the matrix e^{At}.

3.2-8 Use the formula (3.2-18) to find e^{At}, in its simplest form, for the A matrix

$$A = \begin{bmatrix} 0 & 1 \\ -1 & -1 \end{bmatrix}$$

3.2-9 Use the Laplace transform to solve the following ODE, with α as a parameter and zero initial conditions. Reduce the solution to its simplest form (i.e., one trig. function, not two).

$$\dot{y} + y = \sin(10t + \alpha)U_{-1}(t), \qquad 0 \leq \alpha \leq \pi/2$$

Plot a few graphs (e.g., in MATLAB) of the solution, for $0 \leq t \leq 5$ s, and use these to explain the effect of different values of α. Suggest a practical situation that this model describes.

3.2-10 (a) Put the following ODE into state-space form and solve the state equations by Laplace transform.

$$\ddot{y} + 2\dot{y} + 25y = 10\sin(\omega_1 t)U_{-1}(t)$$

(b) Construct a plot of the amplitude of the particular solution, y_p, as ω_1 is varied from 1 to 20 rad/s.

Section 3.3

3.3-1 Given the following A and B matrices, use Cramer's rule to find the transfer function $X_2(s)/U(s)$.

$$A = \begin{bmatrix} 1 & 0 & 1 \\ 0 & 0 & 2 \\ -1 & -3 & -2 \end{bmatrix} \qquad B = \begin{bmatrix} 0 \\ 1 \\ 1 \end{bmatrix}$$

3.3-2 Use the Laplace transform to find the step response of the transfer function:

$$\frac{s+\alpha}{s^2+s+1}$$

with alpha as a parameter. Program the answer and obtain plots of the step response for positive and negative alpha. Describe the effect of the zero on the system step response.

3.3-3 Use the Laplace transform to find the step response of the simple-lead transfer function $s\tau/(s\tau+1)$.

3.3-4 Use the Laplace transform to find the unit impulse response of a standard form quadratic-lag transfer function.

3.3-5 (a) Show that the Laplace transform of a periodic function, $f(t) \equiv f(t+kT)$, k = integer, is given by

$$F(s) = \frac{\mathcal{L}[f_1(t)]}{(1-e^{-Ts})}, \quad \text{where } f_1(t) \equiv \begin{cases} f(t) & 0 \le t \le T \\ 0 & \text{elsewhere} \end{cases}$$

(b) Sketch the poles and zeros of $F(s)$, assuming a set for $F_1(s)$.

3.3-6 (a) The transfer function of a zero-order hold is given by:

$$G(s) = \frac{1-e^{-Ts}}{s}$$

Explain the effect of the factor $(1-e^{-Ts})$ and contrast it with the same factor in Problem 3.3-5.

(b) Sketch all of the poles and zeros of $G(s)$.

Section 3.4

3.4-1 Program the second-order ABM formula, Equation (3.4-14), as an M-file. Use it to integrate the Van der Pol equation (Example 3.4-1) and perform an execution speed versus accuracy comparison with the RK4 integration.

3.4-2 Simulate the Lorenz equations:

$$\dot{x} = 10(y-x)$$
$$\dot{y} = (r-z)x - y$$
$$\dot{z} = xy - 8z/3$$

using the format of Example 3.4-1. Investigate values of the parameter r around 24.06 and 166, and describe the behavior in detail using phase portraits and waveforms.

Section 3.5

3.5-1 (a) Program the transport aircraft model in Section 3.5.

(b) Check your model using the data in Table 3.6-1; calculate the weighted sum of squares of the derivatives for each test case.

(c) Devise an iterative algorithm, using derivative information, to minimize the sum of squares by varying just one of the variables: throttle, elevator, or alpha.

3.5-2 (a) Program the F-16 model given in Section 3.5.

(b) Make a plot of CM(alpha,el) with "el" as a parameter.

(c) Plot CZ(alpha, beta, el) to best display its 3-D nature.

(d) Make a driver program for your model in part (a) and obtain the test case results given in Table 3.5-2.

Section 3.6

3.6-1 (a) With the transport aircraft model from Problem 3.5-1, use the TRIM.m program to reproduce the steady-state trim conditions given in Table 3.6-1.

(b) Use the trim program to find out how steeply the aircraft (in clean configuration, with $x_{cg} = 0.25\bar{c}$) can climb for a range of speeds from 200 to 500 ft/s, at sea level. Compute the rate of climb (ROC) for each speed and determine the speed at which the ROC is a maximum.

3.6-2 (a) Devise a trim algorithm or use the program in the Appendix B1 to trim the F-16 model. Duplicate some longitudinal trims from Tables 3.6-2 and 3.6-3.

3.6-3 (a) Derive Equation (3.6-3) for the pitch attitude in terms of the flight-path angle.

(b) Derive Equation (3.6-5).

3.6-4 Modify the trim program used in Problem 3.6-1 to trim the transport aircraft for a prescribed angle of attack by varying V_T. Derive a trim condition for $\alpha = 15°$ at 10,000 ft.

3.6-5 Use the trim and time-history programs to duplicate the results of Examples 3.6-3 and 3.6-4.

Section 3.7

3.7-1 Given the nonlinear state equations,

$$\dot{x}_1 = x_1^3 - x_2^2 + 8$$
$$\dot{x}_2 = x_1 x_2$$

(a) find all of the singular (equilibrium) points.

(b) linearize the equations and find the algebraic "A matrix."

(c) find the numerical A matrix and its eigenvalues (by hand) at each singular point, and describe the type of perturbed behavior that you would expect near each point.

3.7-2 Program the MATLAB linearization algorithm given in Section 3.7 and add a calculation of the B matrix. Use this to confirm the results of Example 3.7-1.

3.7-3 Write a program to compute the matrices $E^{-1}A$ and $E^{-1}B$ for the decoupled longitudinal equations given in Section 2.5 from the stability derivatives. Test it on the transport aircraft model and compare the results with those given in Example 3.7-1.

3.7-4 (a) Derive the result given in Equation 3.7-4.

(b) Incorporate Equation 3.7-4 in the linearization program and compare its performance with the original algorithm, on the transport aircraft model.

Section 3.8

3.8-1 Find the eigenvalues and eigenvectors of the transport aircraft A matrix given in Example 3.8-5. Normalize the eigenvectors and describe what variables are chiefly involved in each mode.

3.8-2 Run linear and nonlinear time-history simulations of a step throttle-input to the transport aircraft model using the data of Example 3.8-5. Compare the various speed and altitude responses and confirm the points made in Example 3.8-5 about the transfer functions.

3.8-3 (a) Obtain magnitude and phase Bode plots for the transport aircraft throttle-to-speed transfer function using the data of Example 3.8-5. (b) Repeat (a) for the throttle-to-altitude transfer function. Explain how the features of the plots match the transfer function factors, and identify all asymptotes.

3.8-4 (a) Use the transport aircraft dynamics in Example 3.8-5 to find the Bode magnitude and phase plots of the elevator to pitch-rate transfer function.

(b) Determine a short-period approximation and show it on the same plots as in part (a).

(c) Repeat (a) using the elevator to pitch-attitude transfer function and explain the difference between the two sets of graphs.

Section 3.9

3.9-1 (a) A unity-feedback control system has the forward-path transfer function $G(s) = 18(s+2)/[s(s+3)(s+6)]$. Calculate, by hand, the steady state error when the reference input is a unit-ramp function.

(b) Confirm the answer to (a) by means of a simulation and plot.

3.9-2 (a) A feedback control system has forward path SISO transfer functions G_1 followed by G_2 in cascade, and a feedback transfer function H. An additive disturbance $D(s)$ is injected between G_1 and G_2. Find the transfer functions from D to the output Y, and to the error $(R - Y)$.

(b) If $G_1 = 10/s$, $G_2 = 1/[s(s+5)]$, and $H = 10(s+.9)/(s+9.0)$, find the error as a function of time when the disturbance is a unit step.

(c) If we are free to redistribute the gain in the forward path, how can the error be reduced?

3.9-3 (a) A unity feedback control system has $G_c = k(s+z)/s$ and $G_p = 10/(s+10)$. Determine the compensator parameters k and z to achieve dominant closed-loop poles with $\zeta = 1/\sqrt{5}$, and the highest possible error constant.

(b) This system controls the azimuth rotation of a radar antenna. The antenna is tracking a target with a velocity vector $[2000, 0, 0]^T$ starting from an initial position of $[-10000, 2000, 0]^T$ at $t = 0$ (coordinate origin at the radar, right-handed with x and y in the horizontal plane, and y pointing to the closest approach point of the target). Use MATLAB "lsim" to obtain a plot of the tracking error as the target goes past the radar. Use a state-space model and calculate an initial condition vector to avoid a large transient.

3.9-4 Repeat Example 3.9-1 and obtain your own Nyquist plot. Calculate, by hand, the value of K at the stability boundary.

3.9-5 (a) A feedback control system has the loop transfer function:

$$GH(s) = \frac{K}{(s+1)(s+2)(s+4)}$$

Sketch the Nyquist D-contour and the Nyquist plot, and label all of the significant corresponding points.

(b) Solve, by hand, the equation $Im[GH(j\omega)] = 0$, to find the value of K that gives neutral stability.

3.9-6 A feedback control system has the loop transfer function:

$$GH(s) = \frac{K(s+6)}{(s)(s+4)(s^2+4s+8)}$$

(a) Make a rough sketch of the expected root-locus plot.

(b) Calculate, by hand, any real-axis breakaway and entry points, and the angles and real-axis intersection points of any asymptotes.

(c) Use any available commercial software to get an accurate root-locus plot.

3.9-7 Redesign Example 3.9-2 with root locus, to try to get dominant poles with $\zeta = 1/\sqrt{2}$ and p/z not greater than 10.

3.9-8 Given the loop transfer function $G(s)H(s) = K(s+1)/[s(s-1)]$, with $K > 0$, (a) draw the D-contour and Nyquist plot and identify corresponding points on each; (b) calculate the value of K that gives marginal stability; (c) find the gain and phase margins when $K = 2$.

3.9-9 (a) Show that the polar plot of the transfer function $G(s) = (s+z)/(s+p)$ is a semicircle on the real axis as diameter.
  (b) Derive the expression for the maximum phase-lead or phase-lag, the frequency at which this occurs, and the gain at this frequency.
  (c) Sketch the polar plot for the phase-lead case.

3.9-10 Design a lead compensator for the unity-feedback control system in Example 3.9-2 [forward-path transfer function $100/(s^2(s+10))$]. Use a lead compensator with a pole-to-zero ratio of 10. Design for the largest possible loop gain consistent with a gain margin of at least 12 dB and (a) a 30 degree phase margin; (b) a 45 degree phase margin. Derive the state equations with the compensator included (as in Example 3.9-2), close the loop with the appropriate gain matrix, and compare the step responses of these two designs.

CHAPTER 4

AIRCRAFT DYNAMICS AND CLASSICAL CONTROL DESIGN

4.1 INTRODUCTION

In the previous chapters we have developed mathematical tools, realistic aircraft models, and algorithms for performing flight simulation and flight control design. Before we attempt to use all of these tools, models, and algorithms, we must have a clear idea of their applicability and the rationale and design goals for automatic flight control systems. Some idea of the history of the development of automatic flight controls is helpful in this respect.

Historical Perspective

The success of the Wright brothers in achieving the first powered flight in December 1903 has been attributed to both their systematic design approach (they built and used a wind tunnel), and the emphasis they placed on making their aircraft controllable by the pilot rather than inherently stable. However, the difficulties of controlling the early aircraft and the progress toward longer flight times led quickly to the development of an automatic control system. Thus, in 1912 an autopilot was developed by the Sperry Gyroscope Company and tested on a Curtiss flying boat. By 1914 the "Sperry Aeroplane Stabilizer" had reached such a state of development that a public flying demonstration was given in which the mechanic walked along the wing while the pilot raised his hands from the controls.

World War I (1914–1918) provided the impetus for great progress in aircraft design. However, a human pilot was perfectly capable of providing the normal stabilizing and control functions for the aircraft of this era, and the time was not ripe for rapid developments in automatic control. The small perturbation theory of aircraft dynamics had been developed (Bryan, 1911) and in the 1920s stability derivatives were measured and calculated, and the theory was confirmed by flight-tests. Little practical use was made of the theory because even the problem of finding the roots of a quartic equation was difficult at the time. Development of autopilots continued, using gyroscopes as the reference sensor and pneumatic servomechanisms to position the control surfaces. A Sperry autopilot also helped Wiley Post to fly around the world in less than eight days in 1933.

In the late 1930s classical control theory began to develop. The need to design stable telephone repeater amplifiers with closely controlled gain led to the work of Black in "regeneration theory" and to Nyquist's frequency-domain stability criterion. The same stimuli also led to Bode's complex-frequency-domain theory for the relationships between gain and phase, and his logarithmic plots of gain and phase. World War II (1939–1945) led to further developments in control theory because of the need for radar tracking, and the development of servomechanisms for positioning guns and radar antennas. Once again wartime spurred improvements in aircraft design. The large expansion of the speed-altitude envelope and the need to carry and dispose of large payloads led to large variations in the aircraft dynamics, thus creating a need to analyze the dynamic behavior. Larger aircraft required power-boosted control surfaces, and developments in hydraulic servomechanisms resulted. Also, the need to fly at night and in bad weather conditions led to developments in radio-navigation aids and a need to couple these aids to the autopilot. Thus, in 1947, a U.S. Air Force C-53 made a transatlantic flight, including takeoff and landing, completely under the control of an autopilot.

By the late 1940s the concepts of frequency response and transfer functions had become more generally known and the first analog computers were becoming available. The root-locus technique, published by W. R. Evans in 1948, was a major development in analyzing and designing control systems (it is even more useful in the computer age!). Analyses of the stability and performance of aircraft under automatic control began to be performed more commonly by the aircraft companies. The aircraft altitude-speed envelope was being expanded rapidly by the first jet fighters and by a series of research aircraft (the "X" series in the United States).

The rocket-powered Bell X-1 aircraft made its first flight in January 1946; in October 1947 it achieved supersonic flight, and in August 1949 an altitude of nearly 72,000 ft was reached. The envelope was extended farther by the next generation of X-planes, X-1A through X-1D. After reaching Mach 2.44 and 75,000 ft altitude, *inertia coupling* (see Sections 1.5 and 4.5) caused the X-1A to spin around all three axes, almost killing the pilot, Major Charles Yeager. Inertia coupling effects were encountered because these aircraft had the basic form of a modern jet fighter with short stubby wings, most of the mass concentrated along the longitudinal axis, and relatively small tail surfaces for directional stability. Before the problem was fully understood, a number of aircraft of the period suffered inertia coupling effects, sometimes with disastrous results. These included the X-2 and X-3 and the F-100 jet fighter during the course of its production program in 1953.

Many other factors besides inertia coupling contributed to the need for a more analytical approach to aircraft stability and control problems. The changes in aircraft mass properties, together with the need to reduce the area of the aerodynamic surfaces (for lower drag at high speed), caused changes in the natural modes of the aircraft, so that they were no longer easily controllable by the pilot. In addition, the damping of the natural modes tended to decrease as the altitude limits of the airplanes were expanded; these factors made it more important to predict the frequency and damping of the modes analytically. Also, the expansion of the aircraft speed-altitude envelope meant that much greater variations in the dynamics of the aircraft were encountered.

Power-boosted or fully powered control surfaces were introduced because of the increasing aerodynamic loads associated with greater performance and larger aircraft, and because they could eliminate the many hours of flight-test needed to balance the control surfaces carefully. Properly balanced control surfaces were previously necessary to provide a suitable feel to the pilot's controls. With power-boosted controls the feel could be modified with springs and bobweights, and with fully powered irreversible controls the feel could be provided completely artificially. Thus, the "handling qualities" of the aircraft could be adjusted to be satisfactory over a very wide envelope. Power-boosted controls also made possible the use of *stability augmentation*, in which signals from angular rate sensors could be fed to the control surface actuators to modify the natural modes of the aircraft. In addition, they facilitated the use of more complex autopilots.

The year 1949 saw the first flight of the de Havilland Comet, the aircraft that essentially defined the modern jet transport aircraft. In the early 1950s the problems of supersonic flight up to Mach 3 and beyond were beginning to be investigated. The Lockheed X-7 unmanned rocket plane was built to provide a testbed for a ramjet engine. During a five-year test program beginning in 1951, it also provided information on high-speed aerodynamics, aerothermodynamics, special fuels, and special materials. Data from programs such as this undoubtedly contributed to the design of aircraft such as the F-104 and the SR-71. The X-15 rocket plane, which first flew in 1959, expanded the envelope for manned flight to beyond Mach 6 and above 300,000 ft. This aircraft was equipped with a Honeywell-designed adaptive control system that provided three-axis stability augmentation, and a transition from aerodynamic control to reaction control as the aerodynamic controls became ineffective at high altitude.

In the early 1960s small fighter aircraft were approaching Mach 2 speeds; a French Mirage achieved Mach 2.3, and later an F-4 Phantom made a record-breaking Mach 2.4 flight. In the civil aviation field, this was the time of the Boeing 707 and Douglas DC8 passenger jets, and the development of the Aerospatiale-British Aerospace Concorde SST. The digital computer was beginning to have a major impact on engineering, the techniques of numerical analysis assumed greater importance, and this stimulated the growth of modern control theory in the mid-1960s.

A great deal of hypersonic aerodynamics knowledge was gained from the X-15 program and from hypersonic wind tunnel studies in the late 1950s. The X-20 (Dyna-Soar) vehicle, to be built by Boeing under a 1960 contract, was to be a rocket-launched unpowered glider that would gather data to solve the problems of pilot-controlled reentry from orbit. The final design was a unique V-shaped vehicle with a thick wing and upturned wing tips. Although the program was canceled before completion of the first vehicle, it pioneered the technology for the U.S. space shuttle. Later, the unmanned ASSET (1963–1965) and PRIME (X-23A; 1966–1967) vehicles provided flight data on structures, materials, control systems, and other technologies for maneuvering reentry. This was followed in 1969 and the early 1970s by the X-24 manned, blunt lifting-body vehicles. These provided data on the low-speed characteristics of maneuverable reentry vehicles, including stability characteristics, pilot experience for comparison with simulators, man-vehicle interface data, and much control system information.

Because of the digital computer the 1970s saw great strides in computational fluid dynamics, structural and flutter (structural divergence) analysis, simulation of complex dynamical systems, and the application of guidance and control theory in real-time onboard digital computers. Simulation techniques made possible realistic pilot training on the ground, and the automatic flight control system on board an aircraft allowed the dynamic behavior of an entirely different aircraft to be simulated. Thus, space-shuttle pilots trained on a Gulfstream-II aircraft that simulated the feel of the space shuttle.

In the 1970s, flight-control technology advances allowed the F-16 aircraft to be designed for "relaxed static stability" and all-electric (full "fly-by-wire") control. Previous aircraft had used "high-authority" electrical control superimposed on the basic electrohydraulic system (e.g., the F-111), or as in the case of the Concorde, an electrical system with mechanical backup. The processing of the electrical signals for the automatic flight control systems was still in analog rather than digital form.

The 1980s saw the flight testing of aircraft with additional aerodynamic control surfaces that provided direct-lift control or direct sideforce control (such as the AFTI F-16 and the Grumman forward-swept wing X-29A aircraft) and with digital flight control systems (e.g., McDonnell F-15E and F-18). The AFTI F-16 aircraft allowed the use of sideforce control through a ventral fin, and direct-lift control through the combination of the horizontal tail and wing leading-edge flaps. The *decoupled motions* provided by this control were evaluated for possible use in combat situations. The X-29A research aircraft is unstable in pitch (-35 percent static margin at low speed) and has three-surface pitch control (canards, wing flaperons, and strake flaps). The flight control system is a triply redundant digital system (three digital processors with "voting" to eliminate a faulty channel) with analog backup for each processor. These aircraft raise interesting multivariable control problems for modern control theory.

The U.S. space shuttle made its first flight in March 1981. There was also a resurgence of interest in hypersonic flight during the 1980s. Single-stage-to-orbit vehicles were studied, including the British HOTOL (horizontal takeoff and landing) unmanned satellite launch vehicle and the U.S. TAV (Trans-Atmospheric Vehicle)—fully reusable rapid-turnaround vehicles for manned reconnaisance, weapon delivery, and delivery of large payloads to orbit. These were followed in the United States by the NASP (National Aerospace Plane) study contracts on a manned single-stage-to-orbit vehicle. Other studies looked at boosted vehicles; these included the French HERMES vehicle (similar to the space shuttle, manned), and a number of U.S. BGVs (boost-glide vehicles).

Many lessons were learned about the control of hypersonic vehicles. The trajectories must be carefully controlled because the frictional heating in the atmosphere can create temperatures of a few thousand degrees Fahrenheit at critical points on the vehicle. A change in flight conditions can cause localized changes in the airflow, from laminar to turbulent flow, and this can lead to a rapid increase in temperature at some point on the surface of the vehicle. Manual control is difficult or not feasible in most flight phases and, if attempted, would limit the performance. The trajectory can be controlled by feedback comparison with a precomputed reference trajectory, or with real-time trajectory-prediction calculations (as in the case of the space shuttle).

There can be large uncertainties in the aerodynamic coefficients of the hypersonic vehicles, and this complicates the design of the automatic control systems and limits their performance. The control systems must be adapted (gain-scheduled, or self-adaptive) in flight, to allow for the wide variations in vehicle dynamics over the large flight envelope. If fixed "scheduling" is used, difficulties are encountered in sensing the flight conditions. External probes sensing "air data" (dynamic pressure and Mach) can only be used at low Mach numbers, and the air data must be derived from the navigation system and a stored model of the atmosphere. The real atmosphere can show large, unpredictable variations in density; therefore, the control systems must be designed to tolerate these variations.

The era of true "aerospace" vehicles introduces many new challenges for the control engineer. He or she must now think in terms of guidance and control, algorithms and simulation, and numerical methods and digital implementation. Many relatively new analytical techniques are required, including numerical optimization, analysis of sensitivity and robustness to parameter variations, adaptive techniques, and multivariable control. Furthermore, the control engineer can no longer work in isolation; many other technologies will be closely integrated into a design, and constraints will be imposed on the designs from a variety of sources (e.g., structural, thermal, propulsion, energy management and performance, and human factors).

The Need for Automatic Control Systems

The evolution of modern aircraft created a need for power-driven aerodynamic control surfaces and automatic-pilot control systems, as described in the preceding subsection. In addition, the widening performance envelope created a need to augment the stability of the aircraft dynamics over some parts of the envelope. This need for stability augmentation is now described in more detail.

Figure 4.1-1 shows the altitude-Mach envelope of a modern high-performance aircraft; the boundaries of this envelope are determined by a number of factors. The low-speed limit is set by the maximum lift that can be generated (the alpha limit in the figure), and the high-speed limit follows a constant dynamic pressure contour (because of structural limits, including temperature). At the higher altitudes the speed becomes limited by the maximum engine thrust (which has fallen off with altitude). The altitude limit imposed on the envelope is where the combination of airframe and engine characteristics can no longer produce a certain minimum rate of climb (this is the "service ceiling").

INTRODUCTION   259

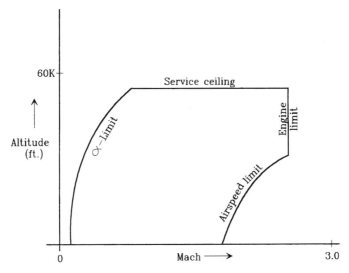

Figure 4.1-1   Aircraft altitude-Mach envelope.

The aircraft envelope covers a very wide range of dynamic pressure. For example, in the landing phase the dynamic pressure may be as low as 50 psf, whereas at Mach 1.2 at sea level the dynamic pressure is 2150 psf. Large variations in dynamic pressure cause correspondingly large variations in the coefficients of the dynamic equations. Other factors also contribute to changes in the aircraft dynamics. The basic aerodynamic coefficients change with Mach number and as functions of the aerodynamic angles, and the mass properties change with different payloads and changing fuel load.

Because of the large changes in aircraft dynamics, a dynamic mode that is stable and adequately damped in one flight condition may become unstable, or at least inadequately damped, in another flight condition. A lightly damped oscillatory mode may cause a great deal of discomfort to passengers or make it difficult for the pilot to control the trajectory precisely. These problems are overcome by using feedback control to modify the aircraft dynamics. The aircraft motion variables are sensed and used to generate signals that can be fed into the aircraft control-surface actuators, thus modifying the dynamic behavior. This feedback must be adjusted according to the flight condition. The adjustment process is called *gain scheduling* because, in its simplest form, it involves only changing the amount of feedback as a function of a "scheduling" variable. These scheduling variables will normally be measured dynamic pressure and/or Mach number.

In the case of low-performance aircraft with relatively narrow envelopes and control surfaces that are not power driven, an unsatisfactory dynamic mode must be corrected by modifying the basic design. As in the case of the high-performance aircraft, this requires an understanding of the dynamic modes and their dependence on the aerodynamic coefficients and aerodynamic derivatives.

260 AIRCRAFT DYNAMICS AND CLASSICAL CONTROL DESIGN

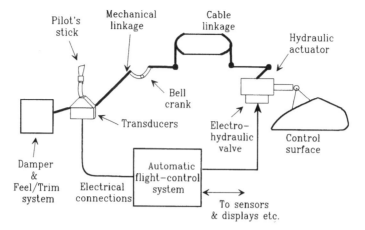

Figure 4.1-2   An electromechanical control system.

Figure 4.1-2 shows how a fully powered aircraft control system might be implemented with mechanical, hydraulic, and electrical components. Because the control surfaces are fully power driven there is no force or motion feedback to the pilot's control stick. This is called an *irreversible* control system, and bob weights and springs (or electrical or hydraulic devices) must be added to the control stick to provide some "feel" to the pilot. The stick and rudder pedals are shown linked to the actuators by a combination of mechanical links and bell cranks, and control wires. The control surfaces are driven by a hydraulic servomechanism that has a follow-up action; that is, the high-power output shaft is driven until its position corresponds to the position of the low-power input shaft.

Augmentation signals are conveniently added to the system of Figure 4.1-2 by electrical means. The signals from rate gyros (angular-rate measuring devices), accelerometers, the air-data computer, and other sources are processed by the flight-control computer. The electrical output of the flight-control computer (converted to analog form) is used to drive electrohydraulic valves, and these superimpose additional motion on the hydromechanical control system.

The Functions of the Automatic Control Systems

The descriptions and analyses of aircraft modes in Chapters 3 and 4 show that they can be divided into different categories. One category includes modes that involve mainly the rotational degrees of freedom; these are the short-period, roll, and dutch roll modes. Their natural frequencies (or time constants, if purely exponential) are determined by the moments of inertia of the aircraft and the moments generated by the aerodynamic surfaces; their damping is determined by the rate-dependent aerodynamic moments. The remaining modes (phugoid and spiral) involve changes in the flight path and are much slower modes. The phugoid mode involves the translational degrees of freedom and is dependent on the aerodynamic forces of lift and drag, and their variation with speed. The spiral mode depends on aerodynamic moments, but only weak aerodynamic forces are involved.

The responsiveness of an aircraft to maneuvering commands is determined in part by the speed of the rotational modes. The frequencies of these modes tend to be sufficiently high that a pilot would find it difficult or impossible to control the aircraft if the modes were lightly damped or unstable. Therefore, it is necessary to provide automatic control systems to give these modes suitable damping and natural frequencies. Such control systems are known as *stability augmentation systems* (SAS). If the augmentation system is intended to control the mode and to provide the pilot with a particular type of response to the control inputs, it is known as a *control augmentation system* (CAS). An example of this is a normal-acceleration CAS, in which the pilot's inputs are intended to control the acceleration generated along the negative z-axis.

The slow modes (phugoid and spiral) are controllable by a pilot. But since it is undesirable for a pilot to have to pay continuous attention to controlling these modes, an automatic control system is needed to provide "pilot relief." An *autopilot* is an automatic control system that provides both pilot relief functions and special functions such as automatic landing.

The common types of SAS, CAS, and the autopilot functions can be listed as follows:

SAS	CAS	Autopilots
Roll damper	Roll rate	Pitch attitude hold
Pitch damper	Pitch rate	Altitude hold
Yaw damper	Normal acceleration	Speed/Mach hold
	Lateral/directional	Automatic landing
		Roll-angle hold
		Turn coordination
		Heading hold/VOR hold

These control systems are described and illustrated by numerical examples in Sections 4.4 through 4.7.

4.2 AIRCRAFT RIGID-BODY MODES

In this section algebraic expressions for the rigid-body modes will be derived so that their dependence on the stability derivatives and on the flight conditions can be examined, and so that conditions for stability can be deduced. When "lat-long" decoupling occurs, it becomes feasible to manipulate the aircraft transformed state equations algebraically. Both the longitudinal and lateral-directional dynamics are still fourth-order, so the modes are obtained from the roots of a fourth-order characteristic polynomial. Algebraic solution of a quartic equation is not practicable, but with some simplifying assumptions based on knowledge of the stability derivatives and the physics of flight, this problem can be bypassed.

Algebraic Derivation of Longitudinal Transfer Functions and Modes

The coefficient matrices for the decoupled longitudinal state equations are given in (2.6-29). The SISO transfer functions can be derived very easily by applying Cramer's rule to the Laplace transformed state equations, as follows. The matrix $(sE - A)$ is given by

$$(sE - A) = \begin{bmatrix} s(V_{T_e} - Z_{\dot{\alpha}}) - Z_\alpha & -(V_{T_e} + Z_q) & -Z_V + X_{T_V}\sin(\alpha_e + \alpha_T) & g_D \sin \gamma_e \\ -sM_{\dot{\alpha}} - M_\alpha - M_{T_\alpha} & s - M_q & -(M_V + M_{T_V}) & 0 \\ -X_\alpha & 0 & s - X_V - X_{T_V}\cos(\alpha_e + \alpha_T) & g_D \cos \gamma_e \\ 0 & -1 & 0 & s \end{bmatrix}$$

(4.2-1)

and the B matrix is

$$B = \begin{bmatrix} Z_{\delta e} & -X_{\delta t}\sin(\alpha_e + \alpha_T) \\ M_{\delta e} & M_{\delta t} \\ X_{\delta e} & X_{\delta t}\cos(\alpha_e + \alpha_T) \\ 0 & 0 \end{bmatrix}$$

If, for example, the $q/\delta e$ transfer function is required, the δ_e column of B must be substituted for the q column of $|sE - A|$. The transfer function is

$$\frac{q}{\delta_e} = \frac{1}{|sE - A|} \begin{vmatrix} s(V_{T_e} - Z_{\dot{\alpha}}) - Z_\alpha & Z_{\delta e} & -Z_V + X_{T_V}\sin(\alpha_e + \alpha_T) & g_D \sin \gamma_e \\ -sM_{\dot{\alpha}} - M_\alpha - M_{T_\alpha} & M_{\delta e} & -(M_V + M_{T_V}) & 0 \\ -X_\alpha & X_{\delta e} & s - X_V - X_{T_V}\cos(\alpha_e + \alpha_T) & g_D \cos \gamma_e \\ 0 & 0 & 0 & s \end{vmatrix}$$

(4.2-2)

It is evident from inspection of the determinant that this transfer function is of the form

$$\frac{q(s)}{\delta_e(s)} = \frac{s(b_2 s^2 + b_1 s + b_0)}{a_4 s^4 + a_3 s^3 + a_2 s^2 + a_1 s + a_0} \quad (4.2\text{-}3)$$

Expressions for the numerator and denominator coefficients can be derived in a straightforward way by expanding the determinants. However, the coefficients are complicated functions of the dimensional derivatives, and are tedious to evaluate without a digital computer. This is a feasible method of deriving transfer functions from the stability derivatives, but it relies on lat-long decoupling and provides very little insight. We will now examine various approximations that lead to transfer functions that are simple enough to provide some insight into the dynamic behavior.

Consider the decoupled longitudinal dynamics; a time-history simulation in Section 3.6 showed that it was possible to excite separately the short-period and phugoid modes. In the phugoid case speed and theta varied, with alpha and q almost constant; while in the short-period case alpha, q, and theta varied, with speed constant. This implies additional decoupling in the dynamic equations that will now be investigated.

Returning to the longitudinal coefficient matrices (2.6-29), with state and control vectors:

$$x = [\alpha \; q \; v_T \; \theta]^T \qquad u = [\delta_e \; \delta_t]^T \qquad (4.2\text{-}4)$$

partition the state equations as:

$$\begin{bmatrix} E_1 & 0 \\ 0 & I \end{bmatrix} \dot{x} = \begin{bmatrix} A_{11} & A_{12} \\ A_{21} & A_{22} \end{bmatrix} x + \begin{bmatrix} B_{11} & B_{12} \\ B_{21} & B_{22} \end{bmatrix} u \qquad (4.2\text{-}5)$$

with $E_1(2 \times 2)$, $A_{ij}(2 \times 2)$, and $B_{ij}(2 \times 1)$. Now if it is to be possible for v_T and θ to vary, without significant changes in α and q, the submatrix A_{12} must introduce very little coupling from the second set of equations into the first. An examination of the appropriate terms of the matrix $E^{-1}A$ in Example 3.7-2 shows that this is the case in that particular example. More generally, the a_{14} term is null when the flight-path angle is zero, and a_{24} is identically zero. The a_{23} term is insignificant when the tuck derivative and the thrust derivative M_{T_V} are negligible. The tuck derivative is negligible at low Mach numbers, and the thrust derivative is often negligible because the thrust vector passes close to the cg. When the a_{13} term is expanded in terms of dimensionless derivatives, components due to $2g_D/V_{T_e}$, C_{L_V}, and C_{T_V} are found. The gravity term is small at normal airspeeds, and the variation of lift coefficient with airspeed is negligible at low Mach numbers. The thrust derivative depends on the type of propulsion, but it is found to be multiplied by $\sin(\alpha_e + \alpha_T)$ and is then usually small under normal flight conditions. In summary, the conditions for decoupling in the A matrix include small flight path angle, small angle of attack, and low Mach number.

If the control stick is held fixed, and there are no feedback control systems operating, the input u is null and we can ignore the B matrix. The eigenvalues of the A matrix then yield *stick-fixed modes* of the aircraft. Here we will look for decoupling in the B matrix and find a transfer function for the decoupled equations. The B_{12} block in (4.2-5) includes the variation of the thrust and pitching moment coefficients with throttle changes. The thrust coefficient term is multiplied by $\sin(\alpha_e + \alpha_T)$ and may often be neglected; the pitching moment term is negligible when the thrust vector passes close to the cm.

We will now neglect the A_{12} and B_{12} terms in (4.2-5), and extract the alpha and pitch-rate equations from the complete dynamics to obtain a short-period transfer function.

The Short-Period Approximation

The short-period approximation obtained from (4.2-5) is

$$\begin{bmatrix} V_{T_e} - Z_{\dot{\alpha}} & 0 \\ -M_{\dot{\alpha}} & 1 \end{bmatrix} \begin{bmatrix} \dot{\alpha} \\ \dot{q} \end{bmatrix} = \begin{bmatrix} Z_\alpha & V_{T_e} + Z_q \\ M_\alpha & M_q \end{bmatrix} \begin{bmatrix} \alpha \\ q \end{bmatrix} + \begin{bmatrix} Z_{\delta_e} \\ M_{\delta_e} \end{bmatrix} \delta_e, \qquad (4.2\text{-}6)$$

where, for compactness, M_α will be assumed to include M_{T_α}. The transfer function matrix is given by

$$C(sE - A)^{-1}B = \frac{C}{\Delta_{sp}}\begin{bmatrix} (s - M_q)Z_{\delta e} + (V_{T_e} + Z_q)M_{\delta e} \\ (sM_{\dot\alpha} + M_\alpha)Z_{\delta e} + \left[s(V_{T_e} - Z_{\dot\alpha}) - Z_\alpha\right]M_{\delta e} \end{bmatrix},$$

where C is the appropriate coupling matrix for α or q, and Δ_{sp} is the short-period characteristic polynomial:

$$\Delta_{sp} = (V_{T_e} - Z_{\dot\alpha})s^2 - \left[Z_\alpha + (V_{T_e} - Z_{\dot\alpha})M_q + (V_{T_e} + Z_q)M_{\dot\alpha}\right]s$$
$$+ M_q Z_\alpha - (V_{T_e} + Z_q)M_\alpha \qquad (4.2\text{-}7)$$

The individual transfer functions are:

$$\frac{\alpha}{\delta_e} = \frac{Z_{\delta e} s + (V_{T_e} + Z_q) M_{\delta e} - M_q Z_{\delta e}}{\Delta_{sp}} \qquad (4.2\text{-}8)$$

$$\frac{q}{\delta_e} = \frac{\left[(V_{T_e} - Z_{\dot\alpha}) M_{\delta e} + Z_{\delta e} M_{\dot\alpha}\right]s + M_\alpha Z_{\delta e} - Z_\alpha M_{\delta e}}{\Delta_{sp}} \qquad (4.2\text{-}9)$$

The short-period mode is normally complex, so comparing the denominator with the quadratic standard form (3.3-16) gives:

$$\omega_{n_{sp}}^2 = \frac{M_q Z_\alpha - M_\alpha(V_{T_e} + Z_q)}{V_{T_e} - Z_{\dot\alpha}} \qquad (4.2\text{-}10a)$$

$$-2\zeta_{sp}\omega_{n_{sp}} = M_q + \frac{M_{\dot\alpha}(V_{T_e} + Z_q) + Z_\alpha}{V_{T_e} - Z_{\dot\alpha}} \qquad (4.2\text{-}10b)$$

The derivatives Z_q and $Z_{\dot\alpha}$ are normally small compared to V_{T_e} and will be dropped from these equations. Then when the dimensionless derivatives are substituted, and the approximation $(C_D \ll C_{L_\alpha})$ is used, the results are:

$$\omega_{n_{sp}}^2 = \frac{\bar{q}S\bar{c}}{J_y}\left[-C_{m_\alpha} - \frac{\rho S\bar{c}}{4m}C_{m_q}C_{L_\alpha}\right] \qquad (4.2\text{-}11)$$

$$\zeta_{sp} = \frac{1}{4\sqrt{2}}\left[\frac{(\rho S\bar{c})\bar{c}^2}{J_y}\right]^{1/2} \frac{\left[-C_{m_q} - C_{m_{\dot\alpha}} + \frac{2J_Y}{m\bar{c}^2}C_{L_\alpha}\right]}{\left[-C_{m_\alpha} - \frac{\rho S\bar{c}}{4m}C_{m_q}C_{L_\alpha}\right]^{1/2}} \qquad (4.2\text{-}12)$$

In the natural frequency formula the term $(\rho S\bar{c}/4m)$ is a mass ratio, and is typically on the order of 0.001. However, $C_{m_q}C_{L_\alpha}$ may be quite large and so, depending on

C_{m_α}, we may not be able to neglect this term. If the pitch damping (C_{m_q}) is small and the term is neglected, the square of the natural frequency is given by pitch-stiffness divided by the pitching inertia ($-\bar{q}S\bar{c}C_{m_\alpha}/J_y$). This is analagous to a spring-mass torsional oscillator. It also shows that *the short-period frequency is proportional to the square-root of dynamic pressure*, within the limits of this approximation. The fact that damping creeps into the natural frequency formula is a consequence of the approximations used earlier.

The inertia ratio ($2J_Y/m\bar{c}^2$) that occurs in the damping formula is equal to twice the square of the quantity: pitching radius of gyration over mean chord. This may be quite large and the C_{L_α} term is likely to dominate the numerator. The denominator is the same expression that occurred in the natural frequency formula. If the mass-ratio term is not dominant, then the variation of the short period damping ratio is determined by the square root of the inertia ratio $[(\rho S\bar{c})\bar{c}^2/J_y]$, and is therefore *proportional to the square root of the air density*.

It must be emphasized again that the above results are only valid at low Mach numbers where the stability derivatives are reasonably constant. Also, the above analysis assumed a damped, oscillatory short-period mode; different behavior will be illustrated later.

The Phugoid Approximation

Approximations for the natural frequency and damping of the phugoid mode will be developed by extending the approach used to derive the short-period results. Refer again to (4.2-5) and assume that only the phugoid mode has been excited. If the derivatives $\dot{\alpha}$ and $\dot{q}$ are then neglected, the first pair of equations reduce to algebraic equations that act as a constraint on the remaining differential equations in the phugoid variables. Therefore, we have

$$0 = A_{11} \begin{bmatrix} \alpha \\ q \end{bmatrix} + A_{12} \begin{bmatrix} v_T \\ \theta \end{bmatrix}$$

$$\begin{bmatrix} v_T \\ \dot{\theta} \end{bmatrix} = A_{21} \begin{bmatrix} \alpha \\ q \end{bmatrix} + A_{22} \begin{bmatrix} v_T \\ \theta \end{bmatrix}$$

When the algebraic equations are used to eliminate α and q from the differential equations, the following equations for the phugoid variables are obtained:

$$\begin{bmatrix} \dot{v}_T \\ \dot{\theta} \end{bmatrix} = \left(A_{22} - A_{21} A_{11}^{-1} A_{12} \right) \begin{bmatrix} v_T \\ \theta \end{bmatrix} \qquad (4.2\text{-}13)$$

In order to evaluate the coefficient matrix we will make the usual assumption that $\gamma_e = 0$. This greatly simplifies the derivation, but as we will see later, γ has a significant effect on the phugoid mode. Equation (4.2-13) now becomes

$$\begin{bmatrix} \dot{v}_T \\ \dot{\theta} \end{bmatrix} =$$

$$\begin{bmatrix} X'_V - \dfrac{X_\alpha \left[M_q \left(Z_V - X_{T_V} \sin(\alpha_e + \alpha_T) \right) - \left(V_{T_e} + Z_q \right) \left(M_V + M_{T_V} \right) \right]}{\Delta_p} & -g_D \\ \dfrac{M_\alpha \left(Z_V - X_{T_V} \sin(\alpha_e + \alpha_T) \right) - Z_\alpha \left(M_V + M_{T_V} \right)}{\Delta_p} & 0 \end{bmatrix} \begin{bmatrix} v_T \\ \theta \end{bmatrix}, \quad (4.2\text{-}14)$$

where

$$X'_V = X_V + X_{T_V} \cos(\alpha_e + \alpha_T)$$

$$\Delta_p = M_q Z_\alpha - M_\alpha \left(V_{T_e} + Z_q \right)$$

The characteristic equation can now be found from $|sI - A|$, and a comparison with the quadratic standard form gives the following expressions for the phugoid natural frequency and damping:

$$\omega_{n_p}^2 = g_D \frac{M_\alpha (Z_V - X_{T_V} \sin(\alpha_e + \alpha_T)) - Z_\alpha (M_V + M_{T_V})}{M_q Z_\alpha - M_\alpha (V_{T_e} + Z_q)} \quad (4.2\text{-}15a)$$

$$2\zeta_P \omega_{n_p} = -X'_V + \frac{X_\alpha \left[M_q (Z_V - X_{T_V} \sin(\alpha_e + \alpha_T)) - (V_{T_e} + Z_q)(M_V + M_{T_V}) \right]}{M_q Z_\alpha - M_\alpha (V_{T_e} + Z_q)}$$

$$(4.2\text{-}15b)$$

These expressions are considerably more complicated than those for the short-period mode; nevertheless, some conclusions can be drawn from them.

Consider the expression for the phugoid frequency, and for simplicity neglect the thrust derivatives and Z_q. Then insert dimensionless derivatives, with $C_{m_e} = 0$ and $C_{D_e} \ll C_{L_\alpha}$; the result is

$$\omega_{n_p}^2 = \frac{2g_D}{\bar{c}} \frac{C_{m_\alpha}(2C_{L_e} + C_{L_V}) - C_{L_\alpha} C_{m_V}}{C_{m_q} C_{L_\alpha} + \dfrac{4m}{\rho S \bar{c}} C_{m_\alpha}} \quad (4.2\text{-}16)$$

This is the equation that will be used to calculate the phugoid frequency, but the variation with flight conditions can be illustrated as follows. The numerator of the second fraction contains the compressibility effects C_{L_V} and C_{m_V}, and the equilibrium lift C_{L_e}. The denominator is the same as a term in the short-period equation (4.2-11), except that here $4m/(es\bar{c})\rho S\bar{c} (\approx 10^3)$ is the reciprocal of the previous mass ratio. The mass ratio times pitch-stiffness term will often dominate the denominator, and then if we also neglect the compressibility terms in the numerator, we get a very simple expression for the phugoid frequency:

AIRCRAFT RIGID-BODY MODES

$$\omega_{n_p}^2 \approx \frac{2g_D}{\bar{c}} \frac{\rho S \bar{c}}{4m}(2C_{L_e}) = \frac{2g_D^2}{V_{T_e}^2} \frac{\bar{q} S C_{L_e}}{m g_D} \qquad (4.2\text{-}17a)$$

In level flight, with a small angle of attack, the lift is approximately equal to the weight and this equation reduces to

$$\omega_{n_p} \approx \frac{g_D}{V_T} \sqrt{2} \qquad (4.2\text{-}17b)$$

Therefore, phugoid frequency is inversely proportional to airspeed, other things being equal. For a given speed, at higher altitude, alpha will be bigger and so the thrust will provide a larger component of the total vertical force, and a smaller aerodynamic lift component will be needed. Therefore, according to (4.2-17a), the frequency will be lower at higher altitude, other things being equal. The result given in Equation (4.2-17b) was found by F. W. Lanchester in 1908 and can be derived for large-amplitude motion from energy considerations.

It is more difficult to derive simple expressions for the damping of the phugoid and furthermore, in the next subsection, the damping equation (4.2-15b) is shown to be quite inaccurate. Nevertheless it is still worthwhile to examine this equation to understand what factors influence the phugoid damping. The second term in the equation is often much smaller than the first, and analyzing only the first term gives

$$2\zeta_p \omega_{n_p} \approx -\left[X_V + X_{T_V} \cos(\alpha_e + \alpha_T)\right]$$

The dimensional derivatives on the right-hand side contain the equilibrium values of drag and thrust, and we will substitute the steady-state condition (2.4-1a) for these quantities, thus

$$2\zeta_p \omega_{n_p} \approx \frac{-2g_D \sin\gamma_e}{V_{T_e}} + \frac{\bar{q} S}{m V_{T_e}}\left[C_{D_V} - C_{T_V} \cos(\alpha_e + \alpha_T)\right]$$

Now consider the level-flight case; use (4.2-17a) to substitute for ω_{n_p}, and equate lift to weight,

$$\zeta_p = \frac{1}{2\sqrt{2}} \frac{\left[C_{D_V} - C_{T_V} \cos(\alpha_e + \alpha_T)\right]}{C_{L_e}} \qquad (4.2\text{-}18)$$

The phugoid mode involves changes in speed, and this equation shows that the damping depends on the changes in drag and thrust with speed. The speed damping derivative C_{D_V} is small until the transonic drag rise begins, and then usually negative in the supersonic regime. Therefore, (4.2-18) indicates the possibility of an unstable phugoid (negative damping) in the supersonic regime, depending on the way in which thrust varies with Mach. Roskam (1979) provides (approximate) comparative analyses of the derivative C_{T_V} for jets, propeller aircraft, rocket aircraft, and unpowered aircraft. However, we should remember that even Equation (4.2-15b) does not necessarily give very accurate results for the phugoid damping. More accurate numerical results given in the next subsection show that at subsonic speeds, *the phugoid damping ratio increases with airspeed and decreases with altitude*. Example 4.2-2 shows, in addition, that the damping decreases rapidly with flight-path angle.

Accuracy of the Short-Period and Phugoid Approximations

The short-period approximation almost always gives a good approximation for the α and q response to elevator inputs with constant throttle setting, and it will play an important role in the numerical designs in this chapter. The phugoid approximation usually gives good accuracy for the period of the phugoid oscillation but not for the damping ratio. These facts are borne out by the transport aircraft model.

The dimensional-derivative evaluation program used in Example 3.7-1 was extended to calculate the short-period and phugoid properties from (4.2-7) and (4.2-14), respectively. Thus, the characteristic roots (or the frequencies and damping ratios) could be calculated for the transport aircraft from any given set of steady-state flight conditions. The program also calculated the matrix $E^{-1}A$ (as used in Example 3.7-1) so that "exact" dynamic modes could be obtained from this matrix using an eigenvalue program. The flight conditions were level flight at sea level, with different airspeeds and cg positions. Table 4.2-1 shows the results of these calculations. An asterisk in the table indicates characteristic roots instead of period and damping ratio.

TABLE 4.2-1: Accuracy of Short-Period and Phugoid Formulae

Airspeed/cg	Calculation	T_{SP}	ζ_{SP}	T_P	ζ_P
200, 0.25	Approximate	7.44	0.555	32.3	0.102
	Exact	7.33	0.565	32.7	−0.129
400, 0.25	Approximate	3.73	0.551	63.5	0.064
	Exact	3.72	0.551	63.6	0.035
600, 0.25	Approximate	2.48	0.551	96.5	0.112
	Exact	2.48	0.551	96.6	0.099
400, 0.30	Approximate	4.04	0.598	65.4	0.067
	Exact	4.04	0.524	65.5	0.033
400, 0.40	Approximate	5.04	0.744	74.1	0.083
	Exact	5.02	0.652	74.3	0.036
400, 0.50	Approximate	(−0.523,	−1.33)*	476	0.691
	Exact	(−0.810 ± j0.200)*		476	0.630
400, 0.55	Approximate	(−1.70,	−0.158,	−0.158,	0.128)*
	Exact	(−1.44,	0.100,	−0.150 ± j0.123)*	

The first three sets of entries show the effect of varying airspeed; the last four sets show the effect of moving the cg position farther aft with speed held constant. The short-period approximation is seen to be a very good approximation for the first five cases. The phugoid approximation gives accurate results for the period; the damping ratio is quite inaccurate but the accuracy appears to improve when the period is large. Note that the phugoid mode is unstable at low airspeed (200 ft/s).

When the cg is moved aft the short-period roots move onto the real axis, and then one real root moves toward the phugoid roots. The short-period and phugoid approximations break down and one real root moves into the right-half plane. At the same time a new oscillatory mode appears that has a phugoid-like period with a short-period damping ratio. This mode is sometimes known as the *third oscillatory mode*, and it is characteristic of a statically unstable airplane. Also, the fact that one real root becomes unstable signals an exponential instability in pitch (a pitch "departure") rather than an oscillatory instability. This is the kind of instability that might be intuitively associated with the loss of positive pitch stiffness.

In the example, the stability boundary for the aft-cg location occurs when the cg lies between $0.501\,\bar{c}$ and $0.502\,\bar{c}$. It is evident that the characteristic equation of the short-period approximation cannot be used as an accurate means of calculating this cg position. However, the condition for a single real root to move into the right-half plane can be derived quite easily from the complete longitudinal dynamics, as we now show.

Pitch Stability

Sections 2.2 and 2.4 described the concept of positive pitch stiffness, and pointed out that positive stiffness was not sufficient to guarantee stability of the longitudinal motion. The stability of the longitudinal motion will now be investigated by means of a dynamic analysis.

The characteristic polynomial of the decoupled longitudinal dynamics can be obtained from the determinant $|sE - A|$, with the E and A matrices as given in (2.6-29). The constant term in the characteristic polynomial is equal to the product of the roots, and therefore the constant term will vanish when a real root reaches the origin, as the pitch-stability limit is reached. This constant term is obtained by putting $s = 0$ in $|sE - A|$, and therefore the stability boundary is given by $|A| = 0$. If the determinant obtained from (2.6-29) is expanded about the (4, 2) element, with $\gamma_e = 0$, the result is

$$0 = |A| = \begin{vmatrix} Z_\alpha & Z_V - X_{T_V}\sin(\alpha_e + \alpha_T) & 0 \\ M_\alpha + M_{T_\alpha} & M_V + M_{T_V} & 0 \\ X_\alpha & X_V + X_{T_V}\cos(\alpha_e + \alpha_T) & -g_D \end{vmatrix}$$

$$= -g_D \begin{vmatrix} Z_\alpha & Z_V - X_{T_V}\sin(\alpha_e + \alpha_T) \\ M_\alpha + M_{T_\alpha} & M_V + M_{T_V} \end{vmatrix}$$

or,

$$Z_\alpha(M_V + M_{T_V}) - (M_\alpha + M_{T_\alpha})(Z_V - X_{T_V}\sin(\alpha_e + \alpha_T)) = 0$$

When the dimensionless derivatives are substituted into this equation, the factors $(\bar{q}S/m)$, $(\bar{q}S\bar{c}/J_y)$, and $(1/V_{T_e})$ are removed, and the equilibrium condition $(C_{M_e} + C_{M_{T_e}}) = 0$ is applied, the stability boundary becomes

$$\left(C_{D_e} + C_{L_\alpha}\right)\left(C_{mV} + C_{m_{T_V}}\right) \\ - \left(C_{m_\alpha} + C_{m_{T_\alpha}}\right)\left[2C_{L_e} + C_{L_V} + \left(2C_{T_e} + C_{T_V}\right)\sin(\alpha_e + \alpha_T)\right] = 0$$

This equation can be simplified by using (2.4-1b) to get the following relationship for steady-state level flight:

$$2C_{T_e}\sin(\alpha_e + \alpha_T) + 2C_{L_e} = 2mg_D/(\bar{q}S) \equiv 2C_{W_e},$$

where C_W ($\approx C_L$) is the aircraft weight made dimensionless in the usual way. Substituting this result into the stability boundary condition, we get

$$(C_{D_e} + C_{L_\alpha})(C_{mV} + C_{m_{T_V}}) - (C_{m_\alpha} + C_{m_{T_\alpha}})\left[2C_{W_e} + C_{L_V} + C_{T_V}\sin(\alpha_e + \alpha_T)\right] = 0$$

To further simplify the expression, neglect the drag coefficient compared to the lift-curve slope, and let the thrust and aerodynamic moment derivatives be included in a single derivative. Then,

$$0 = C_{L_\alpha}C_{mV} - C_{m_\alpha}\left[2C_{W_e} + C_{L_V} + C_{T_V}\sin(\alpha_e + \alpha_T)\right] \quad (4.2\text{-}19)$$

This condition still holds when the last two terms on the right are negligible and, knowing that C_{W_e} and C_{L_α} are always positive, we can deduce that the condition for pitch stability is:

$$C_{m_\alpha} < \frac{C_{L_\alpha}C_{mV}}{2C_{W_e} + C_{L_V} + C_{T_V}\sin(\alpha_e + \alpha_T)} \quad (4.2\text{-}20)$$

When the tuck derivative is zero, (4.2-20) reduces to the static stability condition (see, for example (2.4-19)). When the aircraft has an unstable tuck ($C_{mV} < 0$) at high subsonic Mach numbers, a greater low-speed static margin is required to maintain pitch stability at those Mach numbers. Roskam (1979) points out that the pitch divergence of most subsonic jet-transports is rather slow, and not necessarily objectionable.

Algebraic Derivation of Lateral-Directional Transfer Functions

The lateral-directional coefficient matrices are given by (2.6-31). We will eliminate the E matrix by dividing the first lateral equation by V_{T_e}; the characteristic polynomial is then

AIRCRAFT RIGID-BODY MODES

$$|sI - A| = \begin{vmatrix} s - \dfrac{Y_\beta}{V_{T_e}} & -\dfrac{g_D \cos\theta_e}{V_{T_e}} & -\dfrac{Y_p}{V_{T_e}} & 1 - \dfrac{Y_r}{V_{T_e}} \\ 0 & s & -c\gamma_e/c\theta_e & -s\gamma_e/c\theta_e \\ -L'_\beta & 0 & s - L'_p & -L'_r \\ -N'_\beta & 0 & -N'_p & s - N'_r \end{vmatrix} \quad (4.2\text{-}21)$$

and the B matrix is

$$B = \begin{bmatrix} \dfrac{Y_{\delta a}}{V_{T_e}} & \dfrac{Y_{\delta r}}{V_{T_e}} \\ 0 & 0 \\ L'_{\delta a} & L'_{\delta r} \\ N'_{\delta a} & N'_{\delta r} \end{bmatrix} \quad (4.2\text{-}22)$$

Cramer's rule can now be used to find any particular transfer function. It is usual to make the lateral-directional equations manageable by assuming level flight ($\gamma_e = 0$). Then, for example, the aileron-to-roll-rate transfer function is

$$\frac{p(s)}{\delta_a(s)} = \frac{1}{|sI - A|} \begin{vmatrix} s - \dfrac{Y_\beta}{V_{T_e}} & -\dfrac{g_D \cos\theta_e}{V_{T_e}} & \dfrac{Y_{\delta a}}{V_{T_e}} & 1 - \dfrac{Y_r}{V_{T_e}} \\ 0 & s & 0 & 0 \\ -L'_\beta & 0 & L'_{\delta a} & -L'_r \\ -N'_\beta & 0 & N'_{\delta a} & s - N'_r \end{vmatrix}, \quad (4.2\text{-}23)$$

which is of the form

$$\frac{p(s)}{\delta_a(s)} = \frac{ks(s^2 + 2\zeta_\phi \omega_\phi s + \omega_\phi^2)}{a_4 s^4 + a_3 s^3 + a_2 s^2 + a_1 s + a_0} \quad (4.2\text{-}24)$$

The subscript ϕ has been used on the numerator quadratic because the same factor appears in the roll-angle transfer function, and the notation is in common use.

Once again, the polynomial coefficients are complicated functions of the dimensional derivatives, but some simplifications are possible. If the sideforce and yawing effects of the ailerons are neglected (i.e., neglect $Y_{\delta a}/V_{T_e}$ and $N'_{\delta a}$) the determinant in (4.2-23) has a simple expansion about the third column. Then assuming that $Y_r/V_{T_e} \ll 1$, the numerator of (4.2-24) can be written as

$$ks(s^2 + 2\zeta_\phi \omega_\phi s + \omega_\phi^2) = sL'_{\delta a}\left[s^2 - s(N'_r + Y_\beta/V_{T_e}) + (N'_\beta + Y_\beta N'_r/V_{T_e})\right] \quad (4.2\text{-}25)$$

When the aircraft has negligible *roll-yaw coupling*, the quadratic factor on the right-hand side of (4.2-25) also appears in the lateral-directional characteristic polynomial.

This is shown in the next subsection. The resulting cancellation leaves a particularly simple expression for the aileron-to-roll-rate transfer function.

The lateral/directional characteristic equation does not separate into factors that clearly define each mode. Approximations will be derived that may describe an individual mode reasonably well, but they must be checked for applicability in any given case. Nevertheless, these approximations do provide useful insight into the dynamic behavior, and they will be derived for this reason. We start with the dutch roll approximation.

The Dutch Roll Approximation

The dihedral derivative C_{l_β} determines the amount of rolling in the dutch roll mode, and when this derivative is small, the mode will consist mainly of sideslipping and yawing. The dihedral derivative tends to be large in modern swept-wing aircraft and so it will be neglected only for the purpose of deriving the traditional "three-degrees-of-freedom dutch roll approximation." A more modern approximation will then be given.

The coefficient of the roll angle in the beta-dot equation is the gravity term in the characteristic determinant (4.2-21). When this element is neglected the determinant has a simple reduction about the second column. The reduction of the subsequent third-order determinant can be further simplified if the terms Y_p/V_{T_e} and Y_r/V_{T_e} can be dropped (Y_p is often zero, and $Y_r/V_{T_e} \ll 1$). The cross-derivative term N'_p (yawing moment due to roll rate) is also often negligible. The dihedral derivative then no longer appears in the characteristic polynomial, which is given by

$$|sI - A| = s\left(s - L'_p\right)\left[s^2 - s\left(N'_r + Y_\beta/V_{T_e}\right) + \left(N'_\beta + Y_\beta N'_r/V_{T_e}\right)\right] \quad (4.2\text{-}26)$$

This polynomial has a root at the origin and at $s = L'_p$, which respectively approximate the spiral pole and the roll subsidence pole. The quadratic factor contains the dutch roll poles, and it exactly matches the numerator quadratic of the roll-rate transfer function (4.2-25). Therefore, an approximation to the aileron-to-roll-rate transfer function (4.2-24) is given by

$$\frac{p(s)}{\delta_a(s)} = \frac{L'_{\delta a}}{(s - L'_p)} \quad (4.2\text{-}27)$$

Equation (4.2-26) gives the dutch roll approximations as:

$$\omega_{n_d}^2 = N'_\beta + \left(Y_\beta/V_{T_e}\right) N'_r$$
$$\zeta_d = -\left(N'_r + Y_\beta/V_{T_e}\right) / \left(2\omega_{n_d}\right) \quad (4.2\text{-}28a)$$

A more recent approximation (Ananthkrishnan and Unnikrishnan, 2001) is

$$\omega_{n_d}^2 = N'_\beta + \frac{Y_\beta}{V_{T_e}} N'_r + \frac{g_D}{V_{T_e}} \frac{L'_\beta}{L'_p} - \left(L'_\beta + \frac{Y_\beta}{V_{T_e}} L'_r\right) \frac{N'_p}{L'_p}, \quad (4.2\text{-}28b)$$

whose first two terms agree with (4.2-28a). The damping equation in (4.2-28a) is unchanged, and so improved accuracy in the damping calculation will only come via the more accurate natural frequency.

We will now substitute dimensionless stability derivatives into the traditional dutch-roll formulae and examine the dependence on flight conditions. The derivative N'_β is given by

$$N'_\beta = \frac{N_\beta + (J'_{XZ}/J'_Z) L_\beta}{1 - J'^2_{XZ}/(J'_X J'_Z)} \qquad (4.2\text{-}29)$$

The stability-axes cross-product of inertia J'_{XZ} varies rapidly with the equilibrium angle of attack, typically changing from a small positive value at low alpha to a much larger negative value at high alpha. This larger value is still relatively small compared to J'_Z, so the primed derivatives are normally quite close to their unprimed values. It is possible for N_β to decrease and even change sign at high alpha, but then the linear equations are unlikely to be valid.

If we simply use the unprimed derivative N_β in the formula for the dutch roll frequency, and then substitute the dimensionless derivatives, we obtain

$$\omega^2_{n_d} = \frac{\bar{q} S b}{J'_Z} \left[C_{n_\beta} + \frac{\rho S b}{4m} C_{Y_\beta} C_{n_r} \right] \qquad (4.2\text{-}30)$$

The C_{n_r} term is usually negligible compared to C_{n_β}, and this equation shows that the dutch roll frequency is proportional to the square root of dynamic pressure, assuming constant C_{n_β}. Therefore, at constant altitude, the *frequency increases in proportion to the airspeed*, and for a given speed the *frequency decreases with altitude*.

When unprimed derivatives are substituted into the damping formula, followed by dimensionless derivatives and the natural frequency expression from (4.2-28a), the damping ratio is given by

$$\zeta_d = -\frac{1}{4} \left[\frac{(\rho S b) b^2}{2 J'_Z} \right]^{1/2} \frac{C_{n_r} + (2 J'_Z/mb^2) C_{Y_\beta}}{\left[C_{n_\beta} + (\rho S b/4m) C_{n_r} C_{Y_\beta} \right]^{1/2}} \qquad (4.2\text{-}31)$$

This expression indicates that the dutch roll damping is independent of dynamic pressure. *It will be proportional to the square root of density* since the second term of the denominator is usually negligible.

The dutch roll natural frequency formula tends to be quite accurate if the dihedral derivative is small, although the damping formula is not. This is illustrated in Example 4.2-1. Finally, note that the approximation to the roll subsidence pole, $s = L'_p$, is not very accurate, and a more accurate approximation will be derived next.

The Spiral and Roll Subsidence Approximations

The rolling and spiral modes usually involve very little sideslip. The rolling mode is almost pure rolling motion around the x-stability axis, and the spiral mode consists of yawing motion with some roll. It is common for the spiral mode to be unstable, and the motion then consists of increasing yaw and roll angles in a tightening downward spiral.

These facts allow approximations to be devised by modifying the $\dot{\beta}$ equation and leaving the moment equations unchanged. Sideforce due to sideslip is eliminated from the equation, $\dot{\beta}$ is neglected, and the gravity force is balanced against the force component associated with yaw rate. Thus, in the characteristic determinant (4.2-21) the term $(s - Y_\beta/V_{T_e})$ is eliminated, and the Y_p/V_{T_e} term is again neglected. Because the gravity force is intimately involved in the spiral mode, the mode is dependent on flight-path angle. Unfortunately, the assumption of level flight is needed to allow a reasonably simple analysis and will therefore be used here. The effect of flight-path angle will be investigated numerically in Example 4.2-2. When the simplified determinant is expanded the following second-order characteristic equation is obtained:

$$N'_\beta s^2 + \left(L'_\beta N'_p - L'_p N'_\beta - L'_\beta g_D/V_{T_e}\right) s + \left(L'_\beta N'_r - N'_\beta L'_r\right) g_D/V_{T_e} = 0 \quad (4.2\text{-}32)$$

Equation (4.2-32) normally has two real roots, corresponding to the roll subsidence pole and the spiral pole. Also, the spiral time constant is normally very much greater than the roll time constant. Under these circumstances, if we divide through (4.2-32) by the coefficient of s^2, the coefficient of s (i.e., the negative of the sum of the roots) yields the (negative of the) roll subsidence root. The constant term (i.e., the product of the roots) can then be used to obtain the spiral root. Therefore, we have the further approximations,

Roll Time Constant (τ_R):

$$1/\tau_R \approx -L'_p\left[1 - \frac{L'_\beta N'_p}{N'_\beta L'_p}\right] - \frac{L'_\beta}{N'_\beta}\frac{g_D}{V_{T_e}} \quad (4.2\text{-}33)$$

Spiral Time Constant (τ_S):

$$\tau_S = \frac{L'_\beta \left(N'_p - g_D/V_{T_e}\right) - L'_p N'_\beta}{\left(L'_\beta N'_r - N'_\beta L'_r\right) g_D/V_{T_e}} \quad (4.2\text{-}34)$$

Note that a negative value for the time constant will simply mean an unstable exponential mode. A slightly more accurate formula for τ_s is given in Ananthkrishnan (2001), but the difference is usually negligible.

If we once again neglect the primes and substitute dimensionless derivatives in the roll time-constant equation, we obtain

$$1/\tau_R = -\rho V_{T_e}\frac{Sb^2}{4J'_x}C_{l_p}\left[1 - \frac{C_{l_\beta}C_{n_p}}{C_{n_\beta}C_{l_p}}\right] - \frac{g_D}{V_{T_e}}\frac{J'_Z}{J'_x}\frac{C_{l_\beta}}{C_{n_\beta}} \quad (4.2\text{-}35)$$

In this equation the first term is usually much greater than the second. Therefore the equation indicates that *the roll time constant will vary inversely as the product of density and speed.*

The same procedure can be followed with the spiral time-constant equation (4.2-34). The result is

$$\tau_s = \frac{\dfrac{V_{T_e}}{g_D}\left(C_{l_\beta}C_{n_p} - C_{n_\beta}C_{l_p}\right) - \dfrac{1}{\rho V_{T_e}}\dfrac{4J'_z}{b^2 S}C_{L_\beta}}{\left(C_{l_\beta}C_{n_r} - C_{n_\beta}C_{l_r}\right)} \qquad (4.2\text{-}36)$$

This equation indicates that, at high speed, the spiral time constant is proportional to speed, and at low speed it will tend to become inversely proportional to the product of speed and density.

Spiral Stability

The condition for a pole at the origin is given by $|A| = 0$, and in the case of the lateral dynamics this normally represents the spiral pole becoming neutrally stable. From the characteristic determinant (4.2-21), we obtain

$$|A| = \begin{vmatrix} \dfrac{Y_\beta}{V_{T_e}} & \dfrac{(g_D \cos\theta_e)}{V_{T_e}} & \dfrac{Y_p}{V_{T_e}} & -1 + \dfrac{Y_r}{V_{T_e}} \\ 0 & 0 & c\gamma_e/c\theta_e & s\gamma_e/c\theta_e \\ L'_\beta & 0 & L'_p & L'_r \\ N'_\beta & 0 & N'_p & N'_r \end{vmatrix}$$

When the determinant is expanded the spiral stability boundary is found to be given by

$$\left(L'_\beta N'_r - N'_\beta L'_r\right)\cos\gamma_e + \left(L'_p N'_\beta - L'_\beta N'_p\right)\sin\gamma_e = 0 \qquad (4.2\text{-}37)$$

This equation shows that spiral stability is dependent on flight-path angle, as noted earlier.

Accuracy of the Lateral Mode Approximations

The accuracy of the lateral-modes formulae is often quite good apart from the dutch roll damping. The spiral time constant is also accurately predicted when this mode is unstable. This accuracy will be demonstrated in the following example using a model of a business jet in a cruising flight condition.

Example 4.2-1: Lateral Modes of a Business Jet. The following lateral-directional data for a business jet are taken from Roskam (1979).

Flight Condition:
$W = 13,000$ lbs, $h = 40,000$ ft ($\rho = 0.000588$ slug/ft^3),
$V_T = 675$ ft/s, $\gamma = 0$ deg, $\alpha = 2.7$ deg,
$J_x = 28,000$; $J_z = 47,000$; $J_{xz} = 1,350$ slug-ft^2 (body axes)

Geometrical Data: $S = 232$ ft^2, $b = 34.2$ ft

Stability Derivatives:
$C_{y_\beta} = -0.730$, $C_{y_p} = 0$, $C_{y_r} = +0.400$
$C_{l_\beta} = -0.110$, $C_{l_p} = -0.453$, $C_{l_r} = +0.163$
$C_{n_\beta} = +0.127$, $C_{n_p} = +0.008$, $C_{n_r} = -0.201$

A short program was written to convert the moments of inertia to stability axes, calculate the elements of the decoupled A matrix, and evaluate the approximate equations for the modal characteristics (from (4.2-28a) and (4.2-32), (4.2-33), (4.2-34)). Some intermediate results are:

Stability-Axes Moments of Inertia:
$J'_X = 27,915$ $J'_Z = 47,085$ $J'_{XZ} = 450.0$

Dimensional Derivatives:
$Y_\beta = -56.14$, $Y_p = 0$, $Y_r = 0.7793$
$L_\beta = -4.188$, $L_p = -0.4369$, $L_r = 0.1572$
$N_\beta = 2.867$, $N_p = 0.004575$, $N_r = -0.1149$

Primed Dimensional Derivatives:
$L'_\beta = -4.143$, $L'_p = -0.4369$, $L'_r = 0.1554$
$N'_\beta = 2.800$, $N'_p = -0.002469$, $N'_r = -0.1124$

The full A matrix was calculated from (4.2-21) so that an eigenvalue program could be used to determine the modes "exactly." The exact and approximate results are as follows:

Dutch Roll Mode:
exact: $\omega_n = 1.682$ rad/s, $\zeta = 0.0373$
Equations (4.2-28a): $\omega_n = 1.676$ rad/s, $\zeta = 0.0584$

Roll Subsidence Mode:
exact: $\tau_R = 1.976$s
Equation (4.2-32): $\tau_R = 1.960$s
Equation (4.2-33): $\tau_R = 1.957$s

Spiral mode:
exact: $\tau_S = 978.5$s
Equation (4.2-32): $\tau_S = 976.7$s
Equation (4.2-34): $\tau_S = 978.7$s

These results are in remarkably good agreement, apart from the dutch roll damping. ∎

Mode Variation from the Nonlinear Model

It is not very realistic to use a fixed set of stability derivatives to show the variation of the modal characteristics with flight conditions. Therefore, as a final example we will use the completely numerical approach to calculate the modes of the nonlinear F-16 model at different flight conditions. The modes will only be calculated accurately since the numerical linearization is set up to produce the state-equation coefficient matrices, not the stability derivatives. The variation of the modes with flight-path angle will also be determined, since this could not easily be done with the approximate formulae.

Example 4.2-2: Mode Dependence from the Nonlinear Model. The nonlinear F-16 model allows a realistic examination of the dependence of the modes on flight conditions, since it is not built from a fixed set of aerodynamic derivatives. The following results were obtained by trimming and numerically linearizing the model at the desired flight condition, and then using an eigenvalue program to determine the modes from the full thirteen-state A matrix. Virtually identical results could be obtained by using the decoupled lat-long matrices.

The effect of flight-path angle was investigated by trimming the model according to the second set of conditions in Table 3.6-3 (502 ft/s, $h = 0$ ft, cg = 0.3 $\bar{c}$) but with different values of γ. The modes are shown in Table 4.2-2. It is evident from these results that the "rotational" modes are almost independent of γ. Weak but consistent trends are visible in the dutch roll and roll subsidence modes, and in the short period. Overall, the properties of the rotational modes are remarkably constant, considering the nature of the tabular aerodynamic data and the numerical processing (trimming and linearization) required to obtain them. The "flight-path" modes, phugoid and spiral, are strongly influenced by the flight-path angle. The spiral time constant initially increases as the flight-path angle increases, becomes infinite as the stability boundary is approached, and then decreases with flight-path angle when the mode is unstable. The phugoid period is only weakly affected by γ but increases as γ increases. Phugoid damping is more strongly affected; it decreases with increasing γ and the phugoid becomes unstable at a quite modest flight-path angle.

TABLE 4.2-2: Effect of Flight-Path Angle on F-16 Modes

γ	−5	0	5	10	15	20	deg
T_D	1.934	1.933	1.934	1.937	1.941	1.946	s
ζ_D	0.1346	0.1353	0.1360	0.1366	0.1371	0.1375	
τ_S	55.33	77.91	133.0	461.9	−312.3	−117.0	s
τ_R	0.2777	0.2777	0.2775	0.2772	0.2766	0.2760	s
T_{SP}	3.281	3.277	3.273	3.269	3.266	3.262	s
ζ_{SP}	0.6277	0.6279	0.6281	0.6282	0.6283	0.6283	
T_P	79.60	80.05	80.93	82.39	84.36	86.82	s
ζ_P	0.1297	0.09751	0.06557	0.03396	0.00227	−0.0298	

TABLE 4.2-3: Effect of Speed and Altitude on F-16 Modes

Alt./speed (dyn. pres)	50k, 900 (160)	50k, 600 (71)	0, 900 (963)	0, 367 (160)	ft, ft/s lbs/ft^2
T_D	2.365	2.735	1.143	2.396	s
ζ_D	0.06480	0.07722	0.1272	0.1470	
τ_S	179.2	138.7	122.1	73.52	s
τ_R	1.050	2.230	0.1487	0.4160	s
T_{SP}	4.507	u/s	2.372	4.023	s
ζ_{SP}	0.2615	u/s	0.8175	0.5735	
T_P	102.1	u/s	183.4	56.93	s
ζ_P	0.005453	u/s	0.3242	0.06240	

In Table 4.2-3 the model is trimmed in level flight with various combinations of speed and altitude to illustrate the effect of these two variables on the modes. The cg position is again at 0.3 $\bar{c}$. The flight conditions have been chosen to compare different speeds at the same altitude, the same speed at different altitudes, high and low dynamic pressures at the same altitude, and the same dynamic pressure at two greatly different altitudes. The first trim condition (50,000 ft, 900 ft/s) corresponds to 0.93 Mach and is therefore strictly outside the valid Mach range of the model; this is also true to a lesser extent for the third case (0.81 Mach). We do not have a model that includes compressibility effects, and we will simply consider this example as illustrating the variation of the modes when compressibility is not important. The second trim condition (50,000 ft, 600 ft/s) corresponds to full throttle, while the first case (higher speed) corresponds to only 0.765 throttle. Therefore, a dive and climb maneuver would be needed to get from the second to the first flight condition. The longitudinal dynamics are unstable in the second case. In the fourth flight condition trial-and-error adjustment of the speed was used to make the dynamic pressure the same as the first case.

The tabulated results show that, as expected, the dutch roll has almost the same period at two widely different speed/altitude combinations with the same dynamic pressure. They also show the expected increase in period with altitude (at constant speed), and the decrease in period with airspeed (at constant altitude). The dutch roll damping does tend to be independent of dynamic pressure and to decrease with altitude, as predicted by the theory.

The spiral time constant is expected to vary directly with V_T if the third numerator term in (4.2-36) is negligible, and to vary as $V_T/\bar{q}$ if that term is dominant. The results indicate that the actual variation is somewhere in between these two trends. This is not unexpected because the F-16 has swept wings, and C_{l_β} can be expected to play a significant part in (4.2-36).

The time constant of the roll subsidence mode is approximately proportional to $V_T/\bar{q}$, as predicted. The short-period mode also shows the expected trends; the period is roughly the same at the two equal dynamic pressure conditions, and is much smaller at the high dynamic pressure condition. As predicted, the damping is much more strongly affected by altitude than by dynamic pressure.

In the case of the phugoid period the two sea-level results show that the sixfold increase in dynamic pressure causes an increase in the period of 3.2 times (compared to the prediction of $\sqrt{6}$). At constant dynamic pressure the period increases with altitude, as expected. The phugoid damping also shows the expected trend, increasing with airspeed and decreasing with altitude. ∎

4.3 THE HANDLING-QUALITIES REQUIREMENTS

Background

Control-law design can only be performed satisfactorily if a set of design requirements or performance criteria is available. In the case of control systems for piloted aircraft, generally applicable quantitative design criteria are very difficult to obtain. The reason for this is that the ultimate evaluation of a human-operator control system is necessarily subjective and, with aircraft, the pilot evaluates the aircraft in different ways depending on the type of aircraft and phase of flight. For example, in a dynamic maneuvering situation the pilot may be concerned mainly with the control forces that must be exerted and the resulting six-degrees-of-freedom translational and angular accelerations. In a task requiring precision tracking the pilot's evaluation will be more influenced by visual cues and the response of the aircraft to turbulence.

Also, a pilot's opinion of the *handling qualities* of an aircraft is inevitably influenced by factors other than the obvious control-system considerations of response to control inputs and response to disturbance inputs (e.g., turbulence). He or she will be influenced by the ergonomic design of the cockpit controls, the visibility from the cockpit, the weather conditions, the mission requirements, and physical and emotional factors. The variability introduced by all these factors can only be reduced by averaging test results over many flights and many pilots.

A systematic approach to handling-qualities evaluation is available through *pilot opinion rating* scales such as the Cooper Harper scale (Cooper and Harper, 1969). This rating scale is shown in Table 4.3-1. Once a rating scale like this has been established it is possible to begin correlating the pilot opinion rating with the properties of the aircraft dynamic model, and hence derive some analytical specifications that will guarantee good handling qualities. Although this may seem simple in principle, it has proven remarkably difficult to achieve in practice, and after many years of handling-qualities research it is still not possible to precisely specify design criteria for control systems intended to modify the aircraft dynamics. A survey and a large bibliography covering twenty-five years of handling-qualities research has been given by Ashkenas (1984). The "background information and user guides" for the military flying qualities specifications MIL-F-8785B and MIL-F-8785C (Chalk et al., 1969; Moorhouse and Woodcock, 1982) also provide much useful information.

We first consider some possible ways in which requirements for dynamic response may be specified. The aircraft model may be linearized in a particular flight condition

TABLE 4.3-1: Pilot Opinion Rating and Flying Qualities Level

Aircraft Characteristics	Demands on Pilot in Selected Task or Required Operation	Pilot Rating	Flying Qualities Level
Excellent; highly desirable	Pilot compensation not a factor for desired performance	1	
Good; negligible deficiencies	as above	2	1
Fair; some mildly unpleasant deficiencies	Minimal pilot compensation required for desired performance	3	
Minor but annoying deficiencies	Desired performance requires moderate pilot compensation	4	
Moderately objectionable deficiencies	Adequate performance requires considerable pilot compensation	5	2
Very objectionable but tolerable deficiencies	Adequate performance requires extensive pilot compensation	6	
Major deficiencies	Adequate performance not attainable with maximum tolerable pilot compensation Controllability not in question	7	
Major deficiencies	Considerable pilot compensation required for control	8	3
Major deficiencies	Intense pilot compensation required to retain control	9	
Major deficiencies	Control will be lost during some portion of required operation	10	

and the poles and zeros, or frequency response, of a particular transfer function compared with a specification. Alternatively, certain time responses may be derived from the nonlinear model, in a particular flight condition, and be compared with specifications. Yet another alternative is to model the human operator as an element in a closed control loop containing the aircraft dynamics, and determine what requirements are placed on the operator if the closed-loop control is to have a satisfactory command or disturbance response. All of these techniques have been, or are being, considered by workers in the field, and we will examine some of the ideas in more detail.

Pole-Zero Specifications

Suppose that lat-long decoupling is assumed and the pitch axis is considered. In addition, assume linear dynamic behavior. Then if a transfer function shows that the dynamic response is dominated by a single pair of complex poles (e.g., the short-period poles), the pilot's opinion of the aircraft handling qualities should correlate with the position of these poles. A number of studies have provided data to link pole positions to pilot opinion rating.

In one of the early studies, O'Hara (1967) produced iso-opinion contours for the location of the short-period poles; these were plotted on axes of undamped natural frequency versus damping ratio. They showed that the most satisfactory pilot-opinion rating corresponded to poles inside a closed contour bounded by about 2.4 and 3.8 rad/s, and by damping ratios of about 0.4 and 1.0, with its center at about 3.0 rad/s and $\zeta = 0.65$. This and other similar results form the basis of current pole-position handling-qualities criteria.

Unfortunately for the pole-position criterion, even if the decoupling and linearity assumptions are justified, there are at least two reasons why this approach may not work well. The first is that transfer function zeros are also important (they have a strong effect on step response overshoot). Second, the aircraft and control system dynamics may include quite a lot of poles that contribute significantly to the time response. Pilots are very sensitive to additional dynamics, and the difficulties of specifying requirements on more than just a single pair of poles quickly become prohibitive. The problem of transfer function zeros will be considered first.

The short-period elevator-to-pitch-rate transfer function (4.2-9) plays an important role in the pilot's assessment of the longitudinal-axis flying qualities. In this transfer function the $Z_{\delta e}$, $Z_{\dot{\alpha}}$, and Z_q terms can usually be neglected, with the following result:

$$\frac{q}{\delta_e} = \frac{Z_\alpha M_{\delta e}\left(sV_{T_e}/Z_\alpha - 1\right)}{V_{T_e}s^2 - \left(Z_\alpha + V_{T_e}M_q + V_{T_e}M_{\dot{\alpha}}\right)s + M_q Z_\alpha - V_{T_e}M_\alpha} \quad (4.3\text{-}1)$$

In the handling-qualities literature the dimensional derivative $L_\alpha (\equiv \partial L/\partial \alpha \approx -mZ_\alpha$, $C_D \ll C_{L_\alpha})$ is often used instead of Z_α, and the time constant associated with the transfer function zero is given the symbol T_{θ_2} (T_{θ_1} is associated with the phugoid mode). Therefore, we see that

$$T_{\theta_2} = -V_{T_e}/Z_\alpha \approx mV_{T_e}/L_\alpha \quad (4.3\text{-}2)$$

This time constant is also often expressed in terms of the aircraft load factor response to angle of attack, n_α. Aircraft *load factor*, n, is defined as lift (L) divided by the weight (W), and n_α is the gradient of this quantity with respect to alpha [$n_\alpha = (\partial L/\partial \alpha)/W$]. Therefore, we have

$$T_{\theta_2} = V_{T_e}/(g_D n_\alpha) \quad (4.3\text{-}3)$$

The position of the pitch-rate transfer function zero has been shown to correlate with pilot-opinion ratings of the flying qualities (Chalk, 1963). Shomber and Gertsen

(1967) derived iso-opinion curves involving the short-period frequency and damping, T_{θ_2}, and n_α. When n_α was less than 15 g/rad they found that pilot opinion correlated well with $1/(\omega_n T_{\theta_2})$ and ζ, with the optimum conditions being around $1/(\omega_n T_{\theta_2}) = 0.45$, $\zeta = 0.7$. When n_α was greater than 15, they found that the correlation was with n_α/ω_n (i.e., T_{θ_2} no longer fixed) and ζ, with the optimum conditions near $n_\alpha/\omega_n = 10$, $\zeta = 0.7$. The current military flying-qualities requirements (see later) specify the short-period natural frequency in terms of n_α, and there is still a division of opinion over the importance of T_{θ_2} versus n_α.

The lateral-directional dynamics have proved somewhat less critical than the longitudinal dynamics from the point of view of handling qualities. The normally required changes in the aircraft trajectory can be achieved by a combination of rolling and pitching. O'Hara (1967) used iso-opinion curves to show that lateral dynamics would receive a good rating if the maximum roll acceleration was appropriate to the roll time constant. Both of these quantities are transfer function parameters. Regardless of these studies the current military requirements provide only specifications for the roll time constant and the time to reach a given roll angle. The latter quantity must be obtained from a flight test or a nonlinear simulation.

The dutch roll mode is an unwanted complication in this simple picture; it should be fast and adequately damped (see later) so that the airplane will quickly reorient itself after a directional disturbance. Ideally, the dutch roll should have very little involvement in the lateral dynamics and should therefore almost cancel out of the lateral transfer functions. This requires that quantities ω_ϕ and ζ_ϕ for the complex zeros [see (4.2-24)] should coincide with ω_d and ζ_d for the dutch roll poles. The ratio ω_ϕ/ω_d is the most important quantity in this respect, and iso-opinion curves of ω_ϕ/ω_d versus ζ_d have been plotted (Ashkenas, 1966).

As might be expected, the optimum value of ω_ϕ/ω_d is close to unity for a stable dutch roll. However, there is a subtlety in these results; it can be shown that favorable yaw is generated in a turn when $\omega_\phi/\omega_d > 1$, and the converse is true. We will refer to this again in connection with the lateral directional control augmentation system in Section 4.5.

Finally, consider the case of highly augmented aircraft, where the control systems contribute a number of poles and zeros in addition to those associated with the basic rigid-body transfer functions. Specifications placed on poles and zeros quickly become unmanageable and, as in the case of control system design, one must resort to frequency response techniques. One way in which frequency-domain ideas have been applied to handling qualities specifications is described in the next subsection.

Frequency-Response Specifications

In general the goal of an aircraft control system design should be to produce dominant closed-loop poles that resemble the basic rigid-body poles, with satisfactory damping and natural frequency. The effect of the additional dynamics resulting from the control system compensation networks, and possibly the lower-frequency flexible modes, can be allowed for by determining an "equivalent low-order system" (Craig and Ashkenas, 1971; Hodgkinson, 1979; Bischoff, 1981; Gentry, 1982).

In this concept the coefficients are determined for a low-order transfer function that matches the frequency response of the actual transfer function, over a limited frequency range. The gain and phase are matched simultaneously by adjusting the coefficients of the low-order transfer function to minimize a cost function of the form

$$\text{COST} = \frac{20}{n} \sum_{i=1}^{i=n} \left[\Delta G(\omega_i)^2 + \frac{\Delta P(\omega_i)^2}{57.3} \right] \qquad (4.3\text{-}4)$$

Here n is the number of discrete frequencies (ω_i) used, $\Delta G(\omega_i)$ is the difference in gain (in decibels) between the transfer functions at the frequency ω_i, and $\Delta P(\omega_i)$ is the difference in phase (in degrees) at ω_i. The frequency range used is nominally 0.3 to 10 rad/s, and 20 to 30 discrete frequencies are needed. The upper frequency limit is based on the maximum control frequencies that pilots have been observed to use. The lower limit is based on observations that pilots do not provide continuous closed-loop control at very low frequencies, and the value given does not provide for matching the phugoid mode. The cost function is minimized with a multivariable search routine, in the same way that we obtained steady-state trim in Chapter 3.

The stick-force-to-pitch-rate transfer function is typically used to evaluate the longitudinal dynamics. To compare a particular aircraft with both the short-period and phugoid specifications in the military flying-qualities specifications, the assumed form of this transfer function is

$$\frac{q}{F_s} = \frac{K(s + 1/T_{\theta_1})(s + 1/T_{\theta_2})e^{-\tau s}}{\left(s^2 + 2\zeta_p \omega_{n_p} s + \omega_{n_p}^2\right)\left(s^2 + 2\zeta_{sp} \omega_{n_{sp}} s + \omega_{n_{sp}}^2\right)} \frac{\text{rad/s}}{\text{lb}} \qquad (4.3\text{-}5)$$

Here the subscripts p and sp indicate, respectively, the phugoid and short-period modes. The frequency range for matching the transfer functions should be extended down to about 0.01 rad/s when the phugoid is included.

The term $e^{-\tau s}$ is included in the low-order model to provide an equivalent time delay for matching high-frequency effects from, for example, actuator modes, structural modes and mode filters, and noise filters. The time-delay term contributes only a phase shift to the transfer function; this is consistent with the fact that the phase variations from high-frequency dynamics extend over a larger frequency range than gain variations. The current military requirements suggest that for level-1 handling qualities, the maximum allowable value of the equivalent time delay should be 10.0 ms.

If a good fit to an equivalent low-order system is obtained (e.g., a cost of 10.0 or less), the pole-zero criteria can be applied to this equivalent system. If low values of the cost function cannot be obtained, other criteria must be used.

Another example of frequency-domain specifications applied to aircraft control systems (but not directly to handling qualities) is the military standard requirement document for the design, installation, and test of flight control systems (*MIL-F-9490*, 1975). This provides stability criteria by specifying the minimum gain and phase margins that must be achieved in any actuator path, with all other feedback paths closed. Typical values are 6 dB gain margin and 30 degree phase margin.

Time-Response Specifications

Placing handling-qualities requirements on the time response has the advantage that a time response can readily be obtained from the full nonlinear model dynamics. It does, however, raise the problems of what type of test input to apply and which output variable to observe. In the case of the longitudinal dynamics, it is natural once again to specify requirements on the pitch-rate response. However, fighter aircraft control systems are normally designed to give the pilot control over pitch rate at low speed and normal acceleration (acceleration measured along the body negative z-axis) at high speed. This gives direct control over the variable that stresses the pilot. The two control schemes must be smoothly blended together (see, e.g., Toles, 1985).

Efforts to develop time-response criteria have mostly been linked to the decoupled longitudinal dynamics and have made use of the short-period approximation. They have attempted to define an envelope inside which the pitch rate, angle of attack, or normal acceleration response to an elevator step input should lie. As early as 1963 a step-response envelope for angle of attack was derived from the short-period iso-opinion curves (Rynaski, 1985). Envelope criteria have been published for the pitch-rate response of an SST and of the space shuttle (see Rynaski, 1985).

A time-history envelope criterion, called $C^*(t)$ ("C-star"), was published in 1965 (Tobie et al., 1966), and is still in use. The C^* criterion uses a linear combination of pitch rate and normal acceleration at the pilot's station:

$$C^*(t) = a_{n_p} + 12.4q, \qquad (4.3\text{-}6)$$

where a_{n_p} is the normal acceleration in g's (approximately zero g's = level flight) and q is the pitch rate in radians per second. The envelope for the C^* criterion is shown in Figure 4.3-1. If the response $C^*(t)$ to an elevator step input falls inside the envelope, level-1 flying qualities on the pitch axis will hopefully be obtained. A more recent time-domain criterion than C^* relates pilot opinion ratings to target tracking error and time on target for a step target tracking task (Onstott and Faulkner, 1978).

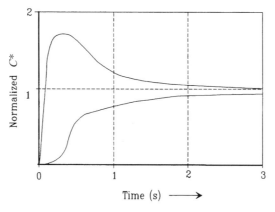

Figure 4.3-1 The C-star envelope.

The cited envelope criteria often give conflicting results and may disagree with the pilot ratings for specific aircraft. Pitch-rate responses having large overshoots and poor settling times have often corresponded to good pilot-opinion ratings. It is known that for fighter aircraft air-combat modes a pitch-rate overshoot is required for good gross acquisition of targets, and a deadbeat pitch-rate response is required for good fine tracking. Rynaski (1985) has argued that angle of attack should be the basic response variable, and it appears that the angle-of-attack response corresponding to good handling qualities may be more like a good conventional step response (i.e., small overshoot and fast nonoscillatory settling).

A time-response criterion, called D^* (or coordination perception parameter), has been devised for the lateral-directional response (Kisslinger and Wendle, 1971). The idea is similar to C^* in that the coordination perception parameter is a blend of lateral acceleration and sideslip angle, and envelope limits for acceptable performance are specified.

Requirements Based on Human Operator Models

For certain types of control tasks it is possible to model a human operator with linear differential equations or a transfer function. An example of such a task is a compensatory tracking task with a random input, that is, a control task in which the operator uses only tracking *error* information to track an unpredictable target. This information may be presented by instruments such as a pilot's artificial horizon display. The human operator model consists of the transfer function, and an added nonanalytic output signal called the *remnant*. The purpose of the remnant is to account for the discrepancies between experimental results with a human operator and analyses using the model. The transfer function model is often given the name *human operator describing function* (not to be confused with the describing function of nonlinear control theory).

The human operator transfer function model for the compensatory tracking task is usually assumed to be

$$Y(s) = \frac{K_p e^{-ds}(\tau_\ell s + 1)}{(\tau_i s + 1)(\tau_n s + 1)} \qquad (4.3\text{-}7)$$

In this transfer function the pure delay, d, may be taken to represent the motor-control functions in the cerebellum and the neuromuscular delay, while the lag τ_n models the mechanical properties of the muscles and limbs. It is known that the speed of response is severely limited by the delay term rather than the lag, and the latter is neglected in many applications. The gain, K_p, lead time constant, τ_ℓ, and lag time constant, τ_i, represent the capability of the human operator to optimize his or her control of a given task. Thus, the operator may use lag compensation to achieve high gain and fine control in some low-bandwidth tasks, or lead compensation to achieve high bandwidth.

This model has been applied to aircraft piloting tasks, and hypotheses (the *adjustment rules*) have been developed for the way in which the adaptive parameters will be "chosen" by the pilot (McRuer et al., 1965). It is also used as the basis of a transfer function method of assessing flying qualities (Neal and Smith, 1970). Interesting examples of the transfer function model applied to a pilot controlling roll angle are given in Etkin (1972) and Roskam (1979).

In the Neal–Smith method the model (4.3-7) is used in conjunction with the aircraft stick-force-to-pitch-attitude transfer function, in a closed pitch-attitude control loop. It is assumed that the human pilot adjusts the lead, lag, and gain, so that the *droop* and *peak-magnification* of the closed-loop frequency response are minimized, as shown in Figure 4.3-2. Therefore, this process is duplicated with the models, the lag τ_n is neglected, and the delay is taken as $d = 0.3$ s. The lead and lag time constants are adjusted, according to the adjustment rules, to optimize the closed-loop frequency response. When this has been done, the maximum lead or lag provided by the pilot model is determined, together with the value of the peak magnification. The pilot opinion rating is then determined from a plot like that shown in Figure 4.3-3.

A later development than the transfer function model of the human operator is the *optimal control model* (OCM), attributable mainly to Baron, Kleinman, and Levison (Kleinman et al., 1970, p. 358). It uses a state-variable formulation and optimal control theory and is based on the assumption that "a well-motivated, well-trained human operator behaves in a near optimal manner, subject to his inherent limitations and constraints and his control task." A description of this model is outside the scope of this chapter, since it has not found its way directly into flying qualities specifications. More information can be found in the book by Sheridan and Ferrill (1974) and in the references cited. A summary of work in human operator modeling, with a fairly comprehensive bibliography, has been given by Gerlach (1977).

Human operator modeling applied to a pilot performing compensatory tracking tasks has now accumulated quite a long history, and attention has turned to modeling the human operator performing other piloting tasks. In a modern fighter aircraft the workload involved in operating all of the different systems (flight control, navigation, radar, weapons, etc.) can be overwhelming, and modeling the human decision-making

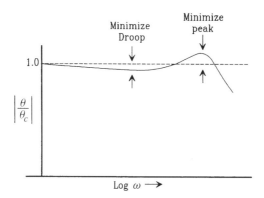

Figure 4.3-2 Closed-loop frequency response for Neal–Smith criterion.

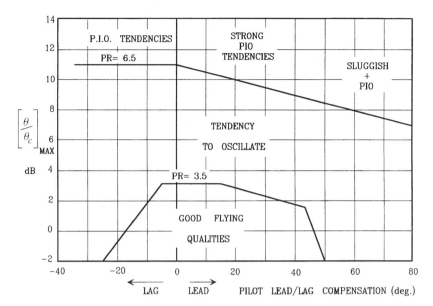

Figure 4.3-3 Neal–Smith evaluation chart.

process has become important. A survey of the relationship of flying qualities specifications to task performance, and the use of pilot models, has been given by George and Moorhouse (1982).

Other Requirements

The preceding subsections described ways in which the dynamic response of an aircraft and its control systems can be characterized, and how these may lead to handling qualities criteria. There are a number of other requirements that must be satisfied for an aircraft to receive a good handling-qualities rating. Some of these have no direct effect on control system design, but they are "inside the loop" that is closed by the pilot. They will be briefly described because of their importance.

One of the more important characteristics of the pilot's controls is the *control feel*, that is, the force and deflection characteristics of the control stick during a particular maneuver. Aircraft whose control surfaces are not power boosted require much careful balancing of the control surfaces, and the addition of a mass and springs to the control stick, in order to obtain satisfactory feel. Aircraft with fully powered, irreversible controls require an artificial-feel system.

Artificial feel may take the form of centering springs, an electromechanical damper, and, for longitudinal control, a mechanical or hydraulic system that provides a stick reaction force proportional to the normal acceleration in g's during a pull-up maneuver. Iso-opinion studies have shown that the amount of stick force per g is quite critical and there is an associated optimum value of stick deflection. Stick force per g requirements are given in the military aircraft specifications, in addition to the control forces required in various flight phases.

Another factor that influences a pilot's opinion of handling qualities, particularly in the landing phase, is speed stability. The aircraft response to a speed disturbance is an exponential change, and this response will typically be rated as satisfactory if it is stable with a time constant of less than about 50 s. An unstable exponential response may be acceptable under some conditions, provided that the time constant is greater than about 25 s.

The Military Flying-Qualities Specifications

In the preceding subsections we attempted to convey some idea of the difficulty of specifying analytical performance criteria for the dynamic behavior of piloted aircraft. The civil and military aviation authorities of various countries are also faced with this problem. In general, their requirements documents are not very analytical and do not provide any way out of our difficulty. However, the U.S. "Military Specification for the Flying Qualities of Piloted Airplanes" (*MIL-F-8785C*, 1980) does provide some analytical specifications that must be met by U.S. military aircraft. A background document and user guide, containing much useful information and a large bibliography, is also available (Chalk et al., 1969). These documents are readily available, and only the mode specifications of *MIL-F-8785C* will be summarized here. (Note that *MIL-F-8785C* has now been superseded by *MIL 1797*, which contains additional information, but this document has limited circulation.)

The military specification defines airplane classes, flight phases, and flying qualities levels, so that different modes can be specified for the various combinations. These are defined in Table 4.3-2; the flying qualities levels are linked to the Cooper–Harper ratings as shown in Table 4.3-1. The specifications for the aircraft modes are as follows.

Phugoid Specifications The military specification dictates that for the different levels of flying qualities, the damping ζ_p and natural frequency ω_{n_p} of the phugoid mode will satisfy the following requirements:

Level 1: $\zeta_p \geq 0.04$
Level 2: $\zeta_p \geq 0.0$
Level 3: $T_{2_p} \geq 55.0$ s

In the level-3 requirement the mode is assumed to be unstable, and T_2 denotes the time required for the mode to double in amplitude. For an exponentially growing sinusoidal mode this time is given by

$$T_2 = \log_e 2/(-\zeta\omega_n) \qquad (\zeta \text{ has negative values})$$

These requirements apply with the pitch control free or fixed; they need not be met transonically in certain cases.

TABLE 4.3-2: Definitions—Flying Qualities Specifications

Airplane Classes

Class I: Small, light airplanes.
Class II: Medium weight, low-to-medium-maneuverability airplanes.
Class III: Large, heavy, low-to-medium-maneuverability airplanes.
Class IV: High-maneuverability airplanes.

Flight Phases

Category A: Nonterminal flight phases generally requiring rapid maneuvering.
Category B: Nonterminal flight phases normally accomplished using gradual maneuvers without precision tracking, although accurate flight-path control may be required.
Category C: Terminal flight phases normally accomplished using gradual maneuvers and usually requiring accurate flight-path control.

Flying Qualities Levels

Level 1: Flying qualities adequate for the mission flight phase.
Level 2: Flying qualities adequate to accomplish the mission flight phase, but some increase in pilot workload or degradation in mission effectiveness exists.
Level 3: Flying qualities such that the airplane can be controlled safely, but pilot workload is excessive, or mission effectiveness is inadequate, or both.

Short-Period Specifications The short-period requirements are specified in terms of the natural frequency and damping of the "short-period mode" of the equivalent low-order system (as defined earlier). The adequacy of the equivalent system approximation is to be judged by the procuring agency. Table 4.3-3a shows the requirements on the equivalent short-period damping ratio ζ_{sp}.

The requirements on equivalent undamped natural frequency ($\omega_{n_{sp}}$) are given in Table 4.3-3b and are specified indirectly, in terms of the quantity $\omega_{n_{sp}}^2/(n/\alpha)$. The denominator ($n/\alpha$) of this term is the aircraft load-factor response to angle of attack in g's per radian, as explained in the subsection on pole-zero specifications.

TABLE 4.3-3a: Short-Period Damping Ratio Limits

	Cat. A & C Flight Phases		Cat. B Flight Phases	
Level	Minimum	Maximum	Minimum	Maximum
1	0.35	1.30	0.30	2.00
2	0.25	2.00	0.20	2.00
3	0.15*	no limit	0.15*	no limit

*May be reduced at altitude > 20,000 ft with approval.

TABLE 4.3-3b: Limits on $\omega_{n_{sp}}^2/(n/\alpha)$

Level	Cat. A Phases		Cat. B Phases		Cat. C Phases	
	Min.	Max.	Min.	Max.	Min.	Max.
1	0.28 $\omega_n \geq 1.0$	3.60	0.085	3.60	0.16 $\omega_n \geq 0.7$	3.60
2	0.16 $\omega_n \geq 0.6$	10.0	0.038	10.0	0.096 $\omega_n \geq 0.4$	10.0
3	0.16	no lim	0.038	no lim	0.096	No limit

There are some additional limits on the minimum value of n/α and the minimum value of ω_n, for different classes of airplane in category C.

Roll Mode Specifications The maximum allowable value of the roll-subsidence mode time constant is given in Table 4.3-4. In addition to these time-constant specifications there is a comprehensive set of requirements on the time required to achieve various (large) changes in roll angle following an abrupt roll command. For example, for air-to-air combat (a flight phase within category A, for class IV airplanes) the minimum allowable time to achieve a certain roll angle depends on airspeed, but for level-1 flying qualities may be as short as 1.0 s for 90 degree roll and 2.8 s for 360 degree roll.

Spiral Mode Specifications The spiral mode is allowed to be unstable, but limits are placed on the minimum time for the mode to double in amplitude, as shown in Table 4.3-5. These requirements must be met following a roll-angle disturbance of up to 20 degrees from trimmed for zero-yaw-rate wings-level flight, with the cockpit controls free.

Dutch Roll Mode Specifications The frequency ω_{n_d} and damping ratio ζ_d of the dutch roll mode must exceed the minimum values given in Table 4.3-6. Note that the quantity $\zeta \omega_n$ is the s-plane real-axis coordinate of the roots, and ω_n is the radial

TABLE 4.3-4: Max. Roll-Mode Time Constant (sec.)

Flight Phase Cat.	Class	Level		
		1	2	3
A	I IV	1.0	1.4	no
	II III	1.4	3.0	lim.
B	All	1.4	3.0	10
C	I, II-C, IV	1.0	1.4	no
	II-L, III	1.4	3.0	lim.

TABLE 4.3-5: Spiral Mode, Minimum Doubling Time

Flight Phase Category	Level 1	Level 2	Level 3
A & C	12 s	8 s	4 s
B	20 s	8 s	4 s

TABLE 4.3-6: Dutch Roll Mode Specifications

Level	Flight Phase Cat.	Class	min ζ_d	min $\zeta_d \omega_{n_d}$	min ω_{n_d}
1	A	I, IV	0.19	0.35	1.0
		II, III	0.19	0.35	0.4
	B	all	0.08	0.15	0.4
	C	I, II-C, IV	0.08	0.15	1.0
		II-L, III	0.08	0.15	0.4
2	all	all	0.02	0.05	0.4
3	all	all	0.02	no lim.	0.04

distance from the origin for complex roots. Therefore, these requirements define an area of the s-plane in which the dutch roll roots must lie.

The lower limit on ζ_d is the larger of the two values that come from the table, except that a value of 0.7 need not be exceeded for class III. Also, class III airplanes may be exempted from some of the minimum ω_d requirements. Airplanes that have a large amount of roll-yaw coupling, as measured by the ratio of the maximum roll angle to the maximum value of sideslip in a dutch roll oscillation, are subject to a more stringent requirement on $\zeta_d \omega_{n_d}$ (see *MIL-F-8785C*).

The military requirements document specifies dynamic response mainly through the pole-zero requirements. These have been summarized here so that the reader may evaluate some of the controller designs described later. Much additional information covering other aspects of flying qualities is available in the requirements document, and it is essential reading for anyone with other than a casual interest in this field.

4.4 STABILITY AUGMENTATION

Most high-performance commercial and military aircraft require some form of stability augmentation. Some military aircraft are actually unstable and would be virtually

292 AIRCRAFT DYNAMICS AND CLASSICAL CONTROL DESIGN

impossible to fly without an automatic control system. The SAS typically uses sensors to measure the body-axes angular rates of the vehicle, and feeds back processed versions of these signals to servomechanisms that drive the aerodynamic control surfaces. In this way an aerodynamic moment proportional to angular velocity and its derivatives can be generated and used to produce a damping effect on the motion. If the basic mode is unstable, or if it is desired to change both damping and natural frequency independently, additional feedback signals will be required, as we will see.

Stability augmentation systems are conventionally designed separately for the longitudinal dynamics and the lateral-directional dynamics, and this is made possible by the decoupling of the aircraft dynamics in most flight conditions. In the next two subsections aircraft model dynamics will be used to describe the design of the various augmentation systems.

Pitch-Axis Stability Augmentation

The purpose of a pitch SAS is to provide satisfactory natural frequency and damping for the short-period mode. This mode involves the variables alpha and pitch rate; feedback of these variables to the elevator actuator will modify the frequency and damping. Figure 4.4-1 shows the arrangement; if the short-period mode is lightly damped but otherwise adequate, only pitch-rate feedback is required. If the frequency and damping are both unsatisfactory or the mode is unstable, alpha feedback is necessary. The phugoid mode will be largely unaffected by this feedback. Outer feedback control loops will often be closed around the pitch SAS to provide, for example, autopilot functions. Automatic adjustment of the augmentation (inner) loop feedback gains may be arranged when the outer feedback loops are engaged, so that the overall performance is optimal.

A physical understanding of the effect of alpha feedback follows from the explanation of pitch stiffness in Chapter 2. A statically unstable aircraft has a pitching moment curve with a positive slope over some range(s) of alpha. If perturbations in alpha are sensed and fed back to the elevator servo to generate a restoring pitching moment, the slope of the pitching moment curve can be made more negative in the region around the operating angle of attack. Furthermore, the overall pitching moment curve and the trimmed elevator deflection will not be affected, thus preserving

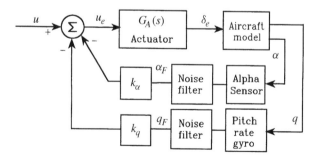

Figure 4.4-1 Pitch-axis stability augmentation.

the trim-drag and maneuverability characteristics that the designer built into the basic airplane design.

The angle of attack measurement may be obtained from the pitot-static air-data system, or a small "wind vane" mounted on the side of the aircraft forebody and positioned (after much testing and calibration) to measure alpha over a wide range of flight conditions. Two sensors may be used, on opposite sides of the aircraft, to provide redundancy and possibly to average out measurement errors caused by sideslipping. In addition, it may be necessary to compute (in real time) a "true" angle of attack from the "indicated angle of attack," airspeed, and Mach number, in order to relate the freestream angle of attack of the airframe to the direction of the flowfield at the sensor position. The signal from the alpha sensor is usually noisy because of turbulence, and a noise filter is used to reduce the amount of noise injected into the control system.

Alpha feedback is avoided if possible because of the difficulty of getting an accurate, rapidly responding, noise-free measurement, and because of the vulnerability of the sensor to mechanical damage. Noise from the alpha sensor can make it difficult to achieve precise pointing (e.g., for targeting), so the amount of alpha feedback is normally restricted.

The pitch-rate sensor is normally a mechanical gyroscopic device, arranged to measure the (inertial) angular rate around the pitch axis. The location of the gyro must be chosen very carefully to avoid picking up the vibrations of the aircraft structure. At a node of an idealized structural oscillation there is angular motion but no displacement, and at an antinode the converse is true. Thus the first choice for the rate gyro location is an antinode corresponding to the most important structural mode. Flight tests must then be used to adjust the position of the gyros. A bad choice of gyro locations can adversely affect handling qualities or, in extreme cases, cause oscillations in the flight control systems (AFWAL-TR-84-3105, 1984). The gyro filter shown in Figure 4.4-1 is usually necessary to remove noise and/or cancel structural mode vibrations.

The sign convention that has been adopted in this book (see Chapter 3) means that a positive elevator deflection leads to a negative pitching moment. Therefore, for convenience, a phase-reversal will be included between the elevator actuator and the control surface in each example, so that the positive-gain root-locus algorithm can be used for design.

Example 4.4-1: The Effects of Pitch-Rate and Alpha Feedback. The longitudinal (four-state) Jacobian matrices for the F-16 model in the nominal flight condition in Table 3.6-3 are:

$$A = \begin{bmatrix} v_T & \alpha & \theta & q \\ -1.9311E-02 & 8.8157E+00 & -3.2170E+01 & -5.7499E-01 \\ -2.5389E-04 & -1.0189E+00 & 0.0000E+00 & 9.0506E-01 \\ 0.0000E+00 & 0.0000E+00 & 0.0000E+00 & 1.0000E+00 \\ 2.9465E-12 & 8.2225E-01 & 0.0000E+00 & -1.0774E+00 \end{bmatrix}$$

$$B = \begin{bmatrix} \delta_e \\ 1.7370E-01 \\ -2.1499E-03 \\ 0.0000E+00 \\ -1.7555E-01 \end{bmatrix} \qquad (1)$$

$$C = \begin{bmatrix} 0.000000E+00 & 5.729578E+01 & 0.000000E+00 & 0.000000E+00 \\ 0.000000E+00 & 0.000000E+00 & 0.000000E+00 & 5.729578E+01 \end{bmatrix} \begin{matrix} \alpha \\ q \end{matrix}$$

The single input is the elevator deflection, δ_e, in degrees, and the two outputs are the appropriate feedback signals: alpha and pitch rate. The entries in the C matrix are the conversions to units of degrees, for consistency with the input.

Either of the two SISO transfer functions obtained from the coefficient matrices will exhibit the dynamic modes for this flight condition; the elevator-to-alpha transfer function is

$$\frac{\alpha}{\delta_e} = \frac{-0.1232(s+75.00)(s+0.009820 \pm j0.09379)}{(s-0.09755)(s+1.912)(s+0.1507 \pm j0.1153)} \qquad (2)$$

Unlike the transfer functions for stable cg positions (e.g., $x_{CG} = 0.3\,\bar{c}$) in Chapter 3, this transfer function does not exhibit the usual phugoid and short-period poles. The pole at $s \approx .098$ indicates an unstable exponential mode with a time constant of about 10 s. The complex-conjugate pole pair corresponds to an oscillatory mode with a period of 33 s and damping ratio of 0.79; this is like a phugoid period with a short-period damping ratio. This mode is the "third oscillatory mode" of the statically unstable airplane (see Section 4.2).

The modes described above obviously do not satisfy the requirements for good handling qualities, and providing continuous control of the unstable mode would be a very demanding job for a pilot. We will now show that alpha and pitch-rate feedback together will restore stability and provide virtually complete control of the position of the short-period poles.

The configuration shown in Figure 4.4-1 will be used, with an alpha filter but, for simplicity, no pitch-rate filter. The actuator and alpha-filter models are taken from the original F-16 model report (Nguyen et al., 1979), and are both simple-lag filters with time constants $\tau_a = 1/20.2$ s and $\tau_F = 0.1$ s, respectively. The aircraft state-space model (1) augmented with these models, is

$$\dot{x} = \left[\begin{array}{cccc:c:c} & & & & \vdots & \vdots & 0 \\ & & A & & \vdots & -B & \vdots & 0 \\ & & & & \vdots & & \vdots & 0 \\ & & & & \vdots & & \vdots & 0 \\ \hdashline 0 & 0 & 0 & 0 & \vdots & -20.2 & \vdots & 0 \\ 0 & 10.0 & 0 & 0 & \vdots & 0 & \vdots & -10.0 \end{array}\right] \begin{bmatrix} v_T \\ \alpha \\ \theta \\ q \\ -- \\ x_a \\ x_F \end{bmatrix} + \begin{bmatrix} 0 \\ 0 \\ 0 \\ 0 \\ ---- \\ 20.2 \\ 0 \end{bmatrix} u_e$$

(3a)

$$y = \begin{bmatrix} \alpha \\ q \\ -- \\ \alpha_F \end{bmatrix} = \begin{bmatrix} & C & : & 0 & & 0 \\ & & : & 0 & & 0 \\ \hline 0 & 0 & 0 & 0 & : & 0 & 57.29578 \end{bmatrix} x \qquad (3b)$$

Notice that the original state equations are still satisfied, and that the original δ_e input is now connected to the actuator state x_a through the phase reversal. The actuator is driven by a new input, u_e. Also, the α filter is driven by the α state of the aircraft dynamics, and an additional output has been created so that the filtered signal α_F is available for feedback. These state equations could also have been created by simulating the filters as part of the aircraft model and running the linearization program again. In the rest of this chapter the augmented matrices will be created by the MATLAB "series" command, as used in Chapter 3.

The state equations (3) can now be used to obtain the loop transfer functions needed for root-locus design. In the case of the innermost (alpha) loop, we already know that the α-loop transfer function will consist of Equation (2) with the two lag filters in cascade, and the effect of the feedback k_α can be anticipated using a sketch of the pole and zero positions. The goal of the alpha feedback is to pull the unstable pole, at $s = 0.098$, back into the left-half s-plane. Let the augmented coefficient matrices in Equation (3) be denoted by aa, ba, and ca. Then the following MATLAB commands can be used to obtain the root locus:

```
k= logspace(-2,1,2000);
r= rlocus(aa,ba,ca(3,:),0,k);     % 3rd row of C
plot(r)
grid on
axis([-20,1,-10,10])
```

Figures 4.4-2a and b show the root-locus plot for the inner loop on two different scales. The expanded scale near the origin (Figure 4.4-2b) shows that the effect of the alpha feedback is to make the loci from the third-mode poles come together on the real axis (near $s = -0.2$). The branch going to the right then meets the locus coming from the unstable pole, and they leave the real axis to terminate on the complex zeros near the origin. This provides a pair of closed-loop poles that correspond to a phugoid mode. The left branch from the third mode poles meets the locus from the pole at $s = -1.9$, and they leave the axis near $s = -1$ to form a short-period mode. Alpha feedback has therefore produced the anticipated effect: the aircraft is stable with conventional longitudinal modes.

The larger-scale plot (Figure 4.4-2a) shows that as the magnitude of the alpha feedback is increased, the frequency of the new short-period poles increases and they move toward the right-half plane. The movement toward the right-half plane is in accordance with the constant net damping rule, and the filter and actuator poles moving left. A slower (less expensive) actuator would place the actuator pole closer to the origin and cause the short period poles to have a lower frequency at a given damping ratio. The position of the short period poles for $k_\alpha = 0.5$ is $(-0.70 \pm j2.0)$. At this position the natural frequency is about 2.2 rad/s, which

296 AIRCRAFT DYNAMICS AND CLASSICAL CONTROL DESIGN

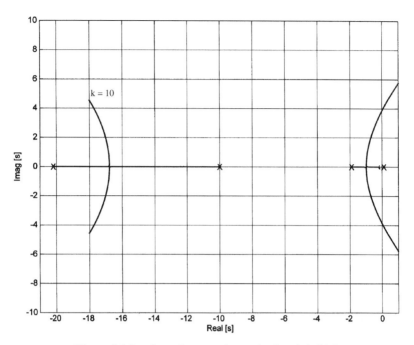

Figure 4.4-2a Inner-loop root-locus plot for pitch SAS.

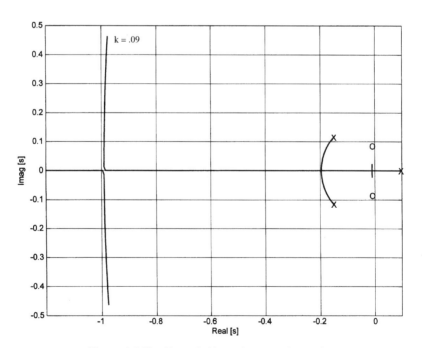

Figure 4.4-2b Expanded inner-loop root-locus plot.

is acceptable according to the flying qualities requirements, but the damping ratio ($\zeta = 0.33$) is quite low.

A root-locus plot will now show the effect of varying k_q, with k_α fixed at 0.5. The following MATLAB commands can be used:

```
acl= aa - ba*kα*ca(3,:);           % Choose kα
%[z,p,k]= ss2zp(acl,ba,ca(2,:),0)  % q/u transf. fn
r= rlocus(acl,ba,ca(2,:),0);
plot(r)
```

The q/u transfer function with $k_\alpha = 0.5$ and $k_q = 0$ is

$$\frac{q}{u} = \frac{203.2s(s+10.0)(s+1.027)(s+0.02174)}{(s+20.01)(s+10.89)(s+0.6990 \pm j2.030)(s+0.008458 \pm j0.08269)} \quad (4)$$

Note that the zeros of this transfer function are the $1/T_{\theta_1}$ and $1/T_{\theta_2}$ unaugmented open-loop zeros, with the addition of a zero at $s = -10$. This zero has appeared because of the MIMO dynamics (two outputs, one input). It originally canceled the alpha-filter pole out of the pitch-rate transfer function, but the inner-loop feedback has now moved the alpha-filter pole to $s = -10.89$.

Figure 4.4-3 shows the root-locus plot for variable k_q. The phugoid poles move very slightly, but are not visible on the plot. The short-period poles follow a circular

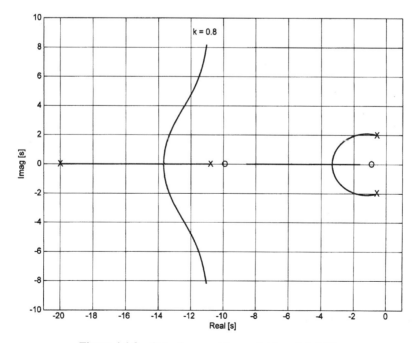

Figure 4.4-3 Outer-loop root-locus plot for pitch SAS.

arc around $s = -1$ (roughly constant natural frequency) as the pitch-rate feedback is increased. The poles become real for quite low values of k_q and, with larger values, a new higher-frequency oscillatory mode is created by the filter and actuator poles. Such a mode would be objectionable to the pilot, and we look for lower values of k_q that make the short period poles match the flying qualities requirements, with no additional oscillatory mode. The value $k_q = 0.25$ places the short-period poles at $s = -2.02 \pm j1.94$. This corresponds to a natural frequency of 2.8 rad/s and a damping ratio of $\zeta = 0.72$. The corresponding closed-loop transfer function for pitch rate is given by

$$\frac{q}{u} = \frac{203.2s(s + 10.0)(s + 1.027)(s + 0.02174)}{(s + 16.39)(s + 11.88)(s + 2.018 \pm j1.945)(s + 0.008781 \pm j0.06681)} \quad (5)$$

The original actuator pole has moved from $s = -20.2$ to $s = -16.39$, and the α-filter pole has moved from $s = -10$ to $s = -11.88$. Apart from these factors, this transfer function is very similar to the stable-cg transfer function in Example 3.8-3 but with improved short-period pole positions. ∎

Example 4.4-1 shows that alpha feedback stabilizes the unstable short-period mode and determines its natural frequency, while the pitch rate feedback mainly determines the damping. The amount of alpha feedback needed to get a satisfactory natural frequency was 0.5 degrees of elevator deflection per degree of alpha. The alpha signal is noisy and sometimes unreliable, and this large amount of alpha feedback is preferably avoided. In the second root-locus plot it can be seen that, as the pitch-rate feedback is varied, the locus of the short-period poles circles around the $1/T_{\theta_2}$ zero. Therefore, by moving the zero to the left a higher natural frequency can be achieved, or the same natural frequency can be achieved with less alpha feedback. This will be demonstrated in the next example.

Example 4.4-2: A Pitch-SAS Design. The coefficient matrices aa, bb, cc, from Example 4.4-1 are used again here, and the alpha-feedback gain will be reduced to $k_\alpha = 0.1$. A lag compensator with a pole at $s = -1$ and a zero at $s = -3$ will be cascaded with the plant to effectively move the $1/T_{\theta_2}$ zero to $s = -3$. The MATLAB commands are:

```
acl= aa - ba*0.1*ca(3,:);          % Close alpha loop, Kα=.1
qfb= ss(acl,ba,ca(2,:),0);         % SISO system for q f.b.
z=3; p=1;
lag= ss(-p,1,z-p,1);               % Lag compensator
csys= series(lag,qfb);             % Cascade Comp. before plant
[a,b,c,d]= ssdata(csys);
k= logspace(-2,0,2000);
r= rlocus(a,b,c,d,k);
plot(r)
grid on
axis([-20,1,-10,10])
```

The root-locus plot is the same shape as Figure 4.4-3, and when the pitch-rate feedback gain is $k_q = 0.2$, the closed-loop transfer function is

$$\frac{q}{u} = \frac{203.2s(s+10.0)(s+1.027)(s+0.0217)(s+3)}{(s+18.02)(s+10.3)(s+1.025)(s+1.98 \pm j2.01)(s+0.0107 \pm j0.0093)} \quad (1)$$

When the pole and zero close to $s = -1$ are canceled out, this transfer function is essentially the same as in Example 4.4-1 except that there is a zero at $s = -3$ instead of $s = -1$. This zero can be replaced by a zero at $s = -1$ once again, by placing the lag compensator in the feedback path. However, a zero at $s = -1$ produces a much bigger overshoot in the step response than the zero at $s = -3$. Therefore the flying qualities requirements on T_{θ_2} should be checked (see Section 4.3) to obtain some guidance on the position of the zero.

This example shows that the same short-period mode, as in Example 4.4-1, can be achieved with much less alpha feedback and less pitch-rate feedback. Also, the transfer function (1) shows that no additional modes are introduced. A dynamic compensator is the price paid for this. Section 4.3 shows that the $1/T_{\theta_2}$ zero will move with flight conditions, and so the compensator parameters may have to be changed with flight conditions. ∎

Lateral-Directional Stability Augmentation/Yaw Damper

Figure 4.4-4 shows the most basic augmentation system for the lateral-directional dynamics. Body-axis roll rate is fed back to the ailerons to modify the roll-subsidence

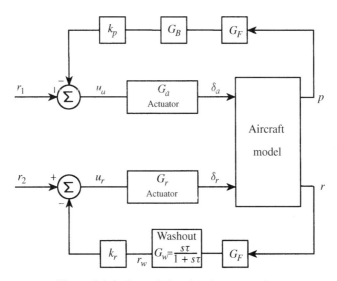

Figure 4.4-4 Lateral-directional augmentation.

mode, and yaw rate is fed back to the rudder to modify the dutch roll mode (yaw-damper feedback). The lateral (rolling) motion is not, in general, decoupled from the yawing and sideslipping (directional) motions. Therefore, the augmentation systems will be analyzed with the aid of the multivariable state equations (two inputs, ailerons and rudder, and two or more outputs), as implied by the figure. This analysis will be restricted to the simple feedback scheme shown in the figure; in a later section additional feedback couplings will be introduced between the roll and yaw channels.

The purpose of the yaw-damper feedback is to use the rudder to generate a yawing moment that opposes any yaw rate that builds up from the dutch roll mode. This raises a difficulty; in a coordinated steady-state turn the yaw rate has a constant nonzero value (see Table 3.6-3, and also the subsection on turn coordination) which the yaw-rate feedback will try to oppose. Therefore, with the yaw-damper operating, the pilot must apply larger than normal rudder pedal inputs to overcome the action of the yaw damper and coordinate a turn. This has been found to be very objectionable to pilots. A simple control-systems solution to the problem is to use "transient rate-feedback," in which the feedback signal is differentiated (approximately) so that it vanishes during steady-state conditions. The approximate differentiation can be accomplished with a simple first-order high-pass filter (see Table 3.3-1), called a "washout filter" in this kind of application.

In Figure 4.4-4, G_W is the washout filter, the transfer function G_a represents an equivalent transfer function for differential actuation of the left and right ailerons, and G_r is the rudder actuator. The transfer functions G_F represent noise filtering and any effective lag at the output of the roll-rate and yaw-rate gyros, and G_B is a *bending mode filter*. The bending mode filter is needed because the moments generated by the ailerons are transmitted through the flexible-beam structure of the wing, and their effect is sensed by the roll-rate gyro in the fuselage. The transfer function of this path corresponds to a general low-pass filtering effect, with resonances occurring at the bending modes of the wing. Because the wing bending modes are relatively low in frequency, they can contribute significant phase shift, and possibly gain changes, within the bandwidth of the roll-rate loop. The bending-mode filter is designed to compensate for these phase and gain changes.

To understand the purpose of the roll-rate feedback, consider the following facts. In Section 4.2 the variation of the roll time constant with flight conditions was analyzed, and in Chapter 2 the change of aileron effectiveness with angle of attack was described. These effects cause large, undesirable variations in aircraft roll performance that result in the pilot flying the aircraft less precisely. Closed-loop control of roll rate is used to reduce the variation of roll performance with flight conditions.

While the roll time constant is a feature of the linear small-perturbation model and gives no indication of the maximum roll rate or time to roll through a large angle, it is relevant to the initial speed of response and control of smaller-amplitude motion. Figure 4.4-5 shows a plot of the reciprocal of the F-16 roll time constant versus alpha, and shows that this time constant may become unacceptably slow at high angles of attack. The plot was derived by trimming the F-16 model in straight and level flight at sea level, with the nominal cg position, over a range of speeds. At angles of attack

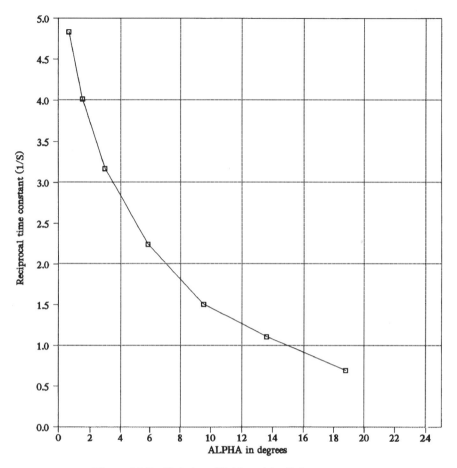

Figure 4.4-5 Variation of F-16 model roll time-constant.

greater than about 20 degrees the roll pole coupled with the spiral pole to form a complex pair.

Landing approach takes place at a relatively high angle of attack, and the roll-rate feedback may be needed to ensure good roll response. Also, satisfactory damping of the dutch roll mode is particularly important during landing approach in gusty crosswind conditions. Our F-16 model does not include flaps and landing gear, so the design of the augmentation loops will simply be illustrated on a low-speed, low-altitude flight condition. If we take the F-16 model dynamics at zero altitude, with the nominal cg position and an airspeed of 205.0 ft/s (alpha = 18.8°), the roll pole is real and quite slow ($\tau = 1.44$ s), and the dutch roll is very lightly damped ($\zeta = 0.2$). The state equations can be found by linearization, and a five-state set of lateral-directional equations can be decoupled from the full thirteen-state set. The coefficient matrices are found to be:

$$A = \begin{bmatrix} \beta & \phi & \psi & p & r \\ -0.13150 & 0.14858 & 0.0 & 0.32434 & -0.93964 \\ 0.0 & 0.0 & 0.0 & 1.0 & 0.33976 \\ 0.0 & 0.0 & 0.0 & 0.0 & 1.0561 \\ -10.614 & 0.0 & 0.0 & -1.1793 & 1.0023 \\ 0.99655 & 0.0 & 0.0 & -0.0018174 & -0.25855 \end{bmatrix}$$

$$B = \begin{bmatrix} \delta_a & \delta_r \\ 0.00012049 & 0.00032897 \\ 0.0 & 0.0 \\ 0.0 & 0.0 \\ -0.1031578 & 0.020987 \\ -0.0021330 & -0.010715 \end{bmatrix} \quad (4.4\text{-}1)$$

$$C = \begin{bmatrix} 0.0 & 0.0 & 0.0 & 57.29578 & 0.0 \\ 0.0 & 0.0 & 0.0 & 0.0 & 57.29578 \end{bmatrix} \begin{matrix} p \\ r \end{matrix}$$

$$D = \begin{bmatrix} 0 & 0 \\ 0 & 0 \end{bmatrix}$$

The null column in the A matrix shows that the state ψ is not coupled back to any other states, and it can be omitted from the state equations when designing an augmentation system. The C matrix has been used to convert the output quantities to degrees, to match the control surface inputs. The transfer functions of primary interest are:

$$\frac{p}{\delta_a} = \frac{-5.911(s - 0.05092)(s + 0.2370 \pm j1.072)}{(s + 0.06789)(s + 0.6960)(s + 0.4027 \pm j2.012)} \quad (4.4\text{-}2)$$

$$\frac{r}{\delta_a} = \frac{-0.1222(s + 0.4642)(s + 0.3512 \pm j4.325)}{(s + 0.06789)(s + 0.6960)(s + 0.4027 \pm j2.012)} \quad (4.4\text{-}3)$$

$$\frac{p}{\delta_r} = \frac{+1.202(s - 0.05280)(s - 2.177)(s + 1.942)}{(s + 0.06789)(s + 0.6960)(s + 0.4027 \pm j2.012)} \quad (4.4\text{-}4)$$

$$\frac{r}{\delta_r} = \frac{-0.6139(s + 0.5078)(s + 0.3880 \pm j1.5439)}{(s + 0.06789)(s + 0.6960)(s + 0.4027 \pm j2.012)} \quad (4.4\text{-}5)$$

The dutch roll poles are not canceled out of the p/δ_a transfer function by the complex zeros. Therefore, coupling exists between the rolling and yawing motions, and the dutch roll mode will involve some rolling motion. These transfer functions validate the decision to use the MIMO state equations for the analysis. At lower angles of attack the dutch roll poles will typically be largely canceled out of the p/δ_a transfer function, leaving only the roll-subsidence and spiral poles.

The two roll-rate transfer functions, given above, contain NMP zeros close to the origin. This is because gravity will cause the aircraft to begin to sideslip as it rolls. Then, if the dihedral derivative C_{l_β} is negative (positive roll stiffness), the aircraft will have a tendency to roll in the opposite direction. This effect will be more pronounced in a slow roll when the sideslip has a chance to build up.

The rudder-to-roll-rate transfer function has another NMP zero farther away from the origin, corresponding to faster-acting non-minimum-phase effects. A positive deflection of the rudder directly produces a positive rolling moment (see Table 3.5-1) and a negative yawing moment. The negative yawing moment rapidly leads to positive sideslip, which will in turn produce a negative rolling moment if the aircraft has positive roll stiffness. This effect tends to cancel the initial positive roll, and the NMP zero is the transfer function manifestation of these competing effects.

Example 4.4-3: A Roll Damper/Yaw Damper Design. In Figure 4.4-4 the aileron and rudder actuators will be taken as simple lags with a corner frequency of 20.2 rad/s (as in the original model), and the bending mode filter will be omitted. The coefficient matrices for the plant will be (4.4-1) with the ψ state removed, and denoted by ap, bp, cp, dp. Positive deflections of the control surfaces lead to negative values for the principal moments (Table 3.5-1) so, in order to use the positive gain root-locus for design, we will insert a phase reversal at the output of the control-surface actuators (in the C matrix). The aileron and rudder actuators will be combined into one two-input, two-output state model, and cascaded with the plant as follows:

```
aa= [-20.2 0; 0 -20.2];    ba= [20.2 0; 0 20.2];      % Actuator
ca= [-1 0; 0 -1];          da= [0 0; 0 0];            % SIGN CHANGE
actua= ss(aa,ba,ca,da);                               % u1= δ_a, u2= δ_r
plant= ss(ap,bp,cp,dp);           % x1=beta, x2=phi, x3=p, x4=r
sys1 = series(actua,plant);       % y1=p, y2=r (degrees)
```

The washout filter will be incorporated in a two-input, two-output model, with the first input-output pair being a direct connection:

```
aw= [-1/τ_w];        bw= [0 1/τ_w];        % τ_w to be defined
cw= [0; -1];         dw= [1 0; 0 1];       % y1=p y2=washed-r
wash= ss(aw,bw,cw,dw);
sys2= series(sys1,wash);    % x1=wash, x2=beta,.., x6=ail, x7=rdr
```

The washout filter time constant is a compromise; too large a value is undesirable since the yaw damper will then interfere with the entry into turns. The following root-locus design plots can also be used to show that too small a value will reduce the achievable dutch roll damping (see Problem 4.4-3). The time constant is normally of the order of 1 s, and $\tau_W = 1.0$ s is used here.

Experience shows that the roll-damping loop is the less critical loop, and it is conveniently closed first. The p/u_a transfer function is the same as (4.4-2) with an additional pole at $s = -20.2$ and the static loop sensitivity changed to 119 (i.e., 20.2 times the original value of 5.91). The MATLAB commands to obtain a root-locus plot, and to close the loop are:

```
[a,b,c,d]= ssdata(sys2);
k= linspace(0,.9,3000);
r= rlocus(a,b(:,1),c(1,:),0,k);
plot(r)                         % Roll channel root locus
grid on
axis([-12,1,-5,5])
```

Figure 4.4-6 is the root-locus plot for positive k_p. It shows that the feedback has had the desired effect of speeding up the roll-subsidence pole, which moves to the left in the s-plane and eventually combines with the actuator pole to form a complex pair. The spiral pole (not visible) moves a little to the right toward the NMP zero at $s = 0.05$, and the dutch roll poles change significantly as they move toward the open-loop complex zeros. If the feedback gain is made too high in this design, it will be found to be excessive at lower angles of attack. Furthermore, a high value will simply cause the aileron actuators to reach their rate and deflection limits more rapidly, as they become less effective at the higher angles of attack. A feedback gain of $k_p = 0.2$ puts the roll subsidence pole at $s = -1.37$, which is about twice as fast as the open-loop value. This is a suitable starting value for investigating the effect of closing the yaw damper loop:

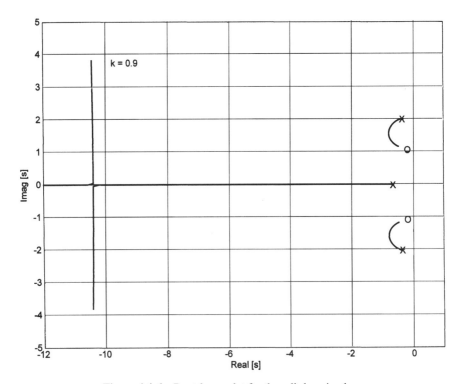

Figure 4.4-6 Root-locus plot for the roll-damping loop.

```
ac11= a - b(:,1)*k_p*c(1,:);            % Close roll loop
[z,p,k1]= ss2zp(ac11,b(:,2),c(2,:),0)   % Yaw tr. fn. + wash
r= rlocus(ac11,b(:,2),c(2,:),0,k);
plot(r)                                 % Yaw channel root locus
```

The transfer function r_W/u_r (with $k_p = 0.2$) is

$$\frac{r_W}{u_r} = \frac{12.40s(s+18.8)(s+0.760)(s+0.961 \pm j0.947)}{(s+1)(s+18.9)(s+1.37)(s+0.0280)(s+20.2)(s+0.752 \pm j1.719)} \quad (1)$$

A root-locus plot for closing the yaw-rate loop through the feedback gain k_r is shown in Figure 4.4-7. Although not shown in the figure, one of the actuator poles is effectively canceled by the zero at $s = -18.8$; the remaining actuator pole moves to the right to meet the roll pole and form a new complex pair. As the magnitude of k_r is increased, the spiral pole moves slightly closer to the washout zero at the origin, and the washout pole moves toward the zero at $s = -0.76$. At first the dutch roll poles move (approximately) around an arc of constant natural frequency and increasing damping ratio toward the complex zeros. After k_r reaches about 3.5, the natural frequency begins to decrease and the damping ratio tends to remain constant.

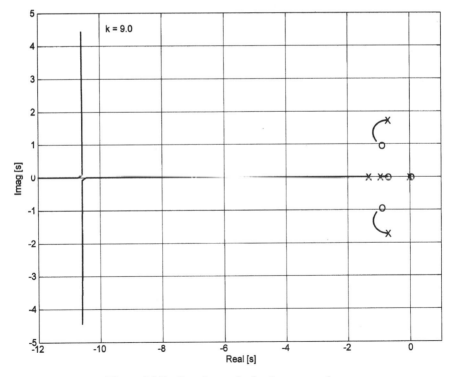

Figure 4.4-7 Root-locus plot for the yaw-rate loop.

This feedback gain was considered to be the optimum value for the dutch roll poles, and so the yaw-rate loop was closed:

```
acl2= a - b*[.2 0; 0 3.5]*c;
[z,p,k1]= ss2zp(acl2,b(;,1),c(1,:),0)     % c.l. roll-rate t.f.
```

The principal transfer functions were found to be:

$$\frac{p}{r_1} = \frac{119.4(s+17.4)(s-0.0502)(s+3.74)(s+0.262 \pm j0.557)}{(s+18.7)(s+17.7)(s+0.0174)(s+3.29)(s+0.861)(s+1.18 \pm j1.33)} \tag{2}$$

$$\frac{r}{r_2} = \frac{12.4(s+18.8)(s+1.00)(s+0.760)(s+0.961 \pm j0.947)}{(s+17.7)(s+18.7)(s+3.29)(s+0.861)(s+0.0174)(s+1.18 \pm j1.33)}, \tag{3}$$

where r_1 and r_2 are the roll-rate and yaw-rate reference inputs, as shown in Figure 4.4-4.

Transfer functions (2) and (3) show that the dutch roll poles, and the washout pole (at $s = -0.861$), do not cancel out of the p/r_1 transfer function, so there is still strong coupling between the roll and yaw channels. The dutch roll natural frequency and damping ($\omega_n = 1.78$ rad/s, $\zeta = 0.67$) are now satisfactory, but the appearance of the relatively slow washout pole in the lateral dynamics may mean that the roll response is not much improved. Since we no longer have a simple dominant poles situation, a time response simulation is needed to assess the design. Before this is undertaken, the effect of a higher gain in the roll-rate loop will be considered.

If the roll-rate loop is closed, with $k_p = 0.4$, the roll subsidence pole moves out to $s = -3.08$, and the zero in the yaw-rate loop transfer function (1) moves from $s = -0.76$ to $s = -3.40$. This causes different behavior in the root-locus plot for the yaw-rate loop, as shown in Figure 4.4-8. The washout pole now moves to the left instead of the right. A comparison of Figures 4.4-7 and 4.4-8 shows that the price paid for this potential improvement in roll response is that the maximum dutch roll frequency is reduced. If the yaw-rate loop is closed with $k_r = 1.3$, to obtain the highest possible damped frequency for the dutch roll poles, the closed-loop transfer functions are

$$\frac{p}{r_1} = \frac{119.4(s+19.27)(s+1.74)(s-0.0507)(s+0.334 \pm j0.787)}{(s+19.25)(s+17.4)(s+0.00767)(s+2.82)(s+1.57)(s+0.987 \pm j0.984)} \tag{4}$$

$$\frac{r}{r_2} = \frac{12.40(s+1.00)(s+17.1)(s+3.40)(s+0.486 \pm j0.459)}{(s+19.25)(s+17.4)(s+0.00767)(s+2.82)(s+1.57)(s+0.987 \pm j0.984)} \tag{5}$$

The dutch roll frequency has decreased to $\omega_n = 1.39$ rad/s, and the damping has increased to $\zeta = 0.71$; these values still represent good flying qualities (see Table 4.3-6). An improvement in the roll response should have been obtained since the slow

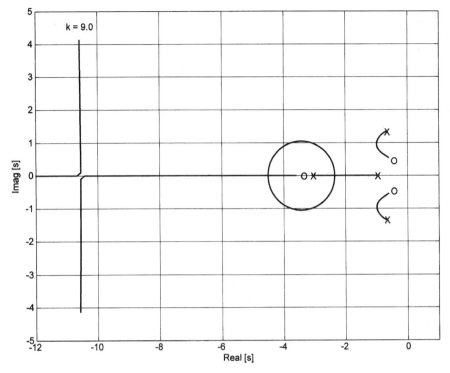

Figure 4.4-8 Alternate yaw-rate root-locus.

washout pole is nearly canceled by the zero at $s = -1.74$, and the roll-subsidence pole (at $s = -2.82$) may now dominate the roll response. Note the way in which one actuator pole almost cancels out of each transfer function. Also, in the yaw-rate response, note the zero at $s = -1$ that originally canceled the washout pole. The transfer functions still show significant roll-yaw coupling.

The roll response of this design can only be assessed with a simulation, and because of the presence of the slow spiral pole in the transfer functions, a doublet pulse should be used as the input. The time responses were obtained by closing the yaw-rate and roll-rate loops with the feedback gains above ($k_p = 0.4$, $k_r = 1.3$) and using the commands:

```
acl2= a - b*[.4 0; 0 1.3]*c;         % Close roll & yaw
t= [0:.02:10];                        % 501 points for plot
u= [-1.8*ones(1,51),1.8*ones(1,50),zeros(1,400)]';   % Doublet
[y,x]= lsim(acl2,b(:,1),c(1,:),0,u,t);   % Linear simulation
plot(t,y,t,u)
grid on
```

Figure 4.4-9 compares the roll-rate response of the open-loop dynamics (augmented with the actuators) with the closed-loop response. The doublet input is negative for 1 s, positive for 1 s, then zero, with unit amplitude in the open-loop case. In the

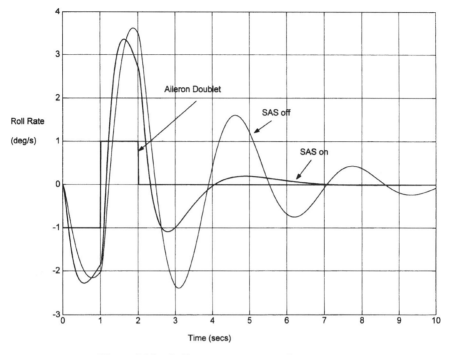

Figure 4.4-9 Roll-rate response to an aileron doublet.

closed-loop case the overall gain is different, and the doublet was adjusted to 1.8 degrees so that the responses were of similar amplitude. The figure exhibits the major improvement in the dutch roll damping and the small but significant improvement in the roll-rate speed of response. ∎

This example indicates the difficulties of multivariable design when significant cross-coupling is present in the dynamics. It also shows the difficulty of obtaining a good roll response at low dynamic pressure and high alpha. The design could be pursued further by investigating the effect of changing the washout time constant and using compensation networks, such as a phase lead, in the yaw-rate feedback loop. As pointed out earlier, increasing the bandwidth of the control loops may simply lead to saturation of the control-surface actuators, and the limitations of the basic aircraft must be considered first.

4.5 CONTROL AUGMENTATION SYSTEMS

When an aircraft is under manual control (as opposed to autopilot control) the stability augmentation systems of the preceding section are, in most cases, the only automatic flight control systems needed. But in the case of high-performance military aircraft, where the pilot may have to maneuver the aircraft to its performance limits and perform tasks such as precision tracking of targets, specialized control augmentation systems (CAS) are needed. Flight control technology has advanced to the point where the flight control system (FCS) can provide the pilot with selectable "task-tailored control laws." For example, although the role of a fighter aircraft has changed to include launching missiles from long range, the importance of the classical dogfight is still recognized. A dogfight places a premium on high maneuverabilty and "agility" (ability to change maneuvers quickly) in the aircraft and a control system that allows the pilot to take advantage of this maneuverability. In this situation a suitable controlled variable for the pitch axis is the *normal acceleration* of the aircraft. This is the component of acceleration in the negative direction of the body-fixed z-axis. It is directly relevant to performing a maximum-rate turn and must be controllable up to the structural limits of the airframe, or the pilot's physical limits. Therefore, for a dogfight, a "g-command" control system is an appropriate mode of operation of the flight control system. Other reasons for using this type of system will be described when we come to consider an example.

Another common mode of operation for a pitch-axis control augmentation system is as a pitch-rate command system. When a situation requires precise tracking of a target, by means of a sighting device, it has been found that a deadbeat response to pitch-rate commands is well suited to the task. Control of pitch rate is also the preferred system for approach and landing. Systems have been designed (Toles, 1985) which blend together the control of pitch rate and normal acceleration.

With respect to lateral/directional control, the most prevalent control augmentation system is a roll-rate command system. This system may be designed to roll the aircraft around its own velocity vector rather than the body axis. The reasons for this are described in the following sections.

Pitch-Rate Control Augmentation Systems

Figure 4.5-1 is a block diagram of a pitch-rate CAS. Type-0 control is not very satisfactory because the control inputs to the plant may be quite large (e.g., several degrees of elevator deflection) while the gains in the error channel are not usually very high and entail large control errors. Therefore, proportional-plus-integral compensation is

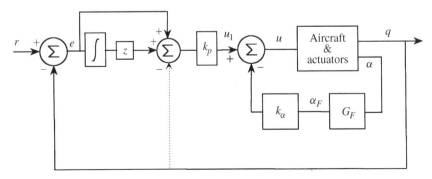

Figure 4.5-1 Pitch-rate control augmentation.

310   AIRCRAFT DYNAMICS AND CLASSICAL CONTROL DESIGN

used to provide more precise control. Inner-loop alpha feedback is used, as in Example 4.4-1, when the pitch stiffness is inadequate.

The proportional path of the PI compensator can be replaced by an equivalent inner-loop pitch-rate feedback shown as a dotted line. This leaves the closed-loop poles unchanged but removes the PI zero from the closed-loop transfer function, thus reducing step-response overshoot (see Equation 3.9-24). It will be shown to be convenient to keep the PI zero while performing root-locus design.

The design of the pitch-rate CAS will now be illustrated by an example. It will be shown that the design can be performed on the short-period dynamics, but some caution must be used.

Example 4.5-1: A Pitch-Rate CAS Design. The F-16 longitudinal dynamics corresponding to the nominal flight condition in Table 3.6-3 will be used once again. The A, B, C coefficient matrices are given in Example 4.4-1. These equations do not exhibit a short-period mode, but the α and q equations are only loosely coupled to v_T and θ and can be extracted as in Section 4.2. The final design will be verified on the complete dynamics. The elevator actuator and α-filter dynamics will be those used in Section 4.4, and a sign change will be incorporated at the actuator output.

The design procedure will be to close the alpha loop, then inspect the actuator to pitch-rate transfer function and choose a position for the PI zero that is likely to yield a satisfactory root-locus plot. This procedure will be illustrated by MATLAB statements.

We first define the plant matrices, cascade the actuator and filter, and close the alpha feedback loop:

```
ap=[-1.0189 .90506; .82225 -1.0774];   % x1= alpha  x2=q
bp=[-2.1499E-3; -1.7555E-1];           % Elevator input
cp=[57.29578  0; 0  57.29578];         % y1= alpha, y2= q
dp=[0  0];
sysp= ss(ap,bp,cp,dp);                 % Plant
sysa= ss(-20.2,20.2,-1,0);             % Actuator & SIGN CHANGE
[sys1]= series(sysa,sysp);             % Actuator then Plant
sysf= ss(-10,[10 0],[1; 0],[0 0; 0 1]);  % Alpha Filter
[sys2]= series(sys1,sysf);             % Actuator+Plant+Filter
[a b c d]= ssdata(sys2);               % Extract a,b,c,d
acl= a - b*[k_\alpha  0]*c;            % Close Alpha-loop
[z,p,k]= ss2zp(acl,b,c(2,:),0)         % q/u1 transf. fn.
```

The filter has been defined with two inputs and two outputs, and one input-output pair is a direct connection so that q is available as output-2. When the inner-loop feedback gain k_α is chosen, the zeros, poles, and gain of the q/u_1 transfer function will be calculated.

The final design will be relatively slow unless the integrator pole can be moved well to the left or made to coincide with a zero. Some trial designs show that this demands a smaller amount of alpha feedback than that used in Example 4.4-1; this will be demonstrated by comparing two different values of k_α.

Consider first the situation with $k_\alpha = 0.20$. The q/u_1 transfer function is then given by

$$\frac{q}{u_1} = \frac{203.2(s + 10.0)(s + 1.029)}{(s + 10.38)(s + 20.13)(s + 0.8957 \pm j1.152)}$$

The behavior of the outer-loop root locus with the added PI compensator can now be anticipated. As k_p is varied, the integrator pole will move toward the zero at -1.029; the compensator zero should be placed to the left of this zero, and the short-period poles will circle around the compensator zero. The following commands will add the PI compensator and plot the root loci:

```
sys3= ss(acl,b,c,[0;0]);           % Alpha-loop closed
sysi= ss(0,3,1,1);                 % PI= (s+3)/s
sys4= series(sysi,sys3);           % x1=alpha-f, , ,x5= PI
[aa,bb,cc,dd]= ssdata(sys4);
k= linspace(0,.9,1000);
r= rlocus(aa,bb,cc(2,:),0,k);
plot(r)
axis=([-16,0,-8,8])
grid on
```

The root locus is shown in Figure 4.5-2 for a compensator zero at $s = -3.0$. When k_p reaches about 0.5 the filter and actuator poles form a second complex pair, the integrator pole has moved to $s = -0.91$, and the short-period poles are at $s = -3.2 \pm j3.4$. Increasing k_p causes the second complex pair to quickly become less damped, while the integrator pole moves only slightly farther left. If the amount of alpha feedback is reduced, the integrator pole can be moved closer to the zero at $s = -1.029$ before the second complex pole pair appears, while maintaining a satisfactory short-period pair.

The alpha feedback was eventually reduced to $k_\alpha = 0.08$, and the compensator zero was retained at $s = -3.0$ with the intention of causing the short-period poles to pass near $s = -4 \pm j3$ ($\omega_n = 5, \zeta = 0.8$). The root-locus plot was the same shape as Figure 4.5-2. With $k_p = 0.5$ the slow integrator pole reached $s = -1.02$ and stopped moving left, the short-period poles reached $s = -3.4 \pm j3$, and the actuator and filter poles were still short of combining to form a complex pair. The closed-loop (unity-feedback) transfer function was

$$\frac{q}{r} = \frac{101.6(s + 3.00)(s + 10.0)(s + 1.029)}{(s + 10.7)(s + 13.7)(s + 1.02)(s + 3.43 \pm j3.03)}$$

This was considered to be a promising design and the closed-loop step-response was simulated with the following code:

```
acl2= aa - bb*0.5*cc(2,:);         % close outer loop
sys= ss(acl2,0.5*bb,cc(2,:),0);    % unity feedback
step(sys,3)
```

312 AIRCRAFT DYNAMICS AND CLASSICAL CONTROL DESIGN

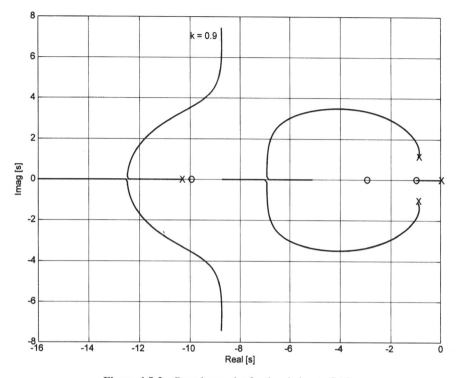

Figure 4.5-2 Root-locus plot for the pitch-rate CAS.

Figure 4.5-3 (heavy curve) shows the step response. This response has a fast rise time and a large overshoot (almost 20 percent) and does not satisfy the "deadbeat" requirement. The other curve shows the pitch-rate step response when the compensator zero is removed. The rise time is now longer, but the settling time is about the same and the overshoot is only about 2 percent. This is potentially a good design and we will move on to apply the same feedback gains to the complete longitudinal dynamics.

When the feedback gains $k_\alpha = 0.08$, and $k_p = 0.5$ are used on the full dynamics given in Example 4.4-1, the closed-loop transfer function is

$$\frac{q}{r} = \frac{304.8(s + 10.0)(s + 1.027)(s + 0.02174)s}{(s + 10.75)(s + 13.67)(s + 1.016)(s + 3.430 \pm j3.032)(s + 0.02173)s}$$

Observe that this transfer function contains the subset of poles and zeros given by the short-period approximation, and that the phugoid mode has degenerated to two real poles with this small amount of alpha feedback. Also, the phugoid poles are canceled by zeros, and so would play no part in the pitch-rate response in this case.

This example illustrates some of the features of a pitch-rate CAS. An actual design can only be optimized by careful comparison with the flying qualities requirements,

CONTROL AUGMENTATION SYSTEMS   313

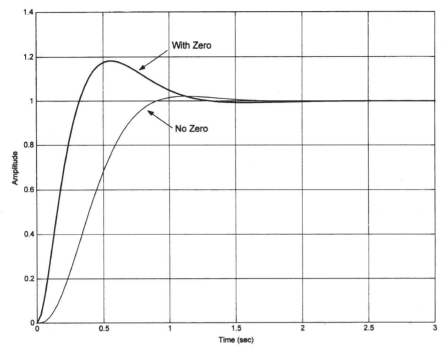

Figure 4.5-3 Step-response of the pitch-rate CAS.

piloted simulation, and flight test. During the design process nominal designs must be performed at several points throughout the speed-altitude envelope, and the feedback gains will be a function of some "scheduling" parameters such as dynamic pressure.

■

Normal Acceleration Control Augmentation Systems

In a fighter aircraft, if an accelerometer is placed close to the pilot's station, aligned along the body z-axis, and used as the feedback sensor for control of the elevator, the pilot has precise control over his z-axis g-load during high-g maneuvers. If 1 g is subtracted from the accelerometer output, the control system will hold the aircraft approximately in level flight with no control input from the pilot. If the pilot blacks out from the g-load, and relaxes any force on the control stick, the aircraft will return to 1 g flight. Other useful features of this system are that the accelerometer output contains a component proportional to alpha and can inherently stabilize an unstable short-period mode, and the accelerometer is an internal sensor that is less noisy and more reliable than an alpha sensor.

The normal acceleration a_n at a point P, fixed in the aircraft body, is defined to be the component of acceleration at P in the negative-z direction of the body axes. Following (1.2-13a), the component of **G** can be expressed in terms of the aircraft pitch and roll angles with respect to the NED system (ignoring the oblateness of the Earth). The accelerometer output is then proportional to the specific force:

$$f_n = a_n + |\mathbf{G}|\cos\theta\cos\phi \qquad (4.5\text{-}1a)$$

If the measurement is expressed in g units, the ratio $|\mathbf{G}|/g_D$ is very close to unity near to the Earth's surface, so that

$$f_n \approx a_n + \cos\theta\cos\phi \qquad g \text{ units} \qquad (4.5\text{-}1b)$$

In level flight, at small angles of attack, the feedback signal for the control system is

$$(f_n - 1) \approx a_n \qquad g \text{ units} \qquad (4.5\text{-}1c)$$

This normal acceleration is approximately zero in steady level flight; it is often called the "incremental" normal acceleration, and given the symbol n_z when in g units. Note that the component of acceleration along the lift axis, in steady-state flight with α, β, and ϕ small, can be written (see Equations (2.5-29)) in terms of the load factor as $(n - \cos\theta)$ g-units. At small alpha the lift direction is nearly coincident with the body negative z-axis, and,

$$a_n \approx n - \cos\theta \qquad g \text{ units} \qquad (4.5\text{-}1d)$$

or,

$$f_n \approx n, \quad \theta \text{ small} \qquad g \text{ units} \qquad (4.5\text{-}1e)$$

Therefore, the accelerometer measurement is an approximate measurement of load factor, under the above conditions.

If the accelerometer is on the body x-axis, at a distance x_a forward of the aircraft cg, and the aircraft is not rolling and yawing, the transport acceleration at that point is (see (1.2-13b)),

$$a_n = -\left(\mathbf{a}^z_{CM/i} - \dot{Q}x_a\right)/g_D \qquad g\text{-units}$$

If this equation is included in the nonlinear aircraft model then, in steady level-flight, the normal acceleration is close to zero and numerical linearization will yield a linear equation for a_n. A linear equation can also be obtained algebraically by finding the increment in the aerodynamic and thrust forces due to perturbations in the state and control variables, and this involves the Z-derivatives (see, for example, McRuer et al., 1973). For the nonlinear F-16 model numerically linearized at the nominal flight condition in Table 3.6-3, the output equation for normal acceleration at the cg ($x_a = 0$) is found to be

$$a_n = 0.003981 v_T + 15.88\alpha + 1.481 q + 0.03333 \delta_e, \qquad (4.5\text{-}2)$$

where α and q are in radians, and δ_e is in degrees.

The normal acceleration in (4.5-2) depends on v_T, α, and q (the quantities that define the longitudinal aerodynamic forces) and on elevator deflection, which produces aerodynamic forces directly. This direct feed term was also noted in Example 3.7-2 and leads to a transfer function of relative degree zero. Note that a_n is insensitive to the pitch attitude when θ is small.

The elevator-to-normal-acceleration transfer function corresponding to (4.5-2) can be found from the Jacobian matrices and is

$$\frac{a_n}{\delta_e} = \frac{0.03333(s - 0.003038)(s + 0.01675)(s + 6.432)(s - 13.14)}{(s - 0.09756)(s + 1.912)(s + 0.1507 \pm j0.1153)} \quad (4.5\text{-}3)$$

This transfer function has the same poles that were noted in Example 4.4-1. Because of the NMP zero at $s = 13.14$, the normal acceleration response to a negative step elevator-command (aircraft nose-up) will be an initial negative acceleration, quickly followed by the expected positive normal acceleration.

The physical explanation for the non-minimum-phase behavior is that when the elevator control surface is deflected trailing edge upward to produce a positive normal acceleration, this creates a downward increment of force on the tail. The result is that the cg of the aircraft may drop momentarily during the pitch-up, so the normal acceleration may briefly become negative before it builds up positively. At the pilot's station ahead of the cg, the normal acceleration also depends on the pitch angular acceleration about the cg, so only a positive normal acceleration may be felt in a pitch-up. Table 4.5-1 shows the elevator-to-normal-acceleration transfer function zeros for a range of accelerometer positions, from the cg forward. The zeros close to the origin do not change significantly from the positions given in (4.5-3), and only the static loop sensitivity and the remaining zeros are shown.

Table 4.5-1 shows that as the accelerometer position is moved forward, the NMP zero moves out toward infinity and the static loop sensitivity decreases, thus keeping the transfer function dc gain constant. Eventually the static loop sensitivity changes sign and a zero comes in from infinity along the negative real axis, finally combining with the other real zero to form a complex pair. At a position near 6.1 ft forward of the cg the non-minimum-phase effect disappears, and this point corresponds to an

TABLE 4.5-1: Transfer Function Zeros versus Accelerometer Position

x_a (ft)	Static-Loop-Sensitivity and Numerator Factors
0	0.03333 $(s + 6.432)$ $(s - 13.14)$
5	0.006042 $(s + 9.171)$ $(s - 50.82)$
6	0.0005847 $(s + 10.68)$ $(s - 450.7)$
6.1	0.00004005 $(s + 10.90)$ $(s - 6448.2)$
7	-0.004872 $(s + 14.73)$ $(s + 39.23)$
15	-0.04852 $(s + 3.175 \pm j6.925)$

"instantaneous center of rotation" when an elevator input is suddenly applied. Note that in the case of the real F-16 aircraft, the pilot's station is approximately 15 ft ahead of the cg and is therefore not close to the instantaneous center of rotation.

It is also important to place the accelerometer close to a node of the most important fuselage bending mode. If this is not done, structural oscillations will be coupled into the rigid-body control system and may degrade the handling qualities or even lead to an "aeroservoelastic" limit-cycle oscillation (see, e.g., AFWAL-TR-84-3105). Inevitably, the design of a normal-acceleration control system to achieve good handling qualities is difficult and can require a good deal of refinement based on flight test results. A control system that has a good normal acceleration step response may have a pitch-rate response with a very large overshoot, and conversely, a reduction in the pitch-rate overshoot may lead to a sluggish normal acceleration response. The C^* criterion is appropriate for initial evaluation of the control system, since it is based on a blend of normal acceleration at the pilot's station and pitch rate.

Finally, note that an accelerometer is an internal (within the fuselage) sensor, with higher reliability and lower noise than the external alpha sensor. However, both accelerometers and alpha sensors are typically employed on modern fighter aircraft, and this reduces the dependence on the alpha sensor. A disadvantage of normal acceleration feedback is that the gain of the transfer function (4.5-3) varies widely with dynamic pressure. Accelerometer noise may become a problem if, at low dynamic pressure, the gain has to be greatly boosted to achieve a desirable closed-loop response. We will now investigate the features of the normal acceleration CAS by means of a design example.

Example 4.5-2: A Normal Acceleration CAS Design. The configuration shown in Figure 4.5-4 will be used. The dynamics will be the same as Example 4.5-1, but an output equation for normal acceleration must be determined. Therefore, using numerical linearization of the F-16 model with the accelerometer 15 ft forward of the cg (i.e., at the pilot's station), and the nominal flight condition from Table 3.6-3, the output equation is found to be

$$a_n = 0.0039813 v_T + 16.262\alpha + 0.97877q - 0.048523\delta_e \quad (1)$$

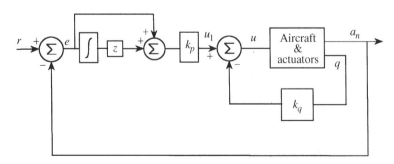

Figure 4.5-4 Normal-acceleration control augmentation.

In (1) the dependence on v_T is quite weak, so the states V_T and θ will be dropped, with the final results checked on the complete dynamics. The MATLAB commands to cascade the actuator with the plant, and close the pitch-rate loop are:

```
ap=[-1.0189 .90506; .82225 -1.0774];   % x1= alpha x2=q
bp=[-2.1499E-3; -1.7555E-1];           % Elevator input
cp=[0  57.29578; 16.262 .97877];       % y1=q y2= an
dp=[0; -0.048523];
sysp= ss(ap,bp,cp,dp);                 % Plant
sysa= ss(-20.2,20.2,-1,0);             % Actuator, SIGN CHANGE
[sys1]= series(sysa,sysp);             % Actuator then Plant
[a,b,c,d]= ssdata(sys1);               % a_n/u transfer fn.
acl= a - b*[0.4 0]*c;                  % Close q loop
[z,p,k]= ss2zp(acl,b,c(2,:),d);        % a_n/u_1 transfer fn.
```

The plant transfer function from elevator actuator to normal acceleration is found to be

$$\frac{a_n}{u} = \frac{.9802(s + 3.179 \pm j6.922)}{(s + 20.20)(s + 1.911)(s + 0.1850)} \quad (2)$$

The effect of the inner-loop pitch-rate feedback is to speed up the two slow poles, and at quite low gain the pole from $s = -1.911$ combines with the actuator pole to form a complex pair. Speeding up these poles is desirable for a fast time response, but as noted previously, the amount of pitch-rate feedback is limited by practical considerations (pickup of structural noise). The value $k_q = 0.4$ (degrees of elevator deflection per degree per second of pitch rate) is in line with our past experience (0.25 to 0.5) and leads to the following closed-loop transfer function.

$$\frac{a_n}{u_1} = \frac{0.9802(s + 3.179 \pm j6.922)}{(s + 13.78)(s + 7.661)(s + 0.8601)} \quad (3)$$

The outer-loop root locus with the added PI compensator can now be anticipated. The compensator pole (at $s = 0$) will combine with the pole at $s = -0.8601$ to form a complex pair, and these poles will move toward the complex zeros of (3). This behavior will be modified depending on the position of the compensator zero. If the compensator zero is well to the left of $s = -0.8601$, these branches will be deflected only slightly to the left before landing on the complex zeros. At the same time the pole at $s = -7.661$ will move toward the compensator zero, creating a potential slow-pole problem. The complex zeros are not well damped and it is difficult to achieve fast, well-damped, complex poles together with a fast real pole.

The alternative is to place the compensator zero close to the pole at $s = -0.8601$ so that this pole is effectively canceled. The loci for the short-period poles will then break away from the real axis somewhere closer to the pole at $s = -7.661$ before proceeding to the complex zeros. Some trial and error shows that this approach leads to a better time response, and it will be followed here. In practice, the sensitivity of the poles to gain variations, noise pickup, and possible advantages of additional compensator poles and zeros would have to be considered.

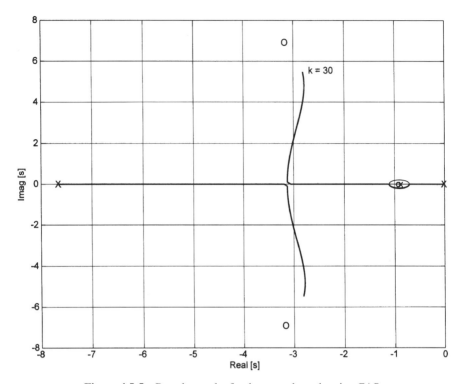

Figure 4.5-5 Root-locus plot for the normal acceleration CAS.

Figure 4.5-5 shows the outer-loop root locus (i.e., k_p varied) when the PI compensator zero is placed at $s = -0.9$ (to demonstrate that exact cancellation is not required). The effect of the imperfect cancellation is visible near $s = -0.9$, and the locus of the short-period poles shows that satisfactory damping and natural frequency can be achieved without the use of an additional lead compensator.

The short-period poles should be made well damped because the compensator zero can be anticipated to cause an overshoot in the closed-loop step response. When the root locus is calibrated with a few values of k_p, a value $k_p = 5$ puts the short-period poles at $s = -3.00 \pm j2.18$ ($\omega_n = 3.7$, $\zeta = 0.81$). The closed-loop transfer function is then

$$\frac{a_n}{r} = \frac{4.901(s + 0.9000)(s + 3.179 \pm j6.922)}{(s + 20.28)(s + 0.9176)(s + 3.000 \pm j2.180)} \tag{4}$$

Figure 4.5-6 shows the closed-loop step response corresponding to this transfer function and the normalized C^* response. The a_n response is fast and well damped; the initial rate of rise is particularly fast because of the pitch acceleration component of the response. The rate limitations of the elevator actuator would modify this response slightly. The associated pitch-rate response (not illustrated) shows an overshoot of approximately 100 percent, but the C^* response falls almost exactly in

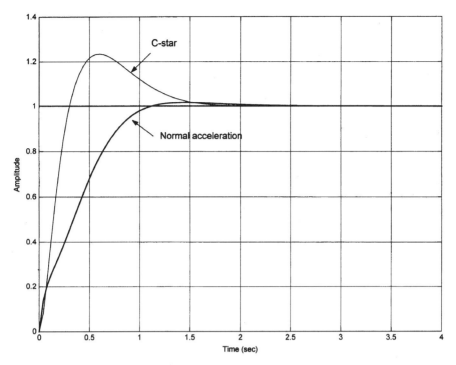

Figure 4.5-6 Normal acceleration CAS; step and C-star responses.

the middle of the level-1 envelope (see Section 4.3). The C^* values were computed by adding the component $12.4q$ to the normal acceleration output equation in the closed-loop Jacobian matrices. An initial time-response run was performed so that the steady-state value of C^* could be determined for use in normalizing the response.

The closed-loop transfer function obtained by applying the same feedback gains to the complete longitudinal dynamics (i.e., phugoid included) is

$$\frac{a_n}{r} = \frac{4.901(s+0.900)(s+3.175 \pm j6.925)(s+0.01685)(s-0.003139)}{(s+20.28)(s+0.9194)(s+3.000 \pm j2.186)(s+0.01637)(s-0.003219)} \tag{5}$$

Notice that this transfer function contains, to a very good approximation, the poles and zeros of (4), thereby justifying the use of the short-period approximation. In (5) the phugoid mode is degenerate (two real poles) and one pole is unstable, whereas in Section 4.4 a stable phugoid was achieved with the basic stability augmentation system. This is because the normal acceleration equation (1) contains a component due to v_T, and this component is being fed back in a positive sense (positive δ_e gives positive v_T). The phugoid mode is almost canceled by the transfer function zeros in this case, and the unstable pole is very slow. Nevertheless an unstable phugoid is undesirable.

The problem can be avoided by retaining some inner-loop alpha feedback and using less gain in the normal acceleration loop. An alternative possibility is to modify the feedback signal by subtracting $\cos\theta \cos\phi$ from the accelerometer output, as in (4.5-1b), to remove the gravity component. If this is done the feedback signal will contain a θ component that will be in the correct sense to provide a stabilizing effect on the phugoid mode. This control system would hold a steady climb or dive with no control stick deflection, and need little input in a coordinated turn. It would probably prove objectionable to pilots. ∎

Lateral/Directional Control Augmentation

The roll/yaw stability augmentation system described in Section 4.4 is adequate for most aircraft, but for aircraft that must maneuver rapidly at high angles of attack, a more refined lateral-directional control augmentation system is required. The lateral aerodynamic control surfaces (ailerons and differential elevator) tend to cause the aircraft to roll about its longitudinal axis, and at high alpha, this can lead to some highly undesirable effects.

Consider the effect of a rapid 90 degree body-axis roll at high alpha. It is easy to visualize that the angle of attack will be converted immediately, and almost entirely, to a sideslip angle. This is referred to as *kinematic coupling* of alpha and beta. Because of this rapid elimination of the angle of attack, the body-axis roll is counterproductive. The most important purpose of a roll is to initiate a turn, which is then achieved by using angle of attack to produce the lift that will subsequently generate the required centripetal acceleration.

The sideslip created by kinematic coupling is referred to as *adverse sideslip* because it will tend to oppose the roll (remember that C_{l_β} is normally negative; a right roll will generate positive beta through kinematic coupling and hence a negative rolling moment). The sideslip will exist until the aircraft has yawed into the wind once more, and then if the angle of attack must be reestablished, the result will be an inefficient turn entry. Most modern fighters therefore use automatic control systems designed to roll the aircraft about the stability x-axis, thus maintaining the initial angle of attack.

Finally, large sideslip angles are undesirable for several important reasons. The effectiveness of the aerodynamic control surfaces may be greatly reduced; directional stability may be lost so that, in some cases, aircraft have been known to tumble (end over end). Even if directional stability is maintained, a large sideforce can be developed that may possibly break the vertical tail.

Another important effect that occurs during a roll is *inertia coupling*. Suppose that the aircraft has been designed to roll around the stability x-axis, with no sideslip. Then the transformations in Section 2.3 can be used to determine the body-axes roll and yaw rates that result in a stability-axes roll-rate P_s with zero yaw rate R_s. The relevant equations are

$$P_s = P \cos\alpha + R \sin\alpha \qquad (4.5\text{-}4a)$$

$$0 = R_s = -P \sin\alpha + R \cos\alpha \qquad (4.5\text{-}4b)$$

or,

$$R = P \tan \alpha \tag{4.5-4c}$$

When alpha is positive R and P must have the same sign, and if alpha is large, body-axes yaw rates comparable to the body-axes roll rate must be generated. Therefore, in a rapid high-alpha roll, gyroscopic (inertia coupling) effects will generate a significant body-axes pitching moment. Euler's equations of motion, Equations (1.5-6), illustrate the inertia-coupling effects when the cross-products of inertia can be neglected. Using these equations, the pitching moment, M_{IC}, due to inertia coupling is given by

$$M_{IC} = \dot{Q} J_Y = (J_Z - J_X) PR \tag{4.5-5}$$

For modern fighter aircraft with stubby wings and engine(s) on or near the longitudinal axis, the moment of inertia, J_X, is usually small compared to J_Z (while J_Z and J_Y are comparable in magnitude). Therefore, a rapid roll (right or left) about the stability x-axis, at large positive alpha, can produce a strong nose-up pitching moment. To avoid a "pitch departure" the pitch-axis control augmentation system must cause the horizontal tail to generate an opposing aerodynamic moment. At high alpha it may be difficult to obtain the neccessary aerodynamic pitching moment because of the horizontal tail stalling. Even when adequate pitching moment is available, the required yawing moment may be unachievable because the rudder is blanketed by the wings. Conventional aircraft therefore have greatly degraded roll response at high alpha, and furthermore, the control systems must often be designed to limit the commanded roll rate to avoid a pitch departure.

Figure 4.5-7 illustrates the essential features of a lateral-directional CAS for a modern fighter aircraft; compensation networks, limiters, and so on, are added as necessary. The aileron-control channel is the same as that shown in Figure 4.4-4 for the lateral-directional SAS, except that the aileron-actuator input now has a cross-connection to the rudder actuator via an alpha-dependent gain (also Mach dependent in general). This cross-connection, known as the *aileron-rudder interconnect* (ARI), may be implemented hydromechanically on some aircraft or electrically on others. Its purpose is to provide the component of yaw rate necessary to achieve a stability-axis roll.

The ARI gain must be determined, as a function of alpha and Mach number, to achieve the exact amount of yaw rate required to satisfy the constraint equation (4.5-4c). The gain is typically estimated from the known aerodynamic data, and adjusted using nonlinear simulation. We can avoid this by incorporating the constraint $R = P \tan \alpha$ in our steady-state trim program, and trimming the aircraft for a "steady-state" roll (see Section 3.6). Table 4.5-2 shows an abridged set of trim data for different roll rates and two different pitch rates; the angular units are all in degrees. The trim program has driven the lateral acceleration a_y (along the body y-axis), to essentially zero (about $10^{-6} g$'s), with a small sideslip angle.

The table shows that angle of attack is almost independent of the roll rate, but it is dependent on pitch rate. Therefore, the second half of the table is for a pitch rate

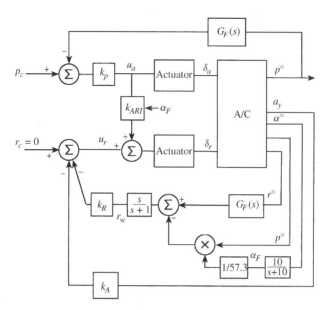

Figure 4.5-7 A lateral-directional CAS.

of 5 deg/s and serves to provide data for a higher alpha condition (6.5 degrees). Pitch rates of 10, 15, and 20 deg./s were used to provide additional data; maximum engine thrust is reached in between the last two conditions. The table indicates that for a stability-axis roll under the conditions shown (i.e., $M = 0.45$, etc.), the required ratio of rudder deflection to aileron deflection is -0.33 at $\alpha = 2.2°$ and 0.12 at $\alpha = 6.5°$.

TABLE 4.5-2: Trim Conditions for Determining ARI Gain

		$cg = 0.35\bar{c}$	$V_T = 502$	$h = 0$	$\bar{q} = 300$	$M = 0.450$	
	P	α	β	a_y	ail	rdr	rdr/ail
	10	2.12	-0.012	0	-0.813	0.269	-0.331
	20	2.12	-0.023	0	-1.63	0.537	-0.329
	45	2.15	-0.050	0	-3.66	1.20	-0.328
$Q = 0$	60	2.17	-0.065	0	-4.88	1.60	-0.328
	90	2.22	-0.091	0	-7.32	2.38	-0.325
	120	2.28	-0.112	0	-9.76	3.14	-0.322
	180	2.39	-0.139	0	-14.6	4.63	-0.317
	10	6.53	0.012	0	-0.835	-0.0948	0.114
	20	6.52	0.022	0	-1.67	-0.195	0.117
$Q = 5$	45	6.46	0.046	0	-3.76	-0.445	0.118
	60	6.41	0.058	0	-5.01	-0.595	0.119
	90	6.28	0.071	0	-7.52	-0.897	0.119
	180	5.87	0.048	0	-15.0	-1.83	0.122

Using the additional data for other angles of attack showed that the ratio of rudder to aileron deflection, k_{ARI}, was a good fit to the straight line:

$$k_{ARI} = 0.13\alpha - 0.7, \qquad (\alpha \text{ in degrees}) \tag{4.5-6}$$

In a practical design the effect of Mach number must also be determined, and a two-dimensional lookup-table might be constructed for k_{ARI}. Because of time and space limitations, (4.5-6) will be used here, and the design example will not involve large variations of Mach number.

The ARI alone would be an open-loop attempt to achieve a stability-axis roll and, to improve on this, feedback control is used to drive the lateral acceleration to zero (as in Table 4.5-2). Figure 4.5-7 shows how lateral acceleration is fed back and compared with a null reference input, and the error signal is used to drive the rudder actuator. This is also known as a *turn-coordination* scheme and can be used in autopilot systems to respond to radio-navigation steering signals or relieve the pilot of the need to coordinate turns.

Like the normal acceleration CAS, lateral acceleration feedback suffers from a wide variation of sensitivity. High values of feedback gain are needed at low speed, and this may cause problems with accelerometer noise. At low speed ($M < 0.3$) sideslip-angle feedback is normally used instead of lateral acceleration, but has the disadvantage that a beta sensor is less reliable than an accelerometer.

The inner feedback loop in the rudder channel provides dutch roll damping by feeding back an approximation to the stability-axis yaw rate (Equation (4.5-4b)) to the rudder. Thus, the filtered alpha signal, converted to radians (as necessary), is used as an approximation to $\sin\alpha$, multiplied by the roll rate and subtracted from the yaw rate. The stability-axis yaw rate is washed out so that it operates only transiently and does not contribute to a control error when a steady yaw rate is present. Note that, according to (2.5-29), the yaw-rate feedback is equivalent to a combination of beta and beta-dot feedback.

When necessary the pilot can still sideslip the airplane, because rudder inputs are applied directly to the rudder actuator. The control system will tend to reject this disturbance input, so the desirable effect of limiting the sideslipping capability will be achieved.

A practical lateral-directional CAS, based on the concept above, will be a complex system involving gain scheduling (with angle of attack and dynamic pressure or Mach), multipliers and limiters, and discrete switching (to change the control laws automatically at the alpha limits). It is a particularly good illustration of the fact that aircraft control systems incorporate many nonlinear and time-varying effects, and that the "tuning" of a design is often done by trial and error, using computer simulation as a tool, together with piloted simulation and flight tests. An example of a lateral-directional CAS design based on Figure 4.5-7 will now be given.

Example 4.5-3: A Lateral-Directional CAS Design. This design will be performed on the F-16 model, in the nominal flight condition of Table 3.6-3 (level flight at sea level, $\alpha = 2.115°$), and will follow Figure 4.5-7. The lateral accelerometer is at the aircraft cg, and the coefficient matrices found by linearizing the aircraft model lateral-directional dynamics are, in MATLAB format,

324 AIRCRAFT DYNAMICS AND CLASSICAL CONTROL DESIGN

```
ap= [-3.2201E-01  6.4040E-02  3.6382E-02  -9.9167E-01;    % x1=β
      0.0          0.0         1.0         3.6928E-02;    % x2=φ
     -3.0649E+01   0.0        -3.6784E+00   6.6461E-01;   % x3=p
      8.5395E+00   0.0        -2.5435E-02  -4.7637E-01 ]; % x4=r

bp= [ 2.9506E-04   8.0557E-04;
      0.0          0.0                                    % input-1 = δₐ
     -7.3331E-01   1.3154E-01;                            % input-2 = δᵣ
     -3.1865E-02  -6.2017E-02 ];

cp= [-5.0249E+00   0.0        -8.1179E-03   1.1932E-01;   % y1= ay
      0.0          0.0         5.7296E+01   0.0;          % y2= p
      0.0          0.0         0.0          5.7296E+01 ]; % y3= r

dp= [ 4.6043E-03   1.2571E-02;
      0.0          0.0;
      0.0          0.0 ];
```

The control-surface actuator dynamics will be the same as Example 4.4-3. The filtered alpha signal is fixed at the trim value, and ARI and roll-rate feedback equations are linearized around this value:

```
kari= .13*2.115 - 0.7;
aa= [-20.2 0; 0 -20.2];              % Two Actuators
ba= [20.2 0; 20.2*kari 20.2];  % Inp-1= Ail., Inp-2=ARI & rdr
ca= [-1 0; 0 -1]; da= [0 0; 0 0];    % SIGN CHANGE in C
actua= ss(aa,ba,ca,da);
plant= ss(ap,bp,cp,dp);              % x1=beta, x2=phi, x3=p, x4=r
sys1 = series(actua,plant);          % x5= aileron  x6= rudder
```

The washout filter has a time constant of 1 s and is included in a three-input, three-output state-space model with direct connections for the p and a_y signals. This model is cascaded at the output of the plant:

```
km= 2.115/57.3;                      % Multiply p by alpha in rads.
aw= [-1]; bw= [0 -km 1];             % Washout filter
cw= [0; 0; -1];                      % outputs ay,p,rw
dw= [1 0 0; 0 1 0; 0 -km 1];         % inputs ay,p,r
wash= ss(aw,bw,cw,dw);
sys2= series(sys1,wash);             % x1=wash x2=beta, etc
[a,b,c,d]= ssdata(sys2);             % Complete augmented system
```

The ARI affects only the B matrix, and when the poles and zeros of the principal transfer functions are checked it is found, as expected, that the effect of the ARI is to move only the zeros of transfer functions from the u_a input. The open-loop transfer function from actuator input to roll-rate, with the ARI connected, is given by:

$$\frac{p}{u_a} = \frac{913.4(s + 0.4018 \pm j2.945)(s - 0.002343)}{(s + 0.4235 \pm j3.064)(s + 3.616)(s + 0.01433)(s + 20.20)} \quad (1)$$

The roll subsidence pole is at $s = -3.615$, the spiral pole is stable at $s = -0.01433$, and the dutch roll poles are lightly damped and almost cancel out of this transfer function. Positive u_a inputs $(-\delta_a)$ will initially produce a positive roll rate, and the "slow" NMP zero indicates that this will disappear as the spiral trajectory becomes established.

A root-locus plot for the roll-rate loop showed that the dutch roll poles moved toward the canceling zeros. The spiral pole moved toward the NMP zero at $s = 0.0023$, and the roll-subsidence pole joined with the actuator pole to form a high-frequency complex pair, whose damping decreased as the feedback gain was increased. A fast roll-rate response was desired so it was decided to allow this complex pair, but keep them well damped. Roll-rate gains close to those used in Example 4.4-3 were tried and the gain $k_p = 0.2$ was chosen, which produced a damping ratio of about 0.7 for the complex pair and a stable spiral mode. The roll-rate loop was closed and the transfer function r_w/u_r was found with the following commands:

```
acl= a - b*[0 .2 0; 0 0 0]*c;      % close roll loop
[z,p,k]= ss2zp(acl,b(:,2),c(3,:),0)   % rw/ur transf. fn.
```

giving:

$$\frac{r_w}{u_r} = \frac{77.40s(s + 0.1030 \pm j0.2272)(s + 11.84 \pm j10.10)}{(s+1)(s+20.2)(s+0.0027)(s+0.4323 \pm j2.976)(s+11.90 \pm j10.70)} \tag{2}$$

A root-locus plot for this loop shows that the dutch roll poles have their highest natural frequency and good damping when $k_r = 0.8$, and this gain was used as the initial gain for investigating the lateral-acceleration feedback loop:

```
acl= a - b*[0 .2 0; 0 0 .8]*c;     % Close roll & yaw loops
[z,p,k]= ss2zp(acl,b(:,2),c(1,:),0)   % ay/uy transfer fn.
```

After removing an approximate cancellation of two complex pairs of poles and zeros, the lateral acceleration transfer function is

$$\frac{a_y}{u_r} = \frac{-0.2539(s+1)(s-4.157)(s+4.00)(s-0.0002)}{(s+16.54)(s+2.168)(s+0.0026)(s+1.701 \pm j1.486)} \tag{3}$$

The NMP zero at $s = 4.157$ is analogous to the NMP zero in the elevator-to-normal-acceleration transfer function. Ignoring for the moment the "slow" NMP zero at $s = 0.0002$, the Laplace transform final-value theorem shows that a positive-step u_r (negative rudder deflection) produces positive lateral acceleration, but the NMP zero at $s = 4.157$ indicates that this acceleration will initially be negative. The explanation is that negative rudder deflection immediately produces a negative sideforce contribution from the tail, but then, as negative sideslip builds up, the sideforce at the cg will become positive.

If the dihedral derivative is negative (positive stiffness) the aircraft will next begin to roll right, and negative sideforce will again occur as its weight starts a positive sideslipping motion. This is the cause of the "slow" NMP zero at $s = 0.0002$. The purpose of the lateral acceleration feedback is to cancel the sideslip that is causing the short-term lateral acceleration, so lateral acceleration must be fed back negatively to u_r. The following commands will give the root-locus plot for the lateral acceleration feedback, with the roll and yaw-rate loops closed:

```
k= linspace(0,100,2000);
r= rlocus(acl,b(:,2),c(1,:),0,k);
plot(r)
grid on
axis([-23.5,.5,-12,12)
```

The root-locus plot, Figure 4.5-8, shows that increasing the lateral acceleration feedback causes the dutch roll poles to circle around in the left-half s-plane, before terminating in the right-half plane on the NMP zero and at infinity. Increasing the inner-loop yaw-rate feedback causes the dutch roll poles to circle farther to the left in the s-plane and allows more lateral-acceleration feedback to be used. However, using large amounts of lateral-acceleration feedback creates a slow real pole by pulling the washout pole back to the right (it was moved left by the rate feedback) and makes the dutch roll pole positions quite sensitive to gain changes.

A conservative choice, $k_r = 0.8$, $k_a = 10$, was made for the yaw rate and acceleration feedback gains. The relevant closed-loop transfer functions are:

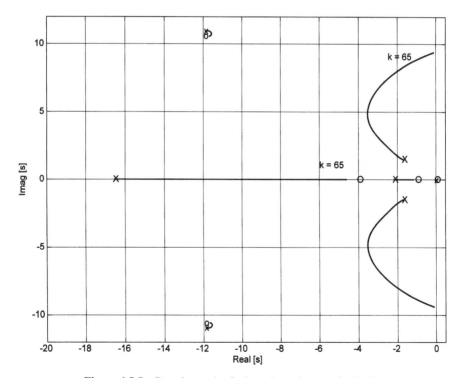

Figure 4.5-8 Root-locus plot for lateral-acceleration feedback.

$$\frac{p}{p_C} = \frac{182.7(s+13.10)(s+2.428 \pm j2.243)(s+1.538)(s-0.002347)}{(s+13.42)(s+2.386 \pm j2.231)(s+1.575)(s+0.002116)(s+11.78 \pm j10.96)} \quad (3)$$

$$\frac{a_y}{r_C} = \frac{-0.2539(s-4.157)(s+4.00)(s+11.92 \pm j10.58)(s+1.00)(s-0.0001965)}{(s+13.42)(s+2.386 \pm j2.231)(s+1.575)(s+0.002116)(s+11.78 \pm j10.96)} \quad (4)$$

A number of poles and zeros can be canceled out of the transfer functions, and there is good decoupling between the two channels. The static loop sensitivity of the first transfer function has changed because the feedback gain k_p has been moved into the forward path, as shown in Figure 4.5-6. Note that the dutch roll mode is satisfactory, and the spiral mode is stable but with an increased time constant.

4.6 AUTOPILOTS

Most of the flying-qualities specifications do not apply directly to autopilot design. In the case of pilot-relief autopilot modes, the autopilot must be designed to meet specifications on steady-state error and disturbance rejection, with less emphasis on dynamic response. In addition, special consideration must be given to the way in which the autopilot is engaged and disengaged, so that uncomfortable or dangerous transient motions are not produced. For example, the altitude-hold autopilot that we will design could not be engaged directly at a few hundred feet below the commanded altitude. Otherwise the result would be a very steep climb, possibly leading to a stall if the engine thrust was not increased.

On the other hand, navigation-coupled autopilot modes must be designed to have a dynamic response that is appropriate to their function. For example, in an automatic terrain-following mode an autopilot must track a randomly changing input of quite wide bandwidth, without significant overshoots in its response. A number of autopilot designs will now be illustrated using the transport-aircraft and F-16 dynamic models.

Pitch-Attitude Hold

This autopilot is normally used only when the aircraft is in wings-level flight. The controlled variable is θ ($\theta = \gamma + \alpha$), and the sensor is an attitude reference gyro (which provides an error signal proportional to the deviation from a preset orientation in inertial space). The controller does not hold the flight-path angle, γ, constant because the angle of attack changes with flight conditions. Thus, if thrust is increased, alpha will tend to decrease and the aircraft will climb; and as aircraft weight decreases (as fuel is burned), alpha will decrease, also causing a gradual climb. Similarly, a preset climb will gradually level out as decreasing air density causes alpha to increase. Because of these characteristics the pitch-attitude-hold autopilot is not very important in its own right. However, the same feedback configuration is used in the inner loops of other autopilots, such as altitude hold and automatic landing.

The block diagram of an attitude-hold autopilot is shown in Figure 4.6-1. Dynamic compensation, $G_c(s)$, is necessary if a small steady-state error and good transient

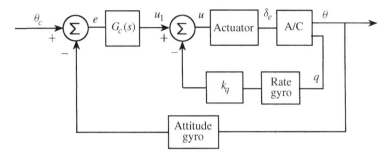

Figure 4.6-1 A pitch-attitude autopilot.

response are required. Inner-loop rate feedback is used to provide additional design freedom and to promote good short-period damping. If the principles are investigated by using only the short-period approximation for the aircraft dynamics, adding an integrator to obtain pitch from pitch rate, and a lag model for the elevator actuator, some root-locus sketches will show that the pitch attitude feedback reduces the damping of the short-period mode and eventually makes it unstable. Pitch attitude is one of the variables involved in the phugoid mode, and an analysis using the complete pitch dynamics will show that the pitch-attitude feedback increases the phugoid damping, and eventually produces two stable real poles. An accurate analysis of the effect on the phugoid mode requires that the altitude state also be included in the plant dynamics.

Two design examples will be given, with and without a dynamic compensator, and these designs will be used later as parts of more complex autopilots. The first example will be for a high-altitude cruise condition, and the second for a landing condition.

Example 4.6-1: A Simple Pitch-Attitude-Hold Autopilot. This example will demonstrate the basic characteristics with no dynamic compensation, so G_c will be simply a gain k_θ. We will also neglect the dynamics of the gyros. The dynamics of the transport-aircraft model in a level-flight cruise condition at 25,000 ft, 500 ft/s true airspeed, and $x_{cg} = 0.25\bar{c}$, are given by:

$$A = \begin{bmatrix} v_T & \alpha & \theta & q & h \\ -0.0082354 & 18.938 & -32.170 & 0.0 & 5.9022E-05 \\ -0.00025617 & -0.56761 & 0.0 & 1.0 & 2.2633E-06 \\ 0.0 & 0.0 & 0.0 & 1.0 & 0.0 \\ 1.3114E-05 & -1.4847 & 0.0 & -0.47599 & -1.4947E-07 \\ 0.0 & -500.00 & 500.00 & 0.0 & 0.0 \end{bmatrix}$$

(1)

$B^T = [0 \quad 0 \quad 0 \quad -0.019781 \quad 0]$ (single input δ_e)

$C = \begin{bmatrix} 0 & 0 & 57.296 & 0 & 0 \\ 0 & 0 & 0 & 57.296 & 0 \end{bmatrix} \begin{matrix} \theta \\ q \end{matrix} \qquad D = \begin{bmatrix} 0 \\ 0 \end{bmatrix}$

These plant matrices will be renamed ap, bp, and so on, and augmented with a simple-lag elevator-actuator model of time-constant 0.1 s. The plant sign change needed to make positive pitch-rate correspond to positive elevator deflection will be incorporated in the actuator dynamics. The design procedure will yield values for k_q and k_θ. The MATLAB commands are:

```
plant= ss(ap,bp,cp,dp);
aa= [-10]; ba= [10];            % Actuator
ca= [-1]; da= [0];              % sign change for plant
actua= ss(aa,ba,ca,da);
sys1 = series(actua,plant);
[a,b,c,d]= ssdata(sys1);
```

The transfer function from δ_e to θ is found to be

$$\frac{\theta}{\delta_e} = \frac{-1.133(s+0.5567)(s+0.01897)(s+1.666E-4)}{(s+0.5234 \pm j1.217)(s+0.002471 \pm j0.08988)(s+1.892E-4)} \quad (2)$$

All of the modes are stable, but the complex modes are quite lightly damped ($\zeta_{sp} = 0.395$, $\zeta_p = 0.027$) in this flight condition. The altitude pole is almost canceled by a zero, but omitting the altitude state will cause a noticeable error in the phugoid parameters. The effect of pitch-attitude feedback on this transfer function can be deduced from the root-locus rules. The altitude pole will move to the nearby zero, and the phugoid poles will move to the real axis and eventually terminate on the two remaining zeros. When the effect of the actuator pole is accounted for, the short-period poles must move toward the right-half plane (approaching 60 degree asymptotes). Thus, the short-period mode becomes less well damped as the phugoid damping increases.

The steady-state pitch-attitude error can be minimized by making the compensator gain as large as possible. A simple design procedure is to fix k_θ and then use a root-locus plot to adjust k_q for best short-period damping. If the damping is more than adequate then k_θ can be increased further. The MATLAB commands are:

```
acl= a - b*[kθ  0]*c;           % Choose kθ
k= linspace(0,10,1000);
r= rlocus(acl,b,c(2,:),0,k);    % Root locus for kq
plot(r)
```

Figure 4.6-2 shows the root-locus plot for k_q, when $k_\theta = 4.0$ (elevator degrees per degree of pitch). All of the poles except the short-period poles are on the real axis, and the damping of the short-period poles passes through a maximum as k_q varies. The upper branch of the loci will move upward and to the right as k_θ is increased, thus reducing the maximum damping that can be attained.

The maximum short-period damping in Figure 4.6-2 is more than adequate, and a gain ($k_q = 2.5$) corresponding to lower damping and reduced natural frequency ($\zeta = 0.64$, $\omega_n = 3.12$) was selected. The gains $k_q = 2.5$ and $k_\theta = 4.0$, and the following MATLAB commands, will give the closed-loop transfer function:

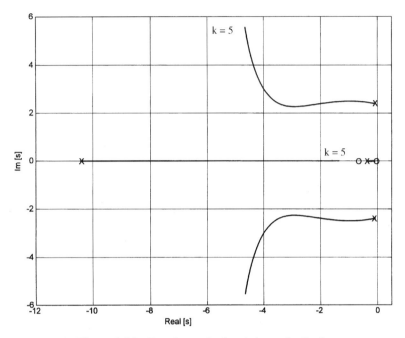

Figure 4.6-2 Root locus plot for pitch-rate feedback.

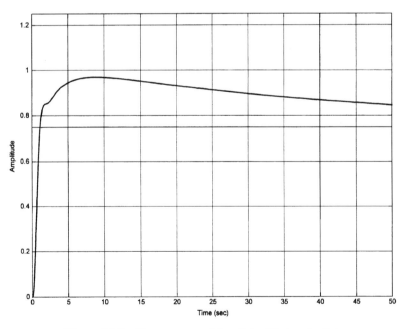

Figure 4.6-3 Step-response of the pitch-attitude autopilot.

```
acl= a - b*[4 2.5]*c;            % Close both loops
[z,p,k]= ss2zp(acl,4*b,c(1,:),0) % Closed-loop, Unity fb
```

The closed-loop transfer function is

$$\frac{\theta}{\theta_c} = \frac{45.33(s+0.5567)(s+0.01897)(s+1.666E-4)}{(s+1.999 \pm j2.389)(s+6.646)(s+0.3815)(s+0.02522)(s+1.718E-4)} \tag{3}$$

The altitude pole is almost canceled by a zero, but the cancellation of the degenerate phugoid poles is less exact and they are readily apparent in the step response, shown in Figure 4.6-3. The step response also has a large steady-state error and eventually settles at about 0.77.

This design has the disadvantage that as k_θ is increased to reduce the steady-state error, large values of k_q (i.e., $k_q > 2.5°$ of elevator per degree per second of pitch rate) must be used to obtain adequate damping of the short-period poles. This is likely to cause problems with rate-sensor noise or structural mode feedback. ∎

In the next example dynamic compensation will be used to provide fast-responding, more precise control of pitch attitude, so that the controller can be used for the flare and touchdown of an automatic landing system.

Example 4.6-2: A Pitch-Attitude Hold with Dynamic Compensation.

When the transport aircraft model is trimmed with landing gear and flaps deployed, at $V_T = 250$ ft/s, $h = 50$ ft, $\gamma = -2.5°$, and $x_{cg} = 0.25\bar{c}$, the dynamics are described by:

$$A = \begin{bmatrix} -3.8916E-02 & 1.8992E+01 & -3.2139E+01 & 0.0000E+00 \\ -1.0285E-03 & -6.4537E-01 & 5.6129E-03 & 1.0000E+00 \\ 0.0000E+00 & 0.0000E+00 & 0.0000E+00 & 1.0000E+00 \\ 8.0847E-05 & -7.7287E-01 & -8.0979E-04 & -5.2900E-01 \end{bmatrix} \begin{matrix} v_T \\ \alpha \\ \theta \\ q \end{matrix}$$

(1)

$$B^T = [0 \; 0 \; 0 \; -0.010992] \quad (\delta_e)$$

$$C = \begin{bmatrix} 0 & 0 & 57.296 & 0 \\ 0 & 0 & 0 & 57.296 \end{bmatrix} \begin{matrix} \theta \\ q \end{matrix} \quad D = \begin{bmatrix} 0 \\ 0 \end{bmatrix}$$

For simplicity the altitude state has been omitted, since its effect on the design is negligible. Once again the plant matrices will be renamed ap, bp, and so on, and the same actuator dynamics as Example 4.6-1 will be used:

```
plant= ss(ap,bp,cp,dp);
aa= [-10]; ba= [10];                    % Actuator
ca= [-1]; da= [0];                      % sign change for plant
actua= ss(aa,ba,ca,da);
sys1 = series(actua,plant);             % Actuator & Plant
[a,b,c,d]= ssdata(sys1);
acl= a - b*[0 kq]*c;                    % Close Pitch-rate fb
%[z,p,k]= ss2zp(acl,b,c(1,:),0)
qclosed= ss(acl,b,c(1,:),0);            % SISO system for theta
```

The pitch-rate feedback gain, k_q, will be limited to a smaller value than Example 4.6-1 for the reasons mentioned there. A gain $k_q = 1.0$ results in a short-period damping ratio of $\zeta_{sp} = 0.74$; the elevator-input-to-pitch-attitude transfer function is then

$$\frac{\theta}{u_1} = \frac{6.298(s + 0.6112)(s + 0.07305)}{(s + 0.9442 \pm j0.8674)(s + 0.01836 \pm j0.1328)(s + 9.288)} \quad (2)$$

and this value of k_q will be used for the rest of the design. A PI compensator will be used to remove the steady-state pitch error. The PI zero will be placed between the zeros at $s = -0.07$ and $s = -0.6$, so that the PI pole will move toward $s = -0.07$. The phugoid poles will move toward the real axis between the other two zeros and become heavily damped. The following commands can be used to add the PI compensator, determine the gain and phase margins, close the pitch attitude loop, and test the step response:

```
zero= ?                                 % Choose a PI zero position
picomp= ss([0],[zero],[1],[1]);         % PI Compensator
syspi = series(picomp,qclosed);         % PI comp & system
[a,b,c,d]= ssdata(syspi);
k1= ?                                   % Choose Proportional Gain
% margin(a,k1*b,c,0);                   % Gain & Phase Margins
acl= a - b*k1*c;                        % Close Pitch Loop
closd= ss(acl,k1*b,c,0);                % Scale b for unity feedback
step(closd,50)                          % Cl.-loop step response
```

Some trial and error with the step response led to a PI zero at $s = -0.2$ and $k_1 \approx 2$. The resulting step response still exhibits an overshoot with a small short-period oscillation superimposed on a well-damped phugoid oscillation, which takes a long time to settle. A large increase in loop gain should reduce the residues in the slow poles, but will degrade the short-period damping unless an additional compensator is used.

A root-locus sketch using transfer function (2), with the PI pole and zero added, shows that a phase-lead zero to the left of the short-period poles should pull the poles to the left and allow higher loop-gain for a given damping. The phase-lead compensator was given a pole to zero ratio of 10 (the maximum recommended), and the pole frequency was adjusted to maximize the gain and phase margins while progressively raising the loop gain. The relevant code is:

```
pole= ?                         % Choose a pole position
lead= ss(-pole,pole,-.9,1);     % Lead compensator
sysall= series(lead,syspi);     % PI + Lead + Plant
[a,b,c,d]= ssdata(sysall);
k1= ?                           % Choose Proportional Gain
margin(a,k1*b,c,0);             % Gain & Phase Margins
```

The compensator

$$G_c = 40 \, \frac{s+0.2}{s} \, \frac{s+1.4}{s+14} \qquad (3)$$

gives a phase margin of 66.8 degrees at 0.33 Hz and a gain margin of 21 dB at 1.75 Hz. The phase and gain margin plots with this compensator are shown in Figure 4.6-4. The closed-loop pitch attitude transfer function, with unity feedback, is given by

$$\frac{\theta}{\theta_c} = \frac{251.9(s+0.6112)(s+0.07305)(s+1.40)(s+0.20)}{(s+2.121 \pm j1.762)(s+0.2717 \pm j0.1516)(s+0.06335)(s+4.170)(s+16.19)} \qquad (4)$$

In the transfer function (4) the short-period mode has increased in frequency [compared to (2)], and the phugoid mode has increased in frequency and become more

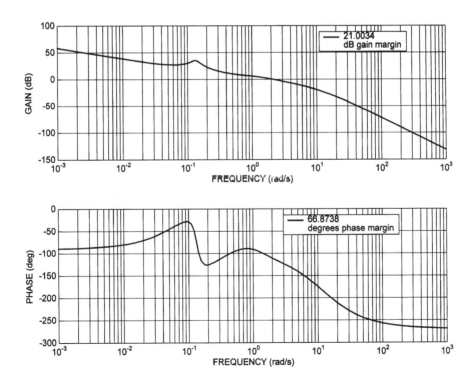

Figure 4.6-4 Bode plots for the pitch-attitude autopilot.

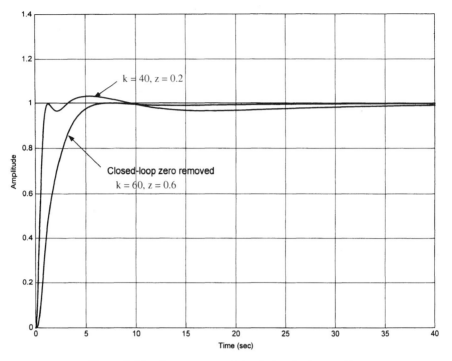

Figure 4.6-5 Step-response of the pitch-attitude autopilot.

damped. The step response, shown in Figure 4.6-5, has a fast rise-time but is slow to settle and contains an undesirable undershoot at about 2 s. If the PI compensator closed-loop zero is removed, by the modification shown in Figure 3.9-9, the rise-time will be slower but a response that resembles that of a dominant complex pair can be obtained. This is also shown in Figure 4.6-5. ∎

Altitude Hold/Mach Hold

Altitude hold is an important pilot-relief mode; it allows an aircraft to be held at a fixed altitude in an air-route corridor, to meet air-traffic control requirements. The sensed altitude is normally the *pressure altitude*, that is, altitude computed in the air-data computer from external pressure measurements. In a modern passenger aircraft the altitude hold will typically hold the aircraft well within ± 200 ft and provide a warning signal if the deviation exceeds ± 100 ft. The system will have *limited authority* over the horizontal control surfaces and will again warn the pilot if the control limits have been reached. These situations will often occur, for example, in rapidly rising air currents deflected upward by mountain ranges ("mountain waves"). A modern system may also have an "easy-on" or "fly-up, fly-down" feature that allows the autopilot to take the aircraft to an assigned altitude without exceeding certain rate-of-climb and pitch-attitude limits (e.g., 2000 to 3000 ft/min, 20 degrees pitch attitude).

The Mach-hold autopilot is chiefly used on commercial passenger jets during climb and descent. During a climb the throttles may be set at a fairly high power level, and feedback of Mach number to the elevator will be used to achieve a constant-Mach climb. The speed will vary over the range of altitude, but the constant Mach number will provide the best fuel efficiency. Similarly, a descent will be flown at constant Mach with the throttles near idle. At the cruising altitude, control of both the throttle and elevator will be used to provide altitude-hold and speed-hold, for pilot relief and efficient cruising. In the following example an altitude-hold design will be illustrated.

Example 4.6-3: An Altitude-Hold Autopilot Design. The altitude-hold configuration is shown in Figure 4.6-6, where G_c is a compensator and G_F is the effective lag of the pressure-altitude measurement. In the interest of simplicity the altitude-sensor lag will be omitted from this example. Again for simplicity, the basic pitch-attitude autopilot from Example 4.6-1 will be used to provide the inner loops of the design, and the compensator G_c will still allow good altitude control to be achieved. The first design goal will be to achieve a high loop gain, for good rejection of low-frequency (lf) altitude disturbances and small altitude error. Second, an altitude response that is deadbeat and relatively slow will be required for energy efficiency and passenger comfort.

Altitude is one of the state variables, and by adding an appropriate row to the C matrix in Example 4.6-1, the transfer function from the pitch-attitude command to altitude can be determined. The altitude feedback has a strong effect on the phugoid poles and a relatively weak effect on the short-period poles. Therefore, the damping of the short-period mode will initially be set close to the desired final value. Thus, based on the experience of Example 4.6-1, the pitch-rate and pitch-attitude feedback loops will be closed by gains $k_q = 2.5$ and $k_p = 3.0$. Starting from "sys1" in Example 4.6-1, we have,

```
[a b c d]= ssdata(sys1)        % Actuators & Plant
acl= a - b*[3 2.5]*c,          % Close k_p and k_q loops
ch= [0 0 0 0 1 0]              % C matrix for Altitude
[z,p,k]= ss2zp(acl,3*b,ch,0)   % Transfer fn. h/θ_c
```

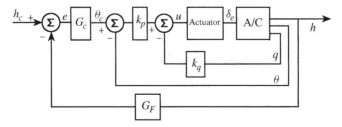

Figure 4.6-6 An altitude-hold autopilot.

The transfer function from θ_c to h, with unity feedback of θ, is:

$$\frac{h}{\theta_c} = \frac{168.4(s + 0.002264)}{(s + 2.261 \pm j1.936)(s + 6.170)(s + 0.3333)(s + 0.02750)(s + 1.731E - 4)} \quad (1)$$

with a short-period damping ratio of 0.76.

A root-locus sketch shows that the poles from $s = -0.028$ and $s = -0.333$ will break away from the real axis to form phugoid poles. The phugoid poles will move toward the right-half plane, while the short-period poles and the pole from $s = -6.17$ move left. A phase-lead compensator with its zero close to $s = -0.333$ will improve the gain and phase margins. A compensator pole-to-zero ratio of 8 was chosen (a compromise between noise accentuation and too little phase lead), and the pole frequency was adjusted for the best margins each time the gain was raised. A deadbeat step response was found to require a large phase margin ($\approx$ 70 degrees). The commands were:

```
sys2= ss(acl,3*b,ch,0);
pole= ?                           % Choose Lead-Comp. Pole
lead= ss(-pole,pole,-.875,1);     % Lead-compensator
sys3= series(lead,sys2);          % Cascade with c.l. system
[a,b,c,d]= ssdata(sys3);
k= ?                              % Choose loop gain
margin(a,k*b,c,0)                 % Check Margins
```

and the lead compensator

$$G_c = \frac{s + 0.3}{3(s + 2.4)} \quad (2)$$

gives gain and phase margins of 13.3 dB and 71.2 degrees, respectively. Unfortunately, the lead compensator reduces the lf loop gain. The transfer function (1) has an lf gain of about 72.8 dB (or 4380) and the compensator (2) reduces this by 27.6 dB (i.e., 1/24). The final loop gain of 45.2 dB (or 182) would allow a steady-state altitude error of 1 ft per 183 ft, a rather poor performance. The performance can be improved by adding a lag compensator that boosts the lf gain, while adding negligible phase lag in the frequency range of the lead compensator. The same effect can be achieved by using a PI compensator to make the altitude control loop type-1, and placing the PI zero close to the origin. A simple lag compensator has the advantage that it can be implemented with passive components (see Table 3.3-1), provided that the time constant is not too large. Modern electronics has diminished this advantage but, for an analog design, a lag compensator is still simpler and more reliable than a PI compensator, and its use will be illustrated here.

Practical considerations limit the maximum time constant of an analog lag compensator to about 100 s (pole at $s = -0.01$). If the compensator zero is chosen to give a large lf gain increase, then it will be found that in the closed-loop transfer function, the slow poles (from $s = -0.0275$ and $s = -0.01$) have relatively large residues (i.e., do not cancel with zeros). If the lag compensator zero is placed near $s = -0.05$

(i.e., an lf gain increase of 5) these slow poles will have a relatively small effect on the closed-loop time response. There will also be less phase lag in the frequency range where the lead compensator is to be added. Therefore, the lag-compensator zero was placed at $s = -0.05$. The commands were:

```
lag= ss(-.01,.01,4,1);
sys4= series(lag,sys3);
[a,b,c,d]= ssdata(sys4);
margin(a,.3333*b,c,0);
acl= a - .3333*b*c;
closed= ss(acl,.3333*b,c,0);
step(closed,30)
```

The gain and phase margins with both compensators are 13.1 dB and 65.7 degrees, respectively. The lf loop gain is 913, which is adequate. Figure 4.6-7 shows the phase and gain margin Bode plots.

The closed-loop altitude transfer function is

$$\frac{h}{h_c} = \frac{56.14(s + 0.30)(s + 0.050)(s + 0.002264)}{(s + 6.29)(s + 2.75 \pm j2.03)(s + 0.673 \pm j0.604)(s + 0.267)(s + 0.053)(s + 0.00224)} \tag{3}$$

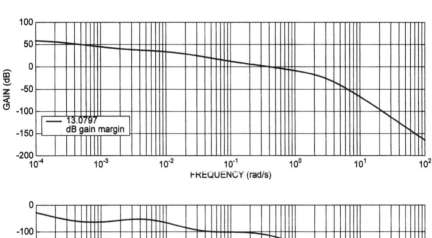

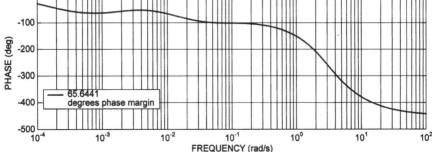

Figure 4.6-7 Bode plots for the altitude-hold autopilot.

AIRCRAFT DYNAMICS AND CLASSICAL CONTROL DESIGN

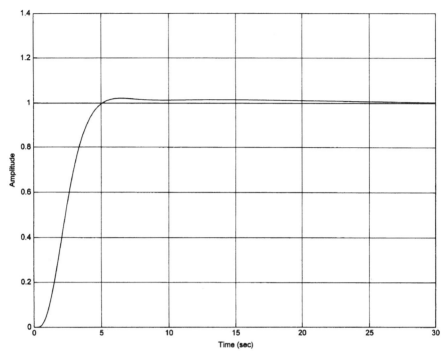

Figure 4.6-8   Step response of the altitude-hold autopilot.

A comparison with transfer function (1) shows that the fast poles have not moved significantly, the three slowest poles essentially cancel out of the transfer function, and a new complex pair has been created.

The step response is shown in Figure 4.6-8. The effect of the slow poles is visible as a small, slowly decaying, displacement from the final value. The steady-state error will be negligible because of the high value of the lf loop gain. The response is essentially deadbeat and is considerably slower than a pitch-axis response. It could be slowed down further by reduction of the loop gain or by using additional lag compensation, but is considered to be satisfactory. As pointed out earlier, it is obvious that this autopilot would not normally be directly engaged with a large altitude error.

The reader may wish to consider repeating this design for cruising conditions at, say, 35,000 ft, to determine the need for scheduling of the controller gains. ∎

Automatic Landing Systems

In Section 4.1 we referred to the need for automatic control in situations where controlling the trajectory of an air vehicle was too difficult a task for a human pilot. A particular case of this is the landing phase in conditions such as bad weather or limited visibility. Landing in limited visibility may be achieved by providing the pilot with instruments to determine the aircraft's position relative to a reference trajectory, but a landing in more difficult conditions requires full automatic control with the pilot playing only a supervisory role.

Automatic control of the longitudinal trajectory requires simultaneous control of engine thrust and pitch attitude because, for example, using only the elevator to attempt to gain altitude may result in a loss of speed and an eventual stall. If the landing speed is such that the aircraft is on the "back side" of the power curve (see Section 3.6), the throttle controls altitude and the elevator controls airspeed (increased power causes a gain in altitude, down-elevator causes a gain in speed).

An aircraft is normally *reconfigured* for landing and takeoff by deploying wing leading and trailing-edge devices (slats, flaps) so that the wing effectively has more camber and area. This provides more lift at low speed, and increased drag; the wing is thereby optimized for a low-speed landing. The reconfiguration has the effect of moving the minimum of the power curve to lower speed. Thus, most aircraft do not operate on the back side of the power curve, although naval aviators are routinely taught to fly in this regime for aircraft-carrier operations. The reconfigured wing and extended landing gear produce a strong nose-down moment, which in turn leads to a trim with a large amount of "up" elevator. We will see this effect in our transport-aircraft model in the following example, and these conditions play a role in determining the elevator size and deflection limit during the aircraft design.

A typical automatic landing system uses a radio beam directed upward from the ground at 3 degrees, with equipment onboard the aircraft to measure the angular deviation from the beam and compute the perpendicular displacement of the aircraft from the *glide path*. Additional equipment is used to provide azimuth information, so that the aircraft can be lined up with the runway. The glide path must usually be intercepted at, at least, 3000 ft altitude (over the outer marker), and the aircraft will descend with an airspeed of 130 to 150 knots (220 to 253 ft/s) under automatic control.

Figure 4.6-9 shows an elevation view of a descending trajectory with velocity V_A (in the tangent frame) and flight-path angle γ. The reference trajectory has an angle γ_R, and the radio beam equipment is at the position Q. Assume that the aircraft passes through the radio beam at point P and time t_1, and that the descent is too gradual (as

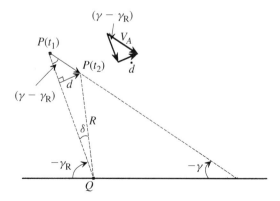

Figure 4.6-9 Glide-slope geometry for autoland.

shown). The resulting positive glide-path deviation that builds up is denoted by d. The automatic landing equipment measures the angular deviation δ and the range R and calculates d from

$$d = R \sin \delta \qquad (4.6\text{-}1)$$

An onboard automatic control system is used to maneuver the aircraft so that d is driven back to zero.

To design a control system we must relate d to the aircraft trajectory. The geometry of the figure shows that the derivative of d is given by

$$\dot{d} = V_A \sin(\gamma - \gamma_R) \qquad (4.6\text{-}2)$$

Therefore, $d(t)$ can be derived by integrating this equation with the aircraft state equations, with the initial condition $d(t_1) = 0$ applied at the time t_1 at which the aircraft intersects the glide path. Note that when d is computed from (4.6-1), the sensitivity of d to flight-path changes will depend on the range R. This effect will be assumed to be compensated for in the onboard computer, so that an automatic control system can be designed for some nominal value of the range. The design of the longitudinal control system for automatic landing will now be presented as an example.

Example 4.6-4: Longitudinal Control for Automatic Landing. Figure 4.6-10 is a block diagram of the auto-land control system. The transport aircraft model in the landing configuration will be used. The throttle servo and engine response will be modeled by a single 5 s lag, and the elevator servo by a 0.1 s lag, as shown; sensor lags have been neglected. The compensators that must be designed are G_1 and G_2, and the pitch-attitude controller will be taken from Example 4.6-2.

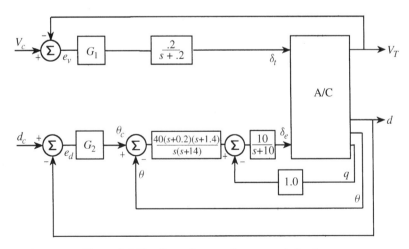

Figure 4.6-10 Control system for automatic landing.

Equation (4.6-2) was added to the transport aircraft model, with γ_R as a model input and d as an additional state. The model was trimmed with gear down and landing flap settings for the conditions $V_T = 250$ ft/s, $\gamma = -2.5°$, $x_{cg} = 0.25\ \bar{c}$, and $h = 750$ ft. The A and B Jacobian matrices for this flight condition are

$$ap = \begin{array}{c} \\ \\ \\ \\ \\ \\ \end{array} \begin{array}{cccccc} v_T & \alpha & \theta & q & h & d \\ \left[\begin{array}{cccccc} -0.038580 & 18.984 & -32.139 & 0 & 1.3233E-4 & 0 \\ -0.0010280 & -0.63253 & 0.0056129 & 1.0 & 3.7553E-6 & 0 \\ 0 & 0 & 0 & 1.0 & 0 & 0 \\ 7.8601E-5 & -0.75905 & -0.00079341 & -0.51830 & -3.0808E-7 & 0 \\ -0.043620 & -249.76 & 249.76 & 0 & 0 & 0 \\ 0 & -250.00 & 250.00 & 0 & 0 & 0 \end{array}\right] \end{array};$$

$$bp = \begin{array}{c} \\ \\ \\ \\ \end{array} \begin{array}{cc} \delta_t & \delta_e \\ \left[\begin{array}{cc} 10.100 & 0 \\ -1.5446E-4 & 0 \\ 0 & 0 \\ 0.024656 & -0.010770 \\ 0 & 0 \\ 0 & 0 \end{array}\right] \end{array};$$

The classical design procedure will be, as usual, to close one loop at a time. In this example there are four loops to be closed. Each loop closure changes both the poles and zeros for the other loop transfer functions. Preferably, the loops should be closed in a sequence that minimizes the number of design iterations. The pitch-attitude control loops are inner loops that we might logically expect to close first. If the effects of thrust and speed on pitching moment are not strong, then the pitch-attitude loops will be affected only by the change in angle of attack with speed. When the pitch attitude loop is closed, changes in speed will cause changes in the angle of attack, and therefore in the flight-path angle and in $\dot{d}$. The pitch attitude control thus determines the interaction of the speed loop on the d loop. We will close the pitch attitude loops first, using the controller from Example 4.6-2. The d control loop cannot hold the required trajectory without closing the speed (auto-throttle) loop, but the speed loop can function independently of the d loop. Therefore, the speed loop should logically be closed next.

Following Example 4.6-2, the pitch-attitude controller can be applied with the following commands:

```
cp= [ 0 0 57.29578 0 0 0; 0 0 0 57.29578 0 0];    % theta & q
dp= [0, 0];
plant= ss(ap,bp(:,2),cp,dp);         % Elev. input
actua= ss(-10,10,-1,0)               % Change sign at output
sys1 = series(actua,plant);          % 1 i/p, 2 o/p
```

```
[a,b,c,d]= ssdata(sys1);
acl= a - b*[0 1]*c;                 % Close q-loop, k_q=1
qclosed= ss(acl,b,c(1,:),0);        % SISO
lead= ss(-14,14,-.9,1);             % Lead compensator
sys2= series(lead,qclosed);
picomp = ss(0,0.2,1,1);             % PI compensator
sys3= series(picomp,sys2);
[a,bt,c,d]= ssdata(sys3);
acl= a - bt*40*c(1,:);              % Close theta loop
```

The closed-loop zero of the PI compensator has been retained because additional zeros provide more phase lead when compensating the outer loops. The transfer function from throttle to speed can now be found:

```
cvt= [1 0 0 0 0 0 0 0];             % C-matrix for VT
bth= [bp(:,1);0;0;0];               % B-matrix for Throttle i/p
[z,p,k]= ss2zp(acl,bth,cvt,0);      % Get poles and zeros
```

After canceling some very close pole-zero pairs, the throttle-to-speed transfer function is

$$\frac{v_t}{\delta_t} \approx \frac{10.10(s + 0.2736 \pm j0.1116)(s + 0.001484)}{(s + 0.2674 \pm j0.1552)(s + 0.0002005)(s + 0.06449)} \tag{1}$$

A root-locus sketch of transfer function (1) (not shown) shows that the pole from $s = -0.06$ will move left to meeet the pole from the throttle servo (when added), and they will break away from the real axis to approach 90 degree asymptotes. Given the slow response of the throttle servo and engine lag we may try to speed up the auto-throttle loop with a phase-lead compensator, although this may cause the throttle servo to saturate frequently. Adding the throttle servo, and closing the loop with no additional gain, shows an infinite gain margin, and a phase margin of about 10 degrees. The low-frequency loop-gain is about 60 dB and the resulting small steady-state error will be acceptable. Therefore, a lead compensator was chosen to improve the phase margin of this loop.

The compensator

$$G_1(s) = \frac{10(s+1)}{s+10} \tag{2}$$

gives a phase margin of about 60 degrees and retains the same lf loop gain. The speed control loop can now be closed:

```
th2vt = ss(acl,bth,cvt,0);          % SISO, throttle to speed
servo = ss(-.2,.2,1,0);             % Throttle servo & Eng. lag
ut2vt = series(servo,th2vt);
splead= ss(-10,10,-.9,1);           % Phase lead
compsp= series(splead,ut2vt);       % Compensator & plant
[a,b,c,d]= ssdata(compsp);
acl= a - b*10*c;                    % Close auto-throttle loop
```

With the speed loop closed, the d/θ_c transfer function for the final loop closure is found from:

```
btheta= [40*bt; 0; 0];      % B-matrix for unity f.b. theta-loop
cd    = [0 0 0 0 0 1 0 0 0 0];  % C-matrix for d output
[z,p,k]= ss2zp(acl,btheta,cd,0)  % d/θc transfer fn.
```

After removing some canceling pole-zero pairs, the transfer function reduces to

$$\frac{d}{\theta_c} = \frac{675.2(s+1.40)(s+0.20)}{(s+2.021 \pm j1.770)(s+0.2725 \pm j0.1114)(s+4.409)(s+16.16)(s+0.001475)} \quad (3)$$

Notice that the zeros were created by the compensators in the pitch attitude controller. A sketch of the root-locus plot will show that the poles at $s \approx (-0.27 \pm j0.11)$ move into the right-half plane as the loop gain is increased, and the margin command shows that with unity feedback the gain and phase margins would be negative. In addition, a Bode plot shows that the lf gain levels out at 69 dB (2818). Using the margin command to optimize a phase-lead compensator, we find that $2(s+0.6)/(s+6)$ will provide a phase margin of 58.4 degrees and a gain margin of 11.1 dB, but the lf gain is then reduced to 2818/5, or 563. It is desirable to follow the glide slope closely when near the ground; therefore, we will make the controller type-1 by adding a PI compensator, as well as the phase lead.

When a PI controller and lead compensator are cascaded with transfer function (3), a root-locus sketch shows that the integrator pole and the pole from $s = -0.0015$ will circle to the left in the s-plane, to terminate eventually on the PI zero and the zero at $s = -0.20$. By placing the PI zero near to the zero at $s = -0.2$ and, using high gain, we will hope to get small residues in these slow poles and avoid a very sluggish closed-loop response. Note that the zero at $s = -0.20$ from the pitch attitude controller is now partly responsible for determining the speed of response of the d loop.

The compensator,

$$G_2(s) = 1.0 \times \frac{(s+0.18)(s+0.5)}{s(s+5)} \quad (4)$$

was derived by examining the effect of the PI zero on the closed-loop poles and zeros and the step response. The lead compensator was adjusted to obtain a compromise between the gain and phase margins, and these were, respectively, 14.7 dB and 51.6 degrees. Figures 4.6-11a and b show the Bode plots; note that the pole at $s = -0.0015$ causes the low-frequency phase lag to approach -180 degrees (like a type-2 system) before the lead compensation begins to take effect.

When some close poles and zeros are canceled, the principal closed-loop transfer functions are

$$\frac{v_T}{v_c} \approx \frac{20.20(s+1)}{(s+7.627)(s+1.280 \pm j0.9480)} \quad (5)$$

$$\frac{d}{d_c} \approx \frac{677.0(s+1.40)(s+0.50)(s+0.20)(s+0.180)}{(s+16.2)(s+5.16 \pm j1.65)(s+1.38 \pm j1.69)(s+0.292 \pm j0.348)(s+0.179 \pm j0.0764)} \quad (6)$$

344 AIRCRAFT DYNAMICS AND CLASSICAL CONTROL DESIGN

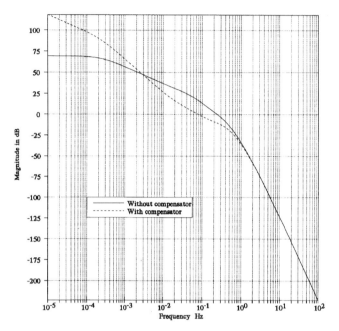

Figure 4.6-11a Bode magnitude plot for the automatic-landing controller.

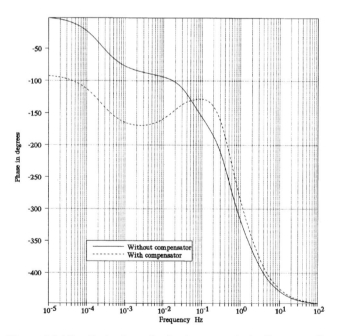

Figure 4.6-11b Bode phase plot for the automatic-landing controller.

Note that in the d transfer function, the slowest pair of complex poles is close to terminating on the zeros at $s = -0.18$ and $s = -0.20$. The step responses could be evaluated by a linear simulation using the closed-loop state equations. Instead a nonlinear simulation of the glide-path descent will be illustrated in Section 4.7. ∎

Roll-Angle Hold Autopilots

In its simplest form, as a wing leveler, the roll-angle autopilot has a history going back to the experiments of Elmer Sperry (see Section 4.1). A sensor incorporating an attitude reference, such as a gyroscope, is used to sense deviations from a reference attitude in the aircraft $y-z$ plane. Feedback of the deviation signal to the ailerons can then be used to control the roll angle of the aircraft. The autopilot will hold the wings level and thus provide a pilot-relief function for long flights and eliminate the danger of the pilot being caught unaware in a coordinated spiral motion toward the ground.

If the aircraft is held at some attitude other than wings-level, additional control systems must be used to control sideslip and pitch rate, so that a coordinated turning motion is produced. Depending on the commanded pitch rate, the aircraft may gain or lose altitude in a turn. If a means of varying the roll reference is provided, the aircraft can be steered in any direction by a single control. These control systems can provide the inner loops for other autopilots that allow an aircraft to fly on a fixed compass heading, or follow a radio-navigational beam in the presence of crosswinds. Such systems will be described later.

Figure 4.6-12 shows a block diagram of a roll-angle hold autopilot. High-performance aircraft virtually always have available a roll-rate gyro for use by a SAS or CAS, and this can be used to provide inner-loop rate damping for the autopilot. If the roll-rate gyro is not available, then for good performance, a compensator is needed in the roll-angle error path. There is usually no requirement for precise tracking of roll-angle commands, so type-0 roll angle control can be used. By the same token, the *velocity error* due to straight roll-rate feedback (i.e., no washout) is not important, particularly since the roll rate is not usually sustained for very long.

If the aircraft has strong roll-yaw coupling, the roll-angle-to-ailerons feedback must be considered as part of a multivariable design, as in Sections 4.4 and 4.5. This is often not the case, and in the lateral transfer function, the poles associated with the directional controls are approximately canceled by zeros. The transfer function for the

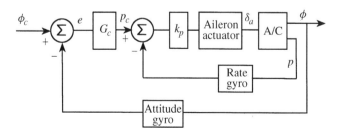

Figure 4.6-12 A roll-angle control system.

346 AIRCRAFT DYNAMICS AND CLASSICAL CONTROL DESIGN

roll-angle loop is then determined by the roll-subsidence pole, the spiral pole, and the actuator and compensator (if any) poles. If roll-rate feedback is used, in conjunction with the roll-angle feedback, there is good control over the position of the closed-loop poles and quite large amounts of feedback can be used. A roll-angle autopilot design will now be illustrated.

Example 4.6-5: A Roll-Angle-Hold Autopilot. This example will use the controller subroutine from the lateral-directional CAS in Example 4.5-3, and with the same flight conditions. In Figure 4.6-12 the dynamics of the gyros will be neglected. With $k_p = 0.2$, the closed-loop transfer function from the roll-rate command, p_c, to the roll angle in Figure 4.6-12 is found to be

$$\frac{\phi}{p_C} = \frac{182.7(s + 13.09)(s + 2.429 \pm j2.241)(s + 1.540)}{(s + 13.42)(s + 2.386 \pm j2.231)(s + 1.575)(s + 0.002116)(s + 11.78 \pm j10.96)} \quad (1)$$

or, approximately,

$$\frac{\phi}{P_c} = \frac{182.7}{(s + 11.78 \pm j10.96)(s + 0.002116)} \quad (2)$$

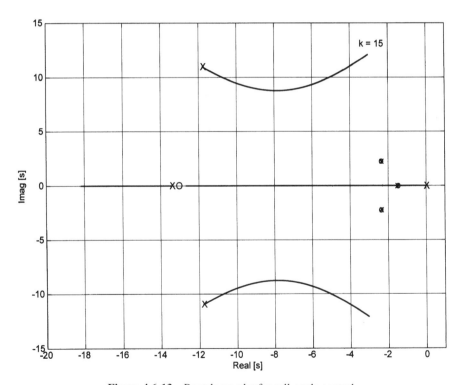

Figure 4.6-13 Root-locus plot for roll-angle control.

In this transfer function the complex pole pair arose from the actuator pole and the roll-subsidence pole, and the real pole is the spiral pole. The spiral pole is close to the origin, and approximates an integration between the roll-rate and the roll angle. When the roll-angle feedback loop is closed, the spiral pole moves to the left and the complex poles move to the right. The root-locus plot is shown in Figure 4.6-13.

A proportional gain (for G_c) of $k_\phi = 5.0$ gave the complex poles a damping ratio of $\zeta = 0.71$ (at $s = -8.88 \pm j8.93$), and the real pole was at $s = -5.4$. The roll-angle control loop is well damped but unrealistically fast. The commanded attitude will be more tightly controlled in the steady state, but the aileron actuators may be driven into rate limiting if abrupt roll-angle commands are applied. This control system will be used in the next subsection in a nonlinear simulation. ∎

Turn Coordination and Turn Compensation

A coordinated turn is defined as zero lateral acceleration of the aircraft cg (i.e., zero component of inertial acceleration on the body y-axis). In a symmetrical aircraft the components of acceleration in the plane of symmetry need not be zero, and so the coordinated turn need not be a steady-state condition. In an asymmetrical aircraft the sideslip angle may not be exactly zero in a coordinated turn because of, for example, asymmetric thrust or the effects of the angular momentum of spinning rotors. Turn coordination is desirable for passenger comfort and, in a fighter aircraft, it allows the pilot to function more effectively. In addition, by minimizing sideslip, it maintains maximum aerodynamic efficiency and also minimizes undesirable aerodynamic loading of the structure. Automatic turn coordination is also useful for a remotely piloted vehicle performing video surveillance or targeting.

In a coordinated turn, level or otherwise, the aircraft maintains the same pitch and roll attitude with respect to the reference coordinate system, but its heading changes continuously at a constant rate. Therefore, the Euler-angle rates $\dot\phi$ and $\dot\theta$ are identically zero, and the rate $\dot\psi$ is the turn rate. The Euler kinematical equations (1.3-21) show that, under these conditions, the body-axes components of the angular velocity are

$$P = -\dot\psi \sin\theta$$
$$Q = \dot\psi \sin\phi \cos\theta \quad (4.6\text{-}3)$$
$$R = \dot\psi \cos\phi \cos\theta$$

If the aircraft is equipped with angular-rate control systems on each axis these rates can be computed, and then they can be used as the controller commands to produce a coordinated turn. In level flight, with small sideslip, the turn coordination constraint is given by Equation (3.6-7):

$$\tan\phi = \frac{\dot\psi}{g_D} \frac{V_T}{\cos\theta} \quad (4.6\text{-}4)$$

348 AIRCRAFT DYNAMICS AND CLASSICAL CONTROL DESIGN

If $\cos\theta \approx 1.0$ then, for a specified turn rate $\dot\psi$, the required pitch and yaw rates can be calculated and the roll rate can be neglected. This produces a quite satisfactory level turn.

Alternative coordination schemes include feedback of sideslip or lateral acceleration to the rudder, or computing just a yaw-rate command as a function of measured roll-angle (see Blakelock, 1965 for details). If, in addition, a pitch-rate command is calculated from the above equations as a function of roll angle, the turn can be held level. This is referred to as "turn compensation" (Blakelock, 1965); it can also be achieved by using altitude feedback to the elevator. An example of turn coordination is given in Example 4.7-5.

Autopilot Navigational Modes

Automatic navigation is an important autopilot function for both military and civil aircraft, and the most important systems will be briefly summarized. A *heading-hold* autopilot is designed to hold the aircraft on a given compass heading. The conventional method of implementing this autopilot is to close an additional yaw-angle feedback loop around the roll-angle control system (including turn compensation) that was illustrated above. Figure 4.6-14 shows the arrangement. The transfer function relating heading angle to roll angle uses the linearized equation obtained from (4.6-4), when ϕ is small and $\cos\theta \approx 1$. Note that the transfer function gain is inversely proportional to speed. An investigation of the root locus for the heading angle loop, and the effects of flight conditions, will be left to the reader (see also Blakelock, 1965 and Roskam, 1979).

A *VOR-hold* (VHF Omni Range) autopilot is an autopilot designed to home on an omnidirectional radio beacon. The heading-angle hold system (including turn compensation, etc.) is used to implement this autopilot, and Figure 4.6-15 shows how this is done. The transfer function derived from the geometry of the beam-following is similar to that derived for the automatic-landing longitudinal control system. The system normally requires proportional plus integral compensation, and possibly lead compensation also. Again, it is left to the reader to investigate further (Blakelock, 1965; Roskam, 1979).

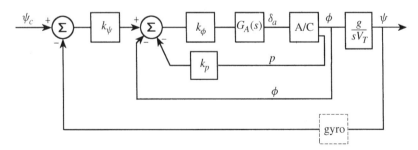

Figure 4.6-14 A heading-hold control system.

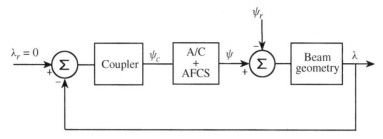

Figure 4.6-15 A VOR-hold autopilot.

A specialized military autopilot that is particularly interesting is a *terrain-following terrain-avoidance* (TFTA) autopilot. This system uses the aircraft's radar or a separate radar carried underneath the aircraft (as in the LANTIRN system). The radar provides guidance commands to fly at constant height (e.g., 100 to 400 ft) above the Earth's surface at high speed. The fly-up, fly-down commands are usually applied to a g-command control system as described in Section 4.5, and the lateral-directional guidance commands are applied to a roll-angle steering control system as in Example 4.7-5.

4.7 NONLINEAR SIMULATION

The linear designs illustrated in previous sections are only the first stage in the design of complete aircraft control systems. At the second stage the control systems must be evaluated on a nonlinear model of the aircraft, with larger amplitude maneuvers and over a larger portion of the envelope. To perform this evaluation nonlinear control-system elements must be modeled (e.g., any multipliers or nonlinear calculations in the control system equations, and rate-limiting and deflection limiting in the control surface actuators). Actuator performance will be strongly affected by the aerodynamic loads on the control surface, further complicating the nonlinear behavior. During this second stage the nonlinear simulation can be done with preprogrammed commands, or with a desktop flight simulator with no cockpit, rudimentary controls, and limited video display capabilities. At a later stage the nonlinear simulation will be performed with a cockpit mock-up and out-of-the-window video displays, and used for piloted evaluation of the aircraft. This will eventually be followed by pilot training in the simulator.

In nonlinear simulation it is highly desirable to separate the control systems equations from the equations of motion and from the aerodynamic database. If this is done errors are easier to find, different controller designs can be substituted easily, there is less chance of corrupting unrelated computer code, and not all of the code has to be recompiled when changes are made. In a big organization different groups of people are responsible for the aerodynamic database and the control systems, and this partitioning of the computer software is very appropriate. The state-space formulation greatly facilitates the achievement of this objective. State variables that are needed

350 AIRCRAFT DYNAMICS AND CLASSICAL CONTROL DESIGN

for the controllers can be numbered independently of state variables needed for the equations of motion, or any other equation set, and at compilation time or during code interpretation, all of the state variables can be placed in one large array for numerical solution purposes. This is like a parallel operation on all of the state variables simultaneously, and there is no question of different variables being of different age. In this section examples of nonlinear simulation will be provided using the transport aircraft in MATLAB code and the F-16 model in Fortran.

Example 4.7-1: Pitch-Rate CAS Nonlinear Simulation. In this example the pitch-rate controller designed in Example 4.5-1 will be converted to a subprogram (i.e., a function subprogram or a subroutine subprogram) that can be linked with a nonlinear aircraft model, an integration routine, and a driver program, to perform flight simulation. This subprogram must have access to the output variables of the aircraft model and the controller command inputs (e.g., as formal parameters, as "global" variables, or through a "common" allocation of memory). A set of controller state variables that can be appended to the array of aircraft state variables must be used. The output of this subprogram is an array of derivatives, and they must be appended to the array of derivatives from the aircraft model.

The first statements in the controller subprogram are specific to the programming language, and must define the variable types and how they are to be passed to and from the subprogram. The rest of the subprogram is essentially independent of the programming language, and consists of the matrix state equations of Example 4.5-1 translated into individual state equations, as in Table 3.3-1. The aircraft model has thirteen state variables so here the controller states will be numbered from fourteen on and appended to the aircraft state vector. In Fortran, the code is:

```
subroutine FC(time,x,xdot)              ! x in, x-dot out
dimension x(*), xdot(*)                 ! assumed-size arrays
common/controls/thtl,el,ail,rdr,qcom    ! controls & commands
common/output/an,alat,ax,qbar,amach,vt,alpha,theta,q  ! from a/c

el= -x(14)                              ! actuator state-->elevator
call f16(time,x,xdot)                   ! Aircraft model
xdot(15)= 10.0*(alpha-x(15) )           ! Alpha filter
xdot(16)= qcom - q                      ! PI integrator input
u= 1.5*x(16) -.5*q - .08*x(15)          ! Control law
xdot(14)= 20.2*(u- x(14) )              ! Elevator actuator
return
end
```

Values have already been assigned to the state variables when this subroutine is called, but the longitudinal control inputs must be assigned before the aircraft model equations can be executed. Therefore, the elevator control is assigned to the actuator state before calling the aircraft model. The lateral-directional controls and states will not be changed from their trim values. Throttle commands and the pitch-rate command "qcom" will be assigned in a separate subprogram. Control surface rate and deflection limits are not modeled; this will be done in later examples. Note that this controller and the aircraft model can be numerically linearized when linked together, and the Jacobian matrices will agree very accurately with the closed-loop matrices in Example 4.5-1. This provides a check for correct operation.

Accurate initialization will allow the longitudinal dynamics to be exercised without waiting for an initial transient to die out. The alpha-filter state should be initialized with the trim value of alpha, and the elevator state with the trim value of elevator deflection (both in degrees). The initial value of the error integrator state can be calculated as follows. From Figure 4.5-1 we see that the steady-state (no integrator input) elevator deflection is given by:

$$-\delta_e = k_p z x_i - k_\alpha \alpha$$

When the trim values are inserted in this equation, the trim value of the integrator output is found to be $x_i = 0.6186$. The aircraft trim-data file can now be augmented with the initial conditions for the three controller states.

A simulation of the F-16 aircraft model, with this controller, will now be used to illustrate some points about controller design. The following discrete-time subprogram was used to provide simulation commands.

```
subroutine DISCRETE(time,TS,x,xdot,)
dimension x(*), xdot(*)
common/controls/thtl,el,ail,rdr,qcom

if (time .lt. 10.0) then
    qcom= 0.0
else if (time .lt. 20.0) then
    qcom= 8.65
    thtl= 1.0
else if (time .lt. 50.0) then
    qcom= 0.0
else
    qcom= 10.0
end if
return
end
```

Figure 4.7-1 shows the vertical-plane trajectory produced by the commands, and Figure 4.7-2 shows the pitch-rate response. The aircraft is given full throttle and a pitch-up command at $t = 10$ s; the pitch rate command has been adjusted to bring the aircraft vertical at $t = 20$ s. At $t = 50$ s the airspeed has fallen to about 300 ft/s and the altitude is approximately 16,000 ft; therefore, the flight conditions are greatly different from the controller design conditions (sea level and 502 ft/s). The pitch-rate response has a large overshoot at $t = 50$ s because of the off-nominal design conditions. The aircraft dynamics change with flight conditions as described in Section 4.2, and in a practical controller design, the gain coefficients would be "*gain-scheduled*" as functions of dynamic pressure and/or Mach number. The time of flight was 100 s for the trajectory shown, and the lateral-directional dynamics did not become significantly involved in the motion. The sideslip angle peaked up to about

352 AIRCRAFT DYNAMICS AND CLASSICAL CONTROL DESIGN

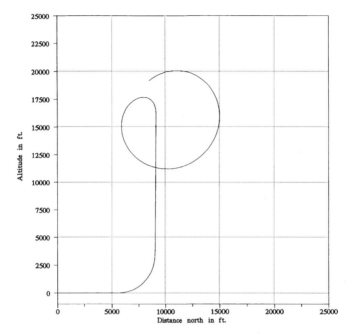

Figure 4.7-1 Aircraft trajectory with pitch-rate CAS.

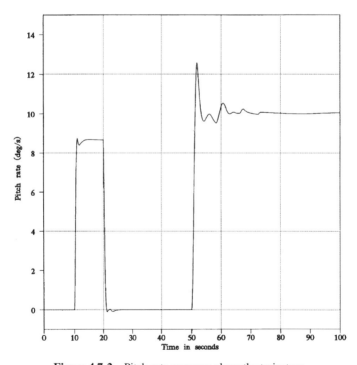

Figure 4.7-2 Pitch-rate response along the trajectory.

0.1 degrees after the pitch-over command at $t = 50$ s but then returned to very small values. The angle of attack reached a peak of approximately 15 degrees at $t = 55$ s.

■

Example 4.7-2: Lateral-Directional CAS Nonlinear Simulation. In this example the lateral-directional controller designed in Example 4.5-3 is programmed for nonlinear simulation and used, together with the pitch-rate controller from Example 4.7-1, to provide complete 6-DOF control. The controller code is:

```
subroutine FC(time,x,xd)
dimension x(*), xd(*)
real m
common/controls/thtl,el,ail,rdr,pcom,qcom,rcom
common/output/an,ay,ax,qbar,m,alpha,beta,phid,thtad,
   & pd,qd,rd                     ! d means degree units
el = -x(1)                        ! actuator state -> el
ail= -x(4)                        ! : : : : :    -> ail
rdr= -x(5)                        ! : : : : :    -> rdr
call f16(time,x(7),xd(7))          ! aircraft dynamics
xd(3)= qcom - qd                   ! error integrator
u = 1.5*x(3) - .5*qd - .08*x(2)    ! pitch control law
xd(1)= 20.2*(u-x(1))               ! elevator actuator
xd(2)= 10.0*( alpha - x(2) )       ! alpha filter
ua   = 0.2*(pcom-pd)               ! roll control law
xd(4)= 20.2*( ua - x(4) )          ! aileron actuators
ari  = (0.13*x(2) - 0.7)*ua        ! ARI
rs   = rd - pd*x(2)/57.3           ! yaw-rate feedback
xd(6)= rs - x(6)                   ! washout
err= rcom +/- .8*xd(6) - 10.0*ay   ! yaw control law
xd(5)= 20.2*( err + ari - x(5) )   ! rudder actuator
return
end
```

This time the controller states are numbered first and the aircraft states are appended to these. The nominal flight condition of Table 3.6-3 was used, and the six compensator states were included in the trim data file. The actuator states must be set to the trimmed values of the corresponding aircraft controls, and the alpha-filter state to the value of alpha in degrees. The other controller states can be set to zero since the rest of the controller is linear.

A nonlinear simulation was chosen that would exercise the ARI through high-alpha and fast roll rates, yet be easily preprogrammed for non-real-time simulation. The trajectory chosen was a pull-up into a vertical loop, with a 180 degree roll at the top of the loop, and continuing into a second vertical loop. The preprogrammed commands were:

```
subroutine DISCRETE(time,ts,x,xd)
dimension x(*),xd(*)
common/controls/thtl,el,ail,rdr,pcom,qcom,rcom

if (time .lt. 5.0) then
```

354 AIRCRAFT DYNAMICS AND CLASSICAL CONTROL DESIGN

```
    qcom= 0.0
    pcom= 0.0
    rcom= 0.0
else if (time .lt. 15.0) then
    qcom= 15.0                          ! Pull up at 15 deg/s
    thtl= 1.0                           ! at full throttle
else if (time .lt. 17.0) then
    pcom= 150.0                         ! rolling for 2 sec.
else
    pcom= 0.0
end if
return
end
```

The alpha-filter and actuator states were initialized exactly with the correct initial conditions, but the other controller states were left uninitialized, so the first 5 s of the flight were used to let any transients die away. Full throttle and a 15 deg/s pitch rate command are applied at $t = 5$ s, and then a roll-rate command pulse is applied between 15 s and 17 s. The desired roll rate is therefore 90 deg/s, but because of the finite error of the type-0 roll-rate loop, the rate command had to be adjusted by trial and error to achieve the 180 degree roll.

An elevation view of the trajectory is shown in Figure 4.7-3a. The first loop corresponds to a normal acceleration of about 4 g, and the aircraft speed decreases

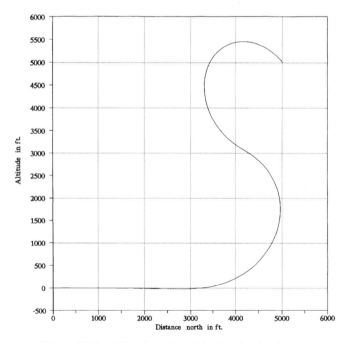

Figure 4.7-3a Elevation view of the simulated trajectory.

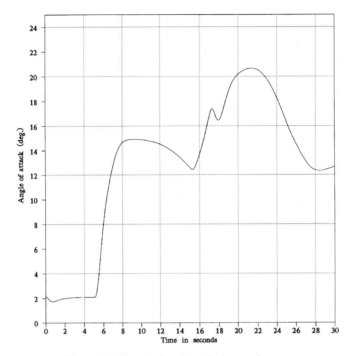

Figure 4.7-3b Angle of attack along trajectory.

roughly linearly from 500 ft/s at 5.0 s to 270 ft/s at 24 s (near the top of the second loop). Figures 4.7-3*b* and *c* show angle of attack, roll attitude, and pitch attitude. Alpha increases rapidly as the loop is started, remains roughly constant to provide the centripetal acceleration while the pitch attitude is between 45 and 90 degrees, and then starts to fall off as gravity helps to provide the centripetal acceleration. During the second loop alpha rises to a larger peak, because the airspeed has dropped considerably by then.

The roll angle of 0 degrees suddenly becomes a roll angle of 180 degrees as the aircraft passes through the vertical attitude condition, and this wings-level attitude is held until the roll is started at 15 s. The attitude angles are computed by integrating the angular rates (state derivatives), not from trigonometric functions, so the roll angle may contain multiple 360 degree ambiguities, depending on how the angular rates behave.

Figure 4.7-3*d* shows the fast roll-rate response and the corresponding yaw rate that is generated by the ARI. The pitch rate undergoes a positive perturbation during the roll, and this is due to the nose-up pitching moment generated by inertia coupling. Figure 4.7-3*e* shows the positive increment in elevator deflection that is generated by the longitudinal control system to counteract the inertia-coupling moment. As stated earlier, a major problem when rolling rapidly at still higher angles of attack is that the longitudinal control surfaces may be unable to generate a large enough nose-down moment.

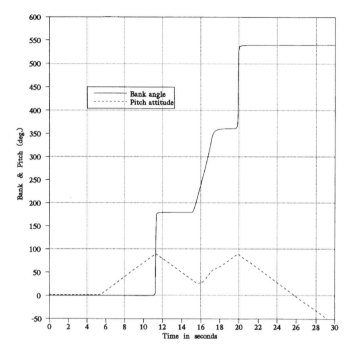

Figure 4.7-3c Roll and pitch angles along trajectory.

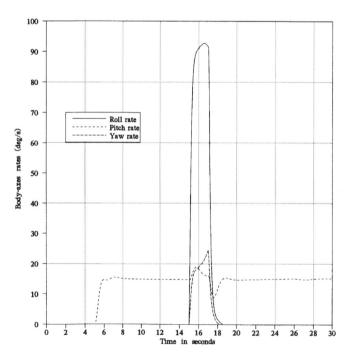

Figure 4.7-3d Body-axes angular rates along trajectory.

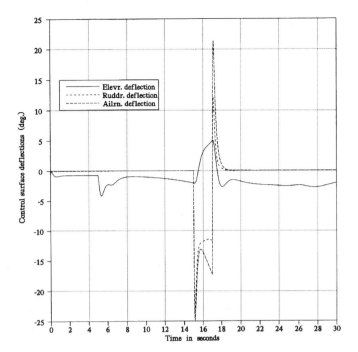

Figure 4.7-3e Control-surface deflections along trajectory.

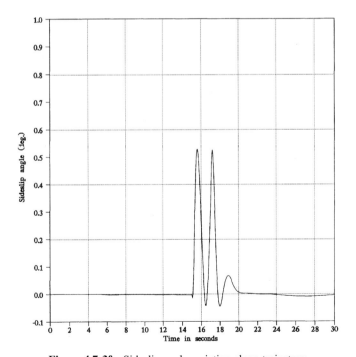

Figure 4.7-3f Sideslip angle variation along trajectory.

In this example the elevator deflections are quite small, but the aileron and rudder deflections are large. This is due to the combination of high demanded roll-rate and low aileron effectiveness (because of the high alpha and relatively low dynamic pressure). It is also partly due to the fact that while the rudder is generating the required yaw rate, it is also generating a rolling moment that opposes the aileron rolling moment. The large peak deflections are due to the instantaneous demand for the high roll rate. Note that the aileron and rudder deflection rates may have reached or exceeded the capabilities of their actuators; this concern is addressed later.

Figure 4.7-3f shows that the control system has done an excellent job of keeping the sideslip angle small during this demanding maneuver. The sideslip excursions are biased positively, that is, toward adverse beta. This is desirable in general; a combination of adverse and proverse beta tends to excite the dutch roll mode. The ARI gain is quite critical, and the values used are close to optimal. Larger values will produce a single negative and single positive beta excursion that are more nearly symmetrical about zero but considerably larger in magnitude. Figure 4.7-3g shows the variation of airspeed with time. The maneuvers have caused the speed to fall continuously until the top of the second loop is passed, despite full throttle being used throughout. The twofold reduction in speed will cause a fourfold reduction in dynamic pressure.

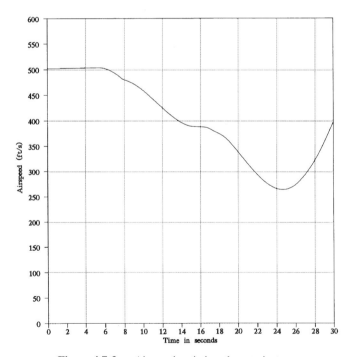

Figure 4.7-3g Airspeed variation along trajectory.

Finally, note that the performance may appear satisfactory for these flight conditions, but the design must be evaluated at other altitude/speed combinations. Gain scheduling with Mach number will probably be required, and much more comprehensive simulation is necessary before the design can be considered practical. ∎

Example 4.7-3: Simulation of Automatic Landing. This example will simulate longitudinal control for an automatic landing of the transport aircraft model using the longitudinal controller designed in Example 4.6-4. The MATLAB controller code can be constructed quite easily from the code in Example 4.6-4, with the help of Figure 4.6-10 and Table 3.3-1:

```
% GLIDE.M Glide-Slope Controller for Ex. 4.7-3
function [xd]= glide(time,x,u)
global xd
u(2)   = -x(8);                      % set elevator
u(1)   = x(11);                      % set throttle
[xd]   = transp(time,x,u);           % call aircraft
xd(14) = 0.0 - x(7);                 % d-loop error
dpi    = xd(14) + .18*x(14);         % PI compensation
xd(13) = dpi - 5.0*x(13);            % Phase Lead
thcom  = xd(13) + .5*x(13);          % theta command

xd(9)  = thcom - 57.29578*x(3);      % Pitch error
tpi    = xd(9) + 0.2*x(9);           % PI integrator
xd(10) = tpi - 14.0*x(10);           % Phase Lead
qcom   = 40.0*(xd(10) + 1.4*x(10));  % Pitch-rate command
qerr   = qcom - 57.29578*x(4);       % q error
xd(8)  = 10.0*(qerr-x(8));           % El. actuator

ev     = 250 - x(1);                 % speed error
xd(12) = ev - 10.0*x(12);            % lead compensator
ut     = 10.0*(ev - 9.0*x(12));      % lead comp.
xd(11) = 0.2*(ut - x(11));           % throttle lag
```

and the sequence of .M files involved in the simulation is:

```
NLSIM.M   - Nonlinear simulation from Chap. 3
RK4       - Fourth-order Runge-Kutta from Ch. 3
GLIDE.M   - Controller routine above
TRANSP.M  - Transport-Aircraft model from Ch. 3
ADC.M     - Atmosphere model for Transp. Aircraft
```

The chosen initial conditions were level flight at $V_T = 250$ ft/s, $h = 1500$ ft, with flaps and gear deployed, and $x_{cg} = 0.25$. An initial-condition data file can be obtained by using TRIM.M (Chapter 3), as follows. In the steady state all of the integrator inputs in Figure 4.6-10 are zero, and it is easy to write algebraic equations for all of the controller variables in terms of the aircraft states and controls. When these equations are included in the cost function, the trim program will produce an initial condition file for all fourteen controller and aircraft states (see Ex. 4.7-4). In the transport aircraft

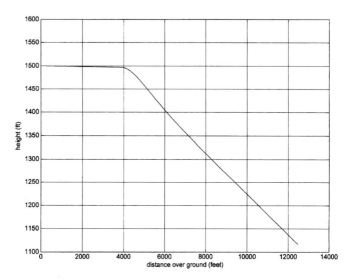

Figure 4.7-4a Automatic landing; elevation profile.

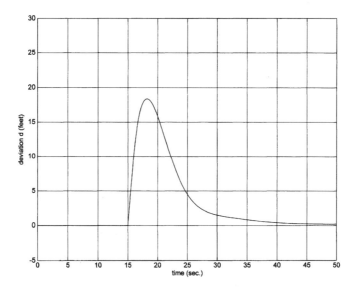

Figure 4.7-4b Automatic landing; deviation from glide path.

model the reference flight-path angle was programmed to change from zero to -2.5 degrees at $t = 15$ s, to represent glide-path capture.

Figure 4.7-4a shows the trajectory for 50 s of simulated flight; the aircraft starts out in level flight with no transient because the controller was accurately initialized. Figure 4.7-4b shows the deviation from the glide path. Figure 4.7-4c shows the behavior of alpha, pitch attitude, and elevator deflection, and Figures 4.7-4d and e

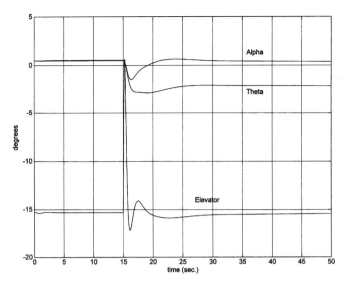

Figure 4.7-4c Automatic landing; controlled variables.

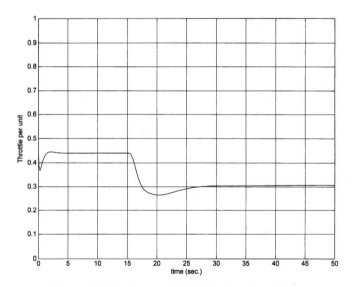

Figure 4.7-4d Automatic landing; throttle variation.

show, respectively, the corresponding variation of throttle position and airspeed. It is evident that the airplane is driven onto the glide path quickly and smoothly, without large excursions in pitch attitude. Airspeed is held very nearly constant and the throttle is changed smoothly and gently. Because of the tight control the elevator shows some rapid excursions, which could cause rate-limiting in a real actuator.

362 AIRCRAFT DYNAMICS AND CLASSICAL CONTROL DESIGN

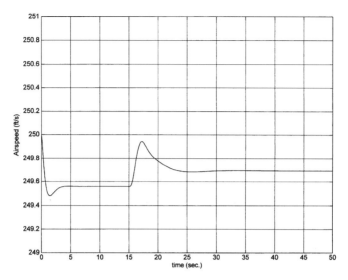

Figure 4.7-4e   Automatic landing; airspeed variation.

The final component of this design is the automatic "flare" control that makes the aircraft begin to level out as the altitude approaches zero, and touch down with an acceptably small rate of descent. This is described in the following subsection.

Flare Control

At an altitude of about 50 ft above the runway the automatic landing system must start to reduce the rate of descent of the aircraft, achieve the correct pitch attitude for landing, and begin to reduce the airspeed. This portion of the trajectory is called the landing flare, and the geometry of the flare is illustrated in Figure 4.7-5. On the glide path the aircraft is descending at a rate of 10 ft/s or greater and will hit the ground

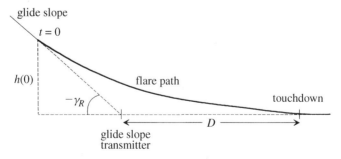

Figure 4.7-5   Landing-flare geometry.

hard if the flare is not executed. The rate of descent must be reduced to less than about 2.0 ft/s by touchdown. The pitch attitude angle will depend on the airspeed and will be only a few degrees for a large jetliner; military aircraft may land with large pitch angles to make use of aerodynamic braking. Altitude rate ($\dot{h} = V_T \sin \gamma$) is a natural choice for the controlled variable since it determines the impact, can be derived in the radar altimeter, makes the control system independent of ground effect and wind disturbances, and involves control of one less plant integration than altitude.

Modern digital-controller-based automatic landing systems can yaw the aircraft to deal with crosswinds while lining up the runway, de-crab the aircraft and dip a wing to keep the lateral velocity component small, and level the wings immediately before touchdown.

Example 4.7-4: Automatic Flare Control. In this example we will use altitude-rate from the aircraft model as the controlled variable, and switch from the glide-path controller to the flare controller when the altitude reaches 50 ft. The speed loop will continue to operate with the same command input. Switching from one controller to another can cause large transients in the aircraft states, which is disconcerting and dangerous so close to the ground. To avoid this, the flare controller must be initialized with the final conditions on the glide slope, and then commanded to go smoothly to the new altitude rate. Here we have used only a simple controller with one additional state, $x(15)$, which is initialized to zero. Thus, the following code shows only a PI compensator and a gain, with a step command of -2 ft/s, for the altitude-rate controller. Some logic (variable "MODE") is used to ensure that control does not momentarily switch back to the glide-path controller if integration errors or transients cause a fluctuation in altitude at changeover. The airspeed on the glide path was chosen to give the aircraft a slightly pitched-up attitude at touchdown, and was 235 ft/s (139 kts). The controller code is as follows:

```
              % FLARE.M   Glide-Slope & Flare Controller
function [xd]= flare(time,x,u);
global   xd mode
u(2)     = -x(8);                      % set elevator
u(1)     = x(11);                      % set throttle
[xd]     = transp(time,x,u);           % call aircraft
h        = x(5);                       % altitude
vcom=235;                              % commanded speed
if time<1
    mode=0;                            % glideslope mode
end
if h>50 & mode==0                      % d-controller
   xd(14)= 0.0 - x(7);                 % integrate d-error
   dpi= xd(14) + .18*x(14);            % PI comp.
   xd(13)= dpi - 5.0*x(13);            % lead compensation
   thcom = xd(13) + 0.5*x(13);         % theta command
   xd(15)=0.;                          % for flare controller
elseif h<=0 | mode==2                  % roll-out mode
```

```
      mode=2;
      xd=zeros(1,15);
      thcom=0;
   elseif h<=50 | mode==1        % flare controller
      mode=1;                    % lock out other modes
      hdot=-2;                   % sink-rate command
      xd(15)= ( hdot - xd(5) );  % PI integrator
      thcom= .1*( xd(15) + .5*x(15) );  % C. L. Zero at -.5
      xd(13)=0;
   end
   xd(9)  = thcom - 57.29578*x(3);   % integrate pitch error
   tpi    = xd(9) + 0.2*x(9);        % PI compensation
   xd(10) = tpi - 14.0*x(10);        % phase-lead comp.
   qcom   = 40* ( xd(10) + 1.4*x(10) );  % pitch-rate command
   qerr   = qcom - 57.29578*x(4);    % pitch-rate error
   xd(8)  = 10.0 * (qerr - x(8) );   % Elevator actuator

   ev     = vcom - x(1);             % Autothrottle code
   xd(12) = ev - 10.0*x(12);         % phase-lead pole
   ut     = 10.0*(ev - 9.0*x(12) );  % phase-lead zero
   xd(11) = 0.2*(ut - x(11));        % throttle servo
```

and the cost function used with the trim program was:

```
% Cost Function for 3-DOF Aircraft
function [f]=ssland(s);
global x u gamma
u(1)= s(1);            % throttle
u(2)= s(2);            % elevator
x(2)= s(3);            % alpha
x(3)= x(2)+ gamma;     % theta
x(8)= -u(2);
x(10)= x(8)/(40*1.4);
x(9)= 14*x(10)/.2;
x(13)= 57.29578*x(3)/.5;
x(14)= 5*x(13)/.18;
x(11)= u(1);
x(12)= x(11)/10;
x(15)= 0.0;            % flare control state
time= 0.0;
[xd]=transp(time,x,u);
f= xd(1)^2 + 100*xd(2)^2 + 10*xd(4)^2;
```

A nonlinear simulation of the controller, with the transport aircraft model, was started from an initial state corresponding to an altitude of 300 ft on the glide-path simulation of Example 4.7-3. Figures 4.7-6a through c show the trajectory, vertical speed, and pitch attitude. The vertical speed and pitch attitude have reached the desired values well before touchdown. The pitch-attitude change is faster than necessary; the small well-damped oscillation in pitch would be barely noticeable to the passengers.

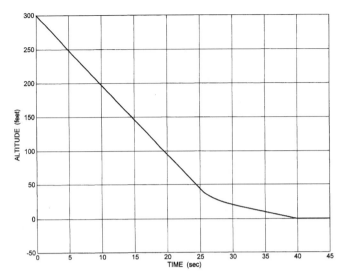

Figure 4.7-6a Automatic landing; glideslope and flare trajectory.

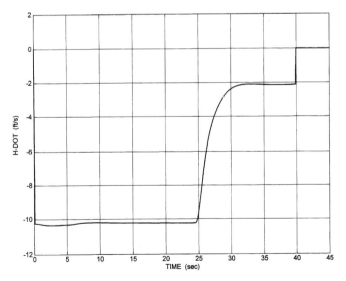

Figure 4.7-6b Altitude-rate during flare.

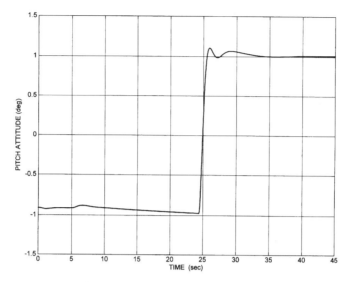

Figure 4.7-6c Pitch-attitude during flare.

Example 4.7-5: A Roll-Angle-Steering Control System. The controller routine for the lateral-directional CAS is easily modified to include the roll-angle feedback and the turn compensation. The controller and command routines are shown below.

```
      subroutine FC(time,x,xd)
      dimension x(*), xd(*)
      real m
      common/controls/thtl,el,ail,rdr,bank,qcom,rcom
      common/output/an,ay,ax,qbar,m,alpha,beta,phid,thtad,
    &    pd,qd,rd
      el = -x(1)
      ail= -x(4)
      rdr= -x(5)
      call f16(time,x(7),xd(7))
      qcom= rd*tan(phid/57.29578)       ! turn compensation
      xd(3)= qcom - qd
      u= 1.5*x(3) - .5*qd - .08*x(2)
      xd(1)= 20.2*(u-x(1))
      xd(2)= 10.0*( alpha - x(2) )
      ua   = 1.0*(bank-phid-.2*pd)      ! bank angle control law
      xd(4)= 20.2*( ua - x(4) )
      ari  = (0.13*x(2) - 0.7)*ua
      rs   = rd - pd*x(2)/57.3
      xd(6)= rs - x(6)
      err= rcom - .8*xd(6) - 10.0*ay
      xd(5)= 20.2*( err + ari - x(5) )
```

```
      return
      end

      subroutine DISCRETE(time,ts,x,xd)
      dimension x(*),xd(*)
      common/controls/thtl,el,ail,rdr,bank,qcom,rcom
      if (time .ge. 20.0) Then
        bank= -70.0
      else if (time .ge. 5.0) Then
        bank= 70.0
      else
        bank = 0.0
        qcom = 0.0
        rcom = 0.0
      end if
      return
      end
```

The same initial condition data used for the lateral-directional CAS were used with this controller, and a simulation was flown using the discrete-time commands shown above. Figure 4.7-7a shows the ground track of the aircraft in response to these commands. The altitude decreased by about 600 ft during the 30 s simulation because of the open-loop turn compensation and the finite control error of the lateral-directional

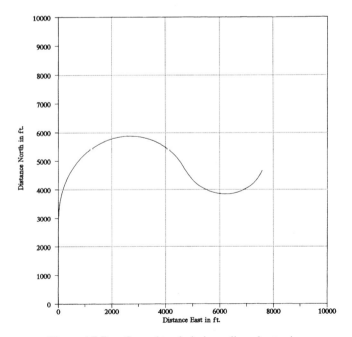

Figure 4.7-7a Ground track during roll-angle steering.

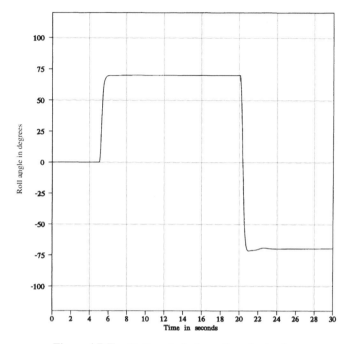

Figure 4.7-7b Roll angle during roll-angle steering.

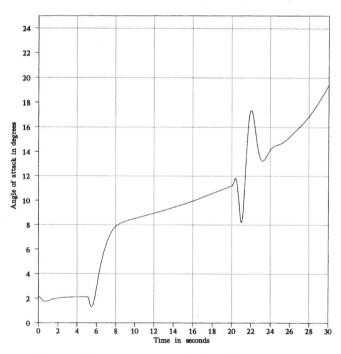

Figure 4.7-7c Angle of attack during roll-angle steering.

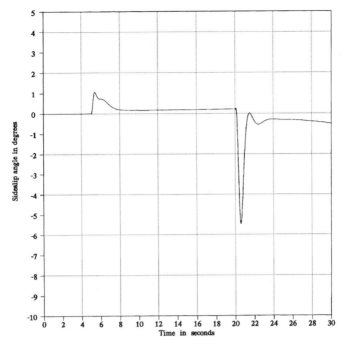

Figure 4.7-7d Sideslip angle during roll-angle steering.

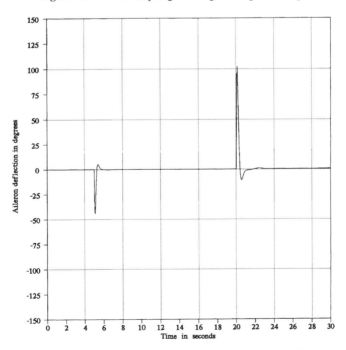

Figure 4.7-7e Aileron deflection during roll-angle steering.

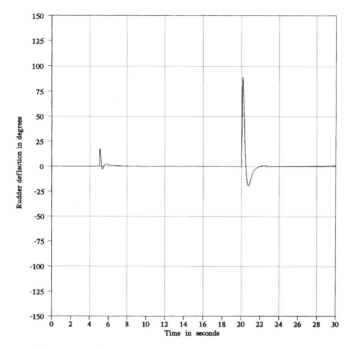

Figure 4.7-7f Rudder deflection during roll-angle steering.

control systems. The speed decreased by about 200 ft/s during the simulation, because of the maneuvers.

Figures 4.7-7*b*, *c*, and *d* show, respectively, the fast well-damped roll-angle response, the angle of attack, and the sideslip angle. The aileron and rudder deflections are shown in Figures 4.7-7*e* and *f*; these show short-duration deflections that are well beyond the limits of the control surfaces and raise the following important points.

First, the simulation results may be unrealistic if the control-surface rate and deflection limits are not modeled. Second, control system limiting will be caused by the abrupt large-amplitude commands and the high gains that have been used in the roll-angle control. This is not necessarily a problem if the system response is still acceptable, since the fastest possible roll response may be desired, and the high gains also provide a small control error for low-amplitude inputs. Third, the airplane handling qualities are the most important consideration, and in this situation the stick prefilter and the maximum roll rate of the airframe will play a major part in determining the pilot's opinion of the roll performance. ■

Example 4.7-6: Simulation of a Controller with Limiters. Here we deal with "hard-limiting" as distinct from merely nonlinear behavior. To simulate limiting behavior the control variables, state variables, and state derivatives must be modified when the limiting occurs. Other variables are essentially dummy variables that are assigned according to the states and controls. Rate-limiting can be simulated by simply "clamping" the appropriate state derivative when its maximum value is reached. State-variable limits must be dealt with by modifying their derivatives also. When a state variable reaches a limit, a nonzero derivative is allowed only if it is in the direction that takes the state variable off the limit. This is related to the integrator wind-up problem described in Chapter 3. In this example, the code used for Example 4.7-2 has been modified by the addition of rate and deflection limits, and is shown below. In-line comments have been added to explain the details.

The behavior of the controller, with limiters, is illustrated by a time-history simulation similar to that of Example 4.7-2, and the discrete-time commands are shown below. The integration step size was reduced to 1 ms in this simulation, to capture the action of the limiters with sufficient accuracy. This can make the execution quite slow.

```
subroutine FC(time,x,xd)
dimension x(*), xd(*)
real m
common/controls/thtl,el,ail,rdr,pcom,qcom,rcom
common/output/an,ay,ax,qbar,m,alpha,beta,phid,thtad,pd,qd,
& rd
data erl,edl,arl,adl,rrl,rdl/60.0,25.0,80.0,21.5,120.0,30.0/
el = -x(1)
ail= -x(4)
rdr= -x(5)
call f16(time,x(7),xd(7))
xd(2)= 10.0*( alpha - x(2) )
xd(3)= qcom - qd
u= 1.5*x(3) - .5*qd - .08*x(2)
xd(1)= 20.2*(u-x(1))

if( abs(xd(1)) .gt. erl) then
    xd(1)= sign(erl, xd(1))          ! Elevator rate limit
end if
if(x(1) .gt. edl) then
  x(1)= edl                          ! Elevator +deflection limit
  it(xd(1) .gt. 0.0) xd(1)= 0.0      ! Stop integrating positively
  if(xd(3) .lt. 0.0) xd(3)= 0.0      ! clamp error integrator
else if (x(1) .lt. -edl) then
    x(1)= -edl                       ! Elevator -deflection limit
    if(xd(1) .lt. 0.0) xd(1)= 0.0    ! stop integrating negatively
    If(xd(3) .gt. 0.0) xd(3)= 0.0    ! clamp error integrator
else
    continue
end if

ua    = 0.2*(pcom-pd)
xd(4)= 20.2*( ua - x(4) )
if( abs(xd(4)) .gt. arl) then
    xd(4)= sign(arl, xd(4))          ! Aileron rate limit
end if
```

```
      if(x(4) .gt. adl) then
         x(4)= adl                          ! Aileron deflection limit
         if(xd(4) .gt. 0.0) xd(4)= 0.0
      else if (x(4) .lt. -adl) then
         x(4)= -adl
         if(xd(4) .lt. 0.0) xd(4)= 0.0
      else
         continue
      end if

      temp= ua
      if (abs(temp) .gt. adl) then
           temp= sign(adl,temp)             ! limit ARI to aileron limit
      end if
      ari = (0.13*x(2) - 0.7)*temp
      rs  = rd - pd*x(2)/57.3
      xd(6)= rs - x(6)
      err= rcom - .8*xd(6) - 10.0*ay
      xd(5)= 20.2*( err + ari - x(5) )

      if( abs(xd(5)) .gt. rrl) then
          xd(5)= sign(rrl, xd(5))           ! Rudder rate limit
      end if
      if(x(5) .gt. rdl) then
         x(5)= rdl
         if(xd(5) .gt. 0.0) xd(5)= 0.0
      else if (x(5) .lt. -rdl) then
         x(5)= -rdl
         if(xd(5) .lt. 0.0) xd(5)= 0.0
      else
         continue
      end if
      return
      end

      subroutine DISCRETE(time,ts,x,xd)
      parameter (ll=20)
      dimension x(*),xd(*)
      common/controls/thtl,el,ail,rdr,pcom,qcom,rcom
      common/output/op(ll)
      if (time .lt. 5.0) then
        qcom= 0.0
        pcom= 0.0
        rcom= 0.0
      else if (time .lt. 12.0) then
        qcom= 20.0
        thtl= 1.0
      else if (time .lt. 16.0) then
        qcom= 0.0
        pcom= 300.0
      else
        pcom = 0.0
        qcom = 0.0
```

```
end if
return
end
```

A pull-up at 20 deg/s is simulated, so that the angle of attack attains quite large values. When alpha reaches about 23 degrees (Figure 4.7-8a) a large roll-rate command is applied; this causes the aileron and rudder actuators to saturate almost immediately (Figure 4.7-8b). The directional controls are then unable to control the sideslip tightly, and some of the angle of attack is rapidly converted to sideslip (Figure 4.7-8c) through kinematic coupling. As the roll and yaw rates build up (Figure 4.7-8d), the inertia coupling moment becomes strong. The elevator deflection then goes from a small negative value to its positive limit, as it tries to oppose the inertia coupling moment. Figure 4.7-8e shows the roll-angle variation, Figure 4.7-8f shows pitch attitude, and Figure 4.7-8g shows the decrease in airspeed during the maneuvers.

In the flight condition illustrated, the available roll and yaw rates are insufficient to cause a pitch departure (due to inertia coupling), but the elevator saturation means that there is no longitudinal control available for 2 or 3 s. For this aircraft, pitch departure appears to be a problem only at very low dynamic pressure and high alpha, and more details can be found in Nguyen et al. (1979). These types of problems are usually solved by using command limiters to limit the roll rate that the pilot can command or

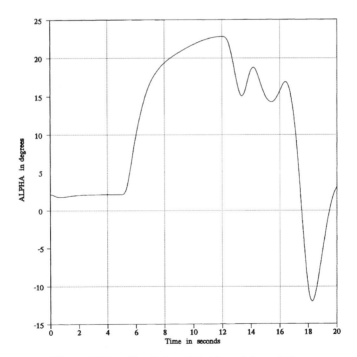

Figure 4.7-8a Simulation of limiting; alpha variation.

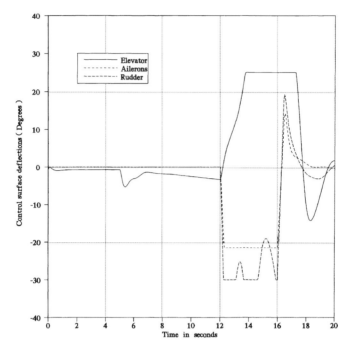

Figure 4.7-8b Simulation of limiting; control surface deflections.

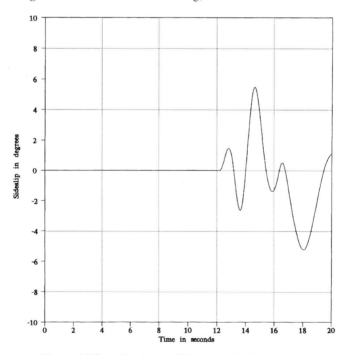

Figure 4.7-8c Simulation of limiting; sideslip variation.

NONLINEAR SIMULATION 375

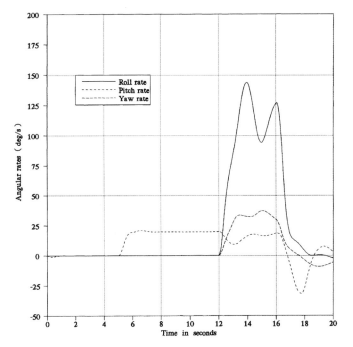

Figure 4.7-8d Simulation of limiting; aircraft angular rates.

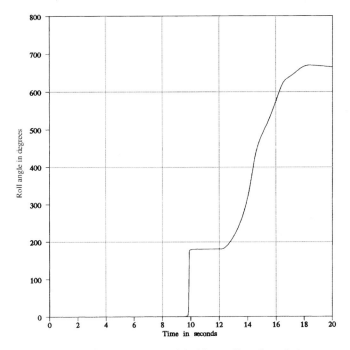

Figure 4.7-8e Simulation of limiting; roll angle variation.

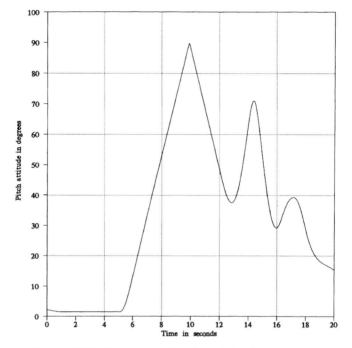

Figure 4.7-8f Simulation of limiting; pitch-attitude variation.

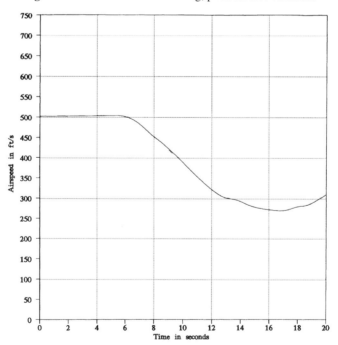

Figure 4.7-8g Simulation of limiting; airspeed variation.

the angle of attack that can be reached through the longitudinal controls. The limiting values must be made functions of the flight conditions, and the design process is a lengthy one involving much nonlinear simulation. ■

4.8 SUMMARY

In this chapter we have described the effect of flight conditions on the aircraft modes, presented some background in handling qualities and control design criteria, and described the purpose and design requirements of a large number of commonly used control systems. The design examples are quite realistic, having been performed on nonlinear aircraft models that are quite accurately representative of two very different types of aircraft. An infinite number of variations on the designs is possible and some of these are suggested in the chapter problems. Time and space limitations have not allowed the control designs to be gain-scheduled over the aircraft envelope and to be evaluated thoroughly in terms of the handling-qualities requirements, but the necessary cababilities have been developed.

The state-space formulation of modern control has provided an exceptionally convenient framework for the software and the use of classical design techniques. It should be evident that a primary requirement for successful design of aircraft control systems is an understanding of the physics of flight, and that interpreting the results of simulations is a vital aspect of this. Classical control theory fits extremely well into this picture because it relates very closely to the physics of the problems, and usually provides clues to the modifications needed to make the design successful. In the following chapters modern design techniques will be introduced. These techniques will come into their own in situations that we found difficult to handle up to this point, such as shaping the closed-loop time response when a number of poles and zeros all contribute significantly.

REFERENCES

AFWAL-TR-84-3105. *Proceedings of the Aeroservoelastic Specialists Meeting*, 2–3 October 1984, vols. I and II, USAF Flight Dynamics Laboratory, Wright-Patterson AFB, Ohio.

Ananthkrishnan, N., and S. Unnikrishnan. "Literal Approximations to Aircraft Dynamic Modes." *Journal of Guidance, Control, and Dynamics* 24, no. 6 (November–December 2001): 1196–1203.

Ashkenas, I. L. "Some Open and Closed-Loop Aspects of Airplane Lateral-Directional Handling Qualities." *AGARD Report 533*, 1966.

———. "Twenty-five Years of Handling Qualities Research." *AIAA Journal of Aircraft* 21, no. 5 (May 1984): 289–301.

Bischoff, D. E. "Longitudinal Equivalent Systems Analysis of Navy Tactical Aircraft." AIAA paper 81-1775, *AIAA Guidance and Control Conference*, Albuquerque, NM, 19–21 August 1981.

Blakelock, J. H. *Automatic Control of Aircraft and Missiles*. New York: Wiley, 1965.

Bryan, G. H. *Stability in Aviation*. London: Macmillan, 1911.

Chalk, C. R. "Fixed-Base Simulator Investigation of the Effects of L_α and True Speed on Pilot Opinion Rating of Longitudinal Flying Qualities." *USAF ASD-TDR-63-399*, Wright-Patterson AFB, Ohio, November 1963.

Chalk, C. R., T. P. Neal, and T. M. Harris. *Background Information and User Guide for MIL-F-8785B (ASG)*. Cornell Aeronautical Laboratories Inc., Buffalo, N.Y., August 1969 (AFFDL-TR-69-72; AD860856).

Cooper, G. E., and R. P. Harper, Jr. "The Use of Pilot Rating in the Evaluation of Aircraft Handling Qualities." *NASA Technical Note D-5153*. Washington, D.C.: NASA, 1969.

Craig, S. J., and I. L. Ashkenas. *Background Data and Recommended Revisions for MIL-8785B (ASG), "Military Specification—Flying Qualities of Piloted Airplanes."* Systems Technology Inc. Technical Report TR-189-1, Hawthorne, Calif., March 1971.

Etkin, B. *Dynamics of Atmospheric Flight*. New York: Wiley, 1972.

Gentry, T. A. "Guidance for the Use of Equivalent Systems with MIL-F-8785C." Paper 82–1355, *AIAA Atmospheric Flight Mechanics Conference*, San Diego, 1982.

George, F. L., and D. J. Moorhouse. "Relationship of the Flying Qualities Specification to Task Performance." *AFFTC/NASA-Dryden/AIAA Workshop Flight Testing to Identify Pilot Workload and Pilot Dynamics*. Edwards AFB, Calif., 19–21 January 1982.

Gerlach, O. H. "Developments in Mathematical Models of Human Pilot Behavior." *Aeronautical Journal* (July 1977): 293–305.

Hodgkinson, J. "Equivalent Systems Approach for Flying Qualities Specification." MCAIR 79-017, presented at *SAE Aerospace Control and Guidance Systems Committee Meeting*, Denver, Colo. 7–9 March 1979.

Kisslinger, R. L., and M. J. Wendle. "*Survivable Flight Control System Interim Report*," No. 1, "*Studies Analysis and Approach, Supplement for Control Criteria Studies*." Technical Report AFFDL-TR-71-20, suppl. 1, Wright-Patterson AFB, Ohio, May 1971.

Kleinman, D. L., S. Baron, and W. H. Levison. "An Optimal Control Model of Human Response. Part I: Theory and Validation; Part II: Prediction of Human Performance in a Complex Task. *Automatica* 6 (1970): 357–369.

McRuer, D. T., D. Graham, E. Krendel, and W. Reisiner, Jr. *Human Pilot Dynamics in Compensatory Systems*. USAF, Wright-Patterson AFB, Ohio, AFFDL-TR-65-15, 1965.

McRuer, D., I. Ashkenas, and D. Graham. *Aircraft Dynamics and Automatic Control*. Princeton, N.J.: Princeton University Press, 1973.

MIL-F-8785C. "U.S. Dept. of Defense Military Specification: Flying Qualities of Piloted Airplanes." 5 November 1980.

MIL-F-9490D. "Flight Control Systems: Design, Installation, and Test of, Piloted Aircraft, General Specification for." USAF, June 1975.

Moorhouse, D. J., and R. J. Woodcock. "Background Information and User Guide for MIL-F-8785C." *AFWAL-TR-81-3109 (AD-A119-421)*. Wright-Patterson AFB, Ohio, July 1982.

Neal, T. P., and R. E. Smith. "An In-flight Investigation to Develop Control System Design Criteria for Fighter Airplanes." *AFFDL-TR-70-74*, vols. I and II, Air Force Flight Dynamics Laboratory, Wright-Patterson AFB, Ohio, December 1970.

Nguyen, L. T., et al. "Simulator Study of Stall/Post-Stall Characteristics of a Fighter Airplane with Relaxed Longitudinal Static Stability." *NASA Tech. Paper 1538*. Washington, D.C.: NASA December 1979.

O'Hara, F. "Handling Criteria." *Journal of the Royal Aeronautical Society* 71, no. 676 (1967): 271–291.

Onstott, E. D., and W. H. Faulkner. "Production Evaluation and Specification of Closed-loop and Multi-axis Flying Qualities." *AFFDL-TR-78-3*, Wright-Patterson AFB, Ohio, February 1978.

Roskam, J. *Airplane Flight Dynamics and Automatic Flight Controls*. Lawrence, Kans.: Roskam Aviation and Engineering Corp., 1979.

Rynaski, E. G. "Flying Qualities in the Time Domain." AIAA paper 85-1849, *AIAA Guidance, Navigation and Control Conference*, Snowmass, Colo., August 1985.

Sheridan, T. B., and W. R. Ferrell. *Man-Machine Systems: Information, Control, and Decision Models of Human Operator Performance*. Cambridge, Mass.: MIT Press, 1974.

Shomber, H. A., and W. M. Gertsen. "Longitudinal Handling Qualities Criteria: An Evaluation." *Journal of Aircraft* 4, no. 4 (1967): 371–76.

Tobie, H. N., E. M. Elliot, and L. G. Malcom. "A New Longitudinal Handling Qualities Criterion." *Proceedings of the National Aerospace Electronics Conference*; Dayton, Ohio, 16–18 May 1966, pp. 93–99.

Toles, R. D. "Application of Digital Multi-input/Multi-output Control Law with Command Path Blended Gains for the Pitch Axis Task-tailored Air Combat Mode of the AFTI/F-16." *Proceedings of the 1985 American Control Conference*, Boston, June 1985, p. 1475.

PROBLEMS

Section 4.2

4.2-1 Use the results of Section 4.2 to write a program that will calculate the damping and natural frequency of the phugoid and short-period modes from the dimensionless longitudinal stability derivatives. Use it to check the results given in Table 4.2-1.

4.2-2 Write a program to determine the lateral directional modes from the appropriate dimensionless derivatives, and use it to check the results of Example 4.2-1. Determine both the approximate values from the equations in Section 4.2 and accurate values from the eigenvalues of the coefficient matrices.

Section 4.4

4.4-1 When the F-16 model is trimmed for level flight at 30,000 ft and 820 ft/s, the dynamic pressure and angle of attack are the same as those of the nominal sea-level condition in Table 3.6-3. The A and B matrices are given by

$$A = \begin{bmatrix} v_T & \alpha & \theta & q \\ -5.9172E-03 & 8.8482E+00 & -3.2170E+01 & -3.5136E-01 \\ -9.5423E-05 & -6.2426E-01 & 0.0000E+00 & 9.6439E-01 \\ 0.0000E+00 & 0.0000E+00 & 0.0000E+00 & 1.0000E+00 \\ 2.4638E-11 & 8.2290E-01 & 0.0000E+00 & -6.6011E-01 \end{bmatrix}$$

$$B = \begin{bmatrix} \delta_e \\ 1.7391E-01 \\ -1.3172E-03 \\ 0.0 \\ -1.7569E-01 \end{bmatrix}$$

Explain how and why these dynamic equations differ from those in Example 4.4-1. Augment these dynamics with the same alpha filter and elevator actuator as in the example, and find α and q feedback gains that will yield closed-loop longitudinal dynamics as close as possible to those obtained in Example 4.4-1.

4.4-2 Using the same flight conditions as in Problem 4.4-1, the lateral-directional dynamics of the F-16 model are

$$A = \begin{bmatrix} \beta & \phi & p & r \\ -.19729 & .039205 & .036657 & -.99645 \\ 0 & 0 & 1.0 & .036878 \\ -30.666 & 0 & -2.2538 & .40705 \\ 8.5461 & 0 & -.015556 & -.29186 \end{bmatrix}$$

$$B = \begin{bmatrix} \delta_a & \delta_r \\ 1.8078E-04 & 4.9356E-04 \\ 0 & 0 \\ -0.73388 & 0.13165 \\ -0.031891 & -0.062067 \end{bmatrix}$$

Determine and identify the modes. Find suitable feedback gains for basic lateral-directional stability augmentation, as in Example 4.4-3. Compare the roll-rate response to an aileron doublet with the response shown in Example 4.4-3.

4.4-3 Repeat Example 4.4-3 with washout time constants of 0.5 s and 2.0 s and describe the difference in behavior and the effect on the dutch roll dampling.

4.4-4 Repeat Example 4.4-2 using the F-16 dynamics given in Problem 4.4-1. Is the lag compensator still useful?

Section 4.5

4.5-1 Repeat the pitch-CAS design (Example 4.5-1) using the F-16 dynamics given in Problem 4.4-1. Does the position of the PI zero need to be changed? Could this design be made to work, over at least the low-Mach part of the envelope, by changing only the proportional gain. Is it possible to say anything about the need for gain-scheduling the controller? Determine the step response with and without the PI zero.

4.5-2 Repeat the design of the normal-acceleration CAS in Example 4.5-2, but with the accelerometer placed at the aircraft c.g. Attempt to obtain a step response that is fast but well damped and similar to that shown in Figure 4.5-6.

4.5-3 Use the MATLAB program NLSIM from Chapter 3 to simulate some maneuvers with the lateral-directional CAS from Example 4.5-3. In the state-equations MATLAB function, use linear state equations for the longitudinal and lateral-directional dynamics but model the nonlinear ARI and yaw-rate feedback. Compare the results with and without the ARI operating.

Section 4.6

4.6-1 Redesign the pitch-attitude hold in Example 4.6-1, using the short-period approximation, with an additional integrator to produce pitch from pitch rate. Evaluate the design with a step-response simulation performed on the full dynamics.

4.6-2 Redesign the pitch-attitude hold in Example 4.6-2 with the PI zero placed at $s = -0.1$; compare the step response with that given in the text. Can this design be performed using the short-period approximation?

4.6-3 Attempt to design a Mach-hold autopilot using the same dynamics as in Example 4.6-3 and Figure 4.6-6, but with Mach as the controlled variable. Show that the throttle to Mach transfer function contains an NMP zero that makes this difficult or impossible. What is the way out of this difficulty?

4.6-4 Redesign the d-loop of the glide-slope controller in Example 4.6-4 with the PI zero at $s = -0.1$. Also attempt to design the lead compensator so that the elevator is less active during the acquisition of the glide slope. Compare the simulation results with those given in the text.

4.6-5 Modify Example 4.6-5 to use PI control and lead compensation (if necessary) so that zero steady-state roll-angle error can be achieved.

Section 4.7

4.7-1 Use NLSIM to repeat the pitch CAS nonlinear simulation in Example 4.7-1 but, instead of the nonlinear F-16 model, use the linear state equations for the nominal flight condition. Show graphs for comparison with those in Example 4.7-1, and discuss the differences.

4.7-2 Reproduce the glide-path and flare simulations described in Examples 4.7-3 and 4.7-4. Add code to simulate rate limiting in the elevator actuator, and determine how low the maximum elevator slew-rate can be before the results are significantly changed.

CHAPTER 5

MODERN DESIGN TECHNIQUES

5.1 INTRODUCTION

Modern control theory has made a significant impact on the aircraft industry in recent years. Bryson (e.g., 1985; Ly et al., 1985) pioneered in applying it to aircraft control. Boeing (Gangsaas et al., 1986) has implemented control systems designed using modern techniques, for instance, in the Boeing 767 autopilot. Honeywell has promoted modern robust design (Doyle and Stein, 1981). Linear quadratic methods were used by General Dynamics in the control system of the AFTI/F-16 (AFWAL-TR-84-308, 1984).

Therefore, in aircraft control systems design it is essential to have an understanding of modern control theory. Unfortunately, the traditional modern design techniques based on state-variable feedback that are available in current texts are not suitable for aircraft control design. This is due to several things, one of which is their dependence on selecting large numbers of design parameters—namely, the performance index weighting matrices. Any design method for aircraft controls should eliminate the need for this trial-and-error selection. Thus the techniques in Ly et al. (1985), Gangsaas et al. (1986), Davison and Ferguson (1981), and Moerder and Calise (1985) all rely on modified design techniques that use output feedback or order-reduction techniques in conjunction with the minimization of a nonstandard performance index.

In the remainder of the book we focus on modern design techniques that are suitable for use in aircraft control. Included are such approaches as eigenstructure assignment, model following, dynamic inversion, LQG/LTR, and LQ output-feedback design. Each of these techniques has its proponents, and each has its advantages and disadvantages, as we will attempt to demonstrate. We will focus on output-feedback design, with performance criteria that are more general than the usual integral-quadratic form. Using this approach it is straightforward to design controllers that have a sensible structure from the point of view of the experience within the aircraft industry, without the trial-and-error selection of a large number of design parameters.

Our strategy in the next chapters will be different than in the first part of the book due to the different character of the material to be covered. We will first develop each modern design technique and then present examples showing how it is used in aircraft controls. In several instances we will consider the same examples presented in Chapter 4; this will afford an opportunity to contrast the classical and modern approaches to design. We now discuss some basic philosophy of modern control design.

Limitations of Classical Control

In Chapter 4 we showed how to design aircraft control systems using classical control techniques. The essence of classical design was *successive loop closure* guided by a good deal of intuition and experience that assisted in selecting the control system structure. For instance, we knew it was desirable to provide inner rate-feedback loops around a plant to reduce the effect of plant parameter variations. In conjunction with this we used standard compensator structures designed to approximate derivative action to stabilize the system, or integral action to eliminate steady-state error.

The one-loop-at-a-time design approach was aided by such tools as root locus, Bode and Nyquist plots, and so on, that enabled us to visualize how the system dynamics were being modified. However, the design procedure became increasingly difficult as more loops were added and did not guarantee success when the dynamics were multivariable, that is, when there were multiple inputs, multiple outputs, or multiple feedback loops.

Philosophy of Modern Control

Two concepts are central to modern control system design. The first is that *the design is based directly on the state-variable model,* which contains more information about the system than the input–output (black box) description. The state-variable model was introduced into system theory, along with matrix algebra, by R. Kalman (1958, 1960). Since we have already seen how to extract state equations from the nonlinear aircraft dynamics and use them for analysis, we are at this point in a good position to use them for control design.

The second central concept is *the expression of performance specifications in terms of a mathematically precise performance criterion* which then yields *matrix equations for the control gains.* These matrix equations are solved using readily available computer software. The classical successive-loop-closure approach means that the control gains are selected individually. In complete contrast, solving matrix equations allows *all the control gains to be computed simultaneously* so that all loops are closed at the same time. This simultaneous design means that we will have greater insight into the design freedom than is possible when the system has more than one input and/or output, or multiple control loops. Moreover, using modern control theory we are able to design control systems more quickly and directly than when using classical techniques.

As in classical control, we are able with modern techniques to select the structure of the control system using the intuition developed in the aircraft industry. Thus it is straightforward to include washout circuits, integral control, and so on. The key to this is the use of *output feedback* design techniques, introduced in Section 5.3 and used throughout the chapter.

The modern control formulation means that the trial and error of one-loop-at-a-time design disappears. Instead, the fundamental engineering decision is *the selection of a suitable performance criterion*. Let us now discuss some important design problems and their associated performance criteria.

Fundamental Design Problems

Pole-Placement/Eigenvector Assignment. Modern control techniques are available for assigning the poles in multi-input/multi-output (MIMO) systems to desired locations *in one step* by solving equations for the feedback gains. These are called *pole-placement techniques*. Once we move away from classical one-loop-at-a-time design and obtain the capability to compute all the feedback gains simultaneously, it will become clear that in the MIMO case *it is possible to do more than simply assign the poles*. In fact, the closed-loop eigenvectors may also be selected within limits.

Desirable pole locations for aircraft design may be found in the military flying qualities specifications (see Section 4.3). However, while discussing flying-qualities requirements, we noted that the time response depends not only on the pole locations—it also depends on the zeros of the individual single-input/single-output (SISO) transfer functions, or equivalently on the eigenvectors (see the discussion on system modes in Chapter 3). Thus the capability of modern controls design to select both the closed-loop poles and eigenvectors is relevant in aircraft design.

In this design approach, the performance criterion is to achieve specified pole locations and eigenvectors. We will discuss pole-placement/eigenstructure assignment in Section 5.2.

Regulator Problem. A fundamental design problem is the *regulator problem,* where it is necessary to regulate the outputs of the system to zero while ensuring that they exhibit desirable time-response characteristics. The regulator problem is important in the design of stability augmentation sytems and autopilots.

Stable regulation of systems implies closed-loop stability, but using modern control we may do more than simply ensure stability. To exercise our design freedom, we select as our performance criterion an *integral-squared performance index* (*PI*) similar to those used in classical design (D'Azzo and Houpis, 1988). That is, the squares of the states and inputs are integrated to obtain the PI. The control gains that minimize the PI are found by solving matrix equations using computer programs. Note that if the integral of the squares of the states is made small, then in some sense the states themselves are forced to stay near zero. Selecting different weighting coefficients in the PI for the various state components results in different time-domain behavior in the closed-loop system. Thus modern control regulator design is fundamentally a *time-domain design technique* useful in shaping the closed-loop response. This is in contrast to classical controls, where most techniques are in the frequency domain.

We will discuss the regulator problem in Section 5.3. A deficiency of the traditional approach to modern regulator design using state feedback and the standard quadratic PI is that, to obtain suitable responses, one must select a large number of design parameters—namely, the PI weighting matrices. To avoid such trial-and-error approaches, we use modified PIs (Section 5.5) that are more suitable for aircraft control system design.

Tracker Problem. Another fundamental design problem is the *tracker problem*, where it is desired for an aircraft to follow or track a command signal. The command may be either constant or time varying. This is also referred to as the *servodesign problem*. The tracker problem is important in the design of control augmentation systems, where, for instance, the command signal may be a desired pitch rate or normal acceleration command. The tracker problem also relies on the selection of an integral-squared PI. However, now it is desired to keep the outputs not at zero, but near the reference command signals.

The modern control technique we will use for tracker design is not the standard one—it has been modified in several respects to make it more suitable for the purpose of aircraft control design. Specifically, we select a general PI that can easily be modified to attain different performance objectives. The result is a convenient technique for aircraft control design that does not involve the trial-and-error tuning of large numbers of parameters. This design approach is described in Sections 5.4 and 5.5.

Model Following. An important approach to control design is *model following*, where it is desired for the aircraft to perform like an ideal model with desirable flying qualities. We have seen in Chapter 4 that one way to specify good flying qualities is to prescribe a low-order model (e.g., with one zero and a complex pole pair) whose response the closed-loop system should match. In this design technique the performance criterion is some measure of the difference between the model and controlled aircraft responses. We cover model-following design in Section 5.6, showing how to design controllers that make the aircraft behave like the model.

As another application, we show that model-following design offers a very straightforward approach to the design of an automatic flare control system. Note that in flare control, it is desired for the aircraft to follow an exponential path to a smooth touchdown—here, the model is just the desired trajectory.

Dynamic Inversion. The aircraft is fundamentally a nonlinear system. Unfortunately, it is very difficult to design control systems directly for nonlinear systems. Therefore, most aircraft control systems are based on linear systems design at a given operating point. This requires that different controllers be designed for different operating points. Using gain scheduling, the linear controllers are then combined to provide control effectiveness over an operating envelope. One technique that can be applied to nonlinear systems is *dynamic inversion*. This is based on the technique of *feedback linearization*. Dynamic inversion control requires that the controller have a full model of the nonlinearities of the aircraft. Using modern high-power computers, this is possible. In Section 5.8 we cover dynamic inversion, which has gained in popularity in recent years for aircraft control design.

Robust Design. It is important to incorporate notions from classical control theory into modern design. Particularly vital is the frequency-domain approach to robustness analysis. However, it is well known that in a multivariable system, individual gain and phase margins between different pairs of inputs and outputs mean little from the point of view of overall robustness. Therefore, in Chapter 6 we generalize frequency-domain robustness analysis techniques to MIMO systems using the concept of the *singular value*. There, we also present the linear quadratic regulator/loop-transfer recovery (LQG/LTR) technique, which has recently gained popularity in aircraft controls design.

Observers, Kalman Filter, and Regulators. In Chapter 6 we also cover the design of observers and the Kalman filter, which are dynamical systems that estimate the full state from measurements of the outputs. By using feedback of these state estimates in conjunction with the observer dynamics, we are able to design a dynamic linear quadratic regulator, which is just a compensator similar to those obtained using classical techniques. In the modern approach, however, a convenient design method for *multivariable systems* is achieved by solving matrix equations to guarantee specified performance.

Digital Control. The control systems of modern aircraft are implemented on digital microprocessors. Examples are the F-15E, F18, and late models of the F-16. The advantages of digital control include the ability to implement complicated multiloop control systems, reprogram the controller gains and structure (e.g., for gain scheduling), obtain redundancy for failure tolerance, and use digital signal processing techniques to filter the control signals.

Digital control design introduces some new problems, such as the need to account for the delays associated with the sampling and hold processes and control computation. Also important is the development of design techniques that overcome the drawbacks associated with z-plane design, where the need arises for extreme accuracy in placing the poles within the unit circle. Implementation problems include accounting for actuator saturation and the effects of computer finite word length, roundoff error, overflow, and so on. We discuss the basics of digital control in Chapter 7.

5.2 ASSIGNMENT OF CLOSED-LOOP DYNAMICS

Classical design techniques such as root locus and Bode analysis are directly applicable only to SISO systems. Using such techniques a single feedback gain may be selected to place the closed-loop poles to guarantee desirable time responses and robustness qualities. In the case of multiple inputs and outputs, or multiple control loops, the classical techniques require successive closures of individual loops and involve a significant amount of trial and error.

In the MIMO case it is possible to do more than simply place the poles. This extra freedom is difficult to appreciate from the point of view of classical control theory due to the successive SISO design approach. In this section we want to show how modern control theory can be used in the multivariable case to place the poles as well as to *take advantage of the extra freedom arising from multiple inputs* to assign the closed-loop eigenvectors. This is important since, as we saw in Sections 3.8 and 4.3, the time response of a multivariable system depends not only on the poles but also on the zeros of the individual SISO transfer functions, or equivalently on the eigenvectors.

The *eigenstructure assignment* technique discussed in this section offers good insight and is especially useful for the design of *decoupling controllers,* as we will show in an example. As far as obtaining suitable time responses for multivariable systems goes, linear quadratic approaches like those in Sections 5.3 to 5.7 and Section 6.5 have generally been found more appropriate in the aircraft industry.

We will now discuss some basic feedback concepts from the point of view of modern control theory.

State Feedback and Output Feedback

We have shown that the linearized equations of motion of an aircraft may be written in the state-space form

$$\dot{x} = Ax + Bu \tag{5.2-1}$$

$$y = Cx, \tag{5.2-2}$$

with $x(t) \in \mathbf{R}^n$ the state, $u(t) \in \mathbf{R}^m$ the control input, and $y(t) \in \mathbf{R}^p$ the measured output.

Let us select a feedback control input of the form

$$u = -Kx + v, \tag{5.2-3}$$

where $v(t)$ is an auxiliary input which might be, for instance, the pilot's command and K is an $m \times n$ gain matrix to be determined. This is called a *state-variable feedback* since all of the state components are fed back. The feedback gain K is an $m \times n$ matrix of scalar control gains.

Substituting the control into the state equation yields the closed-loop system

$$\dot{x} = (A - BK)x + Bv \tag{5.2-4}$$

The closed-loop plant matrix is $(A - BK)$, and we would like to select the feedback gain K for good closed-loop performance.

It is a fundamental result of modern control theory that if the system is *controllable*, all of the closed-loop poles may be assigned to desired locations by selection of K. Controllability means that the control input $u(t)$ independently affects all the system modes. It can be tested for by examining the controllability matrix (Kailath, 1980)

$$U = \begin{bmatrix} B & AB & A^2B & \cdots & A^{n-1}B \end{bmatrix} \quad (5.2\text{-}5)$$

The system is controllable if U has full rank of n, that is, if U has n linearly independent columns. This is equivalent to the nonsingularity of the gramian UU^T, which is a square $n \times n$ matrix whose determinant can be evaluated.

In the next subsection we will see that if there is more than one control loop, corresponding to more than one control gain, we cannot only place the poles, but also to a certain extent select the eigenvectors.

Unfortunately, in aircraft control systems it is usually not possible or economically feasible to measure all the states accurately. It is possible to design a dynamic observer or Kalman filter to provide estimates $\hat{x}(t)$ of the states $x(t)$, and then use *feedback of the estimates* by modifying (5.2-3) to read $u = -K\hat{x} + v$. Indeed, we do discuss this approach in Section 6.4, since we need it to cover the LQG/LTR robust design technique in Section 6.5. However, since the aircraft dynamics are nonlinear, all the parameters of any linear observer would need to be gain-scheduled. This is inconvenient if the order of the observer is large.

Therefore, to obtain realistic aircraft control schemes, we should feed back not the entire state $x(t)$, but only the measurable outputs $y(t)$. The *output feedback* control law is

$$u = -Ky + v = -KCx + v, \quad (5.2\text{-}6)$$

which on substitution into (5.2-1) yields the closed-loop system

$$\dot{x} = (A - BKC)x + Bv \quad (5.2\text{-}7)$$

Now the closed-loop plant matrix is $(A - BKC)$. The output feedback matrix K is an $m \times p$ matrix of scalar gains. Thus since p is generally less than n, there are fewer scalar control gains to select in output-feedback design than in state-feedback design.

An important advantage of output feedback, as we will see, is that it allows us to *incorporate a compensator of desired form into the feedback system.* In aircraft control, there is a wealth of experience that often dictates the form of the control system. For instance, a washout filter may be required, or a PI controller may be needed for zero steady-state error.

Unlike the state feedback case, there is no convenient test to determine for a given system if the closed-loop poles may be independently assigned using output feedback. Pole placement using output feedback is more difficult to accomplish than using state feedback. The basic thrust of this chapter is to investigate the selection of the output feedback gain matrix K to obtain desirable closed-loop characteristics. Note that this will involve more than simply placing the poles, since desirable time responses depend on the poles as well as the zeros of the individual SISO transfer functions. These zeros can also be influenced using feedback if there is more than one input and output.

The gain matrix K is $m \times n$ for state feedback and $m \times p$ for output feedback. In the MIMO case there could be many gain elements, each corresponding to a feedback path. In classical control theory the individual gains must be separately selected using trial-and-error successive loop closure design. By contrast, using modern controls design *all the elements of K are selected simultaneously*. Thus all the feedback loops in a complicated control system can be closed at the same time with a modern control approach.

We will now discuss the selection of K to yield desired closed-loop poles and eigenvectors—that is, to assign the closed-loop eigenstructure. Both state feedback and output feedback will be considered. First, let us recall the importance of the eigenvectors in the system response.

Modal Decomposition

We have discussed the importance of the system modes in Sections 3.8 and 4.2. Let us now carry that discussion a bit further. Let λ_i be an eigenvalue with right eigenvector v_i and left eigenvector w_i, so that

$$Av_i = \lambda_i v_i, \qquad w_i^T A = \lambda_i w_i^T \qquad (5.2\text{-}8)$$

Since $y = Cx$, we may use the results of Section 3.8 to write the output as

$$y(t) = \sum_{i=1}^{n} \left(w_i^T x_0\right) Cv_i e^{\lambda_i t} + \sum_{i=1}^{n} Cv_i \int_0^t e^{\lambda_i(t-\tau)} w_i^T Bu(\tau)\, d\tau \qquad (5.2\text{-}9)$$

The initial condition is $x(0) = x_0$. This equation is valid when the Jordan form of matrix A is diagonal.

From this equation we may note that Cv_i is a direction in the output space associated with λ_i, while the influence of the control input $u(t)$ on eigenvalue λ_i is determined by $w_i^T B$.

If $Cv_i = 0$, motion in the direction v_i cannot be observed in the output and we say that λ_i is *unobservable*. If $w_i^T B = 0$, the control input $u(t)$ can never contribute to the motion in the direction v_i and we say that λ_i is *uncontrollable*.

Clearly, we may affect the coupling between the inputs, states, and outputs by selecting the vectors v_i and w_i in the closed-loop system; that is, we can influence the *numerators* of the individual SISO transfer functions as well as the poles. To see this clearly, examine the transfer function

$$H(s) = C(sI - A)^{-1}B \qquad (5.2\text{-}10)$$

Let the Jordan matrix J and matrix of eigenvectors M be

$$J = \text{diag}\{\lambda_i\}, \qquad M = [\, v_1 \quad v_2 \quad \cdots \quad v_n \,]^T, \qquad (5.2\text{-}11)$$

so that

$$M^{-1} = [\, w_1 \quad w_2 \quad \cdots \quad w_n \,]^T \qquad (5.2\text{-}12)$$

Now recall that $A = MJM^{-1}$ and use the fact that $(QP)^{-1} = P^{-1}Q^{-1}$ for any two compatible square matrices P and Q, to write (5.2-10) as

$$H(s) = CM(sI - J)^{-1}M^{-1}B,$$

or, since $(sI - J)^{-1}$ is diagonal with elements like $1/(s - \lambda_i)$, as

$$H(s) = \sum_{i=1}^{n} \frac{Cv_i w_i^T B}{s - \lambda_i} \qquad (5.2\text{-}13)$$

This equation gives the partial fraction expansion of $H(s)$ in terms of the eigenstructure of A when A is diagonalizable.

Several things may be said at this point. First, if λ_i is unobservable or uncontrollable, its contribution to the partial fraction expansion of $H(s)$ is zero. In this case we say that the state-space description (5.2-1), (5.2-2) is *not minimal*. Second, the terms Cv_i and $w_i^T B$ determine the residues of the poles, and hence the *zeros* of the individual SISO transfer functions in the $p \times m$ matrix $H(s)$. By proposing a technique for selecting the closed-loop eigenvectors by feedback, we are therefore proposing a method of shaping the time response that goes beyond what is possible using only pole placement.

In the next subsections we will show how to design feedback gains that achieve desired closed-loop eigenvectors using both state feedback and output feedback. Meanwhile, in the following example we give some insight on desirable eigenvectors from the point of view of aircraft behavior, recalling some results from Sections 3.8 and 4.2. We also show that the eigenvectors may be selected to obtain *decoupling between the system modes*.

Example 5.2-1: Selecting Eigenvectors for Decoupling. In Sections 3.8 and 4.2 we studied the aircraft longitudinal and lateral modes, showing which states are involved in each one. To ensure that the controlled aircraft exhibits suitable flying qualities, we should take care to design the control system so that this basic modal structure is preserved (Sobel and Shapiro, 1985; Andry et al., 1983).

In this example we idealize the findings of Sections 3.8 and 4.2 a bit. That is, we make more categorical statements about the mode couplings in order to obtain concrete design objectives.

a. Longitudinal Axis. Assuming that the state equations are augmented by a simple lag elevator actuator model, in the linearized perturbed longitudinal equations of an aircraft the state can be taken as

$$x = \begin{bmatrix} \alpha \\ q \\ \theta \\ v_T \\ \delta_e \end{bmatrix} \quad (1)$$

The states are angle of attack, α, pitch rate, q, pitch angle, θ, total velocity, v_T, and elevator actuator state, δ_e, the elevator deflection. We have ordered the states this way to make the upcoming discussion clearer [see (2)].

The short-period mode is due primarily to a coupling of energy between α and q. The phugoid mode is due primarily to a coupling between θ and v_T. It is desirable for the forward velocity to be unaffected by short-period oscillations, while the phugoid oscillations should have no influence on angle of attack.

To achieve this behavior, we could select the closed-loop eigenvectors as

$$v_1 = \begin{bmatrix} 1 \\ -1 \\ 0 \\ 0 \\ x \end{bmatrix} + j \begin{bmatrix} -1 \\ 1 \\ 0 \\ 0 \\ x \end{bmatrix}, \quad v_3 = \begin{bmatrix} 0 \\ 0 \\ 1 \\ -1 \\ x \end{bmatrix} + j \begin{bmatrix} 0 \\ 0 \\ -1 \\ 1 \\ x \end{bmatrix}, \quad (2)$$

with v_2 and v_4 the complex conjugates of v_1 and v_3, respectively. Components whose values we do not care about are denoted by x. Then one oscillatory mode, the one with directions specified by v_1 and v_2, will involve the first two components of the state vector but will not inject energy into components 3 and 4. This will be a "good" short-period mode. Similarly, the phugoid mode, described by the eigenvectors v_3 and v_4, will involve components 3 and 4, but will not affect components 1 and 2.

A good design will have these closed-loop eigenvectors. The closed-loop poles should also be specified to attain the desired frequency and damping of the short-period and phugoid modes. The former will be defined by λ_1 and λ_2, while the latter will be determined by λ_3 and λ_4.

It may be necessary to modify the A matrix to a form that involves the nondimensional time to make the modal coupling of the eigenvectors more apparent (McRuer et al., 1973). That is, the modified A matrix should have the eigenvectors in (2).

b. Lateral Axis. In the linearized perturbed lateral equations of an aircraft the state can be taken as

$$x = \begin{bmatrix} r \\ \beta \\ p \\ \phi \\ \delta_r \\ \delta_a \end{bmatrix}, \qquad (3)$$

where we have assumed first-order lags for the rudder and aileron actuators. The states are yaw rate, r, sideslip angle, β, roll rate, p, bank angle, ϕ, and the actuator states, δ_r, the rudder deflection, and δ_a, the aileron deflection.

Roll commands should not excite the dutch roll mode. Thus let us associate r and β in the closed-loop system with the dutch roll mode (see Section 4.2). Then the roll subsidence mode, which involves p, should not influence r and β. Similarly, the dutch roll oscillation should have no effect on roll rate or bank angle. Desirable eigenvectors to achieve this decoupling between modes are given by

$$v_1 = \begin{bmatrix} 1 \\ -1 \\ 0 \\ 0 \\ x \\ x \end{bmatrix} + j \begin{bmatrix} -1 \\ 1 \\ 0 \\ 0 \\ x \\ x \end{bmatrix}, \qquad v_3 = \begin{bmatrix} 0 \\ 0 \\ 1 \\ x \\ x \\ x \end{bmatrix}, \qquad (4)$$

with v_2 the complex conjugate of v_1 and x denoting entries whose values we are not concerned about. The closed-loop poles should also be selected for desirable time response: λ_1 and λ_2 for the dutch roll and λ_3 for the roll subsidence mode. The desired closed-loop spiral mode may be selected as λ_4. ∎

Eigenstructure Assignment by Full State Feedback

Now that we have seen what the eigenvectors mean from the point of view of the aircraft behavior, we will discuss the assignment of both the closed-loop poles and eigenvectors, first by full state feedback and then in the next subsection by output feedback. This represents an extension of classical control theory in several ways. First, we are able to deal in a natural fashion with MIMO systems, *selecting all the control gains simultaneously for suitable performance*. Second, we will be able to use the extra freedom in systems with more than one input and output to assign the eigenvectors as well as the poles, thus directly influencing the *zeros* of the individual SISO transfer functions. Third, we will be able to address the problem of *decoupling of the modes* through considerations like those in Example 5.2-1.

Matrix Equation for Eigenstructure Assignment.

For ease of presentation we will assume that B and C have full rank m and p, respectively. Our discussion will be based on the polynomial matrix

$$C(s) = [sI - A \quad B], \qquad (5.2\text{-}14)$$

with s a complex variable and A and B the system plant and input matrices.

In this subsection we follow Moore (1976) and consider full state feedback of the form

$$u = -Kx \qquad (5.2\text{-}15)$$

Under the influence of this control input the closed-loop system becomes

$$\dot{x} = (A - BK)x \qquad (5.2\text{-}16)$$

To select K so that a desired eigenvalue λ_i and associated eigenvector v_i are assigned to the closed-loop system, suppose that we can find a vector $u_i \in \mathbf{R}^m$ to satisfy the equation

$$[\lambda_i I - A \quad B] \begin{bmatrix} v_i \\ u_i \end{bmatrix} = 0 \qquad (5.2\text{-}17)$$

Now, choose the feedback gain K to satisfy

$$K v_i = u_i \qquad (5.2\text{-}18)$$

Using the last two equations, we may write

$$0 = (\lambda_i I - A)v_i + Bu_i \qquad (5.2\text{-}19)$$

$$0 = [\lambda_i I \quad (A - BK)] v_i, \qquad (5.2\text{-}20)$$

so that according to (5.2-8), v_i is assigned as a closed-loop eigenvector for eigenvalue λ_i.

As Cv_i was shown in the preceding subsection to be a direction in the output space $\mathbf{R}^p$ associated with v_i, so u_i is the associated direction in the input space $\mathbf{R}^m$. That is, motions of $u(t)$ in the direction of u_i will cause motions of $x(t)$ in the direction of v_i, resulting in motions of $y(t)$ in the direction of Cv_i.

To complete the picture, suppose that n eigenvalues λ_i and associated eigenvectors v_i are chosen, and that in each case we have found a vector u_i that satisfies (5.2-17). Then we may define K by

$$K[v_1 \quad v_2 \quad \cdots \quad v_n] = [u_1 \quad u_2 \quad \cdots \quad u_n] \qquad (5.2\text{-}21)$$

or by appropriate definition of the matrices V and U,

$$KV = U \tag{5.2-22}$$

Then, for each value of $i = 1, \ldots, n$ (5.2-20) will hold, so that each λ_i will be assigned as a closed-loop pole with associated eigenvector v_i. This is the design technique for eigenstructure assignment using full state feedback. It remains only to discuss a few points.

Design Considerations. Since, by definition, the closed-loop eigenvectors must be linearly independent, it is necessary to select v_i as linearly independent vectors. Then (5.2-22) may be solved for K to give

$$K = UV^{-1} \tag{5.2-23}$$

Another issue is that the closed-loop system and feedback gain must be real and not complex. Thus if a complex closed-loop pole λ_i is selected, it is also necessary to select as a closed-loop pole its complex conjugate λ_i^*. Moreover, if v_i is to be the closed-loop eigenvector associated with a complex pole λ_i, then in order for (5.2-21) to have a real solution K, it is necessary to select v_i^* (i.e., the complex conjugate of v_i) as the eigenvector for λ_i^*.

To see that under these circumstances (5.2-22) indeed has a real solution K, note first that if u_i solves (5.2-19) for a given λ_i and v_i, then u_i^* solves the equation for their complex conjugates. Therefore, if $u_i = u_R + ju_I$ and $v_i = v_R + jv_I$, then to assign the desired eigenstructure K must satisfy

$$K\,[v_R + jv_I \quad v_R - jv_I] = [u_R + ju_I \quad u_R - ju_I] \tag{5.2-24}$$

Postmultiplying both sides of this equation by

$$M = \frac{1}{2}\begin{bmatrix} 1 & -j \\ 1 & j \end{bmatrix},$$

it is seen that this equation is equivalent to the real equation

$$K\,[v_R \quad v_I] = [u_R \quad u_I], \tag{5.2-25}$$

which clearly has as a solution a real gain matrix K. Thus, if v_i is complex, then to obtain a real value for K it is only necessary to use not v_i and v_i^* (respectively u_i and u_i^*) in (5.2-21) but the real and imaginary parts of v_i (respectively u_i).

Finally, we must investigate the conditions for existence of a solution to (5.2-17). It is unfortunately not usually possible to specify independently an arbitrary λ_i and v_i and obtain a solution u_i to this equation. Indeed, assuming that λ_i is not an open-loop pole, we have

$$v_i = -(\lambda_i I - A)^{-1} B u_i \tag{5.2-26}$$

Thus, for the existence of a solution u_i, the desired v_i must be a linear combination of the m columns of the linear operator

$$L_i = (\lambda_i I - A)^{-1} B \qquad (5.2\text{-}27)$$

Since B has full rank m by assumption, the matrix L_i also has rank m. Thus v_i must lie in an m-dimensional subspace of $\mathbf{R}^n$ that depends on the choice of λ_i. This means that we have *m degrees of freedom in selecting the closed-loop eigenvector v_i* once λ_i has been selected.

This last point is the crucial difference between classical SISO design and multivariable eigenstructure assignment. If $m = 1$, which corresponds to the single-input case, then eigenvector v_i has only one degree of freedom once the desired eigenvalue λ_i has been selected; that is, there is no additional freedom to choose the eigenvector. However, in the multi-input case where $m > 1$, we can have additional freedom to specify the internal structure of the closed-loop system by selecting m degrees of freedom of v_i arbitrarily. In the preceding subsection we have seen the importance of this in terms of design performance.

The successive-loop-closure approach of classical control, where only one feedback gain is selected at a time, obscures the extra design freedom arising from multiple inputs. In modern control, where all gains are selected simultaneously, this freedom is clearly revealed.

Design Procedure. The following design procedure for eigenstructure assignment is suggested. For a desired closed-loop pole/vector pair of λ_i and v_i^d, solve the equation

$$\begin{bmatrix} \lambda_i I - A & B \\ d & 0 \end{bmatrix} \begin{bmatrix} v_i \\ u_i \end{bmatrix} = \begin{bmatrix} 0 \\ v_i^d \end{bmatrix} \qquad (5.2\text{-}28)$$

for u_i and the *achievable eigenvector* v_i. Repeat for $i = 1, \ldots, n$ to select n closed-loop poles. If the v_i's are not linearly independent, modify the choices for λ_i and/or v_i^d and repeat. Finally, determine the required state-variable feedback gain K using (5.2-23).

The design matrix D may be chosen for several different design objectives:

1. $D = I$. This is the case where the desired vectors v_i^d are eigenvectors (as in Example 5.2-1).
2. $D = C$. This is the case where the desired vectors are directions in the output space $\mathbf{R}^p$, so that we desire $Cv_i = v_i^d$.
3. If certain components of v_i are of no concern (see Example 5.2-1), the corresponding columns of D should be selected as zero. The remaining columns should be selected as columns from the $d \times d$ identify matrix, with d the number of rows of D. The elements of v_i that they multiply should be as specified by the elements of v_i^d. We illustrate further in Example 5.2-3.

We have seen that (5.2-28) may not have an exact solution v_i, u_i. It is necessary to find a solution so that (1) the first n equations hold exactly (i.e., (5.2-19) must hold exactly), and (2) the second block equation $Dv_i = v_i^d$ holds as closely as possible (then our design objectives are most closely matched). Subroutine LLBQF in the IMSL library (IMSL) allows us to do this. It gives a least-squares solution to the second equation in the sense that $\|Dv_i - v_i^d\|^2$ is minimized over all possible v_i for which there exists a u_i that satisfies (5.2-19) (where $\|w\|$ is the Euclidean norm of vector w).

An interactive design technique is suggested wherein:

1. Given the desired λ_i and v_i^d, (5.2-28) is solved for the pair v_i, u_i meeting the requirements above.
2. The achievable eigenvector v_i is compared with the desired eigenvector and if it is unsatisfactory either v_i^d or λ_i may be modified and step 1 repeated.

Eigenstructure Assignment by Output Feedback

In an aircraft control system, all of the states are not generally available for measurement. Instead, only selected outputs are available for control purposes. It is not difficult to modify the eigenstructure assignment technique so that the admissible controls are of the form

$$u = -Ky, \qquad (5.2\text{-}29)$$

with output $y(t) \in \mathbf{R}^p$ given by (5.2-2) (Srinathkumar, 1978). In this case, we will show that p eigenvalues may easily be assigned, with m degrees of freedom in the choice of the associated eigenvectors.

Matrix Equation for Eigenstructure Assignment. In the case of output feedback the closed-loop system is

$$\dot{x} = (A - BKC)x \qquad (5.2\text{-}30)$$

and it is only necessary to replace (5.2-18) by

$$KCv_i = u_i \qquad (5.2\text{-}31)$$

Then, according to (5.2-17),

$$0 = (\lambda_i I - (A - BKC))v_i, \qquad (5.2\text{-}32)$$

so that v_i is assigned as a closed-loop eigenvector for eigenvalue λ_i. In this case, (5.2-21) for K is replaced by

$$KC[\,v_1 \quad v_2 \quad \cdots \quad v_r\,] = [\,u_1 \quad u_2 \quad \cdots \quad u_r\,], \qquad (5.2\text{-}33)$$

where r is the number of closed-loop eigenvalues selected.

If $r = p$ and the vectors Cv_i are linearly independent, we may define

$$V = [v_1 \quad v_2 \quad \cdots \quad v_r], \qquad U = [u_1 \quad u_2 \quad \cdots \quad u_r] \qquad (5.2\text{-}34)$$

and solve for K using

$$K = U(CV)^{-1} \qquad (5.2\text{-}35)$$

Thus, it is clear what is lost by using incomplete state information for feedback purposes, for we can in general no longer assign n poles arbitrarily.

Extensions. What we have demonstrated is a technique for assigning by output feedback p closed-loop poles, with m degrees of freedom in specifying the components of each associated closed-loop eigenvector (where m is the number of inputs).

If $m \leq p$, so that the number of inputs is greater than or equal to the number of outputs, the technique just presented is suitable. However, if $m > p$, we must use the technique on the "dual" system. That is, (A, B, C) is replaced by $A^T, C^T, B^T)$ and the design is performed to find K^T. In this case we may assign m closed-loop poles, with p degrees of freedom in assigning the associated eigenvectors.

A problem with eigenstructure assignment using output feedback is that it is not possible to tell what happens to the $n - p$ poles that are not assigned. Indeed, some of them may become unstable, even though the original plant was stable. If this occurs, or if some closed-loop poles are too lightly damped, the design should be repeated using different values for λ_i or v_i^d. Generally, it is found that if one does not ask for too much in terms of modifying the original plant behavior, that is, if most of the desired closed-loop poles are not too different from the open-loop poles, instability of the unassigned poles is not a problem (as long as they are open-loop stable).

Srinathkumar shows that it is possible to assign an almost arbitrary set of $\min(n, m + p - 1)$ eigenvalues, but we will not go into details here. Kwon and Youn (1987) show that it may be possible to assign $m + p$ poles in some examples.

The next concept is quite important, so we will illustrate it by an exercise.

Example 5.2-2: Eigenstructure Assignment Using Dynamic Regulator.

We have shown how to select constant feedback gains to assign the closed-loop eigenstructure. However, it is possible to obtain a desired modal structure by using a dynamic compensator. This exercise shows how to design a dynamic compensator for eigenstructure assignment.

Consider the plant

$$\dot{x} = Ax + Bu \qquad (1)$$

$$y = Cx \qquad (2)$$

with the regulator dynamics defined by

$$\dot{z} = Fz + Gy \qquad (3)$$

$$u = Hz + Jy \qquad (4)$$

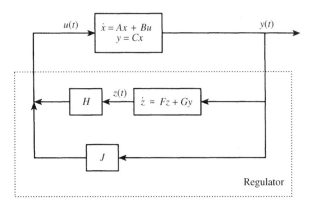

Figure 5.2-1 Plant with regulator.

This corresponds to the situation in Figure 5.2-1. Matrices F, G, H, and J are to be selected to yield a desired closed-loop eigenstructure.

Show that by defining the augmented plant, input, and output matrices

$$\begin{bmatrix} A & 0 \\ 0 & 0 \end{bmatrix}, \quad \begin{bmatrix} B & 0 \\ 0 & I \end{bmatrix}, \quad \begin{bmatrix} C & 0 \\ 0 & I \end{bmatrix} \tag{5}$$

and the gain matrix

$$K = \begin{bmatrix} -J & -H \\ -G & -F \end{bmatrix} \tag{6}$$

the problem of determining F, G, H, and J to yield desired closed-loop poles λ_i and eigenvectors v_i^d may be solved by using the techniques of this section to determine K.

Note: A problem with this approach is that the regulator matrix F cannot be guaranteed stable. An alternative approach to regulator/observer design is given in Andry et al. (1984). ∎

Example 5.2-3: Eigenstructure Design of Longitudinal Pitch Pointing Control. This example is taken from Sobel and Shapiro (1985). A linearized model of the short-period dynamics of an advanced (CCV-type) fighter aircraft is given. These dynamics are augmented by elevator and flaperon actuator dynamics given by the simplified model $20/(s+20)$ so that the state vector is

$$x = \begin{bmatrix} \alpha \\ q \\ \gamma \\ \delta_e \\ \delta_f \end{bmatrix}, \tag{1}$$

where the state components are, respectively, angle of attack, pitch rate, flight-path angle, elevator deflection, and flaperon deflection. The control inputs are elevator and flaperon commands so that

$$u = \begin{bmatrix} \delta_{e_c} \\ \delta_{f_c} \end{bmatrix} \quad (2)$$

The plant and control matrices are

$$A = \begin{bmatrix} -1.341 & 0.9933 & 0 & -0.1689 & -0.2518 \\ 43.223 & -0.8693 & 0 & -17.251 & -1.5766 \\ 1.341 & 0.0067 & 0 & 0.1689 & 0.2518 \\ 0 & 0 & 0 & -20 & 0 \\ 0 & 0 & 0 & 0 & -20 \end{bmatrix}$$

$$B = \begin{bmatrix} 0 & 0 \\ 0 & 0 \\ 0 & 0 \\ 20 & 0 \\ 0 & 20 \end{bmatrix} \quad (3)$$

and the open-loop eigenvalues are

$$\left. \begin{array}{l} \lambda_1 = 5.452 \\ \lambda_2 = -7.662 \end{array} \right\} \text{unstable short-period mode}$$

$$\lambda_3 = 0.0 \quad \text{pitch-attitude mode} \quad (4)$$

$$\lambda_4 = -20 \quad \text{elevator actuator mode}$$

$$\lambda_5 = -20 \quad \text{flaperon actuator mode}$$

The measured output available for control purposes is

$$y = \begin{bmatrix} q \\ n_{zp} \\ \gamma \\ \delta_e \\ \delta_f \end{bmatrix}, \quad (5)$$

where n_{zp} is normal acceleration at the pilot's station. The altitude rate $\dot{h}$ is obtained from the air-data computer and the flight-path angle is then computed using

$$\gamma = \frac{\dot{h}}{TAS} \quad (6)$$

400 MODERN DESIGN TECHNIQUES

with TAS the true airspeed. The control surface deflections are measured using linear variable differential transformers (LVDT). The relation between $y(t)$ and $x(t)$ is given by

$$y = \begin{bmatrix} 0 & 1 & 0 & 0 & 0 \\ 47.76 & -0.268 & 0 & -4.56 & 4.45 \\ 0 & 0 & 1 & 0 & 0 \\ 0 & 0 & 0 & 1 & 0 \\ 0 & 0 & 0 & 0 & 1 \end{bmatrix} x \equiv Cx \quad (7)$$

Since there are five outputs and two control inputs, we may place all the closed-loop poles as well as assign the eigenvectors within two-dimensional subspaces. This roughly corresponds to selecting two components of each eigenvector arbitrarily.

The desired closed-loop short-period poles are chosen to meet military specifications for category A, level-1 flight (Mil. Spec. 1797, 1987) (see Section 4.3). Thus, the desired short-period damping ratio and frequency are 0.8 and 7 rad/s, respectively.

For stability, we specify that the desired closed-loop pitch-attitude mode should decay exponentially with a time constant of 1, so that the pole should be at $s = -1$. The actuator poles should be near -20; however, selecting repeated poles can yield problems with the design algorithm. The desired eigenvalues are thus selected as

$$\begin{aligned} \lambda_1 &= -5.6 + j4.2 \\ \lambda_2 &= -5.6 - j4.2 \end{aligned} \biggr\} \text{ short-period mode}$$

$$\lambda_3 = -1.0 \quad \text{pitch-attitude mode} \quad (8)$$

$$\lambda_4 = -19.0 \quad \text{elevator actuator mode}$$

$$\lambda_5 = -19.5 \quad \text{flaperon actuator mode}$$

In pitch pointing, the control objective is to allow pitch-attitude control while maintaining constant flight-path angle. To achieve this we select the desired closed-loop eigenvectors to decouple pitch-rate and flight-path angle. Thus, an attitude command should be prevented from causing a significant flight-path change. The desired closed-loop eigenvectors are shown in Table 5.2-1, where x denotes elements of no concern to us. Recall that α and q are associated with the short-period mode.

We now discuss the design procedure and the selection of the D matrix in the design equation (5.2-28). We must determine the vectors v_i and u_i for use in (5.2-33) to solve for the feedback gain matrix K. To accomplish this, first consider the desired structure of the short-period mode. According to Table 5.2-1, the required short-period eigenvectors have two "don't care" entries. Define v_1^d in terms of the required eigenvector as

ASSIGNMENT OF CLOSED-LOOP DYNAMICS

$$v_1^d = \begin{bmatrix} 1 & 0 & 0 & 0 & 0 \\ 0 & 1 & 0 & 0 & 0 \\ 0 & 0 & 1 & 0 & 0 \end{bmatrix} \left(\begin{bmatrix} 1 \\ -1 \\ 0 \\ x \\ x \end{bmatrix} + j \begin{bmatrix} -1 \\ 1 \\ 0 \\ x \\ x \end{bmatrix} \right) = \begin{bmatrix} 1 \\ -1 \\ 0 \end{bmatrix} + j \begin{bmatrix} -1 \\ 1 \\ 0 \end{bmatrix} \quad (9)$$

to be the desired vector associated with $\lambda_1 = -5.6 + j4.2$, and select D as the 3×6 coefficient matrix in (9). Then (5.2-28) may be solved for v_1 and u_1. Then the vectors associated with $\lambda_2 = \lambda_1^*$ are $v_2 = v_1^*$, $u_2 = u_1^*$. The achievable eigenvectors v_1 and v_2 associated with the short-period mode are shown in Table 5.2-1.

To determine whether the results to this point are satisfactory, the achievable eigenvectors v_1 and v_2 are compared with the desired eigenvectors. They are satisfactory since there is no coupling to state component 3. Note that although we attempted to select three components of the eigenvectors knowing that there are only two degrees of freedom in this selection, we have nevertheless been fortunate in attaining our design objectives. Had we not been so lucky, it would have been necessary to try different desired eigenvectors, or else slightly different values for the closed-loop poles.

Moving on to the desired structure of λ_3, examine Table 5.2-1 to define

$$v_3^d = \begin{bmatrix} 0 & 1 & 0 & 0 & 0 \\ 0 & 0 & 1 & 0 & 0 \end{bmatrix} \begin{bmatrix} x \\ 0 \\ 1 \\ x \\ x \end{bmatrix} = \begin{bmatrix} 0 \\ 1 \end{bmatrix} \quad (10)$$

to be the desired vector associated with $\lambda_3 = -1.0$ and select D as the 2×6 coefficient matrix in (10). Then (5.2-28) may be solved for v_3 and u_3. The result is the achievable eigenvector v_3 shown in Table 5.2-1; again, it is suitable.

TABLE 5.2-1: Desired and Achievable Eigenvectors

Desired Eigenvectors						Achievable Eigenvectors				
$\begin{bmatrix} 1 \\ -1 \\ 0 \\ x \\ x \end{bmatrix} + j \begin{bmatrix} -1 \\ 1 \\ 0 \\ x \\ x \end{bmatrix}$	$\begin{bmatrix} x \\ 0 \\ 1 \\ x \\ x \end{bmatrix}$	$\begin{bmatrix} x \\ x \\ x \\ 1 \\ x \end{bmatrix}$	$\begin{bmatrix} x \\ x \\ x \\ x \\ 1 \end{bmatrix}$	$\begin{matrix} \alpha \\ q \\ \gamma \\ \delta_e \\ \delta_f \end{matrix}$	$\begin{bmatrix} -0.93 \\ 1 \\ 0 \\ -5.13 \\ 8.36 \end{bmatrix} + j \begin{bmatrix} 1 \\ -9.5 \\ 0 \\ 0.129 \\ -5.16 \end{bmatrix}$	$\begin{bmatrix} -1 \\ 0 \\ 1 \\ -2.8 \\ 3.23 \end{bmatrix}$	$\begin{bmatrix} -0.051 \\ 1.07 \\ -0.006 \\ 1 \\ 0 \end{bmatrix}$	$\begin{bmatrix} 0.01 \\ 0.06 \\ -0.014 \\ 0 \\ 1 \end{bmatrix}$
α/q short period	γ	δ_e	δ_f			α/q short period	γ	δ_e	δ_f

402 MODERN DESIGN TECHNIQUES

To design for the desired structure of λ_4, examine Table 5.2-1 to define

$$v_4^d = [0 \ 0 \ 0 \ 1 \ 0] \begin{bmatrix} x \\ x \\ x \\ 1 \\ x \end{bmatrix} = 1 \tag{11}$$

to be the desired vector associated with $\lambda_4 = -19.0$ and select D as the 1×6 coefficient matrix in (11). Then (5.2-28) may be solved for v_4 and u_4. The results are in the table. Similar procedures apply for λ_5.

Now that all the requisite vectors v_i and u_i, $i = 1, 2, 3, 4, 5$, have been computed, they are used, along with the C matrix from (7), to solve for the feedback gain using (5.2-33). The result is

$$K = \begin{bmatrix} -0.931 & -0.149 & -3.25 & -0.153 & 0.747 \\ 0.954 & 0.210 & 6.10 & 0.537 & -1.04 \end{bmatrix} \tag{12}$$

To check the design, a computer simulation was performed. The closed-loop system was excited with an initial condition of 0.2 rad in angle of attack. Note from

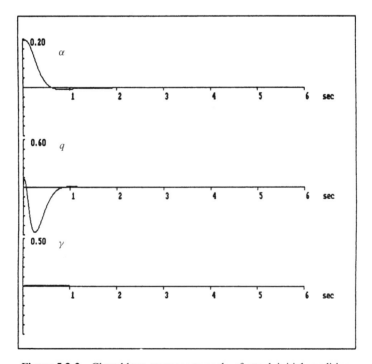

Figure 5.2-2 Closed loop response to angle-of-attack initial condition.

Figure 5.2-2 that this excited the short-period mode but had negligible effect on the flight-path angle. ∎

5.3 LINEAR QUADRATIC REGULATOR WITH OUTPUT FEEDBACK

Our objective in this section is to show how to use modern techniques to design stability augmentation systems (SAS) and autopilots. This is accomplished by regulating certain states of the aircraft to zero while obtaining desirable closed-loop response characteristics. It involves the problem of stabilizing the aircraft by placing the closed-loop poles at desirable locations.

Using classical control theory, we were forced to take a one-loop-at-a-time approach to designing multivariable SAS and autopilots. In this section we will select a performance criterion that reflects our concern with closed-loop stability and good time responses, and then derive matrix equations that may be solved for *all the control gains simultaneously.* These matrix equations are solved using digital computer programs (see Appendix B). This approach thus closes all the loops simultaneously and results in a simplified design strategy for MIMO systems or SISO systems with multiple feedback loops.

Once the performance criterion has been selected, the control gains are explicitly computed by marix design equations, and closed-loop stability will generally be guaranteed. This means that *the engineering judgment in modern control enters into the selection of the performance criterion.* Different criteria will result in different closed-loop time responses and robustness properties.

We assume the plant is given by the linear time-invariant state-variable model

$$\dot{x} = Ax + Bu \qquad (5.3\text{-}1)$$

$$y = Cx, \qquad (5.3\text{-}2)$$

with $x(t) \in \mathbf{R}^n$ the state, $u(t) \in \mathbf{R}^m$ the control input, and $y(t) \in \mathbf{R}^p$ the measured output. The controls will be output feedbacks of the form

$$u = -Ky, \qquad (5.3\text{-}3)$$

where K is an $m \times p$ matrix of constant feedback coefficients to be determined by the design procedure. Since the regulator problem only involves stabilizing the aircraft and inducing good closed-loop time responses, $u(t)$ will be taken as a pure feedback with no auxiliary input (see Section 5.2).

As we will see in Section 5.4, output feedback will allow us to design aircraft controllers of any desired structure. This is one reason for preferring it over full state feedback.

In the regulator problem, we are interested in obtaining good time responses as well as in the stability of the closed-loop system. Therefore, we will select a performance criterion *in the time domain.* Let us now present this criterion.

Quadratic Performance Index

The objective of state regulation for the aircraft is to drive any initial condition error to zero, thus guaranteeing stability. This may be achieved by selecting the control input $u(t)$ to minimize a quadratic *cost* or *performance index* (*PI*) of the type

$$J = \tfrac{1}{2} \int_0^\infty \left(x^\mathrm{T} Q x + u^\mathrm{T} R u \right) dt, \tag{5.3-4}$$

where Q and R are symmetric positive semidefinite *weighting matrices*. Positive semidefiniteness of a square matrix M (denoted $M \geq 0$) is equivalent to all its eigenvalues being non-negative, and also to the requirement that the quadratic form $x^\mathrm{T} M x$ be non-negative for all vectors x. Therefore, the definiteness assumptions on Q and R guarantee that J is non-negative and lead to a sensible minimization problem. This quadratic PI is a vector version of an integral-squared PI of the sort used in classical control (D'Azzo and Houpis, 1988).

To understand the motivation for the choice of (5.3-4), consider the following. If the square root $\sqrt{M}$ of a positive semidefinite matrix M is defined by

$$M = \sqrt{M}^\mathrm{T} \sqrt{M}, \tag{5.3-5}$$

we may write (5.3-4) as

$$J = \tfrac{1}{2} \int_0^\infty \left(\left\| \sqrt{Q} x \right\|^2 + \left\| \sqrt{R} u \right\|^2 \right) dt, \tag{5.3-6}$$

with $\|w\|$ the Euclidean norm of a vector w (i.e., $\|w\|^2 = w^\mathrm{T} w$). If we are able to select the control input $u(t)$ so that J takes on a minimum finite value, certainly the integrand must become zero for large time. This means that both the linear combination $\sqrt{Q} x(t)$ of the states and the linear combination $\sqrt{R} u(t)$ of the controls must go to zero. In different designs we may select Q and R for different performance requirements, corresponding to specified functions of the state and input. In particular, if Q and R are both chosen nonsingular, the entire state vector $x(t)$ and all the controls $u(t)$ will go to zero with time if J has a finite value.

Since a bounded value for J will guarantee that $\sqrt{Q} x(t)$ and $\sqrt{R} u(t)$ go to zero with time, this formulation for the PI is appropriate for the regulator problem, as any initial condition errors will be driven to zero.

If the state vector $x(t)$ consists of capacitor voltages $v(t)$ and inductor currents $i(t)$, then $\|x\|^2$ will contain terms like $v^2(t)$ and $i^2(t)$. Similarly, if velocity $s(t)$ is a state component, $\|x\|^2$ will contain terms like $s^2(t)$. Therefore, the minimization of the PI (5.3.4) is a generalized *minimum energy* problem. We are concerned with minimizing the energy in the states without using too much control energy.

The relative magnitudes of Q and R may be selected to trade off requirements on the smallness of the state against requirements on the smallness of the input. For instance, a larger control-weighting matrix R will make it necessary for $u(t)$ to be smaller to ensure that $\sqrt{R} u(t)$ is near zero. We say that a larger R *penalizes* the controls more, so that they will be smaller in norm relative to the state vector. On the other hand, to make $x(t)$ go to zero more quickly with time, we may select a larger Q.

As a final remark on the PI, we will see that the positions of the closed-loop poles depend on the choices for the weighting matrices Q and R. That is, Q and R may be chosen to yield good time responses in the closed-loop system.

Let us now derive matrix design equations that may be used to solve for the control gain K that minimizes the PI. The result will be the design equations in Table 5.3-1. Software to solve these equations for K is described in Appendix B.

Solution of the LQR Problem

The LQR problem with output feedback is the following. Given the linear system (5.3-1), (5.3-2), find the feedback coefficient matrix K in the control input (5.3-3) that minimizes the value of the quadratic PI (5.3-4). In contrast with most of the classical control techniques given in earlier chapters, this is a *time-domain* design technique.

By substituting the control (5.3-3) into (5.3-1) the closed-loop system equations are found to be

$$\dot{x} = (A - BKC)x \equiv A_c x \qquad (5.3\text{-}7)$$

The PI may be expressed in terms of K as

$$J = \tfrac{1}{2} \int_0^\infty x^T \left(Q + C^T K^T R K C \right) x \, dt \qquad (5.3\text{-}8)$$

The design problem is now to select the gain K so that J is minimized subject to the dynamical constraint (5.3-7).

This *dynamical* optimization problem may be converted into an equivalent *static* one that is easier to solve as follows. Suppose that we can find a constant, symmetric, positive-semidefinite matrix P so that

$$\frac{d}{dt}\left(x^T P x\right) = -x^T \left(Q + C^T K^T R K C \right) x \qquad (5.3\text{-}9)$$

Then J may be written as

$$J = \tfrac{1}{2} x^T(0) P x(0) - \tfrac{1}{2} \lim_{t \to \infty} x^T(t) P x(t) \qquad (5.3\text{-}10)$$

Assuming that the closed-loop system is asymptotically stable so that $x(t)$ vanishes with time, this becomes

$$J = \tfrac{1}{2} x^T(0) P x(0) \qquad (5.3\text{-}11)$$

If P satisfies (5.3-9), we may use (5.3-7) to see that

$$-x^T \left(Q + C^T K^T R K C\right) x = \frac{d}{dt}\left(x^T P x\right) = \dot{x}^T P x + x^T P \dot{x} \qquad (5.3\text{-}12)$$
$$= x^T \left(A_c^T P + P A_c\right) x$$

Since this must hold for all initial conditions, and hence for all state trajectories $x(t)$, we may write

$$g \equiv A_c^T P + P A_c + C^T K^T R K C + Q = 0 \qquad (5.3\text{-}13)$$

If K and Q are given and P is to be solved for, this is called a *Lyapunov equation*. (A Lyapunov equation is a symmetric linear matrix equation. Note that the equation does not change if its transpose is taken.)

In summary, for any fixed feedback matrix K if there exists a constant, symmetric, positive-semidefinite matrix P that satisfies (5.3-13), and if the closed-loop system is stable, the cost J is given in terms of P by (5.3-11). This is an important result in that the $n \times n$ auxiliary matrix P is independent of the state. Given a feedback matrix K, P may be computed from the Lyapunov equation (5.3-13). Then only the initial condition $x(0)$ is required to compute the closed-loop cost under the influence of the feedback control $u = -Ky$ *before we actually apply it.*

It is now necessary to use this result to compute the gain K that minimizes the PI. By using the trace identity

$$\text{tr}(AB) = \text{tr}(BA) \qquad (5.3\text{-}14)$$

for any compatibly dimensioned matrices A and B (with the trace of a matrix the sum of its diagonal elements), we may write (5.3-11) as

$$J = \tfrac{1}{2}\text{tr}(PX), \qquad (5.3\text{-}15)$$

where the $n \times n$ symmetric matrix X is defined by

$$X \equiv x(0)x^T(0) \qquad (5.3\text{-}16)$$

It is now clear that the problem of selecting K to minimize (5.3-8) subject to the dynamical constraint (5.3-7) on the states is equivalent to the *algebraic* problem of selecting K to minimize (5.3-15) subject to the constraint (5.3-13) on the auxiliary matrix P.

To solve this modified problem, we use the Lagrange multiplier approach (Lewis, 1986) to modify the problem yet again. Thus adjoin the constraint to the PI by defining the Hamiltonian

$$\mathcal{H} = \text{tr}(PX) + \text{tr}(gS) \qquad (5.3\text{-}17)$$

with S a symmetric $n \times n$ matrix of Lagrange multipliers that still needs to be determined. Then our constrained optimization problem is equivalent to the simpler problem of minimizing (5.3-17) without constraints. To accomplish this we need only set the partial derivatives of $\mathcal{H}$ with respect to all the independent variables P, S, and K equal to zero. Using the facts that for any compatibly dimensioned matrices A, B, and C and any scalar y,

$$\frac{\partial}{\partial B} \text{tr}(ABC) = A^T C^T \tag{5.3-18}$$

and

$$\frac{\partial y}{\partial B^T} = \left[\frac{\partial y}{\partial B}\right]^T, \tag{5.3-19}$$

the necessary conditions for the solution of the LQR problem with output feedback are given by

$$0 = \frac{\partial \mathcal{H}}{\partial S} = g = A_c^T P + P A_c + C^T K^T R K C + Q \tag{5.3-20}$$

$$0 = \frac{\partial \mathcal{H}}{\partial P} = A_c S + S A_c^T + X \tag{5.3-21}$$

$$0 = \frac{1}{2}\frac{\partial \mathcal{H}}{\partial K} = RKCSC^T - B^T P S C^T \tag{5.3-22}$$

The first two of these are Lyapunov equations and the third is an equation for the gain K. If R is positive definite (i.e., all eigenvalues greater than zero, which implies nonsingularity; denoted $R > 0$) and CSC^T is nonsingular, then (5.3-22) may be solved for K to obtain

$$K = R^{-1} B^T P S C^T \left(C S C^T\right)^{-1} \tag{5.3-23}$$

To obtain the output feedback gain K minimizing the PI (5.3-4), we need to solve the three coupled equations (5.3-20), (5.3-21), and (5.3-23). This situation is quite strange, for to find K we must determine along the way the values of two auxiliary and apparently unnecessary $n \times n$ matrices, P and S. These auxiliary quantities may, however, not be as unnecessary as it appears, for note that the optimal cost may be determined directly from P and the initial state by using (5.3-11).

The Initial Condition Problem. Unfortunately, the dependence of X in (5.3-16) on the initial state $x(0)$ is undesirable, since it makes the optimal gain dependent on the initial state through Equation (5.3-21). In many applications $x(0)$ may not be known. This dependence is typical of output-feedback design. We will see at the end of this chapter that in the case of state feedback it does not occur. Meanwhile, it is usual (Levine and Athans, 1970) to sidestep this problem by minimizing not the PI (5.3-4) but its *expected value*, that is, $E\{J\}$. Then (5.3-11) and (5.3-16) are replaced by

$$E\{J\} = \tfrac{1}{2}E\left\{x^{\mathrm{T}}(0)Px(0)\right\} = \tfrac{1}{2}\mathrm{tr}(PX), \tag{5.3-24}$$

where the symmetric $n \times n$ matrix

$$X \equiv E\left\{x(0)x^{\mathrm{T}}(0)\right\} \tag{5.3-25}$$

is the initial autocorrelation of the state. It is usual to assume that nothing is known of $x(0)$ except that it is uniformly distributed on a surface described by X. That is, we assume that the actual initial state is unknown, but that it is nonzero with a certain expected Euclidean norm. For instance, if the initial states are assumed to be uniformly distributed on the unit sphere, then $X = I$, the identity. This is a sensible assumption for the regulator problem, where we are trying to drive arbitrary nonzero initial states to zero.

The design equations for the LQR with output feedback are collected in Table 5.3-1 for convenient reference. We will now discuss their solution for K.

Determining the Optimal Feedback Gain

The importance of this modern LQ approach to control design is that the matrix equations in Table 5.3-1 are used to solve for all the $m \times p$ elements of K at once. This

TABLE 5.3-1: LQR with Output Feedback

System Model

$\dot{x} = Ax + Bu$

$y = Cx$

Control

$u = -Ky$

Performance Index

$J = \tfrac{1}{2}E\left[\int_0^\infty (x^{\mathrm{T}}Qx + u^{\mathrm{T}}Ru)dt\right]$

with

$Q \geq 0, \quad R > 0$

Optimal Gain Design Equations

$$0 = A_c^{\mathrm{T}}P + PA_c + C^{\mathrm{T}}K^{\mathrm{T}}RKC + Q \tag{5.3-26}$$

$$0 = A_c S + S A_c^{\mathrm{T}} + X \tag{5.3-27}$$

$$K = R^{-1}B^{\mathrm{T}}PSC^{\mathrm{T}}(CSC^{\mathrm{T}})^{-1} \tag{5.3-28}$$

where

$A_c = A - BKC, \quad X = E\left\{x(0)x^{\mathrm{T}}(0)\right\}$

Optimal Cost

$$J = \tfrac{1}{2}\mathrm{tr}(PX) \tag{5.3-29}$$

corresponds to closing all the feedback loops simultaneously. Moreover, as long as certain reasonable conditions (to be discussed) on the plant and PI weighting matrices hold, *the closed loop system is generally guaranteed to be stable*. In view of the trial-and-error successive-loop-closure approach used in stabilizing multivariable systems using classical approaches, this is quite important.

The equations for P, S, and K are coupled nonlinear matrix equations in three unknowns. It is important to discuss some aspects of their solution for the optimal feedback gain matrix K.

Numerical Solution Techniques. There are three basic numerical techniques for determining the optimal output-feedback gain K. First, we may use a numerical optimization routine such as the simplex algorithm in Nelder and Mead (1964) and Press et al. (1986). This algorithm would use only (5.3-26) and (5.3-29). For a given value of K, it would solve the Lyapunov equation for P and then use P in the second equation to determine $E\{J\}$. Based on this, it would vary the elements of K to minimize $E\{J\}$. The Lyapunov equation may be solved using, for instance, subroutine ATXPXA (Bartels and Stewart, 1972). See also the NASA controls design package ORACLS (Armstrong, 1980).

A second approach for computing K is to use a gradient-based routine such as Davidon–Fletcher–Powell (Press et al., 1986). This routine would use all of the design equations in Table 5.3-1. For a given value of K, it would solve the two Lyapunov equations to find the auxiliary matrices P and S. Then it would use the third design equation in the form (5.3-22). Note that if P satisfies the first Lyapunov equation, then $g = 0$ so that [see (5.3-17)] $E\{J\} = \frac{1}{2}E\{H\}$ and $\partial E\{J\}/\partial K = \frac{1}{2}\partial E\{\mathcal{H}\}/\partial K$. Thus the third design equation gives the gradient of $E\{J\}$ with respect to K, which would be used by the routine to update the value of K.

Finally, an iterative solution algorithm was presented in Moerder and Calise (1985). It is given in Table 5.3-2. It was shown in Moerder and Calise (1985) that the algorithm converges to a local minimum for J if the following conditions hold.

Conditions for Convergence of the LQ Solution Algorithm:

1. There exists a gain K such that A_c is stable. If this is true, we call the system (5.3-1)/(5.3-2) *output stabilizable*.
2. The output matrix C has full row rank p.
3. Control weighting matrix R is positive definite. This means that all the control inputs should be weighted in the PI.
4. Q is positive semidefinite and $(\sqrt{Q}, A)$ is *detectable*. That is, the observability matrix polynomial

$$O(s) \equiv \begin{bmatrix} sI - A \\ -\sqrt{Q} \end{bmatrix} \qquad (5.3\text{-}30)$$

has full rank n for all values of the complex variable s not contained in the left-half plane (Kailath, 1980).

TABLE 5.3-2: Optimal Output Feedback Solution Algorithm

1. Initialize:

    Set $k = 0$.

    Determine a gain K_0 so that $A - BK_0C$ is asymptotically stable.

2. kth iteration:

    Set $A_k = A - BK_kC$.

    Solve for P_k and S_k in

    $$0 = A_k^T P_k + P_k A_k + C^T K_k^T R K_k C + Q$$
    $$0 = A_k S_k + S_k A_k^T + X$$

    Set $J_k = \frac{1}{2}\text{tr}(P_k X)$.

    Evaluate the gain update direction

    $$\Delta K = R^{-1} B^T PSC^T (CSC^T)^{-1} - K_k$$

    Update the gain by

    $$K_{k+1} = K_k + \alpha \Delta K$$

    where α is chosen so that

    $A - BK_{k+1}C$ is asymptotically stable

    $$J_{k+1} \equiv \tfrac{1}{2}\text{tr}(P_{k+1} X) \leq J_k$$

    If J_{k+1} and J_k are close enough to each other, go to 3.

    Otherwise, set $k = k + 1$ and go to 2.

3. Terminate:

    Set $K = K_{k+1}$, $J = J_{k+1}$.

    Stop.

If these conditions hold, the algorithm finds an output-feedback gain that stabilizes the plant and minimizes the PI. The detectability condition means that any unstable system modes must be observable in the PI. Then if the PI is bounded, which it is if the optimization algorithm is successful, signals associated with the unstable modes must go to zero as t becomes large, that is, they are stabilized in the closed-loop system.

Initial Stabilizing Gain. Since all three algorithms for solving the matrix equations in Table 5.3-1 for K are iterative in nature, a basic issue for all of them is the selection of an initial stabilizing output-feedback gain K_0. That is, to start the algorithms, it is necessary to provide a K_0 such that $(A - BK_0C)$ is stable. See, for instance, Table 5.3-2.

One technique for finding such a gain is given in Broussard and Halyo (1983). Another possibility is to use the eigenstructure assignment techniques of the preceding section to determine an initial gain for the LQ solution algorithm. We could even select a stabilizing gain using the classical techniques of Chapter 4 and then use modern design techniques to tune the control gains for optimal performance.

A quite convenient technique for finding an initial stabilizing gain K_0 is discussed in Section 5.5. This involves finding a full $m \times n$ state-variable feedback matrix and then zeroing the entries that are not needed in the $m \times p$ output-feedback matrix for the given measured outputs. Note that there are many techniques for finding a full state feedback that stabilizes a system given A and B (see Section 5.7 and Lewis, 1986).

Iterative Design. Software that solves for the optimal output-feedback gain K is described in Appendix B. Given good software, design using the LQ approach is straightforward. A design procedure would involve selecting the *design parameters Q and R*, determining the optimal gain K, and simulating the closed-loop response and frequency-domain characteristics. If the results are not suitable, different matrices Q and R are chosen and the design is repeated. Good software makes a design iteration take only a few minutes.

This approach introduces the notion of *tuning the design parameters Q and R for good performance.* In the next two sections we will present sensible techniques for obtaining suitable PI weighting matrices Q and R that do not depend on individually selecting all of their entries.

Example 5.3-1 will illustrate these notions.

Selection of the PI Weighting Matrices

Once the PI weighting matrices Q and R have been selected, the determination of the optimal feedback gain K is a formal procedure relying on the solution of nonlinear coupled matrix equations. Therefore, the engineering judgment in modern LQ design appears in the selection of Q and R. There are some guidelines for this which we will now discuss.

Observability in the Choice of Q. For stabilizing solutions to the output-feedback problem, it is necessary for $(\sqrt{Q}, A)$ to be detectable. The detectability condition basically means that Q should be chosen so that all unstable states are weighted in the PI. Then, if J is bounded so that $\sqrt{Q}x(t)$ vanishes for large t, the open-loop unstable states will be forced to zero through the action of the control. This means exactly that the unstable poles must have been stabilized by the feedback control gain.

A stronger condition than detectability is *observability*, which amounts to the full rank of $O(s)$ for all values of s. Observability is easier to check than detectability since it is equivalent to the full rank n of the *observability matrix*

$$O \equiv \begin{bmatrix} \sqrt{Q} \\ \sqrt{Q}A \\ \vdots \\ \sqrt{Q}A^{n-1} \end{bmatrix}, \qquad (5.3\text{-}31)$$

which is a constant matrix and so easier to deal with than $O(s)$. In fact, O has full rank n if and only if the observability gramian $O^T O$ is nonsingular. Since the gramian is an $n \times n$ matrix, its determinant is easily examined using available software (e.g., singular-value decomposition/condition number [IMSL]). The observability of $(\sqrt{Q}, A)$ means basically that *all* states are weighted in the PI.

From a numerical point of view, if $(\sqrt{Q}, A)$ is observable, a positive definite solution P to (5.3-26) results; otherwise, P may be singular. Since P helps determine K through (5.3-28), it is found that if P is singular, it may result in some zero-gain elements in K. That is, if $(\sqrt{Q}, A)$ is not observable, the LQ algorithm can refuse to close some of the feedback loops.

This observability condition amounts to a restriction on the selection of Q, and is a drawback of modern control (see Example 5.3-1). In Section 5.5 we will show how to avoid this condition by using a modified PI.

The Structure of Q. The choice of Q can be confronted more easily by considering the performance objectives of the LQR. Suppose that a *performance output*

$$z = Hx \qquad (5.3\text{-}32)$$

is required to be small in the closed-loop system. For instance, in an aircraft lateral regulator it is desired for the sideslip angle, yaw rate, roll angle, and roll rate to be small (see Example 5.3-1). Therefore, we might select $z = [\beta \ r \ \phi \ p]^T$. Once $z(t)$ has been chosen, the performance output matrix H may be formally written down.

The signal $z(t)$ may be made small by LQR design by selecting the PI

$$J = \tfrac{1}{2} \int_0^\infty \left(z^T z + u^T R u \right) dt, \qquad (5.3\text{-}33)$$

which amounts to using the PI in Table 5.3-1 with $Q = H^T H$, so that Q may be computed from H. That is, by weighting *performance outputs* in the PI, Q is directly given.

Maximum Desired Values of $z(t)$ and $u(t)$. A convenient guideline for selecting Q and R is given in Bryson and Ho (1975). Suppose that the performance output (5.3-32) has been defined so that H is given. Consider the PI

$$J = \tfrac{1}{2} \int_0^\infty \left(z^T \bar{Q} z + u^T R u \right) dt \qquad (5.3\text{-}34)$$

Then, in Table 5.3-1 we have $Q = H^T \bar{Q} H$. To select $\bar{Q}$ and R, one might proceed as follows, using the *maximum allowable deviations* in $z(t)$ and $u(t)$.

Define the maximum allowable deviation in component $z_i(t)$ of $z(t)$ as z_{iM} and the maximum allowable deviation in component $u_i(t)$ of the control input $u(t)$ as u_{iM}. Then $\bar{Q}$ and R may be selected as $\bar{Q} = \text{diag}\{q_i\}$, $R = \text{diag}\{r_i\}$, with

$$q_i = \frac{1}{z_{iM}^2}, \qquad r_i = \frac{1}{r_{iM}^2} \tag{5.3-35}$$

The rationale for this choice is easy to understand. For instance, as the allowed limits z_{iM} on $z_i(t)$ decrease, the weighting in the PI placed on $z_i(t)$ increases, which requires smaller excursions in $z_i(t)$ in the closed-loop system.

Implicit Model Following. The implicit model-following design technique in Section 5.6 shows how to select Q and R so that the closed-loop system behaves like a prescribed ideal model. The ideal model may be selected according to flying-qualities requirements (see Section 4.3). It should be selected so that its poles and zeros correspond to the desired closed-loop time-response characteristics.

Asymptotic Properties of the LQR. Consider the PI

$$J = \tfrac{1}{2} \int_0^\infty \left(x^T Q x + \rho u^T R u \right) dt, \tag{5.3-36}$$

where ρ is a scalar design parameter. There are some quite nice results that describe the asymptotic performance of the LQR as ρ becomes small and as ρ becomes large (Kwakernaak and Sivan, 1972; Harvey and Stein, 1978; Grimble and Johnson, 1988).

These results detail the asymptotic closed-loop eigenstructure of the LQR and are of some assistance in selecting Q and R. Unfortunately, they are only well developed for the case of full state-variable feedback, where $C = I$ and all the states are allowed for feedback. Thus they are appropriate in connection with the discussion in Section 5.7.

Example 5.3-1: LQR Design for F-16 Lateral Regulator. In Chapter 4 we designed a roll damper/yaw damper for a low-speed flight condition of the F-16. Successive loop closures were used to perform the design using the root-locus approach. In this example we should like to demonstrate the power of the LQ design equations in Table 5.3-1 by designing a lateral regulator.

In our approach we will select the design parameters Q and R in the table and then use the design equations there to close all the feedback loops simultaneously by computing K. The objective is to design a closed-loop controller to provide for the function of a lateral SAS as well as the closure of the roll-attitude loop. This objective involves the design of two feedback channels with multiple loops, but it is straightforward to deal with using modern control techniques. The simplicity of MIMO design using the LQR will be evident.

a. Aircraft State Equations. We used the F-16 linearized lateral dynamics at the nominal flight condition in Table 3.6-3 ($V_T = 502$ ft/s, 300 psf dynamic pressure, cg at 0.35 $\bar{c}$), retaining the lateral states sideslip, β, bank angle, ϕ, roll rate, p, and yaw rate, r. Additional states δ_a and δ_r are introduced by the aileron and rudder actuators

$$\delta_a = \frac{20.2}{s + 20.2} u_a, \qquad \delta_r = \frac{20.2}{s + 20.2} u_r \qquad (1)$$

A washout filter

$$r_w = \frac{s}{s + 1} r \qquad (2)$$

is used, with r the yaw rate and r_w the washed-out yaw rate. The washout filter state is denoted x_w. Thus the entire state vector is

$$x = \begin{bmatrix} \beta & \phi & p & r & \delta_a & \delta_r & x_w \end{bmatrix}^T \qquad (3)$$

The full-state-variable model of the aircraft plus actuators, washout filter, and control dynamics is of the form

$$\dot{x} = Ax + Bu, \qquad (4)$$

with

$$A = \begin{bmatrix} -0.3220 & 0.0640 & 0.0364 & -0.9917 & 0.0003 & 0.0008 & 0 \\ 0 & 0 & 1 & 0.0037 & 0 & 0 & 0 \\ -30.6492 & 0 & -3.6784 & 0.6646 & -0.7333 & 0.1315 & 0 \\ 8.5396 & 0 & -0.0254 & -0.4764 & -0.0319 & -0.0620 & 0 \\ 0 & 0 & 0 & 0 & -20.2 & 0 & 0 \\ 0 & 0 & 0 & 0 & 0 & -20.2 & 0 \\ 0 & 0 & 0 & 57.2958 & 0 & 0 & -1 \end{bmatrix}$$

$$B = \begin{bmatrix} 0 & 0 \\ 0 & 0 \\ 0 & 0 \\ 0 & 0 \\ 20.2 & 0 \\ 0 & 20.2 \\ 0 & 0 \end{bmatrix} \qquad (5)$$

The control inputs are the rudder and aileron servo inputs so that

$$u = \begin{bmatrix} u_a \\ u_r \end{bmatrix} \qquad (6)$$

and the output is

$$y = \begin{bmatrix} r_w \\ p \\ \beta \\ \phi \end{bmatrix} \qquad (7)$$

Thus, $y = Cx$ with

$$C = \begin{bmatrix} 0 & 0 & 0 & 57.2958 & 0 & 0 & -1 \\ 0 & 0 & 57.2958 & 0 & 0 & 0 & 0 \\ 57.2958 & 0 & 0 & 0 & 0 & 0 & 0 \\ 0 & 57.2958 & 0 & 0 & 0 & 0 & 0 \end{bmatrix} \qquad (8)$$

The factor of 57.2958 converts radians to degrees. The feedback control will be output feedback of the form $u = Ky$, so that K is a 2×4 matrix. That is, we will select eight feedback gains.

For this system the open-loop dutch roll mode has poles at $-0.4425 \pm j3.063$ and so has insufficient damping. The spiral mode has a pole at -0.01631.

b. LQR Output Feedback Design. For the computation of the feedback gain K, it is necessary to select PI weighting matrices Q and R in Table 5.3-1. Then the software described in Appendix B is used to compute the optimal gain K using the design equations in the table. Out philosophy for selecting Q and R follows.

First, let us discuss the choice of Q. It is desired to obtain good stability of the dutch roll mode, so that β^2 and r^2 should be weighted in the PI by factors of q_{dr}. To obtain stability of the roll mode, which in closed-loop will consist primarily of p and ϕ, we may weight p^2 and ϕ^2 in the PI by factors of q_r. We do not care about δ_a and δ_r, so it is not necessary to weight them in the PI; the control weighting matrix R will prevent unreasonably large control inputs. Thus, so far we have

$$Q = \text{diag}\{q_{dr}, q_r, q_r, q_{dr}, 0, 0, 0\} \qquad (9)$$

We do not care directly about x_w; however, it is necessary to weight it in the PI. This is because omitting it would cause problems with the observability condition. A square root of Q in (9) is

$$\sqrt{Q} = \begin{bmatrix} \sqrt{q_{dr}} & \sqrt{q_r} & \sqrt{q_r} & \sqrt{q_{dr}} & 0 & 0 & 0 \end{bmatrix} \qquad (10)$$

Consequently, the observability matrix (5.3-31) has a right-hand column of zero; hence the system is unobservable. This may be noted in simpler fashion by examining the A matrix in (5), where the seventh state x_w is seen to have no influence on the states that are weighted in (9). To correct this potential problem, we chose

$$Q = \text{diag}\{q_{dr}, q_r, q_r, q_{dr}, 0, 0, 1\} \qquad (11)$$

As far as the R matrix goes, it is generally satisfactory to select it as

$$R = \rho I, \tag{12}$$

with I the identity matrix and ρ a scalar design parameter.

Now the design equations in Table 5.3-1 were solved using the software described in Appendix B for several choices of ρ, q_{dr}, q_r. After a few trials, we obtained a good result using $\rho = 0.1, q_{dr} = 50, q_r = 100$. For this selection the optimal feedback gain was

$$K = \begin{bmatrix} -0.56 & -0.44 & 0.11 & -0.35 \\ -1.19 & -0.21 & -0.44 & 0.26 \end{bmatrix} \tag{13}$$

The resulting closed-loop poles were at

$$\begin{aligned} s &= -3.13 \pm j0.83 & \text{dutch roll mode } (r, \beta) \\ &\ -0.82 \pm j0.11 & \text{roll mode } (p, \phi) \\ &\ -11.47 \pm j17.18, -15.02 \end{aligned} \tag{14}$$

To verify the design a simulation was performed. The initial state was selected as $x(0) = [1\ 0\ 0\ 0\ 0\ 0\ 0]^T$; that is, we chose $\beta(0) = 1$. Figure 5.3-1 shows the results. Part a shows the dutch roll mode and part b the roll mode. Note that the responses correspond to the poles in (14), where the dutch roll is the faster mode. Compare to the results of Example 4.4-3.

This design has two deficiencies. First, it uses eight feedback gains in (13). This is undesirable for two reasons: (1) it requires the gain scheduling of all eight gains, and (2) the control system has no structure. That is, all outputs are fed back to both inputs; zeroing some of the gains would give the controller more structure in terms of feeding back certain outputs to only one or the other of the inputs.

The second deficiency is that it was necessary to juggle the entries of Q to obtain a good solution. Actually, due to our weighting of β^2 and r^2 by q_{dr}, and ϕ^2 and p^2 by q_r, the design was fairly straightforward and took about half an hour in all. It was, however, necessary to weight the washout filter state x_w, which is not obvious without considering the observability question.

In Section 5.5 we will show how to overcome both of these deficiencies: the former using "constrained output feedback" and the latter using time weighting like t^k in the PI.

c. Effect of Weighting Parameters.
It is interesting to examine more closely the effects of the design parameters, namely, the entries of the PI weighting matrices Q and R. Using the same Q as above, we show the sideslip response in Figure 5.3-2a for control weightings of $\rho = 0.1, 0.5,$ and 1. Increased control weighting in the PI generally suppresses the control signals in the closed-loop system; that is, less control effort is allowed. As less control effort is allowed, the control is less

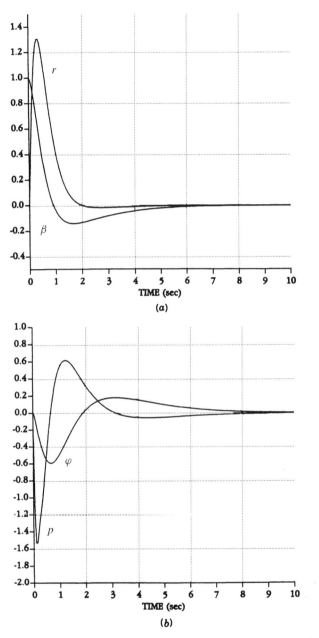

Figure 5.3-1 Closed-loop lateral response: (a) dutch roll states β and r; (b) roll mode states ϕ and p.

418 MODERN DESIGN TECHNIQUES

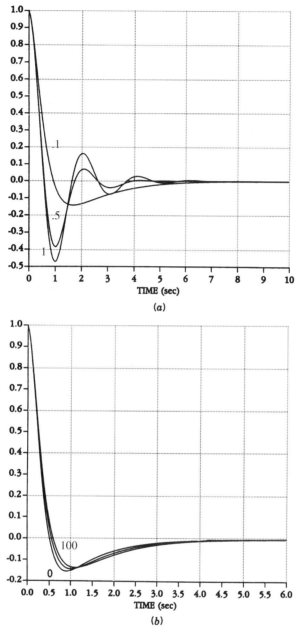

Figure 5.3-2 Effect of PI weighting parameters: (*a*) sideslip as a function of ρ (ρ = 0.1, 0.5, 1); (*b*) sideslip as a function of q_{dr} (q_{dr} = 0, 50, 100).

effective in controlling the modes. Indeed, according to the figure, as ρ increases the undershoot in β increases. Moreover, with increasing ρ the control is also less effective in suppressing the undesirable oscillations in the dutch roll mode which were noted in the open-loop system.

As far as the effect of the dutch roll weighting q_{dr} goes, examine Figure 5.3-2b, where $\rho = 0.1$ and $q_r = 100$ as in part a, but the sideslip response is shown for $q_{dr} = 0, 50$, and 100. As q_{dr} increases, the undershoot decreases, reflecting the fact that increased weighting on β^2 in the PI will result in smaller excursions in β in closed-loop.

One last point is worth noting. The open-loop system is stable; therefore, it is clear that it is detectable, since all the unstable modes are observable for any choice of Q (there are no unstable modes). Thus, the design would work if we omitted the weighting on x_w^2 in the Q matrix (although, it turns out, the closed-loop poles are not as good). In general, however, the detectability condition is difficult to check in large systems that are open-loop unstable; thus the observability condition is used instead. Failing to weight an undetectable state can lead to some zero elements of K, meaning that some feedback loops are not closed. Thus, to guarantee that this does not occur, Q should be selected so that $(\sqrt{Q}, A)$ is observable.

To avoid all this discussion on observability, we may simply use a modified nonstandard PI with weighting like t^k. Such a PI is introduced in Section 5.5 and leads to a simplified design procedure.

d. Gain Scheduling. For implementation on an aircraft, the control gains in (13) should be gain-scheduled. To accomplish this, the nonlinear aircraft equations are linearized at several equilibrium flight conditions over the desired flight envelope to obtain state-variable models like (4) with different A and B matrices. Then the LQR design is repeated for those different systems.

A major advantage of LQR design will quickly be apparent, for once the control structure has been selected, it takes only a minute or two to run the software to find the optimal gains for a new A and B using the design equations in Table 5.3-1. Note that the optimal gains for one point in the gain schedule can be used as initial stabilizing gains in the LQ solution algorithm for the next point.

It is important, however, to be aware of an additional consideration. The optimal gains at each gain-scheduling point should guarantee *robust stability and performance*; that is, they should guarantee stability and good performance at points *near* the design equilibrium point. Such robust stability can be verified after the LQ design by using multivariable frequency-domain techniques. These techniques are developed in Section 6.2, where the remarks on robustness to plant parameter variations are particularly relevant to gain scheduling. ■

5.4 TRACKING A COMMAND

In aircraft control we are often interested not in regulating the state near zero, which we discussed in the preceding section, but in *following a nonzero reference command signal.* For example, we may be interested in designing a control system for optimal step-response shaping. This reference-input tracking or *servodesign* problem is important in the design of command augmentation systems (CAS), where the reference command may be, for instance, desired pitch rate or normal acceleration. In this section and the next we cover tracker design.

It should be mentioned that the *optimal* linear quadratic (LQ) tracker of modern control is not a causal system (Lewis, 1986). It depends on solving an "adjoint" system of differential equations backward in time, and so is impossible to implement. A suboptimal "steady-state" tracker using full state-variable feedback is available, but it offers no convenient structure for the control system in terms of desired dynamics such as PI control, washout filters, and so on. Thus, there have been problems with using it in aircraft control.

Modified versions of the LQ tracker have been presented in Davison and Ferguson (1981) and Gangsaas et al. (1986). There, controllers of desired structure can be designed since the approaches are output-feedback based. The optimal gains are determined numerically to minimize a PI with, possibly, some constraints.

It is possible to design a tracker by first designing a regulator using, for instance, Table 5.3-1. Then some feedforward terms are added to guarantee perfect tracking (Kwakernaak and Sivan, 1972). The problem with this technique is that the resulting tracker has no convenient structure and often requires derivatives of the reference command input. Moreover, servosystems designed using this approach depend on knowing the dc gain exactly. If the dc gain is not known exactly, the performance deteriorates. That is, the design is *not robust* to uncertainties in the model.

Here we discuss an approach to the design of tracking control systems which is more useful in aircraft control applications (Stevens et al., 1991). This approach will allow us to design a servo control system that has any structure desired. This structure will include a unity-gain outer loop that feeds the performance output back and subtracts it from the reference command, thus defining a tracker error $e(t)$ which should be kept small (see Figure 5.4-1). It can also include compensator dynamics

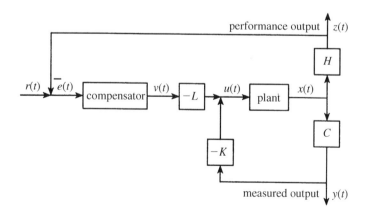

Figure 5.4-1 Plant with compensator of desired structure.

such as a washout filter or an integral controller. The control gains are chosen to minimize a quadratic performance index (PI). We are able to give explicit design equations for the control gains (see Table 5.4-1), which may be solved using the software described in Appendix B.

A problem with the tracker developed in this section is the need to select the design parameters Q and R in the PI in Table 5.4-1. There are some intuitive techniques available for choosing these parameters (see Section 5.3); however, in Section 5.5 we will show how modified PIs may be used to make the selection of Q and R almost transparent, yielding tracker design techniques that are very convenient for use in aircraft control systems design. We will show, in fact, that *the key to achieving required performance using modern design strategies is in selecting an appropriate PI.*

Tracker with Desired Structure

In aircraft control design there is a wealth of experience and knowledge that dictates in many situations what sort of compensator dynamics yield good performance from the point of view of both the control engineer and the pilot. For example, a washout circuit may be required, or it may be necessary to augment some feedforward channels with integrators to obtain a steady-state error of exactly zero.

The control system structures used in classical aircraft design also give good *robustness properties.* That is, they perform well even if there are disturbances or uncertainties in the system. Thus, the multivariable approach developed here usually affords this robustness. Formal techniques for verifying closed-loop robustness for multivariable control systems are given in Chapter 6.

Our approach to tracker design allows controller dynamics of any desired structure and then determines the control gains that minimize a quadratic PI over that structure. Before discussing the tracker design, let us recall from Chapter 3 how compensator dynamics may be incorporated into the aircraft state equations.

A dynamic compensator of prescribed structure may be incorporated into the system description as follows.

Consider the situation in Figure 5.4-1, where the plant is described by

$$\dot{x} = Ax + Bu \qquad (5.4\text{-}1)$$

$$y = Cx, \qquad (5.4\text{-}2)$$

with state $x(t)$, control input $u(t)$, and $y(t)$ the *measured output* available for feedback purposes. In addition,

$$z = Hx \qquad (5.4\text{-}3)$$

is a *performance output,* which must track the given *reference input* $r(t)$. The performance output $z(t)$ is not generally equal to $y(t)$.

It is important to realize that for perfect tracking it is necessary to have as many control inputs in vector $u(t)$ as there are command signals to track in $r(t)$ (Kwakernaak and Sivan, 1972).

The dynamic compensator has the form

$$\dot{w} = Fw + Ge$$
$$v = Dw + Je, \quad (5.4\text{-}4)$$

with state $w(t)$, output $v(t)$, and input equal to the *tracking error*

$$e(t) = r(t) - z(t) \quad (5.4\text{-}5)$$

F, G, D, and J are known matrices chosen to include the desired structure in the compensator.

The allowed form for the plant control input is

$$u = -Ky - Lv, \quad (5.4\text{-}6)$$

where the constant gain matrices K and L are to be chosen in the control design step to result in satisfactory tracking of $r(t)$. This formulation allows for both feedback and feedforward compensator dynamics.

As we have seen in Chapter 3, these dynamics and output equations may be written in augmented form as

$$\frac{d}{dt}\begin{bmatrix} x \\ w \end{bmatrix} = \begin{bmatrix} A & 0 \\ -GH & F \end{bmatrix}\begin{bmatrix} x \\ w \end{bmatrix} + \begin{bmatrix} B \\ 0 \end{bmatrix}u + \begin{bmatrix} 0 \\ G \end{bmatrix}r \quad (5.4\text{-}7)$$

$$\begin{bmatrix} y \\ v \end{bmatrix} = \begin{bmatrix} C & 0 \\ -JH & D \end{bmatrix}\begin{bmatrix} x \\ w \end{bmatrix} + \begin{bmatrix} 0 \\ J \end{bmatrix}r \quad (5.4\text{-}8)$$

$$z = [H \quad 0]\begin{bmatrix} x \\ w \end{bmatrix}, \quad (5.4\text{-}9)$$

and the control input may be expressed as

$$u = -[K \quad L]\begin{bmatrix} y \\ v \end{bmatrix} \quad (5.4\text{-}10)$$

Note that in terms of the augmented plant/compensator state description, the admissible controls are represented as a *constant output feedback* $[K \ L]$. In the augmented description, all matrices are known except the gains K and L, which need to be selected to yield acceptable closed-loop performance.

A comment on the compensator matrices F, G, D, and J is in order. Often, these matrices are completely specified by the structure of the compensator. Such is the case, for instance, if the compensator contains integrators. However, if it is desired to include a washout or a lead-lag, it may not be clear exactly how to select the time constants. In such cases, engineering judgment will usually give some insight. However, it may sometimes be necessary to go through the design to be proposed, and then if required, return to readjust F, G, D, and J and reperform the design.

LQ Formulation of the Tracker Problem

By redefining the state, the output, and the matrix variables to streamline the notation, we see that the augmented equations (5.4-7)–(5.4-9) that contain the dynamics of both the aircraft and the compensator are of the form

$$\dot{x} = Ax + Bu + Gr \qquad (5.4\text{-}11)$$

$$y = Cx + Fr \qquad (5.4\text{-}12)$$

$$z = Hx \qquad (5.4\text{-}13)$$

In this description, let us take the state $x(t) \in \mathbf{R}^n$, control input $u(t) \in \mathbf{R}^m$, reference input $r(t) \in \mathbf{R}^q$, performance output $z(t) \in \mathbf{R}^q$, and measured output $y(t) \in \mathbf{R}^p$. The admissible controls (5.4-10) are proportional output feedbacks of the form

$$u = -Ky = -KCx - KFr \qquad (5.4\text{-}14)$$

with constant gain K to be determined. This situation corresponds to the block diagram in Figure 5.4-2. Since K is an $m \times p$ matrix, we intend to close all the feedback loops simultaneously by computing K.

Using these equations the closed-loop system is found to be

$$\dot{x} = (A - BKC)x + (G - BKF)r$$

$$\equiv A_c x + B_c r \qquad (5.4\text{-}15)$$

In the remainder of this subsection, we will use the formulation (5.4-11)–(5.4-14), assuming that the compensator, if required, has already been included in the system

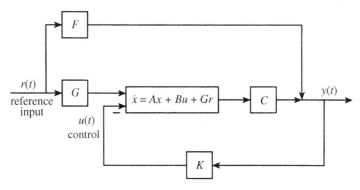

Figure 5.4-2 Plant/feedback structure.

dynamics and demonstrating how to select the constant output-feedback gain matrix K using LQ techniques.

Our formulation differs sharply from the traditional formulations of the optimal tracker problem (Kwakernaak and Sivan, 1972; Lewis, 1986). Note that (5.4-14) includes both feedback and feedforward terms, so that both the closed-loop poles *and compensator zeros* may be affected by varying the gain K (see Example 5.4-1). Thus, we should expect better success in shaping the step response than by placing only the poles.

Since the performance specifications of aircraft are often given in terms of time-domain criteria (Mil. Spec. 1797, 1987) (see Section 4.3) and these criteria are closely related to the step response, we will assume henceforth that the reference input $r(t)$ is a step command with magnitude r_0. Designing for such a command will yield suitable time-response characteristics. Although our design is based on step-response shaping, it should be clearly realized that the resulting control system, if properly designed, will give good time responses for *any arbitrary reference command signal $r(t)$*.

Let us now formulate an optimal control problem for selecting the control gain K to guarantee tracking of $r(t)$. Then we will derive the design equations in Table 5.4-1, which are used to determine the optimal K. These equations are solved using software like that described in Appendix B.

The Deviation System. Denote steady-state values by overbars and deviations from the steady-state values by tildes. Then the state, output, and control deviations are given by

$$\tilde{x}(t) = x(t) - \bar{x} \tag{5.4-16}$$

$$\tilde{y}(t) = y(t) - \bar{y} = C\tilde{x} \tag{5.4-17}$$

$$\tilde{z}(t) = z(t) - \bar{z} = H\tilde{x} \tag{5.4-18}$$

$$\tilde{u}(t) = u(t) - \bar{u} = -KCx - KFr_0 - (-KC\bar{x} - KFr_0) = -KC\tilde{x}(t)$$

or

$$\tilde{u} = -K\tilde{y} \tag{5.4-19}$$

The tracking error $e(t) = r(t) - z(t)$ is given by

$$e(t) = \tilde{e}(t) + \bar{e} \tag{5.4-20}$$

with the error deviation given by

$$\tilde{e}(t) = e(t) - \bar{e} = (r_0 - Hx) - (r_0 - H\bar{x}) = -H\tilde{x}$$

or

$$\tilde{e} = -\tilde{z} \tag{5.4-21}$$

Since in any acceptable design the closed-loop plant will be asymptotically stable, A_c is nonsingular. According to (5.4-15), at steady state

$$0 = A_c \bar{x} + B_x r_0, \tag{5.4-22}$$

so that the steady-state response $\bar{x}$ is

$$\bar{x} = -A_c^{-1} B_c r_0 \tag{5.4-23}$$

and the steady-state error is

$$\bar{e} = r_0 - H\bar{x} = \left(1 + HA_c^{-1} B_c\right) r_0 \tag{5.4-24}$$

To understand this expression, note that the closed-loop transfer function from r_0 to z [see (5.4-15) and (5.4-13)] is

$$H(s) = H(sI - A_c)^{-1} B_c \tag{5.4-25}$$

The steady-state behavior may be investigated by considering the dc value of $H(s)$ (i.e., $s = 0$); this is just $-HA_c^{-1} B_c$, the term appearing in (5.4-24).

Using (5.4-16), (5.4-19), and (5.4-23) in (5.4-15) the closed-loop dynamics of the state deviation are seen to be

$$\dot{\tilde{x}} = A_c \tilde{x} \tag{5.4-26}$$

$$\tilde{y} = C\tilde{x} \tag{5.4-27}$$

$$\tilde{z} = H\tilde{x} = -\tilde{e} \tag{5.4-28}$$

and the control input to the deviation system (5.4-26) is (5.4-19). Thus the step-response shaping problem has been converted to a *regulator problem* for the deviation system

$$\dot{\tilde{x}} = A\tilde{x} + B\tilde{u} \tag{5.4-29}$$

Again, we emphasize the difference between our approach and traditional ones (e.g., Kwakernaak and Sivan, 1972). Once the gain K in (5.4-19) has been found, the control for the plant is given by (5.4-14), which inherently has both feedback and feedforward terms. Thus no extra feedforward term need be added to make $\bar{e}$ zero.

Performance Index. To make the tracking error $e(t)$ in (5.4-20) small, we propose to attack two equivalent problems: the problem of regulating the error deviation $\tilde{e}(t) = -\tilde{z}(t)$ to zero, and the problem of making the steady-state error $\bar{e}$ small.

Note that we do not assume a type-1 system, which would force $\bar{e}$ to be equal to zero. This can be important in aircraft control, where it may not be desirable to force the system to be of type-1 by augmenting all control channels with integrators.

This augmentation complicates the servo structure. Moreover, it is well known from classical control theory that suitable step responses may often be obtained without resorting to inserting integrators in all the feedforward channels.

To make both the error deviation $\tilde{e}(t) = -H\tilde{x}(t)$ and the steady-state error $\bar{e}$ small, we propose selecting K to minimize the performance index (PI)

$$J = \tfrac{1}{2}\int_0^\infty \left(\tilde{e}^T\tilde{e} + \tilde{u}^T R\tilde{u}\right) dt + \tfrac{1}{2}\bar{e}^T V \bar{e}, \tag{5.4-30}$$

with $R > 0$, $V \geq 0$, design parameters. The integrand is the standard quadratic PI with, however, a weighting V included on the steady-state error. Note that the PI weights the control *deviations* and not the controls themselves. If the system is of type 1, containing integrators in all the feedforward paths, then V may be set to zero since the steady-state error is automatically zero.

Making the error deviation $\tilde{e}(t)$ small improves the transient response, while making the steady-state error $\bar{e}(t)$ small improves the steady-state response. If the system is of type 0, these effects involve a trade-off, so that then there is a design trade-off involved in selecting the size of V.

We can generally select $R = rI$ and $V = vI$, with r and v scalars. This simplifies the design since now only a few parameters must be tuned during the interactive design process.

According to (5.4-21), $\tilde{e}^T\tilde{e} = \tilde{x}^T H^T H \tilde{x}$. Referring to Table 5.3-1, therefore, it follows that the matrix Q there is equal to $H^T H$, where H is known. That is, weighting the error deviation in the PI has already shown us how to select the design parameter Q, affording a considerable simplification.

The problem we now have to solve is how to select the control gains K to minimize the PI J for the deviation system (5.4-29). Then the tracker control for the original system is given by (5.4-14).

We should point out that the proposed approach is suboptimal in the sense that minimizing the PI does not necessarily minimize a quadratic function of the total error $e(t) = \bar{e} + \tilde{e}(t)$. It does, however, guarantee that both $\tilde{e}(t)$ and $\bar{e}$ are small in the closed-loop system, which is a design goal.

Solution of the LQ Tracker Problem

It is now necessary to solve for the optimal feedback gain K that minimizes the PI. The design equations needed are now derived. They appear in Table 5.4-1.

By using (5.4-26) and a technique like the one used in Section 5.3 (see Problems), the optimal cost is found to satisfy

$$J = \tfrac{1}{2}\tilde{x}^T(0) P \tilde{x}(0) + \tfrac{1}{2}\bar{e}^T V \bar{e}, \tag{5.4-31}$$

with $P \geq 0$ the solution to

$$0 = g \equiv A_c^T P + P A_c + Q + C^T K^T RKC, \tag{5.4-32}$$

with $Q = H^T H$ and $\bar{e}$ given by (5.4-24).

In our discussion of the linear quadratic regulator we assumed that the initial conditions were uniformly distributed on a surface with known characteristics. While this is satisfactory for the regulator problem, it is an unsatisfactory assumption for the tracker problem. In the latter situation the system starts at rest and must achieve a given final state that is dependent on the reference input, namely (5.4-23). To find the correct value of $\tilde{x}(0)$, we note that since the plant starts at rest (i.e., $x(0) = 0$), according to (5.4-16),

$$\tilde{x}(0) = -\bar{x}, \tag{5.4-33}$$

so that the optimal cost (5.4-31) becomes

$$J = \tfrac{1}{2}\bar{x}^T P \bar{x} + \tfrac{1}{2}\bar{e}^T V \bar{e} = \tfrac{1}{2}\mathrm{tr}(PX) + \tfrac{1}{2}\bar{e}^T V \bar{e}, \tag{5.4-34}$$

with P given by (5.4-32), $\bar{e}$ given by (5.4-24), and

$$X \equiv \bar{x}\bar{x}^T = A_c^{-1} B_c r_0 r_0^T B_c^T A_c^{-T}, \tag{5.4-35}$$

with $A_c^{-T} = (A_c^{-1})^T$.

The optimal solution to the unit-step tracking problem, with (5.4-11) initially at rest, may now be determined by minimizing J in (5.4-34) over the gains K, subject to the constraint (5.4-32) and Equations (5.4-24) and (5.4-35).

This algebraic optimization problem can be solved by any well-known numerical method (see Press et al., 1986; Söderström, 1978). A good approach for a fairly small number ($mp \leq 10$) of gain elements in K is the Simplex minimization routine (Nelder and Mead, 1964). To evaluate the PI for each fixed value of K in the iterative solution procedure, one may solve (5.4-32) for P using subroutine ATXPXA (Bartels and Stewart, 1972) and then employ (5.4-34). Software for determining the optimal control gains K is described in Appendix B.

Design Equations for Gradient-Based Solution.

As an alternative solution procedure, one may use gradient-based techniques (e.g., the Davidon–Fletcher–Powell algorithm [Press et al., 1986]), which are generally faster than non-gradient-based approaches.

To find the gradient of the PI with respect to the gains, define the Hamiltonian

$$\mathcal{H} = \tfrac{1}{2}\mathrm{tr}(PX) + \tfrac{1}{2}\mathrm{tr}(gS) + \tfrac{1}{2}\bar{e}^T V \bar{e}, \tag{5.4-36}$$

with S a Lagrange multiplier. Now, using the basic matrix calculus identities,

$$\frac{\partial Y^{-1}}{\partial x} = -Y^{-1} \frac{\partial Y}{\partial x} Y^{-1} \tag{5.4-37}$$

$$\frac{\partial UV}{\partial x} = \frac{\partial U}{\partial x} V + U \frac{\partial V}{\partial x} \tag{5.4-38}$$

$$\frac{\partial y}{\partial x} = \mathrm{tr}\left[\frac{\partial y}{\partial z} \cdot \frac{\partial z^T}{\partial x}\right] \tag{5.4-39}$$

we may proceed as in the preceding section, with, however, a little more patience due to the extra terms (see the Problems!), to obtain the necessary conditions for a solution given in Table 5.4-1.

To find K by a gradient minimization algorithm, it is necessary to provide the algorithm with the values of J and $\partial J/\partial K$ for a given K. The value of J is given by the expression in Table 5.4-1 for the optimal cost. To find $\partial J/\partial K$ given K, solve (5.4-40)–(5.4-41) for P and S. Then since these equations hold, $\partial J/\partial K = \partial \mathcal{H}/\partial K$, which may be found using (5.4-42). These equations should be compared to those in Table 5.3-1. Note that the dependence of X on the gain K [see (5.4-45)] and the presence of $\bar{e}$ in the PI have resulted in extra terms being added in (5.4-42).

TABLE 5.4-1: LQ Tracker with Output Feedback

System Model

$$\dot{x} = Ax + Bu + Gr$$

$$y = Cx + Fr$$

$$z = Hx$$

Control

$$u = -Ky$$

Performance Index

$$J = \tfrac{1}{2}\int_0^\infty \left(\tilde{x}^T Q\tilde{x} + \tilde{u}^T R\tilde{u}\right) dt + \tfrac{1}{2}\bar{e}^T V\bar{e}, \quad \text{with } Q = H^T H$$

Optimal Output Feedback Gain

$$0 = \frac{\partial \mathcal{H}}{\partial S} = A_c^T P + PA_c + Q + C^T K^T RKC \tag{5.4-40}$$

$$0 = \frac{\partial \mathcal{H}}{\partial P} = A_c S + SA_c^T + X \tag{5.4-41}$$

$$0 = \frac{\partial \mathcal{H}}{\partial K} = RKCSC^T - B^T PSC^T + B^T A_c^{-T}\left(P + H^T VH\right)\bar{x}\bar{y}^T$$
$$- B^T A_c^{-T} H^T V r_0 \bar{y}^T \tag{5.4-42}$$

with r a unit step of magnitude r_0 and

$$\bar{x} = -A_c^{-1} B_c r_0 \tag{5.4-43}$$

$$\bar{y} = C\bar{x} + Fr_0 \tag{5.4-44}$$

$$X = \overline{xx}^T = A_c^{-1} B_c r_0 r_0^T B_c^T A_c^{-T} \tag{5.4-45}$$

where

$$A_c = A - BKC, \quad B_c = G - BKF$$

Optimal Cost

$$J = \tfrac{1}{2}\text{tr}(PX) + \tfrac{1}{2}\bar{e}^T V\bar{e}$$

Determining the Optimal Feedback Gain

The issues in finding the optimal output-feedback gain K in the tracker problem of Table 5.4-1 are the same as those discusssed in connection with the regulator problem of Table 5.3-1. They are: choice of Q to satisfy detectability, choice of solution technique, finding an initial stabilizing gain, and iterative design by tuning Q and R.

We emphasize that there are only a few design parameters in our approach, namely, r and v (since we can can generally select $R = rI$, $V = vI$). Thus it is not difficult or time consuming to come up with good designs. Much of the simplicity of our approach derives from the fact that ^-Q in the PI is equal to $H^T H$, which is known.

Let us now illustrate the servo design procedure by an example.

Example 5.4-1: Normal Acceleration CAS. In Chapter 4 we designed a normal acceleration CAS using classical control theory. In that example, successive loop closures were used with root-locus design to obtain the feedback gains. Here we will show that using the LQ design equations in Table 5.4-1 we can close all the loops simultaneously. Thus the design procedure is more straightforward. We will also demonstrate that using LQ design, *the algorithm automatically selects the zero of the compensator for optimal performance.*

a. Control System Structure. The normal acceleration control system is shown in Figure 5.4-3, where r is a reference step input in g's and $u(t)$ is the elevator actuator voltage. An integrator has been added in the feedforward path to achieve zero steady-state error. The performance output that should track the reference command r is $z = n_z$, so that the tracking error is $e = r - n_z$. The state and measured output are

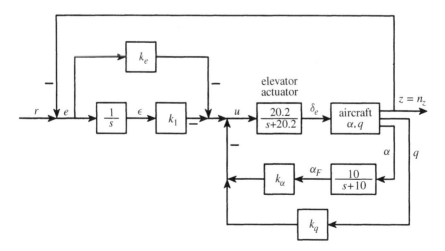

Figure 5.4-3 G-command system.

430 MODERN DESIGN TECHNIQUES

$$x = \begin{bmatrix} \alpha \\ q \\ \delta_e \\ \alpha_F \\ \epsilon \end{bmatrix}, \quad y = \begin{bmatrix} \alpha_F \\ q \\ e \\ \epsilon \end{bmatrix}, \quad (1)$$

with $\epsilon(t)$ the integrator output and α_F the filtered measurements of angle of attack.

Exactly as in Chapter 4, we linearized the F-16 dynamics about the nominal flight condition in Table 3.6-3 (502 ft/s, level flight, dynamic pressure of 300 psf, $x_{cg} = 0.35\,\bar{c}$) and augmented the dynamics to include the elevator actuator, angle-of-attack filter, and compensator dynamics. The result is

$$\dot{x} = Ax + Bu + Gr \quad (2)$$

$$y = Cx + Fr \quad (3)$$

$$z = Hx, \quad (4)$$

with

$$A = \begin{bmatrix} -1.01887 & 0.90506 & -0.00215 & 0 & 0 \\ 0.82225 & -1.07741 & -0.17555 & 0 & 0 \\ 0 & 0 & -20.2 & 0 & 0 \\ 10 & 0 & 0 & -10 & 0 \\ -16.26 & -0.9788 & 0.04852 & 0 & 0 \end{bmatrix}$$

$$B = \begin{bmatrix} 0 \\ 0 \\ 20.2 \\ 0 \\ 0 \end{bmatrix} \quad G = \begin{bmatrix} 0 \\ 0 \\ 0 \\ 0 \\ 1 \end{bmatrix} \quad (5a)$$

$$C = \begin{bmatrix} 0 & 0 & 0 & 57.2958 & 0 \\ 0 & 57.2958 & 0 & 0 & 0 \\ -16.26 & -0.9788 & 0.04852 & 0 & 0 \\ 0 & 0 & 0 & 0 & 1 \end{bmatrix} \quad F = \begin{bmatrix} 0 \\ 0 \\ 1 \\ 0 \end{bmatrix} \quad (5b)$$

$$H = [\,16.26 \quad 0.9788 \quad -0.04852 \quad 0 \quad 0\,] \quad (5c)$$

The factor of 57.2958 is added to convert angles from radians to degrees.

The control input is

$$u = -Ky = -\begin{bmatrix} k_\alpha & k_q & k_e & k_I \end{bmatrix} y = -k_\alpha \alpha_F - k_q q - k_e e - k_I \epsilon \quad (6)$$

It is desired to select the four control gains to guarantee a good response to a step command r. Note that k_α and k_q are feedback gains, while k_e and k_I are feedforward gains.

Note that the proportional-plus-integral compensator is given by

$$k_e + \frac{k_I}{s} = k_e \frac{s + k_I/k_e}{s}, \tag{7}$$

which has a zero at $s = -k_I/k_e$. Since the LQ design algorithm will select all four control gains, it will *automatically select the optimal location for the compensator zero*.

b. Performance Index and Determination of the Control Gains. Due to the integrator, the system is of type 1. Therefore, the steady-state error $\bar{e}$ is automatically equal to zero. A natural PI thus seems to be

$$J = \tfrac{1}{2}\int_0^\infty \left(\tilde{e}^2 + \rho \tilde{u}^2\right) dt \tag{8}$$

with ρ a scalar weighting parameter. Since $\tilde{e} = H\tilde{x}$, this corresponds to the PI in Table 5.4-1 with

$$Q = H^T H = \begin{bmatrix} 264 & 16 & 1 & 0 & 0 \\ 16 & 1 & 0 & 0 & 0 \\ 1 & 0 & 0 & 0 & 0 \\ 0 & 0 & 0 & 0 & 0 \\ 0 & 0 & 0 & 0 & 0 \end{bmatrix} \tag{9}$$

This is, unfortunately, not a suitable Q matrix since (H, A) is not observable in open loop. Indeed, according to Figure 5.4-3, observing the first two states α and q can never give information about ϵ in the open-loop configuration (where the control gains are zero). Thus, the integrator state is unobservable in the PI. Since the integrator pole is at $s = 0$, (H, A) is undetectable (unstable unobservable pole), so that any design based on (9) would, in fact, yield a value for the integral gain of $k_I = 0$.

We will show in Section 5.5 a very convenient way to correct problems like this. There we will introduce a time weighting of t^k into the PI. In the meantime, to correct the observability problem here, let us select

$$Q = \begin{bmatrix} 264 & 16 & 1 & 0 & 0 \\ 16 & 1 & 0 & 0 & 0 \\ 1 & 0 & 0 & 0 & 0 \\ 0 & 0 & 0 & 0 & 0 \\ 0 & 0 & 0 & 0 & 1 \end{bmatrix}, \tag{10}$$

where we include a weighting on $\epsilon(t)$ to make it observable in the PI.

Now, we selected $\rho = 1$ and solved the design equations in Table 5.4-1 for the optimal control gain K using the software described in Appendix B. For this Q and ρ the feedback matrix was

$$K = [0.006 \quad -0.152 \quad 1.17 \quad 0.996] \tag{11}$$

and the closed-loop poles were

$$s = -1.15 \pm j0.69$$
$$-1.60, -9.98, -19.54 \tag{12}$$

These yield a system that is not fast enough; the complex pair is also unsuitable in terms of flying-qualities requirements.

After repeating the design using several different Q and ρ, we decided on

$$Q = \begin{bmatrix} 264 & 16 & 1 & 0 & 0 \\ 16 & 60 & 0 & 0 & 0 \\ 1 & 0 & 0 & 0 & 0 \\ 0 & 0 & 0 & 0 & 0 \\ 0 & 0 & 0 & 0 & 100 \end{bmatrix}, \tag{13}$$

$\rho = 0.01$. The decreased control weighting ρ has the effect of allowing larger control effort and so speeding up the response. The increased weighting on the integrator output $\epsilon(t)$ has the effect of forcing n_z to its final value of r more quickly, hence also speeding up the response. The increased weighting on the second state component q has the effect of regulating excursions in $\tilde{q}(t)$ closer to zero, and hence of providing increased damping.

With this Q and ρ the control matrix was

$$K = [-1.629 \quad -1.316 \quad 18.56 \quad 77.6] \tag{14}$$

and the closed-loop poles were at

$$s = -2.98 \pm j3.17,$$
$$-19.31 \pm j4.64$$
$$-5.91 \tag{15}$$

The closed-loop step response is shown in Figure 5.4-4; it is fairly fast with an overshoot of 6 percent. Note the hump in the initial response due to the non-minimum-phase zero. Further tuning of the elements of Q and R could provide less overshoot, a faster response, and a smaller gain for the angle-of-attack feedback. (It is worth noting that we will obtain a far better response with more reasonable gains in Example 5.5-2, where we use a PI with time-dependent weighting like t^k.)

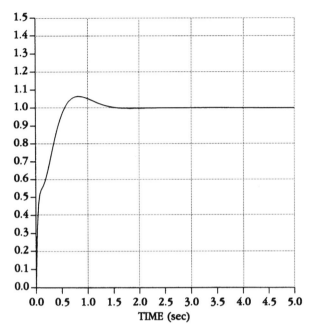

Figure 5.4-4 Normal acceleration step response.

According to (7), the compensator zero has been placed by the LQ algorithm at

$$s = -\frac{k_I}{k_e} = -4.18 \tag{16}$$

Using the software described in Appendix B, the entire design, including determining K for different choices of Q and ρ until a suitable design was reached, took about 30 minutes.

c. Discussion. We can now emphasize an important aspect of modern LQ design. As long as $Q \geq 0$, $R > 0$, and $(\sqrt{Q}, A)$ is observable, the closed-loop system designed using Table 5.4-1 is generally stable. Thus, the LQ theory has allowed us to tie the control system design to some *design parameters* which may be tuned to obtain acceptable behavior—namely, the elements of weighting matrices Q and R. Using the software described in Appendix B, for a given Q and R the optimal gain K is easily found. If it is not suitable in terms of time responses and closed-loop poles, the elements of Q and R can be changed and the design repeated. The importance of this is that for admissible Q and R, *closed-loop stability is guaranteed.*

A disadvantage of the design equations in Table 5.4-1 is the need to try different Q and R until suitable performance is obtained, as well as the need for (H, A) to be observable. In Section 5.5 we will introduce a different PI with time weighting of t^k which eliminates these deficiencies.

434 MODERN DESIGN TECHNIQUES

Another point needs to be made. Using the control (6)/(3) in (2) yields the closed-loop plant

$$\dot{x} = (A - BKC)x + (G - BKF)r \qquad (17)$$

whence the closed-loop transfer function from $r(t)$ to $z(t)$ is

$$H(s) = H(sI - (A - BKC))^{-1}(G - BKF) \qquad (18)$$

Note that the transfer function numerator depends on the optimal gain K. That is, this scheme uses optimal positioning of *both the poles and zeros* to attain step-response shaping.

d. Selection of Initial Stabilizing Gain. In order to initialize the algorithm that determines the optimal K by solving the design equations in Table 5.4-1, it is necessary to find an initial gain that stabilizes the system. In this example we simply selected gains with signs corresponding to the static loop sensitivity of the individual transfer functions, since this corresponds to negative feedback. The static loop sensitivities from u to α and from u to q are negative, so negative gains were chosen for these loops. The initial gain used was

$$K = [-0.1 \quad -0.1 \quad 1 \quad 1] \qquad (19)$$

■

5.5 MODIFYING THE PERFORMANCE INDEX

Modern control theory affords us the ability to close all the feedback loops simultaneously by solving matrix equations for the gain matrix K. With a sensible problem formulation, it also *guarantees the stability of the closed-loop system.* These two fundamental properties make modern design very useful for aircraft control systems. One should recall the difficulty in guaranteeing closed-loop stability in multiloop control systems using one-loop-at-a-time design (Chapter 4).

An additional important advantage is as follows. The standard aircraft control system structures used in classical design have been developed to yield good *robustness properties.* That is, they yield good performance even if there are disturbances in the systems, or modeling inaccuracies such as plant parameter variations or high-frequency unmodeled dynamics (e.g., flexible aircraft modes). Since the approach described here allows these standard structures to be incorporated into the control system, it generally yields robust control systems. We will discuss procedures for formally verifying robustness in Chapter 6.

In the LQ regulator design method of Section 5.3 and the LQ tracker design method of Section 5.4, it was necessary to select the performance index (PI) weighting matrices Q and R as design parameters. Moreover, it was necessary to satisfy an observability property in selecting Q. There are some good approaches that give guidance in selecting Q, such as Bryson's approach (see Section 5.3). Note also that in Table 5.4-1, $Q = H^T H$, where H is known. However, due to the observability requirement the design parameters Q and R do not necessarily correspond to actual performance objectives.

In this section we show how to modify the PI to considerably simplify the selection of the weighting matrices Q and R in Table 5.4-1. The observability of $(\sqrt{Q}, A)$ will be unnecessary. The PIs shown in this section correspond to actual performance objectives and involve only a few design parameters, even for systems with many states and many control gains to determine. These facts, coupled with the capability already demonstrated of employing a compensator with any desired structure, will result in a powerful and convenient approach to the design of multivariable aircraft control systems.

A wide range of performance objectives may be attained by using modifications of the PI. We will consider several modifications, all of which are useful depending on the performance objectives. The important concept to grasp is that *the key to obtaining suitable closed-loop behavior using LQ design lies in selecting a suitable PI for the problem at hand.* At the end of the section we present several examples in aircraft control design to demonstrate this issue as well as the directness of the approach.

We will again be concerned with the system-plus-compensator

$$\dot{x} = Ax + Bu + Gr \qquad (5.5\text{-}1)$$

We are trying to determine controls that are static output feedbacks of the form

$$u = -Ky \qquad (5.5\text{-}2)$$

with

$$y = Cx + Fr \qquad (5.5\text{-}3)$$

the measured output and

$$z = Hx \qquad (5.5\text{-}4)$$

the performance output, which is to track the reference command r. If we are interested in regulation and not tracking, then G and F do not appear in the equations and z is not defined.

Constrained Feedback Matrix

In many applications it is desired for certain elements of the feedback gain matrix K to be zero to avoid coupling between certain output/input pairs. Zeroing certain gains allows us to specify the detailed structure of the control system. For instance, it may be desired that the error in channel 1 of the controller not be coupled to the control input in channel 2. Zeroing some gains also simplifies the gain-scheduling problem by reducing the number of nonzero gains requiring tabulation. This is called *constrained* output-feedback design.

Gain Element Weighting. Certain elements k_{ij} of K can be made small simply by weighting them in the performance index, that is, by selecting a PI like

$$J = \tfrac{1}{2} \int_0^\infty \left(\tilde{x}^T Q \tilde{x} + \tilde{u}^T R \tilde{u} \right) dt + \sum_i \sum_j g_{ij} k_{ij}^2 \qquad (5.5\text{-}5)$$

Gain-element weight g_{ij} is chosen large to make the (i, j)-th element k_{ij} of the feedback matrix K small in the final design. Then, in implementing the controller, the small elements of K may simply be set to zero.

The design problem is now to minimize

$$J = \tfrac{1}{2}\text{tr}(PX) + \sum_i \sum_j g_{ij} k_{ij}^2, \qquad (5.5\text{-}6)$$

with P satisfying the matrix equation in Table 5.3-1 or Table 5.4-1, as appropriate. This may be accomplished by using the equations in Table 5.3-1 (if we are interested in regulation) or Table 5.4-1 (if we are interested in tracking) to numerically minimize the PI, but with the extra term involving the gain weighting that appears in (5.5-6) (Moerder and Calise, 1985).

Computing an Initial Stabilizing Gain. The iterative algorithms that solve the design equations in Tables 5.3-1 and 5.4-1 require initial stabilizing gains. Unfortunately, stabilizing output-feedback gains can be complicated to find in large multivariable systems. A few ways to find K_0 so that $(A - BK_0C)$ is stable were mentioned in Section 5.3 and Example 5.4-1d. Gain element weighting can be used to solve the problem of determining an initial stabilizing output-feedback gain, as we now see.

There are many techniques for finding a stabilizing *state-variable* feedback given the plant system matrix A and control matrix B (see Kailath, 1980, as well as Section 5.7). That is, it is straightforward to find a K_0 so that $(A - BK_0)$ is stable. Routines that perform this are available in standard software packages such as ORACLS (Armstrong, 1980). Unfortunately, for flight control purposes, state-feedback design is unsuitable for reasons such as those we have discussed. However, suppose that an $m \times n$ stabilizing state-feedback gain has been found. Then, to determine an $m \times p$ output-feedback gain, it is only necessary to weight in the PI the elements of the state feedback matrix that do not correspond to measured outputs. The algorithm will then provide a suitable output feedback gain matrix by driving these elements to zero.

Gain Element Fixing. There is an attractive alternative to gain element weighting for fixing gain matrix elements. If a numerical technique such as Simplex (Press et al.,

1986) is used to determine the optimal control by varying K and directly evaluating J, we may simply fix certain elements of K and not allow the Simplex to vary them. This allows the fixed elements to be retained at any desired (possibly nonzero) value and takes far fewer computations than gain element weighting, especially if many elements of K are fixed.

If, on the other hand, a gradient-based routine such as Davidon–Fletcher–Powell (Press et al., 1986) is used in conjunction with the design equations in Tables 5.3-1 or 5.4-1, it is easy to modify the gradient $\partial J/\partial K$ to leave certain elements of K fixed. Indeed, to fix element (i, j) of K, one need only set element (i, j) of $\partial J/\partial K$ equal to zero.

These approaches require fewer operations than the gain weighting approach based on (5.5-5) and are incorporated in the software described in Appendix B, which is called program LQ. Illustrations of control design using constrained output feedback are provided in the examples.

Derivative Weighting

As we will soon show in an example, it is often convenient to weight in the PI not the states themselves but their derivatives. This is because rates of change of the states can in some design specifications be more important than the values of the states. For instance, elevator rate of change has a closer connection with required control energy than does elevator deflection. To accomodate such situations, we may consider the PI

$$J = \tfrac{1}{2} \int_0^\infty \dot{\tilde{x}}^T Q \dot{\tilde{x}}\, dt \qquad (5.5\text{-}7)$$

One way to formulate this optimization problem is to convert this PI to one that weights the states and inputs but has a state/input cross-weighting term (simply substitute (5.4-29) into J). This optimization problem is solved in Lewis (1986).

An alternative (see the Problems) is to minimize

$$J = \tfrac{1}{2}\mathrm{tr}\left[P\dot{\tilde{x}}(0)\dot{\tilde{x}}^T(0) \right], \qquad (5.5\text{-}8)$$

with P the solution to

$$A_c^T P + P A_c + Q = 0 \qquad (5.5\text{-}9)$$

Again, any optimization technique may be used. More details on this formulation may be found in Quintana et al. (1976).

In the step-response shaping problem, the value of the initial state derivative vector to use in (5.5-8) is easy to determine since $x(0) = 0$ and $\tilde{x}$ is a constant so that according to (5.4-16) and (5.4-15)

$$\dot{\tilde{x}}(0) = B_c r_0 \qquad (5.5\text{-}10)$$

Time-Dependent Weighting

One final form of the PI remains to be discussed. A step response that is apparently good (i.e., fast, with acceptable overshoot and settling time) may contain a slow pole(s) with small residue, so that the response creeps for a long time as it nears its final value. The quadratic performance criterion penalizes small errors relatively lightly and so does not tend to suppress this kind of behavior.

Thus, in the spirit of the classical (ITAE, ISTSE, etc.) performance indices (D'Azzo and Houpis, 1988) we define a PI that contains a time-weighted component:

$$J = \tfrac{1}{2} \int_0^\infty \left(t^k \tilde{x}^T P \tilde{x} + \tilde{x}^T Q \tilde{x} \right) dt \tag{5.5-11}$$

If we are interested in including a control-weighting term $\tilde{u}^T R \tilde{u}$ in (5.5-11) and in using the output feedback (5.5-2), we may add the term $C^T K^T R K C$ (since $\tilde{u}^T R \tilde{u} = \tilde{x}^T C^T K^T R K C \tilde{x}$) to the appropriate state-weighting matrix P or Q, depending on whether we wish to multiply the control weighting term by t^k. For instance, if the control-input term is not to be weighted by t^k, the PI (5.5-11) takes on the form

$$J = \tfrac{1}{2} \int_0^\infty \left[t^k \tilde{x}^T P \tilde{x} + \tilde{x}^T \left(Q + C^T K^T R K C \right) \tilde{x} \right] dt \tag{5.5-12}$$

If it is desired to have the control weighting multiplied by t^k, the term $C^T K^T R K C$ should be added to P instead of Q.

Whether or not the control effort should be time-weighted is a matter for experiment with the particular design. The time-varying weighting in the PI places a heavy penalty on errors that occur late in the response and is thus very effective in suppressing the effect of a slow pole as well as in eliminating lightly damped settling behavior.

Due to the factor t^k, the optimal gain K that minimizes J is time varying. However, to obtain useful designs we will determine the suboptimal solution that assumes a time-invariant control gain K. Note that time-varying gains would be very difficult to gain-schedule.

We may successively integrate by parts (see the Problems) to show that the value of (5.5-12) for a given value of K is given by successively solving the nested Lyapunov equations

$$0 = g_0 \equiv A_c^T P_0 + P_0 A_c + P$$

$$0 = g_1 \equiv A_c^T P_1 + P_1 A_c + P_0$$

$$\vdots \tag{5.5-13}$$

$$0 = g_{k-1} \equiv A_c^T P_{k-1} + P_{k-1} A_c + P_{k-2}$$

$$0 = g_k \equiv A_c^T P_k + P_k A_c + k! P_{k-1} + Q + C^T K^T R K C$$

Then

$$J = \tfrac{1}{2}\tilde{x}^T(0) P_k \tilde{x}(0) = \tfrac{1}{2}\bar{x}^T P_k \bar{x} = \tfrac{1}{2}\text{tr}(P_k X) \qquad (5.5\text{-}14)$$

A minimization routine such as Simplex (Nelder and Mead, 1964; Press et al., 1986) can be used to find the optimal gains using (5.5-13) and (5.5-14) to evaluate the PI for a specified value of the gain K.

Alternatively, to use a faster gradient-based routine, we may determine the gradient of J with respect to K. To do so, define the Hamiltonian

$$\mathcal{H} = \tfrac{1}{2}\text{tr}(P_k X) + \tfrac{1}{2}\text{tr}(g_0 S_0) + \ldots + \tfrac{1}{2}\text{tr}(g_k S_k), \qquad (5.5\text{-}15)$$

where $S_i \geq 0$ are matrices of undetermined Lagrange multipliers. Then, by differentiating $\mathcal{H}$ with respect to all variables, necessary conditions for a minimum may be found (see the Problems). These design equations for the LQ tracker with time weighting are summarized in Table 5.5-1.

To use a gradient-based optimization routine such as Davidon–Fletcher–Powell (Press et al., 1986), we may proceed as follows. For a given K, solve the nested Lyapunov equations for P_i and S_i. Since the g_i are then all zero, (5.5-15) shows that $J = \mathcal{H}$. Then (5.5-23) gives the gradient of J with respect to K, which is used by the gradient-based routine to find the updated value of K.

If it is desired to use LQ *regulator* design (as opposed to tracker design, that is, Table 5.3-1) with time-dependent weighting, one need only set $X = I$ (assuming that $E\{x(0)x^T(0)\} = I$) and $\bar{x} = 0$ in the tracker design equations of Table 5.5-1.

Software to determine the optimal value of K given the design parameters k, Q, and R (for both the regulator and tracker) is described in Appendix B. It is called program LQ.

A combination of derivative and time-dependent weighting occurs in the PI:

$$J = \tfrac{1}{2}\int_0^\infty \left(t^k \dot{\tilde{x}}^T P \dot{\tilde{x}} + \dot{\tilde{x}}^T Q \dot{\tilde{x}}\right) dt \qquad (5.5\text{-}16)$$

The optimal gains in this situation may be determined by minimizing

$$J = \tfrac{1}{2}\dot{\tilde{x}}^T(0) P_k \dot{\tilde{x}}(0) = \tfrac{1}{2} r_0^T B_c^T P_k B_c r_0 \qquad (5.5\text{-}17)$$

subject to (5.5-13) with $R = 0$.

A Fundamental Design Property

We now mention a fact of key importance in connection with time-dependent weighting. We will be very concerned to use PIs that are sensible from a design point of view. That is, we will not be content to select P and Q in Table 5.5-1 as $n \times n$ matrices and juggle their entries until a suitable design occurs. This sort of approach is one of the fundamental flaws of modern LQ design.

TABLE 5.5-1: LQ Tracker with Time-Weighted PI

System Model
$$\dot{x} = Ax + Bu + Gr$$
$$y = Cx + Fr$$
Control
$$u = -Ky$$
Performance Index
$$J = \tfrac{1}{2}\int_0^\infty \left[t^k \tilde{x}^T P \tilde{x} + \tilde{x}^T(Q + C^T K^T RKC)\tilde{x}\right] dt$$
Optimal Output Feedback Control
$$0 = g_0 \equiv A_c^T P_0 + P_0 A_c + P$$
$$0 = g_1 \equiv A_c^T P_1 + P_1 A_c + P_0$$
$$\vdots \qquad (5.5\text{-}21)$$
$$0 = g_{k-1} \equiv A_c^T P_{k-1} + P_{k-1} A_c + P_{k-2}$$
$$0 = g_k \equiv A_c^T P_k + P_k A_c + k! P_{k-1} + Q + C^T K^T RKC$$

$$0 = A_c S_k + S_k A_c^T + X$$
$$0 = A_c S_{k-1} + S_{k-1} A_c^T + k! S_k$$
$$0 = A_c S_{k-2} + S_{k-2} A_c^T + S_{k-1} \qquad (5.5\text{-}22)$$
$$\vdots$$
$$0 = A_c S_0 + S_0 A_c^T + S_1$$
$$0 = \frac{\partial \mathcal{H}}{\partial K} = RKC S_k C^T - B^T(P_0 S_0 + \cdots + P_k S_k)C^T + B^T A_c^{-T} P_k \bar{x} \bar{y}^T \qquad (5.5\text{-}23)$$
with r a unit step of magnitude r_0 and
$$\bar{x} = -A_c^{-1} B_c r_0 \qquad (5.5\text{-}24)$$
$$\bar{y} = C\bar{x} + F r_0 \qquad (5.5\text{-}25)$$
$$X = \bar{x}\bar{x}^T = A_c^{-1} B_c r_0 r_0^T B_c^T A_c^{-T} \qquad (5.5\text{-}26)$$
where
$$A_c = A - BKC, \quad B_c = G - BKF$$
Optimal Cost
$$J = \tfrac{1}{2}\text{tr}(P_k X)$$

A sensible PI is one of the form
$$J = \tfrac{1}{2}\int_0^\infty (t^k \tilde{e}^T \tilde{e} + r \tilde{u}^T \tilde{u}) dt, \qquad (5.5\text{-}18)$$

where according to Section 5.4, the error deviation is given by
$$\tilde{e} = -H\tilde{x}, \qquad (5.5\text{-}19)$$

with $z = Hx$ the performance output. This PI corresponds to our desire to make the error small without too much control energy. Since $\tilde{e}^T\tilde{e} = \tilde{x}^T H^T H \tilde{x}$, it amounts to using the PI in Table 5.5-1 with $Q = 0$, $R = rI$, and $P = H^T H$.

However, if (H, A) is not observable and if $k = 0$, there may be problems with any LQ design (Lewis, 1986). Specifically, in this case the Lyapunov equation

$$A_c^T P + P A_c + H^T H + C^T K^T R K C = 0 \quad (5.5\text{-}20)$$

may not have a positive-definite solution P. This could result in some of the feedback gains being set to zero in the LQ optimal solution.

To correct this, we could add a term like $\tilde{x}^T Q \tilde{x}$ in the PI, with $(\sqrt{Q}, A)$ observable. This, however, is exactly what we are trying to avoid, since it will give us all of the elements of Q as design parameters that should be varied until a suitable K results. To avoid this counterintuitive approach, we need only select $k > 0$ in the PI in Table 5.5-1. To see why, consider the case $Q = 0$ and examine Table 5.5-1. Note that even if $(\sqrt{P}, A)$ is not observable, $\left[k! P_{k-1} \right)^{1/2}, A \right]$ may be observable for some $k > 0$. If so, the last Lyapunov equation in (5.5-21) will have a positive definite solution P_k, which will correct the observability problem. That is, by using time weighting, the LQ observability problem is corrected. We will illustrate this point in Example 5.5-2.

Example 5.5-1: Constrained Feedback Control for F-16 Lateral Dynamics.
In Example 5.3-1 we showed how to design a lateral stability augmentation system for an F-16. The resulting gain matrix K had eight nonzero entries. It would be desirable to avoid gain scheduling such a large number of gains, as well as to avoid feedback from roll rate and bank angle to rudder, and from washed-out yaw-rate and sideslip to aileron. That is, the gain matrix should have the form

$$K = \begin{bmatrix} 0 & x & 0 & x \\ x & 0 & x & 0 \end{bmatrix} \quad (1)$$

This *constrained output feedback* regulator is quite easy to design using the techniques just discussed. Indeed, select a PI of the form (5.5-5) with $g_{11} = 1000$, $g_{13} = 1000$, $g_{22} = 1000$, $g_{24} = 1000$ in order to weight the unwanted entries of $K = [k_{ij}]$. Then the algorithm of Table 5.3-1, with the modified equation (5.5-6) used to evaluate the PI in a numerical minimization scheme, yields the feedback gain matrix

$$K = \begin{bmatrix} -1\text{E}-3 & -0.55 & 1\text{E}-3 & -0.49 \\ -1.14 & -1\text{E}-3 & 0.05 & 1\text{E}-3 \end{bmatrix} \approx \begin{bmatrix} 0 & -0.55 & 0 & -0.49 \\ -1.14 & 0 & 0.05 & 0 \end{bmatrix} \quad (2)$$

The same Q and R were used as in Example 5.3-1. The resulting closed-loop poles are

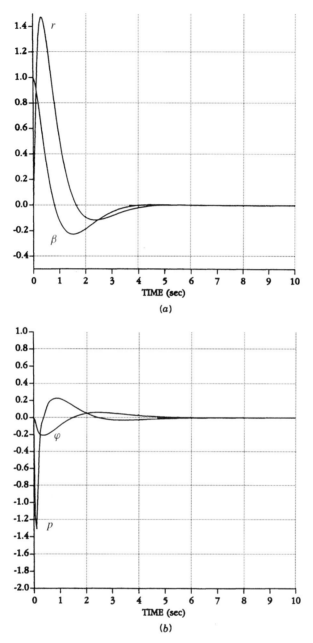

Figure 5.5-1 Closed-loop lateral response. (*a*) dutch roll states β and r, (*b*) spiral and roll subsidence states ϕ and p.

$$s = -1.16 \pm j0.99 \quad \text{dutch roll mode } (r, \beta)$$
$$-0.79 \quad \text{spiral mode}$$
$$-7.42 \quad \text{roll subsidence mode}$$
$$-11.54 \pm j19.51, -12.27 \tag{3}$$

Note that the spiral and roll subsidence modes now consist of two real poles so that the complex roll mode is absent. The closed-loop response is shown in Figure 5.5-1. It should be compared to the response obtained in Example 5.3-1 as well as in examples in Chapter 4.

An alternative design technique is simply to use the option in program LQ of instructing the program to leave certain elements of K fixed at zero during the minimization procedure. ∎

Example 5.5-2: Time-Dependent Weighting Design of Normal Acceleration CAS.
In Example 5.4-1 we designed a normal acceleration CAS. A deficiency with that approach was the need to check for the observability of $(\sqrt{Q}, A)$; there, unobservability led us to weight the integrator output in Q. In this example we show how to avoid the observability issue by using time-dependent weighting in the PI.

The aircraft and controller dynamics are the same as in Example 5.4-1. Here, however, we will select the time-weighted PI

$$J = \tfrac{1}{2} \int_0^\infty \left(t^2 \tilde{e}^2 + \rho \tilde{u}^2 \right) dt, \tag{1}$$

which is entirely sensible from a performance point of view and contains only one design parameter to be tuned. This corresponds to the PI in Table 5.5-1 with $P = H^T H, Q = 0, R = \rho$.

Selecting $\rho = 0.05$ and using program LQ we obtained the control gains

$$K = [-0.847 \quad 0.452 \quad 1.647 \quad 8.602], \tag{2}$$

the closed-loop poles

$$s = -1.90 \pm j2.58$$
$$-2.35$$
$$-13.88 \pm j3.12, \tag{3}$$

and the step response shown in Figure 5.5-2. It is much better than the result of Example 5.4-1, and was obtained without juggling the elements of the Q matrix or worrying about observability issues. By using time weighting in the PI, we have formulated a design problem that has only one design parameter that needs to be

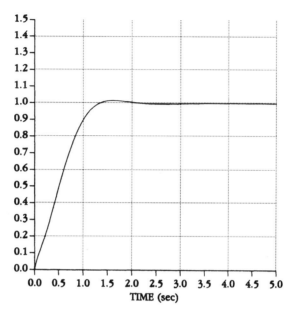

Figure 5.5-2 Normal acceleration step response.

varied, namely, the control weighting, ρ. This entire design took 5 minutes. Contrast to Example 4.5-3. ∎

Example 5.5-3: Pitch-Rate Control System Using LQ Design.

In this example we reconsider pitch-rate control system design using LQ techniques. The approach to be used here should be compared to the classical approach used in Chapter 4. It will be demonstrated how two of the PIs just developed can simplify the control system design, since they have *only one design parameter that must be tuned to obtain good performance*. This LQ technique is therefore in sharp contrast to the classical approach, where we had to vary all three elements of the gain matrix in successive loop-closure design. It is also in contrast to the traditional modern LQ approaches, where all the elements of the PI weighting matrices must generally be tuned to obtain good performance and where the observability properties of the PI must be considered in selecting the state weighting matrix.

Since we are using a modern LQ-based approach, a sensible formulation of the problem should result in closed-loop stability for all selections of the design parameter. This is an extremely important property of modern control design techniques and in complete contrast to classical techniques, where stability in multiloop systems can be difficult to achieve.

a. Aircraft and Control System Dynamics. The pitch control system is shown in Figure 5.5-3, where the control input is elevator actuator voltage $u(t)$ and r is a

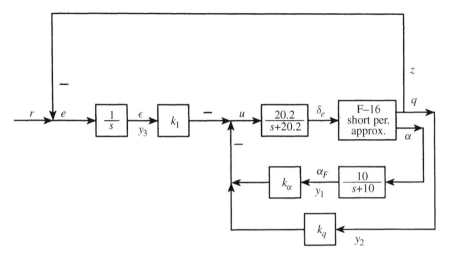

Figure 5.5-3 Pitch-rate control system.

reference step input corresponding to the desired pitch command. Thus the performance output, $z(t)$, is the pitch rate, q. The measured outputs $y(t)$ are pitch, q, and angle of attack, α; however, since α measurements are quite noisy, a low-pass filter with a cutoff frequency of 10 rad/s is used to provide filtered measurements α_F of the angle of attack. To ensure zero steady-state error an integrator was added in the feedforward channel; this corresponds to the compensator dynamics. The integrator output is ϵ.

We used the short-period approximation to the F-16 dynamics linearized about the nominal flight condition in Table 3.6-3 (502 ft/s, 0 ft altitude, level flight, with the cg at 0.35 $\bar{c}$). Thus the basic aircraft states of interest are α and q. An additional state is introduced by the elevator actuator. The elevator deflection is δ_e.

The states and outputs of the plant plus compensator are

$$x = \begin{bmatrix} \alpha \\ q \\ \delta_e \\ \alpha_F \\ \epsilon \end{bmatrix}, \quad y = \begin{bmatrix} \alpha_F \\ q \\ \epsilon \end{bmatrix} \qquad (1)$$

and the system dynamics are described by

$$\dot{x} = Ax + Bu + Gr \qquad (2)$$

$$y = Cx + Fr \qquad (3)$$

$$z = Hx \qquad (4)$$

with

$$A = \begin{bmatrix} -1.01887 & 0.90506 & -0.00215 & 0 & 0 \\ 0.82225 & -1.07741 & -0.17555 & 0 & 0 \\ 0 & 0 & -20.2 & 0 & 0 \\ 10.0 & 0 & 0 & -10 & 0 \\ 0 & -57.2958 & 0 & 0 & 0 \end{bmatrix}$$

$$B = \begin{bmatrix} 0 \\ 0 \\ 20.2 \\ 0 \\ 0 \end{bmatrix} \quad G = \begin{bmatrix} 0 \\ 0 \\ 0 \\ 0 \\ 1 \end{bmatrix}$$

$$C = \begin{bmatrix} 0 & 0 & 0 & 57.2958 & 0 \\ 0 & 57.2958 & 0 & 0 & 0 \\ 0 & 0 & 0 & 0 & 1 \end{bmatrix} \quad F = \begin{bmatrix} 0 \\ 0 \\ 0 \end{bmatrix}$$

$$H = \begin{bmatrix} 0 & 57.2958 & 0 & 0 & 0 \end{bmatrix}$$

The factor of 57.2958 is added to convert angles from radians to degrees.

The control input is

$$u = -Ky = -\begin{bmatrix} k_\alpha & k_q & k_I \end{bmatrix} y = -k_\alpha \alpha_F - k_q q - k_I \epsilon \tag{5}$$

It is required to select the feedback gains to yield good closed-loop response to a step input at r, which corresponds to a single-input/multi-output design problem.

Now consider two LQ designs based on two different performance indices. The modified PIs introduced in this section will mean that we do not need to worry about observability issues and that *only one design parameter will appear*. This is significant in view of the fact that there are five states and three control gains to find.

Since the integrator makes the system type 1, the steady-state error $\bar{e}$ is equal to zero and

$$e(t) = \tilde{e}(t) \tag{6}$$

Thus the PI term involving $\bar{e}$ in Section 5.4 is not required.

b. Time-Dependent Weighting Design. Consider the PI

$$J = \tfrac{1}{2} \int_0^\infty \left(t^2 \tilde{e}^2 + \rho \tilde{u}^2 \right) dt \tag{7}$$

This is a natural PI that corresponds to the actual performance requirements of keeping the tracking error small without using too much control energy, and also has the

MODIFYING THE PERFORMANCE INDEX

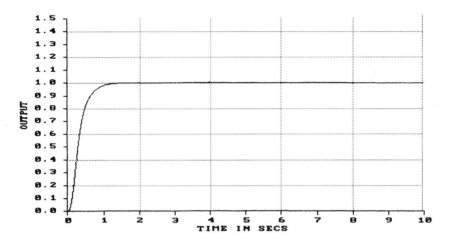

Figure 5.5-4 Pitch-rate step response using time-dependent weighting design.

important advantage of requiring the adjustment of only one design parameter ρ. It amounts to using $P = H^T H$, $Q = 0$, $R = \rho$ in Table 5.5-1.

Program LQ was used to solve the design equations in Table 5.5-1 for several values of ρ. A good step response was found with $\rho = 1$, which yielded optimal gains of

$$K = [-0.046 \quad -1.072 \quad 3.381] \tag{8}$$

closed-loop poles of $s = -8.67 \pm j9.72, -9.85, -4.07$, and -1.04, and the step response in Figure 5.5-4. Compare to the results of Example 4.5-1.

c. Derivative Weighting Design. Since elevator actuator *rate* has a stronger intuitive connection to "control activity" than does elevator displacement, let us illustrate derivative weighting by repeating the design. Select the PI

$$J = \tfrac{1}{2} \int_0^\infty \left(pt^2 e^2 + \dot{\delta}_e^2 \right) dt \tag{9}$$

Since $e(t) = \dot{\epsilon}(t)$, this may be written

$$J = \tfrac{1}{2} \int_0^\infty \left(pt^2 \dot{\epsilon}^2 + \dot{\delta}_e^2 \right) dt, \tag{10}$$

with $\epsilon(t)$ and $\delta_e(t)$ the deviations in the integrator output and elevator deflection. This is exactly the derivative weighting PI (5.5-16) with $P = \text{diag}\{0, 0, 0, 0, p\}$ and $Q = \text{diag}\{0, 0, 1, 0, 0\}$.

It should be emphasized that we have again been careful to formulate the problem in such a way that only one design parameter, namely, p, needs to be adjusted in the iterative design phase.

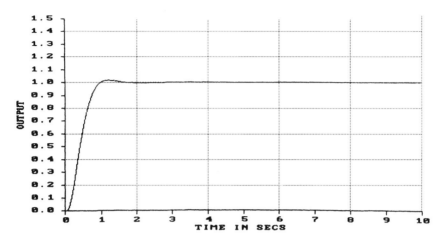

Figure 5.5-5 Pitch-rate step response using derivative weighting design.

The software described in Appendix B was used to minimize (5.5-17) subject to (5.5-13) for several values of p. The weight $p = 10$ led to a good step response, as shown in Figure 5.5-5. The feedback gain matrix was

$$K = [-0.0807 \quad -0.475 \quad 1.361] \tag{11}$$

and the closed-loop poles were at $s = -3.26 \pm j2.83, -1.02, -10.67,$ and -14.09. These poles are virtually identical to those obtained in Example 4.5-1. Compare the design process in this example with the design process in that example. ∎

Example 5.5-4: Multivariable Wing Leveler. In this example, we will illustrate a multi-input/multi-output (MIMO) design using the LQ approach developed in this chapter. This example should be compared with Chapter 4, where we designed a two-input/two-output roll damper/yaw damper using classical control by successive loop closures.

a. Control System Structure. The control system shown in Figure 5.5-6 is meant to hold the aircraft's wings level while providing yaw damping by holding washed-out yaw rate, r_w, at zero. It is a two-channel system. In the upper channel there is an outer-loop unity-gain feedback of bank angle, ϕ, with an inner-loop feedback of roll rate, p. This channel has a PI compensator to make the system type 1 to achieve zero steady-state bank angle error. The control input for the upper channel is aileron deflection, δ_a. The lower channel has a feedback of washed-out yaw rate, r_w; in this channel the control input is rudder deflection, δ_r.

The reference command is $r_c = \begin{bmatrix} r_\phi & r_r \end{bmatrix}^T$. The tracking control system should hold ϕ at the commanded value of r_ϕ, and r_w at the commanded value of r_r, which is equal to zero. To hold the wings level, r_ϕ is set equal to zero, although it could be any commanded bank angle. The tracking error is $e = \begin{bmatrix} e_\phi & e_r \end{bmatrix}^T$ with

MODIFYING THE PERFORMANCE INDEX 449

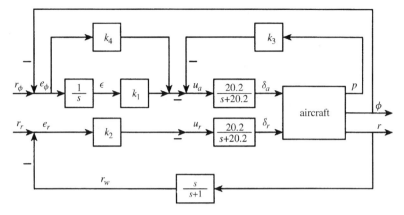

Figure 5.5-6 Wing-leveler lateral control system.

$$e_\phi = r_\phi - \phi$$
$$e_r = r_r - r_w \tag{1}$$

b. State Equations for Aircraft and Control Dynamics. As in Example 5.3-1, we used the F-16 linearized lateral dynamics at the nominal flight condition in Table 3.6-3 ($V_T = 502$ ft/s, 300 psf dynamic pressure, cg at 0.35 $\bar{c}$) retaining the lateral states sideslip, β, bank angle, ϕ, roll rate, p, and yaw rate, r. Additional states δ_a and δ_r are introduced by the aileron and rudder actuators. The washout filter state is called x_w. We denote by ϵ the output of the controller integrator in the upper channel. Thus the entire state vector is

$$x = [\beta \quad \phi \quad p \quad r \quad \delta_a \quad \delta_r \quad x_w \quad \epsilon]^T \tag{2}$$

The full state-variable model of the aircraft plus actuators, washout filter, and control dynamics is of the form

$$\dot{x} = Ax + Bu + Gr_c, \tag{3}$$

with

$$A = \begin{bmatrix} -0.3220 & 0.0640 & 0.0364 & -0.9917 & 0.0003 & 0.0008 & 0 & 0 \\ 0 & 0 & 1 & 0.0037 & 0 & 0 & 0 & 0 \\ -30.6492 & 0 & -3.6784 & 0.6646 & -0.7333 & 0.1315 & 0 & 0 \\ 8.5395 & 0 & -0.0254 & -0.4764 & -0.0319 & -0.0620 & 0 & 0 \\ 0 & 0 & 0 & 0 & -20.2 & 0 & 0 & 0 \\ 0 & 0 & 0 & 0 & 0 & -20.2 & 0 & 0 \\ 0 & 0 & 0 & 57.2958 & 0 & 0 & -1 & 0 \\ 0 & -1 & 0 & 0 & 0 & 0 & 0 & 0 \end{bmatrix}$$

$$B = \begin{bmatrix} 0 & 0 \\ 0 & 0 \\ 0 & 0 \\ 0 & 0 \\ 20.2 & 0 \\ 0 & 20.2 \\ 0 & 0 \\ 0 & 0 \end{bmatrix} \quad G = \begin{bmatrix} 0 & 0 \\ 0 & 0 \\ 0 & 0 \\ 0 & 0 \\ 0 & 0 \\ 0 & 0 \\ 0 & 0 \\ 1 & 0 \end{bmatrix} \quad (4)$$

The performance output that should follow the reference input $[r_\phi \quad r_r]^\mathrm{T}$ is

$$z = \begin{bmatrix} \phi \\ r_w \end{bmatrix} = \begin{bmatrix} 0 & 1 & 0 & 0 & 0 & 0 & 0 & 0 \\ 0 & 0 & 0 & 57.2958 & 0 & 0 & -1 & 0 \end{bmatrix} x = Hx, \quad (5)$$

where the factor 57.2958 converts radians to degrees. According to the figure, if we define the measured output as

$$y = \begin{bmatrix} \epsilon \\ e_r \\ p \\ e_\phi \end{bmatrix} = Cx + Fr_c \quad (6)$$

with

$$C = \begin{bmatrix} 0 & 0 & 0 & 0 & 0 & 0 & 0 & 1 \\ 0 & 0 & 0 & -57.2958 & 0 & 0 & 1 & 0 \\ 0 & 0 & 1 & 0 & 0 & 0 & 0 & 0 \\ 0 & -1 & 0 & 0 & 0 & 0 & 0 & 0 \end{bmatrix}$$

$$F = \begin{bmatrix} 0 & 0 \\ 0 & 1 \\ 0 & 0 \\ 1 & 0 \end{bmatrix} \quad (7)$$

the control input $u = [u_a \quad u_r]^\mathrm{T}$ may be expressed as

$$u = -Ky \quad (8)$$

with

$$K = \begin{bmatrix} k_1 & 0 & k_3 & k_4 \\ 0 & k_2 & 0 & 0 \end{bmatrix} \quad (9)$$

The control gains, k_i, must now be determined for satisfactory closed-loop response. Therefore, this is an output-feedback design problem exactly of the form addressed in this chapter. Note that some of the entries of K must be constrained to zero to yield the desired control structure shown in Figure 5.5-6.

c. LQ Output Feedback Design. To guarantee tracking by $z(t)$ of the reference command $r_c(t)$, we may select the PI

$$J = \tfrac{1}{2}\int \left(t^2 \tilde{x}^T P \tilde{x} + \tilde{u}^T \tilde{u}\right) dt + \tfrac{1}{2} v \bar{e}^T \bar{e} \tag{10}$$

with $\tilde{x}(t)$ and $\tilde{u}(t)$ the state and control deviations defined in Section 5.4 and $\bar{e}$ the steady-state error. Although the integrator in the upper control channel guarantees that $\bar{e}_\phi$ will be zero, the steady-state error weighting v is required to ensure that $\bar{e}_r$ is small. Note that v is a scalar.

The design equations for K using this PI are given in Table 5.5-1, with, however, the extra terms from Table 5.4-1 added to (5.5-23) due to the steady-state error weighting v. Thus K is easily determined using program LQ.

Several attempts were made to obtain suitable closed-loop behavior using different values for v and P. Finally, it was found that good behavior was obtained with $v = 10$ and P selected to weight the states β, ϕ, p, r, and ϵ, as well as the cross-term in ϕr. That is,

$$p_{11} = p_{22} = p_{33} = p_{44} = p_{88} = 100, \qquad p_{24} = p_{42} = 10 \tag{11}$$

The motivation for the p_{24} cross-weighting is that after a few design attempts with different P, it was found that there were always several barely stable and badly damped complex pole pairs in the closed-loop system. The p_{24}, p_{42} cross-weighting penalizes the dutch roll mode, which was one of the ones yielding problems. The motivation for selecting p_{88} weighting is that good results are generally obtained if the integrator output is weighted.

Using the final selection of v and P, the control gains were found to be

$$k_1 = 15.04, \qquad k_2 = 0.1822,$$
$$k_3 = -5.348, \qquad k_4 = 22.52, \tag{12}$$

yielding closed-loop poles of

$$-0.72 \pm j3.03$$
$$-1.12 \pm j0.07$$
$$-2.43, -5.05 \tag{13}$$
$$-15.3, -19.4$$

d. Simulation. The closed-loop response to a reference command of $r_\phi = 1, r_r = 0$ is shown in Figure 5.5-7. The transient response and steady-state errors are both quite

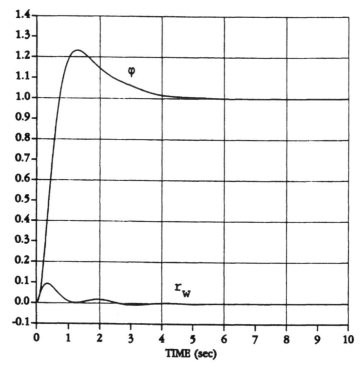

Figure 5.5-7 Closed-loop response to a command of $r_\phi = 1$, $r_r = 0$. Bank angle ϕ (rad) and washed-out yaw rate (rad/s).

satisfactory. This is despite the presence of an underdamped pole pair at $-0.72 \pm j3.03$. One should recall the discussion in Chapter 4, where the strong coupling between the aircraft roll and yaw channels was emphasized. Despite this, Figure 5.5-7 shows that we have been quite successful in decoupling the yaw rate from the bank angle. ∎

Example 5.5-5: Glide-Slope Coupler. A glide-slope coupler is part of an automatic landing system—it guides an aircraft down a predetermined flight path to the end of a runway. At the end of the descent another control system, the automatic flare control (Example 5.6-1), is switched in to cause the aircraft to flare to a landing.

In this example we design a glide-slope coupler for the longitudinal dynamics of a medium-sized transport aircraft. Our approach should be compared to the frequency-domain approach in Example 4.6-4. See also Blakelock (1965).

a. Aircraft Dynamics. The important inputs are both elevator and throttle for this problem, since both are needed to fly down a glide path in a coordinated manner. Exactly as in Example 4.6-4, the longitudinal dynamics of the aircraft were linearized

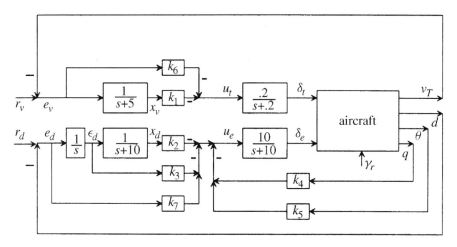

Figure 5.5-8 Glide-slope coupler.

about a velocity of $V_T = 250$ ft/s with the cg at $0.25\ \bar{c}$ and including throttle and elevator actuators. The state and control input are

$$x = [v_T \ \alpha \ \theta \ q \ \delta_t \ \delta_e]^T, \qquad u = [u_t \ u_e]^T, \qquad (1)$$

with v_T the deviation from trim velocity. The dynamics are described by

$$\dot{x} = Ax + Bu, \qquad (2)$$

where A and B may be found by referring to Example 4.6-4. (In finding the A and B in (2) from the matrices in Example 4.6-4, note our selection of states.)

At this point it is worthwhile to examine Figure 5.5-8, which we are starting to construct.

b. Glide-Slope Geometry. The glide-slope geometry is discussed in Example 4.6-4. The commanded or reference flight path angle $-\gamma_r$ is generally 2.5 degrees. The perpendicular distance from the glide path is $d(t)$.

Our control objectives in the glide-slope coupler are to regulate to zero the off-glide-path distance, d, and the deviation from trim velocity, v_T. Then the aircraft will remain on the glide path with the nominal velocity of $V_T = 250$ ft/s. To accomplish this, the two control inputs are throttle and elevator. The outputs available for feedback are pitch rate, q, pitch angle, θ, v_T, and d, which is available from measurements taken from the ground.

The component of velocity perpendicular to the glide path is given by

$$\dot{d} = V_T \sin(\gamma - \gamma_r) \approx V_T(\gamma - \gamma_r) \qquad (3)$$

when $(\gamma - \gamma_r)$ is small. We will assume that the velocity deviation v_T is small, and take V_T in (3) as the trim velocity 250 ft/s. To follow the glide path, we require $\dot{d} = 0$, so that the flight path angle γ should be equal to γ_r. Then the aircraft will descend at an angle of $\gamma_r = -2.5°$.

In terms of variables in the state vector in (1), we may use $\gamma = \theta - \alpha$ to write

$$\dot{d} = V_T\theta - V_T\alpha - \frac{V_T}{57.2958}\gamma_r = V_T\theta - V_T\alpha - 4.3633\gamma_r, \tag{4}$$

with θ and α in radians and γ_r in degrees. Therefore, we may include the off-glide-path distance d as a state in (1) by redefining

$$x = [v_T \quad \alpha \quad \theta \quad q \quad d \quad \delta_t \quad \delta_e]^T \tag{5}$$

c. Control System Structure. Our objective is to regulate v_T and d to zero. Thus, we may define the performance output as

$$z = \begin{bmatrix} v_T \\ d \end{bmatrix} = Hx \tag{6}$$

Now examine Figure 5.5-8, which we have drawn to show that this may be considered as a tracking problem with reference commands r_v and r_d of zero. The tracking error is $e = [e_v \quad e_d]^T$ with

$$\begin{aligned} e_v &= r_v - v_T \\ e_d &= r_d - d \end{aligned} \tag{7}$$

To obtain zero steady-state error in $v_T(t)$ and $d(t)$, we could add integrators in each of the forward error paths. However, according to the open-loop dynamics in Example 4.6-4 there are already several poles near the origin. Adding more poles near the origin makes the problem of stabilization more difficult.

Since we are more concerned about keeping d exactly zero, let us only add an integrator in the forward path corresponding to the tracking error in d. We can then obtain a small enough error in v_T without a forward-path integrator by using weighting of the steady-state error, as we will soon see.

An additional consideration for including a forward-path integrator in the d channel is the following. Note from (4) and Figure 5.5-8 that the commanded glide-path angle γ_r acts as a constant disturbance of magnitude $-2.5°$ into the system. The disturbance affects $\dot{d}$. To reject this constant disturbance, we need a type-1 system with respect to d, which requires the integrator in the d feedforward path.

We can gain considerable insight by having root-locus design techniques in mind during a design by modern control. Thus to pull the closed-loop poles into the left-half plane, we may add compensator zeros in the left-half plane. To implement the compensators without pure differentiation, we should add poles relatively far in the left-half plane, where they will not appreciably affect the root locus. Thus let us propose a lead compensator in each forward channel (see Figure 5.5-8).

The compensators we propose are of the form

$$\frac{w_v}{e_v} = \frac{k_1}{s+5} + k_6 = k_6 \frac{s + (5 + k_1/k_6)}{s+5} \tag{8}$$

$$u_t = -w_v$$

and

$$\frac{w_d}{e_d} = \frac{k_2}{s(s+10)} + \frac{k_3}{s} + k_7$$

$$= k_7 \frac{s^2 + (10 + k_3/k_7)s + (k_2 + 10k_3)/k_7}{s(s+10)} \tag{9}$$

$$u_e = -w_d$$

The important point to note is that by varying the control gains, we may adjust *both the compensator gain and its zeros.* Thus the LQ optimization routine can adjust *the zeros of the compensators,* presumably inducing lead compensation where it is required. We have selected the throttle compensator pole at $s = -5$ and the distance compensator pole at $s = -10$; however, any poles far to the left compared to the aircraft poles would suffice.

As we have seen in Example 4.6-4, selecting multiple control gains by classical techniques requires a successive-loop-closure approach. We hope to show that finding suitable gains using modern control theory is far easier, given a sensible problem formulation.

To formulate the controller so that the gains may be determined by our output-feedback LQ approach, note that state variable representations of (8) and (9) are given by

$$\dot{x}_v = -5x_v + e_v = -5x_v - v_T + r_v \tag{10}$$

$$u_t = -k_1 x_v - k_6 e_v = -k_1 x_v - k_6(-v_T + r_v) \tag{11}$$

and

$$\dot{\epsilon}_d = e_d = -d + r_d \tag{12}$$

$$\dot{x}_d = -10 x_d + \epsilon_d \tag{13}$$

$$u_e = -k_2 x_d - k_3 \epsilon_d - k_7 e_d = -k_2 x_d - k_3 \epsilon_d - k_7(-d + r_d) \tag{14}$$

The dynamical equations (4), (10), (12), and (13) may be incorporated into the system description by defining the augmented state

$$x = [v_T \quad \alpha \quad \theta \quad q \quad d \quad \delta_t \quad \delta_e \quad x_v \quad x_d \quad \epsilon_d]^T \tag{15}$$

Then the augmented system is described by

$$\dot{x} = Ax + Bu + Gr \tag{16}$$

with

$$A = \begin{bmatrix} -0.04 & 19.0096 & -32.1689 & 0 & 0 & 10.1 & 0 & 0 & 0 & 0 \\ -0.001 & -0.64627 & 0 & 1 & 0 & 0 & 0 & 0 & 0 & 0 \\ 0 & 0 & 0 & 1 & 0 & 0 & 0 & 0 & 0 & 0 \\ 0 & -0.7739 & 0 & -0.529765 & 0 & 0.02463 & -0.011 & 0 & 0 & 0 \\ 0 & -250 & 250 & 0 & 0 & 0 & 0 & 0 & 0 & 0 \\ 0 & 0 & 0 & 0 & 0 & -0.2 & 0 & 0 & 0 & 0 \\ 0 & 0 & 0 & 0 & 0 & 0 & -10 & 0 & 0 & 0 \\ -1 & 0 & 0 & 0 & 0 & 0 & 0 & -5 & 0 & 0 \\ 0 & 0 & 0 & 0 & 0 & 0 & 0 & 0 & -10 & 1 \\ 0 & 0 & 0 & 0 & -1 & 0 & 0 & 0 & 0 & 0 \end{bmatrix}$$

$$\tag{17}$$

$$B = \begin{bmatrix} 0 & 0 & 0 \\ 0 & 0 & 0 \\ 0 & 0 & 0 \\ 0 & 0 & 0 \\ 0 & 0 & -4.3633 \\ 0.2 & 0 & 0 \\ 0 & 10 & 0 \\ 0 & 0 & 0 \\ 0 & 0 & 0 \\ 0 & 0 & 0 \end{bmatrix}, \quad G = \begin{bmatrix} 0 & 0 \\ 0 & 0 \\ 0 & 0 \\ 0 & 0 \\ 0 & 0 \\ 0 & 0 \\ 0 & 0 \\ 1 & 0 \\ 0 & 0 \\ 0 & 1 \end{bmatrix}$$

To incorporate the constant disturbance γ_r required in (4), we have defined an augmented input

$$u' = \begin{bmatrix} u^T & \gamma_r \end{bmatrix}^T = \begin{bmatrix} u_i & u_e & \gamma_r \end{bmatrix}^T \tag{18}$$

Inputs such as γ_r which are not actual controls, nor reference signals $r(t)$ in the usual tracking system sense, are called *exogenous inputs*. Although they play the role of disturbances in the system, they are crucial in guaranteeing the desired system behavior. Indeed, were we to ignore γ_r, the glide-slope coupler would always make the aircraft fly a horizontal path!

It should be clearly understood that for the design of the control system, only the control input $u(t)$ is used. The full input $u'(t)$ will be required only in the simulation

state, where γ_r will be set equal to -2.5 degrees to obtain the desired landing approach behavior.

In (16)/(17) the reference input is defined as

$$r = [r_v \quad r_d]^T, \tag{19}$$

which is zero for the glide-slope coupler.

The equations (11) and (14) may be incorporated by defining a new measured output as

$$y = [x_v \quad x_d \quad \epsilon_d \quad q \quad \theta \quad e_v \quad e_d]^T \tag{20}$$

Then

$$y = Cx = Fr \tag{21}$$

with

$$C = \begin{bmatrix} 0 & 0 & 0 & 0 & 0 & 0 & 0 & 1 & 0 & 0 \\ 0 & 0 & 0 & 0 & 0 & 0 & 0 & 0 & 1 & 0 \\ 0 & 0 & 0 & 0 & 0 & 0 & 0 & 0 & 0 & 1 \\ 0 & 0 & 0 & 57.2958 & 0 & 0 & 0 & 0 & 0 & 0 \\ 0 & 0 & 57.2958 & 0 & 0 & 0 & 0 & 0 & 0 & 0 \\ -1 & 0 & 0 & 0 & 0 & 0 & 0 & 0 & 0 & 0 \\ 0 & 0 & 0 & 0 & -1 & 0 & 0 & 0 & 0 & 0 \end{bmatrix}$$

$$F = \begin{bmatrix} 0 & 0 \\ 0 & 0 \\ 0 & 0 \\ 0 & 0 \\ 0 & 0 \\ 1 & 0 \\ 0 & 1 \end{bmatrix} \tag{22}$$

Now, according to Figure 5.5-8, the control vector $u(t)$ is given by the output feedback

$$u = \begin{bmatrix} u_t \\ u_e \end{bmatrix} = -\begin{bmatrix} k_1 & 0 & 0 & 0 & 0 & k_6 & 0 \\ 0 & k_2 & k_3 & k_4 & k_5 & 0 & k_7 \end{bmatrix} y = -Ky, \tag{23}$$

which has some elements constrained to zero.

According to (6), we may write

$$z = \begin{bmatrix} v_T \\ d \end{bmatrix} = \begin{bmatrix} 1 & 0 & 0 & 0 & 0 & 0 & 0 & 0 & 0 \\ 0 & 0 & 0 & 0 & 1 & 0 & 0 & 0 & 0 \end{bmatrix} x = Hx \quad (24)$$

At this point we have succeeded in casting the glide slope coupler design problem into the formulation required in Tables 5.4-1 and 5.5-1.

It is important to understand the construction of the matrices in (17), (22), and (24), for this problem formulation stage is one of the most important phases in the LQ design technique.

d. PI and Control Design.

The other important phase in LQ design is the selection of an appropriate PI. Since the loop gain around the velocity loop is not of type 1, we will require weighting of the steady-state error to force v_T to go to zero at steady state. Thus let us propose the PI

$$J = \tfrac{1}{2} \int_0^\infty \left(q t^2 \tilde{e}^T \tilde{e} + \tilde{u}^T \tilde{u} \right) dt + \tfrac{1}{2} \bar{v} \bar{e}^T \bar{e} \quad (25)$$

The motivation for the weighting t^2 follows. Weighting $\tilde{e}$ in the PI makes practical sense since we want it to vanish. However, $\tilde{e}^T \tilde{e} = \tilde{x}^T H^T H \tilde{x}$, and (H, A) is not observable. In fact, the compensator states are not observable through $z = Hx$. An LQ design without the weighting t^2 would, therefore, fail. To correct the situation, we could weight the entire state in the PI by using a term like $\tilde{x}^T Q \tilde{x}$. However, this would give us too many design parameters (i.e., the elements of Q) and lead to a counterintuitive situation.

We prefer to work with sensible PIs, and in this situation we want to retain the weighting of $\tilde{e}(t)$, which is the variable of direct concern to us. Therefore, we use t^2 weighting to correct the observability problem. See the discussion preceding Example 5.5-1.

With t^2 weighting, a large value of the scalar q will result in a closed-loop system that is too fast. After several design iterations, it was found that suitable values for the PI design parameters were $q = 0.001$, $v = 100$. We employed program LQ to solve for the optimal gain K using the design equations of Table 5.5-1, including the steady-state error weighting from Table 5.4-1. We selected the option of fixing seven of the gain elements to zero as required by (23).

With $q = 0.001$, $v = 100$, the optimal control gains were

$$K = \begin{bmatrix} 2.598 & 0 & 0 & 0 & 0 & -0.9927 & 0 \\ 0 & 583.7 & -58.33 & -2.054 & -1.375 & 0 & 6.1 \end{bmatrix} \quad (26)$$

and the closed-loop poles were at

$$-0.27 \pm j1.01$$

$$-0.36 \pm j0.49$$
$$-0.37 \pm j0.09 \tag{27}$$
$$-1.18, -4.78, -8.38, -10.09$$

Thus, the slowest time constant is $1/0.27 \approx 4$ s.

e. Simulation and Discussion. A simulation of the glide-slope coupler appears in Figure 5.5-9. The aircraft was initialized in level flight at 1500 ft. The glide-slope coupler was switched on as the aircraft crossed through the glide path.

For simulation purposes, we used the exogenous input $\gamma_r = -2.5°$ (the desired glide-path angle) and reference commands of $r_v = 0, r_d = 0$. Altitude h was added as a state using the equation for vertical velocity

$$\dot{h} = V_T \sin \gamma \approx V_T (\theta - \alpha), \tag{28}$$

with V_T assumed to be the trim velocity of 250 ft/s.

According to the altitude plot in Figure 5.5-9a, after a small transient lasting about 20 s, the aircraft settles down onto the glide path and follows it down. Touchdown occurred at 137.5 s. Figure 5.5-9b shows the off-glide-path error d.

Figure 5.5-9c shows angle of attack and pitch angle. Note that after the transient, the flight-path angle is given by $\gamma = \theta - \alpha = -2.5°$. Since in the descending configuration the aircraft is no longer at the original trim condition, a small angle of attack α of -0.18 degrees remains at steady state. The final pitch angle θ is -2.68 degrees.

According to Figure 5.5-9d, the velocity deviation v_T settles out at 0.29 ft/s. This is a consequence of the fact that there is no integrator in the forward e_v path in Figure 5.5-8. Thus the steady-state velocity on the glide path is $V_T = 250.29$ ft/s; this is very suitable from a performance point of view. The smallness of the steady-state deviation despite the fact that the v_T loop is of type 0 is a consequence of the steady-state error weighting v in the PI (25).

Finally, the elevator and throttle control efforts δe and δt are shown in Figure 5.5-9e. Note the coordinated control achieved in this two-input system using the LQ approach. Since the descent down the glide path does not represent the original trim condition, the steady-state values of the control efforts are not zero. Intuitively, less throttle is required to maintain 250 ft/s if the aircraft is descending.

Figure 5.5-9 shows that as the aircraft passes through the glide path, the elevator is pushed forward and the throttle is cut. As a result, the angle of attack and pitch angle decrease. After a slight positive position error, d, and an initial increase in velocity, v_T, further control effort stabilizes the aircraft on the glide path.

It is interesting to note the fundamental mechanism behind the glide slope coupler. Namely, we regulate $\dot{d}$ in (3)–(4) to zero so that $\gamma = \gamma_r = -2.5°$. Then, according to (28), $\dot{h} = V_T \sin \gamma_r$, the appropriate descent rate to stay on the glide path.

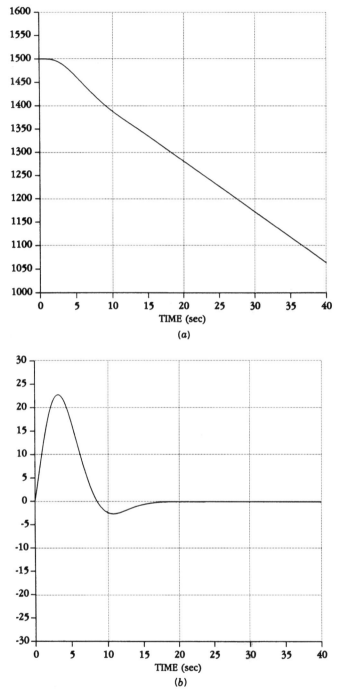

Figure 5.5-9 Glide-slope coupler responses: (*a*) altitude h (ft); (*b*) off-glide path distance *d* (ft).

MODIFYING THE PERFORMANCE INDEX 461

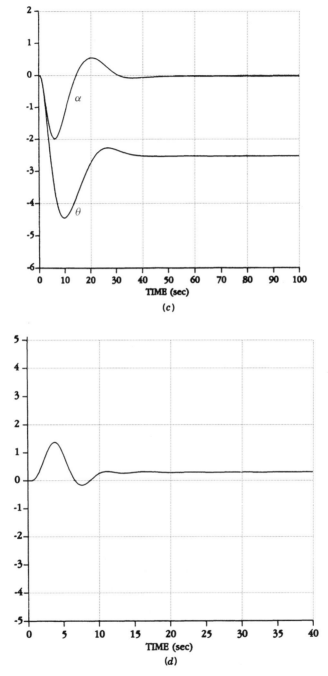

Figure 5.5-9 (*continued*) (*c*) Angle of attack α and pitch angle θ (deg); (*d*) velocity deviation v_T (ft/s).

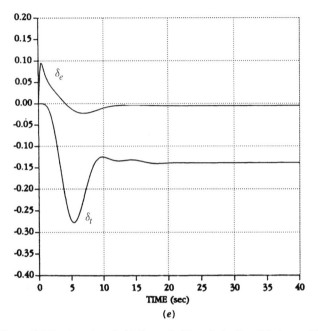

Figure 5.5-9 (*continued*) (*e*) Control efforts δ_e (rad) and δ_i (per unit).

With the optimal gains in (26), according to (8) the velocity channel compensator is

$$\frac{w_v}{e_v} = -0.9927 \frac{s + 2.38}{s + 5}, \tag{29}$$

which is a lead compensator as anticipated. The zeros in the d channel compensator could similarly be found. It is important to note that our formulation has resulted in the *compensator zeros being selected in an optimal fashion.* This is an improvement over root-locus design, where the zeros are determined using the engineering judgment that actually only applies for single-input/single-output systems.

It should be mentioned that determining an initial stabilizing gain K_0 for program LQ is not easy. In this example, we used the root-locus techniques from Chapter 4 to find the initial gain. Other approaches were discussed earlier in the subsection entitled "Constrained Feedback Matrix." ∎

5.6 MODEL-FOLLOWING DESIGN

In Section 4.3 we discussed flying qualities and gave the military flying-qualities specifications for the various aircraft modes. These desirable flying qualities could be viewed as constituting an *ideal model with good performance* which we would like to reproduce in the actual aircraft. For instance, to obtain good longitudinal performance we could select suitable short-period and phugoid poles from the flying-qualities specifications tabulated in Section 4.3. Then we could determine a state-variable realization of an ideal model with this behavior (see Stern and Henke, 1971). Finally, we could design a control system to make the actual aircraft behave like this ideal model.

This approach to control system design is the powerful model-following design technique. In this section we show how to design controllers that make the aircraft behave like a desired model. We will discuss two fundamentally different sorts of model-following control, "explicit" and "implicit," which result in controllers of different structure (Armstrong, 1980; Kreindler and Rothschild, 1976; O'Brien and Broussard, 1978).

Explicit Model-Following Control

Regulation with Model-Following Behavior. First, we will consider the regulator problem, where the objective is to drive the plant state to zero. Then we will treat the more difficult tracker or servo problem, where the plant is to follow a reference command with behavior like the prescribed model. Let the plant be described in state-variable form by

$$\dot{x} = Ax + Bu \tag{5.6-1}$$

$$y = Cx \tag{5.6-2}$$

$$z = Hx, \tag{5.6-3}$$

with state $x(t) \in \mathbf{R}^n$ and control input $u(t) \in \mathbf{R}^m$. The measured output $y(t)$ is available for feedback purposes.

A model is prescribed with dynamics

$$\underline{\dot{x}} = \underline{A}\underline{x} \tag{5.6-4}$$

$$\underline{z} = \underline{H}\underline{x}, \tag{5.6-5}$$

where the model matrix $\underline{A}$ reflects a system with desirable handling qualities such as speed of response, overshoot, and so on. The model states suitable for feedback purposes are given by

$$\underline{y} = \underline{C}\underline{x} \tag{5.6-6}$$

Model quantities will be denoted by underbars or the subscript m.

Notice that the model has no reference input, since we are considering the regulator problem here. That is, the plant should have the same unforced response as the model, which translates into suitable locations of the poles.

It is desired to select the plant control $u(t)$ so that the plant performance output $z(t)$ matches the model output $\underline{z}(t)$, for then the plant will exhibit the desirable time response of the model. That is, we should like to minimize the *model mismatch error*

$$e = \underline{z} - z = \underline{H}\underline{x} - Hx \tag{5.6-7}$$

To achieve this control objective, let us select the performance index

$$J = \tfrac{1}{2} \int_0^\infty \left(e^T Q e + u^T R u \right) dt, \tag{5.6-8}$$

with $Q > 0$ (to ensure that all components of the error vanish) and $R > 0$.

We can cast this model-matching problem into the form of the regulator problem whose solution appears in Table 5.3-1 as follows.

Define the augmented state $x' = \begin{bmatrix} x^T & \underline{x}^T \end{bmatrix}^T$ and the augmented system

$$\dot{x}' = \begin{bmatrix} A & 0 \\ 0 & \underline{A} \end{bmatrix} x' + \begin{bmatrix} B \\ 0 \end{bmatrix} u \equiv A'x' + B'u \tag{5.6-9}$$

$$y' = \begin{bmatrix} y \\ \underline{y} \end{bmatrix} = \begin{bmatrix} C & 0 \\ 0 & \underline{C} \end{bmatrix} x' \equiv C'x', \tag{5.6-10}$$

so that

$$e = \begin{bmatrix} -H & \underline{H} \end{bmatrix} x' \equiv H'x' \tag{5.6-11}$$

Then the PI (5.6-8) may be written

$$J = \tfrac{1}{2} \int_0^\infty \left((x')^T Q' x' + u^T R u \right) dt, \tag{5.6-12}$$

with

$$Q' = \begin{bmatrix} H^T Q H & -H^T Q \underline{H} \\ -\underline{H}^T Q H & \underline{H}^T Q \underline{H} \end{bmatrix} \tag{5.6-13}$$

At this point it is clear that the design equations of Table 5.3-1 can be used to select $u(t)$ if the primed quantities A', B', C', Q' are used there. The conditions for convergence of the algorithm in Table 5.3-2 require that (A', B', C') be output stabilizable and $(\sqrt{Q'}, A')$ be detectable. Since the model matrix $\underline{A}$ is certainly stable, the block diagonal form of A' and C' shows that output stabilizability of the plant (A, B, C) is required. The second condition requires detectability of the plant (H, A).

The form of the resulting output-feedback control law is quite interesting. Indeed, the optimal feedback is of the form

$$u = -K'y' \equiv -\begin{bmatrix} K_p & K_m \end{bmatrix} y' = -K_p y - K_m \underline{y} \tag{5.6-14}$$

Thus not only the plant output but also the *model* output is required. That is, the model acts as a *compensator* to drive the plant states to zero in such a fashion that the performance output $z(t)$ follows the model output $\underline{z}(t)$.

Tracking with Model-Following Behavior.
Unfortunately, while the model-following regulator problem has a direct solution that is easy to obtain, the model-following *tracker* problem is not so easy. In this situation, we should like the plant (5.6-1)–(5.6-3) to behave like the model

$$\dot{\underline{x}} = \underline{A}\underline{x} + \underline{B}r \tag{5.6-15}$$

$$\underline{z} = \underline{H}\underline{x}, \tag{5.6-16}$$

which is *driven by the reference input* $r(t)$. The approach above yields

$$\dot{x}' = \begin{bmatrix} A & 0 \\ 0 & \underline{A} \end{bmatrix} x' + \begin{bmatrix} B \\ 0 \end{bmatrix} u + \begin{bmatrix} 0 \\ \underline{B} \end{bmatrix} r \equiv A'x' + B'u + G'r \tag{5.6-17}$$

and thus the derivation in Section 5.3 results in a PI that contains a term in $r(t)$, for which the determination of the optimal feedback gains is not easy (Lewis, 1986).

A convenient technique for designing a practical tracker is the *command generator tracker (CGT)* technique, where the tracking problem is converted into a regulator problem (Franklin et al., 1986). In this approach, a generator system is assumed for the reference input. We will apply it here.

Thus, suppose that for some initial conditions the reference command $r(t)$ satisfies the differential equation

$$r^{(d)} + a_1 r^{(d-1)} + \cdots + a_d r = 0 \tag{5.6-18}$$

for a given degree d and set of coefficients a_i. Most command signals of interest satisfy such an equation. For instance, the unit step of magnitude r_0 satisfies

$$\dot{r} = 0 \tag{5.6-19}$$

with $r(0) = r_0$, while the ramp (velocity command) with slope v_0 satisfies

$$\ddot{r} = 0 \tag{5.6-20}$$

with $r(0) = 0, \dot{r}(0) = v_0$. We call (5.6-18) the *command generator system*.

Define the command generator characteristic polynomial as

$$\Delta(s) = s^d + a_1 s^{d-1} + \cdots + a_d \tag{5.6-21}$$

Then denoting d/dt in the time domain by D, we may write

$$\Delta(D)r = 0 \tag{5.6-22}$$

Multiplying the augmented dynamics (5.6-17) by $\Delta(D)$ results in the modified system

$$\dot{\xi} = A'\xi + B'\mu, \tag{5.6-23}$$

where the modified state and control input are

$$\xi = \Delta(D)x' = (x')^{(d)} + a_1(x')^{(d-1)} + \cdots + a_d x' \quad (5.6\text{-}24)$$

$$\mu = \Delta(D)u = u^{(d)} + a_1 u^{(d-1)} + \cdots + a_d u \quad (5.6\text{-}25)$$

The reason for these manipulations is that because of (5.6-22), the reference command $r(t)$ does not appear in (5.6-23). Let us partition ξ as

$$\xi = \begin{bmatrix} \xi_p \\ \xi_m \end{bmatrix} \quad (5.6\text{-}26)$$

Applying $\Delta(D)$ to the model mismatch error (5.6-7) results in

$$\Delta(D)e = [-H \quad H]\xi = H'\xi \quad (5.6\text{-}27)$$

This may be expressed in terms of state variables using the observability canonical form (Kailath, 1980), which for scalar $e(t)$ and $d = 3$, for instance, is

$$\dot{\epsilon} = \begin{bmatrix} 0 & 1 & 0 \\ 0 & 0 & 1 \\ -a_3 & -a_2 & -a_1 \end{bmatrix} \epsilon + \begin{bmatrix} 0 \\ H' \end{bmatrix} \xi \equiv F\epsilon + \begin{bmatrix} 0 \\ H' \end{bmatrix} \xi \quad (5.6\text{-}28)$$

$$e = [1 \quad 0 \quad 0]\epsilon, \quad (5.6\text{-}29)$$

where $\epsilon(t) = \begin{bmatrix} e & \dot{e} & \cdots & e^{(d-1)} \end{bmatrix}^T$ is the vector of the error and its first $d-1$ derivatives.

Collecting all the dynamics (5.6-23)–(5.6-28) into one system yields

$$\frac{d}{dt}\begin{bmatrix} \epsilon \\ \xi \end{bmatrix} = \begin{bmatrix} F & \begin{matrix} 0 \\ H' \end{matrix} \\ 0 & A' \end{bmatrix} \begin{bmatrix} \epsilon \\ \xi \end{bmatrix} + \begin{bmatrix} 0 \\ B' \end{bmatrix} \mu \quad (5.6\text{-}30)$$

Let us now note what we have achieved. Using the command generator polynomial $\Delta(s)$, we have prefiltered the augmented state, control input, and error to obtain a system (5.6-30) *that is not driven by the reference input* $r(t)$. Using this system we may now perform an LQ *regulator* design, since if its state goes to zero, the tracking error $e(t)$ vanishes. That is, by performing a regulator design (using Table 5.3-1) for (5.6-30), we may design a *tracker* control system that causes the original plant to follow the reference command with performance like that of the ideal model.

For the regulator design, we will take the outputs available for feedback in (5.6-30) as

$$v = \begin{bmatrix} I & 0 & 0 \\ 0 & C & 0 \\ 0 & 0 & \underline{C} \end{bmatrix} \begin{bmatrix} \epsilon \\ \xi_p \\ \underline{\xi}_m \end{bmatrix} \quad (5.6\text{-}31)$$

To achieve small error without using too much control energy, we may select the PI (5.6-8) (with $u(t)$ replaced by $\mu(t)$). According to (5.6-29), the error is given in terms of the state of (5.6-30) by

$$e = h \begin{bmatrix} \epsilon \\ \xi \end{bmatrix}, \quad (5.6\text{-}32)$$

with $h = [1 \ 0 \ \cdots \ 0]$ the first row of the identity matrix. Therefore, in the PI we should weight the state of (5.6-30) using

$$Q' = h^T Q h \quad (5.6\text{-}33)$$

Since the observability canonical form is always observable, the augmented system (5.6-30) is detectable if the plant (H, A) and the model $(\underline{H}, \underline{A})$ are both detectable.

Now, by applying the equations of Table 5.3-1 to the system (5.6-30) with outputs (5.6-31) and PI weights Q' and R, we may compute the control gains in the control law

$$\mu = -\begin{bmatrix} K_\epsilon & K_p & K_m \end{bmatrix} \begin{bmatrix} \epsilon \\ C\xi_p \\ \underline{C}\underline{\xi}_m \end{bmatrix} \quad (5.6\text{-}34)$$

or

$$\Delta(s)u = -K_\epsilon \epsilon - K_p C \Delta(s) x - K_m \underline{C} \Delta(s) \underline{x} \quad (5.6\text{-}35)$$

To determine the optimal tracking control input $u(t)$ for the original system, write this as

$$\Delta(s)(u + K_p y + K_m \underline{y}) = -K_\epsilon \epsilon \equiv -[K_d \ \cdots \ K_2 \ K_1] \begin{bmatrix} e \\ \dot{e} \\ \vdots \\ e^{(d-1)} \end{bmatrix} \quad (5.6\text{-}36)$$

Thus we obtain the transfer function

$$\frac{u + K_p y + K_m \underline{y}}{e} = -\frac{K_1 s^{d-1} + \cdots + K_{d-1} s + K_d}{s^d + a_1 s^{d-1} + \cdots + a_d}, \quad (5.6\text{-}37)$$

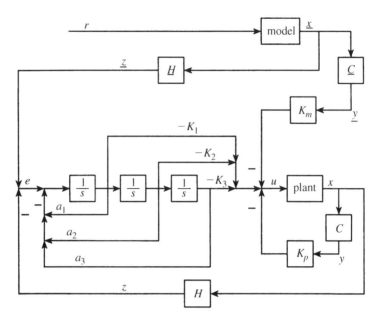

Figure 5.6-1 Explicit model-following command generator tracker for $d = 3$.

which may be implemented in reachability canonical form (Kailath, 1980) to obtain the control structure shown in Figure 5.6-1.

CGT Structure. The structure of this *model-following command generator tracker (CGT)* is very interesting. It consists of an output feedback K_p, a feedforward compensator that is nothing but the reference model with a gain of K_m, and an additional feedforward filter in the error channel that guarantees perfect tracking. Note that if $d = 1$ so that $r(t)$ is a unit step, the error filter is a PI controller. If $d = 2$ so that $r(t)$ is a ramp, the error filter consists of two integrators, resulting in a type-2 system that gives zero steady-state error. What this means is the CGT design *automatically adds the compensator of appropriate structure to guarantee that the system has the correct type for perfect tracking.*

It is extremely interesting to note that the augmented state description (5.6-30) is nothing but the state description of Figure 5.6-1. It should be emphasized that this technique is extremely direct to apply. Indeed, given the prescribed model and the command generator polynomial $\Delta(s)$, the system (5.6-30)/(5.6-31) may be written down immediately, and the equations in Table 5.3-1 used to select the feedback gains.

A word on the command generator assumption (5.6-22) is in order. In point of fact, for aircraft applications $r(t)$ is usually the pilot's command input. For control system design it is not necessary to determine the actual coefficients a_i that describe the pilot command, although this is one approach (Kreindler and Rothschild, 1976). Instead, the performance objectives should be taken into account to select $\Delta(s)$. For instance, if it desired for the aircraft to follow a position command, we may select the command generator $\dot{r} = 0$. On the other hand, if the aircraft should follow a rate (velocity) command, we may select $\ddot{r} = 0$. Then when the actual command input $r(t)$ is applied (which may be neither a unit step nor a unit ramp), the aircraft will exhibit the appropriate closed-loop behavior.

Implicit Model-Following Control

We will now discuss a formulation that results in a radically different sort of control scheme. In explicit model following, which is also called *model in the system control*, the model explicitly appeared in the controller as a feedforward compensator. On the other hand, implicit model following, also called *model in the performance index*, is a completely different approach in which the model does not appear in the control structure. Indeed, implicit model following can be viewed simply as a technique for selecting the weighting matrices in the PI in a meaningful way (see Armstrong, 1980, and Kreindler and Rothschild, 1976).

Suppose that the performance output $z(t)$ of the plant prescribed by (5.6-1)–(5.6-3) is required to follow the model given by

$$\dot{z} = \underline{A}z \qquad (5.6\text{-}38)$$

The model matrix $\underline{A}$ has poles corresponding to desirable handling qualities of the plant, such as may be found in Mil. Spec. 1797 (1987) and Stern and Henke (1971).

When the control objective is met, the performance output will satisfy the differential equation (5.6-38). Thus we may define an error by

$$e = \dot{z} - \underline{A}z \qquad (5.6\text{-}39)$$

This is a different sort of error than we have seen before.

To make $e(t)$ small without using too much control energy, we may choose $u(t)$ to minimize the PI (5.6-8). Since $\dot{z} = HAx + HBu$, this becomes

$$J = \tfrac{1}{2} \int_0^\infty \left[(HAx + HBu - \underline{A}Hx)^T Q (HAx + HBu - \underline{A}Hx) + u^T Ru \right] dt \qquad (5.6\text{-}40)$$

or

$$J = \tfrac{1}{2} \int_0^\infty \left(x^T Q' x + 2 x^T W u + u^T R' u \right) dt, \qquad (5.6\text{-}41)$$

where

$$Q' = (HA - \underline{A}H)^T Q (HA - \underline{A}H)$$
$$W = (HA - \underline{A}H)^T QHB, \qquad R' = \left(B^T H^T QHB + R \right) \qquad (5.6\text{-}42)$$

The additional term in W is a *cross-weighting* between $u(t)$ and $x(t)$.

In Table 5.3-1 we have given the LQ regulator design equations to determine the optimal output feedback gains for the case $W = 0$. By using techniques like those in that derivation (see the Problems), we may derive the modified design equations for the case of $W \neq 0$. They are

$$0 = A_c^T P + P A_c + Q + C^T K^T R K C - W K C - C^T K^T W^T \quad (5.6\text{-}43)$$

$$0 = A_c S + S A_c^T + X \quad (5.6\text{-}44)$$

$$0 = R K C S C^T - (P B + W)^T S C^T, \quad (5.6\text{-}45)$$

where

$$A_c = A - B K C \quad (5.6\text{-}46)$$

The optimal cost is still given by

$$J = \tfrac{1}{2}\operatorname{tr}(PX) \quad (5.6\text{-}47)$$

To find the optimal output-feedback gains in

$$u = -Ky \quad (5.6\text{-}48)$$

for implicit model following, it is only necessary to solve these design equations using Q', W, and R'. For this, a technique like that in Table 5.3-2 may be used. Alternatively, algorithms such as the Simplex or Davidon–Fletcher–Powell may be employed.

Note that implicit model following in the regulator case is nothing but a convenient technique for selecting the PI weighting matrices Q', R' (and W) to guarantee desirable behavior, since the right-hand sides of (5.6-42) are known. Indeed, it is reasonable to select $R = \rho I$ and $Q = I$.

It is possible to design a tracking control system using implicit model following by using the CGT approach. However, this system has an undesirable structure from the point of view of aircraft control since it generally requires derivatives of the performance output $z(t)$.

Example 5.6-1: Automatic Flare Control by Model-Following Design.

Model-following design may be used to design a control system that makes the aircraft behave like an ideal model (Kreindler and Rothschild, 1976). Such a model may be constructed using the military flying-qualities requirements discussed in Section 4.3 so that it has good performance. However, this is not the only use for model-following design in aircraft controls.

In this example, we complete the design of the automatic landing system that was begun in Example 5.5-5. There, we constructed a glide-slope coupler whose function is to conduct an aircraft down a glide path toward the runway. Here we will show that explicit model-following design may be used to design the automatic flare control system whose function is to cause the aircraft to flare gently to a touchdown.

a. Determining the Reference Model. The control system is basically an altitude hold system with a time-varying reference or commanded altitude $\underline{h}(t)$. A gentle flare is described by an exponential, so that the commanded altitude should obey the differential equation

$$\underline{\dot{h}} = -\frac{1}{\tau}\underline{h} + r, \qquad \text{initial condition } \underline{h}(0) = \underline{h}_0, \tag{1}$$

where τ and $\underline{h}_0$ are chosen for the desired flare characteristics. Equation (1) is the reference model (see 5.6-15)

$$\begin{aligned}\underline{\dot{x}} &= \underline{Ax} + \underline{B}r \\ \underline{z} &= \underline{Hx},\end{aligned} \tag{2}$$

with $\underline{A} = -1/\tau$, $\underline{B} = 1$, $\underline{H} = 1$. Thus $\underline{z} = \underline{x} = \underline{h}$. The model reference input is $r(t)$, which is equal to the constant value of zero in this example. Then

$$\underline{h}(t) = \underline{h}_0 e^{-t/\tau} \tag{3}$$

The geometry of the flare path shown in Figure 5.6-2 may be used to determine the flare time constant τ and initial altitude $\underline{h}_0$ (see Blakelock, 1965). In Example 5.5-5 we designed a glide-slope coupler for a total velocity of $V_T = 250$ ft/s. Thus, on the glide path the rate of descent is

$$\dot{h} = V_t \sin(-2.5°) \approx -V_T \frac{2.5}{57.2958} = -10.91 \text{ ft/s} \tag{4}$$

The flare control system is turned on at time $t = 0$ shown in the figure. Therefore, for (1) we obtain $\underline{\dot{h}}(0) = -10.91$, and

$$\underline{h}_0 = -\tau\underline{\dot{h}}(0) = 10.91\tau \text{ ft} \tag{5}$$

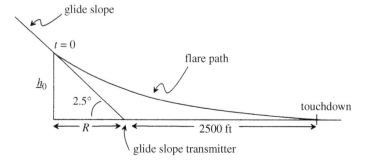

Figure 5.6-2 Flare-path geometry.

The distance R is thus given by

$$R = \frac{h_0}{\tan(2.5°)} \approx \underline{h}_0 \times \frac{57.2958}{2.5} = V_T \tau \text{ ft} \tag{6}$$

If it is desired to touch down 2500 ft beyond the glide-slope transmitter, and if we assume that $\underline{h}(t)$ given in (1) vanishes in 4τ seconds, then

$$4\tau V_T = R + 2500 = V_T \tau + 2500 \tag{7}$$

or

$$3\tau V_T = 2500, \tag{8}$$

so that

$$\tau = 3.333 \text{ s} \tag{9}$$

This yields the reference model

$$\dot{\underline{h}} = -0.3\underline{h} + r, \qquad \underline{h}_0 = 36.37 \text{ ft}, \tag{10}$$

with reference input $r(t)$ taking the constant value of zero.

b. Basic Aircraft and Controller. The flare control system is shown in Figure 5.6-3. For small flight-path angles the aircraft altitude is given by

$$\dot{h} = V_T \sin\gamma = V_T \sin(\theta - \alpha) \approx V_T\theta - V_T\alpha, \tag{11}$$

which is the same as the equation in Example 5.5-5 for $\dot{d}$ (with d the off-glide path distance) without the term in γ_r. What this means is that an altitude-hold system is given by the lower d-hold channel in the glide-slope coupler in Figure 5.5-8, with d

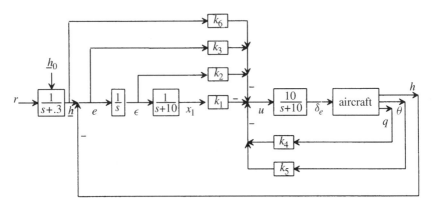

Figure 5.6-3 Automatic flare control system.

replaced everywhere by h and γ_r removed. Indeed, the control gains in that system were used as an initial stabilizing guess in the LQ design for this example.

In this example we want to illustrate the model-following design procedure for the h control channel only. A complete design would include a second velocity-hold channel exactly as in the glide-scope coupler.

We used low-velocity longitudinal Jacobians for a medium-sized transport linearized about $V_T = 250$ ft/s, cg $= 0.25\,\bar{c}$, as in Example 5.5-5. For the flare control system h-channel, we may use the short-period approximation, with, however, θ retained due to the need to compute the altitude using (11).

The state of the aircraft plus the lead compensator is

$$x = [\alpha \quad \theta \quad q \quad h \quad \delta_e \quad x_1]^T \tag{12}$$

with x_1 the compensator state (see Figure 5.6-3). The performance output is

$$z = h = [0 \quad 0 \quad 0 \quad 1 \quad 0 \quad 0]x = Hx \tag{13}$$

and the control input $u(t)$ is the elevator servo command. According to the figure, the measured outputs corresponding to the aircraft and the lead compensator are

$$y = [x_1 \quad q \quad \theta]^T \tag{14}$$

c. Explicit Model-Following Control. We should like the reference output $z(t)$ to follow the model altitude $\underline{h}(t)$ given by (2)/(10). Since the model's reference input $r(t)$ has the constant value of zero, $r(t)$ satisfies the differential equation

$$\dot{r} = 0, \tag{15}$$

so that the command generator polynomial (5.6-21) is given by

$$\Delta(s) = s \tag{16}$$

The model mismatch latitude error (5.6-7) is given by

$$e = \underline{h} - h \tag{17}$$

Therefore, the observability canonical form realization (5.6-28) is

$$\dot{\epsilon} = [-H \quad H]\xi = [0 \quad 0 \quad 0 \quad -1 \quad 0 \quad 0 \mid 1]\xi, \tag{18}$$

with $\xi(t)$ the modified state $\Delta(s)[\underline{x}^T \quad x^T]^T$.

According to (18), $F = 0$ in the augmented system (5.6-30). Thus, we are required to incorporate an integrator in the control system (see (5.6-37) and Figure 5.6-1). This we have already done in Figure 5.6-3.

The overall dynamics of the modified system (5.6-30) are given by

$$\dot{X} = AX + Bu$$
$$y = CX, \tag{19}$$

with X the augmented state that contains the basic aircraft and compensator dynamics, the model dynamics (10), and the integrator required by (18). For convenience, we will order the states differently than in (5.6-30), taking

$$X = \begin{bmatrix} \alpha & \theta & q & h & \delta_e & x_1 & \epsilon & \underline{h} \end{bmatrix}^T \tag{20}$$

According to Figure 5.6-3, the outputs are

$$Y = \begin{bmatrix} x_1 & \epsilon & e & q & \theta & \underline{h} \end{bmatrix}^T \tag{21}$$

The model state $\underline{h}$ is included as an output due to the development leading to (5.6-37). With this structure, the plant matrices are given by

$$A = \begin{bmatrix} -0.64627 & 0 & 1 & 0 & 0 & 0 & 0 & 0 \\ 0 & 0 & 1 & 0 & 0 & 0 & 0 & 0 \\ -0.7739 & 0 & -0.52977 & 0 & -0.011 & 0 & 0 & 0 \\ -250 & 250 & 0 & 0 & 0 & 0 & 0 & 0 \\ 0 & 0 & 0 & 0 & -10 & 0 & 0 & 0 \\ 0 & 0 & 0 & 0 & 0 & -10 & 1 & 0 \\ 0 & 0 & 0 & -1 & 0 & 0 & 0 & 1 \\ 0 & 0 & 0 & 0 & 0 & 0 & 0 & -0.3 \end{bmatrix} \tag{22}$$

$$B = \begin{bmatrix} 0 \\ 0 \\ 0 \\ 0 \\ 10 \\ 0 \\ 0 \\ 0 \end{bmatrix} \tag{23}$$

$$C = \begin{bmatrix} 0 & 0 & 0 & 0 & 0 & 1 & 0 & 0 \\ 0 & 0 & 0 & 0 & 0 & 0 & 1 & 0 \\ 0 & 0 & 0 & -1 & 0 & 0 & 0 & 1 \\ 0 & 0 & 57.2958 & 0 & 0 & 0 & 0 & 0 \\ 0 & 57.2958 & 0 & 0 & 0 & 0 & 0 & 0 \\ 0 & 0 & 0 & 0 & 0 & 0 & 0 & 1 \end{bmatrix} \tag{24}$$

Then, according to Figure 5.6-3, the control input $u(t)$ is given by

$$u = -Ky = (k_1 x_1 + k_2 \epsilon + k_3 e + k_4 q + k_5 \theta + k_6 \underline{h}) \tag{25}$$

The control structure shown in Figure 5.6-3 and described here is nothing but the structure required for model following according to Figure 5.6-1.

d. PI and LQ Control Gain Design. Although the explicit model-following design technique discussed in this section involves using the LQ *regulator* design equations from Table 5.3-1 on the augmented system (5.6-30), we have found that the results are generally better using LQ *tracker design with time-weighted PI*. Thus, we used the design equations in Table 5.5-1 with the auxiliary matrices

$$G = [0 \quad 0 \quad 0 \quad 0 \quad 0 \quad 0 \quad 0 \quad 1]^T$$
$$F = [0 \quad 0 \quad 0 \quad 0 \quad 0 \quad 0]^T \tag{26}$$
$$H = [0 \quad 0 \quad 0 \quad -1 \quad 0 \quad 0 \quad 0 \quad 1]^T,$$

which were determined from Figure 5.6-3. (Note the redefinition of the matrix H.)

The PI was selected as

$$J = \tfrac{1}{2} \int_0^\infty \left(qt^2 \tilde{e}^2 + \tilde{u}^2 \right) dt \tag{27}$$

It is important to note that a sensible formulation of the problem has resulted in the appearance of *only one design parameter, q, in the PI*. Thus we will not be faced with tuning many design parameters in an effort to obtain suitable responses. In view of the fact that there are eight states and six control gains to determine, this is quite significant. No steady-state error weighting is used in the PI since the plant is of type 1.

After several design iterations using the software of Appendix B to solve for K using the design equations in Table 5.5-1, we decided on $q = 0.001$ and obtained the control gains

$$K = [593.4 \quad -59.30 \quad 6.154 \quad -0.56 \quad -1.00 \quad -0.01852] \tag{28}$$

The closed-loop poles were at

$$\begin{aligned} &-0.15 \pm j0.23 \\ &-0.15 \pm j1.02 \\ &-0.30, -0.92 \\ &-9.43, -10.22 \end{aligned} \tag{29}$$

Note that the model pole of $s = -0.3$ has not moved since it is uncontrollable.

e. Simulation and Discussion. The controlled flare is shown in Figure 5.6-4—it matches the desired flare $\underline{h}(t)$ very well. To obtain this graph it is necessary to

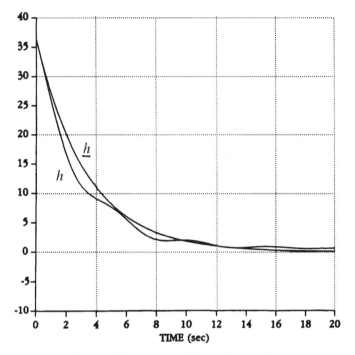

Figure 5.6-4 Controlled flare, altitude in feet.

use initial conditions $x(0)$ corresponding to the equilibrium state on the glide slope from Example 5.5-5. The flight-path angle γ is shown in Figure 5.6-5a. Shown in Figure 5.6-5b is the elevator command, δ_e; in examining this figure recall that upward elevator deflection (i.e., back stick) is defined as negative.

The poles in (29) are quite slow and there is one badly damped pair. However, the time responses are acceptable. This is because the flare control system is engaged with the aircraft on the glide path, so that there are no sudden reference command changes to excite the underdamped mode. Moreover, the flare is gentle, so that the time scale of the desired motion is on the order of the time scale of the closed-loop poles.

Although the control gain from the model state $\underline{h}$ to elevator servo command u is small, it plays a very important function. As may be seen in Blakelock (1965), the tendency of the flare control system without model state feedforward is to lag behind the desired response. This results in a flare that is always below the desired path and requires a modification in the design flare time constant τ. The feedforward of h corrects this problem in a simple manner.

Using the gains in (28), the compensator in the forward error channel of Figure 5.6-3 has the transfer function

$$\frac{k_1}{s(s+10)} + \frac{k_2}{s} + k_3 = \frac{6.154(s+0.364)}{s+10}, \tag{30}$$

MODEL-FOLLOWING DESIGN 477

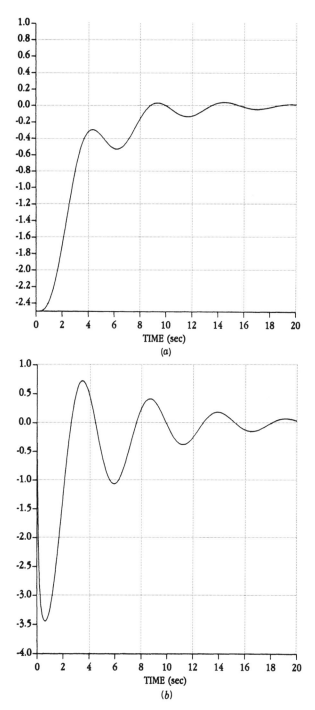

Figure 5.6-5 Aircraft response during controlled flare: (*a*) flight-path angle γ (deg); (*b*) elevator command δ_e (deg).

where the pole at $s = 0$ has been canceled by a zero at $s = 0$ to yield a simplified compensator. Thus, there is no integrator in the feedforward path, and the model-following behavior does not rely on the system being of type 1. The ratio of the zero to the poles in the lead compensator is excessive, and the design may be repeated using, for instance, a compensator pole at $s = -5$ instead of $s = -10$ (and no integrator). ∎

5.7 LINEAR QUADRATIC DESIGN WITH FULL STATE FEEDBACK

In the previous sections of this chapter we have seen how to design control systems using a variety of modern control techniques that rely only on measuring a system *output*. These output-feedback approaches are very suitable for aircraft control design since they allow us to design a compensator with any desired dynamical structure. This cannot be accomplished using full state feedback.

In this section we intend to explore full state-variable feedback in the linear quadratic regulator (LQR) for the insight it provides. That is, for the system

$$\dot{x} = Ax + Bu, \qquad (5.7\text{-}1)$$

with $x \in \mathbf{R}^n, u \in \mathbf{R}^m$, we want to examine control laws of the form

$$u = -Kx, \qquad (5.7\text{-}2)$$

which result in the closed-loop system

$$\dot{x} = (A - BK)x \equiv A_c x \qquad (5.7\text{-}3)$$

In the previous sections we defined the measurable output

$$y = Cx \qquad (5.7\text{-}4)$$

and restricted ourselves to controls of the form

$$u = -Ky = -KCx \qquad (5.7\text{-}5)$$

Here we plan to examine the simplifications in the control design equations that come about when $C = I$. As we will see, we can draw some conclusions that will give more insight into modern control theory.

The Relevance of State Feedback

Although all the states are seldom measurable in aircraft control systems, we have several objectives for looking at state-variable feedback design in this section. First, it is clear that state feedback is just the special case of output feedback with $C = I$. That is, it assumes that all the states can be measured. Thus the theory for state-variable feedback will tell us *the best performance that we can expect* in the closed-loop system by using static output feedback, where all of the states are not available as measurements.

Second, the output-feedback design equations in Tables 5.3-1, 5.4-1, and 5.5-1 are not the LQR equations with which the reader may be familiar. We would like to show how they relate to the more traditional Riccati equation.

If all the states are involved in the feedback, there are some very powerful stability results of which the reader should be aware. Indeed, under some reasonable assumptions it is possible to *guarantee the stability of the closed-loop system* using the optimal LQ state feedback gain. Similar theoretical results for output feedback have not yet been discovered.

Finally, we will need state feedback in Chapter 6 when we discuss dynamic regulators and LQG/LTR robust design. A limitation of state feedback is that all the states are not generally available, but only the outputs are measured. However, we can design a full state feedback $u = -Kx$, and then a dynamic observer to estimate the states from the measured outputs. Then the state *estimates* $\hat{x}$ may be fed back, instead of the states themselves, in a control law such as $u = -K\hat{x}$. The combination of state feedback plus an observer is called a *dynamic regulator*. It is a compensator of the sort used in classical control, but it is easy to design for multivariable systems, overcoming a deficiency of the classical approach, where multiloop and MIMO systems are hard to deal with.

The Riccati Equation and Kalman Gain

By setting $C = I$ all of our work in Sections 5.3 and 5.4 applies to state feedback. That is, all the work of deriving the control design equations for state feedback has already been done. Let us see how the LQR design equations simplify in the case of full state feedback.

To regulate the performance output

$$z = Hx \qquad (5.7\text{-}6)$$

to zero, let us select the PI

$$J = \tfrac{1}{2} \int_0^\infty \left(x^\mathrm{T} Q x + u^\mathrm{T} R u \right) dt, \qquad (5.7\text{-}7)$$

with $Q = H^\mathrm{T} H \geq 0$, $R > 0$.

The output feedback gain K in (5.7-5) that minimizes the PI may be found using the design equations in Table 5.3-1. To obtain the optimal state feedback in (5.7-2), we may simply set $C = I$ in the table. The results are

$$0 = A_c^\mathrm{T} P + P A_c + Q + K^\mathrm{T} R K \qquad (5.7\text{-}8)$$

$$0 = A_c S + S A_c^\mathrm{T} + X \qquad (5.7\text{-}9)$$

$$K = R^{-1} B^\mathrm{T} P S S^{-1}, \qquad (5.7\text{-}10)$$

where the initial state autocorrelation is

$$X = E\left\{x(0)x^T(0)\right\} \tag{5.7-11}$$

The problems in computing the output feedback gains include the need to know X and the selection of an initial stabilizing gain K_0 for the algorithm in Table 5.3-2. Moreover, although we gave conditions for the convergence to a local minimum of the algorithm in that table, little is known about the necessary and sufficient conditions for the existence of an output-feedback gain that satisfies the design equations and stabilizes the plant.

All of these problems vanish in the case of state feedback, as we will now show. According to (5.7-10),

$$K = R^{-1}B^T P, \tag{5.7-12}$$

that is, the solution S to (5.7-9) *is not needed to solve for the optimal state feedback gain*. The gain K is called the *Kalman gain*. Using (5.7-12) in (5.7-8) yields

$$0 = A_c^T P + PA_c + Q + PBR^{-1}B^T P \tag{5.7-13}$$

or, according to (5.7-3),

$$0 = \left(A - BR^{-1}B^T P\right)^T P + P\left(A - BR^{-1}B^T P\right)$$
$$+ Q + PBR^{-1}B^T P$$
$$0 = A^T P + PA + Q - PBR^{-1}B^T P \tag{5.7-14}$$

This matrix quadratic equation is called the *algebraic Riccati equation (ARE)*. It is named after Count J. F. Riccati, who used a related equation in the study of heat flow (Riccati, 1724). Since the equation is equal to its own transpose (verify!), the solution P is symmetric ($P = P^T$).

Since S is not needed to find the optimal state-feedback gain K, this gain does not depend on X in (5.7-9). That is, contrary to the case with output feedback, to compute the optimal state feedback gains *no information about the initial state $x(0)$ is needed*. Thus, it is not required to take expected values of the PI as we did in Section 5.3. Therefore, according to the development in that section, the optimal cost is given by

$$J = \tfrac{1}{2}x^T(0)Px(0) \tag{5.7-15}$$

The state-feedback LQR is summarized in Table 5.7-1.

Setting $C = I$ has allowed us to replace the solution of three coupled matrix equations by the solution of *one nonlinear matrix equation* for P. Then the Kalman gain is given in terms of P by (5.7-17). The importance of this is that there are many good techniques for solving the Riccati equation using *standard software packages* (e.g., ORACLS [Armstrong, 1980], MATRIX, [1989], PC-MATLAB [Moler et al.,

TABLE 5.7-1: LQR with State Feedback

System Model

$$\dot{x} = Ax + Bu$$

Control

$$u = -Kx$$

Performance Index

$$J = \tfrac{1}{2}\int_0^\infty \left(x^T Q x + u^T R u\right) dt$$

Optimal LQ Design Equations

ALGEBRAIC RICCATI EQUATION (ARE)

$$0 = A^T P + PA + Q - PBR^{-1}B^T P \qquad (5.7\text{-}16)$$

KALMAN GAIN

$$K = R^{-1}B^T P \qquad (5.7\text{-}17)$$

Optimal Cost

$$J = \tfrac{1}{2}x^T(0)Px(0)$$

1987], and IMSL [1980]). On the other hand, the specialized software for solving the output-feedback problem in Tables 5.3-1, 5.4-1, or 5.5-1 can be used to solve the full state-feedback problem by setting $C = I$.

Guaranteed Closed-Loop Stability

The theory for the LQ regulator with state feedback is well developed. In fact, the next stability result is so fundamental that we set it apart as a theorem (Lewis, 1986). The notion of detectability was introduced while discussing Table 5.3-2. We say that (A, H) is detectable if there exists an L so that $A - LH$ is stable; this amounts to the observability of the unstable modes of A. We say that (A, B) is *stabilizable* if there exists a feedback gain K such that $A_c = A - BK$ is stable. This amounts to the controllability of the unstable modes of A.

Theorem. Let H be any matrix so that $Q = H^T H$. Suppose that (H, A) is detectable. Then (A, B) is stabilizable if and only if:

(a) There exists a unique positive semidefinite solution P to the Riccati equation, and

(b) The closed-loop system (5.7-3) is asymptotically stable if the Kalman gain K is computed using (5.7-17) in terms of this positive semidefinite solution P. ∎

This result is at the heart of modern control theory. Exactly as in classical control, it allows us to examine *open-loop* properties (i.e., detectability and stabilizability) and draw conclusions about the closed-loop system. As long as (H, A) is detectable, so that all the unstable modes appear in the PI, and (A, B) is stabilizable, so that the control $u(t)$ has sufficient influence on the system, the LQ regulator using state feedback will *guarantee* a stable closed-loop system. A similar easily understandable result has not yet been discovered for output feedback.

Detectability is implied by the stronger condition of observability, which is easy to check by verifying that the observability matrix has full rank n (see Section 5.3). Stabilizability is implied by controllability, which is easy to check by verifying that the controllability matrix has full rank n (see Section 5.2). Thus the controllability of (A, B) and the observability of (H, A) guarantee closed-loop stability of the LQ regulator with state feedback.

This theorem, coupled with the availability of good software for solving the ARE, means that it is always straightforward to find a state-variable feedback gain K that stabilizes any stabilizable plant, no matter how many inputs or outputs it has.

Since output feedback amounts to a partial state feedback, it is clear that if the conditions of the theorem do not hold, we should not expect to be able to stabilize the plant using any output feedback. (Unless time-dependent weighting of the form t^k is used in the PI to avoid the observability requirement; see Section 5.5). Thus in the case of output-feedback design these conditions should hold *as a minimum*. In fact, we saw that the algorithm of Table 5.3-2 requires the detectability of $(\sqrt{Q}, A)$ and the output stabilizability of the system. Output stabilizability is a stronger condition than stabilizability.

In the case of a full state feedback, it is possible in simple examples to give a direct correlation between the PI weighting matrices and the closed-loop poles. Let us investigate this connection for systems obeying Newton's laws.

Example 5.7-1: LQR with State Feedback for Systems Obeying Newton's Laws.

In this example we will see that in the case of full state feedback for simple systems, there is a direct connection between the PI weights and the closed-loop damping ratio and natural frequency.

Systems obeying Newton's laws may be described by the state equation

$$\dot{x} = \begin{bmatrix} 0 & 1 \\ 0 & 0 \end{bmatrix} x + \begin{bmatrix} 0 \\ 1 \end{bmatrix} u = Ax + Bu, \tag{1}$$

where the state is $x = [d \quad v]^T$, with $d(t)$ the position and $v(t)$ the velocity, and the control $u(t)$ is an acceleration input. Indeed, note that (1) says nothing other than $\ddot{d} = u$, or $a = F/m$.

Let the PI be

$$J = \tfrac{1}{2} \int_0^\infty \left(x^T Q x + u^2 \right) dt, \tag{2}$$

with $Q = \text{diag}\{q_d^2, q_v\}$. In this example, we will see the effect of q_d and q_v. Note that it is not useful to include a separate control weighting r, since only the ratios q_d^2/r and q_v/r are important in J.

LINEAR QUADRATIC DESIGN WITH FULL STATE FEEDBACK 483

Since the Riccati solution P is symmetric, we may assume that

$$P = \begin{bmatrix} p_1 & p_2 \\ p_2 & p_3 \end{bmatrix} \tag{3}$$

for some scalars p_1, p_2, p_3 to be determined. Using A, B, Q, and $r = 1$ in the Riccati equation in Table 5.7-1 yields

$$0 = \begin{bmatrix} 0 & 0 \\ 1 & 0 \end{bmatrix} \begin{bmatrix} p_1 & p_2 \\ p_2 & p_3 \end{bmatrix} + \begin{bmatrix} p_1 & p_2 \\ p_2 & p_3 \end{bmatrix} \begin{bmatrix} 0 & 1 \\ 0 & 0 \end{bmatrix} + \begin{bmatrix} q_d^2 & 0 \\ 0 & q_v \end{bmatrix}$$

$$- \begin{bmatrix} p_1 & p_2 \\ p_2 & p_3 \end{bmatrix} \begin{bmatrix} 0 & 0 \\ 0 & 1 \end{bmatrix} \begin{bmatrix} p_1 & p_2 \\ p_2 & p_3 \end{bmatrix} \tag{4}$$

The reader should verify that this may be multiplied out to obtain the three scalar equations

$$0 = -p_2^2 + q_d^2 \tag{5a}$$

$$0 = p_1 - p_2 p_3 \tag{5b}$$

$$0 = 2p_2 - p_3^2 + q_v \tag{5c}$$

Solving these equations in the order (5a), (5c), (5b) gives

$$p_2 = q_d \tag{6a}$$

$$p_3 = \sqrt{2}\sqrt{q_d + \frac{q_v}{2}} \tag{6b}$$

$$p_1 = q_d \sqrt{2}\sqrt{q_d + \frac{q_v}{2}}, \tag{6c}$$

where we have selected the signs that make P positive definite.

According to Table 5.7-1, the Kalman gain is equal to

$$K = R^{-1} B^T P = [0 \quad 1] \begin{bmatrix} p_1 & p_2 \\ p_2 & p_3 \end{bmatrix} = [p_2 \quad p_3] \tag{7}$$

Therefore,

$$K = \left[q_d \quad \sqrt{2}\sqrt{q_d + \frac{q_v}{2}} \right] \tag{8}$$

It should be emphasized that in the case of state feedback, we have been able to find an *explicit expression* for K in terms of the PI weights. This is not possible for output feedback.

Using (8), the closed-loop system matrix is found to be

$$A_c = (A - BK) = \begin{bmatrix} 0 & 1 \\ -q_d & -\sqrt{2}\sqrt{q_d + \dfrac{q_v}{2}} \end{bmatrix} \quad (9)$$

Therefore, the closed-loop characteristic polynomial is

$$\Delta_c(s) = |sI - A_c| = s^2 + 2\zeta\omega s + \omega^2, \quad (10)$$

with the optimal natural frequency ω and damping ratio ζ given by

$$\omega = \sqrt{q_d}, \quad \zeta = \frac{1}{\sqrt{2}}\sqrt{1 + \frac{q_v}{2q_d}} \quad (11)$$

It is now clear how selection of the weights in the PI affects the closed-loop behavior. Note that if no velocity weighting q_v is used, the damping ratio becomes the familiar $1/\sqrt{2}$.

Note that (A, B) is reachable since

$$U = [B \quad AB] = \begin{bmatrix} 0 & 1 \\ 1 & 0 \end{bmatrix} \quad (12)$$

is nonsingular. The observability matrix is

$$O = \begin{bmatrix} \sqrt{Q} \\ \sqrt{Q}A \end{bmatrix} = \begin{bmatrix} q_d & 0 \\ 0 & \sqrt{d_v} \\ 0 & q_d \\ 0 & 0 \end{bmatrix} \quad (13)$$

Therefore, observability is guaranteed if and only if the position weighting q_d is greater than zero. Then the theorem says that we should be able to rely on a stable closed-loop system. Examining (11) makes it clear that this is indeed the case. ∎

5.8 DYNAMIC INVERSION DESIGN

In this chapter we have presented some basic tools of modern control design for linear systems. Since aircraft are inherently nonlinear systems, applying these linear design tools means that one must design several linear controllers, and then gain-schedule them over the operating regime of the aircraft (see Problem 5.4-9). There are alternative techniques that can deal directly with the known nonlinearities of the aircraft dynamics, using these nonlinearities in the controller to improve the system performance. These techniques are generally based on the *feedback* linearization approach (Slotine and Li, 1991) developed by Hunt and Su (1983) and Jacubczyk and Respondek (1980).

In this section we introduce the technique known as *dynamic inversion*, which has grown popular in recent years (Adams and Banda, 1993; Lane and Stengel, 1988; Enns et al., 1994; Tomlin et al., 1995; Wright Laboratory Report 1996). The dynamic inversion controller takes into account the nonlinearities of the aircraft, and thus does not require gain scheduling. As such it is suitable for a wide range of operating conditions, including high-angle-of-attack and hypervelocity design. To simulate the dynamic inversion control scheme, we use the technique of computer simulation for nonlinear dynamical systems given in Section 3.3. For this, we use the MATLAB software (*MATLAB Reference Guide*).

Though dynamics inversion is used for nonlinear systems and shows its true power there, we will start this section with a linear derivation and design example to get a feel for how it works. Then, we will study dynamic inversion controls design and simulation for a nonlinear aircraft.

Dynamic Inversion for Linear Systems

Derivation of Dynamic Inversion Controller. Let the plant be described in state-variable form by

$$\dot{x} = Ax + Bu \tag{5.8-1}$$

$$y = Cx, \tag{5.8-2}$$

with state $x(t) \in \mathbf{R}^n$, control input $u(t) \in \mathbf{R}^m$, and output $y(t) \in \mathbf{R}^p$. The entire state $x(t)$ is available for feedback purposes.

It is assumed that the system is *square,* that is, the number of inputs m is equal to the number of outputs p so that vectors $u(t)$ and $y(t)$ have the same dimension. This often occurs for aircraft systems, since there is often one control actuator per degree of freedom. If this is not the case, we may make some amendments to the following procedure. For instance, in modern high-performance aircraft, there may be more actuators than degrees of freedom (e.g., elevators, horizontal stabilators, and thrust vectoring for longitudinal dynamics). In this event, the control dimension may be reduced to obtain a square system by several techniques, including ganging, pseudo control, and daisy chaining (Wright Laboratory Report 1996). These are all techniques for *allocating control effectiveness* among several redundant actuators.

For a square system, then, it is desired to control the output $y(t)$ so that it follows a desired reference trajectory $r(t)$. Define the tracking error as

$$e(t) = r(t) - y(t) \tag{5.8-3}$$

In dynamic inversion, one differentiates the output $y(t)$ until the control $u(t)$ appears in the expression for the derivative. This is known technically as *input-output feedback linearization* (Slotine and Li, 1991). Taking the first derivative yields

$$\dot{y} = C\dot{x} = CAx + CBu, \qquad (5.8\text{-}4)$$

where $u(t)$ appears if matrix CB is not zero. In this case, since the system is square, so is matrix CB. If CB is nonsingular, then we are done. If $CB = 0$, then we continue to differentiate, obtaining

$$\ddot{y} = C\ddot{x} = CA\dot{x} + CB\dot{u} = CA^2 x + CABu \qquad (5.8\text{-}5)$$

If matrix CAB is nonsingular we are done. If $CAB = 0$, we differentiate again, continuing until the coefficient multiplying $u(t)$ is nonzero.

For aircraft, it is generally the case that CB is nonsingular. This is because of the way in which the control actuators enter into the aircraft dynamics equations, with one actuator for each degree of freedom. Then, we may stop at (5.8-4). Define an auxiliary input $v(t)$ by

$$v = CBu + CAx - \dot{r} \qquad (5.8\text{-}6)$$

so that

$$u = (CB)^{-1}(\dot{r} - CAx + v) \qquad (5.8\text{-}7)$$

Substituting this expression for $u(t)$ into (5.8-4) yields

$$\begin{aligned}\dot{y} &= CAx + CB\left[(CB)^{-1}(\dot{r} - CAx + v)\right] \\ &= CAx + \dot{r} - CAx + v\end{aligned} \qquad (5.8\text{-}8)$$

or

$$\dot{e} = -v \qquad (5.8\text{-}9)$$

The auxiliary input $v(t)$ was selected to make expression (5.8-7) hold in order to cancel the term CAx, and also so that CB does not appear in (5.8-9).

System (5.8-9) is the *error dynamics*. To complete the design it is only necessary to select $v(t)$ so that this system is stable. Due to the way in which $v(t)$ was defined by (5.8-7), the error dynamics have a very simple form; indeed, the error system has p poles at $s = 0$. This means that it is very easy to select $v(t)$ to stabilize this system. A variety of techniques may be used, including robust control, LQG/LTR (see Chapter 6) and other linear system design techniques (Adams and Banda, 1993; Lane and Stengel, 1988; Enns et al., 1994; Tomlin et al., 1995; Wright Laboratory Report 1996).

A simple choice for $v(t)$ is

$$v = Ke \qquad (5.8\text{-}10)$$

Then, one has the closed-loop error dynamics given by

$$\dot{e} = -Ke, \qquad (5.8\text{-}11)$$

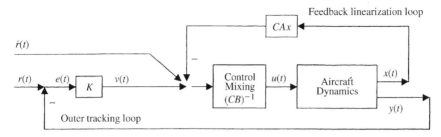

Figure 5.8-1 Dynamic inversion controller.

which is a stable system as long as gain matrix K is positive definite. In practice, one usually selects K diagonal to keep the control channels in the outer loop decoupled. The gain K should be selected so that the closed-loop system satisfied MILSPEC flying-qualities requirements.

The overall dynamic inversion control input is given by

$$u = (CB)^{-1}(\dot{r} + Ke - CAx) \tag{5.8-12}$$

The control scheme given by this is shown in Figure 5.8-1. Note that (5.8-10) is simply an *outer proportional feedback tracking loop,* while (5.8-7) is an *inner control loop* using full state variable feedback. This inner loop is called the *feedback linearization loop.* Its function is to make the system from $v(t)$ to $y(t)$ appear like a linear system with poles at the origin (5.8-9). This greatly simplifies the design of the outer tracking loop. There is also a feedforward term involving $\dot{r}(t)$, which is known as *velocity feedforward.* This greatly improves the tracking accuracy of the closed-loop system.

Note that to implement the dynamic inversion control algorithm (5.8-12) one must know CA and CB. That is, *a model of the aircraft dynamics is actually built into the controller.* This is what makes the outer control loop design so simple. Moreover, full state feedback is required for the inner loop.

This completes the design of the dynamic inversion controller. The full power of this approach will be seen in the next subsection when we apply the technique to nonlinear aircraft systems.

Zero Dynamics. Equation (5.8-11) only gives the error dynamics of the output $y(t)$. The full closed-loop system is obtained by substituting the control (5.8-7) into the state equation (5.8-1). This yields

$$\dot{x} = Ax + B(CB)^{-1}(\dot{r} - CAx + v)$$
$$\dot{x} = [I - B(CB)^{-1}]Ax + B(CB)^{-1}(\dot{r} + v) \tag{5.8-13}$$

The *zero dynamics* are defined as the dynamics of the system when the input $v(t)$ is selected to give an output $y(t)$ equal to zero. Since $y(t) = 0$, then $\dot{y}(t) = 0$, so that (5.8-9) shows that

$$v = -\dot{e} = \dot{y} - \dot{r} = -\dot{r}$$

Substituting this value for $v(t)$ into (5.8-13) yields the zero dynamics

$$\dot{x} = \left[I - B(CB)^{-1}C \right] Ax \equiv A_z x \qquad (5.8\text{-}14)$$

Note that the dimension of the entire state is n, while the dimension of the error dynamics (5.8-11) is $p < n$. The error dynamics are guaranteed stable by the choice of $v(t)$; however, there remain n-p poles that may or may not be stable. These poles are unobservable selecting the output $y(t)$, and so they cannot be moved using the dynamic inversion controller. These n-p poles are exactly given by the zero dynamics. If some of these internal zeros are non-minimum phase, then the closed-loop system designed by dynamic inversion will be unstable.

The poles of the matrix A_z consist of p poles at the origin (namely, the poles of error dynamics system (5.8-9)), plus the n-p internal zeros. Define the operator

$$P \equiv I - B(CB)^{-1}C \qquad (5.8\text{-}15)$$

Note that $P^2 = P$ so that P is a projection. Furthermore $PB \equiv [I - B(CB)^{-1}C]B = 0$ and $CP \equiv C[I - B(CB)^{-1}C] = 0$, so that P is the projection on the nullspace of C along the range of B. Thus, $A_z = PA$ describes those dynamics that are both in the nullspace of C and in range perpendicular of B. These are precisely the modes that are unobservable using the output $y(t)$, and cannot be controlled using the dynamic inversion approach.

Selection of Controlled Variables (CVs). For dynamic inversion to be successful, it is necessary to *select the controlled variable (CV) $y(t)$ so that the zero dynamics are stable*. This may be checked by computing A_z for the selected output matrix C, and finding its poles. If they are not stable, then a new C matrix must be selected. Once a suitable C matrix has been found, the p poles of the error dynamics may be selected using (5.8-10).

The outputs to be controlled in fighter aircraft are usually selected as:

$$\begin{aligned}
&\text{pitch axis CV:} && q + n_{zp}/V_{co} \\
&\text{roll axis CV:} && p + \alpha r &&& (5.8\text{-}16) \\
&\text{yaw axis CV:} && r - \alpha p - (g \sin\phi \cos\theta)/V + k\beta
\end{aligned}$$

Discussion of these controlled variables may be found in Enns et al. (1994). These CVs are suitable for most conventional flight regimes and piloting tasks. They may need modifications for high-α or very-low-speed flight.

The pitch axis CV is motivated by the C-star criterion C^*. See the discussion in Section 4.3 on C^*. One uses the normal acceleration at the pilot's station n_{zp} and not n_z at the center of gravity of the aircraft since the latter yields unstable zero dynamics. The crossover velocity V_{co} of the CV should be selected to match the MILSPEC requirements on n_{zp} and pitch rate q. Even using this CV, dynamic inversion design can destabilize the phugoid mode. This problem may be avoided by adding a small airspeed term to the pitch axis CV (Enns et al., 1994) to obtain

DYNAMIC INVERSION DESIGN

pitch axis CV: $\quad q + n_{zp}/V_{co} + kv_T \quad$ (5.8-17)

The gain k is selected so that the zero dynamics are stable. This is illustrated in the next example. The roll axis and yaw axis CVs generally do not have such a stability problem.

In the next example we show how to select CVs and design a dynamic inversion controller for linearized longitudinal dynamics. To verify the performance of the controller we employ MATLAB, using the technique for computer simulation of systems that was introduced in Section 3.3. This technique applies for linear or nonlinear systems, and also employs directly the actual controller. Once the simulation results are satisfactory, this controller can simply be cut out of the code and programmed into the aircraft computer.

Example 5.8-1: Dynamic Inversion Design for Linear F-16 Longitudinal Dynamics.
Consider the linearized F-16 longitudinal dynamics of Chapter 3. Including an elevator actuator, the states are given by

$$x = [v_T \quad \alpha \quad \theta \quad q \quad \delta_e]^T \quad (1)$$

The control input is the elevator input u_e. The A and B matrices are given by

$$A = \begin{bmatrix} -0.1270 & -235.0000 & -32.2000 & -9.5100 & -0.2440 \\ 0 & -0.9690 & 0 & 0.9080 & -0.0020 \\ 0 & 0 & 0 & 1.0000 & 0 \\ 0 & -4.5600 & 0 & -1.5800 & -0.2000 \\ 0 & 0 & 0 & 0 & -20.2000 \end{bmatrix} \quad B = \begin{bmatrix} 0 \\ 0 \\ 0 \\ 0 \\ 20.2 \end{bmatrix} \quad (2)$$

In Chapter 4 both the normal acceleration n_z at the cg and the normal acceleration n_{zp} at the pilot's station are given. One has

$$n_z = [0.004 \quad 15.88 \quad 0 \quad 1.481 \quad 0.033]\,x \quad (3)$$

$$n_{zp} = [0.004 \quad 16.2620 \quad 0 \quad 0.9780 \quad -0.0485]\,x \quad (4)$$

Computing $C^* = n_{zp} + 12.4q$ as in Chapter 4 yields

$$C^* = [0.004 \quad 16.2620 \quad 0 \quad 13.3780 \quad -0.0485]\,x \quad (5)$$

a. Zero Dynamics for Different Controlled Variables. To find a suitable CV, one may compute the zero dynamics $A_z = [I - B(CB)^{-1}C]\,A$ for the above outputs. Then, the eigenvalues of A_z may be determined. One notes that all these computations are very easy using the MATLAB software.

For $y(t) = n_z$ one obtains

$$13.8346, \quad -6.5717, \quad -0.1242, \quad 0.0633, \quad 0$$

Since the number of inputs and outputs is $p = 1$, one obtains one pole at zero, corresponding to the error dynamics (5.8-9). The remaining poles of A_z are the zero dynamics. Since there are unstable zeros, performing dynamic inversion design using n_z as the CV would destabilize the system.

For $y(t) = n_{zp}$ one obtains the poles of A_z as

$$-3.6652 \pm 7.6280i, \quad -0.1244, \quad 0.0550, \quad 0$$

One has the pole at zero of the error dynamics, plus the internal zeros using this choice for $y(t)$. The situation is better, and there is only one slightly unstable pole in the zero dynamics.

For $y(t) = C^*$ one obtains

$$-56.3022, \quad -2.1389, \quad -0.1252, \quad 0.0325, \quad 0,$$

which still reveals a slightly unstable zero dynamics pole.

Accordingly, none of these would be appropriate choices for the CV for dynamic inversion design, though n_{zp} and C^* are both better than n_z. The trouble is the velocity term in n_{zp} and C^*. One may correct the problem by adding a small airspeed term to n_{zp} or C^* (Enns et al., 1994). After a few tries, we decided on

$$y(t) = C^* - 0.014 v_T = n_{zp} + 12.4q - 0.014 v_T, \tag{6}$$

which yields

$$y(t) = [-.01 \quad 16.2620 \quad 0 \quad 13.3780 \quad -0.0485]\, x = Cx \tag{7}$$

Computing now A_z and finding its eigenvalues yields

$$-56.2243, \quad -2.0209, \quad -0.1423, \quad -0.0758, \quad 0$$

The unstable zero dynamics pole is gone and so we proceed to simulate the dynamic inversion controller.

b. Simulation of the Dynamic Inversion Controller. To simulate the dynamic inversion controller given in Figure 5.8-1 we use a method of computer simulation that is based on the nonlinear state equation $\dot{x} = f(x, u)$. This technique was detailed in Section 3.3 and works for either linear or nonlinear systems. MATLAB has a built-in function that performs Runge-Kutta integration, and is very convenient to use here.

The third-order Runge-Kutta integrator in MATLAB is called *ode23*, and it requires a MATLAB M file containing the nonlinear state dynamics. This M file is given in Figure 5.8-2. The form of this M file is very important. Note that one first computes the dynamic inverse controller (5.8-12), and then computes the state equation derivatives for the aircraft. Thus, the first portion of the M file is exactly the code required to implement the controller on the actual aircraft.

```
% Inverse Dynamics Controller for F16 Linear Dynamics
function xdot=F16LinDynInv(t,x)
global y

%       VT= x(1);        ! True airspeed
%       ALPH= x(2);      ! Angle of Attack in rads.
%       THTA= x(3);      ! Pitch attitude in rads.
%       Q   = x(4);      ! Pitch rate rad/s
%       elev= x(5);      ! elevator actuator

% Inverse Dynamics Controller

    % Model of aircraft
    a=  [ -0.1270   -235.0000   -32.2000   -9.5100   -0.2440
             0        -0.9690        0       0.9080   -0.0020
             0           0            0       1.0000       0
             0        -4.5600        0      -1.5800   -0.2000
             0           0            0          0    -20.2000 ];
    b=  [0 0 0 0 20.20]';
    c=  [ -0.01    16.2620      0    13.3780   -0.0485];
    &  % y=cstar' modifed

    % command input

    r= 1   ;  % check step response
    rdot= 0 ;

    % controller parameters

    K= 10;

    % plant outputs, tracking errors, and control inputs

    y= c*x ;   % y= cstar' modified
    e= r-y ;
    v= K*e ;
    w= rdot - c*a*x + v ;
    u= inv(c*b) * w ;
    thtl=0
    uelev= u ;

% Aircraft State Equations

xdot(1)= -0.1270*x(1)  -235.0*x(2)  -32.3*x(3)  -9.51*x(4)  -0.244*x(5)  +62.8&thtl ;
xdot(2)=                -0.9690*x(2)              +0.908*x(4)  -0.002*x(5)  -0.04*thtl ;
xdot(3)=                                            x(4) ;
xdot(4)=                -4.56*x(2)                -1.58*x(4)  -0.2*x(5) ;
xdot(5)=                                                      -020.2*x(5)  +20.2*uelev ;

xdot=xdot' ;
```

Figure 5-8-2 Dynamic Inversion Controller and simulation code.

It is important to note that the dynamic inversion controller must know the aircraft dynamics A, B, C. That is, the controller must contain a model of the aircraft dynamics.

We selected the desired trajectory $r(t)$ equal to a unit step, since if the step response of the controlled aircraft is suitable, then the controller has a good performance for a wide range of pilot input commands.

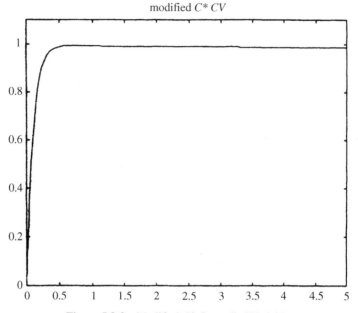

Figure 5.8-3 Modified C^* Controlled Variable.

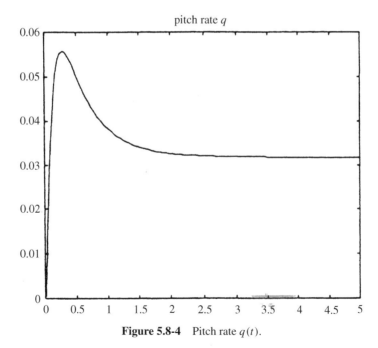

Figure 5.8-4 Pitch rate $q(t)$.

The MATLAB command lines required to run the simulation are given by

```
»[t,x]=ode23('F16LinDynInv',[0 5],[0 0 0 0 0]');
»y=x*cstarmodd';
»plot(t,y)
```

where cstarmodd is the C matrix defined by (7). The second argument in *ode23* specifies the integration time interval 0–5 s, and the third argument specifies zero initial conditions $x(0)$.

The output $y(t)$ is shown in Figure 5.8-3. If the time constant is not suitable according to MILSPEC requirements, one may simply select another value of K in (5.8-12) and repeat the simulation. The pitch rate is shown in Figure 5.8-4. ∎

A Pathological Case. In aircraft control one generally has CB nonsingular in (5.8-4). If $CB = 0$ one may proceed as discussed there. However, in pathological situations it may occur that CB is neither zero nor nonsingular. Then one must proceed as follows.

Differentiating repeatedly one obtains

$$\ddot{y} = C\ddot{x} = CA\dot{x} + CB\dot{u} = CA^2 x + CB\dot{u} + CABu = CA^2 x + C[B \quad AB]\begin{bmatrix}\dot{u}\\u\end{bmatrix}$$

$$\dddot{y} = C\dddot{x} = CA^3 x + C[B \quad AB \quad A^2 B]\begin{bmatrix}\ddot{u}\\\dot{u}\\u\end{bmatrix}$$

Continuing for n steps, with n the number of states, one obtains the n-th derivative of $y(t)$ as

$$y^{(n)}(t) = CA^n x(t) + CU_n \underline{u}(t), \qquad (5.8\text{-}18)$$

where the controllability matrix is

$$U_n = \begin{bmatrix} B & AB & \cdots & A^{n-1}B \end{bmatrix} \qquad (5.8\text{-}19)$$

and $\underline{u}(t)$ is a vector of $u(t)$ and its first $n-1$ derivatives.

Now, if the system is controllable, then U_n has rank n. If in addition the C matrix has rank p, then CU_n has rank p. In this case, though CU_n is not square, it has a right-inverse given by

$$(CU_n)^+ = (CU_n)^T \left[(CU_n)(CU_n)^T\right]^{-1} \qquad (5.8\text{-}20)$$

for note that $(CU_n)(CU_n)^+ = I$, the $p \times p$ identity matrix. Thus, one may define

$$\underline{u} = (CU_n)^+ \left(r^{(n)} - CA^n x + v\right) \qquad (5.8\text{-}21)$$

and substitute into (5.8-18) to obtain

$$y^{(n)}(t) = CA^n x(t) + CU_n \underline{u}(t) = CA^n x(t) + CU_n \left[(CU_n)^+ (r^{(n)} - CA^n x + v) \right],$$

which yields

$$y^{(n)}(t) = r^{(n)} + v \quad (5.8\text{-}22)$$

or

$$e^{(n)} = -v \quad (5.8\text{-}23)$$

This is the error dynamics. It has p poles at the origin.

Selecting now the outer loop structure given by

$$v = K_{n-1} e^{(n-1)} + \cdots + K_0 e \quad (5.8\text{-}24)$$

gives the closed-loop error dynamics

$$e^{(n)} + K_{n-1} e^{(n-1)} + \cdots + K_0 e = 0 \quad (5.8\text{-}25)$$

The gains K_i can be selected to make this system stable. Note that this requires feedforward of the tracking error $e(t)$ and its derivatives.

In this pathological case, the inverse dynamics controller is given by (5.8-21) and (5.8-24). A dynamical system can then be employed to extract the control input $u(t)$ from its derivative vector $\underline{u}(t)$.

Dynamic Inversion for Nonlinear Systems

Since the aircraft is inherently a nonlinear system, we will now discuss dynamic inversion control for nonlinear systems (Slotine and Li, 1991; Enns et al., 1994; Wright Laboratory Report, 1996). Dynamic inversion is one of few control techniques that can directly be extended to nonlinear systems.

Derivation of Dynamic Inversion Controller. Let the plant be described in nonlinear state-variable form by

$$\dot{x} = f(x) + g(x)u \quad (5.8\text{-}26)$$

$$y = h(x), \quad (5.8\text{-}27)$$

with state $x(t) \in R^n$, control input $u(t) \in R^m$, and output $y(t) \in R^p$. The entire state $x(t)$ is available for feedback purposes. It is assumed that the system is *square*, that is, the number of inputs m is equal to the number of outputs p so that vectors $u(t)$ and $y(t)$ have the same dimension.

Note that the system is linear in the control input $u(t)$. This generally holds for aircraft systems, though if it does not, and one has instead the more general state equation

$$\dot{x} = f(x, u),$$

one can use a modified form of the upcoming development (Enns et al., 1994).

To make the system follow a desired trajectory $r(t)$, the tracking error is defined as

$$e(t) = r(t) - y(t) \tag{5.8-28}$$

Differentiate the output to obtain

$$\dot{y} = \frac{\partial h}{\partial x}\dot{x} = \frac{\partial h}{\partial x}f(x) + \frac{\partial h}{\partial x}g(x)u \equiv F(x) + G(x)u \tag{5.8-29}$$

Define now the control input by

$$u = G^{-1}(x)\left[-F(x) + \dot{r} + v\right] \tag{5.8-30}$$

with $v(t)$ an auxiliary input to be defined. Substituting this expression into (5.8-29) yields the error dynamics

$$\dot{e} = -v \tag{5.8-31}$$

Now, any linear design technique, including robust control techniques, LQR/LTR, and so on, can be used to select $v(t)$ to stabilize this linear system with p poles at the origin. One convenient choice is simply

$$v = Ke \tag{5.8-32}$$

with K positive definite.

The overall dynamic inversion controller is given by

$$u = G^{-1}(x)\left[-F(x) + \dot{r} + Ke\right] \tag{5.8-33}$$

This controller is depicted in Figure 5.8-5. It requires full state feedback for the inner loop.

Note that the control $u(t)$ has been selected to make the plant from $v(t)$ to $y(t)$ be simply a linear system with p poles at the origin. This is accomplished by the inner feedback linearization loop, which is now nonlinear. Then, an outer tracking loop is closed to complete the design. Any linear design technique, including robust control, H-infinity, or LQG/LTR, may be used for this outer-loop design.

It is important to note that the control (5.8-33) contains a model of the aircraft dynamics, since it requires $F(x)$ and $G(x)$. Therefore, to implement it, one must know the nonlinear functions in the aircraft equation. In the upcoming example, this amounts to including the nonlinear aircraft functions in the controller, but in practice

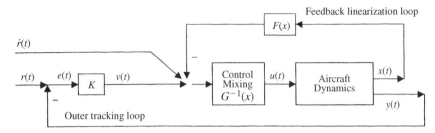

Figure 5.8-5 Nonlinear dynamic inversion controller.

it usually entails including *full lookup-tables* in the controller. This can become cumbersome, but is possible with today's computing systems.

In aircraft systems, $G(x)$ is usually nonsingular. If $G(x)$ is not nonsingular, then one must take more steps to derive the controller. If $G(x) = 0$, one may proceed as in Slotine and Li (1991), repeatedly differentiating $y(t)$ using Lie derivatives.

The *CV*s are selected as detailed for linear systems, though nonlinear versions of the controlled outputs may be used. For instance, for the roll axis one might use the nonlinear version $p \cos \alpha + r \sin \alpha$ (Enns et al., 1994).

In the nonlinear case, it is more difficult to test the selected controlled variables than in the linear case, since one does not have the artifice of the zero dynamics matrix A_z. However, one may linearize the nonlinearities and use a version of the technique presented for linear systems. Specifically, the full closed-loop dynamics are given by

$$\dot{x} = f(x) + g(x)G^{-1}(x)\left[-F(x) + \dot{r} + v\right]$$

$$= \left[I - gG^{-1}\frac{\partial h}{\partial x}\right]f(x) + gG^{-1}\left[\dot{r} + v\right] \quad (5.8\text{-}34)$$

The zero dynamics are given by

$$\dot{x} = \left[I - gG^{-1}\frac{\partial h}{\partial x}\right]f(x) \quad (5.8\text{-}35)$$

These may be linearized to determine the suitability of the *CV* at a specific operating point. Stability may also be checked by simulation. Simply simulate (5.8-35) in MATLAB, selecting different initial conditions and verifying that the state converges to zero in each case. This amounts to plotting a phase portrait of the zero dynamics (Tomlin et al., 1995).

Example 5.8-2: Dynamic Inversion Design for Nonlinear Longitudinal Dynamics.

We now present a nonlinear version of Example 5.8-1. A longitudinal model $\dot{x} = f(x) + g(x)u$ of an aircraft similar to that presented in Chapter 3 is given in Figure 5.8-6. The states are

$$x = [v_T \quad \alpha \quad \theta \quad q \quad \delta_e]^T \tag{1}$$

and the control input is the elevator actuator input u_e.

The normal acceleration is given by

$$n_z = \bar{q}S(C_L \cos\alpha + C_D \sin\alpha)/mg \tag{2}$$

and the normal acceleration at the pilot's station is

$$n_{zp} = n_z + 15\dot{q}/g = \bar{q}S(C_L \cos\alpha + C_D \sin\alpha)/mg + 15M/gI_{yy}, \tag{3}$$

with M the pitching moment (MOM in Figure 5.8-6). The output is selected as

$$y = C^* = n_{zp} + 12.4q \equiv h(x) \tag{4}$$

We aim to apply (5.8-33) to compute the dynamic inversion controller. To do this we must determine

$$F(x) = \frac{\partial h}{\partial x}f(x) \quad \text{and} \quad G(x) = \frac{\partial h}{\partial x}g(x) \tag{5}$$

$f(x)$ and $g(x)$ are easily determined from the nonlinear dynamics and are given in Figure 5.8-7. Finding $\partial h/\partial x$ is tedious and the results are as follows. First, $\partial n_z/\partial x$ is given as

$$\frac{\partial n_z}{\partial v_T} = \rho v_T S(C_L \cos\alpha + C_D \sin\alpha)/mg$$

$$\frac{\partial n_z}{\partial \alpha} = \bar{q}S[(C_D + 4.58)\cos\alpha - 0.515C_L \sin\alpha]/mg \tag{6}$$

$$\frac{\partial n_z}{\partial \theta} = 0, \quad \frac{\partial n_z}{\partial q} = 0, \quad \frac{\partial n_z}{\partial \delta_e} = 0$$

Next, one has $\partial M/\partial x$ given by

$$\frac{\partial M}{\partial v_T} = \rho v_T S\bar{c}C_M + \frac{S\bar{c}^2}{4}\rho q C_{mq}$$

$$\frac{\partial M}{\partial \alpha} = -0.75\bar{q}\bar{c}S$$

$$\frac{\partial M}{\partial \theta} = 0 \tag{7}$$

$$\frac{\partial M}{\partial q} = \frac{S\bar{c}^2\bar{q}}{2v_T}C_{mq}$$

$$\frac{\partial M}{\partial \delta_e} = -0.9\bar{q}\bar{c}S$$

498 MODERN DESIGN TECHNIQUES

```
% Nonlinear Longitudinal Aircraft Model (for small airplane)
% B. Stevens file modified by F. Lewis on 8 May 2000

function xdot=NonLinDynInv(t,x);

% Definition of some constants for the aircraft used for simulation;
    WEIGHT = 2300.0; G=32.2; MASS=WEIGHT/G;
    IYY = 2094.;
    RHO = 2.377E-3;
    S = 175.0;
    CBAR = 4.89;
    CMQ = -12.0;
    RTOD = 57.29578;      ! radians to degrees

    VT=  x(1);            ! True airspeed
    ALPH= x(2);           ! Angle of Attack in rads.
    THTA= x(3);           ! Pitch attitude in rads.
    Q  =  x(4);           ! Pitch rate rad/s
    EL=  x(5);            ! elevator actuator

% Computed control inputs are thtl (throttle) and uelev (elev. act. command)

    GAM= THTA - ALPH;
    CBV= 0.5*CBAR/VT;
    CL = 0.25 + 4.58*ALPH;                  ! Linear lift curve
    CM = 0.015 - 0.75*ALPH - 0.9*x(5);      ! Linear pitching moment
    CD = .038 + .053*CL*CL;                 ! Parabolic drag
    QBAR= 0.5*RHO*VT*VT;                    ! Dynamic pressure
    LIFT= QBAR*S*CL;
    DRAG= QBAR*S*CD;
    MOM = QBAR*S*CBAR*(CM + CBV*CMQ*Q);  ! Added pitch damping
    FT= (338.02 + 1.5651*vt - .00884*vt**2 )* thtl; ! Nonlinear Thrust

% State Equations

    xdot(1)= (FT*cos(ALPH) - DRAG - WEIGHT*sin(GAM) )/MASS;
    xdot(2)= (-FT*sin(ALPH) - LIFT + WEIGHT*cos(GAM))/(MASS*VT) + Q;
    xdot(3)= Q;
    xdot(4)= MOM/IYY;
    xdot(5)= -20.2*x(5) + 20.2*uelev;

% outputs

    nz= QBAR*S*(CL*cos(ALPH) + CD*sin(ALPH)) / (G*MASS) ! Normal accel.
    nzp= nz + 15*MOM/(g*IYY) ! Normal accel. at pilot's station
    cstar= nzp - 12.4q    ! controlled variable
```

Figure 5.8-6 Nonlinear model of aircraft longitudinal dynamics.

Finally, one has

$$\frac{\partial C^*}{\partial x} = \frac{\partial n_z}{\partial x} + \frac{15}{gI_{yy}}\frac{\partial M}{\partial x} + k,$$

where $k = [0 \quad 0 \quad 0 \quad 12.40]^T$. All of these are included finally in Figure 5.8-7.

DYNAMIC INVERSION DESIGN

```
% Nonlinear Longitudinal Dynamic Inversion Controller
  function xdot=NonLinDynInvCtrlr(t,x);

% MODEL OF AIRCRAFT DYNAMICS USED IN CONTROLLER

% Definition of some constants for the aircraft used for simulation;
    WEIGHT = 2300.0; G=32.2; MASS=WEIGHT/G;
    IYY = 2094.;
    RHO = 2.377E-3;
    S = 175.0;
    CBAR = 4.89;
    CMQ = -12.0;
    RTOD = 57.29578;    % radians to degrees

    VT= x(1);       % True airspeed
    ALPH= x(2);     % Angle of Attack in rads.
    THTA= x(3);     % Pitch attitude in rads.
    Q   = x(4);     % Pitch rate rad/s
    EL= x(5);       % elevator actuator

% Computed control inputs are thtl (throttle) and EL (elevator command)
    GAM= THTA - ALPH;
    CBV= 0.5*CBAR/VT;
    CL = 0.25 + 4.58*ALPH;              % Linear lift curve
    CM = 0.015 - 0.75*ALPH - 0.9*x(5);  % Linear pitching moment
    CD = .038 + 0.53*CL*CL;             % Parabolic drag
    QBAR= 0.5*RHO*VT*VT;                % Dynamic pressure
    LIFT= QBAR*S*CL;
    DRAG= QBAR*S*CD;
    MOM = QBAR*S*CBAR*(CM + CBV*CMQ*Q); % Added pitch damping
    FT= (338.02 + 1.5651*vt - .00884*vt**2 )* thtl; % Nonlinear Thrust

% function f(x)
    f1= (FT*cos(ALPH) - DRAG - WEIGHT*sin(GAM) )/MASS;
    f2= (-FT*sin(ALPH) - LIFT + WEIGHT*cos(GAM))/(MASS*VT) + Q;
    f3= Q;
    f4= MOM/IYY;
    f5= -20.2*x(5);

% function g(x)
          g=[0  0  0  0  20.2]' ;
```

Figure 5.8-7 Nonlinear dynamic inversion controller (Part I).

To simulate the dynamic inversion controller, one may write a single MATLAB M file containing both the controller in Figure 5.8-7 and the aircraft dynamics in Figure 5.8-6. The form of this M file will be similar in spirit to that used in Example 5.8-1. This is left for the enterprising reader.

It is very important to note that the dynamic inversion controller in Figure 5.8-7 requires full knowledge of all the nonlinear dynamics of the aircraft. In this example this entails including all the analytic expressions used in the aircraft model. However,

500 MODERN DESIGN TECHNIQUES

```
% function dh/dx
    dnzdvt= RHO*VT*S*(CL*cos(ALPH) + CD*sin(ALPH))/WEIGHT;
    dnzdal= QBAR*S*((CD+4.58)*cos(ALPH) - 0.515*CL*sin(ALPH))/WEIGHT;
    dmdvt= RHO*VT*S*CBAR*CM + (CBAR^2)*S*RHO*Q*CMQ/4;
    dmdal= -0.75*QBAR*CBAR*S;
    dmdq=  (CBAR^2)*S*QBAR*CMQ/(2*VT);
    dmddel= -0.9*QBAR*CBAR*S;
    h1= dnzdvt + 15*dmdvt/(G*IVY);
    h2= dnzdal + 15*dmdal/(G*IVY);
    h3= 0;
    h4=         15dmdq/(G*IVY) + 12.4;
    h5=         15dmddel/(G*IVY);
% COMMAND INPUT
    r= 1;   % check step response
    rdot= 0;
% DYNAMIC INVERSION CONTROL INPUT
    e= r-y;
    Fctrl= h1*f1 + h2*f2 + h3*f3 + h4*f4 + h5*f5 ;
    Gctrl= h5*g(5);
    K= 10;
    uelev= (-Fctrl + rdot + K*e)/Gctrl;
    thtl= 0;
```

Figure 5.8-7 (*continued*) Nonlinear dynamic inversion controller (Part II).

in practice, it generally involves including full aircraft lookup-tables in the controller. In this example, the Jacobian $\partial h/\partial x$ was computed analytically. In practice, one may use a numerical differentiation routine as part of the controller. ∎

5.9 SUMMARY

In this chapter we showed how to use modern control techniques to design multivariable and multiloop aircraft flight control systems. The approach is based on the state-variable model and a mathematical performance criterion selected according to the performance objectives. The matrix of control gains is determined by solving explicit matrix equations using computer software. Using such an approach, all the feedback loops are closed simultaneously to yield the guaranteed performance desired. This is in contrast to the classical techniques of Chapter 4, which relied on trial-and-error successive loop closures to find the control gains individually.

Two basic modern design techniques were covered. In Section 5.2 we discussed eigenstructure assignment techniques that take advantage of the freedom inherent in design for systems with more than one input and/or output to assign the closed-loop poles and eigenvectors. In the remainder of the chapter we covered linear quadratic (LQ) techniques, where the control gains are selected to minimize generalized quadratic performance indices (PIs). Design equations were derived for the control gains minimizing these PIs and listed in tabular form for easy reference. The design equations may be solved for the control gains using software like that described in Appendix B.

In Section 5.5 the thrust was to introduce modified nonstandard PIs allowing LQ designs with only a small number of design parameters that require tuning for suitable performance. The point was made that successful control system design hinges on the selection of a suitable PI.

Our primary thrust was to use output feedback to allow the design of a compensator with any desired structure. The PI was an integral of the squares of the states and control inputs; thus the LQ techniques used in this chapter are *time-domain* techniques.

In Section 5.8 we discussed dynamic inversion design, which results in a controller with an inner feedback linearization loop and an outer tracking loop.

REFERENCES

Adams, R.J., and S. S. Banda. "Robust Flight Control Design Using Dynamic Inversion and Structured Singular Value Synthesis." *IEEE Transactions on Control System Technology* 1, no. 2 (June 1993): 80–92.

AFWAL-TR-84-3008. "AFTI/F-16 Development and Integration Program, DFCS Phase Final Technical Report." General Dynamics, Fort Worth, Tex., December 1984.

Andry, A. N., Jr., E. Y. Shapiro, and J. C. Chung. "Eigenstructure Assignment for Linear Systems." *IEEE Transactions on Aerospace and Electronic Systems,* vol. AES-19, no. 5, pp. 711–729, September 1983.

Andry, A. N., Jr., J. C. Chung, and E. Y. Shapiro. "Modalized Observers." *IEEE Transactions on Automatic Control,* vol. AC-29, no. 7, July 1984, pp. 669–672.

Armstrong, E. S. *ORACLS: A Design System for Linear Multivariable Control.* New York: Marcel Dekker, 1980.

Bartels, R. H., and G. W. Stewart. "Solution of the Matrix Equation $AX + XB = C$." *Communications of the ACM,* 15, no. 9 (September 1972): 820–826.

Blakelock, J. H. *Automatic Control of Aircraft and Missiles.* New York: Wiley, 1965.

Broussard, J., and N. Halyo. "Active Flutter Control using Discrete Optimal Constrained Dynamic Compensators." *Proceedings of the American Control Conference,* June 1983, pp. 1026–1034.

Bryson, A. E., Jr. "New Concepts in Control Theory, 1959–1984." *Journal of Guidance,* 8, no. 4 (July–August 1985): 417–425.

Bryson, A. E., Jr., and Y.-C. Ho. *Applied Optimal Control.* New York: Hemisphere, 1975.

Davison, E. J., and I. J. Ferguson. "The Design of Controllers for the Multivariable Robust Servomechanism Problem using Parameter Optimization Methods." *IEEE Transactions on Automatic Control,* vol. AC-26, no. 1, February 1981, pp. 93–110.

D'Azzo, J. J., and C. H. Houpis. *Linear Control System Analysis and Design.* New York: McGraw-Hill, 1988.

Doyle, J. C., and G. Stein. "Multivariable Feedback Design: Concepts for a Classical/Modern Synthesis." *IEEE Transactions on Automatic Control,* vol. AC-26, no. 1, February 1981, pp. 4–16.

Enns, D., D. Bugajski, R. Hendrick, and G. Stein. "Dynamic Inversion: An Evolving Methodology for Flight Control Design." *International Journal of Control* 59, no. 1 (1994): 71–91.

Franklin, G. F., J. D. Powell, and A. Emami-Naeini. *Feedback Control of Dynamic Systems.* Reading, Mass.: Addison-Wesley, 1986.

Gangsaas, D., K. R. Bruce, J. D. Blight, and U.-L. Ly. "Application of Modern Synthesis to Aircraft Control: Three Case Studies." *IEEE Transactions on Automatic Control,* vol. AC-31, no. 11, November 1986, pp. 995–1014.

Grimble, M. J., and M. A. Johnson. *Optimal Control and Stochastic Estimation: Theory and Applications.* Vol. 1. New York: Wiley, 1988.

Harvey, C. A., and G. Stein. "Quadratic Weights for Asymptotic Regulator Properties." *IEEE Transactions on Automatic Control,* vol. AC-23, no. 3, 1978, pp. 378–387.

Hunt, L. R., R. Su, and G. Meyer. "Global Transformations of Nonlinear Systems." *IEEE Transactions on Automatic Control* 28 (1983): 24–31.

IMSL. *Library Contents Document.* 8th ed., International Mathematical and Statistical Libraries, Inc., NBC Building, 7500 Bellaire Blvd., Houston, TX 77036, 1980.

Jacubczyk, B., and W. Respondek. "On Linearization of Control Systems." *Bulletin Academie Polonaise de Science et Mathematique* 28 (1980): 517–522.

Kailath, T. *Linear Systems.* Englewood Cliffs, N.J.: Prentice-Hall, 1980.

Kalman, R. "Contributions to the Theory of Optimal Control." *Boletin de la Sociedad de Matematica Mexicana* 5 (1958): 102–119.

Kalman, R. E. "A New Approach to Linear Filtering and Prediction Problems." *Transactions of the ASME Journal of Basic Engineering* 82 (1960): 34–35.

Kreindler, E., and D. Rothschild. "Model-following in Linear-quadratic Optimization." *American Institute of Aeronautics and Astronautics Journal* 14, no. 7 (July 1976): 835–842.

Kwakernaak, H., and R. Sivan. *Linear Optimal Control Systems.* New York: Wiley, 1972.

Kwon, B.-H., and M.-J. Youn. "Eigenvalue-generalized Eigenvector Assignment by Output Feedback." *IEEE Transactions on Automatic Control,* vol. AC-32, no. 5, May 1987, pp. 417–421.

Lane, S. H., and R. F. Stengel. "Flight Control Using Non-linear Inverse Dynamics." *Automatica* 24, no. 4 (1988): 471–483.

Levine, W. S., and M. Athans. "On the Determination of the Optimal Constant Output Feedback Gains for Linear Multivariable Systems," *IEEE Transactions on Automatic Control,* vol. AC-15, no. 1, February 1970, pp. 44–48.

Lewis, F. L. *Optimal Control.* New York: Wiley, 1986.

Ly, U.-L., A. E. Bryson, and R. H. Cannon. "Design of Low-order Compensators using Parameter Optimization." *Automatica* 21, no. 3 (1985): 315–318.

MacFarlane, A.G.J. "The Calculation of Functionals of the Time and Frequency Response of a Linear Constant Coefficient Dynamical System." *Quarterly Journal of Mechanical Applied Mathematics* 16, pt. 2 (1963): 259–271.

MATLAB Reference Guide, 1994, The MathWorks, 24 Prime Park Way, Natick, MA 01760.

$MATRIX_x$, Integrated Systems, Inc., 2500 Mission College Blvd., Santa Clara, CA 95054, 1989.

McRuer, D., I. Ashkenas, and D. Graham. *Aircraft Dynamics and Automatic Control.* Princeton N.J.: Princeton University Press, 1973.

Mil. Spec. 1797. "Flying Qualities of Piloted Vehicles." 1987.

Moerder, D. D., and A. J. Calise. "Convergence of a Numerical Algorithm for Calculating Optimal Output Feedback Gains." *IEEE Transactions on Automatic Control,* vol. AC-30, no. 9, September 1985, pp. 900–903.

Moler, C., J. Little, and S. Bangert. *PC-Matlab,* The Mathworks, Inc., 20 North Main St., Suite 250, Sherborn, MA 01770, 1987.

Moore, B. C. "On the Flexibility Offered by State Feedback in Multivariable Systems beyond Closed-loop Eigenvalue Assignment." *IEEE Transactions on Automatic Control* (October 1976): 689–692.

Nelder, J. A., and R. Mead. "A Simplex Method for Function Minimization." *Computing Journal* 7 (1964): 308–313.

O'Brien, M. J., and J. R. Broussard. "Feedforward Control to Track the Output of a Forced Model." *Proceedings of the IEEE Conference on Decision Control,* December 1978.

Press, W. H., B. P. Flannery, S. A. Teukolsky, and W. T. Vetterling. *Numerical Recipes: The Art of Scientific Computing.* New York: Cambridge University Press, 1986.

Quintana, V. H., M. A. Zohdy, and J. H. Anderson. "On the Design of Output Feedback Excitation Controllers of Synchronous Machines." *IEEE Transactions of Power Apparatus Systems,* vol. PAS-95, no. 3, 1976, pp. 954–961.

Riccati, J. F. "Animadversiones in aequationes differentiales secundi gradus." *Actorum Eruditorum quae Lipsiae publicantur, Suppl.* 8 (1724): 66–73.

Shahian, B., and M. Hassul. *Control System Design Using MATLAB.* Englewood Cliffs, N.J.: Prentice-Hall, 1993.

Slotine, J.-J. E., and W. Li. *Applied Nonlinear Control.* Englewood Cliffs, N.J.: Prentice-Hall, 1991.

Sobel, K. M., and E. Y. Shapiro. "Eigenstructure Assignment for Design of Multimode Flight Control Systems." *IEEE Control Systems Magazine* (May 1985): 9–14.

Söderström, T. "On Some Algorithms for Design of Optimal Constrained Regulators." *IEEE Transactions on Automatic Control,* vol. AC-23, no. 6, December 1978, pp. 1100–1101.

Srinathkumar, S. "Eigenvalue/eigenvector Assignment using Output Feedback." *IEEE Transactions on Automatic Control,* vol. AC-23, no. 1, February 1978, pp. 79–81.

Stern, G., and H. A. Henke. "A Design Procedure and Handling-quality Criteria for Lateral-directional Flight Control Systems." *AF-FDL-TR-70-152,* Wright-Patterson AFB, Ohio, May 1971.

Stevens, B. L., F. L. Lewis, and F. Al-Sunni. "Aircraft Flight Controls Design using Output Feedback." *Journal of Guidance, Control, and Dynamics,* 15, no. 1, 238–246. January–February 1992.

Tomlin, C., J. Lygeros, L. Benvenuti, and S. Sastry. "Output Tracking for a Non-minimum Phase Dynamic CTOL Aircraft Model." *Proceedings of the Conference on Decision and Control,* pp. 1867–1872. New Orleans, December 1995.

Wright Laboratory Report WL-TR-96-3099. Wright-Patterson AFB, OH 45433-7562, May 1996.

PROBLEMS

Section 5.2

5.2-1 Eigenstructure Assignment with Full State Feedback. The short-period approximation of an aircraft with the cg far aft might be described by

$$\dot{x} = \begin{bmatrix} -1.10188 & 0.90528 & -0.00212 \\ 4.0639 & -0.77013 & -0.16919 \\ 0 & 0 & -10 \end{bmatrix} x + \begin{bmatrix} 0 & 0 \\ 0 & 1 \\ 10 & 0 \end{bmatrix} u, \quad (1)$$

which includes an elevator actuator of $10/(s+10)$. The state is $x = [\alpha \quad q \quad \delta_e]^T$. An extra control input u_2 has been added to illustrate the extra design freedom available in multivariable systems.

(a) Find the poles.

(b) To conform to flying-qualities specifications, it is desired to assign closed-loop short-period eigenvalues λ_1 and λ_2 of $-2 \pm j2$. The actuator pole does not matter but may be assigned to $s = -15$ to speed up its response. The desired closed-loop eigenvectors are

$$v_1 = v_2^* = [0.20 + j0.35 \quad -0.98 + j0.07 \quad 0]^T,$$
$$v_3 = [0 \quad 0 \quad 1]^T$$

Find the state-feedback gain K in $u = -Kx$ to assign the desired eigenstructure.

5.5-2 Eigenstructure Assignment with Output Feedback. In Problem 5.2-1, a more realistic situation occurs when only measurements of α and q are taken. Then the control is $u = -Ky$ with $y = [\alpha \quad q]^T$. Only two poles may now be assigned. Select desired closed-loop poles as λ_1 and λ_2 in Problem 5.1, with the same eigenvectors v_1 and v_2. Find the required output-feedback gain K. Find the closed-loop poles. What happens to the actuator pole?

5.2-3 In Problem 5.2-1, change the control input to $B = [0 \quad 0 \quad 10]^T$ and use feedback of the output $y = [\alpha \quad q]^T$. Now two poles can be assigned, but there is no freedom in selecting the eigenvectors. Select the desired closed-loop poles $\lambda_1 = \lambda_2^* = -2 + j2$. Find the achievable associated eigenvectors v_1 and v_2. Find the feedback gain K. Find the closed-loop actuator pole.

Section 5.3

5.3-1 Fill in the details in the derivation of the design equations in Table 5.3-1.

5.3-2 Output-Feedback Design for Scalar Systems

(a) Consider the case where $x(t), u(t), y(t)$ are all scalars. Show that the solution S to the second Lyapunov equation in Table 5.3-1 is not needed to determine the output-feedback gain K. Find an explicit solution for P and hence for the optimal gain K.

(b) Repeat for the case where $x(t)$ and $y(t)$ are scalars, but $u(t)$ is an m-vector.

5.3-3 Use (5.3-28) to eliminate K in the Lyapunov equations of Table 5.3-1, hence deriving two coupled nonlinear equations that may be solved for the optimal auxiliary matrices S and P. Does this simplify the solution of the output-feedback design problem?

5.3-4 Software for Output-Feedback Design. Write a program that finds the gain K minimizing the PI in Table 5.3-1 using the Simplex algorithm in (Press et al., 1986). Use it to verify the results of Example 5.3-1. Can you tune the elements of Q and R to obtain better closed-loop responses than the ones given?

5.3-5 For the system

$$\dot{x} = \begin{bmatrix} 0 & 1 \\ 0 & 0 \end{bmatrix} x + \begin{bmatrix} 0 \\ 1 \end{bmatrix} u, \qquad y = [\,1 \quad 1\,]\,x \tag{1}$$

find the output-feedback gain that minimizes the PI in Table 5.3-1 with $Q = I$. Try various values of R to obtain a good response. You will need the software from Problem 5.3-4. The closed-loop step response may be plotted using the software described in Appendix B. (Note that system (1) is nothing but Newton's law, since if $x = [p \quad v]^T$, then $\ddot{p} = u$, where $u(t)$ may be interpreted as an acceleration input F/m.)

5.3-6 Gradient-Based Software for Output-Feedback Design. Write a program that finds the gain K minimizing the PI in Table 5.3-1 using the Davidon–Fletcher–Powell algorithm (Press et al., 1986). Use it to verify the results of Example 5.3-1.

Section 5.4

5.4-1 Derive (5.4-31).

5.4-2 Derive the necessary conditions in Table 5.4-1.

5.4-3 In Example 5.4-1, use the observability matrix to verify that the original proposed value of $Q = H^T H$ has $(\sqrt{Q}, A)$ unobservable while the Q that contains a (5,5) element has $(\sqrt{Q}, A)$ observable.

5.4-4 Software for LQ Output-Feedback Design. Write a program to solve for the optimal gain K in Table 5.4-1 using the Simplex algorithm (Press et al., 1986). Use it to verify Example 5.4-1.

5.4-5 In Example 5.4-1 we used an output with four components. There is an extra degree of freedom in the choice of control gains that may not be needed. Redo the example using the software from Problem 5.4-4, with the output defined as $y = [\alpha_F \quad q \quad \varepsilon]^T$.

5.4-6 To see whether the angle-of-attack filter in Example 5.4-1 complicates the design, redo the example using $y = [\alpha \quad q \quad e \quad \epsilon]^T$.

5.4-7 Redo Example 5.4-1 using root-locus techniques like those in Chapter 4. Based on this, are the gains selected by the LQ algorithm sensible from the point of view of classical control theory?

5.4-8 Gradient-Based Software for LQ Output-Feedback Design. Write a program to solve for the optimal gain K in Table 5.4-1 using the Davidon–Fletcher–Powell algorithms (Press et al., 1986). Use it to verify Example 5.4-1.

5.4-9 Gain Scheduling. To implement a control law on an aircraft, it must be gain-scheduled over the flight envelope where it will be used. In Section 3.5 a software longitudinal model was given for a transport aircraft. In Section 3.6 it was shown how to use a trim program to obtain linearized state-variable models at different trim conditions. Using the trim software, obtain three state-variable models for the short-period approximation at 0 ft altitude for speeds of 170, 220, and 300 ft/s. Redo the normal acceleration CAS in Example 5.4-1 for each of these three state-space models. The result is three sets of control gains, each of which is valid for one of the trim conditions. To implement the gain-scheduled control law, write a simple program that selects between the control gains depending on the actual measured speed of the aircraft. Use linear interpolation between the three gain element values for points between the three equilibrium conditions.

Section 5.5

5.5-1 Show the validity of (5.5-8) and (5.5-9).

5.5-2 Use a technique like that employed in Section 5.3 to derive the expression for the optimal cost in terms of P_k that appears in Table 5.5-1. You will need to successively integrate by parts (MacFarlane 1963).

5.5-3 Derive the necessary conditions in Table 5.5-1.

5.5-4 Software for Output-Feedback LQR Design. Write a program that finds the gain K minimizing the PI in Table 5.3-1 using the Simplex algorithm (Press et al., 1986). Include gain-element weighting using (5.5-6). Use this software to verify the results of Example 5.5-1.

5.5-5 Software for Output-Feedback LQ Tracker Design. Write a program that finds the gain K minimizing the PI in Table 5.4-1 using the Simplex algorithm (Press et al., 1986). Include gain-element weighting using (5.5-6).

5.5-6 In Example 5.4-1 we used an output with four components. There is an extra degree of freedom in the choice of control gains which may not be needed. Using the gain-element weighting software from Problem 5.5-5, redo the example with a large weight on the gain element multiplying $e(t)$ to drive it to zero. Is the performance as good? Try tuning the performance index weights for better performance.

5.5-7 Software for Time-Weighted Output-Feedback Tracker Design. Write a program that finds the gain K minimizing the PI in Table 5.5-1 using the Simplex algorithm (Press et al., 1986). Include gain-element weighting using (5.5-6). Use this software to verify the results of Example 5.5-2. Redo the design using weighting of t^3, t^4. Is there any significant difference from the t^2 case?

5.5-8 Root-Locus Design. Redo Example 5.5-3, finding the control gains using root-locus techniques like those in Chapter 4. Compare this procedure to modern design using software that solves the design equations in Table 5.5-1.

Section 5.6

5.6-1 Derive the implicit-model following design equations (5.6-43)–(5.6-45).

5.6-2 Using the control gains found in the flare control system of Example 5.6-1, determine the compensator zeros in Figure 5.6-3.

5.6-3 A system obeying Newton's laws is described by the state equations

$$\dot{x} = \begin{bmatrix} 0 & 1 \\ 0 & 0 \end{bmatrix} x + \begin{bmatrix} 0 \\ 1 \end{bmatrix} u, \quad y = [1 \quad 1]x$$

The state is $x = [p \quad v]^T$, with $p(t)$ the position and $v(t)$ the velocity.

Using the CGT approach, design an explicit-model following controller that makes the position follow a quadratic input command $r(t) = r_0 + r_1 t + r_2 t^2$.

5.6-4 It is desired to make the scalar plant

$$\dot{x} = x + u, \quad y = x, \quad z = x$$

behave like the scalar model

$$\dot{\underline{x}} = -2\underline{x} + r, \quad \underline{y} = \underline{x}, \quad \underline{z} = \underline{x}$$

with reference input r equal to the unit step. Use explicit model following to design a servosystem:
(a) Draw the controller structure.
(b) Select the control gains using LQR design on the augmented system.

Section 5.7

5.7-1 **Damped Harmonic Oscillator.** The damped harmonic oscillator is described by

$$\dot{x} = \begin{bmatrix} 0 & 1 \\ -\omega_n^2 & -2\zeta\omega_n \end{bmatrix} x + \begin{bmatrix} 0 \\ 1 \end{bmatrix} u,$$

with ζ the damping ratio and ω_n the natural frequency. This system is useful in modeling systems with an oscillatory mode (e.g., short-period mode, fuel slosh).
(a) Repeat Example 5.7-1 for this system.
(b) For several choices of the PI weighting parameters, find the optimal gain and simulate the closed-loop response. (You can check your results using the software written to solve the design equations in Table 5.3-1, 5.4-1, or 5.5-1 by setting $C = I$ there.)

Section 5.8

5.8-1 A basic helicopter model (Shahian and Hassul, 1993) is given by

$$\dot{x} = \begin{bmatrix} -0.4 & 0 & -0.01 \\ 1 & 0 & 0 \\ -1.4 & 9.8 & -0.02 \end{bmatrix} x + \begin{bmatrix} 6.3 \\ 0 \\ 9.8 \end{bmatrix} \delta,$$

where the state is $x = [q \quad \theta \quad v]$, with q = pitch rate, θ = pitch angle, and v = horizontal velocity. The control input is the rotor tilt angle δ.

a. Select different controlled variables as outputs and investigate the stability of the zero dynamics.

b. Select a CV that yields stable zero dynamics. Design the dynamic inversion controller. Simulate using MATLAB.

5.8-2 A nonlinear system is given by

$$\dot{x}_1 = x_1 x_2 + x_3$$
$$\dot{x}_2 = -2x_2 + x_1 u$$
$$\dot{x}_3 = \sin x_1 + 2x_1 x_2 + u$$

a. Select $y(t) = x_1(t)$ as the controlled variable. Investigate the stability of the zero dynamics.

b. Design the dynamic inversion controller. Simulate using MATLAB.

5.8-3 Perform the full simulation in Example 5.8-2. That is, combine the dynamic inversion controller and the aircraft into one M file and plot the outputs using MATLAB as done in Example 5.8-1.

CHAPTER 6

ROBUSTNESS AND MULTIVARIABLE FREQUENCY-DOMAIN TECHNIQUES

6.1 INTRODUCTION

Modeling Errors and Stability Robustness

In the design of aircraft control systems it is important to realize that the rigid-body equations that are the basis for design in Chapters 4 and 5 are only an approximation to the nonlinear aircraft dynamics. An aircraft has flexible modes that are important at high frequencies; we neglected these in our rigid-body design model. These *unmodeled high-frequency dynamics* can act to destabilize a control system that may have quite suitable behavior in terms only of the rigid-body model.

Moreover, as the aircraft changes its equilibrium flight condition, the linearized rigid-body model describing its perturbed behavior changes. This *parameter variation* is a low frequency effect that can also act to destabilize the system. To compensate for this variation, one may determine suitable controller gains for linearized models at several design equilibrium points over a flight envelope. Then, these design gains may be scheduled in computer lookup-tables for suitable controller performance over the whole envelope. For gain scheduling to work, it is essential for the controller gains at each design equilibrium point to guarantee stability for actual flight conditions near that equilibrium point. Thus it is important to design controllers that have *stability robustness,* which is the ability to provide stability in spite of modeling errors due to high-frequency unmodeled dynamics and plant parameter variations.

Disturbances and Performance Robustness

It is often important to account for disturbances such as wind gusts and also for sensor measurement noise. Disturbances can often act to cause unsatisfactory performance in a system that has been designed without taking them into account. Thus, it is important to design controllers that have *performance robustness,* which is the ability to guarantee acceptable performance (in terms, for instance, of percent overshoot, settling time, etc.) even though the system may be subject to disturbances.

Classical Robust Design

In classical control, robustness may be designed into the system from the beginning by providing sufficient gain and phase margin to counteract the effects of inaccurate modeling or disturbances. In terms of the Bode magnitude plot, it is known that the loop gain should be high at low frequencies for performance robustness, but low at high frequencies, where unmodeled dynamics may be present, for stability robustness. The concept of bandwidth is important in this connection, as is the concept of the sensitivity function.

Classical control design techniques are generally in the frequency domain, so they afford a convenient approach to robust design for single-input/single-output (SISO) systems. However, it is well known that the individual gain margins, phase margins, and sensitivities of all the SISO transfer functions in a multivariable or multiloop system have little to do with its overall robustness. Thus there have been problems in extending classical robust design notions to multi-input/multi-output (MIMO) systems.

Modern Robust Design

Modern control techniques provide a direct way to design multiloop controllers for MIMO systems by closing all the loops simultaneously. Performance is guaranteed in terms of minimizing a quadratic performance index (PI) which, with a sensible problem formulation, generally implies closed-loop stability as well. However, all our work in Chapter 5 assumed that the aircraft model is exactly known and that there are no disturbances. In fact, this is rarely the case.

In this chapter we show that the classical frequency-domain robustness measures are easily extended to MIMO systems in a rigorous fashion by using the notion of the *singular value.* In Section 6.2 we develop the *multivariable loop gain and sensitivity* and describe the *multivariable Bode magnitude plot.* In terms of this plot, we present bounds that *guarantee* both robust stability and robust performance for multivariable systems, deriving notions that are entirely analogous to those in classical control.

In Section 6.3 we give a design technique for robust multivariable controllers using modern output-feedback theory, showing how robustness may be guaranteed. The approach is a straightforward extension of classical techniques. To yield both suitable time-domain performance and robustness, an iterative approach is described that is simple and direct using the software described in Appendix B. We illustrate by designing a pitch rate control system that has good performance despite the presence of flexible modes and wind gusts.

A popular modern approach to the design of robust controllers is *linear quadratic Gaussian/loop-transfer recovery* (LQG/LTR). This approach has been used extensively by Honeywell in the design of advanced multivariable aircraft control systems.

LQG/LTR relies on the *separation principle,* which involves designing a full state-variable feedback (as in Section 5.7) and then an *observer* to provide the state estimates for feedback purposes. The result is a dynamic compensator that is similar to those resulting from classical control approaches. The importance of the separation principle is that compensators can be designed for *multivariable systems* in a straightforward manner by solving matrix equations. In Section 6.4 we discuss observers and the Kalman filter. In Section 6.5 we cover LQG/LTR design.

A recent approach to modern robust design is H-infinity design (Francis et al., 1984; Doyle et al., 1989; Kaminer et al., 1990). However, using H-infinity design it is difficult to obtain a controller with a desired structure. For this reason, as well as due to space limitations, we will not cover H-infinity design.

6.2 MULTIVARIABLE FREQUENCY-DOMAIN ANALYSIS

We will deal with system uncertainties, as in classical control, using robust design techniques which are conveniently examined in the frequency domain. To this point, our work in modern control has been in the time domain, since the LQ performance index is a time-domain criterion.

One problem that arises immediately for MIMO systems is that of extending the SISO Bode magnitude plot. We are not interested in making several individual SISO frequency plots for various combinations of the inputs and outputs in the MIMO system and examining gain and phase margins. Such approaches have been tried and may not always yield much insight into the true behavior of the MIMO system. This is due to the coupling that generally exists between *all* inputs and *all* outputs of a MIMO system.

Thus in this section we introduce the *multivariable loop gain and sensitivity* and the *multivariable Bode magnitude plot,* which will be nothing but the plot versus frequency of the *singular values* of the transfer function matrix. This basic tool allows much of the rich experience of classical control theory to be applied to MIMO systems. Thus, we will discover that for robust performance the minimum singular value of the loop gain should be large at low frequencies, where disturbances are present. On the other hand, for robust stability the maximum singular value of the loop gain should be small at high frequencies, where there are significant modeling inaccuracies. We will also see that to guarantee stability despite parameter variations in the linearized model due to operating point changes, the maximum singular value should be below an upper limit.

Sensitivity and Cosensitivity

Figure 6.2-1 shows a standard feedback system of the sort that we have seen several times in our work to date. The plant is $G(s)$, and $K(s)$ is the feedback/feedforward compensator, which can be designed by any of the techniques we have covered. The plant output is $z(t) \in \mathbf{R}^q$, the plant control input is $u(t) \in \mathbf{R}^m$, and the reference input is $r(t) \in \mathbf{R}^q$.

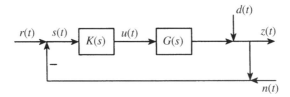

Figure 6.2-1 Standard feedback configuration.

We have mentioned in Section 5.4 that perfect tracking may not be achieved unless the number m of control inputs $u(t)$ is greater than or equal to the number q of performance outputs $z(t)$ (Kwakernaak and Sivan, 1972). Therefore, we will assume that $m = q$ so that the plant $G(s)$ and compensator $K(s)$ are square. This is only a consequence of sensible design, not a restriction on the sorts of plants that may be considered.

We have added a few items to the figure to characterize uncertainties. The signal $d(t)$ represents a *disturbance* acting on the system of the sort appearing in classical control. This could represent, for instance, wind gusts. The *sensor measurement noise* or errors are represented by $n(t)$. Both of these signals are generally vectors of dimension q. Typically, the disturbances occur at low frequencies, say below some ω_d, while the measurement noise $n(t)$ has its predominant effect at high frequencies, say above some value ω_n. Typical Bode plots for the magnitudes of these terms appear in Figure 6.2-2 for the case that $d(t)$ and $n(t)$ are scalars. The reference input is generally also a low-frequency signal (e.g., the unit step).

The tracking error is

$$e(t) \equiv r(t) - z(t) \tag{6.2-1}$$

Due to the presence of $n(t)$, $e(t)$ may not be symbolized in Figure 6.2-1. The signal $s(t)$ is in fact given by

$$s(t) = r(t) - z(t) - n(t) = e(t) - n(t) \tag{6.2-2}$$

Let us perform a frequency-domain analysis on the system to see the effects of the uncertainties on system performance. In terms of Laplace transforms we may write

$$Z(s) = G(s)K(s)S(s) + D(s) \tag{6.2-3}$$

$$S(s) = R(s) - Z(s) - N(s) \tag{6.2-4}$$

$$E(s) = R(s) - Z(s) \tag{6.2-5}$$

Now we may solve for $Z(s)$ and $E(s)$, obtaining the closed-loop transfer function relations (see the Problems)

$$Z(s) = (I + GK)^{-1}GK(R - N) + (I + GK)^{-1}D \tag{6.2-6}$$

$$E(s) = \left[I - (I + GK)^{-1}GK\right]R + (I + GK)^{-1}GKN - (I + GK)^{-1}D \tag{6.2-7}$$

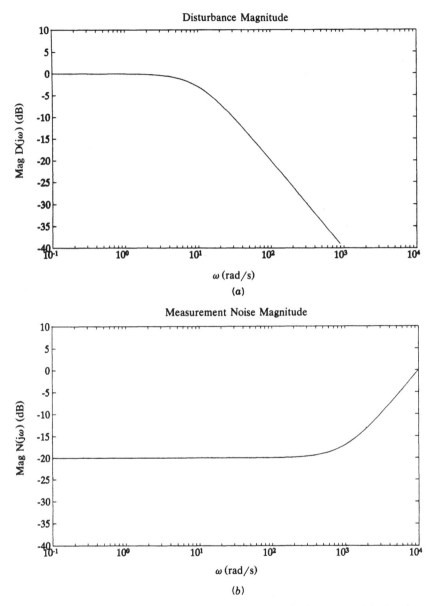

Figure 6.2-2 Typical Bode plots for the uncertain signals in the system: (*a*) disturbance magnitude; (*b*) measurement noise magnitude.

It is important to note that unlike the case for SISO systems, care must be taken to perform the matrix operations in the correct order (for instance, $GK \neq KG$). The multiplications by matrix inverses must also be performed in the correct order.

We can put these equations into a more convenient form. According to the matrix inversion lemma, (6.2-7) may be written as

$$E(s) = (I + GK)^{-1}(R - D) + (I + GK)^{-1}GKN \qquad (6.2\text{-}8)$$

Moreover, since GK is square and invertible, we can write

$$(I + GK)^{-1}GK = \left[(GK)^{-1}(I + GK)\right]^{-1} = \left[(GK)^{-1} + I\right]^{-1}$$
$$= \left[(I + GK)(GK)^{-1}\right]^{-1} = GK(I + GK)^{-1} \qquad (6.2\text{-}9)$$

Therefore, we may finally write $Z(s)$ and $E(s)$ as

$$Z(s) = GK(I + GK)^{-1}(R - N) + (I + GK)^{-1}D \qquad (6.2\text{-}10)$$
$$E(s) = (I + GK)^{-1}(R - D) + GK(I + GK)^{-1}N \qquad (6.2\text{-}11)$$

To simplify things a bit, define the *system sensitivity*

$$S(s) = (I + GK)^{-1} \qquad (6.2\text{-}12)$$

and

$$T(s) = GK(I + GK)^{-1} = (I + GK)^{-1}GK \qquad (6.2\text{-}13)$$

Since

$$S(s) + T(s) = (I + GK)(I + GK)^{-1} = I \qquad (6.2\text{-}14)$$

we call $T(s)$ the *complementary sensitivity*, or in short, the *cosensitivity*. Note that the *return difference*

$$L(s) = I + GK \qquad (6.2\text{-}15)$$

is the inverse of the sensitivity. The *loop gain* is given by $G(s)K(s)$.

These expressions extend the classical notions of loop gain, return difference, and sensitivity to multivariable systems. They are generally square transfer function matrices of dimension $q \times q$. In terms of these new quantities, we have

$$Z(s) = T(s)(R(s) - N(s)) + S(s)D(s) \qquad (6.2\text{-}16)$$
$$E(s) = S(s)(R(s) - D(s)) + T(s)N(s) \qquad (6.2\text{-}17)$$

According to the second equation, to ensure small tracking errors, we must have $S(j\omega)$ small at those frequencies ω where the reference input $r(t)$ and disturbance $d(t)$ are large. This will yield good *disturbance rejection*. On the other hand, for satisfactory *sensor noise rejection*, we should have $T(j\omega)$ small at those frequencies ω where $n(t)$ is large.

Unfortunately, a glance at (6.2-14) reveals that $S(j\omega)$ and $T(j\omega)$ cannot simultaneously be small at any one frequency ω. According to Figure 6.2-2, we should like to have $S(j\omega)$ small at low frequencies, where $r(t)$ and $d(t)$ dominate, and $T(j\omega)$ small at high frequencies, where $n(t)$ dominates.

These are nothing but the multivariable generalizations of the well-known SISO classical notion that a large loop gain $GK(j\omega)$ is required at low frequencies for satisfactory performance and small errors, but a small loop gain is required at high frequencies where sensor noises are present.

Multivariable Bode Plot

These notions are not difficult to understand on a heuristic level. Unfortunately, it is not so straightforward to determine a clear measure for the "smallness" of $S(j\omega)$ and $T(j\omega)$. These are both square matrices of dimension $q \times q$, with q the number of performance outputs $z(t)$ and reference inputs $r(t)$. They are complex functions of the frequency. Clearly, the classical notion of the Bode magnitude plot, which is defined only for *scalar* complex functions of ω, must be extended to the MIMO case.

Some work was done early on using the frequency-dependent eigenvalues of a square complex matrix as a measure of smallness (Rosenbrock, 1974; MacFarlane, 1970; MacFarlane and Kouvaritakis, 1977). However, note that the matrix

$$M = \begin{bmatrix} 0.1 & 100 \\ 0 & 0.1 \end{bmatrix} \qquad (6.2\text{-}18)$$

has large and small components, but its eigenvalues are both at 0.1.

A better measure of the magnitude of square matrices is the *singular value (SV)* (Strang, 1980). Given any matrix M we may write its *singular value decomposition (SVD)* as

$$M = U \Sigma V^*, \qquad (6.2\text{-}19)$$

with $*$ denoting complex conjugate transpose, U and V square unitary matrices (i.e., $V^{-1} = V^*$, the complex conjugate transpose of V), and

$$\Sigma = \begin{bmatrix} \sigma_1 & & & & & & \\ & \sigma_2 & & & & & \\ & & \ddots & & & & \\ & & & \sigma_r & & & \\ & & & & 0 & & \\ & & & & & \ddots & \\ & & & & & & 0 \end{bmatrix}, \qquad (6.2\text{-}20)$$

with $r = \text{rank}(M)$. The singular values are the σ_i, which are ordered so that $\sigma_1 \geq \sigma_2 \geq \cdots \geq \sigma_r$. The SVD may loosely be thought of as the extension to general matrices (which may be nonsquare or complex) of the Jordan form. If M is a function of $j\omega$, so are U, σ_i, and V.

Since $MM^* = U\Sigma V^* V \Sigma^T U^* = U\Sigma^2 U^*$, it follows that the singular values of M are simply the (positive) square roots of the nonzero eigenvalues of MM^*. A similar proof shows that the nonzero eigenvalues of MM^* and those of M^*M are the same.

We note that the M given above has two singular values, $\sigma_1 = 100.0001$ and $\sigma_2 = 0.0001$. Thus, this measure indicates that M has a large and a small component. Indeed, note that

$$\begin{bmatrix} 0.1 & 100 \\ 0 & 0.1 \end{bmatrix} \begin{bmatrix} -1 \\ 0.001 \end{bmatrix} = \begin{bmatrix} 0 \\ 0.0001 \end{bmatrix} \quad (6.2\text{-}21)$$

while

$$\begin{bmatrix} 0.1 & 100 \\ 0 & 0.1 \end{bmatrix} \begin{bmatrix} 0.001 \\ 1 \end{bmatrix} = \begin{bmatrix} 100.0001 \\ 0.1 \end{bmatrix} \quad (6.2\text{-}22)$$

Thus the singular value σ_2 has the *input direction*

$$\begin{bmatrix} -1 \\ 0.001 \end{bmatrix}$$

associated with it for which the output contains the value σ_2. On the other hand, the singular value σ_1 has an associated input direction of

$$\begin{bmatrix} 0.001 \\ 1 \end{bmatrix}$$

for which the output contains the value σ_1.

There are many nice properties of the singular value that make it a suitable choice for defining the magnitude of matrix functions. Among these is the fact that the maximum singular value is an *induced matrix norm*, and norms have several useful attributes. The use of the SVs in the context of modern control was explored in Doyle and Stein (1981) and Safonov et al. (1981).

A major factor is that there are many good software packages that have good routines for computing the singular value (e.g., subroutine LSVDF in IMSL [1980] or Moler et al. [1987]). Thus, plots like those we will present may easily be obtained by writing only a computer program to drive the available subroutines. Indeed, since the SVD uses unitary matrices, its computation is numerically stable. An efficient technique for obtaining the SVs of a complex matrix as a function of frequency ω is given in Laub (1981).

We note that a complete picture of the behavior of a complex matrix versus ω must take into account the magnitudes of the SVs as well as the *multivariable phase*, which may also be obtained from the SVD (Postlethwaite et al., 1981). Thus, complete MIMO generalizations of the Bode magnitude *and* phase plots are available. However, the theory relating to the phase portion of the plot is more difficult to use in a practical design technique, although a MIMO generalization of the Bode gain-phase relation is available (Doyle and Stein, 1981). Therefore, we will only employ plots of the SVs versus frequency, which correspond to the Bode magnitude plot for MIMO systems.

The magnitude of a square transfer function matrix $H(j\omega)$ at any frequency ω depends on the direction of the input excitation. Inputs in a certain direction in the input space will excite only the SV(s) associated with that direction. However, for any input, the magnitude of the transfer function $H(j\omega)$ at any given frequency ω may be bounded above by its *maximum singular value,* denoted $\bar{\sigma}(H(j\omega))$, and below by its *minimum singular value,* denoted $\underline{\sigma}(H(j\omega))$. Therefore, all our results, as well as the plots we will give, need take into account only these two bounding values of "magnitude."

Example 6.2-1: MIMO Bode Magnitude Plots. Here, we consider a simple nonaircraft system to make some points about the singular value plots. Consider the multivariable system

$$\dot{x} = \begin{bmatrix} -1 & -1 & 0 & 0 \\ 1 & -1 & 0 & 0 \\ 0 & 0 & -2 & 6 \\ 0 & 0 & -6 & -2 \end{bmatrix} x + \begin{bmatrix} 1 & 0 \\ 0 & 0 \\ 0 & 1 \\ 0 & 0 \end{bmatrix} u = Ax + Bu \tag{1}$$

$$z = \begin{bmatrix} 1 & 0 & 0 & 0 \\ 0 & 0 & 1 & 0 \end{bmatrix} x = Hx, \tag{2}$$

which as a 2×2 MIMO transfer function of

$$H(s) = H(sI - A)^{-1}B = \frac{M(s)}{\Delta(s)}, \tag{3}$$

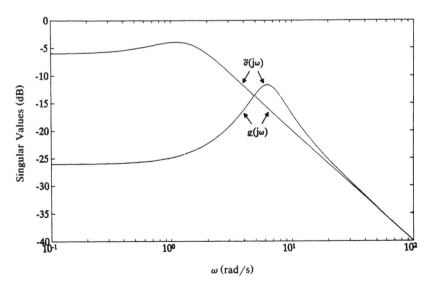

Figure 6.2-3 MIMO Bode magnitude plot of singular values versus frequency.

with

$$\Delta(s) = s^4 + 6s^3 + 50s^2 + 88s + 80$$

$$M(s) = \begin{bmatrix} 1 & 0 \\ 0 & 1 \end{bmatrix} s^3 + \begin{bmatrix} 5 & 0 \\ 0 & 4 \end{bmatrix} s^2 + \begin{bmatrix} 44 & 0 \\ 0 & 6 \end{bmatrix} s + \begin{bmatrix} 40 & 0 \\ 0 & 4 \end{bmatrix} \quad (4)$$

By writing a driver program that calls standard software (e.g., subroutine LSVDF in IMSL, 1980) to evaluate the SVs at closely spaced values of frequency ω, we may obtain the SV plots versus frequency shown in Figure 6.2-3. We call this the *multivariable Bode magnitude plot* for the MIMO transfer function $H(s)$.

Since $H(s)$ is 2×2, it has two singular values. Note that although each singular value is continuous, the maximum and minimum singular values are not. This is due to the fact that the singular values can cross over each other, as the figure illustrates. ∎

Example 6.2-2: Singular Value Plots for F-16 Lateral Dynamics. To illustrate the difference between the singular-value plots and the individual SISO Bode plots of a multivariable system, let us consider the F-16 lateral dynamics of Examples 5.3-1 and 5.5-4. In the latter example, we designed a wing leveler. For convenience, refer to the figure there showing the control system structure. Using the system matrices A and B in that example, which include an integrator in the ϕ channel as well as actuator dynamics and a washout filter, take as the control inputs $u = [u_a \quad u_r]^T$, with u_a the aileron servo input and u_r the rudder servo input. Select as outputs $z = [\epsilon \quad r_w]^T$, with ϵ the integrator output in the ϕ channel and r_w the washed-out yaw rate.

The individual SISO transfer functions in this two-input/two-output open-loop system are

$$H_{11} = \frac{\epsilon}{u_a} = \frac{14.8}{s(s+0.0163)(s+3.615)(s+20.2)} \quad (1)$$

$$H_{12} = \frac{r_w}{u_a}$$
$$= \frac{-36.9s(s+2.237)\left[(s+0.55)^2 + 2.49^2\right]}{(s+0.0163)(s+1)(s+3.165)(s+20.2)\left[s+0.4225)^2 + 3.063^2\right]} \quad (2)$$

$$H_{21} = \frac{\epsilon}{u_r} = \frac{-2.65(s+2.573)(s-2.283)}{s(s+0.0163)(s+3.615)(s+20.2)\left[(s+0.4225)^2 + 3.063^2\right]} \quad (3)$$

$$H_{22} = \frac{r_w}{u_r} = \frac{-0.718s\left[(s+0.139)^2 + 0.446^2\right]}{(s+0.0163)(s+1)(s+20.2)\left[s+0.4225)^2 + 3.063^2\right]} \quad (4)$$

The standard Bode magnitude plots for these SISO transfer functions are shown in Figure 6.2-4. Clearly visible are the resonance due to the dutch roll mode, as well as the integrator in the upper ϕ channel in the figure in Example 5.5-4.

MULTIVARIABLE FREQUENCY-DOMAIN ANALYSIS

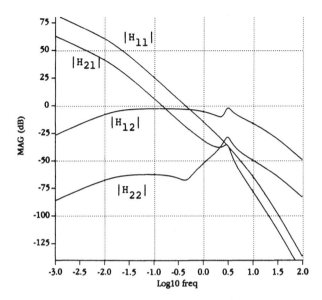

Figure 6.2-4 SISO Bode magnitude plots for F-16 lateral dynamics.

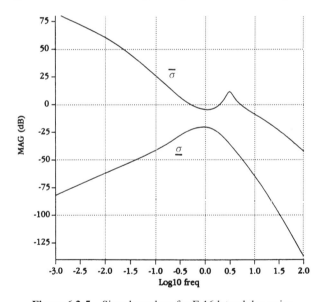

Figure 6.2-5 Singular values for F-16 lateral dynamics.

On the other hand, shown in Figure 6.2-5 are the singular values of this multivariable system. Note that it is not immediately evident how they relate to the SISO plots in Figure 6.2-4. In the next section we will see that bounds for guaranteed robustness are given for MIMO systems in terms of the minimum singular value being large at low frequencies (for performance robustness) and the maximum singular value being small at high frequencies (for stability robustness). The lack of any clear correspondence between Figures 6.2-4 and 6.2-5 shows that these bounds cannot be expressed in terms of the individual SISO Bode plots. ∎

Frequency-Domain Performance Specifications

We have seen how to make a multivariable Bode magnitude plot of a square transfer function matrix. It is now necessary to discuss *performance specifications* in the frequency domain in order to determine what a "desirable" Bode plot means in the MIMO case. The important point is that the low-frequency requirements are generally in terms of the *minimum* singular value being *large*, while the high-frequency requirements are in terms of the *maximum* singular value being *small*.

First, let us point out that the classical notion of *bandwidth* holds in the MIMO case. This is the frequency ω_c for which the loop gain $GK(j\omega)$ passes through a value of 1, or 0 dB. If the bandwidth should be limited due to high-frequency noise considerations, the *largest* SV should satisfy $\bar{\sigma}(GK(j\omega)) = 1$, at the specified cutoff frequency ω_c.

L_2 **Operator Gain.** To relate frequency-domain behavior to time-domain behavior, we may take into account the following considerations (Morari and Zafiriou, 1989). Define the L_2 norm of a vector time function $s(t)$ by

$$\|s\|_2 = \left[\int_0^\infty s^T(t)s(t)dt \right]^{1/2} \tag{6.2-23}$$

This is related to the total energy in $s(t)$ and should be compared to the LQ performance index.

A linear time-invariant system has input $u(t)$ and output $z(t)$ related by the convolution integral

$$z(t) = \int_{-\infty}^{\infty} h(t-\tau)u(\tau)\,d\tau, \tag{6.2-24}$$

with $h(t)$ the impulse response. The L_2 *operator gain,* denoted $\|H\|_2$, of such a system is defined as the smallest value of γ such that

$$\|z\|_2 \leq \gamma \|u\|_2 \tag{6.2-25}$$

This is just the operator norm induced by the L_2 vector norm. An important result is that the L_2 operator gain is given by

$$\|H\|_2 = \max_\omega \left[\bar{\sigma}(H(j\omega)) \right], \tag{6.2-26}$$

with $H(s)$ the system transfer function. That is, $\|H\|_2$ is nothing but the *maximum value* over ω of the maximum singular value of $H(j\omega)$. Thus, $\|H\|_2$ is an *H-infinity norm* in the frequency domain.

This result gives increased importance to $\overline{\sigma}(H(j\omega))$, for if we are interested in keeping $z(t)$ small over a range of frequencies, we should take care that $\overline{\sigma}(H(j\omega))$ is small over that range.

It is now necessary to see how this result may be used in deriving frequency-domain performance specifications. Some facts we will use in this discussion are

$$\underline{\sigma}(GK) - 1 \leq \underline{\sigma}(I + GK) \leq \underline{\sigma}(GK) + 1 \quad (6.2\text{-}27)$$

$$\overline{\sigma}(M) = \frac{1}{\underline{\sigma}(M^{-1})}, \quad (6.2\text{-}28)$$

$$\overline{\sigma}(AB) \leq \overline{\sigma}(A)\overline{\sigma}(B) \quad (6.2\text{-}29)$$

for any matrices A, B, GK, M, with M nonsingular.

Before we begin a discussion of performance specifications, let us note the following. If $S(j\omega)$ is small, as desired at low frequencies, then

$$\overline{\sigma}(S) = \overline{\sigma}\left[(I + GK)^{-1}\right] = \frac{1}{\underline{\sigma}(I + GK)} \approx \frac{1}{\underline{\sigma}(GK)} \quad (6.2\text{-}30)$$

That is, a large value of $\underline{\sigma}(GK)$ guarantees a small value of $\overline{\sigma}(S)$.

On the other hand, if $T(j\omega)$ is small, as is desired at high frequencies, then

$$\overline{\sigma}(T) = \overline{\sigma}\left[GK(I + GK)^{-1}\right] \approx \overline{\sigma}(GK) \quad (6.2\text{-}31)$$

That is, a small value of $\overline{\sigma}(GK)$ guarantees a small value of $\overline{\sigma}(T)$.

This means that specifications that $S(j\omega)$ be small at low frequencies and $T(j\omega)$ be small at high frequencies may equally well be formulated in terms of $\underline{\sigma}(GK)$ being large at low frequencies and $\overline{\sigma}(GK)$ being small at high frequencies. Thus, all of our performance specifications will be in terms of the *minimum and maximum SVs of the loop gain* $GK(j\omega)$. The practical significance of this is that we need only compute the SVs of $GK(j\omega)$, not those of $S(j\omega)$ and $T(j\omega)$. These notions are symbolized in Figure 6.2-6, where it should be recalled that $S + T = I$.

Now, we will first consider low-frequency specifications on the singular value plot, and then high-frequency specifications. According to our discussion relating to (6.2-17), the former will involve the reference input $r(t)$ and disturbances $d(t)$, while the latter will involve the sensor noise $n(t)$.

Low-Frequency Specifications. For low frequencies let us suppose that the sensor noise $n(t)$ is zero so that (6.2-17) becomes

$$E(s) = S(s)(R(s) - D(s)) \quad (6.2\text{-}32)$$

Thus, to keep $\|e(t)\|_2$ small, it is only necessary to ensure that the L_2 operator norm $\|S\|_2$ is small at all frequencies where $R(j\omega)$ and $D(j\omega)$ are appreciable. This may be achieved by ensuring that, at such frequencies, $\overline{\sigma}(S(j\omega))$ is small. As we have just seen, this may be guaranteed if we select

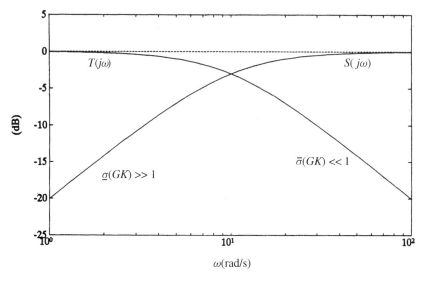

Figure 6.2-6 Magnitude specifications on $S(jw)$, $T(jw)$, and $GK(jw0)$.

$$\underline{\sigma}(GK(j\omega)) \gg 1 \quad \text{for } \omega \leq \omega_d, \tag{6.2-33}$$

where $D(s)$ and $R(s)$ are appreciable for $\omega \leq \omega_d$.

Thus, exactly as in the classical case (Franklin et al., 1986), we are able to specify a low-frequency performance bound that guarantees *performance robustness,* that is, good performance in the face of low-frequency disturbances. For instance, to ensure that disturbances are attenuated by a factor of 0.01, we should ensure $\underline{\sigma}(GK(j\omega))$ is greater than 40 dB at low frequencies $\omega \leq \omega_d$.

At this point it is worth examining Figure 6.2-9, which illustrates the frequency-domain performance specifications we are beginning to derive. Another low-frequency performance bound may be derived from steady-state error considerations. Thus, suppose that $d(t) = 0$ and the reference input is a unit step of magnitude r so that $R(s) = r/s$. Then, according to (6.2-32) and the final value theorem (Franklin et al., 1986), the steady-state error e_∞ is given by

$$e_\infty = \lim_{s \to 0} sE(s) = rS(0) \tag{6.2-34}$$

To ensure that the largest component of e_∞ is less than a prescribed small acceptable value δ_∞, we should therefore select

$$\underline{\sigma}(GK(0)) > \frac{r}{\delta_\infty} \tag{6.2-35}$$

The ultimate objective of all our concerns is to manufacture a compensator $K(s)$ in Figure 6.2-1 that gives desirable performance. Let us now mention two low-frequency considerations that are important in the initial stages of the design of the compensator $K(s)$.

To make the steady-state error in response to a unit step at $r(t)$ exactly equal to zero, we may ensure that there is an integrator *in each path* of the system $G(s)$ so that it is of type 1 (Franklin et al., 1986). Thus, suppose that the system to be controlled is given by

$$\dot{x} = Ax + Bv \qquad (6.2\text{-}36)$$
$$z = Hx$$

To add an integrator to each control path, we may augment the dynamics so that

$$\frac{d}{dt}\begin{bmatrix} x \\ \epsilon \end{bmatrix} = \begin{bmatrix} A & B \\ 0 & 0 \end{bmatrix}\begin{bmatrix} x \\ \epsilon \end{bmatrix} + \begin{bmatrix} 0 \\ I \end{bmatrix}u, \qquad (6.2\text{-}37)$$

with ϵ the integrator outputs (see Figure 6.2-7). The system $G(s)$ in Figure 6.2-1 should now be taken as (6.2-37), which contains the integrators as a precompensator.

Although augmenting each control path with an integrator results in zero steady-state error, in some applications this may result in an unnecessarily complicated compensator. Note that the steady-state error may be made as small as desired without integrators by selecting $K(s)$ so that (6.2-35) holds.

A final concern about the low-frequency behavior of $G(s)$ needs to be addressed. It is desirable in many situations to have $\underline{\sigma}(GK)$ and $\overline{\sigma}(GK)$ close to the same value. Then the speed of the responses will be nearly the same in all channels of the system. This is called the issue of *balancing the singular values at low frequency*. The SVs of $G(s)$ in Figure 6.2-1 may be balanced at low frequencies, as follows.

Suppose that the plant has the state-variable description (6.2-36), and let us add a square constant precompensator gain matrix P, so that

$$v = Pu \qquad (6.2\text{-}38)$$

is the relation between the control input $u(t)$ in Figure 6.2-1 and the actual plant input $v(t)$. The transfer function of the plant plus precompensator is now

$$G(s) = H(sI - A)^{-1}BP \qquad (6.2\text{-}39)$$

As s goes to zero, this approaches

$$G(0) = H(-A)^{-1}BP,$$

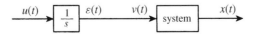

Figure 6.2-7 Plant augmented with integrators.

as long as A has no poles at the origin. Therefore, we may ensure that $G(0)$ has all SVs equal to a prescribed value of γ by selecting

$$P = \gamma \left[H(-A)^{-1} B \right]^{-1}, \qquad (6.2\text{-}40)$$

for then $G(0) = \gamma I$.

The transfer function of (6.2-36) is

$$H(s) = H(sI - A)^{-1} B, \qquad (6.2\text{-}41)$$

whence we see that the required value of the precompensator gain is

$$P = \gamma H^{-1}(0) \qquad (6.2\text{-}42)$$

This is nothing but the (scaled) reciprocal dc gain.

Example 6.2-3: Precompensator for Balancing and Zero Steady-State Error.
Let us design a precompensator for the system in Example 6.2-1 using the notions just discussed. Substituting the values of A, B, and H in (6.2-40) with $\gamma = 1$ yields

$$P = \left[H(-A)^{-1} B \right]^{-1} = \begin{bmatrix} 2 & 0 \\ 0 & 20 \end{bmatrix} \qquad (1)$$

To ensure zero-steady-state error as well as equal singular values at low frequencies, we may incorporate integrators in each input channel along with the gain matrix P by writing the augmented system

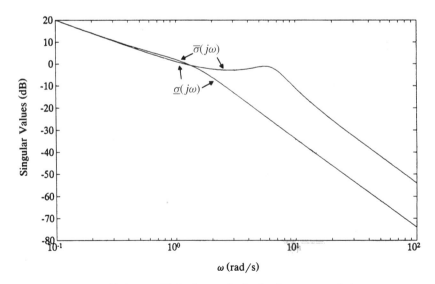

Figure 6.2-8 MIMO Bode magnitude plot for augmented plant.

$$\frac{d}{dt}\begin{bmatrix} x \\ \epsilon \end{bmatrix} = \begin{bmatrix} A & B \\ 0 & 0 \end{bmatrix} \begin{bmatrix} x \\ \epsilon \end{bmatrix} + \begin{bmatrix} 0 \\ P \end{bmatrix} u \qquad (2)$$

The singular-value plots for this plant plus precompensator appear in Figure 6.2-8. At low frequencies there is now a slope of -20 dB/decade as well as equality of $\underline{\sigma}$ and $\overline{\sigma}$. Thus the augmented system is both balanced and of type 1. Compare Figure 6.2-8 to the singular-value plot of the uncompensated system in Figure 6.2-3. The remaining step is the selection of the feedback gain matrix for the augmented plant (2) so that the desired performance is achieved. ∎

High-Frequency Specifications. We now turn to a discussion of high-frequency performance specifications. The sensor noise is generally appreciable at frequencies above some known value ω_n (see Figure 6.2-2). Thus, according to (6.2-17), to keep the tracking error norm $\|e\|_2$ small in the face of measurement noise we should ensure that the operator norm $\|T\|_2$ is small at high frequencies above this value. By (6.2-31) this may be guaranteed if

$$\overline{\sigma}(GK(j\omega)) \ll 1 \qquad \text{for } \omega \geq \omega_n \qquad (6.2\text{-}43)$$

(see Figure 6.2-9). For instance, to ensure that sensor noise is attenuated by a factor of 0.1, we should guarantee that $\overline{\sigma}(GK(j\omega)) < -20$ dB for $\omega \geq \omega_n$.

One final high-frequency robustness consideration needs to be mentioned. It is unusual for the plant model to be exactly known. There are two basic sorts of modeling inaccuracies that concern us in aircraft controls. The first is plant parameter variation due to changes in the linearization equilibrium point of the nonlinear model. This is a low-frequency phenomenon and will be discussed in the next subsection. The

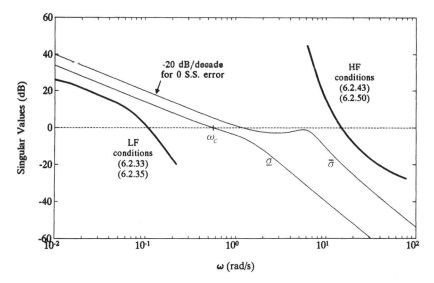

Figure 6.2-9 Frequency-domain performance specifications.

second sort of inaccuracy is due to unmodeled high-frequency dynamics; this we discuss here.

We are assuming a rigid-body aircraft model for the purpose of control design, and in so doing are neglecting flexible and vibrational modes at high frequencies. Thus although our design may guarantee closed-loop stability for the *assumed mathematical model* $G(s)$, stability is not assured for the *actual* plant $G'(s)$ with flexible modes. To guarantee *stability robustness* in the face of plant parameter uncertainty, we may proceed as follows.

The model uncertainties may be of two types. The actual plant model G' and the assumed plant model G may differ by *additive uncertainties* so that

$$G'(j\omega) = G(j\omega) + \Delta G(j\omega), \qquad (6.2\text{-}44)$$

where the unknown discrepancy satisfies a known bound

$$\overline{\sigma}(\Delta G(j\omega)) < a(\omega), \qquad (6.2\text{-}45)$$

with $a(\omega)$ known for all ω.

On the other hand, the actual plant model $G'(s)$ and the assumed plant model $G(s)$ may differ by *multiplicative uncertainties* so that

$$G'(j\omega) = [I + M(j\omega)] G(j\omega), \qquad (6.2\text{-}46)$$

where the unknown discrepancy satisfies a known bound

$$\overline{\sigma}(M(j\omega)) < M(\omega), \qquad (6.2\text{-}47)$$

with $m(\omega)$ known for all ω. We will show several ways of finding the bound $m(\omega)$. In Example 6.2-4 we show how to construct a reduced-order model for the system, which may then be used for control design. There $m(\omega)$ is determined from the neglected dynamics. In Example 6.3-1 we show how $m(\omega)$ may be determined in terms of the aircraft's neglected flexible modes. In the next subsection we show how to determine $m(\omega)$ in terms of plant parameter variations in the linearized model due to operating point changes.

Since we may write (6.2-44) as

$$G'(j\omega) = \left[I + \Delta G(j\omega) G^{-1}(j\omega)\right] G(j\omega) \equiv [I + M(j\omega)] G(j\omega), \qquad (6.2\text{-}48)$$

we will confine ourselves to a discussion of multiplicative uncertainties, following Doyle and Stein (1981).

Suppose that we have designed a compensator $K(s)$ so that the closed-loop system in Figure 6.2-1 is stable. We should now like to derive a frequency-domain condition that guarantees the stability of the *actual* closed-loop system, which contains not $G(s)$, but $G'(s)$, satisfying (6.2-46)/(6.2-47). For this, the multivariable Nyquist condition (Rosenbrock, 1974) may be used.

Thus, it is required that the encirclement count of the map $|I+G'K|$ be equal to the negative number of unstable open-loop poles of $G'K$. By assumption, this number is the same as that of GK. Thus, the number of encirclements of $|I+G'K|$ must remain unchanged for all G' allowed by (6.2-47). This is assured if and only if $|I+G'K|$ remains nonzero as G is warped continuously toward G', or equivalently,

$$0 < \underline{\sigma}\,[I + [I + \epsilon M(s)]\,G(s)K(s)]$$

for all $0 \le \epsilon \le 1$, all $M(s)$ satisfying (6.2-47) and all s on the standard Nyquist contour.

Since G' vanishes on the infinite radius segment of the Nyquist contour, and assuming for simplicity that no indentations are required along the $j\omega$-axis portion, this reduces to the following equivalent conditions:

$$0 < \underline{\sigma}\,[I + G(j\omega)K(j\omega) + \epsilon M(j\omega)G(j\omega)K(j\omega)]$$

for all $0 \le \epsilon \le 1, 0 \le \omega < \infty$, all M,

iff $\qquad 0 < \underline{\sigma}\left[\{I + \epsilon MGK(I+GK)^{-1}\}(I+GK)\right]$

iff $\qquad 0 < \underline{\sigma}\left[I + MGK(I+GK)^{-1}\right]$

all $0 \le \omega < \infty$, and all M,

iff
$$\overline{\sigma}\left[GK(I+GK)^{-1}\right] < \frac{1}{m(\omega)} \qquad (6.2\text{-}49)$$

for all $0 \le \omega < \infty$. Thus stability robustness translates into a requirement that the cosensitivity $T(j\omega)$ be bounded above by the reciprocal of the multiplicative modeling discrepancy bound $m(\omega)$.

In the case of high-frequency unmodeled dynamics, $1/m(\omega)$ is small at high ω, so that according to (6.2-31), we may simplify (6.2-49) by writing it in terms of the loop gain as

$$\overline{\sigma}(GK(j\omega)) < \frac{1}{m(\omega)} \qquad (6.2\text{-}50)$$

for all ω such that $m(\omega) \gg 1$.

This bound for stability robustness is illustrated in Figure 6.2-9.

An example will be useful at this point.

Example 6.2-4: Model Reduction and Stability Robustness. In some situations we have a high-order aircraft model that is inconvenient to use for controller design. Examples occur in engine control and spacecraft control. In such situations, it is possible to compute a reduced-order model of the system which may then be used for controller design. Here we will show a convenient technique for model reduction as well as an illustration of the stability robustness bound $m(\omega)$. The technique described here is from Athans et al. (1986).

a. Model Reduction by Partial-Fraction Expansion. Suppose that the actual plant is described by

$$\dot{x} = Ax + Bu \tag{1a}$$

$$z = Hx, \tag{1b}$$

with $x \in \mathbf{R}^n$. If A is simple with eigenvalues γ_i, right eigenvectors u_i, and left eigenvectors v_i so that

$$Au_i = \lambda_i u_i, \quad v_i^T A = \lambda_i v_i^T, \tag{2}$$

then the transfer function

$$G'(s) = H(sI - A)^{-1} B \tag{3}$$

may be written as the partial-fraction expansion (Section 5.2)

$$G'(s) = \sum_{i=1}^{n} \frac{R_i}{s - \lambda_i}, \tag{4}$$

with residue matrices given by

$$R_i = H u_i v_i^T B \tag{5}$$

If the value of n is large, it may be desirable to find a *reduced-order approximation* to (1) for which a simplified compensator $K(s)$ in Figure 6.2-1 may be designed. Then, if the approximation is a good one, the compensator $K(s)$ should work well when used on the actual plant $G'(s)$.

To find a reduced-order approximation $G(s)$ to the plant, we may proceed as follows. Decide which of the eigenvalues λ_i in (4) are to be retained in $G(s)$. This may be done using engineering judgment, by omitting high-frequency modes, by omitting terms in (4) that have small residues, and so on. Let the r eigenvalues to be retained in $G(s)$ be $\lambda_i, \lambda_2, \ldots, \lambda_r$.

Define the matrix

$$Q = \text{diag}\{Q_i\}, \tag{6}$$

where Q is an $r \times r$ matrix and the blocks Q_i are defined as

$$Q_i = \begin{cases} 1, & \text{for each real eigenvalue retained} \\ \begin{bmatrix} \frac{1}{2} & -\frac{j}{2} \\ \frac{1}{2} & \frac{j}{2} \end{bmatrix}, & \text{for each complex pair retained} \end{cases} \tag{7}$$

Compute the matrices

$$V \equiv Q^{-1} \begin{bmatrix} v_1^T \\ \vdots \\ v_r^T \end{bmatrix} \quad (8)$$

$$U \equiv \begin{bmatrix} u_1 & \cdots & u_r \end{bmatrix} Q \quad (9)$$

In terms of these constructions, the reduced-order system is nothing but a projection of (1) onto a space of dimension r with state defined by

$$w = Vx \quad (10)$$

The system matrices in the reduced-order approximate system

$$\dot{w} = Fw + Gu \quad (11a)$$
$$z = Jw + Du \quad (11b)$$

are given by

$$F = VAU$$
$$G = VB \quad (12)$$
$$J = HU,$$

with the direct-feed matrix given in terms of the residues of the neglected eigenvalues as

$$D = \sum_{i=r+1}^{n} -\frac{R_i}{\lambda_i} \quad (13)$$

The motivation for selecting such a D matrix is as follows. The transfer function

$$G(s) = J(sI - F)^{-1}G + D$$

of the reduced system (11) is given as (verify!)

$$G(s) = \sum_{i=1}^{r} \frac{R_i}{s - \lambda_i} + \sum_{i=r+1}^{n} -\frac{R_i}{\lambda_i} \quad (14)$$

Evaluating $G(j\omega)$ and $G'(j\omega)$ at $\omega = 0$, it is seen that they are equal to dc. Thus the modeling errors induced by taking $G(s)$ instead of the actual $G'(s)$ occur at higher frequencies. Indeed, they depend on the frequencies of the neglected eigenvalues of (1).

To determine the $M(s)$ in (6.2-46) that is induced by the order reduction, note that

$$G' = (I + M)G \tag{15}$$

so that

$$M = (G' - G)G^{-1} \tag{16}$$

or

$$M(s) = \left[\sum_{i=r+1}^{n} \frac{R_i}{\lambda_i} \frac{s}{s - \lambda_i}\right] G^{-1}(s) \tag{17}$$

Then the high-frequency robustness bound is given in terms of

$$m(j\omega) = \bar{\sigma}(M(j\omega)) \tag{18}$$

Note that $M(j\omega)$ tends to zero as ω becomes small, reflecting our perfect certainty of the actual plant at dc.

b. An Example. Let us use an example to illustrate the model-reduction procedure, and show also how to compute the upper bound $m(\omega)$ in (6.2-46)/(6.2-47) on the high-frequency modeling errors thereby induced. To make it easy to see what is going on, we will take a Jordan-form system.

Let there be prescribed the MIMO system

$$\dot{x} = \begin{bmatrix} -1 & 0 & 0 \\ 0 & -2 & 0 \\ 0 & 0 & -10 \end{bmatrix} x + \begin{bmatrix} 1 & 0 \\ 0 & 1 \\ 2 & 0 \end{bmatrix} u = Ax + Bu \tag{19a}$$

$$z = \begin{bmatrix} 1 & 0 & 0 \\ 0 & 1 & 1 \end{bmatrix} x = Cx \tag{19b}$$

The eigenvectors are given by $u_i = e_i$, $v_i = e_i$, $i = 1, 2, 3$, with e_i the i-th column of the 3×3 identity matrix. Thus the transfer function is given by the partial-fraction expansion

$$G'(s) = \frac{R_1}{s+1} + \frac{R_2}{s+2} + \frac{R_3}{s+10}, \tag{20}$$

with

$$R_1 = \begin{bmatrix} 1 & 0 \\ 0 & 0 \end{bmatrix}, \quad R_2 = \begin{bmatrix} 0 & 0 \\ 0 & 1 \end{bmatrix}, \quad R_3 = \begin{bmatrix} 0 & 0 \\ 2 & 0 \end{bmatrix} \tag{21}$$

To find the reduced-order system that retains the poles at $\lambda = -1$ and $\lambda = -2$, define

MULTIVARIABLE FREQUENCY-DOMAIN ANALYSIS   531

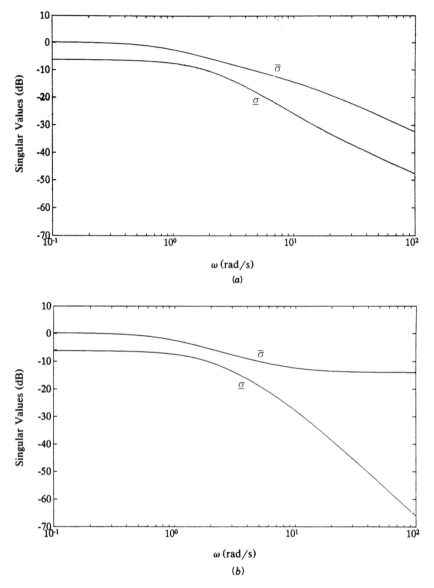

Figure 6.2-10 MIMO Bode magnitude plots of singular values: (*a*) actual plant; (*b*) reduced-order approximation.

$$Q = \begin{bmatrix} 1 & 0 \\ 0 & 1 \end{bmatrix}, \qquad V = \begin{bmatrix} 1 & 0 & 0 \\ 0 & 1 & 0 \end{bmatrix}, \qquad U = \begin{bmatrix} 1 & 0 \\ 0 & 1 \\ 0 & 0 \end{bmatrix} \qquad (22)$$

and compute the approximate system

532 ROBUSTNESS AND MULTIVARIABLE FREQUENCY-DOMAIN TECHNIQUES

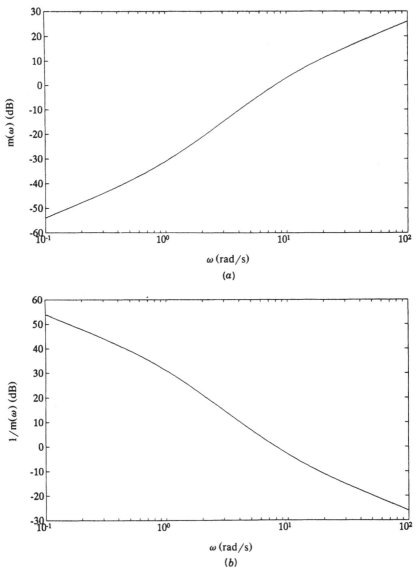

Figure 6.2-11 High-frequency stability-robustness bound: (a) $m(\omega)$; (b) $1/m(\omega)$.

$$\dot{w} = \begin{bmatrix} -1 & 0 \\ 0 & -2 \end{bmatrix} w + \begin{bmatrix} 1 & 0 \\ 0 & 1 \end{bmatrix} u = Fw + Gu \qquad (23a)$$

$$z = \begin{bmatrix} 1 & 0 \\ 0 & 1 \end{bmatrix} w + \begin{bmatrix} 0 & 0 \\ 0.2 & 0 \end{bmatrix} u = Jw + Du \qquad (23b)$$

This has a transfer function of

$$G(s) = \frac{R_1}{s+1} + \frac{R_2}{s+2} + D \qquad (24)$$

Singular-value plots of the actual plant (19) and the reduced-order approximation (23) are shown in Figure 6.2-10.

The multiplicative error is given by

$$M = (G' - G)G^{-1} = \begin{bmatrix} 0 & 0 \\ -\dfrac{0.2s(s+1)}{s+10} & 0 \end{bmatrix}, \qquad (25)$$

whence

$$m(\omega) = \overline{\sigma}(M(j\omega)) = \frac{0.2\omega\sqrt{\omega^2+1}}{\sqrt{\omega^2+100}}, \qquad (26)$$

and the high-frequency bound on the loop gain $GK(j\omega)$ is given by

$$\frac{1}{m(j\omega)} = \frac{5\sqrt{\omega^2+100}}{\omega\sqrt{\omega^2+1}} \qquad (27)$$

This bound is plotted in Figure 6.2-11. Note that the modeling errors become appreciable (i.e., of magnitude 1) at a frequency of 8.0 rad/s. Above this frequency, we should ensure that constraint (6.2-50) on the loop-gain magnitude holds to guarantee stability-robustness. This will be a restriction on any compensator $K(s)$ designed using the reduced-order plant (23). ∎

Robustness Bounds for Plant Parameter Variations

The aircraft is nonlinear, but for controller design we use linearized models obtained at some operating point. In practice, it is necessary to determine linear models at several design operating points over a specified flight envelope, and determine optimal control gains for each one. Then these design control gains are tabulated and scheduled using microprocessors, so that the gains most appropriate for the actual operating point of the aircraft are used in the controller. It is usual to determine which of the design operating points are closest to the actual operating point and use some sort of linear combination of the control gains corresponding to these design points.

It is important for the control gains to stabilize the aircraft at all points near the design operating point for this gain-scheduling procedure to be effective. In passing from operating point to operating point, the parameters of the state-variable model vary. Using (6.2-49), we may design controllers that guarantee robust stability despite plant parameter variations.

Suppose that the nominal perturbed model used for design is

$$\dot{x} = Ax + Bu$$
$$y = Cx \qquad (6.2\text{-}51)$$

which has the transfer function

$$G(s) = C(sI - A)^{-1}B \qquad (6.2\text{-}52)$$

However, due to operating point changes the actual aircraft perturbed motion is described by

$$\dot{x} = (A + \Delta A)x + (B + \Delta B)u$$
$$y = (C + \Delta C)x, \qquad (6.2\text{-}53)$$

where the plant parameter variation matrices are $\Delta A, \Delta B, \Delta C$. It is not difficult to show (see Stevens et al. [1987] and the Problems) that this results in the transfer function

$$G'(s) = G(s) + \Delta G(s),$$

with

$$\Delta G(s) = C(sI - A)^{-1}\Delta B + \Delta C\left(sI - A^{-1}\right)B$$
$$+ C(sI - A)^{-1}\Delta A(sI - A)^{-1}B, \qquad (6.2\text{-}54)$$

where second-order effects have been neglected. Hence (6.2-48) may be used to determine the multiplicative uncertainty bound $m(\omega)$. The cosensitivity $T(j\omega)$ should then satisfy the upper bound (6.2-49) for guaranteed stability in the face of the parameter variations $\Delta A, \Delta B, \Delta C$.

Since $(sI - A)^{-1}$ has a relative degree of at least 1, the high-frequency roll-off of $\Delta G(j\omega)$ is at least -20 dB/decade. Thus, plant parameter variations yield an upper bound for the cosensitivity at low frequencies.

Using (6.2-54) it is possible to design robust controllers over a range of operating points that do not require gain scheduling. Compare with Minto et al. (1990).

6.3 ROBUST OUTPUT-FEEDBACK DESIGN

We should now like to incorporate the robustness concepts introduced in Section 6.2 into the LQ output feedback design procedure for aircraft control systems. This may be accomplished using the following steps.

1. If necessary, augment the plant with added dynamics to achieve the required steady-state error behavior, or to achieve balanced singular values at dc. Use the techniques of Example 6.2-3.
2. Select a performance index, the PI weighting matrices Q and R, and, if applicable, the time weighting factor k in t^k.
3. Determine the optimal output feedback again K using, for instance, Table 5.4-1 or 5.5-1.
4. Simulate the time responses of the closed-loop system to verify that they are satisfactory. If not, select different Q, R, and k and return to step 3.
5. Determine the low-frequency and high-frequency bounds required for performance robustness and stability robustness. Plot the loop gain singular values to verify that the bounds are satisfied. If they are not, select new Q, R, and k and return to step 3.

An example will illustrate the robust output-feedback design procedure.

Example 6.3-1: Pitch-Rate Control System Robust to Wind Gusts and Unmodeled Flexible Mode.
Here we will illustrate the design of a pitch-rate control system that is robust in the presence of vertical wind gusts and the unmodeled dynamics associated with a flexible mode. It would be worthwhile first to review the pitch rate CAS designed in Examples 4.5-1 and 5.5-3.

a. Control System Structure. The pitch-rate CAS system is described in Example 5.5-3. The state and measured output are

$$x = \begin{bmatrix} \alpha \\ q \\ \delta_e \\ \alpha_F \\ \epsilon \end{bmatrix}, \quad y = \begin{bmatrix} \alpha_F \\ q \\ \epsilon \end{bmatrix}, \quad (1)$$

with α_F the filtered angle of attack and ϵ the output of the integrator added to ensure zero steady-state error. The performance output $z(t)$ that should track the reference input $r(t)$ is $q(t)$.

Linearizing the F-16 dynamics about the nominal flight condition in Table 3.6-3 (502 ft/s, level flight, $x_{cg} = 0.35\,\bar{c}$) yields

$$\dot{x} = Ax + Bu + Gr \quad (2)$$
$$y = Cx + Fr \quad (3)$$
$$z = Hx, \quad (4)$$

with the system matrices given in Example 5.5-3.

536 ROBUSTNESS AND MULTIVARIABLE FREQUENCY-DOMAIN TECHNIQUES

The control input is

$$u = -Ky = -\begin{bmatrix} k_\alpha & k_q & k_I \end{bmatrix} y = -k_\alpha \alpha_F - k_q q - k_I \epsilon \tag{5}$$

It is desired to select the control gains to guarantee a good response to a step command r in the presence of vertical wind gusts and the unmodeled dynamics of the first flexible mode.

b. Frequency-Domain Robustness Bounds. According to *Mil. Spec. 1797* (1987), the vertical wind gust noise has a spectral density given in Dryden form as

$$\Phi_w(\omega) = 2L\sigma^2 \frac{1 + 3L^2\omega^2}{(1 + L^2\omega^2)^2}, \tag{6}$$

with ω the frequency in rad/s, σ the turbulence intensity, and L the turbulence scale length divided by true airspeed. Assuming that the vertical gust velocity is a disturbance input that changes the angle of attack, the software described in Chapter 3 can be used to find a control input matrix from gust velocity to x. Then, using stochastic techniques like those in Example 6.4-2, the magnitude of the gust disturbance versus frequency can be found. It is shown in Figure 6.3-1. We took $\sigma = 10$ ft/s and $L = (1700 \text{ ft})/(502 \text{ ft/s}) = 3.49$ s.

Let the transfer function of the rigid dynamics from $u(t)$ to $z(t)$ be denoted by $G(s)$. Then the transfer function including the first flexible mode is given by Blakelock (1965)

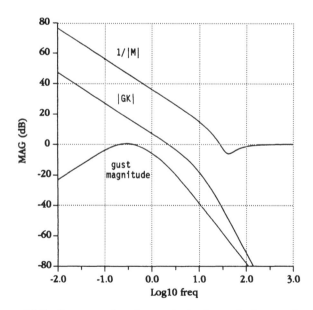

Figure 6.3-1 Frequency-domain magnitude plots and robustness bounds.

$$G'(s) = G(s)F(s), \tag{7}$$

where

$$F(s) = \frac{\omega_n^2}{s^2 + 2\zeta\omega_n s + \omega_n^2}, \tag{8}$$

with $\omega_n = 40$ rad/s and $\zeta = 0.3$. According to Section 6.2, therefore, the multiplicative uncertainty is given by

$$M(s) = F(s) - I = \frac{-s(s + 2\zeta\omega_n)}{s^2 + 2\zeta\omega_n s + \omega_n^2} \tag{9}$$

The magnitude of $1/M(j\omega)$ is shown in Figure 6.3-1.

We should like to perform our controls design using only the rigid dynamics $G(s)$. Then, for performance robustness in the face of the gust disturbance and stability robustness in the face of the first flexible mode, the loop-gain singular values should lie within the bounds implied by the gust disturbance magnitude and $1/|M(j\omega)|$.

c. Controls Design and Robustness Verification.
In Example 5.5-3c we performed a derivative-weighting design and obtained the control gains

$$K = [-0.0807 \quad -0.475 \quad 1.361] \tag{10}$$

The resulting step response is reproduced in Figure 6.3-2, and the closed-loop poles were

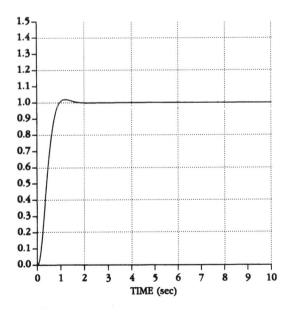

Figure 6.3-2 Optimal pitch-rate step response.

$$s = -3.26 \pm j2.83$$
$$-1.02 \qquad (11)$$
$$-10.67, -14.09$$

To verify that the robustness bounds hold for this design, it is necessary to find the loop-gain $GK(s)$ of the closed-loop system. Thus, in the figure of Example 5.5-3 it is necessary to find the loop transfer function from $e(t)$ around to $e(t)$ (i.e., from $e(t)$ to $-z(t)$). With respect to this loop gain, note that some of the elements in (10) are feedforward gains while some are feedback gains.

The magnitude of $GK(j\omega)$ is plotted in Figure 6.3-1. Note that the robustness bounds are satisfied. Therefore, this design is robust in the presence of vertical turbulence velocities up to 10 ft/s as well as the first flexible mode. ■

6.4 OBSERVERS AND THE KALMAN FILTER

The central theme in Chapter 5 was control design using partial state or output feedback. We saw in Section 5.4 that by using output feedback a compensator of any desired structure may be used, with the feedback gains being selected by modern LQ techniques. Thus, output-feedback design is very suitable for aircraft control. In Section 6.3 we saw how to verify the robustness of the closed-loop system using multivariable Bode plots.

On the other hand, in Section 5.7 we saw that the design equations for full state-variable feedback were simpler than those for output feedback. In fact, in state-variable design it is only necessary to solve the matrix Riccati equation, for which there are many good techniques (ORACLS [Armstrong, 1980], PC-MATLAB [Moler et al., 1987], and MATRIX$_x$ [1989]). By contrast, in output-feedback design it is necessary to solve three coupled nonlinear equations (see Table 5.3-1), which must generally be done using iterative techniques (Moerder and Calise, 1985; Press et al., 1986).

Moreover, in the case of full state feedback, if the system (A, B) is reachable and $(\sqrt{Q}, A)$ is observable (with Q the state weighting in the PI), the Kalman gain is guaranteed to stabilize the plant and yield a global minimum value for the PI. This is a fundamental result of modern control theory, and no such result yet exists for output feedback. The best that may be said is that if the plant is output stabilizable, the algorithm of Table 5.3-2 yields a local minimum for the PI and a stable plant.

Another issue is that the LQ regulator with full state feedback enjoys some important robustness properties that are not guaranteed using output feedback. Specifically, as we will see in Section 6.5, it has an infinite gain margin and 60 percent of phase margin.

Thus, state-feedback design offers some advantages over output feedback if the structure of the compensator is of no concern. Although this is rarely the case in aircraft controls, it is nevertheless instructive to pursue a compensator design technique based on state feedback.

Since all the states are seldom available, the first order of business is to estimate the full state $x(t)$ given only partial information in the form of the measured outputs $y(t)$. This is the *observer design* problem. Having estimated the state, we may then use the *estimate* of the state for feedback purposes, designing a feedback gain *as if* all the states were measurable. The combination of the observer and the state feedback gain is then a dynamic regulator similar to those used in classical control, as we will show in the last portion of this section. In the modern approach, however, it is straightforward to design multivariable regulators with desirable properties by solving matrix equations due to the fundamental *separation principle*, which states that the feedback gain and observer may be designed-separately and then concatenated.

One of our prime objectives in this section and the next is to discuss the linear quadratic Gaussian/loop-transfer recovery (LQG/LTR) technique for control design. This is an important modern technique for the design of robust aircraft control systems. It relies on full state-feedback design, followed by the design of an observer that allows full recovery of the guaranteed robustness properties of the LQ regulator with state feedback.

Of course, observers and filters have important applications in aircraft in their own right. For instance, the angle of attack is difficult to measure accurately; however, using an observer or Kalman filter it is not difficult to estimate the angle of attack very precisely by measuring pitch rate and normal acceleration (see Example 6.4-2).

Observer Design

In aircraft control, all of the states are rarely available for feedback purposes. Instead, only the measured outputs are available. Using modern control theory, if the measured outputs capture enough information about the dynamics of the system, it is possible to use them to *estimate* or *observe* all the states. Then these state estimates may be used for feedback purposes.

To see how a state observer can be constructed, consider the aircraft equations in state-space form

$$\dot{x} = Ax + Bu \tag{6.4-1}$$

$$y = Cx, \tag{6.4-2}$$

with $x(t) \in \mathbf{R}^n$ the state, $u(t) \in \mathbf{R}^m$ the control input, and $y(t) \in \mathbf{R}^p$ the available measured outputs.

Let the estimate of $x(t)$ be $\hat{x}(t)$. We claim that the state observer is a dynamical system described by

$$\dot{\hat{x}} = A\hat{x} + Bu + L(y - C\hat{x}) \tag{6.4-3}$$

or

$$\dot{\hat{x}} = (A - LC)\hat{x} + Bu + Ly \equiv A_0\hat{x} + Bu + Ly \tag{6.4-4}$$

That is, the observer is a system with two inputs, namely, $u(t)$ and $y(t)$, both of which are known.

Since $\hat{x}(t)$ is the state estimate, we could call

$$\hat{y} = C\hat{x} \qquad (6.4\text{-}5)$$

the estimated output. It is desired that $\hat{x}(t)$ be close to $x(t)$. Thus, if the observer is working properly, the quantity $y - \hat{y}$ that appears in (6.4-3) should be small. In fact,

$$\tilde{y} = y - \hat{y} \qquad (6.4\text{-}6)$$

is the *output estimation error*.

It is worth examining Figure 6.4-1, which depicts the state observer. Note that the observer consists of two parts: a *model of the system* involving (A, B, C), and an *error-correcting portion* that involves the output error multiplied by L. We call matrix L the *observer gain*.

To demonstrate that the proposed dynamical system is indeed an observer, it is necessary to show that it manufactures an estimate $\hat{x}(t)$ that is close to the actual state $x(t)$. For this purpose, define the (*state*) *estimation error* as

$$\tilde{x} = x - \hat{x} \qquad (6.4\text{-}7)$$

By differentiating (6.4-7) and using (6.4-1) and (6.4-4), it is seen that the estimation error has dynamics given by

$$\dot{\tilde{x}} = (A - LC)\tilde{x} = A_0\tilde{x} \qquad (6.4\text{-}8)$$

The initial estimation error is $\tilde{x}(0) = x(0) - \hat{x}(0)$, with $\hat{x}(0)$ the initial estimate, which is generally taken as zero.

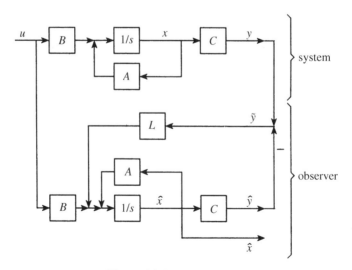

Figure 6.4-1 State observer.

It is required that the estimation error vanish with time for any $\tilde{x}(0)$, for then $\hat{x}(t)$ will approach $x(t)$. This will occur if $A_0 = (A - LC)$ is asymptotically stable. Therefore, as long as we select the observer gain L so that $(A-LC)$ is stable, (6.4-3) is indeed an observer for the state in (6.4-1). The observer design problem is to select L so that the error vanishes suitably quickly. It is a well-known result of modern control theory that the poles of $(A - LC)$ may be arbitrarily assigned to desired locations if and only if (C, A) is observable.

Since, according to Figure 6.4-1, we are injecting the output into the state derivative, L is called an *output injection*. Observers of the sort we are mentioning here are called *output-injection observers*, and their design could be called output-injection design.

It is important to discuss the output-injection problem of selecting L so that $(A - LC)$ is stable, for it is a problem we have already solved under a different guise. The state-feedback control law for system (6.4-1) is

$$u = -Kx, \tag{6.4-9}$$

which results in the closed-loop system

$$\dot{x} = (A - BK)x \tag{6.4-10}$$

The state-feedback design problem is to select K for desired closed-loop properties. We have shown how this may be accomplished in Section 5.7. Thus if we select the feedback gain as the Kalman gain

$$K = R^{-1}B^T P, \tag{6.4-11}$$

with P the positive-definite solution to the algebraic Riccati equation (ARE)

$$0 = A^T P + PA + Q - PBR^{-1}B^T P, \tag{6.4-12}$$

then if (A, B) is reachable and $(\sqrt{Q}, A)$ is observable, the closed-loop system is guaranteed to be stable. The matrices Q and R are design parameters that will determine the closed-loop dynamics, as we have seen in the examples of Chapter 5.

Now, compare (6.4-8) and (6.4-10). They are very similar. In fact,

$$(A - LC)^T = A^T - C^T L^T, \tag{6.4-13}$$

which has the free matrix L^T to the right, exactly as in the state feedback problem involving $(A - BK)$. This important fact is called *duality*, that is, state feedback and output injection are duals. (Note that $A - LC$ and $(A - LC)^T$ have the same poles.)

The important result of duality for us is that *the same theory we have developed for selecting the state-feedback gain may be used to select the output-injection gain L.* In fact, compare (6.4-13) to $(A - BK)$. Now, in the design equations (6.4-11) and (6.4-12) let us replace A, B, and K everywhere they occur by A^T, C^T, and L^T, respectively. The result is

$$L^T = R^{-1}CP$$
$$0 = AP + PA^T + Q - PC^TR^{-1}CP \qquad (6.4\text{-}14)$$

The first of these may be rewritten as

$$L = PC^TR^{-1} \qquad (6.4\text{-}15)$$

We call (6.4-14) the *observer ARE*.

Let us note the following connection between reachability and observability. Taking the transpose of the reachability matrix yields

$$U^T = \begin{bmatrix} B & AB & A^2B & \cdots & A^{n-1}B \end{bmatrix}^T$$
$$= \begin{bmatrix} B^T \\ B^TA \\ \vdots \\ B^T(A^T)^{n-1} \end{bmatrix} \qquad (6.4\text{-}16)$$

However, the observability matrix is

$$V = \begin{bmatrix} C \\ CA \\ \vdots \\ CA^{n-1} \end{bmatrix} \qquad (6.4\text{-}17)$$

Comparing U^T and V, it is apparent that they have the same form. In fact, since U and U^T have the same rank it is evident that (A, B) is reachable if and only if (B^T, A^T) is observable. This is another aspect of duality.

Taking into account these notions, an essential result of output injection is the following. It is the dual of the guaranteed stability using the Kalman gain discussed in Section 5.7. Due to its importance, we formulate it as a theorem.

Theorem. *Let (C, A) be observable and $(A, \sqrt{Q})$ be reachable. Then the error system (6.4-8) using the gain L given by (6.4-15), with P the unique positive definite solution to (6.4-14), is asymptotically stable.* ∎

Stability of the error system guarantees that the state estimate $\hat{x}(t)$ will approach the actual state $x(t)$. By selecting L to place the poles of $(A - LC)$ far enough to the left in the s-plane, the estimation error $\tilde{x}(t)$ can be made to vanish as quickly as desired.

The power of this theorem is that we may treat Q and R as design parameters that may be turned until suitable observer behavior results for the gain computed from the observer ARE. As long as we select Q and R to satisfy the theorem, observer stability is assured. An additional factor, of course, is that software for solving the observer ARE is readily available (e.g., ORACLS [Armstrong, 1980], PC-MATLAB [Moler et al., 1987], and MATRIX$_x$ [1989]).

We have assumed that the system matrices (A, B, C) are exactly known. Unfortunately, in reality this is rarely the case. In aircraft control, for instance, (6.4-1)–(6.4-2) represents a model of a nonlinear system at an equilibrium point. Variations in the operating point will result in variations in the elements of A, B, and C. However, if the poles of $(A - LC)$ are selected far enough to the left in the s-plane (i.e., fast enough), the estimation error will be small despite uncertainties in the system matrices. That is, the observer has some robustness to modeling inaccuracies.

It is worth mentioning that there are many other techniques for the selection of the observer gain L. In the single-output case the observability matrix V is square. Then Ackermann's formula (Franklin et al., 1986) may be used to compute L. If

$$\Delta_0(s) = |sI - (A - LC)| \tag{6.4-18}$$

is the desired observer characteristic polynomial, the required observer gain is given by

$$L = \Delta_0(A) V^{-1} e_n, \tag{6.4-19}$$

with $e_n = [0 \ \cdots \ 0 \ 1]^T$ the last column of the $n \times n$ identity matrix.

A general rule of thumb is that for suitable accuracy in the state estimate $\hat{x}(t)$, the slowest observer pole should have a real part 5 to 10 times larger than the real part of the fastest system pole. That is, the observer time constants should be 5 to 10 times larger than the system time constants.

Example 6.4-1: Observer Design for Double Integrator System.
In Example 5.7-1 we discussed state-feedback design for systems obeying Newton's laws,

$$\dot{x} = \begin{bmatrix} 0 & 1 \\ 0 & 0 \end{bmatrix} x + \begin{bmatrix} 0 \\ 1 \end{bmatrix} u = Ax + Bu, \tag{1}$$

where the state is $x = [d \ v]^T$, with $d(t)$ the position and $v(t)$ the velocity, and the control $u(t)$ is an acceleration input. Let us take position measurements so that the measured output is

$$y = \begin{bmatrix} 1 & 0 \end{bmatrix} x = Cx \tag{2}$$

We should like to design an observer that will reconstruct the full state $x(t)$ given only position measurements. Let us note that simple differentiation of $y(t) = d(t)$ to obtain $v(t)$ is unsatisfactory, since differentiation increases sensor noise. In fact, the observer is a *low-pass* filter that provides estimates while rejecting high-frequency noise. We will discuss two techniques for observer design.

544   ROBUSTNESS AND MULTIVARIABLE FREQUENCY-DOMAIN TECHNIQUES

a. Riccati Equation Design. There is good software available in standard design packages for solving the observer ARE (e.g., ORACLS [Armstrong, 1980] and PC-MATLAB [Moler et al., 1987]). However, in this example we want to solve the ARE analytically to show the relation between the design parameters Q and R and the observer poles.

Selecting $R = 1$ and $Q = \text{diag}\{q_d, q_v^2\}$ with q_d and q_v non-negative, we may assume that

$$P = \begin{bmatrix} p_1 & p_2 \\ p_2 & p_3 \end{bmatrix} \tag{3}$$

for some scalars p_1, p_2, and p_3 to be determined. The observer ARE (6.4-14) becomes

$$0 = \begin{bmatrix} 0 & 1 \\ 0 & 0 \end{bmatrix} \begin{bmatrix} p_1 & p_2 \\ p_2 & p_3 \end{bmatrix} + \begin{bmatrix} p_1 & p_2 \\ p_2 & p_3 \end{bmatrix} \begin{bmatrix} 0 & 0 \\ 1 & 0 \end{bmatrix} + \begin{bmatrix} q_d & 0 \\ 0 & q_v^2 \end{bmatrix}$$
$$- \begin{bmatrix} p_1 & p_2 \\ p_2 & p_3 \end{bmatrix} \begin{bmatrix} 1 & 0 \\ 0 & 0 \end{bmatrix} \begin{bmatrix} p_1 & p_2 \\ p_2 & p_3 \end{bmatrix}, \tag{4}$$

which may be multiplied out to obtain the three scalar equations

$$0 = 2p_2 - p_1^2 + q_d \tag{5a}$$
$$0 = p_3 - p_1 p_2 \tag{5b}$$
$$0 = p_2^2 + q_v^2 \tag{5c}$$

Solving these equations gives

$$p_2 = q_v \tag{6a}$$
$$p_1 = \sqrt{2}\sqrt{q_v + \frac{q_d}{2}} \tag{6b}$$
$$p_3 = q_v \sqrt{2}\sqrt{q_v + \frac{q_d}{2}}, \tag{6c}$$

where we have selected the signs that make P positive definite.

According to (6.4-15), the observer gain is equal to

$$L = \begin{bmatrix} p_1 & p_2 \\ p_2 & p_3 \end{bmatrix} \begin{bmatrix} 1 \\ 0 \end{bmatrix} = \begin{bmatrix} p_1 \\ p_2 \end{bmatrix} \tag{7}$$

Therefore,

$$L = \begin{bmatrix} \sqrt{2}\sqrt{q_v + \frac{q_d}{2}} \\ q_v \end{bmatrix} \tag{8}$$

Using (8), the error system matrix is found to be

$$A_0 = (A - LC) = \begin{bmatrix} -\sqrt{2}\sqrt{q_v + \frac{q_d}{2}} & 1 \\ -q_v & 0 \end{bmatrix} \quad (9)$$

Therefore, the observer characteristic polynomial is

$$\Delta_0(s) = |sI - A_0| = s^2 + 2\zeta\omega s + \omega^2, \quad (10)$$

with the observer natural frequency ω and damping ratio ζ given by

$$\omega = \sqrt{q_v}, \quad \zeta = \frac{1}{\sqrt{2}}\sqrt{1 + \frac{q_d}{2q_v}} \quad (11)$$

It is now clear how selection of Q affects the observer behavior. Note that if $q_d = 0$, the damping ratio becomes the familiar $1/\sqrt{2}$.

The reader should verify that the system is observable and that $(A, \sqrt{Q})$ is reachable as long as $q_v \neq 0$. A comparison with Example 5.7-1, where a state feedback was designed for Newton's system, reveals some interesting aspects of duality.

b. Ackermann's Formula Design. Riccati equation observer design is useful whether the plant has only one or multiple outputs. If there is only one output, we may use Ackermann's formula (6.4-19).

Let the desired observer polynomial be

$$\Delta_0(s) = s^2 + 2\zeta\omega s + \omega^2 \quad (12)$$

for some specified damping ratio ζ and natural frequency ω. Then

$$\Delta_0(A) = A^2 + 2\zeta\omega A + \omega^2 I = \begin{bmatrix} \omega^2 & 2\zeta\omega \\ 0 & \omega^2 \end{bmatrix} \quad (13)$$

$$V = \begin{bmatrix} C \\ CA \end{bmatrix} = I, \quad (14)$$

so that the observer gain is

$$L = \begin{bmatrix} 2\zeta\omega \\ \omega^2 \end{bmatrix} \quad (15)$$

One may verify that the characteristic polynomial of $A_0 = A - LC$ is indeed (12).

c. Simulation. To design an observer with a complex pole pair having damping ratio of $\zeta = 1/\sqrt{2}$ and natural frequency of $\omega = 1$ rad/s, the observer gain was selected as

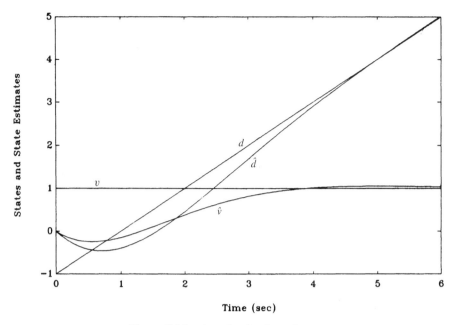

Figure 6.4-2 Actual and estimated states.

$$L = \begin{bmatrix} \sqrt{2} \\ 1 \end{bmatrix} \tag{16}$$

A simulation was performed. The time histories of the actual states and their estimates are shown in Figure 6.4-2. The initial conditions were $d(0) = -1$, $v(0) = 1$ and the input was $u(t) = 0$. The observer was started with initial states of $\hat{d}(0) = 0$, $\hat{v}(0) = 0$.
∎

The Kalman Filter

Throughout Chapter 5 we assumed that the system is exactly known and that no modeling inaccuracies, disturbances, or noises are present. In fact, nature is seldom so cooperative. In Sections 6.2 and 6.3 we showed how to take account of uncertainties in the model and the environment using a robust frequency-domain approach. An alternative is to treat uncertainties using *probability theory*.

In this subsection we develop the Kalman filter, which is based on a probabilistic treatment of process and measurement noises. The Kalman filter is an observer that is used for navigation and other applications that require the reconstruction of the state from noisy measurements. Since it is fundamentally a low-pass filter, it has good noise rejection capabilities. In Example 6.4-2 we show how to use the Kalman filter to estimate the angle of attack in the face of gust disturbances. In Section 6.5 we show

OBSERVERS AND THE KALMAN FILTER

how to use a state-variable feedback and a Kalman filter to design robust aircraft controllers by using the LQG/LTR technique.

We begin with a brief review of probability theory. It is not necessary to follow the derivation to use the Kalman filter: it is only necessary to solve the design equations in Table 6.4-1. Thus, one could skip the review that follows. However, an understanding of the theory will result in more sensible application of the filter. Supplemental references are Gelb (1974) and Lewis (1986b).

A Brief Review of Probability Theory. Suppose that the plant is described by the stochastic dynamical equation

$$\dot{x} = Ax + Bu + Gw \quad (6.4\text{-}20)$$

$$y = Cx + v, \quad (6.4\text{-}21)$$

with state $x(t) = \in \mathbf{R}^n$, control input $u(t) \in \mathbf{R}^m$, and measured output $y(t) \in \mathbf{R}^p$. Signal $w(t)$ is an unknown *process noise* that acts to disturb the plant. It could represent the effects of wind gusts, for instance, or unmodeled high-frequency plant dynamics. Signal $v(t)$ is an unknown *measurement noise* that acts to impair the measurements; it could represent sensor noise.

Since (6.4-20) is driven by process noise, the state $x(t)$ is now also a random process, as is $y(t)$. To investigate average properties of random processes we will require several concepts from probability theory (Papoulis, 1984). The point is that although $w(t)$ and $v(t)$ represent unknown random processes, we do in fact know something about them which can help us in control design. For instance, we may know their average values or total energy content. The concepts we will now define allow us to incorporate this general sort of knowledge into our theory.

Given a random vector $z \in \mathbf{R}^n$, we denote by $f_z(\zeta)$ the *probability density function* (*PDF*) of z. The PDF represents the probability that z takes on a value within the differential region $d\zeta$ centered at ζ. Although the value of z may be unknown, it is quite common in many situations to have a good feel for its PDF.

The *expected value* of a function $g(z)$ of a random vector z is defined as

$$E\{g(z)\} = \int_{-\infty}^{\infty} g(\zeta) f_z(\zeta) d\zeta \quad (6.4\text{-}22)$$

The *mean* or *expected value* of z is defined by

$$E\{z\} = \int_{-\infty}^{\infty} \zeta f_z(\zeta) d\zeta, \quad (6.4\text{-}23)$$

which we will symbolize by $\bar{z}$ to economize on notation. Note that $\bar{z} \in \mathbf{R}^n$.

The covariance of z is given by

$$P_z = E\left\{(z - \bar{z})(z - \bar{z})^\mathrm{T}\right\} \quad (6.4\text{-}24)$$

Note that P_z is an $n \times n$ constant matrix.

An important class of random vectors is characterized by the *Gaussian* or *normal* PDF

$$f_z(\zeta) = \frac{1}{\sqrt{(2\pi)^n |P_z|}} e^{-(\zeta-\bar{z})^T P_z^{-1}(\zeta-\bar{z})/2} \qquad (6.4\text{-}25)$$

In the scalar case $n = 1$ this reduces to the more familiar

$$f_z(\zeta) = \frac{1}{\sqrt{2\pi P_z}} e^{-(\zeta-\bar{z})^2/2P_z}, \qquad (6.4\text{-}26)$$

which is illustrated in Figure 6.4-3. Such random vectors take on values near the mean $\bar{z}$ with greatest probability and have a decreasing probability of taking on values farther away from $\bar{z}$. Many naturally occurring random variables are Gaussian.

If the random vector is a time function, it is called a *random process*, symbolized as $z(t)$. Then the PDF may also be time varying and we write $f_z(\zeta, t)$. One can imagine the PDF in Figure 6.4-3 changing with time. In this situation, the expected value and covariance matrix are also functions of time, so we write $\bar{z}(t)$ and $P_z(t)$.

Many random processes $z(t)$ of interest to us have a time-invariant PDF. These are *stationary* processes and, even though they are random time functions, they have a constant mean and covariance.

To characterize the relation between two random processes $z(t)$ and $x(t)$ we employ the *joint PDF* $f_{zx}(\zeta, \xi, t_1, t_2)$, which represents the probability that $(z(t_1), x(t_2))$ is within the differential area $d\zeta \times d\xi$ centered at (ζ, ξ). For our purposes, we will assume that the processes $z(t)$ and $x(t)$ are *jointly stationary,* that is, the joint PDF is not a function of both times t_1 and t_2, but depends only on the difference $(t_1 - t_2)$.

In the stationary case, the expected value of the function of two variables $g(z, x)$ is defined as

$$E\{q(z(t_1), x(t_2))\} = \int_{-\infty}^{\infty} g(\zeta, \xi) f_{z,x}(\zeta, \xi, t_1 - t_2) d\zeta d\xi \qquad (6.4\text{-}27)$$

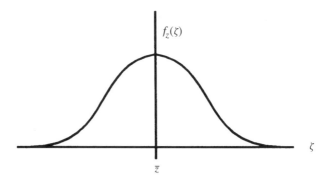

Figure 6.4-3 Gaussian PDF.

In particular, the *cross-correlation matrix* is defined by

$$R_{zx}(\tau) = E\left\{z(t+\tau)x^{\mathrm{T}}(t)\right\} \tag{6.4-28}$$

In the sequel, we will briefly require the cross-correlation matrix of two nonstationary processes, which is defined as

$$R_{zx}(t,\tau) = E\left\{z(t)x^{\mathrm{T}}(\tau)\right\} \tag{6.4-29}$$

Considering $z(t_1)$ and $z(t_2)$ as two jointly distributed random stationary processes, we may define the *autocorrelation function* of $z(t)$ as

$$R_z(\tau) = E\left\{z(t+\tau)z^{\mathrm{T}}(t)\right\} \tag{6.4-30}$$

The autocorrelation function gives us some important information about the random process $z(t)$. For instance,

$$\mathrm{tr}\left[R_z(0)\right] = \mathrm{tr}\left[E\left\{z(t)z^{\mathrm{T}}(t)\right\}\right] = E\left\{\|z(t)\|^2\right\}$$

is equal to the total energy in the process $z(t)$. (In writing this equation recall that for any compatible matrices M and N, $\mathrm{tr}(MN) = \mathrm{tr}(NM)$.)

If

$$R_{zx}(\tau) = 0, \tag{6.4-31}$$

we call $z(t)$ and $x(t)$ *orthogonal*. If

$$R_z(\tau) = P\delta(\tau), \tag{6.4-32}$$

where P is a constant matrix and $\delta(t)$ is the Dirac delta, then $z(t)$ is orthogonal to $z(t+\tau)$ for any $\tau \neq 0$. What this means is that the value of the process $z(t)$ at one time t is unrelated to its value at another time $\tau \neq t$. Such a process is called *white noise*. An example is the thermal noise in an electric circuit, which is due to the thermal agitation of the electrons in the resistors.

Note that $P\delta(0)$ is the covariance of $z(t)$, which is unbounded. We call P a *spectral density matrix*. It is sometimes loosely referred to as a covariance matrix.

Derivation of the Kalman Filter. We may now return to system (6.4-20)/(6.4-21). Neither the initial state $x(0)$, the process noise $w(t)$, nor the measurement noise $v(t)$ is exactly known. However, in practice we may have some feeling for their general characteristics. Using the concepts we have just discussed, we may formalize this general knowledge so that it may be used in control design.

The process noise is due to some sort of system disturbance, such as wind gusts; the measurement noise is due to sensor inaccuracies; and the initial state is uncertain because of our ignorance. Since these are all unrelated, it is reasonable to assume that

$x(0)$, $w(t)$, and $v(t)$ are mutually orthogonal. Some feeling for $x(0)$ may be present in that we may know its mean $\bar{x}_0$ and covariance P_0. We symbolize this as

$$x(0) \sim (\bar{x}_0, P_0) \qquad (6.4\text{-}33)$$

It is not unreasonable to assume that $w(t)$ and $v(t)$ have means of zero, since, for instance, there should be no bias on the measuring instruments. We will also assume that the process noise and measurement noise are white noise processes, so that

$$R_w(\tau) = E\left\{w(t+\tau)w^T(t)\right\} = Q\delta(\tau) \qquad (6.4\text{-}34)$$

$$R_v(\tau) = E\left\{v(t+\tau)v^T(t)\right\} = R\delta(\tau) \qquad (6.4\text{-}35)$$

Spectral density matrices Q and R will be assumed known. (Often, we have a good feeling for the standard deviations of $w(t)$ and $v(t)$.) According to (6.4-30), Q and R are positive semidefinite. We will assume in addition that R is nonsingular. In summary, we will assume that

$$w(t) \sim (0, Q), \quad Q \geq 0 \qquad (6.4\text{-}36)$$

$$v(t) \sim (0, R), \quad R > 0 \qquad (6.4\text{-}37)$$

The assumption that $w(t)$ and $v(t)$ are white may in some applications be a bad one. For instance, wind-gust noise is generally of low frequency. However, suppose that $w(t)$ is not white. Then we can determine a system description

$$\dot{x}_w = A_w x_w + B_w n \qquad (6.4\text{-}38)$$

$$w = C_w x_w + D_w n, \qquad (6.4\text{-}39)$$

which has a white noise input $n(t)$ and output $w(t)$. This is called a *noise-shaping filter*. These dynamics may be combined with the plant (6.4-20)–(6.4-21) to obtain the augmented dynamics

$$\begin{bmatrix} \dot{x} \\ \dot{x}_w \end{bmatrix} = \begin{bmatrix} A & GC_w \\ 0 & A_w \end{bmatrix} \begin{bmatrix} x \\ x_w \end{bmatrix} + \begin{bmatrix} B \\ 0 \end{bmatrix} u + \begin{bmatrix} CD_w \\ B_w \end{bmatrix} n \qquad (6.4\text{-}40)$$

$$y = [C \quad 0] \begin{bmatrix} x \\ x_w \end{bmatrix} + v \qquad (6.4\text{-}41)$$

This augmented system does have a white process noise $n(t)$. A similar procedure may be followed if $v(t)$ is nonwhite. Thus, we can generally describe a plant with nonwhite noises in terms of an augmented system with white process and measurement noises.

The determination of a system (6.4-38)/(6.4-39) that describes nonwhite noise $w(t)$ (or $v(t)$) is based on factoring the spectral density of the noise $w(t)$. For details, see Lewis (1986*b*). We will illustrate the procedure in Example 6.4-2.

We should now like to design an estimator for the stochastic system (6.4-20)/(6.4-21) under the assumptions just listed. We will propose the output-injection observer, which has the form

$$\dot{\hat{x}} = A\hat{x} + Bu + L(y - \hat{y}) \tag{6.4-42}$$

or

$$\dot{\hat{x}} = (A - LC)\hat{x} + Bu + Ly \tag{6.4-43}$$

The time function $\hat{x}(t)$ is the state estimate and

$$\hat{y} = E\{Cx + v\} = C\hat{x} \tag{6.4-44}$$

is the estimate of the output $y(t)$. (This expected value is actually the *conditional mean* given the previous measurements; see Lewis [1986b].)

The estimator gain L must be selected to provide an optimal estimate in the presence of the noises $w(t)$ and $v(t)$. To select L, we will need to define the estimation error

$$\tilde{x}(t) = x(t) - \hat{x}(t) \tag{6.4-45}$$

Using (6.4-20) and (6.4-42), we may derive the error dynamics to be

$$\dot{\tilde{x}} = (A - LC)\tilde{x} + Gw - Lv$$
$$\equiv A_0\tilde{x} + Gw - Lv \tag{6.4-46}$$

Note that the error system is driven by both the process and measurement noise. The output of the error system may be taken as $\tilde{y} = y - \hat{y}$ so that

$$\tilde{y} = C\tilde{x} + v \tag{6.4-47}$$

The *error covariance* is given by

$$P(t) = E\{\tilde{x}\tilde{x}^T\}, \tag{6.4-48}$$

which is time-varying. Thus $\tilde{x}(t)$ is a nonstationary random process. The error covariance is a measure of the *uncertainty* in the estimate. Smaller values for $P(t)$ mean that the estimate is better, since the error is more closely distributed about its mean value of zero if $P(t)$ is smaller.

If the observer is asymptotically stable and $w(t)$ and $v(t)$ are stationary processes, the error $\tilde{x}(t)$ will eventually reach a *steady state* in which it is also stationary with constant mean and covariance. The gain L will be chosen to minimize the *steady-state error covariance* P. Thus the optimal gain L will be a constant matrix of observer gains.

Before determining the optimal gain L, let us compute the mean and covariance of the estimation error $\tilde{x}(t)$. Using (6.4-46) and the linearity of the expectation operator,

$$E\{\dot{\tilde{x}}\} = A_0 E\{\tilde{x}\} + GE\{w\} - LE\{v\}, \qquad (6.4\text{-}49)$$

so that

$$\frac{d}{dt}E\{\tilde{x}\} = A_0 E\{\tilde{x}\} \qquad (6.4\text{-}50)$$

Thus $E\{\tilde{x}\}$ is a deterministic time-varying quantity that obeys a differential equation with system matrix A_0. If $A_0 = A - LC$ is stable, then $E\{\tilde{x}\}$ eventually stabilizes at a steady-state value of zero, since the process and measurement noise are of zero mean. Since

$$E\{\tilde{x}\} = E\{x\} - E\{\hat{x}\} = E\{x\} - \hat{x}, \qquad (6.4\text{-}51)$$

it follows that in this case the estimate $\hat{x}(t)$ approaches $E\{x(t)\}$. Then the estimate is said to be *unbiased*. According also to (6.4-51), the mean of the initial error $\tilde{x}(0)$ is equal to zero if the observer (6.4-43) is initialized to $\hat{x}(0) = \bar{x}_0$, with $\bar{x}_0$ the mean of $x(0)$.

If the process noise $w(t)$ and/or measurement noise $v(t)$ have means that are not zero, then according to (6.4-49), the steady-state value of $E\{\tilde{x}\}$ is not equal to zero. In this case, $\hat{x}(t)$ does not tend asymptotically to the true state $x(t)$, but is offset from it by the constant value $-E\{\tilde{x}\}$. Then the estimates are said to be biased (see the Problems).

To determine the error covariance, note that the solution of (6.4-46) is given by

$$\tilde{x}(t) = e^{A_0 t}\tilde{x}(0) - \int_0^t e^{A_0(t-\tau)} Lv(\tau)\,d\tau + \int_0^t e^{A_0(t-\tau)} Gw(\tau)\,d\tau \qquad (6.4\text{-}52)$$

We will soon require the cross-correlation matrices $R_{v\tilde{x}}(t, t)$ and $R_{w\tilde{x}}(t, t)$. To find them, use (6.4-52) and the assumption that $x(0)$ (and hence $\tilde{x}(0)$), $w(t)$, and $v(t)$ are orthogonal. Thus

$$R_{v\tilde{x}}(t, t) = E\{v(t)\tilde{x}^T(t)\}$$

$$= -\int_0^t E\{v(t)v^T(\tau)\} L^T e^{A_0^T(t-\tau)}\,d\tau \qquad (6.4\text{-}53)$$

Note that

$$R_v(t, \tau) = R\delta(t - \tau) \qquad (6.4\text{-}54)$$

but the integral in (6.4-53) has an upper limit of t. Recall that the unit impulse can be expressed as

$$\delta(t) = \lim_{T \to 0} \frac{1}{T} \Pi\left(\frac{t}{T}\right), \qquad (6.4\text{-}55)$$

where the rectangle function

$$\frac{1}{T}\Pi\left(\frac{t}{T}\right) = \begin{cases} \frac{1}{T}, & |t| < \frac{T}{2} \\ 0 & \text{otherwise} \end{cases} \qquad (6.4\text{-}56)$$

is centered at $t = 0$. Therefore, only half the area of $\delta(t - \tau)$ should be considered as being to the left of $\tau = t$. Hence (6.4-53) is

$$R_{v\tilde{x}}(t, t) = -\tfrac{1}{2}RL^\mathrm{T} \qquad (6.4\text{-}57)$$

Similarly,

$$R_{w\tilde{x}}(t, t) = E\left\{w(t)\tilde{x}^\mathrm{T}(t)\right\}$$
$$= \int_0^t E\left\{w(t)w^\mathrm{T}(\tau)\right\} G^\mathrm{T} e^{A_0^\mathrm{T}(t-\tau)} d\tau \qquad (6.4\text{-}58)$$

or

$$R_{w\tilde{x}}(t, t) = \tfrac{1}{2}QG^\mathrm{T} \qquad (6.4\text{-}59)$$

To find a differential equation for $P(t) = E\left\{\tilde{x}\tilde{x}^\mathrm{T}\right\}$, write

$$\dot{P}(t) = E\left\{\frac{d\tilde{x}}{dt}\tilde{x}^\mathrm{T}\right\} + E\left\{\tilde{x}\frac{d\tilde{x}^\mathrm{T}}{dt}\right\} \qquad (6.4\text{-}60)$$

According to the error dynamics (6.4-46) the first term is equal to

$$E\left\{\frac{d\tilde{x}}{dt}\tilde{x}^\mathrm{T}\right\} = (A - LC)P + \tfrac{1}{2}LRL^\mathrm{T} + \tfrac{1}{2}GQG^\mathrm{T}, \qquad (6.4\text{-}61)$$

where we have used (6.4-57) and (6.4-59). To this equation add its transpose to obtain

$$\dot{P} = A_0 P + PA_0^\mathrm{T} + LRL^\mathrm{T} + GQG^\mathrm{T} \qquad (6.4\text{-}62)$$

What we have derived in (6.4-62) is an expression for the error covariance when the observer (6.4-43) is used with a specific gain L. Given any L such that $(A - LC)$ is stable, we may solve (6.4-62) for $P(t)$, using as initial condition $P(0) = P_0$, with P_0 the covariance of the initial state, which represents the uncertainty in the initial estimate $\hat{x}(0) = \bar{x}_0$.

Clearly, gains that result in smaller error covariances $P(t)$ are better, for then the error $\tilde{x}(t)$ is generally closer to its mean of zero. That is, the error covariance is a measure of the performance of the observer, and smaller covariance matrices are indicative of better observers. We say that P is a measure of the uncertainty in the estimate. (Given symmetric positive semidefinite matrices P_1 and P_2, P_1 is less than P_2 if $(P_2 - P_1) \geq 0$.)

The error covariance $P(t)$ reaches a bounded steady-state value P as $t \to \infty$ as long as A_0 is asymptotically stable. At steady state, $\dot{P} = 0$ so that (6.4-62) becomes the algebraic equation

$$0 = A_0 P + P A_0^T + LRL^T + GQG^T \tag{6.4-63}$$

The steady-state error covariance is the positive (semi)definite solution to (6.4-63). To obtain a constant observer gain, we may select L *to minimize the steady-state error covariance* P. Necessary conditions for L are now easily obtained after the same fashion that the output feedback gain K was obtained in Section 5.3.

Thus define a performance index (PI)

$$J = \tfrac{1}{2} \text{tr}(P) \tag{6.4-64}$$

(Note that $\text{tr}(P)$ is the sum of the eigenvalues of P. Thus a small J corresponds to a small P.) To select L so that J is minimized subject to the constraint (6.4-63), define the Hamiltonian

$$\mathcal{H} = \tfrac{1}{2} \text{tr}(P) + \tfrac{1}{2} \text{tr}(gS), \tag{6.4-65}$$

where

$$g = A_0 P + P A_0^T + LRL^T + GQG^T \tag{6.4-66}$$

and S is an $n \times n$ undetermined (Lagrange) multiplier.

To minimize J subject to the constraint $g = 0$, we may equivalently minimize $\mathcal{H}$ with no constraints. Necessary conditions for a minimum are therefore given by

$$\frac{\partial \mathcal{H}}{\partial S} = A_0 P + P A_0^T + LRL^T + GQG^T = 0 \tag{6.4-67}$$

$$\frac{\partial \mathcal{H}}{\partial P} = A_0^T S + S A_0 + I = 0 \tag{6.4-68}$$

$$\frac{1}{2} \frac{\partial \mathcal{H}}{\partial L} = SLR - SPC^T = 0 \tag{6.4-69}$$

If A_0 is stable, the solution S to (6.4-68) is positive definite. Then, according to (6.4-69),

$$L = PC^T R^{-1} \tag{6.4-70}$$

Substituting this value for L into (6.4-67) yields

$$(A - PC^TR^{-1}C)P + P(A - PC^TR^{-1}C)^T + PC^TR^{-1}CP + GQG^T = 0 \quad (6.4\text{-}71)$$

or

$$AP + PA^T + GQG^T - PC^TR^{-1}CP = 0 \quad (6.4\text{-}72)$$

To determine the optimal observer gain L, we may therefore proceed by solving (6.4-72) for the error covariance P, and then using (6.4-70) to compute L. The matrix quadratic equation (6.4-72) is called the *algebraic (filter) Riccati equation (ARE)*. There are several efficient techniques for solving the ARE for P (e.g., Armstrong, 1980; IMSL, 1980, MATRIX$_x$, 1989, MATLAB [Moler et al., 1987]).

The optimal gain L determined using (6.4-70) is called the *steady-state Kalman gain*, and the observer so constructed is called the *steady-state Kalman filter*. The term *steady state* refers to the fact that although the optimal gain that minimizes $P(t)$ is generally time varying, we have selected the optimal gain that minimizes the *steady-state* error covariance in order to obtain a constant observer gain. Since the gain must eventually be gain-scheduled in actual flight control applications, we require a constant gain to keep the number of parameters to be scheduled within reason.

The design equations for the Kalman filter are collected in Table 6.4-1. A block diagram appears in Figure 6.4-1. The steady-state Kalman filter is the best estimator with constant gains that has the dynamics of the form in the table. Such a filter is said to be *linear*. It can be shown (Lewis, 1986b) that if the process noise $w(t)$ and measurement noise $v(t)$ are *Gaussian*, this is also *the optimal* steady-state estimator of any form.

TABLE 6.4-1: The Kalman Filter

System Model
$\dot{x} = Ax + Bu + Gw$
$y = Cx + v$
$x(0) \sim (\bar{x}_0, P_0), \quad w(t) \sim (0, Q), \quad v(t) \sim (0, R)$

Assumptions
$w(t)$ and $v(t)$ are white noise processes orthogonal to each other and to $x(0)$.

Initialization
$\hat{x}(0) = \bar{x}_0$

Error Covariance ARE
$AP + PA^T + GQG^T - PC^TR^{-1}CP = 0$

Kalman Gain
$L = PC^TR^{-1}$

Estimate Dynamics (Filter Dynamics)
$\dot{\hat{x}} = A\hat{x} + Bu + L(y - C\hat{x})$

556 ROBUSTNESS AND MULTIVARIABLE FREQUENCY-DOMAIN TECHNIQUES

The quantity

$$\tilde{y}(t) = y(t) - \hat{y}(t) = y(t) - C\hat{x}(t) \tag{6.4-73}$$

that drives the filter dynamics in the table is called the residual. For more information on the Kalman filter, see Bryson and Ho (1975), Kwakernaak and Sivan (1972), and Lewis (1986b).

The filter ARE should be compared to the ARE we discussed at the beginning of this section in connection with output-injection design. There, no particular meaning was given to the auxiliary matrix P. In this stochastic setting, we have discovered that it is nothing but the error covariance. Small values of P generally indicate a filter with good estimation performance.

The theorem offered in connection with output-injection observer design also holds here. Thus suppose that (C, A) is observable and $(A, G\sqrt{Q})$ is reachable. Then the ARE has a unique positive definite solution P. Moreover, error system (6.4-46) using the gain Kalman gain L given by (6.4-70), with P the unique positive definite solution to the ARE, is asymptotically stable.

One might be inclined to believe that the less noise in the system, the better. However, the actual situation is quite surprising. For the existence of the Kalman filter it was necessary to assume that $R > 0$, that is, that *the measurement noise corrupts all the measurements*. If there are some noise-free measurements, a more complicated filter known as the *Deyst filter* must be used. Moreover, the assumption that $(A, G\sqrt{Q})$ is reachable means that *the process noise should excite all the states*.

Example 6.4-2: Kalman Filter Estimation of Angle of Attack in Gust Noise.
The short-period approximation to the F-16 longitudinal dynamics is

$$\dot{x} = Ax + B\delta_e + Gw_g, \tag{1}$$

with $x = [\alpha \ q]^T$, α the angle of attack, q the pitch rate, control input δ_e the elevator deflection, and w_g the vertical wind gust disturbance velocity. Using the software described in Chapter 3 to linearize the F-16 dynamics about the nominal flight condition in Table 3.6-3 (true airspeed of 502 ft/s, dynamic pressure of 300 psf, and cg at 0.35 $\bar{c}$), the plant matrices are found to be

$$A = \begin{bmatrix} -1.01887 & 0.90506 \\ 0.82225 & -1.07741 \end{bmatrix}, \quad B = \begin{bmatrix} -0.00215 \\ -0.17555 \end{bmatrix}, \quad G = \begin{bmatrix} 0.00203 \\ -0.00164 \end{bmatrix} \tag{2}$$

The vertical wind gust noise is not white, but according to *Mil. Spec. 1797* (1987) has a spectral density given in Dryden form as

$$\Phi_w(\omega) = 2L\sigma^2 \frac{1 + 3L^2\omega^2}{(1 + L^2\omega^2)^2}, \tag{3}$$

with ω the frequency in rad/s, σ the turbulence intensity, and L the turbulence scale length divided by true airspeed. Taking $\sigma = 10$ ft/s and $L = (1750 \text{ ft})/(502 \text{ ft/s})$

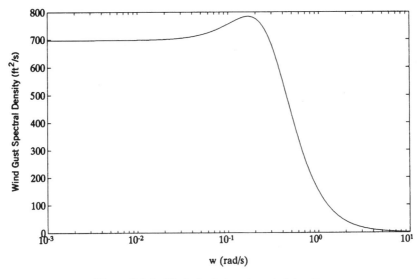

Figure 6.4-4 Vertical wind gust spectral density.

$= 3.49$ s (see *Mil. Spec. 1797* [1987]) the gust spectral density is shown in Figure 6.4-4.

a. Determination of Gust-Shaping Filter. Since w_g is not white, a noise *shaping filter* of the form of (6.4-38), (6.4-39) must be determined by factoring $\Phi_w(s)$ (Lewis, 1986b). Note that

$$\Phi_w(\omega) = 2L\sigma^2 \frac{(1+\sqrt{3}Lj\omega)(1-\sqrt{3}Lj\omega)}{(1+Lj\omega)^2(1-Lj\omega)^2}, \tag{4}$$

so that

$$\Phi_w(s) = H_w(s)H_w(-s) \tag{5}$$

with

$$H_w(s) = \sigma\sqrt{\frac{6}{L}}\frac{s+1/L\sqrt{3}}{L(s+1/L)^2} \tag{6}$$

$$H_w(s) = \sigma\sqrt{\frac{6}{L}}\frac{s+1/L\sqrt{3}}{s^2+2s/L+1/L^2} \tag{7}$$

Now a reachable canonical form realization of $H_w(s)$ (Kailath, 1980) is given by

$$\dot{z} = \begin{bmatrix} 0 & 1 \\ -\frac{1}{L^2} & -\frac{2}{L} \end{bmatrix} z + \begin{bmatrix} 0 \\ 1 \end{bmatrix} w \tag{8}$$

$$w_g = \gamma \left[\frac{1}{L\sqrt{3}} \quad 1 \right] z, \tag{9}$$

where the gain is $\gamma = \sigma\sqrt{g/L}$. Using $\sigma = 10$, $L = 3.49$ yields

$$\dot{z} = \begin{bmatrix} 0 & 1 \\ -0.0823 & -0.5737 \end{bmatrix} z + \begin{bmatrix} 0 \\ 1 \end{bmatrix} w \approx A_w z + B_w w \tag{10}$$

$$w_g = [2.1728 \quad 13.1192]z \equiv C_w z \tag{11}$$

The shaping filter (10)/(11) is a system driven by the *white* noise input $w(t) \sim (0, 1)$ that generates the gust noise $w_g(t)$ with spectral density given by (3).

b. Augmented Plant Dynamics. The overall system, driven by the white noise input $w(t) \sim (0, 1)$ and including an elevator actuator with transfer function $20.2/(s + 20.2)$, is given by (see (6.4-40))

$$\frac{d}{dt}\begin{bmatrix} \alpha \\ q \\ z_1 \\ z_2 \\ \delta_e \end{bmatrix} = \begin{bmatrix} -1.01887 & 0.90506 & 0.00441 & 0.02663 & -0.00215 \\ 0.82225 & -1.07741 & -0.00356 & -0.02152 & -0.17555 \\ 0 & 0 & 0 & 1 & 0 \\ 0 & 0 & -0.0823 & -0.5737 & 0 \\ 0 & 0 & 0 & 0 & -20.2 \end{bmatrix} \begin{bmatrix} \alpha \\ q \\ z_1 \\ z_2 \\ \delta_e \end{bmatrix}$$

$$+ \begin{bmatrix} 0 \\ 0 \\ 0 \\ 0 \\ 20.2 \end{bmatrix} u + \begin{bmatrix} 0 \\ 0 \\ 0 \\ 1 \\ 0 \end{bmatrix} w \tag{12}$$

with $u(t)$ the elevator actuator input. To economize on notation, let us symbolize this augmented system as

$$\dot{x} = Ax + Bu + Gw \tag{13}$$

c. Estimating Angle of Attack. Direct measurements of angle of attack α are noisy and biased. However, pitch rate q and normal acceleration n_z are convenient to measure. Using the software in Chapter 3 it is determined that

$$n_z = 15.87875\alpha + 1.48113q \tag{14}$$

Therefore, let us select the measured output as

$$y = \begin{bmatrix} n_z \\ q \end{bmatrix} = \begin{bmatrix} 15.87875 & 1.48113 & 0 & 0 & 0 \\ 0 & 1 & 0 & 0 & 0 \end{bmatrix} x + v \equiv Cx + v, \tag{15}$$

where $v(t)$ is measurement noise. A reasonable measurement noise covariance is

$$R = \begin{bmatrix} \frac{1}{20} & 0 \\ 0 & \frac{1}{60} \end{bmatrix} \tag{16}$$

Now the algebraic Riccati equation in Table 6.4-1 may be solved using standard available software (e.g., ORACLS [Armstrong, 1980; IMSL, 1980], PC-MATLAB [Moler et al., 1987], MATRIX$_x$ [1989]) to obtain the Kalman gain

$$L = \begin{bmatrix} 0.0375 & -0.0041 \\ -0.0202 & 0.0029 \\ 3.5981 & -0.2426 \\ 1.9061 & -0.2872 \\ 0 & 0 \end{bmatrix}, \tag{17}$$

whence the Kalman filter is given by

$$\dot{\hat{x}} = (A - LC)\hat{x} + Bu + Ly \tag{18}$$

Note that the Kalman gain corresponding to the fifth state δ_e is zero. This is due to the fact that according to (12), the gust noise $w(t)$ does not excite the actuator motor.

To implement the estimator we could use the state formulation (18) in a subroutine, or we could compute the transfer function to the angle-of-attack estimate given by

$$H_\alpha(s) = [1 \quad 0 \quad \cdots \quad 0][sI - (A - LC)]^{-1}[B \quad L] \tag{19}$$

(Note that α is the first component of x.) Then the angle of attack estimate is given by

$$\hat{\alpha}(s) = H(s) \begin{bmatrix} U(s) \\ Y(s) \end{bmatrix}, \tag{20}$$

so that $\alpha(t)$ may be estimated using $u(t)$ and $y(t)$, both of which are known.

Similarly, the estimate of the wind gust velocity $w_g(t)$ may be recovered. ∎

Dynamic Regulator Design Using the Separation Principle

The fundamental approach to regulator and compensator design in this book involves selecting the compensator dynamics using the intuition of classical control and traditional aircraft design. Then the adjustable compensator gains are computed using the output feedback design equations in Table 5.3-1, 5.4-1, or 5.5-1. The advantages of this approach include:

1. Good software for solving the design equations is available (e.g., the Davidon–Fletcher–Powell algorithm [Press et al., 1986]). See Appendix B.

2. General multi-input/multi-output control design is straightforward.
3. All the intuition in classical control design in the aircraft industry can be used to select the compensator structure.
4. Complicated compensator structures are avoided, which is important from the point of view of the pilot's feel for the aircraft and also simplifies the gain-scheduling problem.

However, in complicated modern systems (e.g., aircraft engines) there may be no a priori guidelines for selecting the compensator structure. In this case, a combination of LQ state-feedback and observer/filter design proves very useful for controller design. This combination is known as linear quadratic Gaussian (LQG) design, and is explored next. In Section 6.5 we discuss the LQG/LTR technique for robust design, which has become popular in some aspects of aircraft control.

Linear Quadratic Gaussian Design. The linear quadratic regulator (LQR) and the Kalman filter can be used together to design a dynamic regulator. This procedure is called linear quadratic Gaussian (LQG) design, and will now be described. An important advantage of LQG design is that the compensator structure is given by the procedure, so that it need not be known beforehand. This makes LQG design useful in the control of complicated modern-day systems (e.g., space structures, aircraft engines), where an appropriate compensator structure may not be known.

Suppose that the plant and measured output are given by

$$\dot{x} = Ax + Bu + Gw \qquad (6.4\text{-}74)$$

$$y = Cx + v, \qquad (6.4\text{-}75)$$

with $x(t) \in \mathbf{R}^n$, $u(t)$ the control input, $w(t)$ the process noise, and $v(t)$ the measurement noise. Suppose that the full-state-feedback control

$$u = -Kx + r \qquad (6.4\text{-}76)$$

has been designed, with $r(t)$ the pilot's input command. That is, the state feedback gain K has been selected by some technique, such as the LQR technique in Section 5.7. If the control (6.4-76) is substituted into (6.4-74) the closed-loop system is found to be

$$\dot{x} = (A - BK)x + Br + Gw \qquad (6.4\text{-}77)$$

Full-state-feedback design is attractive because if the conditions in Section 5.7 hold, the closed-loop system is guaranteed stable. Such a strong result has not yet been shown for output feedback. Moreover, using full state feedback all the poles of $(A - BK)$ may be placed arbitrarily as desired. Finally, the state-feedback design equations are simpler than those for output feedback and may be solved using standard available routines (e.g., ORACLS [Armstrong, 1980; IMSL, 1980], PC-MATLAB [Moler et al., 1987], MATRIX$_x$ [1989]). However, the control law (6.4-76) cannot be implemented since all the states are usually not available as measurements.

OBSERVERS AND THE KALMAN FILTER

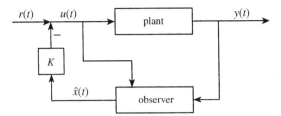

Figure 6.4-5 Regulator design using observer and full state feedback.

Now, suppose that an observer or Kalman filter

$$\dot{\hat{x}} = (A - LC)\hat{x} + Bu + Ly \quad (6.4\text{-}78)$$

has been designed. That is, the filter gain L has been selected by any of the techniques discussed in this section to provide state estimates. Then, since all the states are not measurable and the control (6.4-76) cannot be implemented in practice, we propose to feed back the *estimate* $x(t)$ instead of the actual state $x(t)$. That is, let us examine the feedback law

$$u = -K\hat{x} + r \quad (6.4\text{-}79)$$

The closed-loop structure using this controller is shown in Figure 6.4-5. Due to the fact that the observer is a dynamical system, the proposed controller is nothing but a dynamical regulator of the sort seen in classical control theory. However, in contrast to classical design, the theory makes it easy to design *multivariable regulators with guaranteed stability even for complicated MIMO systems.*

If K is selected using the LQR Riccati equation in Section 5.7 and L is selected using the Kalman filter Riccati equation in Table 6.4-1, this procedure is called LQG design.

We propose to show that using this control:

1. The closed-loop poles are the same as if the full state feedback (6.4-76) had been used.
2. The transfer function from $r(t)$ to $y(t)$ is the same as if (6.4-76) had been used.

The importance of these results is that the state feedback K and the observer gain L may be designed *separately* to yield desired closed-loop plant behavior and observer behavior. This is the *separation principle*, which is at the heart of modern controls design. Two important ramifications of the separation principle are that closed-loop stability is guaranteed and good software is available to solve the matrix design equations that yield K and L.

The Separation Principle. To show the two important results just mentioned, define the estimation error (6.4-45) and examine the error dynamics (6.4-46). In terms of $\tilde{x}(t)$, we may write (6.4-79) as

$$u = -Kx + K\tilde{x} + r, \tag{6.4-80}$$

which, when used in (6.4-74), yields

$$\dot{x} = (A - BK)x + BK\tilde{x} + Br + Gw \tag{6.4-81}$$

Now, write (6.4-81) and (6.4-46) as the augmented system

$$\frac{d}{dt}\begin{bmatrix} x \\ \tilde{x} \end{bmatrix} = \begin{bmatrix} A - BK & BK \\ 0 & A - LC \end{bmatrix}\begin{bmatrix} x \\ \tilde{x} \end{bmatrix} + \begin{bmatrix} B \\ 0 \end{bmatrix}r + \begin{bmatrix} G \\ G \end{bmatrix}w - \begin{bmatrix} 0 \\ L \end{bmatrix}v \tag{6.4-82}$$

$$y = [C \quad 0]\begin{bmatrix} x \\ \tilde{x} \end{bmatrix} + v \tag{6.4-83}$$

This represents the complete dynamics of Figure 6.4-5.

Since the augmented system is block triangular, the closed-loop characteristic equation is

$$\Delta(s) = |sI - (A - BK)| \cdot |sI - (A - LC)| = 0 \tag{6.4-84}$$

That is, the closed-loop poles are nothing but the plant poles that result by choosing K and the desired observer poles that result by choosing L. Thus the state feedback gain K and observer gain L may be selected separately for desirable closed-loop behavior.

The closed-loop transfer function from $r(t)$ to $y(t)$ is given by

$$H_c(s) = [C \quad 0]\begin{bmatrix} A - BK & BK \\ 0 & A - LC \end{bmatrix}^{-1}\begin{bmatrix} B \\ 0 \end{bmatrix},$$

and the triangular form of the system matrix makes it easy to see that

$$H_c(s) = C[sI - (A - BK)]^{-1}B \tag{6.4-85}$$

This, however, is exactly what results if the full state feedback (6.4-76) is used.

Of course, the initial conditions also affect the output $y(t)$. However, since the observer is stable, the effects of the initial error $\tilde{x}(0)$ will vanish with time. The observer poles (i.e., those of $(A - LC)$) should be chosen 5 to 10 times faster than the desired closed-loop plant poles (i.e., those of $(A - BK)$) for good closed-loop behavior.

Discussion. From our point of view, when possible it is usually better to design compensators using output feedback as we have demonstrated in the previous chapters than to use separation principle design. To see why, let us examine the structure of the dynamic compensator in Figure 6.4-5 in more detail.

The control input $u(t)$ may be expressed as

$$U(s) = H_y(s)Y(s) + H_u(s)U(s) + R(s), \tag{6.4-86}$$

where, according to (6.4-79) and (6.4-78), the transfer function from $y(t)$ to $u(t)$ is

$$H_y(s) = -K\,[sI - (A - LC)]^{-1}\,L \qquad (6.4\text{-}87)$$

and the transfer function from $u(t)$ to $u(t)$ is

$$H_u(s) = -K\,[sI - (A - LC)]^{-1}\,B \qquad (6.4\text{-}88)$$

Now, note that the compensator designed by this technique has order equal to the order n of the plant. This means that it has too many parameters to be conveniently gain-scheduled. Moreover, it has no special structure. This means that none of the classical control intuition available in the aircraft industry has been used in its design.

It is possible to design *reduced-order* compensators using the separation principle. Three possible approaches are:

1. Find a reduced-order model of the plant, then design a compensator for this reduced-order model.
2. Design a compensator for the full plant, then reduce the order of the compensator.
3. Design the reduced-order compensator directly from the full-order plant.

One technique for order reduction is the partial-fraction-expansion technique in Example 6.2-3. Other techniques include principal component analysis (Moore, 1982) and the frequency-weighted technique in Anderson and Liu (1989). A very convenient approach is given in Ly et al. (1985).

It is important to realize that although the plant is minimal (i.e., reachable and observable), the LQ regulator may not be. That is, it may have unreachable or unobservable states. A technique for reducing the regulator to minimal form is given in Yousuff and Skelton (1984).

In Section 6.5 we illustrate the design of a LQ regulator in robust design using the LQG/LTR approach.

6.5 LQG/LOOP-TRANSFER RECOVERY

We saw in Sections 6.2 and 6.3 how to use the multivariable Bode plot to design controllers guaranteeing performance robustness and stability robustness using output feedback. In Section 6.4 we discussed the Kalman filter. In this section we propose to cover the linear quadratic Gaussian/loop-transfer recovery (LQG/LTR) design technique for robust controllers. This approach is quite popular in the current literature and has been used extensively by Honeywell and others to design multivariable aircraft flight control systems (Doyle and Stein, 1981; Athans, 1986). It is based on the fact that the linear quadratic regulator (LQR) using state-variable feedback has certain *guaranteed robustness properties*.

Thus, suppose that a state feedback gain K has been computed using the ARE as in Section 5.7. This state feedback cannot be implemented since all of the states are not available as measurements; however, it can be used as the basis for the design of a dynamic LQ regulator by using a Kalman filter to provide state estimates for feedback purposes. We would like to discuss two issues. First, we will show that in contrast to output feedback, state feedback has certain guaranteed robustness properties in terms of gain and phase margins. Then we will see that the Kalman filter may be designed so that the dynamic regulator recovers the desirable robustness properties of full state feedback.

Guaranteed Robustness of the Linear-Quadratic Regulator

We have discussed conditions for performance robustness and stability robustness for the general feedback configuration of the form shown in Figure 6.2-1, where $G(s)$ is the plant and $K(s)$ is the compensator. The linear quadratic regulator using *state feedback* has many important properties, as we have seen Section 5.7. In this subsection we should like to return to the LQR to show that it has certain *guaranteed robustness properties* that make it even more useful (Safonov and Athans, 1977).

Thus, suppose that in Figure 6.2-1, $K(s) = K$, the constant optimal LQ state feedback gain determined using the algebraic Riccati equation (ARE) as in Table 5.7-1. Suppose, moreover, that

$$G(s) = (sI - A)^{-1}B \qquad (6.5\text{-}1)$$

is a plant in state-variable formulation.

For this subsection, it will be necessary to consider the loop gain *referred to the control input* $u(t)$ in Figure 6.2-1. This is in contrast to the work in Section 6.2, where we referred the loop gain to the output $z(t)$, or equivalently to the signal $s(t)$ in the figure. Breaking the loop at $u(t)$ yields the loop gain

$$KG(s) = K(sI - A)^{-1}B \qquad (6.5\text{-}2)$$

Our discussion will be based on the *optimal return difference relation* that holds for the LQR with state feedback (Lewis, 1986a; Grimble and Johnson, 1988; Kwakernaak and Sivan, 1972), namely,

$$\begin{aligned} & \left[I + K(-sI - A)^{-1}B\right]^\mathrm{T} \left[I + K(sI - A)^{-1}B\right] \\ & = I + \frac{1}{\rho}B^\mathrm{T}(-sI - A)^{-\mathrm{T}}Q(sI - A)^{-1}B, \end{aligned} \qquad (6.5\text{-}3)$$

where "$-T$" means the inverse transposal. We have selected $R = I$.

Denoting the i-th singular value of a matrix M as $\sigma_i(M)$, we note that by definition

$$\sigma_i(M) = \sqrt{\lambda_i(M^*M)}, \qquad (6.5\text{-}4)$$

with $\lambda_i(M^*M)$ the ith eigenvalue of matrix M^*M and M^* the complex conjugate transpose of M. Therefore, according to (6.5-3), there results (Doyle and Stein, 1981)

$$\sigma_i[I + KG(j\omega)] = \left[\lambda_i\left[I + \frac{1}{\rho}B^T(-j\omega I - A)^{-T}Q(j\omega I - A)^{-1}B\right]\right]^{1/2}$$

$$= \left[1 + \frac{1}{\rho}\lambda_i\left[B^T(-j\omega I - A)^{-T}Q(j\omega I - A)^{-1}B\right]\right]^{1/2}$$

or

$$\sigma_i[I + KG(j\omega)] = \left[1 + \frac{1}{\rho}\sigma_i^2[H(j\omega)]\right]^{1/2}, \qquad (6.5\text{-}5)$$

with

$$H(s) = H(sI - A)^{-1}B \qquad (6.5\text{-}6)$$

and $Q = H^T H$.

We could call (6.5-5) the *optimal singular-value relation* of the LQR. It is important due to the fact that the right-hand side is known in terms of open-loop quantities *before the optimal feedback gain is found* by solution of the ARE, while the left-hand side is the closed-loop return difference. Thus, exactly as in classical control, we are able to derive properties of the closed-loop system in terms of properties of the open-loop system.

According to this relation, for all ω the minimum singular value satisfies the *LQ optimal singular-value constraint*

$$\underline{\sigma}[I + KG(j\omega)] \geq 1 \qquad (6.5\text{-}7)$$

Thus the LQ regulator always results in a *decreased sensitivity*.

Some important conclusions on the guaranteed robustness of the LQR may now be discovered using the *multivariable Nyquist criterion* (Postlethwaite et al., 1981), which we will refer to as the polar plot of the return difference $I + KG(s)$, where the origin is the critical point (Grimble and Johnson, 1988). (Usual usage is to refer the criterion to the polar plot of the loop gain $KG(s)$, where -1 is the critical point.)

A typical polar plot of $\underline{\sigma}(I + KG(j\omega))$ is shown in Figure 6.5-1, where the optimal singular-value constraint appears as the condition that *all the singular values remain outside the unit disc*. To see how the endpoints of the plots were discovered, note that since $K(sI - A)^{-1}B$ has relative degree of at least 1, its limiting value for $s = j\omega$ as $\omega \to \infty$ is zero. Thus, in this limit, $I + KG(j\omega)$ tends to I. On the other hand, as $\omega \to 0$, the limiting value of $I + KG(j\omega)$ is determined by the dc loop gain, which should be large.

The multivariable Nyquist criterion says that the closed-loop system is stable if none of the singular-value plots of $I + KG(j\omega)$ encircle the origin in the figure.

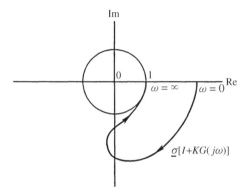

Figure 6.5-1 Typical polar plot for optimal LQ return difference (referred to the plant input).

Clearly, due to the optimal singular-value constraint, no encirclements are possible. This constitutes a proof of the *guaranteed stability* of the LQR.

Multiplying the optimal feedback K by any positive scalar gain k results in a loop gain of $kKG(s)$, which has a minimum singular value plot identical to the one in Figure 6.5-1 except that it is scaled outward. That is, the $\omega \to 0$ limit (i.e., the dc gain) will be larger, but the $\omega \to \infty$ limit will still be 1. Thus the closed-loop system will still be stable. In classical terms, the LQ regulator has *an infinite gain margin*.

The *phase margin* may be defined for multivariable systems as the angle marked "PM" in Figure 6.5-2. As in the classical case, it is the angle through which the polar plot of $\sigma[I + KG(j\omega)]$ must be rotated (about the point 1) clockwise to make the plot go through the critical point.

Figure 6.5-3 combines Figures 6.5-1 and 6.5-2. By using some simple geometry, we may find the value of the angle indicated as 60 degrees. Therefore, due to the LQ singular value constraint, the plot of $\sigma[I + KG(j\omega)]$ must be rotated through at least 60 degrees to make it pass through the origin. The LQR thus has a *guaranteed phase margin of at least 60 degrees.*

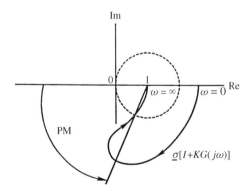

Figure 6.5-2 Definition of multivariable phase margin.

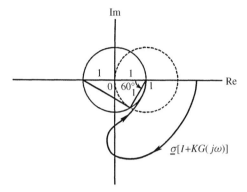

Figure 6.5-3 Guaranteed phase margin of the LQR.

This means that a phase shift of up to 60 degrees may be introduced in any of the m paths in Figure 6.2-1, or in all paths simultaneously as long as the paths are not coupled to each other in the process.

This phase margin is excessive; it is higher than that normally required in classical control system design. This overdesign means that in other performance aspects, the LQ regulator may have some deficiencies. One of these turns out to be that at the crossover frequency (loop gain = 1), the slope of the multivariable. Bode plot is −20 dB/decade, which is a relatively slow attenuation rate (Doyle and Stein, 1981). By allowing a Q weighting matrix in the PI that is not positive semidefinite, it is possible to obtain better LQ designs that have higher roll-off rates at high frequencies (Shin and Chen, 1974; Ohta et al., 1990; Al-Sunni et al., 1991).

A stability robustness bound like (6.2-49) may be obtained for the loop gain referred to the input $u(t)$. It is

$$\overline{\sigma}\left[KG(I+KG)^{-1}\right] < \frac{1}{m(\omega)} \tag{6.5-8}$$

The inverse of this is

$$m(\omega) < \frac{1}{\overline{\sigma}\left[KG(I+KG)^{-1}\right]} = \underline{\sigma}\left[I+(KG)^{-1}\right] \tag{6.5-9}$$

It can be shown (see the Problems) that (6.5-7) implies that

$$\underline{\sigma}\left[I+(KG(j\omega))^{-1}\right] \geq \tfrac{1}{2} \tag{6.5-10}$$

Therefore, *the LQR remains stable for all multiplicative uncertainties in the plant transfer function which satisfy* $m(\omega) < \tfrac{1}{2}$.

Loop-Transfer Recovery

The control design techniques we have discussed in Chapter 5 involve selecting a desirable compensator structure using classical aircraft control intuition. Then the compensator gains are adjusted using output-feedback design for suitable performance. Robustness may be guaranteed using the multivariable Bode plot as shown in Sections 6.2 and 6.3.

However, in some cases, the plant may be so complex that there is little intuition available for selecting the compensator structure. This can be the case, for instance, for a jet engine (Athans et al., 1986). In this event, the technique to be presented in this section may be useful for controller design, since it yields a suitable compensator structure automatically.

Let us examine here the plant

$$\dot{x} = Ax + Bu + Gw \tag{6.5-11}$$

$$y = Cx + v, \tag{6.5-12}$$

with process noise $w(t) \sim (0, M)$ and measurement $n(t) \sim (0, v^2 N)$ both white, $M > 0$, $N > 0$, and v a scalar parameter.

We have seen that the full-state-feedback contol

$$u = -Kx \tag{6.5-13}$$

has some extremely attractive features, including simplified design equations (Section 5.7) and some important guaranteed robustness properties. Unfortunately, *these are not shared by an output-feedback control law,* where the robustness must be checked independently. However, state feedback is usually impossible to use since all the states are seldom available for feedback in any practical application.

According to Figure 6.5-4a, where the plant transfer function is

$$\Phi(s)B = (sI - A)^{-1}B, \tag{6.5-14}$$

the loop gain, breaking the loop at the input $u(t)$, is

$$L_s(s) = K\Phi B \tag{6.5-15}$$

According to Section 6.4, if an observer or Kalman filter is used to produce a state estimate $\hat{x}(t)$, which is then used in the control law

$$u = -K\hat{x}, \tag{6.5-16}$$

the result is a regulator which, due to the separation principle, has the same transfer function as the state-feedback controller. However, it is known that the guaranteed robustness properties of the full-state-feedback controller are generally lost (Doyle, 1978).

In this section we will assume that a state-feedback gain K has already been determined using, for instance, the algebraic Riccati equation design technique in Section 5.7. This K yields suitable robustness properties of $K\Phi B$. We should like to present a technique for designing a Kalman filter that results in a regulator that

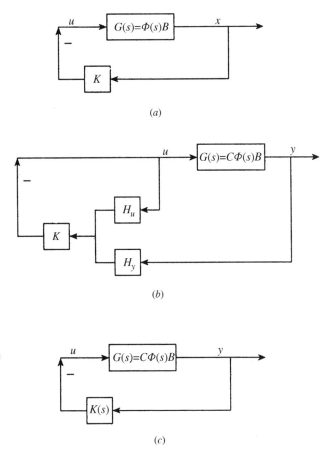

Figure 6.5-4 (a) Loop gain with full state feedback; (b) regulator using observer and estimate feedback; (c) regulator loop gain.

recovers the guaranteed robustness properties of the full-state-feedback control law as the design parameter ν goes to zero. The technique is called LQG/loop-transfer recovery (LQG/LTR), since the loop gain (i.e., loop transfer function) $K\Phi B$ of full state feedback is recovered in the regulator as $\nu \to 0$. As we will see, the key to robustness using a stochastic regulator is in the selection of the noise spectral densities M and N.

Regulator Loop Gain. Using an observer or Kalman filter, the closed-loop system appears in Figure 6.5-4b, where the regulator is given by (Section 6.4)

$$U(s) = -K\,[sI - (A - LC)]^{-1} BU(s) - K\,[sI - (A - LC)]^{-1} LY(s)$$
$$= -H_u(s)U(s) - H_y(s)Y(s) \qquad (6.5\text{-}17)$$

570 ROBUSTNESS AND MULTIVARIABLE FREQUENCY-DOMAIN TECHNIQUES

and L is the observer or Kalman gain. Denoting the observer resolvent matrix as

$$\Phi_0(s) = [sI - (A - LC)]^{-1} \tag{6.5-18}$$

we write

$$H_u = K\Phi_0 B, \quad H_y = K\Phi_0 L \tag{6.5-19}$$

To find an expression for $K(s)$ in Figure 6.5-4c using the regulator, note that $(I + H_u)U = -H_y Y$, so that

$$U = -(I + H_w)^{-1} H_y Y = -K(s)Y \tag{6.5-20}$$

However,

$$(I + H_u)^{-1} K$$
$$= \left[I + K(sI - (A - LC))^{-1} B\right]^{-1} K$$
$$= \left[I - K(sI - (A - BK - LC))^{-1} B\right] K$$
$$= K(sI - (A - BK - LC))^{-1} \left[(sI - (A - BK - LC)) - BK\right]$$
$$= K(sI - (A - BK - LC))^{-1} \Phi_0^{-1},$$

where the matrix inversion lemma was used in the second step. Therefore,

$$K(s) = (I + H_u)^{-1} H_y$$
$$= K [sI - (A - BK - LC)]^{-1} \Phi_0^{-1} \Phi_0 L$$

or

$$K(s) = K [sI - (A - BK - LC)]^{-1} L \equiv K\Phi_r L, \tag{6.5-21}$$

with $\Phi_r(s)$ the regulator resolvent matrix.

We will now show how to make the loop gain (at the input) using the regulator

$$L_r(s) = K(s)G(s) = K\Phi_r LC\Phi B \tag{6.5-22}$$

approach the loop gain $L_s(s) = K\Phi B$ using full state feedback, which is guaranteed to be robust.

Recovery of State-Feedback Loop Gain at the Input. To design the Kalman filter so that the regulator loop gain at the input $L_r(s)$ is the same as the state feedback loop gain $L_s(s)$, we will need to assume that the plant $C\Phi B$ is *minimum phase* (i.e., with stable zeros), with B and C of full rank and $\dim(u) = \dim(y)$. The references for this subsection are Doyle and Stein (1979, 19891); Athans (1986); Stein and Athans (1987); and Birdwell (1989).

Let us propose $G = I$ and the process noise spectral density matrix

$$M = v^2 M_0 + BB^T, \qquad (6.5\text{-}23)$$

with $M_0 > 0$. Then, according to Table 6.4-1,

$$L = PC^T(v^2 N)^{-1} \qquad (6.5\text{-}24)$$

and the Kalman filter ARE becomes

$$0 = AP + PA^T + (v^2 M_0 + BB^T) - PC^T(v^2 N)^{-1} CP \qquad (6.5\text{-}25)$$

According to Kwakernaak and Sivan (1972), if the aforementioned assumptions hold, then $P \to 0$ as $v \to 0$, so that

$$L(v^2 N)L^T = PC^T(v^2 N)^{-1} CP \to BB^T$$

The general solution of this equation is

$$L \to \frac{1}{v} BUN^{-1/2}, \qquad (6.5\text{-}26)$$

with U any unitary matrix.

We claim that in this situation $L_r(s) \to L_s(s)$ as $v \to 0$. Indeed, defining the full-state-feedback resolvent as

$$\Phi_c(s) = (sI - (A - BK))^{-1} \qquad (6.5\text{-}27)$$

we may write

$$\begin{aligned}
L_r(s) &= K(s)G(s) = K[sI - (A - BK \quad LC)]^{-1} LC\Phi B \\
&= K\left[\Phi_c^{-1} + LC\right]^{-1} LC\Phi B \\
&= K\left[\Phi_c - \Phi_c L(I + C\Phi_c L)^{-1} C\Phi_c\right] LC\Phi B \\
&= K\Phi_c L\left[I - (I + C\Phi_c L)^{-1} C\Phi_c L\right] C\Phi B \\
&= K\Phi_c L\left[(I + C\Phi_c L) - C\Phi_c L\right](I + C\Phi_c L)^{-1} C\Phi B \\
&= K\Phi_c L(I + C\Phi_c L)^{-1} C\Phi B \\
&\to K\Phi_c B(C\Phi_c B)^{-1} C\Phi B \\
&= K\Phi B(I + K\Phi B)^{-1} \left[C\Phi B(I + K\Phi B)^{-1}\right]^{-1} C\Phi B \\
&= \left[K\Phi B(C\Phi B)^{-1}\right] C\Phi B = K\Phi B \qquad (6.5\text{-}28)
\end{aligned}$$

The matrix inversion lemma was used in going from line 2 to line 3 and from line 7 to line 8. The limiting value (6.5-26) for L was used at the arrow.

What we have shown is that using $G = I$ and the process noise given by (6.5-23), as $v \to 0$ the regulator loop gain using a Kalman filter approaches the loop gain using full state feedback. This means that as $v \to 0$, all the robustness properties of the full-state-feedback control law are recovered in the stochastic regulator.

The *LQG/LTR design procedure* is thus as follows:

1. Use the control ARE in Table 5.7-1 to design a state feedback gain K with desirable properties. This may involve iterative design varying the PI weighting matrices Q and R.
2. Select $G = I$, process noise spectral density $M = v^2 M_0 + BB^T$, and noise spectral density $v^2 N$ for some $M_0 > 0$ and $N > 0$. Fix the design parameter v and use the Kalman filter ARE to solve for the Kalman gain L.
3. Plot the maximum and minimum singular values of the regulator loop gain $L_r(s)$ and verify that the robustness bounds are satisfied. If they are not, decrease v and return to 2.

A *reduced-order* regulator with suitable robustness properties may be designed by the LQG/LTR approach using the notions at the end of Section 6.4. That is, either a regulator may be designed for a reduced-order model of the plant, or the regulator designed for the full-order plant may then have its order reduced. In using the first approach, a high-frequency bound characterizing the unmodeled dynamics should be used to guarantee stability robustness.

An interesting aspect of the LQR/LTR approach is that the recovery process may be viewed as a *frequency-domain linear quadratic* technique that trades off the smallness of the sensitivity $S(j\omega)$ and the cosensitivity $T(j\omega)$ at various frequencies. These notions are explored in Stein and Athans (1987) and Safonov et al. (1981).

Non-Minimum-Phase Plants and Parameter Variations. The limiting value of $K(s)$ is given by the bracketed term in (6.5-28). Clearly, as $v \to 0$ the regulator *inverts the plant transfer function $C\phi B$*. If the plant is of minimum-phase, with very stable zeros, the LQG/LTR approach generally gives good results. On the other hand, if the plant is non-minimum-phase or has stable zeros with large time constants, the approach can be unsuitable.

In some applications, however, even if the plant is non-minimum-phase, the LQG/LTR technique can produce satisfactory results (Athans, 1986). In this situation, better performance may result if the design parameter v is not nearly zero. If the right-half plane zeros occur at high frequencies where the loop gain is small, the LQG/LTR approach works quite well.

An additional defect of the LQG/LTR approach appears when there are plant parameter variations. As seen in Section 6.2, stability in the presence of parameter variations requires that the loop-gain singular values be below some upper bound at low frequencies. However, this bound is not taken into account in the LQG/LTR derivation. Thus LQG/LTR can yield problems for aircraft control design, where gain scheduling is required. The H-infinity design approach (Francis et al., 1984; Doyle et al., 1989) has been used with success to overcome this problem.

Recovery of Robust Loop Gain at the Output. We have shown that by designing the state feedback first and then computing the Kalman filter gain using a specific choice of noise spectral densities, the stochastic regulator recovers the robustness of the loop gain $K(s)G(s)$ referred to the input $u(t)$ in Figure 6.5-4. However, in Section 6.2 we saw that for a small tracking error the robustness should be studied in terms of the loop gain $G(s)K(s)$ referred to the error, or equivalently to the system *output*.

Here we should like to show how to design a stochastic regulator that recovers a robust loop gain $G(s)K(s)$. This yields a second LQG/LTR design algorithm.

Thus, suppose that we *first design a Kalman filter* with gain L using Table 6.4-1. By duality theory, one may see that the Kalman filter loop gain

$$L_k(s) = C\Phi L \qquad (6.5\text{-}29)$$

enjoys exactly the same guaranteed robustness properties as the state-feedback loop gain $K\Phi B$ that were described earlier in this section.

$$L_r^o(s) = G(s)K(s) = C\Phi BK\Phi_r L \qquad (6.5\text{-}30)$$

Thus, we should like to determine how to design a state-feedback gain K so that $L_r^o(s)$ approaches $C\Phi L$. As we will see, the key to this is in the selection of the PI weighting matrices Q and R in Table 5.7-1.

To determine K, let us propose the PI

$$J = \tfrac{1}{2} \int_0^\infty (x^T Q x + \rho^2 u^T R u)\,dt \qquad (6.5\text{-}31)$$

with

$$Q = \rho^2 Q_0 + C^T C, \qquad (6.5\text{-}32)$$

with $Q_0 > 0$. By using techniques dual to those above, we may demonstrate that as $\rho \to 0$ the state feedback gain determined using Table 5.7-1 approaches

$$K \to \frac{1}{\rho} R^{-1/2} WC, \qquad (6.5\text{-}33)$$

with W a unitary matrix. Using this fact, it may be shown that

$$L_r^o(s) = G(s)K(s) \to C\Phi L \qquad (6.5\text{-}34)$$

The LQR/LTR design technique for loop gain recovery at the output is therefore exactly dual to that for recovery at the input. Specifically, the Kalman gain L is

574 ROBUSTNESS AND MULTIVARIABLE FREQUENCY-DOMAIN TECHNIQUES

first determined using Table 6.4-1 for desired robustness properties. Then Q and R are selected, with Q of the special form (6.5-32). For a small value of ρ, the state-feedback gain K is determined using Table 5.7-1. If the singular-value Bode plots of $L_r^o(s)$ do not show acceptable robustness, ρ is decreased and a new K is determined.

If the plant $C\Phi B$ is minimum-phase, all is well as ρ is decreased. However, if there are zeros in the right-half plane, there could be problems as ρ becomes too small, although with care the LQG/LTR technique often still produces good results for suitable ρ.

Example 6.5-1: LQG/LTR Design of Aircraft Lateral Control System. We will illustrate the loop-transfer recovery technique on a lateral aircraft CAS design. This example should be compared with examples in Chapter 4 and Example 5.5-4. All computations, including solving for the state feedback gains and Kalman filter gains, were carried out very easily using MATLAB (Moler et al., 1987).

a. Control Objective. The tracking control system shown in Figure 6.5-5 is meant to provide coordinated turns by causing the bank angle $\phi(t)$ to follow a desired command while maintaining the sideslip angle $\beta(t)$ at zero. It is a two-channel system with control input $u = [u_\phi \ \ u_\beta]^T$.

The reference command is $r = [r_\phi \ \ r_\beta]^T$. The control system should hold ϕ at the commanded value of r_ϕ and $\beta(t)$ at the commanded value of r_β, which is equal to zero. The tracking error is $e = [e_\phi \ \ e_\beta]^T$ with

$$\begin{aligned} e_\phi &= r_\phi - \phi \\ e_\beta &= r_\beta - \beta \end{aligned} \quad (1)$$

The negatives of the errors appear in the figure since a minus sign appears in $u = -K\hat{x}$ as is standard for LQG design.

b. State Equations of Aircraft and Basic Compensator Dynamics. To obtain the basic aircraft dynamics, the nonlinear F-16 model was linearized at the nominal flight condition in Table 3.6-3 ($V_T = 502$ ft/s, 0 ft altitude, 300 psf dynamic pressure,

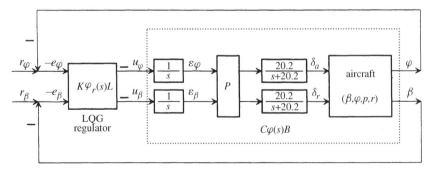

Figure 6.5-5 Aircraft turn coordinator control system.

cg at 0.35 $\bar{c}$) retaining the states sideslip β, bank angle ϕ, roll rate p, and yaw rate r. Additional states δ_a and δ_r are introduced by the aileron and rudder actuators, both of which are modeled as having approximate transfer functions of $20.2/(s + 20.2)$. The aileron deflection is δ_a and the rudder deflection is δ_r.

The singular values versus frequency of the basic aircraft with actuators are shown in Figure 6.5-6. Clearly, the steady-state error will be large in closed-loop since the loop gain has neither integrator behavior nor large singular values at dc. Moreover, the singular values are widely separated at dc, so that they are not balanced.

To correct these deficiencies we may use the techniques of Example 6.2-3. The dc gain of the system is given by

$$H(0) = \begin{bmatrix} -727.37 & -76.94 \\ -2.36 & 0.14 \end{bmatrix} \quad (2)$$

First, the dynamics are augmented by integrators in each control channel. We denote the integrator outputs by ϵ_ϕ and ϵ_β. The singular value plots including the integrators are shown in Figure 6.5-7. The dc slope is now -20 dB/decade, so that the closed-loop steady-state error will be zero. Next, the system was augmented by $P = H^{-1}(0)$ to balance the singular values at dc. The net result is shown in Figure 6.5-8, which is very suitable.

The entire state vector, including aircraft states and integrator states, is

$$x = \begin{bmatrix} \beta & \phi & p & r & \delta_a & \delta_r & \epsilon_\phi & \epsilon_\beta \end{bmatrix}^T \quad (3)$$

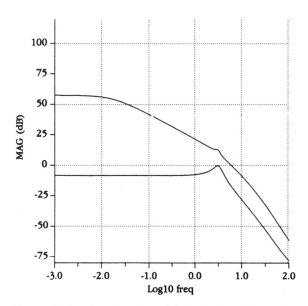

Figure 6.5-6 Singular values of the basic aircraft dynamics.

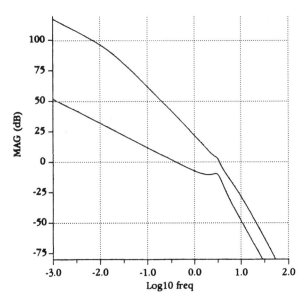

Figure 6.5-7 Singular values of aircraft augmented by integrators.

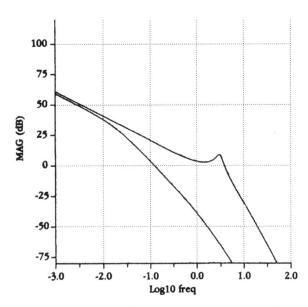

Figure 6.5-8 Singular values of aircraft augmented by integrators and inverse dc gain matrix P.

The full-state-variable model of the aircraft plus actuators and integrators is of the form

$$\dot{x} = Ax + Bu \qquad (4)$$

with

$$A = \begin{bmatrix} -0.3220 & 0.0640 & 0.0364 & -0.9917 & 0.0003 & 0.0008 & 0 & 0 \\ 0 & 0 & 1 & 0.0037 & 0 & 0 & 0 & 0 \\ -30.6492 & 0 & -3.6784 & 0.6646 & -0.7333 & 0.1315 & 0 & 0 \\ 8.5395 & 0 & -0.0254 & -0.4764 & -0.0319 & -0.0620 & 0 & 0 \\ 0 & 0 & 0 & 0 & -20.2 & 0 & -0.01 & -5.47 \\ 0 & 0 & 0 & 0 & 0 & -20.2 & -0.168 & 51.71 \\ 0 & 0 & 0 & 0 & 0 & 0 & 0 & 0 \\ 0 & 0 & 0 & 0 & 0 & 0 & 0 & 0 \end{bmatrix} \qquad (5)$$

$$B = \begin{bmatrix} 0 & 0 \\ 0 & 0 \\ 0 & 0 \\ 0 & 0 \\ 0 & 0 \\ 0 & 0 \\ 1 & 0 \\ 0 & 1 \end{bmatrix} \qquad (6)$$

The output is given by $y = [\phi \quad \beta]^T$, or

$$y = \begin{bmatrix} 0 & 57.2958 & 0 & 0 & 0 & 0 & 0 & 0 \\ 57.2958 & 0 & 0 & 0 & 0 & 0 & 0 & 0 \end{bmatrix} x = Cx, \qquad (7)$$

where the factor of 57.2958 converts radians to degrees. Then

$$e = r - y \qquad (8)$$

c. Frequency-Domain Robustness Bounds.
We now derive the bounds on the loop-gain MIMO Bode magnitude plot that guarantee robustness of the closed-loop system. Consider first the high-frequency bound. Let us assume that the aircraft model is accurate to within 10 percent up to a frequency of 2 rad/s, after which the uncertainty grows without bound at the rate of 20 dB/decade. The uncertainty could be due to actuator modeling inaccuracies, aircraft flexible modes, and so on. This behavior is modeled by

578 ROBUSTNESS AND MULTIVARIABLE FREQUENCY-DOMAIN TECHNIQUES

$$m(\omega) = \frac{s+2}{20} \tag{9}$$

We assume $m(\omega)$ to be a bound on the multiplicative uncertainty in the aircraft transfer function (Section 6.2).

For stability robustness, despite the modeling errors, we saw in Section 6.2 that the loop gain referred to the output should satisfy

$$\overline{\sigma}(GK(j\omega)) < \frac{1}{m(\omega)} = \left|\frac{20}{s+2}\right| \tag{10}$$

when $1/m(\omega) \ll 1$. The function $1/m(\omega)$ is plotted in Figure 6.5-9.

Turning to the low-frequency bound on the closed-loop loop gain, the closed-loop system should be robust to wind gust disturbances. Using techniques like those in Examples 6.3-1, the gust magnitude plot shown in Figure 6.5-10a may be obtained. According to Section 6.2, for robust performance despite wind gusts, the minimum loop-gain singular value $\underline{\sigma}(GK(j\omega))$ should be above this bound.

d. Target Feedback Loop Design. The robustness bounds just derived are expressed in terms of the singular-value plots referred to $e(t)$. To recover the loop gain $GK(j\omega)$ at $e(t)$, or equivalently at the output, the Kalman filter should be designed first, so that we should employ LQG/LTR algorithm 2. Then $C\Phi(s)L$ is the target feedback loop which should be recovered in the state-feedback design phase.

In standard applications of the LQG/LTR technique, the regulator is designed for robustness, but the time responses are not even examined until the design has been

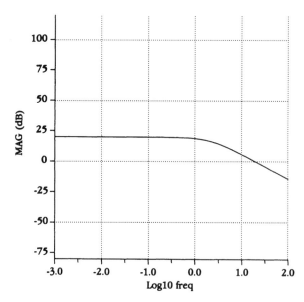

Figure 6.5-9 Multiplicative uncertainty bound $1/m(\omega)$ for the aircraft dynamical model.

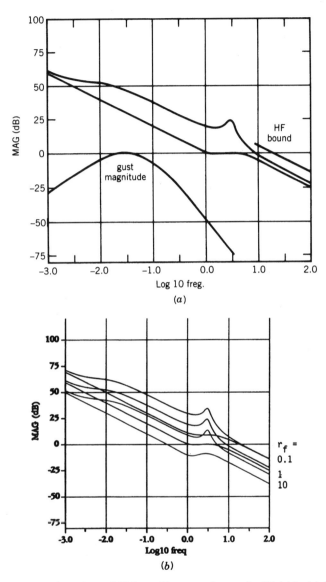

Figure 6.5-10 Singular values of Kalman filter open-loop gain $C\Phi(s)L$: (a) for $r_f = 1$, including robustness bounds; (b) for various values of r_f.

completed. It is difficult to obtain decent time responses using this approach. In this example we should like to emphasize the fact that *it is not difficult to obtain good time responses as well as robustness using LQG/LTR.* It is only necessary to select the Kalman gain L in Table 6.4-1 for good robustness properties as well as *suitable step responses* of the target feedback loop $C\Phi(s)L$, where $\Phi(s) = (sI - A)^{-1}$.

Using MATLAB, the Kalman filter design equations in Table 6.4-1 were solved using

$$Q = \text{diag}\{0.01, 0.01, 0.01, 0.01, 0, 0, 1, 1\}, \tag{11}$$

$R = r_f I$, and various values of r_f. The maximum and minimum singular values of the filter open-loop gain $C\Phi(s)L$ for $r_f = 1$ are shown in Figure. 6.5-10a, which also depicts the robustness bounds. The singular values for several values of r_f are shown in Figure 6.5-10b. Note how the singular-value magnitudes increase as r_f decreases, reflecting improved rejection of low-frequency disturbances. The figures show that the robustness bounds are satisfied for $r_f = 1$ and $r_f = 10$, but that the high-frequency bound is violated for $r_f = 0.1$.

The associated step responses of $C\Phi(s)L$ with reference commands of $r_\phi = 1$, $r_\beta = 0$ are shown in Figure 6.5-11. The response for $r_f = 10$ is unsuitable, while the response for $r_f = 0.1$ is too fast and would not be appreciated by the pilot. On the other hand, the response for $r_f = 1$ shows suitable time of response and overshoot characteristics, as well as good decoupling between the bank angle $\phi(t)$ and the sideslip $\beta(t)$.

Therefore, the target feedback loop was selected as $C\Phi(s)L$ with $r_f = 1$, since this results in a design that has suitable robustness properties and step responses. The corresponding Kalman gain is given by

$$L = \begin{bmatrix} -0.007 & 0.097 \\ 0.130 & -0.007 \\ 0.199 & -0.198 \\ -0.093 & -0.020 \\ -0.197 & -0.185 \\ 1.858 & 1.757 \\ 0.685 & -0.729 \\ 0.729 & 0.684 \end{bmatrix} \tag{12}$$

The Kalman filter poles (e.g., those of $A - LC$) are given by

$$s = -0.002, -0.879, -1.470,$$
$$-3.952 \pm j3.589,$$
$$-7.205, -20.2, -20.2 \tag{13}$$

Although there is a slow pole, the step response is good, so this pole evidently has a small residue.

It is of interest to discuss how the frequency and time responses were plotted. For the frequency response, we used the open-loop system

$$\dot{\hat{x}} = A\hat{x} + Le$$
$$\hat{y} = C\hat{x}, \tag{14}$$

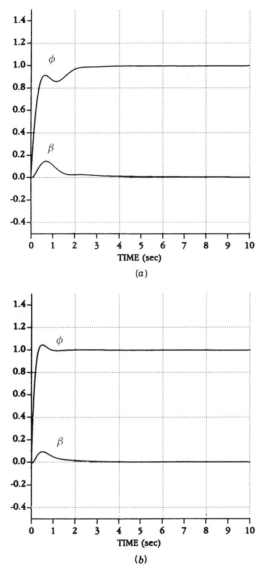

Figure 6.5-11 Step responses of target feedback loop $C\Phi(s)L$: (a) $r_f = 10$; (b) $r_f = 1$.

which has a transfer function of $C\Phi(s)L = C(sI - A)^{-1}L$. A program was written which plots the singular values versus frequency for a system given in state-space form. This yielded Figure 6.5-10.

For the step response, it is necessary to examine the closed-loop sytem. In this case, the loop is closed by using $e = r - \hat{y}$ in (14), obtaining

$$\dot{\hat{x}} = (A - LC)\hat{x} + Lr$$
$$\hat{y} = C\hat{x}$$
(15)

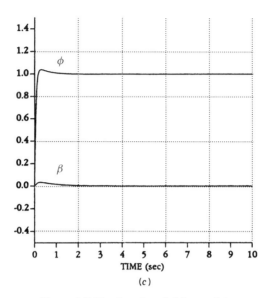

Figure 6.5-11 (*continued*) (c) $r_f = 0.1$.

Using these dynamics in program TRESP (Appendix B) with $r = [1 \quad 0]^T$ produces the step response plot.

A word on the choice for Q is in order. The design parameters Q and R should be selected so that the target feedback loop $C\Phi(s)L$ has good robustness and time-response properties. It is traditional to select $Q = BB^T$, which accounts for the last two diagonal entries of (11). However, in this example it was impossible to obtain good step responses using this selection for Q. Motivated by the fact that the process noise in the aircraft excites the first four states as well, we experimented with different values for Q, plotting in each case the singular values and step responses. After a few iterations, the final choice (11) was made.

e. Loop Transfer Recovery at the Output.

The target feedback loop $C\Phi(s)L$ using $r_f = 1$ has good properties in both the frequency and time domains. Unfortunately, the closed-loop system with LQG regulator has a loop gain referred to the output of $C\Phi(s)BK\Phi_r(s)L$, with the regulator resolvent given by

$$\Phi_r(s) = [sI - (A - LC - BK)]^{-1} \quad (16)$$

On the other hand, LQG/LTR algorithm 2 shows how to select a state-feedback gain K so that the LQG regulator loop gain approaches the ideal loop gain $C\Phi(s)L$. Let us now select such a feedback gain matrix.

Using MATLAB, the LQR design problem in Table 5.7-1 was solved with $Q = C^T C$, $R = \rho^2 I$, and various values of $r_c \equiv \rho^2$ to obtain different feedback gains K. Some representative singular values of the LQG loop gain $C\Phi(s)BK\Phi_r(s)L$ are plotted in Figure 6.5-12, where L is the target-loop Kalman gain (12). Note how

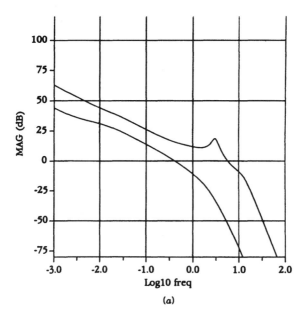

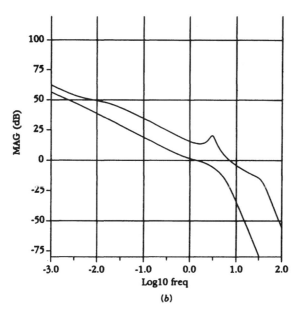

Figure 6.5-12 Singular value plots for the LQG regulator: (a) LQG with $r_c = 10^{-3}$; (b) LQG with $r_c = 10^{-7}$.

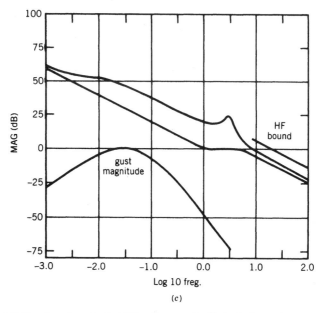

Figure 6.5-12 (*continued*) (*c*) LQG with $r_c = 10^{-11}$, including robustness bounds.

the actual singular values approach the target singular values in Figure 6.5-10a as r_c decreases. A good match is obtained for $r_c = 10^{-11}$.

Figure 6.5-12c also depicts the robustness bounds, which are satisfied for this choice of $r_c = 10^{-11}$. The corresponding step responses are given in Figure 6.5-13. A suitable step response that matches well the target response of Figure 6.5-11b results when $r_c = 10^{-11}$.

It is of interest to discuss how these plots were obtained. For the LQG singular value plots, the complete dynamics are given by

$$\dot{x} = Ax + Bu$$
$$\dot{\hat{x}} = (A - LC)\hat{x} + Bu + Lw \qquad (17)$$
$$u = -K\hat{x},$$

where $w(t) = -e(t)$. These may be combined into the augmented system

$$\begin{bmatrix} \dot{x} \\ \dot{\hat{x}} \end{bmatrix} = \begin{bmatrix} A & -BK \\ 0 & A - LC - BK \end{bmatrix} \begin{bmatrix} x \\ \hat{x} \end{bmatrix} + \begin{bmatrix} 0 \\ L \end{bmatrix} w \qquad (18)$$

$$y = [C \quad 0] \begin{bmatrix} x \\ \hat{x} \end{bmatrix} \qquad (19)$$

LQG/LOOP-TRANSFER RECOVERY 585

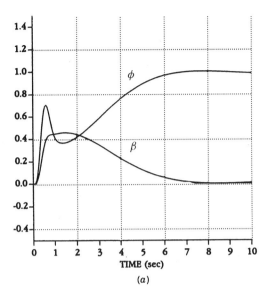

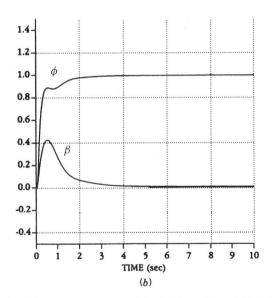

Figure 6.5-13 Closed-loop step responses of the LQG regulator: (*a*) LQG with $r_c = 10^{-3}$; (*b*) LQG with $r_c = 10^{-7}$.

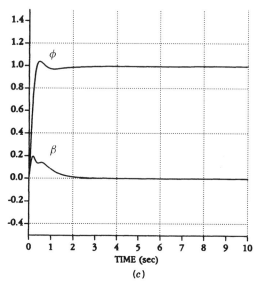

Figure 6.5-13 (*continued*) (*c*) LQG with $r_c = 1-^{-11}$.

which has transfer function $C\Phi(s)BK\Phi_r(s)L$. The singular values are now easily plotted.

For the step responses, the closed-loop system must be studied. To close the loop, set $w = y - r$ in (18) to obtain the closed-loop dynamics

$$\begin{bmatrix} \dot{x} \\ \dot{\hat{x}} \end{bmatrix} = \begin{bmatrix} A & -BK \\ LC & A - LC - BK \end{bmatrix} \begin{bmatrix} x \\ \hat{x} \end{bmatrix} + \begin{bmatrix} 0 \\ -L \end{bmatrix} r \qquad (20)$$

$$y = [C \quad 0] \begin{bmatrix} x \\ \hat{x} \end{bmatrix} \qquad (21)$$

These are used with program TRESP in Appendix B to obtain Figure 6.5-13.

The final LQG regulator is given by the Kalman gain L in (12) and the feedback gain K corresponding to $r_c = 10^{-11}$.

f. Reduced-Order Regulator. The LQG regulator just designed has order $n = 8$, the same as the plant. This is excessive for an aircraft lateral control system. A reduced-order regulator that produces very good results may easily be determined using the partial-fraction-expansion approach in Example 6.2-4, principal component analysis (Moore, 1982), or other techniques. This is easily accomplished using MAT-LAB. The singular value plots and step response using the reduced-order regulator should be examined to verify robustness and suitable performance. ∎

6.6 SUMMARY

In Section 6.2 we extended some classical frequency-domain analysis techniques to multivariable systems using the notion of the singular value. We defined the multivariable loop gain, return difference, and sensitivity, and showed that the multivariable Bode magnitude plot is just the plot of the maximum and minimum singular values of the loop gain versus frequency. To guarantee stability robustness to unmodeled high-frequency dynamics and plant parameter variations, as well as performance robustness in the presence of disturbances, we derived various frequency-domain bounds that the singular values of the loop gain must satisfy.

A convenient robust output-feedback design approach was presented in Section 6.3 that results in acceptable time-domain performance with guaranteed robustness.

In Section 6.4 we covered the design of multivariable observers for estimating the full state of the aircraft model from the measured outputs. We discussed the Kalman filter, showing an example of its use by reconstructing the angle of attack from normal acceleration and pitch-rate measurements in the presence of wind-gust noise. We showed how to use full state feedback and an observer to design a dynamic regulator.

Finally, in Section 6.5 we covered the popular LQG/LTR robust design technique, illustrating with the design of a multivariable lateral control system.

REFERENCES

Al-Sunni, F. M., B. L. Stevens, and F. L. Lewis. "Negative State Weighting in the Linear Quadratic Regulator for Aircraft Control," *Journal of Guidance, Control, and Dynamics*, 15, no. 5 (September–October, 1992) 1279–1281.

Anderson, B.D.O., and Y. Liu. "Controller Reduction: Concepts and Approaches," *IEEE Transactions on Automatic Control*, vol. AC-34, no. 8, August 1989, pp. 802–812.

Armstrong, E. S. *ORACLS: A Design System for Linear Multivariable Control*. New York: Marcel Dekker, 1980.

Athans, M. "A Tutorial on the LQG/LTR Method." *Proceedings of the American Control Conference*, June 1986, pp. 1289–1296.

Athans, M., P. Kapsouris, E. Kappos, and H. A. Spang III. "Linear-quadratic Gaussian with Loop-Transfer Recovery Methodology for the F-100 Engine." *Journal of Guidance, Control, and Dynamics* 9, no. 1 (January–February 1986): 45–52.

Birdwell, J. D. "Evolution of a Design Methodology for LQG/LTR." *IEEE Control Systems Magazine* (April 1989): 73–77.

Blakelock, J. H. *Automatic Control of Aircraft and Missiles*. New York: Wiley, 1965.

Bryson, A. E., Jr., and Y.-C. Ho. *Applied Optimal Control*. New York: Hemisphere, 1975.

Doyle, J. C. "Guaranteed Margins for LQG Regulators." *IEEE Transactions on Automatic Control* (August 1978): 756–757.

Doyle, J. C., and G. Stein. "Robustness with Observers." *IEEE Transactions on Automatic Control*, vol. AC-24, no. 4, August 1979, pp. 607–611.

———. "Multivariable Feedback Design: Concepts for a Classical/Modern Synthesis." *IEEE Transactions on Automatic Control,* vol. AC-26, no. 1, February 1981, pp. 4–16.

Doyle, J. C., K. Glover, P. P. Khargonekar, and B. Francis. "State-space Solutions to Standard H_2 and H_∞ Control Problems." *IEEE Transactions on Automatic Control,* vol. AC-34, no. 8, August 1989, pp. 831–847.

Francis, B., J. W. Helton, and G. Zames. "H_∞ Optimal Feedback Controllers for Linear Multivariable Systems." *IEEE Transactions on Automatic Control,* vol. AC-29, no. 10, October 1984, pp. 888-900.

Franklin, G. F., J. D. Powell, and A. Emami-Naeini. *Feedback Control of Dynamic Systems.* Reading, Mass.: Addison-Wesley, 1986.

Gelb, A., ed. *Applied Optimal Estimation.* Cambridge, Mass.: MIT Press, 1974.

Grimble, M. J., and M. A. Johnson. *Optimal Control and Stochastic Estimation: Theory and Applications.* Vol. 1. New York: Wiley, 1988.

IMSL. *Library Contents Document.* 8th ed. International Mathematical and Statistical Libraries, Inc., 7500 Bellaire Blvd., Houston, TX 77036, 1980.

Kailath, T. *Linear Systems.* Englewood Cliffs, N.J.: Prentice-Hall, 1980.

Kaminer, I., P. P. Khargonekar, and G. Robel. "Design of Localizer Capture and Track Modes for a Lateral Autopilot using H-infinity Synthesis." *IEEE Control Systems Magazine* 10, no. 4 (June 1990): 13–21.

Kwakernaak, H., and R. Sivan. *Linear Optimal Control Systems.* New York: Wiley, 1972.

Laub, A. J. "An Inequality and Some Computations Related to the Robust Stability of Linear Dynamic Systems." *IEEE Transactions on Automatic Control,* vol. AC-24, no. 2, April 1979, pp. 318–320.

———. "Efficient Multivariable Frequency Response Computations." *IEEE Transactions on Automatic Control,* vol. AC-26, no. 2, April 1981, pp. 407–408.

Lewis, F. L. *Optimal Control.* New York: Wiley, 1986a.

———. *Optimal Estimation.* New York: Wiley, 1986b.

Ly, U.-L., A. E. Bryson, and R. H. Cannon. "Design of Low-order Compensators using Parameter Optimization." *Automatica* 21, no. 3 (1985): 315–318.

MacFarlane, A.G.J. "Return-difference and Return-ratio Matrices and Their Use in the Analysis and Design of Multivariable Feedback Control Systems." *Proceedings of the Institute of Electrical Engineering* 117, no. 10 (October 1970): 2037–2049.

MacFarlane, A.G.J., and B. Kouvaritakis. "A Design Technique for Linear Multivariable Feedback Systems." *International Journal of Control* 25 (1977): 837–874.

MATRIX$_x$, Integrated Systems, Inc., 2500 Mission College Blvd., Santa Clara, CA 95054, 1989.

Mil. Spec. 1797. "Flying Qualities of Piloted Vehicles." 1987.

Minto, K. D., J. H. Chow, and J. W. Beseler. "An Explicit Model-matching Approach to Lateral-axis Autopilot Design." *IEEE Control Systems Magazine* 10, no. 4 (June 1990): 22–28.

Moerder, D. D., and A. J. Calise. "Convergence of a Numerical Algorithm for Calculating Optimal Output Feedback Gains." *IEEE Transactions on Automatic Control,* vol. AC-30, no. 9, September 1985, pp. 900–903.

Moler, C., J. Little, and S. Bangert. *PC-Matlab,* The Mathworks, Inc., 20 North Main St., Suite 250, Sherborn, MA 01770, 1987.

Moore, B. C. "Principal Component Analysis in Linear Systems: Controllability, Observability, and Model Reduction." *IEEE Transactions on Automatic Control,* vol. AC-26, no. 1, 1982, pp. 17–32.

Morari, M., and E. Zafiriou. *Robust Process Control.* Englewood Cliffs, N.J.: Prentice Hall, 1989.

Ohta, H., M. Nakinuma, and P. N. Nikiforuk. "Use of Negative Weights in Linear Quadratic Regulator Synthesis." Preprint, 1990.

Papoulis, A. *Probability, Random Variables, and Stochastic Processes.* 2nd ed. New York: McGraw-Hill, 1984.

Postlethwaite, I., J. M. Edmunds, and A.G.J. MacFarlane. "Principal Gains and Principal Phases in the Analysis of Linear Multivariable Systems." *IEEE Transactions on Automatic Control,* vol. AC-26, no. 1, February 1981, pp. 32–46.

Press, W. H., B. P. Flannery, S. A. Teukolsky, and W. T. Vetterling. *Numerical Recipes: The Art of Scientific Computing.* New York: Cambridge University Press, 1986.

Rosenbrock, H. H. *Computer-Aided Control System Design.* New York: Academic Press, 1974.

Safonov, M. G., and M. Athans. "Gain and Phase Margin for Multiloop LQG Regulators." *IEEE Transactions on Automatic Control,* vol. AC-22, no. 2, April 1977, pp. 173–178.

Safonov, M. G., A. J. Laub, and G. L. Hartmann. "Feedback Properties of Multivariable Systems: The Role and Use of the Return Difference Matrix." *IEEE Transactions on Automatic Control,* vol. AC-26, no. 1, February 1981, pp. 47–65.

Shin, V., and C. Chen. "On the Weighting Factors of the Quadratic Criterion in Optimal Control." *International Journal of Control* 19 (May 1974): 947–955.

Stein, G., and M. Athans. "The LQG/LTR Procedure for Multivariable Feedback Control Design." *IEEE Transactions on Automatic Control,* vol. AC-32, no. 2, February 1987, pp. 105–114.

Stevens, B. L., P. Vesty, B. S. Heck, and F. L. Lewis. "Loop Shaping with Output Feedback." *Proceedings of the American Control Conference* (June 1987): 146–149.

Strang, G. *Linear Algebra and Its Applications.* 2d ed. New York: Academic Press, 1980.

Yousuff, A., and R. E. Skelton. "A Note on Balanced Controller Reduction." *IEEE Transactions on Automatic Control,* vol. AC-29, no. 3, March 1984, pp. 254–257.

PROBLEMS

Section 6.2

6.2-1 Derive in detail the multivariable expressions (6.2-16) and (6.2-17) for the performance output and the tracking error.

6.2-2 Prove (6.2-54). You will need to neglect any terms that contain second-order terms in the parameter variation matrices and use the fact that for small X, $(I - X)^{-1} \approx (I + X)$.

6.2-3 Multivariable Closed-Loop Transfer Relations. In Figure 6.2-1, let the plant $G(s)$ be described by

$$\dot{x} = \begin{bmatrix} 0 & 1 & 0 \\ 0 & -3 & 0 \\ 0 & 0 & 0 \end{bmatrix} x + \begin{bmatrix} 0 & 0 \\ 1 & 0 \\ 0 & 1 \end{bmatrix} u, \quad z = \begin{bmatrix} 1 & 0 & 0 \\ 0 & 0 & 1 \end{bmatrix} x$$

and the compensator is $K(s) = 2I_2$.

(a) Find the multivariable loop gain and return difference.

(b) Find the sensitivity and cosensitivity.

(c) Find the closed-loop transfer function from $r(t)$ to $z(t)$, and hence the closed-loop poles.

6.2-4 For the continuous-time system in Example 6.2-1, plot the individual SISO Bode magnitude plots from input 1 to outputs 1 and 2, and from input 2 to outputs 1 and 2. Compare them to the MIMO Bode plot to see that there is no obvious relation. Thus, the robustness bounds cannot be given in terms of the individual SISO Bode plots.

6.2-5 Software for MIMO Bode Magnitude Plot. Write a computer program to plot the Bode magnitude plot for a multivariable system given in state-space form $\dot{x} = Ax + Bu$, $y = Cx + Du$. Your program should read in A, B, C, D. You may use a SVD routine (e.g., IMSL, 1980 and Press et al., 1986) or the technique in Laub (1981). Use the software to verify Examples 6.2-1 and 6.2-2.

6.2-6 Multivariable Bode Plot. For the system in Problem 6.2-3, plot the multivariable Bode magnitude plots for:

(a) the loop gain GK

(b) the sensitivity S and cosensitivity T. For which frequency ranges do the plots for $GK(j\omega)$ match those for $S(j\omega)$? For $T(j\omega)$?

6.2-7 Bode Plots for F-16 Lateral Regulator. Plot the loop gain multivariable Bode magnitude plot for the F-16 lateral regulator designed in Example 5.3-1.

6.2-8 Balancing and Zero Steady-State Error. Find a precompensator for balancing the SVs at low frequency and ensuring zero steady-state error for the system

$$\dot{x} = \begin{bmatrix} 0 & 1 & 0 \\ -2 & -3 & 0 \\ 0 & 0 & -3 \end{bmatrix} x + \begin{bmatrix} 0 & 0 \\ 1 & 0 \\ 0 & 1 \end{bmatrix} u, \quad z = \begin{bmatrix} 1 & 0 & 0 \\ 0 & 0 & 1 \end{bmatrix} x$$

Plot the SVs of the original and precompensated system.

Section 6.3

6.3-1 Model Reduction and Neglected High-Frequency Modes. An unstable system influenced by high-frequency parasitics is given by

$$\dot{x} = \begin{bmatrix} 0 & 1 & 0 \\ 1 & 0 & 1 \\ 0 & 0 & -10 \end{bmatrix} x + \begin{bmatrix} 0 & 0 \\ 0 & 1 \\ 1 & 0 \end{bmatrix} u, \qquad z = [1 \ 0 \ 0] x$$

(a) Use the technique of Example 6.2-4 to find a reduced-order model that neglects the high-frequency mode at $s = 10$ rad/s. Find the bound $m(j\omega)$ on the magnitude of the neglected portion.

(b) Using techniques like those in Sections 5.4 and 5.5, design a servo control system for the reduced-order model. Try a lead compensator whose gains are varied by the LQ algorithm, as used in Example 5.5-5. Verify the step response of the closed-loop system by performing a simulation on the reduced-order system.

(c) Find the loop gain of the closed-loop system and plot its singular values. Do they fall below the bound $1/m(j\omega)$, thus guaranteeing robustness to the neglected mode? If not, return to part (b) and find other gains that do guarantee stability robustness.

(d) Simulate your controller on the full system, including the high-frequency mode. How does the step response look?

(e) A better controller results if high-frequency dynamics are not neglected in the design stage. Design a servo control system for the full third-order system. It may be necessary to use a more complicated controller. Verify the step response of the closed-loop system by performing a simulation. Compare to the results of part (d).

6.3-2 **Gain Scheduling Robustness.** In the problems for Section 5.4 a gain-scheduled normal acceleration CAS was designed for a transport aircraft using three equilibrium points. Using the results at the end of Section 6.2, we want to check the design for robustness to plant parameter variations. Call the systems at the three equilibrium points (A_i, B_i, C_i), $i = 1, 2, 3$.

(a) In Problem 6.2-5 you wrote a program to plot the MIMO Bode magnitude plots for a state-variable system. Note that a state-space realization of $\Delta G(s)$ in (6.2-54) is given by

$$\dot{x} = \begin{bmatrix} A & -\Delta A \\ 0 & A \end{bmatrix} x + \begin{bmatrix} \Delta B \\ b \end{bmatrix} u, \qquad y = [C \ \Delta C] x$$

That is, this system has transfer function of ΔG. Define $\Delta G_{ij}(s)$ as being computed using $\Delta A = A_i - A_j$, $\Delta B = B_i - B_j$, $\Delta C = C_i - C_j$. Use these facts combined with (6.2-48) to obtain low-frequency bounds for robustness to the gain-scheduling plant parameter variations.

(b) Find the loop-gain singular values of your design for the gain-scheduled CAS. Do they fall below the robustness bounds? If not, select new PI weights and try to improve the design. If this fails, you will need to select more closely spaced equilibrium points for the gain-scheduled design.

Section 6.4

6.4-1 Nonzero-Mean Noise. Use (6.4-49) to write down the best estimate for $x(t)$ in terms of the filter state $\hat{x}(t)$ if the process noise $w(t)$ and measurement noise $v(t)$ have nonzero means of $\overline{w}$ and $\overline{v}$, respectively.

6.4-2 Observer for Angle of Attack. In Example 5.5-3 a low-pass filter of $10/(s+10)$ was used to smooth out the angle-of-attack measurements to design a pitch rate CAS. An alternative is to use an observer to reconstruct α. This completely avoids measurements of the angle of attack.

(a) Considering only the 2×2 short period approximation, design an observer that uses measurements of $q(t)$ to provide estimates of $\alpha(t)$. The observer should have $\zeta = 1/\sqrt{2}$ and $\omega_n = 10$ rad/s. Use Ackermann's formula to find the output-injection matrix L.

(b) Delete the α filter in Example 5.5-3, replacing it by the dynamics of the second-order observer just designed. With the new augmented dynamics, perform the LQ design of Example 5.5-3. Compare the performance of this pitch-rate CAS to the one using the α filter.

6.4-3 Dynamic LQ Regulator for Pitch-Rate CAS. In Example 5.5-3 and Problem 6.4-2, output-feedback design was used to build a pitch-rate CAS. In this problem we would like to use LQG theory to perform the design.

(a) Design an observer for α using q measurements, as described in the previous problem.

(b) Neglect the elevator actuator, considering only the 2×2 short period approximation in Example 5.5-3 plus the feedforward-path integrator. Find the state-feedback gain K to place the poles at $\zeta = 1/\sqrt{2}$, $\omega_n = 3.5$ rad/s; this yields good flying qualities for the short-period mode. Use Ackermann's formula, or the design software for Table 5.3-1 with $C = I$.

(c) Using the 2×2 observer and the state feedback K, construct a dynamic pitch-rate CAS. Verify its performance by plotting the step response.

6.4-4 Kalman Filter. Software for solving the Kalman filter ARE is available in Armstrong (1980) and IMSL (1980); also MATRIX$_x$ (1989) and MATLAB (Moler et al., 1987). Alternatively, the Kalman filter gain L can be found using the software for Table 5.3-1 on the dual plant (A^T, C^T, B^T) with $B = I$. Repeat Example 6.4-2 if the wind gusts have a turbulence intensity of 20 ft/s.

Section 6.5

6.5-1 Show that (6.5-7) implies (6.5-10) (see Laub, 1979).

6.5-2 LQG/LTR Design. Note that the state-feedback gain K can be found using the software for Table 5.3-1 with $C = I$. Likewise, the Kalman filter gain L can be found using the software for Table 5.3-1 on the dual plant (A^T, C^T, B^T), with $B = I$.

(a) In Problem 6.4-3(b), plot the loop-gain singular values assuming full state feedback.

(b) Now angle-of-attack measurements are not allowed. Design a Kalman filter for various values of the design parameter v. In each case, plot the closed-loop step response as well as the loop-gain singular values. Compare the step response and the SVs to the case for full state feedback as v becomes small.

CHAPTER 7

DIGITAL CONTROL

7.1 INTRODUCTION

In Chapters 4 through 6 we have shown how to design continuous-time controllers for aircraft. However, with microprocessors so fast, light, and economical, control laws are usually implemented on modern aircraft in digital form. In view of the requirement for gain scheduling of aircraft controllers, digital control schemes are especially useful, for gain scheduling is very easy on a digital computer.

To provide reliability in the event of failures, modern aircraft control schemes are redundant, with two or three control laws for each application. The actual control to be applied is selected by "voting"; that is, there should be good agreement between two out of three controllers. Such schemes are more conveniently implemented on a microprocessor, where the comparison and voting logic resides.

In this chapter we address the design of digital, or discrete-time, controllers, since the design of such controllers involves some extra considerations of which one should be aware. In Section 7.2 we discuss the *simulation* of digital controllers on a digital computer. Then in Sections 7.3 and 7.4 two approaches to digital control *design* are examined. Finally, some aspects of the actual *implementation* are mentioned in Section 7.5.

In the first approach to digital controls design, covered in Section 7.3, we show how to convert an already designed continuous-time controller to a discrete-time controller using, for instance, the bilinear transform (BLT). An advantage of this *continuous controller redesign* approach is that the sample period T does not have to be selected until after the continuous controller has been designed.

Unfortunately, controller discretization schemes based on transformations such as the BLT are approximations. Consequently, the sampling period T must be small to ensure that the digital controller performs like the continuous version from which it was designed. Therefore, in Section 7.4 we show how the design of the continuous-time controller may be *modified* to take into account some properties of the sampling process, as well as computation delays. Discretization of such a modified continuous controller yields a digital control system with improved performance.

In Section 7.5 we discuss some implementation considerations, such as actuator saturation and controller structure.

There are many excellent references on digital control; some of them are listed at the end of the chapter. We will draw most heavily on Franklin and Powell (1980), Åström and Wittenmark (1984), and Lewis (1992).

7.2 SIMULATION OF DIGITAL CONTROLLERS

A digital control scheme is shown in Figure 7.2-1. The plant $G(s)$ is a continuous-time system, and $K(z)$ is the dynamic digital controller, where s and z are, respectively, the Laplace and Z-transform variables (i.e., $1/s$ represents integration and z^{-1} represents a unit time delay). The digital controller $K(z)$ is implemented using software code in a microprocessor.

The hold device in the figure is a digital-to-analog (D/A) converter that converts the discrete control samples u_k computed by the software controller $K(z)$ into the continuous-time control $u(t)$ required by the plant. It is a *data reconstruction* device. The input u_k and output $u(t)$ for a *zero-order hold (ZOH)* are shown in Figure 7.2-2. Note that $u(kT) = u_k$, so that $u(t)$ is continuous from the right. That is, $u(t)$ is updated at times kT. The sampler with sample period T is an analog-to-digital (A/D) converter that takes the samples $y_k = y(kT)$ of the output $y(t)$ that are required by the software controller $K(z)$.

In this chapter we discuss the design of the digital controller $K(z)$. Once the controller has been designed, it is important to *simulate* it before it is implemented to determine if the closed-loop response is suitable. The simulation should provide the response at all times, including times between the samples.

To simulate a digital controller we may use the scheme shown in Figure 7.2-3. There the continuous dynamics $G(s)$ are contained in the subroutine $F(t, x, \dot{x})$; they are integrated using a Runge–Kutta integrator. Note that two time intervals are involved: the sampling period T and the *Runge–Kutta integration period* $T_R \ll T$. T_R should be selected as an integral divisor of T.

Several numerical integration schemes were discussed in Section 3.5. We have found that the Runge–Kutta routines are very suitable, while Adams–Bashforth routines do not give enough accuracy for digital control purposes. This is especially true

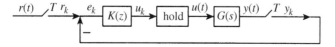

Figure 7.2-1 Digital controller.

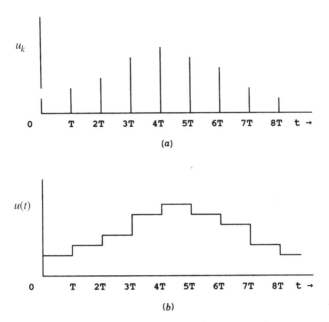

Figure 7.2-2 Data reconstruction using a ZOH: (*a*) discrete control sequence u_k; (*b*) reconstructed continuous signal $u(t)$.

when advanced adaptive and parameter estimation techniques are used. For most purposes, the fixed step size Runge–Kutta algorithm in Appendix B is suitable if T_R is selected small enough. In rare instances it may be necessary to use an adaptive step size integrator such as Runge–Kutta–Fehlburg. In all the examples in this book, the fixed step size version was used.

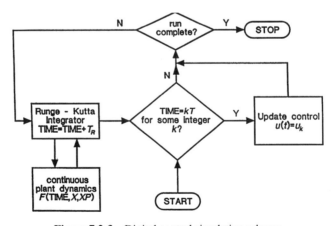

Figure 7.2-3 Digital control simulation scheme.

A driver program that realizes Figure 7.2-3 is given in Figure 7.2-4. It is written in a modular fashion to apply to a wide variety of situations, and calls a Runge–Kutta integration routine such as the one in Appendix B. The Runge–Kutta integrator in turn calls subroutine F(TIME, X, XP) containing the continuous-time dynamics.

The digital controller is contained in subroutine DIG(T, X). Figure 7.2-3 assumes a ZOH; thus, the control input $u(t)$ is updated to u_k at each time kT and then held constant until time $(k+1)T$. The driver program in Figure 7.2-4 performs this.

It is important to realize that this simulation technique provides $x(t)$ as a continuous function of time, even at values *between* the sampling instants (in fact, it provides $x(t)$ at multiples of T_R). This is essential in verifying acceptable *intersample behavior* of the closed-loop system prior to implementing the digital controller on the actual

```
C    DRIVER PROGRAM TO COMPUTE AND SIMULATE DIGITAL CONTROL SCHEME
C    REQUIRES SUBROUTINES:
C      DIG(T,X) FOR DIGITAL CONTROL UPDATE AT SAMPLING INSTANTS
C      RUNKUT(TIME,TR,X,NSTATES) TO INTEGRATE CONTINUOUS DYNAMICS
C      F(TIME,X,XP) TO PROVIDE CONTINUOUS PLANT DYNAMICS
       PROGRAM DIGICON
       REAL X(1)
       COMMON/CONTROL/U(1)
       COMMON/OUTPUT/Y(1)

C    SET RUN TIME, SAMPLING PERIOD, RUNGE KUTTA STEP SIZE
       DATA TRUN,T,TR/5.,0.5,0.01/
C    SET INITIAL PLANT STATE
       DATA X(1)/0./

       TIME= 0.
       N= NINT(TRUN/T)
       NT= NINT(T/TR)

*    DIGITAL CONTROL SIMULATION RUN

       DO 10 K= 0,N-1

C    UPDATE DIGITAL CONTROL INPUT

       CALL DIG(T,X)

C    INTEGRATE CONTINUOUS DYNAMICS BETWEEN SAMPLES

       DO 10 I= 1,NT

C    WRITE TO FILE FOR PLOT

       WRITE(7,*) TIME,X(1),U

10     CALL RUNKUT(TIME,TR,X,1)
       WRITE(7,*) TIME,X(1),U

       STOP
       END
```

Figure 7.2-4 Digital control simulation driver program.

plant. Even though the closed-loop behavior is acceptable at the sample points, with improper digital control system design there can be serious problems between the samples. The basic problem is that a badly designed controller can destroy observability, so that poor intersample behavior is not apparent at the sample points (Lewis, 1992). This simulation scheme allows the intersample behavior to be checked prior to actual implementation.

We will soon present several examples that demonstrate the simulation of digital controllers. First, it is necessary to discuss the design of digital controllers.

7.3 DISCRETIZATION OF CONTINUOUS CONTROLLERS

A digital control design approach that could directly use all of the continuous-time techniques of the previous chapters would be extremely appealing. Therefore, in this section we discuss the design of digital controllers by the redesign of existing continuous controllers. In this approach, the continuous controller is first designed using any desired technique. Then the controller is discretized using, for instance, the bilinear transform, to obtain the digital control law, which is finally programmed on the microprocessor.

An alternative approach to digital control design is given in Lewis (1992). In that approach, it is not necessary to design a continuous-time controller first, but a discrete-time controller is designed *directly* using a sampled version of the aircraft dynamics.

We now show how to discretize a continuous controller to obtain a digital controller. The idea is illustrated by designing a digital PID controller in Example 7.3-1 and a digital pitch rate control system in Example 7.3-2.

Suppose that a continuous-time controller $K^c(s)$ has been designed for the plant $G(s)$ by some means, such as root locus or LQ design. We will discuss two approximate schemes for converting $K^c(s)$ into a discrete-time controller $K(z)$ that can be implemented on a microprocessor. First we discuss the bilinear transformation (BLT) and then the matched pole-zero (MPZ) technique.

The sample period is T seconds, so that the *sampling frequency* is

$$f_s = \frac{1}{T}, \quad \omega_s = \frac{2\pi}{T} \tag{7.3-1}$$

Bilinear Transformation

A popular way to convert a continuous transfer function to a discrete one is the *bilinear transformation (BLT)*, or *Tustin's approximation*. On sampling (Franklin and Powell, 1980) the continuous poles are mapped to discrete poles according to $z = e^{sT}$. As may be seen by series expansion

$$z = e^{sT} \approx \frac{1 + sT/2}{1 - sT/2} \tag{7.3-2}$$

Therefore, to obtain an approximate sampling technique for continuous transfer functions, we may propose inverting this transformation and defining

$$s' = \frac{2}{T}\frac{z-1}{z+1} \tag{7.3-3}$$

An approximate discrete equivalent of the continuous transfer function is then given by

$$K(z) = K^c(s') \tag{7.3-4}$$

We call (7.3-3) the *bilinear transformation,* or BLT.

The BLT corresponds to approximating integration using the trapezoid rule, since if

$$\frac{Y(z)}{U(z)} = \frac{2}{T}\frac{z-1}{z+1} = \frac{2}{T}\frac{1-z^{-1}}{1+z^{-1}}$$

then (recall that z^{-1} is the unit delay in the time domain so that $z^{-1}u_k = u_{k-1}$)

$$u_k = u_{k-1} + \frac{T}{2}(y_k + y_{k-1}) \tag{7.3-5}$$

If the continuous transfer function is

$$K^c(s) = \frac{\prod_{i=1}^{m}(s+t_i)}{\prod_{i=1}^{n}(s+s_i)}, \tag{7.3-6}$$

with the relative degree $r = n - m > 0$, then the BLT yields the approximate discrete equivalent transfer function given by

$$K(z) = \frac{\prod_{i=1}^{m}\left[\frac{2(z-1)}{T(z+1)} + t_i\right]}{\prod_{i=1}^{m}\left[\frac{2(z-1)}{T(z+1)} + s_i\right]}$$

$$K(z) = \left[\frac{T}{2}(z+1)\right]^r \frac{\prod_{i=1}^{m}[(z-1)+(z+1)t_iT/2]}{\prod_{i=1}^{n}[(z-1)+(z+1)s_iT/2]}$$

$$K(z) = \left[\frac{T}{2}(z+1)\right]^r \frac{\prod_{i=1}^{m}[(1+t_iT/2)z - (1-t_iT/2)]}{\prod_{i=1}^{n}[(1+s_iT/2)z - (1-s_iT/2)]} \tag{7.3-7}$$

It can be seen that the poles and finite zeros map to the z-plane according to

$$z = \frac{1+sT/2}{1-sT/2}; \tag{7.3-8}$$

however, the r zeros at infinity in the s-plane map into zeros at $z = -1$. This is sensible since $z = -1$ corresponds to the *Nyquist frequency* ω_N, where $z = e^{j\omega NT} = -1$, so that $\omega_N T = \pi$ or

$$\omega_N = \frac{\pi}{T} = \frac{\omega_s}{2} \tag{7.3-9}$$

This is the highest frequency before folding of $|K(e^{j\omega T})|$ occurs (see Figure 7.4-1). Since the BLT maps the left-half of the s-plane into the unit circle, it maps stable continuous systems $K^c(s)$ into stable discrete $K(z)$.

According to (7.3-7), the BLT gives discretized transfer functions that have a relative degree of zero; that is, the degrees of the numerator and denominator are the same. If

$$K(z) = \frac{b_0 z^n + b_1 z^{n-1} + \cdots + b_n}{z^n + a_1 z^{n-1} + \cdots + a_n} \tag{7.3-10}$$

and $Y(z) = K(z)U(z)$, then the difference equation relation y_k and u_k is

$$y_k = -a_1 y_{k-1} - \cdots - a_n y_{k-n} + b_0 u_k + b_1 u_{k-1} + \cdots + b_n u_{k-n} \tag{7.3-11}$$

and the current output y_k depends on the current input u_k. This is usually an undesirable state of affairs, since it takes some computation time for the microprocessor to compute y_k. Techniques for including the computation time will be discussed later.

If the continuous-time controller is given in the state-space form

$$\dot{x} = A^c x + B^c u$$
$$y = Cx + Du, \tag{7.3-12}$$

one may use the Laplace transform and (7.3-3) to show that the discretized system using the BLT is given by Hanselmann (1987) as

$$x_{k+1} = Ax_k + B_1 u_{k+1} + B_0 u_k$$
$$y_k = Cx_k + Du_k, \tag{7.3-13}$$

with

$$A = \left[I - A^c \frac{T}{2}\right]^{-1} \left[I + A^c \frac{T}{2}\right]$$

$$B_1 = B_0 = \left[I - A^c \frac{T}{2}\right]^{-1} \frac{T}{2} B^c \tag{7.3-14}$$

Note that the discretized system is not a traditional state-space system since x_{k+1} depends on u_{k+1}. Aside from computation time delays, this is not a problem in our applications, since all we require of (7.3-13) is to implement it on a microprocessor. Since (7.3-13) is only a set of difference equations, this is easily accomplished. We illustrate how to discretize a continuous-time controller using the BLT in Examples

7.3-1 and 7.3-2, where we design a digital PID controller and a digital pitch rate controller.

Matched Pole-Zero

The second popular approximation technique for converting a continuous transfer function to a discrete one is the *matched pole–zero (MPZ) method*. Here, both the poles and finite zeros are mapped into the z-plane using the transformation e^{sT}, as follows:

1. If $K^c(s)$ has a pole (or finite zero) at $s = s_i$, then $K(z)$ will have a pole (or finite zero) at

$$z_i = e^{s_i T} \qquad (7.3\text{-}15)$$

2. If the relative degree of $K^c(s)$ is r, so that it has r zeros at infinity, r zeros of $K(z)$ are taken at $z = -1$ by multiplying by the factor $(1+z)^r$.
3. The gain of $K(z)$ is selected so that the dc gains of $K^c(s)$ and $K(z)$ are the same, that is, so that

$$K(1) = K^c(0) \qquad (7.3\text{-}16)$$

An alternative to step 2 is to map only $r - 1$ of the infinite s-plane zeros into $z = -1$. This leaves the relative degree of $K(z)$ equal to 1, which allows one sample period for control computation time. We will call this the *modified MPZ* method.

Thus if

$$K^c(s) = \frac{\Pi_{i=1}^m (s + t_i)}{\Pi_{i=1}^n (s + s_i)} \qquad (7.3\text{-}17)$$

and the relative degree is $r = n - m$, the MPZ discretized transfer function is

$$K(z) = k(z+1)^{r-1} \frac{\Pi_{i=1}^m \left(z - e^{-t_i T}\right)}{\Pi_{i=1}^n \left(z - e^{s_i T}\right)}, \qquad (7.3\text{-}18)$$

where the gain k is chosen to ensure (7.3-16). Note that if $K^c(s)$ is stable, so is the $K(z)$ obtained by the MPZ, since $z = e^{sT}$ maps the left-half s-plane into the unit circle in the z-plane. Although the MPZ requires simpler algebra than the BLT, the latter is more popular in industry.

Digital Design Examples

Now let us show some examples of digital controller design using the BLT and MPZ to discretize continuous controllers.

Example 7.3-1: Discrete PID Controller. Since the continuous PID controller is so useful in aircraft control design, let us demonstrate how to discretize it to obtain a digital PID controller. A standard continuous-time PID controller has the transfer function (Åström and Wittenmark, 1984)

$$K^c(s) = k\left[1 + \frac{1}{T_I s} + \frac{T_D s}{1 + T_D s/N}\right], \qquad (1)$$

where k is the proportional gain, T_I is the integration time constant or "reset" time, and T_D is the derivative time constant. Rather than use pure differentiation, a "filtered derivative" is used that has a pole far left in the s-plane at $s = -N/T_D$. A typical value for N is 3 to 10; it is usually fixed by the manufacturer of the controller.

Let us consider a few methods of discretizing (1) with sample period T seconds.

a. BLT. Using the BLT, the discretized version of (1) is found to be

$$K(z) = k\left[1 + \frac{1}{T_I \dfrac{2(z-1)}{T(z+1)}} + \frac{T_D \dfrac{2(z-1)}{T(z+1)}}{1 + \dfrac{T_D}{N}\dfrac{2(z-1)}{T(z+1)}}\right] \qquad (2)$$

or, on simplifying,

$$K(z) = k\left[1 + \frac{T}{T_{Id}}\frac{z+1}{z-1} + \frac{T_{Dd}}{T}\frac{z-1}{z-\nu}\right] \qquad (3)$$

with the discrete integral and derivative time constants

$$T_{Id} = 2T_I \qquad (4)$$

$$T_{Dd} = \frac{NT}{1 + NT/2T_D} \qquad (5)$$

and the derivative-filtering pole at

$$\nu = \frac{1 - NT/2T_D}{1 + NT/2T_D} \qquad (6)$$

b. MPZ. Using the MPZ approach to discretize the PID controller yields

$$K(z) = k\left[1 + \frac{k_1(z+1)}{T_I(z-1)} + \frac{k_2 N(z-1)}{z - e^{-NT/T_D}}\right], \qquad (7)$$

where k_1 and k_2 must be selected to match the dc gains. At dc, the D terms in (1) and (7) are both zero, so we may select $k_2 = 1$. The dc values of the I terms in (1) and (7) are unbounded. Therefore, to select k_1 let us match the low-frequency gains. At low frequencies, $e^{j\omega T} \approx 1 + j\omega T$. Therefore, for small ω, the I terms of (1) and (7) become

$$K^c(j\omega) = \frac{1}{j\omega T_I}$$

$$K(e^{j\omega T}) \approx \frac{2k_1}{T_I(j\omega T)},$$

and to match them, we require that $k_1 = T/2$.

Thus, using the MPZ the discretized PID controller again has the form (3), but now with

$$T_{Id} = 2T_I \tag{8}$$

$$T_{Dd} = NT \tag{9}$$

$$\nu = e^{-NT/T_D} \tag{10}$$

c. Modified MPZ. If we use the modified MPZ method, then in the I term in (7) the factor $(z + 1)$ does not appear. Then the normalizing gain k_1 is computed to be T. In this case, the discretized PID controller takes on the form

$$K(z) = k\left[1 + \frac{T}{T_{Id}}\frac{1}{z-1} + \frac{T_{Dd}}{T}\frac{z-1}{z-\nu}\right], \tag{11}$$

with

$$T_{Id} = T_I \tag{12}$$

$$T_{Dd} = NT \tag{13}$$

$$\nu = e^{-NT/T_D} \tag{14}$$

Now, there is a control delay of one sample period (T s) in the integral term, which could be advantageous if there is a computation delay.

d. Difference Equation Implementation. Let us illustrate how to implement the modified MPZ PID controller (11) using difference equations, which are easily placed into a software computer program. It is best from the point of view of numerical accuracy in the face of computer round-off error to implement digital controllers as several first- or second-order systems in parallel. Such a parallel implementation may be achieved as follows.

First, write $K(z)$ in terms of z^{-1}, which is the unit delay in the time domain (i.e., a delay of T s, so that, for instance, $z^{-1}u_k = u_{k-1}$), as

$$K(z^{-1}) = k\left[1 + \frac{T}{T_{Id}}\frac{z^{-1}}{1-z^{-1}} + \frac{T_{Dd}}{T}\frac{1-z^{-1}}{1-\nu z^{-1}}\right] \tag{15}$$

(*Note:* there is some abuse in notation in denoting (15) as $K(z^{-1})$; this, we will accept.)

Now, suppose that the control input u_k is related to the tracking error as

$$u_k = K\left(z^{-1}\right) e_k \tag{16}$$

Then, u_k may be computed from past and present values of e_k using auxiliary variables as follows:

$$v_k^I = v_{k-1}^I + \frac{T}{T_{Id}} e_{k-1} \tag{17}$$

$$v_k^D = v v_{k-1}^D + \frac{T_{Dd}}{T}(e_k - e_{k-1}) \tag{18}$$

$$u_k = k\left(e_k + v_k^I + v_k^D\right) \tag{19}$$

The variables v_k^I and v_k^D represent the integral and derivative portions of the PID controller, respectively. For more discussion, see Åström and Wittenmark (1984). ■

Example 7.3-2: Digital Pitch-Rate Controller via BLT. In Example 5.5-3 we designed a pitch-rate control system using LQ output-feedback techniques. Here we demonstrate how to convert that continuous control system into a digital control system. The BLT is popular in industry; therefore, we will use it here.

The continuous controller is illustrated in Figure 7.3-1, where

$$K_1^c(s) = \frac{k_I}{s} \tag{1}$$

$$K_2^c(s) = \frac{10 k_\alpha}{s + 10} \tag{2}$$

$$K_3^c(s) = k_q \tag{3}$$

The most suitable feedback gains in Example 5.5-3 were found using derivative weighting design to be

$$k_I = 1.361, \qquad k_\alpha = -0.0807, \qquad k_q = -0.475 \tag{4}$$

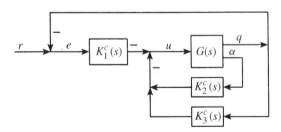

Figure 7.3-1 Continuous pitch-rate controller.

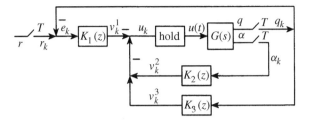

Figure 7.3-2 Digital pitch-rate controller.

A digital control scheme with the same structure is shown in Figure 7.3-2. We have added samplers with period T to produce the samples of pitch rate, q, and angle of attack, α, as well as a hold device to convert the control samples u_k computed by the digital controller back to a continuous-time control input $u(t)$ for the plant. Note that the reference input $r(t)$ must also be sampled.

Since the integrator and alpha smoothing filter are part of the digital controller, the continuous dynamics $G(s)$ in Figure 7.3-2 are given by $\dot{x} = Ax + Bu$, $y = Cx$, with

$$A = \begin{bmatrix} -1.01887 & 0.90506 & -0.00215 \\ 0.82225 & -1.07741 & -0.17555 \\ 0 & 0 & -20.2 \end{bmatrix}, \quad B = \begin{bmatrix} 0 \\ 0 \\ 20.2 \end{bmatrix}$$

$$C = \begin{bmatrix} 57.2958 & 0 & 0 \\ 0 & 57.2958 & 0 \end{bmatrix}, \tag{5}$$

where

$$x = \begin{bmatrix} \alpha \\ q \\ \delta_e \end{bmatrix}, \quad y = \begin{bmatrix} \alpha \\ q \end{bmatrix} \tag{6}$$

Using the BLT, the discrete equivalents to (1)–(3) are found to be

$$K_1(z) = k_1 \frac{z+1}{z-1} \quad \text{with } k_1 = \frac{k_I T}{2} \tag{7}$$

$$K_2(z) = k_2 \frac{z+1}{z-\pi} \quad \text{with } k_2 = \frac{10 k_\alpha T}{10T + 2}, \quad \pi = \frac{1 - 10T/2}{1 + 10T/2} \tag{8}$$

$$K_3(z) = k_q \tag{9}$$

Defining the intermediate signals v_k^1, v_k^2, v_k^3 shown in Fiure 7.3-2 and denoting the unit delay in the time domain by z^{-1}, we may express (7)–(9) in terms of difference equations as follows:

$$e_k = r_k - q_k, \tag{10}$$

$$v_k^1 = k_1 \frac{1+z^{-1}}{1-z^{-1}} e_k$$

or

$$v_k^1 = v_{k-1}^1 + k_1(e_k + e_{k-1}), \tag{11}$$

$$v_k^2 = k_2 \frac{1+z^{-1}}{1-\pi z^{-1}} \alpha_k$$

or

$$v_k^2 = \pi v_{k-1}^2 + k_2(\alpha_k + \alpha_{k-1}), \tag{12}$$

$$v_k^3 = k_q q_k \tag{13}$$

The control samples u_k are thus given by

$$u_k = -\left(v_k^1 + v_k^2 + v_k^3\right) \tag{14}$$

Note the low-pass filtering effects manifested by the averaging of e_k and α_k that occurs in these equations. This will tend to average out any measurement noise.

These difference equations describe the digital controller, and are easily implemented on a microprocessor. First, however, the controller should be simulated. The Fortran subroutine in Figure 7.3-3a may be used with the driver program in

```
C   DIGITAL PITCH RATE CONTROLLER
    SUBROUTINE DIG(IK,T,X)
    REAL X(*), K(2), KI, KA, KQ
    COMMON/CONTROL/U
    COMMON/OUTPUT/AL,Q,UPLOT
    DATA REF, KI,KA,KQ/1., 1.361,-0.0807,-0.475/

    K(1)= KI*T/2
    K(2)= 10*KA*T/(10*T + 2)
    P= (1 - 10*T/2) / (1 + 10*T/2)

    E= REF - Q
    V1= V1   + K(1)*(E + EKM1)
    V2= P*V2 + K(2)*(AL + ALKM1)
    V3= KQ*Q
    U= -(V1 + V2 + V3)
    UPLOT= U

    EKM1= E
    ALKM1= AL

    RETURN
    END
```

(a)

Figure 7.3-3 Digital simulation software: (a) FORTRAN subroutine to simulate digital pitch-rate controller; (continued)

DISCRETIZATION OF CONTINUOUS CONTROLLERS

```
C   CONTINUOUS SHORT PERIOD DYNAMICS

    SUBROUTINE F(TIME,X,XP)
    REAL X(*), XP(1)
    COMMON/CONTROL/U
    COMMON/OUTPUT/AL,Q

    XP(1)= -1.01887*X(1) + 0.90506*X(2) - 0.00215*X(3)
    XP(2)=  0.82225*X(1) - 1.07741*X(2) - 0.17555*X(3)
    XP(3)=                              - 20.2   *X(3) + 20.2*U

    AL   = 57.2958*X(1)
    Q    =                57.2958*X(2)

    RETURN
    END
```
(b)

Figure 7.3-3 (*continued*) (b) subroutine $F(t, x, \dot{x})$ to simulate continuous plant dynamics.

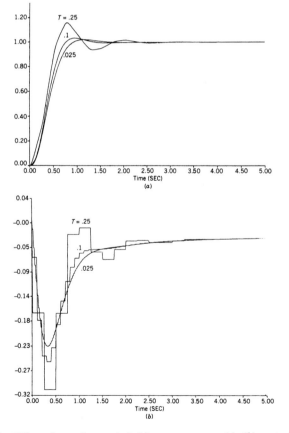

Figure 7.3-4 Effect of sampling period: (a) step response $q(t)$; (b) control input $u(t)$.

Figure 7.2-4 to simulate the digital control law. The subroutine $F(t, x, \dot{x})$ required by the Runge–Kutta integrator for the continuous plant dynamics (5) is given in Figure 7.3-3b.

The step response using this digital controller was plotted for several sampling periods T in Figure 7.3-4. A zero-order hold was used. Note that the step response improves as T becomes small. Indeed, the response for $T = 0.025$ s is indistinguishable from the response using a continuous controller in Example 5.5-3c.

The motivation for selecting $T = 0.025$ s was as follows. The settling time of the continuous controller step response in Example 5.5-3c was $t_s = 1$ s. The settling time is about four times the slowest time constant, which is thus 0.25 s. The sampling period should be selected about one-tenth of this for good performance. ∎

7.4 MODIFIED CONTINUOUS DESIGN

In Section 7.3 we showed how to convert a continuous-time controller to a digital controller using the BLT and MPZ. However, that technique is only an approximate one that gives worse results as the sample period T increases. In this section we show how to *modify the design of the continuous controller* so that it yields a more suitable digital controller. This allows the use of larger sample periods. To do this we will take into account some properties of the zero-order-hold and sampling processes. Using modified continuous design, we are able to design in Example 7.4-1 a digital pitch-rate control system that works extremely well even for relatively large sample periods.

Sampling, Hold Devices, and Computation Delays

We will examine some of the properties of the discretization and implementation processes to see how the continuous controller may be designed in a fashion that will yield an improved digital controller. Specifically, in the design of the continuous controller it is desirable to include the effects of sampling, hold devices, and computation delays.

Sampling and Aliasing. We would like to gain some additional insight on the sampling process (Oppenheim and Schafer, 1975; Franklin and Powell, 1980; Åström and Wittenmark, 1984). To do so, define the Nyquist frequency $\omega_N = \omega_s/2 = \pi/T$, and the sampling frequency $\omega_s = 2\pi/T$ and picture the output $y^*(t)$ of the sampler with input $y(t)$ as the string of impulses

$$y^*(t) = \sum_{k=-\infty}^{\infty} y(t)\delta(t - kT), \qquad (7.4\text{-}1)$$

where $\delta(t)$ is the unit impulse. Since the impulse train is periodic, it has a Fourier series that may be computed to be

$$\sum_{k=-\infty}^{\infty} \delta(t - kT) = \frac{1}{T} \sum_{n=-\infty}^{\infty} e^{jn\omega_s t} \qquad (7.4\text{-}2)$$

MODIFIED CONTINUOUS DESIGN

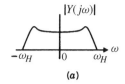

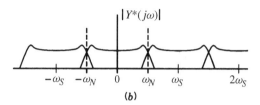

Figure 7.4-1 Sampling in the frequency domain: (a) spectrum of $y(t)$; (b) spectrum of sampled signal $y^*(t)$.

Using this in (7.4-1) and taking the Laplace transform yields

$$Y^*(s) = \frac{1}{T} \int_{-\infty}^{\infty} y(t) \left[\sum_{n=-\infty}^{\infty} e^{jn\omega_s t} \right] e^{-st} \, dt$$

$$Y^*(s) = \frac{1}{T} \sum_{n=-\infty}^{\infty} \int_{-\infty}^{\infty} y(t) e^{-(s-jn\omega_s)t} \, dt$$

$$Y^*(s) = \frac{1}{T} \sum_{n=-\infty}^{\infty} Y(s - jn\omega_s), \qquad (7.4\text{-}3)$$

where $Y(s)$ is the Laplace transform of $y(t)$ and $Y^*(s)$ is the Laplace transform of the sampled signal $y^*(t)$. Due to the factor $1/T$ appearing in (7.4-3), the sampler is said to have a *gain of* $1/T$.

Sketches of a typical $Y(j\omega)$ and $Y^*(j\omega)$ are shown in Figure 7.4-1, where ω_H is the highest frequency contained in $y(t)$. Notice that the digital frequency response is symmetric with respect to ω_N and periodic with respect to ω_s. At frequencies less than ω_N, the spectrum of $Y^*(j\omega)$ has two parts: one part comes from $Y(j\omega)$ and is the portion that should appear. However, there is an additional portion from $Y(j(\omega-\omega_s))$; the "tail" of $Y(j(\omega-\omega_s))$, which contains high-frequency information about $y(t)$, is "folded" back or *aliased* into the lower frequencies of $y^*(j\omega)$. Thus the high-frequency content of $y(t)$ appears at low frequencies and can lead to problems in reconstructing $y(t)$ from its samples.

If $\omega_H < \omega_N$, the tail of $Y^*(j(\omega - \omega_s))$ does not appear to the right of $\omega - \omega_N$ and $y(t)$ can be uniquely reconstructed from its samples by low-pass filtering. This condition is equivalent to

610 DIGITAL CONTROL

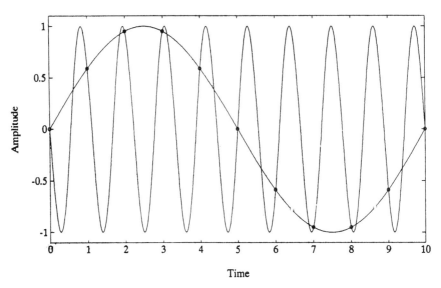

Figure 7.4-2 Example of aliasing in the time domain.

$$\omega_s > 2\omega_H, \quad (7.4\text{-}4)$$

which is the sampling theorem of Shannon that guarantees aliasing does not occur.

It is interesting to see what the sampling theorem means in the time domain. Examining Figure 7.4-2, where we show two continuous signals that have the same samples. If the original signal was the higher-frequency signal, the D/A reconstruction process will produce the lower-frequency signal from the samples of the higher-frequency signal. Thus, aliasing can result in *high-frequency signals being misinterpreted as low-frequency signals.* If the sampling frequency ω_s is greater than twice the highest frequency ω_H appearing in the continuous signal, the problem depicted in the figure does not occur and the signal can be accurately reconstructed from its samples.

Selecting the Sampling Period. For control design, the sampling frequency ω_s must generally be significantly greater than twice the highest frequency of any signal appearing in the system. That is, in control applications the sampling theorem does not usually provide much insight in selecting ω_s. Some guides for selecting the sampling period T are now discussed.

If the continuous-time system has a single dominant complex pole pair with natural frequency of ω, the rise time is given approximately by

$$t_r = \frac{1.8}{\omega} \quad (7.4\text{-}5)$$

It is reasonable to have at least two to four samples per rise time so that the error induced by ZOH reconstruction is not too great during the fastest variations of the continuous-time signal (Åström and Wittenmark, 1984). Then we have $t_r = 1.8/\omega \geq 4T$, or approximately

$$T \leq \frac{1}{2\omega} \tag{7.4-6}$$

However, if high-frequency components are present up to a frequency of ω_H radians and it is desired to retain them in the sampled system, a rule of thumb is to select

$$T \leq \frac{1}{4\omega_H} \tag{7.4-7}$$

These formulae should be used with care, and to select a suitable T it may be necessary to perform digital control designs for several values of T, for each case carrying out a computer simulation of the behavior of the plant under the influence of the proposed controller. Note particularly that using continuous redesign of digital controllers with the BLT or MPZ, even smaller sample periods may be required since the controller discretization technique is only an approximate one.

Zero-Order Hold. The D/A hold device in Figure 7.2-1 is required to reconstruct the plant control input $u(t)$ from the samples u_k provided by the digital control scheme. The zero-order hold (ZOH) is usually used. There, we take

$$u(t) = u(kT) = u_k, \qquad kT \leq t < (k+1)T, \tag{7.4-8}$$

with u_k the k-th sample of $u(t)$. The ZOH yields the sort of behavior in Figure 7.2-2 and has the impulse response shown in Figure 7.4-3. This impulse response may be written as

$$h(t) = u_{-1}(t) - u_{-1}(t-T),$$

with $u_{-1}(t)$ the unit step. Thus the transfer function of the ZOH is

$$G_0(s) = \frac{1 - e^{-sT}}{s} \tag{7.4-9}$$

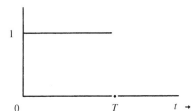

Figure 7.4-3 ZOH impulse response.

To determine the Bode magnitude and phase of $G_0(s)$, write

$$G_0(j\omega) = \frac{1 - e^{-j\omega T}}{j\omega} = e^{-j\omega T/2}\frac{e^{j\omega T/2} - e^{-j\omega T/2}}{j\omega}$$

$$G_0(j\omega) = Te^{-j\omega T/2}\frac{\sin(\omega T/2)}{\omega T/2} = Te^{-j\omega T/2}\operatorname{sinc}\frac{\omega}{\omega_s}, \qquad (7.4\text{-}10)$$

where $\operatorname{sinc} x \equiv (\sin \pi x)/\pi x$. The magnitude and phase of the ZOH are shown in Figure 7.4-4. Note that the ZOH is a low-pass filter of magnitude $T|\operatorname{sinc}(\omega/\omega_s)|$ with a phase of

$$\angle \text{ZOH} = -\frac{\omega T}{2} + \theta = -\frac{\pi\omega}{\omega_s} + \theta, \qquad \theta = \begin{cases} 0, & \sin\dfrac{\omega T}{2} > 0 \\ \pi, & \sin\dfrac{\omega T}{2} < 0 \end{cases} \qquad (7.4\text{-}11)$$

According to (7.4-10), for frequencies ω much smaller than ω_s, the ZOH may be approximated by

$$G_0(s) \approx Te^{-sT/2}, \qquad (7.4\text{-}12)$$

that is, by a pure delay of half the sampling period and a scale factor of T.

As we saw in the digital pitch-rate controller in Example 7.3-1, the performance of the digital controller deteriorates with increasing T, so that sample periods are required which may be too small. (We note that smaller values of T require faster computation to compute u_k; thus a faster, and more expensive, microprocessor may be required for small T.) This deterioration is partly due to the delay introduced by the hold device. We will soon see how to take this delay into account *while designing the continuous controller,* so that discretization yields a digital controller that gives suitable performance for larger values of T.

Computation Delay. If the microprocessor is fast so that the time Δ required to compute the digital control law is negligible, Δ will have little effect when a digital controller is implemented. However, if Δ is appreciable, it can have a deleterious effect on the closed-loop response. Then it may be necessary to account for it.

If $\Delta \leq T$, the computation delay may be accounted for by ensuring that u_k depends only on *previous* values of the outputs. This may often be achieved by using the modified MPZ approach for digital controller design. However, the BLT is more popular and it always yields a u_k that depends on *current* values of the outputs (see (7.3-11)). Moreover, the discrete PID controller (see Example 7.3-1) always has a dependence on the current outputs through the derivative term, even if the modified MPZ is used.

MODIFIED CONTINUOUS DESIGN   613

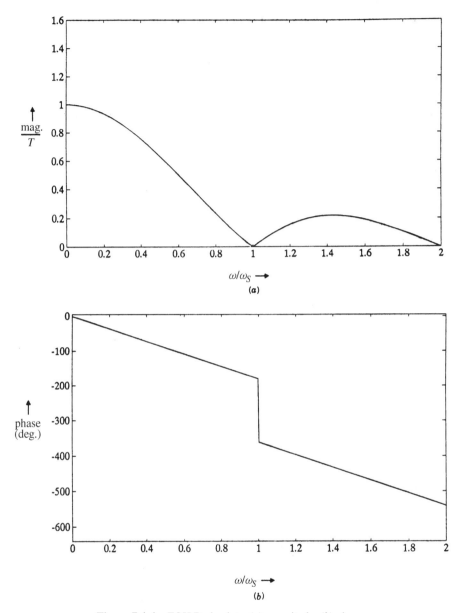

Figure 7.4-4 ZOH Bode plots: (a) magnitude; (b) phase.

If the computation delay is not negligible but is only a fraction of T, it seems inefficient to allow it to cause a delay of a full T seconds in applying the control to the plant. If there is noise present in the system, then using outputs delayed by an entire sample period to compute u_k can, for large sample periods, lead to significant deterioration over using more recent outputs to compute u_k.

Modified Continuous Design Procedures

We will now show how to account for the hold delay and computation delay *while designing the continuous controller $K^c(s)$ for discretization*. Then, when the BLT or MPZ is used to discretize $K^c(s)$, a digital controller $K(z)$ with improved performance will be obtained. We call this approach *modified continuous controller design for discretization*.

A disadvantage of modified continuous design techniques is that the sample period T must be selected prior to the continuous controller design. However, good software makes it easy to redesign the continuous controller with a different value of T. The advantage of the approach is that the effects of the sampling and hold operations, computation delay, and aliasing are apparent *while the continuous design is being performed*. Thus, they may be to some extent compensated for.

Modified continuous design can often allow significantly larger sample periods than direct application of the BLT or MPZ to a continuous controller designed with no consideration that the next step will be conversion to a digital control law. This will be illustrated in Example 7.4-1, where we design a pitch-rate control system by modified continuous design.

Let us discuss aliasing, computation delays, and then the ZOH.

Aliasing. The plant $G(s)$ is generally a low-pass filter. We have seen in Figure 7.4-1 that as long as the sampling frequency ω_s is selected at least twice as large as the plant cutoff frequency ω_H, the effects of aliasing will be small.

However, one type of signal appearing in the closed-loop system that may not be bandlimited is *measurement noise*. High-frequency measurement noise may be aliased down to lower frequencies that are within the plant bandwidth and thus have a detrimental effect on system performance. To avoid this, low-pass *anti-aliasing filters* of the form

$$H_a(s) = \frac{a}{s+a} \tag{7.4-13}$$

may be inserted after the measuring devices and before the samplers. The cutoff frequency a should be selected less than $\omega_N = \omega_s/2$, so that there is good attenuation beyond ω_n rad/s.

If the cutoff frequency of the antialiasing filter is not much higher than the plant cutoff frequency, the filter will affect the closed-loop performance, and it should be appended to the plant *at the design stage* so that the continuous controller is designed taking it into account. See Figure 7.4-5, which represents the actual plant $G(s)$ augmented by various filters, some still to be discussed, that should be taken into account in the design stage.

Computation Delay. The delay associated with a computation time of Δ has a transfer function of

$$G_{\text{comp}}(s) = e^{-s\Delta}, \tag{7.4-14}$$

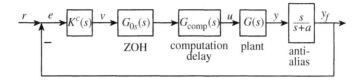

Figure 7.4-5 Modified continuous plant with antialiasing filter and compensation to model hold device and computation delays.

which has a magnitude of 1 and a phase of $-\omega\Delta$ radians. To account for this delay, we may perform the continuous controller design not on the plant $G(s)$, but on $G(s)e^{-s\Delta}$. However, it is awkward to design a controller for a plant whose transfer function is not rational (Franklin et al., 1980). It is more convenient to approximate the delay with a rational transfer function.

For this purpose, we may use Padé approximants to $e^{-s\Delta}$, which match the first few terms of the Taylor series expansion (Su, 1971; Franklin et al., 1986). Table 7.4-1 gives several Padé approximants to $e^{-s\Delta}$. These approximants match the first $n+m+1$ terms of the Taylor series expansion, where n is the denominator degree and m the numerator degree.

To perform a modified continuous design that takes into account the computation delay Δ, it is only necessary to incorporate a Padé approximant $G_{\text{comp}}(s)$ to $e^{-s\Delta}$ of suitable order into the plant as shown in Figure 7.4-5. The continuous controller $K^c(s)$ designed for this modified plant is then discretized using the BLT or MPZ to produce a digital controller $K(z)$.

Notice that the Padé approximants in Table 7.4-1 having finite zeros are *nonminimum-phase*. This is a property of a pure time delay. The advantage of the modified continuous design approach is that the non-minimum-phase nature of the delayed plant manifests itself at the continuous controller design stage, so that the digital controller that results after using the BLT compensates for this problem automatically.

Zero-Order Hold. Finally, let us discuss modified continuous design taking into account the ZOH. Since the sampler has a gain of $1/T$, the sampler plus ZOH has a transfer function of

$$G_{0s}(s) = \frac{1 - e^{-sT}}{sT} \qquad (7.4\text{-}15)$$

Some useful approximants to $G_{0s}(s)$ are given in Table 7.4-2. These have been computed using Padé approximants of e^{-sT}, so they are not strictly speaking Padé approximants, since they only match the first $n + m$ terms of the Taylor series. They are, however, sufficiently accurate for our purposes. Note that the approximants to $G_{0s}(s)$ have unstable zeros. Modified continuous controller design taking into account $G_{0s}(s)$ involves designing a controller for $G(s)G_{0s}(s)$ (see Figure 7.4-5).

TABLE 7.4-1 Padé Approximants to $e^{-s\Delta}$ for Approximation of Computation Delay

$$\frac{1}{1+s\Delta}$$

$$\frac{1-s\Delta/2}{1+s\Delta/2}$$

$$\frac{1}{1+s\Delta+(s\Delta)^2/2}$$

$$\frac{1-s\Delta/3}{1+2s\Delta/3+(s\Delta)^2/6}$$

$$\frac{1-s\Delta/2+(s\Delta)^2/12}{1+s\Delta/2+(s\Delta)^2/12}$$

$$\frac{1}{1+s\Delta+(s\Delta)^2/2+(s\Delta)^3/6}$$

$$\frac{1-s\Delta/4}{1+3s\Delta/4+(s\Delta)^2/4+(s\Delta)^3/24}$$

$$\frac{1-2s\Delta/5+(s\Delta)^2/20}{1+3s\Delta/5+3(s\Delta)^2/20+(s\Delta)^3/60}$$

TABLE 7.4-2 Approximants to $(1 - e^{-sT})/sT$ for Approximation of Hold Delay

$$\frac{1}{1 + sT/2}$$

$$\frac{1 - sT/6}{1 + sT/3}$$

$$\frac{1 - sT/10 + (sT)^2/60}{1 + 2sT/5 + (sT)^2/20}$$

$$\frac{1 - sT/14 + 23(sT)^2/840 - (sT)^3/840}{1 + 3sT/7 + (sT)^2/14 + (sT)^3/120}$$

Implementation. It is important to realize that the antialiasing filter should be implemented using analog circuitry as part of the plant $G(s)$. It should immediately precede the sampler. $G_{\text{comp}}(s)$, on the other hand, is not implemented since it is a model of the computation delay. $G_{0s}(s)$ is implemented by the ZOH and the sampler. $K^c(s)$ is discretized using the BLT or MPZ and becomes the digital controller $K(z)$.

The next example illustrates modified continuous design for discretization.

Example 7.4-1: Digital Pitch-Rate Controller via Modified Continuous Design. In Example 5.5-3 we designed a continuous-time pitch-rate controller. In Example 7.3-2 we showed how to use the BLT to convert that controller into digital form. It was seen that the response was good for $T = 0.025$ s, slightly worse for $T = 0.1$ s, and unacceptable for $T = 0.25$ s.

In this example let us design a modified continuous controller which, on discretization, will yield a better digital controller using larger sample periods than the one of Example 7.3-2. We will select the sampling period in this example to be $T = 0.25$ s.

a. Modified Continuous-Time Plant. To account for the effects of the hold delay we will incorporate a model of the sampling and hold processes into the continuous-time dynamical model of the aircraft as shown in Figure 7.4-5. Let us use a Padé approximant to (7.4-15). Specifically, examining Table 7.4-2, select

$$G_{0s}(s) = \frac{1 - sT/6}{1 + sT/3} = -\frac{1}{2} + \frac{9/2T}{s + 3/T} \qquad (1)$$

According to Figure 7.4-5, the ZOH/sampler approximant should act as a filter on the plant control input $u(t)$. Thus a state-variable representation of $G_{0s}(s)$ is given by

$$\dot{x}_z = -\frac{3}{T}x_z + \frac{9}{2T}v$$

$$u = x_z - \frac{1}{2}v, \qquad (2)$$

where $v(t)$ is the new input shown in Figure 7.4-5. With $T = 0.25$ s this becomes

$$\dot{x}_z = -12x_z + 18v$$
$$u = x_z - 0.5v \tag{3}$$

We should like to propose the same control structure used in Example 5.5-3. There, an angle-of-attack filter and an integrator in the feedforward channel were used. The ZOH/sampler dynamics (3) may be augmented into the system-plus-compensator state equations by defining the augmented state

$$x = \begin{bmatrix} \alpha & q & \delta_e & \alpha_F & \epsilon & x_z \end{bmatrix}^T \tag{4}$$

Then

$$\dot{x} = Ax + Bv + Er \tag{5}$$
$$y = Cx + Fr \tag{6}$$
$$z = Hx \tag{7}$$

with

$$A = \begin{bmatrix} -1.01887 & 0.90506 & -0.00215 & 0 & 0 & 0 \\ 0.82225 & -1.07741 & -0.17555 & 0 & 0 & 0 \\ 0 & 0 & -20.2 & 0 & 0 & 20.2 \\ 10.0 & 0 & 0 & -10 & 0 & 0 \\ 0 & -57.2958 & 0 & 0 & 0 & 0 \\ 0 & 0 & 0 & 0 & 0 & -12 \end{bmatrix},$$

$$B = \begin{bmatrix} 0 \\ 0 \\ -10.1 \\ 0 \\ 0 \\ 18 \end{bmatrix}, \quad E = \begin{bmatrix} 0 \\ 0 \\ 0 \\ 0 \\ 1 \\ 0 \end{bmatrix}$$

$$C = \begin{bmatrix} 0 & 0 & 0 & 57.2958 & 0 & 0 \\ 0 & 57.2958 & 0 & 0 & 0 & 0 \\ 0 & 0 & 0 & 0 & 1 & 0 \end{bmatrix}, \quad F = \begin{bmatrix} 0 \\ 0 \\ 0 \end{bmatrix}$$

$$H = \begin{bmatrix} 0 & 57.2958 & 0 & 0 & 0 & 0 \end{bmatrix}$$

Then, according to Example 5.5-3 the control input $v(t)$ is given by

$$v = -Ky = -\begin{bmatrix} k_\alpha & k_q & k_I \end{bmatrix} y = -k_\alpha \alpha_F - k_q q - k_I \epsilon \tag{8}$$

We are now in a position to perform the controls design to select the control gains.

b. PI and Continuous Controls Design.
To design the continuous-time controller, let us select the PI

$$J = \tfrac{1}{2} \int_0^\infty \left(q_5 t^2 e^2 + \dot{\delta}_e^2 \right) dt \tag{9}$$

that weights elevator rate of change, since this is closely related to actuator energy. Since $e(t) = \dot{\epsilon}(t)$, this may be written

$$J = \tfrac{1}{2} \int_0^\infty \left(q_5 t^2 \dot{\epsilon}^2 + \dot{\delta}_e^2 \right) dt, \tag{10}$$

with $\epsilon(t)$ and $\delta_e(t)$ the deviations in the integrator output and elevator deflection. Thus, this is the PI with derivative weighting discussed in Section 5.5.

Using $q_5 = 5$ and the software described in Appendix B, we computed the optimal gain matrix

$$K = [-0.04238 \quad -0.4098 \quad 0.8426], \tag{11}$$

which gave the closed-loop poles

$$s = -2.40 \pm j4.71$$
$$-1.08, -2.76$$
$$-9.86, -25.80 \tag{12}$$

The closed-loop step response of the continuous-time controller is shown in Figure 7.4-6. Note that it is comparable to the responses shown in Example 5.5-3.

Let us note that the transfer function from $v(t)$ to $q(t)$ contains the approximate ZOH/sampler dynamics described by (1), (3). These include a pole at $s = -12$ which has no significant effect. However, they also include a non-minimum-phase zero at $s = 24$. This zero significantly changes the root locus, and the control gains (11) selected automatically by the LQ approach take this non-minimum-phase zero into account. Indeed, note the delay of approximately $T = 0.25$ s in Figure 7.4-6.

It should also be realized that in contrast to the situation in Example 7.3-2, which relied on the continuous design from Example 5.5-3, the sampling period is now needed to write the continuous dynamics (5), and hence to design the continuous-time controller.

c. Digital Controller.
The modified continuous controller just designed is described by exactly the same equations as in Example 7.3-2, with, however, the modified gain K given in (11). Thus the new digital controller is exactly the same as the one described in that example, though using the modified gains.

To examine the performance of the modified digital controller, we may use the driver program described in Section 7.2, along with the continuous-time aircraft dynamics and the subroutine DIG(IK, T, X) from Example 7.3-2 with the gains in (11). The response for $T = 0.25$ s is shown in Figure 7.4-7. Note that at this design sample

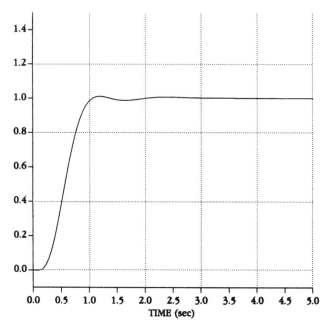

Figure 7.4-6 Step response $q(t)$ using modified continuous-time pitch rate controller.

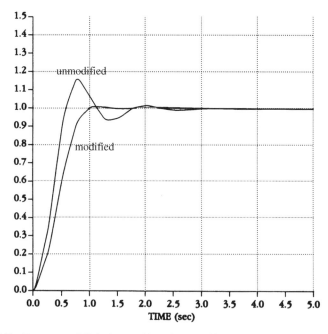

Figure 7.4-7 Response of digital controller using modified and unmodified continuous-time design.

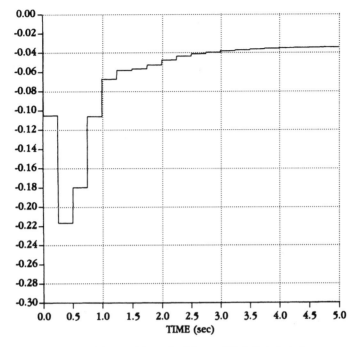

Figure 7.4-8 Control input $u(t)$ required for modified digital pitch rate controller.

period of $T = 0.25$ s, the digital control response is much like the response using the continuous-time controller shown in Figure 7.4-6. It is important to note, however, that, using the digital controller, the delay noted in Figure 7.4-6 does not appear.

For comparison we have also shown in Figure 7.4-7 the unacceptable response from Example 7.3-2 for $T = 0.25$ s. This was the result of using a digital controller, obtained simply by applying the BLT to the unmodified continuous-time controller, which did not take into account the effects of the hold delay.

The control input $u(t)$ required in the modified digital controller with $T = 0.25$ s is shown in Figure 7.4-8. It may be compared to the control signals in Example 7.3-2.

Clearly, the response shown in Figure 7.4-7 obtained using modified continuous design is excellent. It far surpasses the digital control response in Example 7.3-2 for $T = 0.25$ s. Thus we have demonstrated that a sensible technique for taking into account some of the properties of the sample-and-hold process in the design stage of the continuous controller results in improved digital controllers that may be used with larger sample periods T. ∎

7.5 IMPLEMENTATION CONSIDERATIONS

In this chapter we have discussed a design approach for digital controllers that is based on discretizing a continuous-time controller using the BLT or MPZ. It now

DIGITAL CONTROL

behooves us to look at some practical considerations involved with implementing the digital controller. Our discussion will necessarily be brief, giving only an indication of some of the issues. More detail may be discovered in Åström and Wittenmark (1984), Franklin and Powell (1980), Franklin et al. (1980), Phillips and Nagle (1984), Hanselmann (1987), Lewis (1992), and Slivinsky and Borninski (1987). We will mention actuator saturation and windup and controller realization structures.

Actuator Saturation and Windup

Actuator saturation is a problem that occurs in both continuous-time and digital control systems. Since it is easy to protect against by using a digital controller, we have placed it in this section.

A digital controller may be represented in the dynamic state-space form

$$x_{k+1} = Fx_k + Gw_k \tag{7.5-1}$$

$$u_k = Cx_k + Dw_k, \tag{7.5-2}$$

where $x_k \in \mathbf{R}^n$ is the controller state and w_k the controller input, composed generally of the tracking error and the plant measured output.

We have assumed thus far that the plant control input $u_k \in \mathbf{R}^m$ which is computed by the controller can actually be applied to the plant. However, in flight controls the plant inputs (such as elevator deflection δ_e, throttle, and so on) are limited by *maximum* and *minimum* allowable values. Thus, the relation between the *desired plant input* v_k and the *actual plant input* u_k is given by the sort of behavior shown in Figure 7.5-1, where u_H and u_L represent, respectively, the maximum and minimum control effort allowed by the mechanical actuator. Thus, to describe the actual case in an aircraft flight control system, we are forced to include *nonlinear saturation functions* in the control channels as shown in Figure 7.5-2.

Consider the simple case where the controller is an integrator with input w_k and output v_k. Then all is well as long as v_k is between u_L and u_H, for in this region the aircraft input u_k equals v_k. However, if v_k exceeds u_H, then u_k is limited to its maximum value u_H. This in itself may not be a problem. The problem arises if w_k remains positive, for then the integrator continues to integrate and v_k may increase well beyond u_H. Then, when w_k becomes negative, it may take considerable time

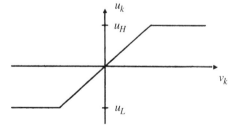

Figure 7.5-1 Actuator saturation function.

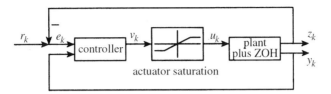

Figure 7.5-2 Flight control system including actuator saturation.

for v_k to decrease below u_H. In the meantime, u_k is held at u_H, giving an incorrect control input to the aircraft. This effect of integrator saturation is called *windup*. It arises because the controllers we design are generally dynamical in nature, which means that they store information or energy.

To correct integrator windup, it is necessary to *limit the state of the controller* so that it is consistent with the saturation effects being experienced by the plant input u_k. This is not difficult to achieve (Åström and Wittenmark, 1984). Indeed, write (7.5-2) in the form

$$0 = u_k - Cx_k - Dw_k,$$

multiply it by L, which will soon be selected, and add it to (7.5-1) to obtain

$$x_{k+1} = (F - LC)x_k + (G - LD)w_k + Lu_k \qquad (7.5\text{-}3)$$

This is the form in which the digital controller should be implemented to avoid windup, as we now argue. A little thought shows that actuator windup occurs in the form (7.5-1) when F is not asymptotically stable. For then, as long as w_k in (7.5-1) is nonzero, x_k will continue to increase. However, by selecting L so that

$$F_0 = F - LC \qquad (7.5\text{-}4)$$

is asymptotically stable, this problem is averted.

A special case occurs when L is selected so that F_0 has al poles at the origin. Then x_k displays *deadbeat behavior*; after n time steps it remains limited to an easily computed value dependent on the values of w_k and u_k (see the Problems).

The *antiwindup gain L* may be selected to place the poles of F_0 arbitrarily if (C, F) is observable. However, as long as (C, F) is *detectable* (i.e., has all its unstable poles observable), windup may be eliminated using this technique.

To complete the design for antiwindup protection, the digital controller should be implemented in the form (7.5-3) and the aircraft control input then selected according to

$$u_k = \text{sat}\,(Cx_k + Dw_k), \qquad (7.5\text{-}5)$$

where the *saturation function* (shown in Figure 7.5-1) is defined for scalars as

$$\text{sat}(v) = \begin{cases} u_H, & v \geq u_H \\ v, & u_L < v < u_H \\ u_L, & v \leq u_L, \end{cases} \tag{7.5-6}$$

with u_H and u_L the maximum and minimum allowable values, respectively. For vectors, the saturation function is defined as

$$\text{sat}(v) = \begin{bmatrix} \text{sat}(v_1) \\ \text{sat}(v_2) \\ \vdots \\ \text{sat}(v_m) \end{bmatrix} \tag{7.5-7}$$

The values of u_H and u_L for each component v_i should be selected to correspond to the actual limits on the components of the plant input u_k.

Note that the limited signal u_k is used in (7.5-3), providing a feedback arrangement in the controller with antiwindup protection. What we have in effect done is include an *observer* with dynamics F_0 in the digital controller. Since F_0 is asymptotically stable, the observer will provide reasonable "estimates" even in the event of saturation.

Where u_k is not saturated, the controller with antiwindup compensation (7.5-3), (7.5-5) is identical to (7.5-1), (7.5-2).

If the controller is given in transfer function form

$$R(z^{-1}) u_k = T(z^{-1}) r_k - S(z^{-1}) w_k, \tag{7.5-8}$$

where r_k is the reference command and z^{-1} is interpreted in the time domain as a unit delay of T seconds, antiwindup compensation may be incorporated as follows.

Select a desired stable observer polynomial $A_0(z^{-1})$ and add $A_0(z^{-1})u_k$ to both sides to obtain

$$A_0 u_k = T r_k - S w_k + (A_0 - R) u_k \tag{7.5-9}$$

A regulator with antiwindup compensation is then given by

$$A_0 v_k = T r_k - S w_k + (A_0 - R) u_k \tag{7.5-10}$$

$$u_k = \text{sat}(v_k) \tag{7.5-11}$$

Example 7.5-1: Antiwindup Compensation for Digital PI Controller. From Example 7.3-1 a general digital PI controller is given by

$$u_k = k \left[1 + \frac{T}{T_I} \frac{1}{z-1} \right] W_k, \tag{1}$$

where we have used design by the modified MPZ to obtain a delay of T seconds in the integrator to allow for computation time. The proportional gain is k and the reset time is T_I; both are fixed in the design stage.

Multiply by z^{-1} and write

$$\left(1 - z^{-1}\right) u_k = k \left[\left(1 - z^{-1}\right) + \frac{T z^{-1}}{T_I}\right] W_k, \quad (2)$$

which is in the transfer function form (7.5-8). The corresponding difference equation form for implementation is

$$u_k = u_{k-1} + k w_k + k \left(-1 + \frac{T}{T_I}\right) w_{k-1} \quad (3)$$

This controller will experience windup problems since the autoregressive polynomial $R = 1 - z^{-1}$ has a root at $z = 1$, making it marginally stable. Thus, when u_k is limited, the integrator will continue to integrate, "winding up" beyond the saturation level.

To correct this problem, select an observer polynomial of

$$A_0\left(z^{-1}\right) = 1 - \alpha z^{-1}, \quad (4)$$

which has a pole at some desirable location $|\alpha| < 1$. The design parameter α may be selected by simulation studies. Then the controller with antiwindup protection (7.5-10)/(7.5-11) is given by

$$\left(1 - \alpha z^{-1}\right) v_k = k \left[1 + (-1 + T/T_I) z^{-1}\right] w_k + (1 - \alpha) z^{-1} u_k \quad (5)$$

$$u_k = \text{sat}(v_k) \quad (6)$$

The corresponding difference equations for implementation are

$$v_k = k w_k + \alpha v_{k-1} + k \left(-1 + \frac{T}{T_I}\right) w_{k-1} + (1 - \alpha) u_{k-1} \quad (7)$$

$$u_k = \text{sat}(v_k) \quad (8)$$

A few lines of Fortran code implementing this digital controller are given in Figure 7.5-3. This subroutine may be used as the control update routine DIG with the digital simulation driver program in Section 7.2.

If $\alpha = 1$, we obtain the special case (2), which is called the *position form* and has no antiwindup compensation.

If $\alpha = 0$, we obtain the *deadbeat antiwindup compensation*

$$v_k = k \left[1 + (-1 + T/T_I) z^{-1}\right] w_k + u_{k-1}, \quad (9)$$

C DIGITAL PI CONTROLLER WITH ANTIWINDUP COMPENSATION

```
      SUBROUTINE CONUP(T)
      REAL K
      COMMON/CONTROL/U
      COMMON/OUTPUT/Z
      COMMON/REF/R
      DATA K,AL,TI,ULOW,UHIGH/ 0.5, 0.2, 5., -0.5, 0.5/

      E= R-Z
      V= K*E + V
      U= AMAX1(ULOW,V)
      U= AMIN1(UHIGH,U)
      V= AL*V + K*(-1 + T/TI)*E + (1-AL)*U

      RETURN
      END
```

Figure 7.5-3 Fortran code implementing PI controller with antiwindup compensation.

with corresponding difference equation implementation

$$v_k = u_{k-1} + kw_k + \left(-1 + \frac{T}{T_I}\right) w_{k-1} \tag{10}$$

If u_k is not in saturation, this amounts to updating the plant control by adding the second and third terms in (10) to u_{k-1}. These terms are, therefore, nothing but $u_k - u_{k-1}$. The compensator with $\alpha = 0$ is thus called the *velocity form* of the PI controller. ∎

Controller Realization Structures

Round-off errors can occur every time an arithmetic operation is performed. Moreover, since all the stable behavior of a discrete system is described by the location of the poles within the unit circle, great accuracy is required in the filter coefficients to obtain desired closed-loop pole locations.

A direct implementation of the digital filter would involve simply writing n difference equations describing (7.5-1)/(7.5-2) and would be virtually guaranteed to have severe numerical problems if n is larger. Specifically, the controller and observer canonical forms (Kailath, 1980) are notoriously unstable numerically. That is, their poles are very sensitive to small variations in their coefficients. It can be shown that the sensitivity to coefficient variations of the impulse response and frequency response is also high in direct implementations (Hanselmann, 1987). For good numerical performance with fixed-point arithmetic, digital filters should be implemented as *cascade or parallel combinations of first- and second-order filters*.

A state-space transformation may be used to place the digital filter into an appropriate form for implementation. To obtain real coefficients, the *real Jordan form* is suitable (Phillips and Nagle, 1984; Hanselmann, 1987; Lewis, 1992). This is a block diagonal form for (7.5-1), (7.5-2) having first- and second-order blocks in cascade and parallel. Corresponding to each real pole there will be first-order blocks, and corresponding to each complex pole there will be second-order blocks.

A form suitable for implementation may also be found by performing a partial fraction expansion (PFE) on the transfer function. A technique for doing this in terms of the eigenstructure is given in Section 5.2. However, a *real PFE* should be found which will have the form (in the case of a simple matrix F)

$$H(z) = D + \sum_{i=1}^{r} H_i(z), \qquad (7.5\text{-}12)$$

where $H_i(z)$ is first-order for real poles and second-order for complex poles. We are therefore concerned with implementing first-order filters of the form

$$H_1\left(z^{-1}\right) = \frac{b_1 z^{-1}}{1 + a_1 z^{-1}} \qquad (7.5\text{-}13)$$

and second-order filters of the form

$$H_2\left(z^{-1}\right) = \frac{b_1 z^{-1} + b_2 z^{-2}}{1 + a_1 z^{-1} + a_2 z^{-2}} \qquad (7.5\text{-}14)$$

To implement $H_1(z^{-1})$, we may write

$$y_k = H_1\left(z^{-1}\right) u_k$$

$$\left(1 + a_1 z^{-1}\right) y_k = b_1 z^{-1} u_k$$

$$y_k = -a_1 y_{k-1} + b_1 u_{k-1}, \qquad (7.5\text{-}15)$$

which is a difference equation that may easily be programmed.

There are many ways to implement the second-order transfer function (Phillips and Nagle, 1984). Among these are *direct forms 1 through 4* (denoted D1, D2, D3, D4) and *cross-coupled forms 1 and 2* (denoted X1, X2).

In Figure 7.5-4 we show D1, D3, and X1. Forms D2, D4, and X2, respectively, are their duals (i.e., all arrows are reversed and the roles of the input and the output are interchanged). In Table 7.5-1 we give a comparison of the number of time-delay elements, multipliers, and summing junctions for each form. Note that D1 and X1 conserve time-delay elements, while D3 conserves summing junctions.

The difference equation implementations of these second-order modules are given in Table 7.5-2, with $y_k = H_2(z^{-1}) u_k$. It is interesting that the difference equations for the X1 module may be written from the complex PFE (and hence, with a little manipulation, from the usual complex Jordan form).

When implementing these modules on a fixed-point microprocessor, it is important to incorporate overflow protection (Slivinsky and Borninski, 1987). When

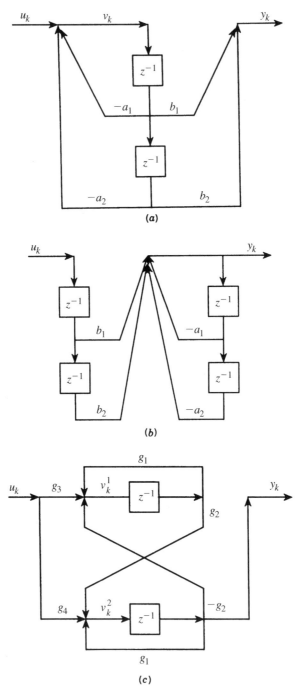

Figure 7.5-4 Implementations of second-order digital filters: (*a*) direct form 1, D1; (*b*) direct form 3, D3; (*c*) cross-coupled form 1, X1.

TABLE 7.5-1 Elements of Second-Order Modules

	Structure		
	D1	D3	X1
Time-delay elements	2	4	2
Multipliers	4	4	6
Summing junctions	2	1	2

TABLE 7.5-2 Difference Equation Implementation of Second-Order Modules

D1:
$$v_k = -a_1 v_{k-1} - a_2 v_{k-2} + u_k$$
$$y_k = b_1 v_{k-1} + b_2 v_{k-2}$$

D3:
$$y_k = -a_1 y_{k-1} - a_2 y_{k-2} + b_1 u_{k-1} + b_2 u_{k-2}$$

X1:
$$v_k^1 = g_1 v_{k-1}^1 - g_2 v_{k-1}^2 + g_3 u_k$$
$$v_k^2 = g_1 v_{k-1}^2 + g_2 v_{k-1}^1 + g_4 u_k$$
$$y_k = v_{k-1}^2$$

where g_i are defined by

$$H_2(z^{-1}) = \frac{Nz^{-1}}{1 + pz^{-1}} + \frac{N^* z^{-1}}{1 + p^* z^{-1}}$$

$$g_1 = -\text{Re}(p)$$
$$g_2 = -\text{Im}(p)$$
$$g_3 = 2\,\text{Im}(N)$$
$$g_4 = 2\,\text{Re}(N)$$

interconnecting them to produce $H(z)$, scaling may be introduced between the modules (Phillips and Nagle, 1984).

7.6 SUMMARY

Since most aircraft control systems are implemented using digital signal processors, in this chapter we have outlined the basics of digital control. In Section 7.2 we discussed how to simulate digital control schemes using a Runge–Kutta integrator on the continuous aircraft dynamics. This approach yields the time responses not only at the sample points, but also between the samples. It is important to check the intersample performance of the closed-loop system before implementing a digital controller on an aircraft, since it can be unsatisfactory even though all is well at the sample points.

In Section 7.3 we gave a design technique for digital controllers that is based on *redesign of an existing continuous-time controller* by discretizing it using approximation techniques like the BLT and MPZ. This results in digital controllers that require small sample periods to work properly. To overcome the requirement for unreasonably small sample periods, in Section 7.4 we showed how to modify the continuous-time controller so that, after discretization, a better digital controller is obtained that works for larger sample periods. This modification allowed the delay properties of the sample-and-hold process to be taken into account.

Finally, in Section 7.5 we mentioned some digital controller implementation considerations. We showed how to design controllers with antiwindup protection to overcome the problems of saturation of the control signals due to control limitations such as elevator deflection stops and throttle maximum limits. We gave some low-order controller structures that allow the implementation of digital controllers with maximum accuracy and efficiency.

REFERENCES

Åström, K. J., and B. Wittenmark. *Computer Controlled Systems.* Englewood Cliffs, N.J.: Prentice-Hall, 1984.

Franklin, G. F., and J. D. Powell. *Digital Control.* Reading, Mass.: Addison-Wesley, 1980.

Franklin, G. F., J. D. Powell, and A. Emami-Naeini. *Feedback Control of Dynamic Systems.* Reading, Mass.: Addison-Wesley, 1980.

Hanselmann, H. "Implementation of Digital Controllers: A Survey." *Automatica* 23, no. 1 (1987): 7–32.

Kailath, T. *Linear Systems.* Englewood Cliffs, N.J.: Prentice Hall, 1980.

Lewis, F. L. *Applied Optimal Control and Estimation.* Englewood Cliffs, N.J.: Prentice Hall, 1992.

Oppenheim, A. V., and R. W. Schafer. *Digital Signal Processing.* Englewood Cliffs, N.J.: Prentice-Hall, 1975.

Phillips, C. L., and H. T. Nagle, Jr. *Digital Control System Analysis and Design.* Englewood Cliffs, N.J.: Prentice-Hall, 1984.

Slivinsky, C., and J. Borninski. "Control System Compensation and Implementation with the TMS32010." In *Digital Signal Processing Applications,* ed. K.-S. Lin. Englewood Cliffs, N.J.: Prentice Hall, 1987.

Su, K. L. *Time-Domain Synthesis of Linear Networks.* Englewood Cliffs, N.J.: Prentice Hall, 1971.

PROBLEMS

Section 7.3

7.3-1 Prove (7.3.13)/(7.3.14) by taking the Laplace transform of $\dot{x} = Ax + Bu$ and then using the BLT.

7.3-2 Digital Pitch-Rate Controller. Design a digital pitch-rate controller (see Example 7.3-2) using the MPZ technique. Simulate the step response for a few sample periods T and compare to the digital controller designed using the BLT.

7.3-3 Digital Normal Acceleration CAS. A normal acceleration CAS was designed in Example 5.4-1.

(a) Using the BLT, design a digital normal acceleration CAS controller. Simulate the time response for various sampling periods.

(b) Repeat using the modified MPZ.

7.3-4 Digital Wing Leveler. A wing leveler was designed in Example 5.5-4.

(a) Using the BLT, design a digital wing leveler. Simulate the time response for various sampling periods.

(b) Repeat using the modified MPZ.

Section 7.4

7.4-1 A Padé approximant for e^{-sT} is

$$G(s) = \frac{1 - 2sT/3 + (sT)^2/6}{1 + sT/3}$$

(a) Use long division to determine how many terms of the Taylor series of e^{-sT} are matched by $G(s)$.

(b) Use $G(s)$ to derive one of the approximants for the ZOH plus sampler shown in Table 7.4-2. How many terms of the Taylor series are matched by this approximant?

7.4-2 Digital Normal Acceleration CAS. A normal acceleration CAS was designed in Example 5.4-1. Using the BLT, design a digital normal acceleration CAS controller. Use modified continuous design, including the hold delay. Use $T = 0.1$ s. Simulate the time response and compare to the results of Problem 7.3-3.

7.4-3 Digital Wing Leveler. A wing leveler was designed in Example 5.5-4. Using the modified MPZ, design a digital wing leveler. Use modified continuous design, including the hold delay. Use $T = 0.1$ s. Simulate the time response and compare to the results of Problem 7.3-4.

Section 7.5

7.5-1 Antiwindup Compensator

(a) Write down the value of the state x_k in the antiwindup controller (7.5-3) for the deadbeat case where F_0 has all poles at the origin. Assume that w_k and u_k are constant and that $k > n$. Note that if F_0 has all poles at the origin, then $F_0^n = 0$, where n is the dimension of F_0.

(b) Repeat for the case where the controller is just an integrator so that $x_{k+1} = x_k + (T/T_I)w_k$, $u_k = \text{sat}(x_k)$. Simplify as far as possible.

7.5-2 Show how to determine the X1 difference equations in Table 7.5-2 directly from the complex Jordan form blocks corresponding to a complex pair of poles.

7.5-3 Antiwindup Protection for Normal Acceleration CAS. In Example 5.4-1 a normal acceleration CAS was designed; it had a PI controller in the feedforward loop. In the problems for Section 7.3 this design was digitized.

(a) Modify the digital normal acceleration CAS to add antiwindup protection.

(b) Now, set limits into the elevator actuator in your simulation program. Obtain time responses with and without the antiwindup protection.

7.5-4 Antiwindup Protection for Pitch-Rate CAS. Repeat the previous problem for the pitch-rate controller in Example 5.5-3, which was digitized in the problems for Section 7.3.

APPENDIX A

F-16 MODEL

This appendix contains the remainder of the data for the F-16 aircraft model given in Chapter 3. The usage of the lookup-tables will be made evident by referring to the aircraft model. These data, Appendix B, and the other programs used in this book can be obtained on a floppy disc, at a nominal cost, from Dr. B. L. Stevens, 1051 Park Manor Terr., Marietta, GA 30064.

Mass Properties

$$\text{Weight (lbs):} \quad W = 20{,}500$$
$$\text{Moments of Inertia (slug-ft}^2\text{):} \quad J_{xx} = 9{,}496$$
$$J_{yy} = 55{,}814$$
$$J_{zz} = 63{,}100$$
$$J_{xz} = 982$$

Wing Dimensions

$$\text{Span} = 30 \text{ ft}$$
$$\text{Area} = 300 \text{ ft}^2$$
$$\text{m.a.c} = 11.32 \text{ ft}$$

Reference CG Location

$$X_{cg} = 0.35 \, \bar{c}$$

Control Surface Actuator Models

	deflection limit	rate limit	time const.
Elevator	±25.0°,	60°/s,	0.0495 s lag
Ailerons	±21.5°,	80°/s,	0.0495 s lag
Rudder	±30.0°,	120°/s,	0.0495 s lag

Engine Angular Momentum

Assumed fixed at 160 slug-ft^2/s

Standard Atmosphere Model

```
      SUBROUTINE ADC(VT,ALT,AMACH,QBAR)   ! air data computer
      DATA R0/2.377E-3/                   ! sea-level density
      TFAC = 1.0 - 0.703E-5 * ALT
      T    = 519.0 * TFAC                 ! temperature
      IF (ALT .GE. 35000.0) T= 390.0
      RHO  = R0 * (TFAC**4.14)            ! density
      AMACH= VT/SQRT(1.4*1716.3*T)        ! Mach number
      QBAR = 0.5*RHO*VT*VT                ! dynamic pressure
C     PS   = 1715.0 * RHO * T             ! static pressure
      RETURN
      END
```

Engine Model

The F-16 engine power response is modeled by a first-order lag (in function PDOT, given below). The rest of the model consists of the throttle gearing (in TGEAR) and the lookup-tables for thrust as a function of operating power level, altitude, and Mach (in THRUST). In the thrust-lookup tables the rows correspond to a Mach number variation from 0 to 1.0 in increments of 0.2, and the columns correspond to altitudes from 0 to 50,000 ft in increments of 10,000 ft. There is a table for each of the power levels "idle," "military," and "maximum." The accompanying linear interpolation algorithm can extrapolate beyond the boundaries of a table, but the results may not be realistic.

```
FUNCTION TGEAR(THTL)    ! Power command v. thtl. relationship
IF(THTL.LE.0.77) THEN
  TGEAR = 64.94*THTL
ELSE
  TGEAR = 217.38*THTL-117.38
END IF
RETURN
END
FUNCTION PDOT(P3,P1)    ! PDOT= rate of change of power
IF (P1.GE.50.0) THEN    ! P3= actual power, P1= power command
  IF (P3.GE.50.0) THEN
    T=5.0
    P2=P1
  ELSE
    P2=60.0
    T=RTAU(P2-P3)
  END IF
ELSE
  IF (P3.GE.50.0) THEN
    T=5.0
    P2=40.0
  ELSE
    P2=P1
    T=RTAU(P2-P3)
  END IF
END IF
PDOT=T*(P2-P3)
RETURN
END
FUNCTION RTAU(DP)       ! used by function PDOT
IF (DP.LE.25.0) THEN
    RTAU=1.0            ! reciprocal time constant
ELSE IF (DP.GE.50.0)THEN
    RTAU=0.1
ELSE
    RTAU=1.9-.036*DP
END IF
RETURN
END
```

```fortran
      FUNCTION THRUST(POW,ALT,RMACH)    ! Engine thrust model
      REAL A(0:5,0:5), B(0:5,0:5), C(0:5,0:5)
      DATA A/
     + 1060.0,   670.0,   880.0,  1140.0,  1500.0,  1860.0,
     +  635.0,   425.0,   690.0,  1010.0,  1330.0,  1700.0,
     +   60.0,    25.0,   345.0,   755.0,  1130.0,  1525.0,
     +-1020.0,  -710.0,  -300.0,   350.0,   910.0,  1360.0,
     +-2700.0, -1900.0, -1300.0,  -247.0,   600.0,  1100.0,
     +-3600.0, -1400.0,  -595.0,  -342.0,  -200.0,   700.0/
C         mil data now
      DATA B/
     +12680.0,  9150.0,  6200.0,  3950.0,  2450.0,  1400.0,
     +12680.0,  9150.0,  6313.0,  4040.0,  2470.0,  1400.0,
     +12610.0,  9312.0,  6610.0,  4290.0,  2600.0,  1560.0,
     +12640.0,  9839.0,  7090.0,  4660.0,  2840.0,  1660.0,
     +12390.0, 10176.0,  7750.0,  5320.0,  3250.0,  1930.0,
     +11680.0,  9848.0,  8050.0,  6100.0,  3800.0,  2310.0/
C         max data now
      DATA C/
     +20000.0, 15000.0, 10800.0,  7000.0,  4000.0,  2500.0,
     +21420.0, 15700.0, 11225.0,  7323.0,  4435.0,  2600.0,
     +22700.0, 16860.0, 12250.0,  8154.0,  5000.0,  2835.0,
     +24240.0, 18910.0, 13760.0,  9285.0,  5700.0,  3215.0,
     +26070.0, 21075.0, 15975.0, 11115.0,  6860.0,  3950.0,
     +28886.0, 23319.0, 18300.0, 13484.0,  8642.0,  5057.0/
C
              H = .0001*ALT
              I = INT(H)
              IF(I.GE.5)I=4
              DH= H-FLOAT(I)
              RM= 5.0*RMACH
              M = INT(RM)
              IF(M.GE.5)M=4
              DM= RM-FLOAT(M)
              CDH=1.0-DH
      S= B(I,M)   *CDH + B(I+1,M)   *DH
      T= B(I,M+1) *CDH + B(I+1,M+1) *DH
      TMIL= S + (T-S)*DM
          IF( POW .LT. 50.0 ) THEN
      S= A(I,M)   *CDH + A(I+1,M)   *DH
      T= A(I,M+1) *CDH + A(I+1,M+1) *DH
      TIDL= S + (T-S)*DM
      THRUST=TIDL+(TMIL-TIDL)*POW*.02
          ELSE
      S= C(I,M)   *CDH + C(I+1,M)   *DH
      T= C(I,M+1) *CDH + C(I+1,M+1) *DH
      TMAX= S + (T-S)*DM
      THRUST=TMIL+(TMAX-TMIL)*(POW-50.0)*.02
          END IF
      RETURN
      END
```

Aerodynamic Data

The aerodynamic data tables and associated interpolation algorithms, given below, will provide values for the body-axes dimensionless aerodynamic coefficients of the F-16 model at arbitrary values of the independent variables. The angle-of-attack range of the tables is from -10 to 45 degrees in 5 degree increments, and the sideslip angle range is from -30 to 30 degrees in either 5 or 10 degree increments. The given interpolation algorithm interpolates linearly between the data points; it will extrapolate beyond the table boundaries, but the results may be unrealistic.

```
      SUBROUTINE DAMP(ALPHA, D)  ! various damping coefficients
      REAL A(-2:9,9),D(9)
      DATA A/
     & -.267,   -.110,    .308,   1.34,    2.08,    2.91,    2.76,
     &  2.05,   1.50,    1.49,   1.83,    1.21,
     &  .882,   .852,    .876,   .958,    .962,    .974,    .819,
     &  .483,   .590,   1.21,   -.493,   -1.04,
     & -.108,  -.108,   -.188,   .110,    .258,    .226,    .344,
     &  .362,   .611,    .529,   .298,   -2.27,
     & -8.80, -25.8,   -28.9,  -31.4,   -31.2,   -30.7,   -27.7,
     & -28.2, -29.0,   -29.8,  -38.3,   -35.3,
     & -.126,  -.026,    .063,   .113,    .208,    .230,    .319,
     &  .437,   .680,    .100,   .447,   -.330,
     & -.360,  -.359,   -.443,  -.420,   -.383,   -.375,   -.329,
     & -.294,  -.230,   -.210,  -.120,   -.100,
     & -7.21,  -.540,   -5.23,  -5.26,   -6.11,   -6.64,   -5.69,
     & -6.00,  -6.20,   -6.40,  -6.60,   -6.00,
     & -.380,  -.363,   -.378,  -.386,   -.370,   -.453,   -.550,
     & -.582,  -.595,   -.637, -1.02,    -.840,
     &  .061,   .052,    .052,  -.012,   -.013,   -.024,    .050,
     &  .150,   .130,    .158,   .240,    .150/
C
      S= 0.2 * ALPHA
      K= INT(S)
      IF(K .LE. -2) K= -1
      IF(K .GE.  9) K=  8
      DA= S - FLOAT(K)
      L = K + INT( SIGN(1.1,DA) )
      DO 1, I= 1,9
  1   D(I)= A(K,I) + ABS(DA) * (A(L,I) - A(K,I))
      END
C
C D1= CXq; D2= CYr; D3= CYp; D4= CZq; D5= Clr; D6= Clp
C D7= Cmq; D8= Cnr; D9= Cnp
```

F-16 MODEL

```
      FUNCTION CX(ALPHA,EL)   ! x-axis aerodynamic force coeff.
      REAL A(-2:9,-2:2)
      DATA A/
     & -.099, -.081, -.081, -.063, -.025,  .044,  .097,
     &  .113,  .145,  .167,  .174,  .166,
     & -.048, -.038, -.040, -.021,  .016,  .083,  .127,
     &  .137,  .162,  .177,  .179,  .167,
     & -.022, -.020, -.021, -.004,  .032,  .094,  .128,
     &  .130,  .154,  .161,  .155,  .138,
     & -.040, -.038, -.039, -.025,  .006,  .062,  .087,
     &  .085,  .100,  .110,  .104,  .091,
     & -.083, -.073, -.076, -.072, -.046,  .012,  .024,
     &  .025,  .043,  .053,  .047,  .040/
C
      S= 0.2 * ALPHA
      K= INT(S)
      IF(K .LE. -2) K= -1
      IF(K .GE.  9) K=  8
      DA= S - FLOAT(K)
      L = K + INT( SIGN(1.1,DA) )
      S= EL/12.0
      M= INT(S)
      IF(M .LE. -2) M= -1
      IF(M .GE.  2) M=  1
      DE= S - FLOAT(M)
      N= M + INT( SIGN(1.1,DE) )
      T= A(K,M)
      U= A(K,N)
      V= T + ABS(DA) * (A(L,M) - T)
      W= U + ABS(DA) * (A(L,N) - U)
      CX= V + (W-V)  * ABS(DE)
      RETURN
      END

      FUNCTION CY(BETA,AIL,RDR)   ! sideforce coefficient
      CY= -.02*BETA + .021*(AIL/20.0) + .086*(RDR/30.0)
      END

      FUNCTION CZ(ALPHA,BETA,EL) ! z-axis force coeff.
      REAL A(-2:9)
      DATA A/  .770,  .241, -.100, -.416, -.731, -1.053,
     & -1.366, -1.646, -1.917, -2.120, -2.248, -2.229/
      S= 0.2 * ALPHA
      K= INT(S)
      IF(K .LE. -2) K= -1
      IF(K .GE.  9) K=  8
      DA= S - FLOAT(K)
      L = K + INT( SIGN(1.1,DA) )
      S= A(K) + ABS(DA) * (A(L) - A(K))
      CZ= S*(1-(BETA/57.3)**2) - .19*(EL/25.0)
      END
```

APPENDIX A

```
      FUNCTION CM(ALPHA,EL)  ! pitching moment coeff.
      REAL A(-2:9,-2:2)
      DATA A/
     &  .205,  .168,  .186,  .196,  .213,  .251,  .245,
     &  .238,  .252,  .231,  .198,  .192,
     &  .081,  .077,  .107,  .110,  .110,  .141,  .127,
     &  .119,  .133,  .108,  .081,  .093,
     & -.046, -.020, -.009, -.005, -.006,  .010,  .006,
     & -.001,  .014,  .000, -.013,  .032,
     & -.174, -.145, -.121, -.127, -.129, -.102, -.097,
     & -.113, -.087, -.084, -.069, -.006,
     & -.259, -.202, -.184, -.193, -.199, -.150, -.160,
     & -.167, -.104, -.076, -.041, -.005/
C    SAME INTERPOLATION AS CX ********************
C
      FUNCTION CL(ALPHA,BETA)   ! rolling moment coeff.
      REAL A(-2:9,0:6)
      DATA A/12*0,
     & -.001, -.004, -.008, -.012, -.016, -.019, -.020,
     & -.020, -.015, -.008, -.013, -.015,
     & -.003, -.009, -.017, -.024, -.030, -.034, -.040,
     & -.037, -.016, -.002, -.010, -.019,
     & -.001, -.010, -.020, -.030, -.039, -.044, -.050,
     & -.049, -.023, -.006, -.014, -.027,
     &  .000, -.010, -.022, -.034, -.047, -.046, -.059,
     & -.061, -.033, -.036, -.035, -.035,
     &  .007, -.010, -.023, -.034, -.049, -.046, -.068,
     & -.071, -.060, -.058, -.062, -.059,
     &  .009, -.011, -.023, -.037, -.050, -.047, -.074,
     & -.079, -.091, -.076, -.077, -.076/
C
      S= 0.2 * ALPHA
      K= INT(S)
      IF(K .LE. -2) K= -1
      IF(K .GE.  9) K=  8
      DA= S - FLOAT(K)
      L = K + INT( SIGN(1.1,DA) )
      S- .2* ABS(BETA)
      M= INT(S)
      IF(M .EQ. 0) M= 1
      IF(M .GE. 6) M= 5
      DB= S - FLOAT(M)
      N= M + INT( SIGN(1.1,DB) )
      T= A(K,M)
      U= A(K,N)
      V= T + ABS(DA) * (A(L,M) - T)
      W= U + ABS(DA) * (A(L,N) - U)
      DUM= V + (W-V) * ABS(DB)
      CL= DUM + SIGN(1.0, BETA)
      RETURN
      END
```

```
      FUNCTION CN(ALPHA,BETA)  ! yawing moment coeff.
      REAL A(-2:9,0:6)
      DATA A/12*0,
     &   .018,   .019,   .018,   .019,   .019,   .018,   .013,
     &   .007,   .004,  -.014,  -.017,  -.033,
     &   .038,   .042,   .042,   .042,   .043,   .039,   .030,
     &   .017,   .004,  -.035,  -.047,  -.057,
     &   .056,   .057,   .059,   .058,   .058,   .053,   .032,
     &   .012,   .002,  -.046,  -.071,  -.073,
     &   .064,   .077,   .076,   .074,   .073,   .057,   .029,
     &   .007,   .012,  -.034,  -.065,  -.041,
     &   .074,   .086,   .093,   .089,   .080,   .062,   .049,
     &   .022,   .028,  -.012,  -.002,  -.013,
     &   .079,   .090,   .106,   .106,   .096,   .080,   .068,
     &   .030,   .064,   .015,   .011,  -.001/
C  NOW USE SAME INTERPOLATION AS CL ********************
C
      FUNCTION DLDA(ALPHA,BETA)  ! rolling mom. due to ailerons
      REAL A(-2:9,-3:3)
      DATA A/-.041, -.052, -.053, -.056, -.050, -.056, -.082,
     &  -.059, -.042, -.038, -.027, -.017,
     &  -.041, -.053, -.053, -.053, -.050, -.051, -.066,
     &  -.043, -.038, -.027, -.023, -.016,
     &  -.042, -.053, -.052, -.051, -.049, -.049, -.043,
     &  -.035, -.026, -.016, -.018, -.014,
     &  -.040, -.052, -.051, -.052, -.048, -.048, -.042,
     &  -.037, -.031, -.026, -.017, -.012,
     &  -.043, -.049, -.048, -.049, -.043, -.042, -.042,
     &  -.036, -.025, -.021, -.016, -.011,
     &  -.044, -.048, -.048, -.047, -.042, -.041, -.020,
     &  -.028, -.013, -.014, -.011, -.010,
     &  -.043, -.049, -.047, -.045, -.042, -.037, -.003,
     &  -.013, -.010, -.003, -.007, -.008/
      S= 0.2 * ALPHA
      K= INT(S)
      IF(K .LE. -2) K= -1
      IF(K .GE.  9) K=  8
      DA= S - FLOAT(K)
      L = K + INT( SIGN(1.1,DA) )
      S= 0.1 * BETA
      M= INT(S)
      IF(M .eq. -3) M= -2
      IF(M .GE.  3) M=  2
      DB= S - FLOAT(M)
      N= M + INT( SIGN(1.1,DB) )
      T= A(K,M)
      U= A(K,N)
      V= T + ABS(DA) * (A(L,M) - T)
      W= U + ABS(DA) * (A(L,N) - U)
      DLDA= V + (W-V)  * ABS(DB)
      RETURN
      END
```

```
      FUNCTION DLDR(ALPHA,BETA) ! rolling moment due to rudder
      REAL A(-2:9,-3:3)      ! use same interpolation as DLDA
      DATA A/ .005,   .017,    .014,    .010,   -.005,   .009,   .019,
     &  .005,  -.000,  -.005,  -.011,   .008,
     &  .007,   .016,   .014,   .014,   .013,   .009,   .012,
     &  .005,   .000,   .004,   .009,   .007,
     &  .013,   .013,   .011,   .012,   .011,   .009,   .008,
     &  .005,  -.002,   .005,   .003,   .005,
     &  .018,   .015,   .015,   .014,   .014,   .014,   .014,
     &  .015,   .013,   .011,   .006,   .001,
     &  .015,   .014,   .013,   .013,   .012,   .011,   .011,
     &  .010,   .008,   .008,   .007,   .003,
     &  .021,   .011,   .010,   .011,   .010,   .009,   .008,
     &  .010,   .006,   .005,   .000,   .001,
     &  .023,   .010,   .011,   .011,   .011,   .010,   .008,
     &  .010,   .006,   .014,   .020,   .000/
C
      FUNCTION DNDA(ALPHA,BETA) ! yawing moment due to ailerons
      REAL A(-2:9,-3:3)      ! use same interpolation as DLDA *
      DATA A/ .001,  -.027,   -.017,   -.013,  -.012,  -.016,   .001,
     &  .017,   .011,   .017,   .008,   .016,
     &  .002,  -.014,  -.016,  -.016,  -.014,  -.019,  -.021,
     &  .002,   .012,   .015,   .015,   .011,
     & -.006,  -.008,  -.006,  -.006,  -.005,  -.008,  -.005,
     &  .007,   .004,   .007,   .006,   .006,
     & -.011,  -.011,  -.010,  -.009,  -.008,  -.006,   .000,
     &  .004,   .007,   .010,   .004,   .010,
     & -.015,  -.015,  -.014,  -.012,  -.011,  -.008,  -.002,
     &  .002,   .006,   .012,   .011,   .011,
     & -.024,  -.010,  -.004,  -.002,  -.001,   .003,   .014,
     &  .006,  -.001,   .004,   .004,   .006,
     & -.022,   .002,  -.003,  -.005,  -.003,  -.001,  -.009,
     & -.009,  -.001,   .003,  -.002,   .001/
C
      FUNCTION DNDR(ALPHA,BETA) ! yawing moment due to rudder
      REAL A(-2:9,-3:3)
      DATA A/ -.018, -.052, -.052, -.052, -.054, -.049, -.059,
     & -.051,  -.030,  -.037,  -.026,  -.013,
     & -.028,  -.051,  -.043,  -.046,  -.045,  -.049,  -.057,
     & -.052,  -.030,  -.033,  -.030,  -.008,
     & -.037,  -.041,  -.038,  -.040,  -.040,  -.038,  -.037,
     & -.030,  -.027,  -.024,  -.019,  -.013,
     & -.048,  -.045,  -.045,  -.045,  -.044,  -.045,  -.047,
     & -.048,  -.049,  -.045,  -.033,  -.016,
     & -.043,  -.044,  -.041,  -.041,  -.040,  -.038,  -.034,
     & -.035,  -.035,  -.029,  -.022,  -.009,
     & -.052,  -.034,  -.036,  -.036,  -.035,  -.028,  -.024,
     & -.023,  -.020,  -.016,  -.010,  -.014,
     & -.062,  -.034,  -.027,  -.028,  -.027,  -.027,  -.023,
     & -.023,  -.019,  -.009,  -.025,  -.010/
C  NOW USE SAME INTERPOLATION AS DLDA ********************
```

APPENDIX B

SOFTWARE

This appendix contains the Fortran code that is required to use the aircraft models given in the text and is not otherwise readily available. For the steady-state trim algorithm (Section B1) we give the basic trimmer subroutine, part of the constraint subroutine, a cost function, and the Simplex minimization algorithm. The user must write a driver program and add additional flight path constraints, as required. In Section B2 a subroutine for numerical linearization is given, and the user need only add a driver program. Software for time-history simulation and control systems design is readily available from other sources and so, in the rest of this appendix, we have given only the Runge–Kutta algorithm that was used for most of the examples.

Appendix A, Appendix B, and the programs used in this book can be obtained on a floppy disc, at a nominal cost, from Dr. B. L. Stevens, 1051 Park Manor Terr., Marietta, GA 30064.

B1. AIRCRAFT STEADY-STATE TRIM CODE

The subroutine "TRIMMER" (below) sets up a function minimization algorithm to determine a steady-state trim condition for either a 6-DOF or 3-DOF (longitudinal only) aircraft model. The subroutine arguments are the number of degrees of freedom (NV) and the "COST" function (which must be declared "EXTERNAL" in the main program). Labeled COMMON storage is used to pass the state and control vectors to and from the main program and the cost function (and, in the case of the control vector, the aircraft model also).

The main program must initialize the state vector according to the trim condition required, the control vector can simply be set to zero initially. It must also set the turn-rate, roll-rate, or pull-up rate, set flags for coordinated turns or stability-axis roll, and pass these through a common block ("CONSTRNT") to the constraint routine. A Simplex routine (given below) is used for function minimization, and it returns the coordinates of the cost function minimum in the simplex vector S. The cost function is then called once more to set the state and control vectors to their final values, and these values are passed through COMMON to the main program to be placed in a data file. Subroutine "SMPLX" can easily be replaced by "ZXMWD" from the IMSL library, or "AMOEBA" from "Numerical Recipes" if desired. The author is indebted to Dr. P. Vesty for this simplex routine.

```
      SUBROUTINE TRIMMER (NV, COST)
      PARAMETER (NN=20, MM=10)
      EXTERNAL COST
      CHARACTER*1 ANS
      DIMENSION S(6), DS(6)
      COMMON/ STATE/ X(NN)
      COMMON/ CONTROLS/ U(MM)
      COMMON/ OUTPUT/ AN,AY,AX,QBAR,AMACH   ! common to aircraft
      DATA RTOD /57.29577951/
      S(1)= U(1)
      S(2)= U(2)
      S(3)= X(2)
      IF(NV .LE. 3) GO TO 10
      S(4)= U(3)
      S(5)= U(4)
      S(6)= X(3)
10    DS(1) = 0.2
      DS(2) = 1.0
      DS(3) = 0.02
      IF(NV .LE. 3) GO TO 20
      DS(4) = 1.0
      DS(5) = 1.0
      DS(6) = 0.02
20    NC= 1000
      WRITE(*,'(1X,A,$)')'Reqd. # of trim iterations (def. = 1000) : '
      READ(*,*,ERR=20)   NC
      SIGMA = -1.0
      CALL SMPLX(COST,NV,S,DS,SIGMA,NC,F0,FFIN)
      FFIN = COST(S)
        IF (NV .GT. 3) THEN
      WRITE(*,'(/11X,A)')'Throttle    Elevator,    Ailerons,    Rudder'
      WRITE(*,'(9X,4(1PE10.2,3X),/)') U(1), U(2), U(3), U(4)
      WRITE(*,99)'Angle of attack',RTOD*X(2),'Sideslip angle',RTOD*X(3)
      WRITE(*,99) 'Pitch angle', RTOD*X(5), 'Bank angle', RTOD*X(4)
      WRITE(*,99) 'Normal acceleration', AN, 'Lateral acceln', AY
      WRITE(*,99) 'Dynamic pressure', QBAR, 'Mach number', AMACH
        ELSE
      WRITE(*,'(/1X,A)')'  Throttle      Elevator      Alpha       Pitch'
      WRITE(*,'(1X,4(1PE10.2,3X))')U(1),U(2),X(2)*RTOD,X(3)*RTOD
      WRITE(*,'(/1X,A)')'Normal acceleration  Dynamic Pressure  Mach '
      WRITE(*,'(5X,3(1PE10.2,7X))') AN,QBAR,AMACH
        END IF
      WRITE(*,99)'Initial cost function ',F0,'Final cost function',FFIN
99    FORMAT(2(1X,A22,1PE10.2))
40    WRITE(*,'(/1X,A,$)') 'More Iterations ? (def= Y) : '
      READ(*,'(A)',ERR= 40) ANS
      IF (ANS .EQ. 'Y'.OR. ANS .EQ. 'y'.OR. ANS .EQ. '/') GO TO 10
      IF (ANS .EQ. 'N'.OR. ANS .EQ. 'n') RETURN
      GO TO 40
      END
```

```
      FUNCTION CLF16 (S)    ! F16 cost function (see text)
      PARAMETER (NN=20)
      REAL S(*), XD(NN)
      COMMON/STATE/X(NN)    ! common to main program
      COMMON/CONTROLS/THTL,EL,AIL,RDR   ! to aircraft
      THTL = S(1)
      EL   = S(2)
      X(2) = S(3)
      AIL  = S(4)
      RDR  = S(5)
      X(3) = S(6)
      X(13)= TGEAR (THTL)
      CALL   CONSTR (X)
      CALL   F (TIME,X,XD)
      CLF16 = XD(1)**2 + 100.0*( XD(2)**2 + XD(3)**2 )
     &      + 10.0*( XD(7)**2 + XD(8)**2 + XD(9)**2 )
      RETURN
      END

      SUBROUTINE CONSTR (X)  ! used by COST, to apply constraints
      DIMENSION X(*)
      LOGICAL COORD, STAB
      COMMON/CNSTRNT/RADGAM,SINGAM,RR,PR,TR,PHI,CPHI,SPHI,COORD,STAB
C     common to main program.
      CALPH = COS(X(2))
      SALPH = SIN(X(2))
      CBETA = COS(X(3))
      SBETA = SIN(X(3))
      IF (COORD) THEN
        ! coordinated turn logic here
      ELSE IF (TR .NE. 0.0) THEN
        ! skidding turn logic here
      ELSE       ! non-turning flight
        X(4) = PHI
        D  = X(2)
        IF(PHI .NE. 0.0) D = -X(2)      ! inverted
        IF( SINGAM .NE. 0.0 ) THEN      ! climbing
        SGOCB = SINGAM / CBETA
        X(5)  = D + ATAN( SGOCB/SQRT(1.0-SGOCB*SGOCB)) ! roc constraint
        ELSE
        X(5)  = D                       ! level
        END IF
        X(7)= RR
        X(8)= PR
        IF (STAB) THEN                  ! stab.-axis roll
          X(9)= RR*SALPH/CALPH
        ELSE
          X(9) = 0.0                    ! body-axis roll
        END IF
      END IF
      RETURN
      END
```

```
      SUBROUTINE SMPLX(FX,N,X,DX,SD,M,Y0,YL)
C This simplex algorithm minimizes FX(X), where X is (Nx1).
C DX contains the initial perturbations in X. SD should be set according
C to the tolerance required; when SD<0 the algorithm calls FX M times
      REAL X(*), DX(*)
      DIMENSION XX(32), XC(32), Y(33), V(32,32)
C
      NV=N+1
      DO 2 I=1,N
      DO 1 J=1,NV
1     V(I,J)=X(I)
2     V(I,I+1)=X(I)+DX(I)
      Y0=FX(X)
      Y(1)=Y0
      DO 3 J=2,NV
3     Y(J)=FX(V(1,J))
      K=NV
4     YH=Y(1)
      YL=Y(1)
      NH=1
      NL=1
      DO 5 J=2,NV
      IF(Y(J).GT.YH) THEN
         YH=Y(J)
         NH=J
      ELSEIF(Y(J).LT.YL) THEN
         YL=Y(J)
         NL=J
      ENDIF
5     CONTINUE
      YB=Y(1)
      DO 6 J=2,NV
6     YB=YB+Y(J)
      YB=YB/NV
      D=0.0
      DO 7 J=1,NV
7     D=D+(Y(J)-YB)**2
      SDA=SQRT(D/NV)
      IF((K.GE.M).OR.(SDA.LE.SD)) THEN
         SD=SDA
         M=K
         YL=Y(NL)
         DO 8 I=1,N
8        X(I)=V(I,NL)
         RETURN
      END IF
      DO 10 I=1,N
      XC(I)=0.0
      DO 9 J=1,NV
9     IF(J.NE.NH) XC(I)=XC(I)+V(I,J)
10    XC(I)=XC(I)/N
      DO 11 I=1,N
11    X(I)=XC(I)+XC(I)-V(I,NH)
```

```
      K=K+1
      YR=FX(X)
      IF(YR.LT.YL) THEN
      DO 12 I=1,N
12    XX(I)=X(I)+X(I)-XC(I)
      K=K+1
      YE=FX(XX)
      IF(YE.LT.YR) THEN
        Y(NH)=YE
        DO 13 I=1,N
13      V(I,NH)=XX(I)
      ELSE
        Y(NH)=YR
        DO 14 I=1,N
14      V(I,NH)=X(I)
      END IF
        GOTO 4
      ENDIF
      Y2=Y(NL)
      DO 15 J=1,NV
15    IF((J.NE.NL).AND.(J.NE.NH).AND.(Y(J).GT.Y2)) Y2=Y(J)
      IF(YR.LT.YH) THEN
        Y(NH)=YR
        DO 16 I=1,N
16  V(I,NH)=X(I)
        IF(YR.LT.Y2) GO TO 4
      ENDIF
      DO 17 I=1,N
17    XX(I)=0.5*(V(I,NH)+XC(I))
      K=K+1
      YC=FX(XX)
      IF(YC.LT.YH) THEN
        Y(NH)=YC
        DO 18 I=1,N
18      V(I,NH)=XX(I)
      ELSE
        DO 20 J=1,NV
        DO 19 I-1,N
19      V(I,J)=0.5*(V(I,J)+V(I,NL))
20      IF(J.NE.NL) Y(J)=FX(V(1,J))
        K=K+N
      ENDIF
      GO TO 4
      END
```

B2. NUMERICAL LINEARIZATION SUBROUTINE

Subroutine JACOB will calculate Jacobian matrices for the set of nonlinear state equations contained in the subroutine F (specified as an argument of JACOB). Subroutine F(TIME, X, XD) should contain "CONTROLS" and "OUTPUT" common blocks as used in the text. The argument FN is a double-precision function used to determine an approximation to each partial derivative that is required.

To calculate the *A, B, C, D* matrices the main program should be designed to call JACOB four times, with FN replaced in turn by each of the partial derivative functions FDX, FDU, YDX, and YDU (given below). The partial derivative functions must be declared "EXTERNAL" in the main program. The vectors X and XD are respectively the state vector and its derivative. The vector V must contain the equilibrium condition and should be replaced by X or U respectively, depending on whether the partial derivatives with respect to X or U are being calculated. The array IO is used to specify the set of integers corresponding to the rows of the Jacobian matrix, and JO is used to specify the set corresponding to the columns. NR and NC are respectively the number of rows and the number of columns in the Jacobian matrix and the linear array ABC contains the columns of the Jacobian matrix, stacked one after the other.

The linearization algorithm chooses smaller and smaller perturbations in the independent variable and compares three successive approximations to the particular partial derivative. If these approximations agree within a certain tolerance, then the size of the perturbation is reduced to determine if an even smaller tolerance can be satisfied. The algorithm terminates successfully when a tolerance TOLMIN is reached or if a tolerance of at least OKTOL can be achieved. If the algorithm does not terminate successfully, then the successive approximations are displayed and the user is asked to decide on the value of the partial derivative.

```
      SUBROUTINE JACOB (FN,F,X,XD,V,IO,JO,ABC,NR,NC)
      DIMENSION X(*),XD(*),V(*),IO(*),JO(*),ABC(*)
      EXTERNAL FN,F
      LOGICAL FLAG, DIAGS
      CHARACTER*1 ANS
      REAL*8 FN,TDV
      DATA DEL,DMIN,TOLMIN,OKTOL /.01, .5, 3.3E-5, 8.1E-4/
C
      DIAGS= .TRUE.
      PRINT '(1X,A,$)', 'DIAGNOSTICS ?   (Y/N, "/"=N) '
      READ(*,'(A)') ANS
      IF (ANS .EQ.'/'.OR. ANS .EQ. 'N' .OR. ANS .EQ. 'n')DIAGS=.FALSE.
      IJ= 1
      DO 40   J=1,NC
      DV= AMAX1( ABS( DEL*V(JO(J)) ), DMIN )
      DO 40   I=1,NR
      FLAG= .FALSE.
1     TOL= 0.1
      OLTOL= TOL
      TDV= DBLE( DV )
      A2= 0.0
```

```
          A1= 0.0
          A0= 0.0
          B1= 0.0
          B0= 0.0
          D1= 0.0
          D0= 0.0
          IF (DIAGS .OR. FLAG) WRITE(*,'(/1X,A8,I2,A1,I2,11X,A12,8X,A5)')
        & 'Element ',I,',',J, 'perturbation','slope'
          DO 20   K= 1,18                  ! iterations on TDV
          A2= A1
          A1= A0
          B1= B0
          D1= D0
          A0= FN(F,XD,X,IO(I),JO(J),TDV)
          Bϕ= AMIN1( ABS(A0), ABS(A1) )
          D0= ABS ( A0 - A1 )
          IF (DIAGS .OR. FLAG) WRITE(*,'(20X,1P2E17.6)') TDV,A0
          IF(K .LE. 2) GO TO 20
          IF (A0 .EQ. A1 .AND. A1 .EQ. A2) THEN
              ANS2= A1
              GO TO 30
          END IF
          IF (A0 .EQ. 0.0) GO TO 25
10        IF( D0 .LE. TOL*B0 .AND. D1 .LE. TOL*B1) THEN
              ANS2= A1
              OLTOL= TOL
              IF(DIAGS .OR. FLAG) WRITE(*,'(1X,A9,F8.7)') 'MET TOL= ',TOL
              IF (TOL .LE. TOLMIN) THEN
                 GO TO 30
              ELSE
                 TOL= 0.2*TOL
                 GO TO 10
              END IF
          END IF
20        TDV= 0.6D0*TDV
25        IF (OLTOL .LE. OKTOL) THEN
             GO TO 30
          ELSE IF (.NOT. FLAG) THEN
             WRITE(*,'(/1X,A)')'NO CONVERGENCE *****'
             FLAG= .TRUE.
             GO TO 1
          ELSE
21           WRITE(*,'(1X,A,$)') 'Enter estimate : '
             READ(*,*,ERR=21) ANS2
             FLAG= .FALSE.
             GO TO 30
          END IF
30        ABC(IJ)= ANS2
          IF (DIAGS) THEN
             PRINT '(27X,A5,1PE13.6)','Ans= ',ANS2
             PAUSE 'Press "enter"'
          END IF
40        IJ= IJ+1
```

```
      RETURN
      END

      DOUBLE PRECISION FUNCTION  FDX(F,XD,X,I,J,DDX)
      REAL*4  XD(*), X(*)
      DOUBLE PRECISION T, DDX, XD1, XD2
      EXTERNAL F
      TIME= 0.0
      T   = DBLE( X(J) )
      X(J)= SNGL( T - DDX )
      CALL  F(TIME,X,XD)
      XD1 = DBLE( XD(I) )
      X(J)= SNGL( T + DDX )
      CALL  F(TIME,X,XD)
      XD2 = DBLE( XD(I) )
      FDX = (XD2-XD1)/(DDX+DDX)
      X(J)= SNGL( T )
      RETURN
      END
```

```
      DOUBLE PRECISION FUNCTION FDU(F,XD,X,I,J,DDU)
      PARAMETER (NIN=10)
      REAL*4  XD(*), X(*)
      COMMON/CONTROLS/U(NIN)
      DOUBLE PRECISION  T, DDU, XD1, XD2
      EXTERNAL F
      TIME= 0.0
      T   = DBLE( U(J) )
      U(J)= SNGL( T - DDU )
      CALL  F(TIME,X,XD)
      XD1 = DBLE( XD(I) )
      U(J)= SNGL( T + DDU )
      CALL  F(TIME,X,XD)
      XD2 = DBLE( XD(I) )
      FDU = (XD2-XD1)/(DDU+DDU)
      U(J)= SNGL( T )
      RETURN
      END
      DOUBLE PRECISION FUNCTION YDX(F,XD,X,I,J,DDX)
      PARAMETER (NOP=20)
      REAL*4  XD(*), X(*)
      COMMON/OUTPUT/Y/(NOP)
      DOUBLE PRECISION  T, DDX, YD1, YD2
      EXTERNAL F
      TIME= 0.0
      T   = DBLE( X(J) )
      X(J)= SNGL( T - DDX )
      CALL  F(TIME,X,XD)
      YD1 = DBLE( Y(I) )
      X(J)= SNGL( T + DDX )
      CALL  F(TIME,X,XD)
      YD2 = DBLE( Y(I) )
      YDX = (YD2-YD1)/(DDX+DDX)
      X(J)= SNGL(T)
      RETURN
      END
      DOUBLE PRECISION FUNCTION YDU(F,XD,X,I,J,DDU)
      PARAMETER (NIN=10, NOP=20)
      REAL*4  XD(*), X(*)
      COMMON/CONTROLS/U(NIN)
      COMMON/OUTPUT/Y(NOP)
      DOUBLE PRECISION  T, DDU, YD1, YD2
      EXTERNAL F
      TIME= 0.0
      T   = DBLE( U(J) )
      U(J)= SNGL( T - DDU )
      CALL  F(TIME,X,XD)
      YD1 = DBLE( Y(I) )
      U(J)= SNGL( T + DDU )
      CALL  F(TIME,X,XD)
      YD2 = DBLE( Y(I) )
      YDU = (YD2-YD1)/(DDU+DDU)
      U(J)= SNGL(T)
      RETURN
      END
```

B3. RUNGE–KUTTA INTEGRATION

This subroutine implements "Runge's fourth-order rule" as described in Chapter 3. Its arguments are the subroutine F containing the nonlinear state equations, the current time TT, the integration time-step DT, the state and state-derivative vectors XX and XD, and the number of state variables NX. Subroutine F should be declared EXTERNAL in the main program unit.

```
      SUBROUTINE RK4(F,TT,DT,XX,XD,NX)
      PARAMETER (NN=30) ! same as main prog.
      REAL*4 XX(*),XD(*),X(NN),XA(NN)
      CALL F(TT,XX,XD)
      DO 1 M=1,NX
      XA(M)=XD(M)*DT
1     X(M)=XX(M)+0.5*XA(M)
      T=TT+0.5*DT
      CALL F(T,X,XD)
      DO 2 M=1,NX
      Q=XD(M)*DT
      X(M)=XX(M)+0.5*Q
2     XA(M)=XA(M)+Q+Q
      CALL F(T,X,XD)
      DO 3 M=1,NX
      Q=XD(M)*DT
      X(M)=XX(M)+Q
3     XA(M)=XA(M)+Q+Q
      TT=TT+DT
      CALL F(TT,X,XD)
      DO 4 M=1,NX
4     XX(M)=XX(M)+(XA(M)+XD(M)*DT)/6.0
      RETURN
      END
```

B4. OUTPUT-FEEDBACK DESIGN

Output-feedback design is not an easy problem. Finding the optimal output-feedback gains to minimize a quadratic performance index (PI)

$$J = \frac{1}{2} \int_0^\infty \left(x^T Q x + u^T R u \right) dt, \tag{B.7.1}$$

involves solving coupled nonlinear matrix design equations of the form (Chapter 5)

$$0 = \frac{\partial H}{\partial S} = A_c^T P + P A_c + C^T K^T R K C + Q \tag{B.7.2}$$

$$0 = \frac{\partial H}{\partial P} = A_c S + S A_c^T + X \tag{B.7.3}$$

$$0 = \frac{1}{2}\frac{\partial H}{\partial K} = RKCSC^{\mathrm{T}} - B^{\mathrm{T}}PSC^{\mathrm{T}}, \tag{B.7.4}$$

where

$$A_c = A - BKC, \qquad X = x(0)x^{\mathrm{T}}(0)$$

In the design of tracking systems, the equations are even worse.

We have used two general approaches to solving such equation sets. In the first, the PI J is computed based on (B.7.2) using

$$J = \tfrac{1}{2}\,\mathrm{tr}(PX) \tag{B.7.5}$$

The simplex routine in Appendix B1 was used to minimize J. In the second approach, a gradient-algorithm (e.g., Davidon–Fletcher–Powell) was used.[*] There, the gradient $\partial J/\partial K$ is computed using all three design equations (B.7.2)–(B.7.4).

---

[*]Press, W. H., B. P. Flannery, S. A. Teukolsky, and W. T. Vetterling, *Numerical Recipes,* New York: Cambridge, 1986.

INDEX

Acceleration:
  angular, 10, 11
  centripetal, 12, 40–41, 49, 52
  Coriolis, 12, 48
  -derivatives, 78, 94, 132, 134
  gravitational, 40–41
  lateral, 285, 321, 323–327, 347–348
  in moving frames, 10–12
  measurement of, 13–15
  normal, 194, 208, 261, 284, 287, 309, 313–320, 429, 443
  roll, 70, 282
  translational, of center of mass, 47, 53
  transport, 12
  vector, 3, 4
Accelerometer:
  equations of, 13–15
  feedback from, 313–320, 323–327, 347–348
    effect on phugoid mode, 319
    transfer function, 315
  position in aircraft, 314–316
Ackermann's formula, 543
Actuator:
  control surface, 260
  limiting/saturation, 304, 349, 371, 622
  models, 294–295, 303, 329
Adams-Bashforth integration, 175
Additive uncertainty, 526
Adjustment rules (human operator model), 285
Aerodynamic angles, *see also* Angle of attack.
  defined, 72
  derivatives of-, 78
  equations for, 74
Aerodynamic center, 62
Aerodynamic data:
  plots, 81 ff
  tables, 98–99
Aerodynamic derivatives, *see* Derivatives
Aerodynamic forces and moments:
  acceleration dependent, 78
  coefficients for, 64, 66, 75 ff
  component buildup, 79
  defined, 74
  rate-dependent, 77
Agility, 309
Aileron:
  -doublet response, 308
  effectiveness, 91–92
  feedback to, 299–303, 320–324, 345–347
  rolling moment, 87
  -rudder interconnect (ARI), 320–323
  sign convention, 186
Aircraft:
  coordinate systems, 72
  dynamic behavior, 208
  electromechanical controls, 260
  envelope, altitude-Mach, 258–259
  flat-Earth equations of motion, 52
  F-16 model, 183
  nonlinear model equations, 107–116
  round-Earth equations of motion, 49
  transport aircraft model, 181
  wing planforms, 69–71

655

Air-data:
  computer, 183
  sensors, 258
Airfoil:
  NACA, 64
  properties, 60–67
Aliasing, 608
Alpha-dot:
  aerodynamic effects, 77–78
  expression for, 78
  state derivative, 109–115
Altitude:
  effect on aerodynamic coefficients, 75–76
  feedback, 335
  -hold autopilot, 334
  pole, 212
  -rate feedback, 363
  state, 206, 215–216
Analog computer simulation, 148–149
Angle of attack:
  aircraft-, 72
  airfoil-, 61
  estimation of, 556
  feedback, 292–298, 310–313
  sensor, 293
Angular momentum, 42–43
  rate of change of-, 44
  of spinning rotors, 46
Angular-motion:
  rigid body-, 42
  state equation, 44
Angular velocity vector, 9–10, 27–28, 42, 49
Anhedral angle, 89
Antiwindup compensation, 225, 623
Area rule for drag reduction, 82
ARI, 320–323
Artificial feel, 256, 260, 287
Aspect ratio, 68–70, 80–81, 83, 133
Atmosphere:
  mathematical model, 634
  standard-, 63, 65, 109
Attitude:
  -hold autopilot, see Autopilots
  pitch, 26–27
Attitude representation, 26–27
Augmentation:
  control-, 261, 308
  stability-, 261, 291
Autocorrelation, 549
Automatic landing systems, 338, 359, 362
Autopilots:
  altitude hold, 334
  automatic landing, 338, 359, 362, 452
  definition and types of-, 261
  flare control, 362, 470
  heading-hold, 348
  Mach hold, 334
  -navigational modes, 348
  pitch attitude hold, 327–331

  roll-angle (bank-angle) hold, 345, 448
  roll-angle steering, 366
  terrain-following, 349
  turn coordination, 347, 574
  VOR-hold, 348
Auto-throttle, 341
Axes, see Coordinate systems
Axial force coefficient, 73, 84

Bandwidth, 168–169
Bank angle, see Roll angle
Bending mode:
  -filter, 300
  of fuselage, 316
Beta-dot, 78, 110–112, 272
  derivatives, 78, 109
  expression for-, 78
  feedback, 323
Bilinear transformation, 598
Bode plots:
  design examples, 240, 242
  for classical design, 165–169
  multivariable, 515
Bryan, G. H., 116

Camber, airfoil-, 61
Ceiling, service-, 258
Center of gravity (cg):
  of aircraft models, 182, 183, 197, 209, 268–269, 315
Center of mass (cm), 3, 13–15, 34, 40–41, 42
  effect on stability, 103–107
  translation of-, 47 ff
Center of pressure (cp), 62, 67
Center of rotation, 316
Centripetal acceleration, see Acceleration
Characteristic equation, 221
Characteristic polynomial, 221
Chord:
  airfoil-, 61
  mean aerodynamic-, 64, 68, 77
Command generator tracker, 465
Command limiters, 321, 373
Companion form matrix, 150, 164
Compensation:
  design of-, 228 ff
  networks, 161
  phase-lead/phase-lag, 168, 228–230, 232, 236, 237–245
  proportional plus derivative (PD), 228
  proportional plus integral (PI), 229–230, 234, 602
  rate-feedback, 228–229
Compensator with desired structure, 421
Compressibility effects, 65, 67, 84, 91, 95, 132
Computation delay, 612
Constrained feedback design, 435
Control augmentation systems (CAS):
  definition and types of-, 261, 308

INDEX

lateral-directional, 320–327, 448
normal-acceleration (g-command), 313–320, 429, 443
pitch-rate, 309–313, 350, 444, 535, 604, 617
stability-axis-roll, 320
Control feel, *see* Handling qualities
Control surface:
  effectiveness, 92, 95, 129, 135
  sign convention, 186
Control system:
  electromechanical, 260
  feel, 260
  irreversible, 209, 260
  multi-input/multi-output (MIMO), 144
  need for-, 258
  single-input/single-output (SISO), 144
  type, 223
Controllability, 388, 542
Controlled variables, 488
Convolution integral, 152
Cooper-Harper scale, *see* Handling qualities
Coordinated turn, *see* Turn
Coordinate rotations, *see* Rotation matrix
  -Earth-related, 39–40
Coordinate systems:
  aircraft-, 37, 72
  Earth-centered, Earth-fixed (ECEF), 36
  Earth-centered-inertial (ECI), 36
  north-east-down (NED), 36
  tangent-plane, 37
  vehicle-carried, 37
Coriolis:
  acceleration, 12, 48–49, 52
  equation of, 10, 11, 28, 29
Corner frequency, 165
Cost function, 191–195, 283, 359, 364
Coupling:
  inertia-, 45, 255, 320
  numerators, 220
Cramer's rule, 271
Cross-correlation, 549
Crossover frequency:
  gain, 237–238
  phase, 237–238
Cross product:
  definition, 6
  matrix, 22, 44, 49, 50
Cross-wind force, 73
C-star,
  criterion, 284
  response, 316, 318–319

Damped frequency, 166
Damping, *see* Mode
  aerodynamic, 77
  critical, 171
  derivatives, 77
  ratio, 164
Data tables (aerodynamic), 98

DC gain, 158, 161, 524
Deadbeat observer, 625
Decoupling:
  in aircraft dynamics, 110, 114, 115, 119, 122, 126, 390, 401
  in LQR design, 452
Derivative of an array, 29
Derivatives (aerodynamic):
  acceleration-, 78
  damping-, 77
  dimensionless, 128–137
  force dimensional-, 120
  moment dimensional, 124
Derivative weighting in LQR design, 437, 447
Describing function, 237
Detectability, 409, 481
Deviation system, 424
Deyst filter, 556
Differentiation of vectors, 4, 9–10
Digital control:
  realization structures, 626
  techniques, 386, 595
Dihedral:
  angle, 89
  derivative, 91, 136, 272
  effect, 89
Dimensional derivative, *see* Derivatives
Dimensionless derivative, *see* Derivatives
Direction angles, 5
Direction cosine matrix, *see* Rotation matrix
Direction cosines, 5
Discretization of continuous controller, 598
Disturbance inputs:
  to aircraft models, 222, 279
  to control systems, 217, 221, 222, 280, 327, 335, 509
Dominant pole-pair, 171
Dot product, 6, 22
Downwash *see* Horizontal tail
Drag:
  -coefficient of aircraft, 79–83
  -coefficient of airfoil section, 66
  definition, 62
  -divergence Mach number, 65, 81
  induced, 70, 90
  polar, 83
  transonic rise, 69, 81
  types of, 79–80
D-star criterion, 285
Dutch-roll:
  approximation, 272–273
  accuracy of-, 275
  dependence on flight conditions, 273
  eigenvectors, 211, 392
  frequency and damping formulae, 272–273
  -mode described, 211
Dynamic inversion, 484
  for linear systems, 485
  for nonlinear systems, 494

Dynamic pressure:
  definition, 63
  effects, 63–65, 75, 91, 95, 97, 109
  variation over envelope, 259

Earth:
  angular rotation rate of, 36
  frames and coordinate systems for-, 37
  -mass gravitational constant, 36
  shape and gravitation of, 34
Efficiency:
  factor for aircraft tail, 95
  factor for induced drag, 80
Eigenvalues:
  definition, 20
  of F-16 model, 209–211
  of rotation matrix, 24–25
Eigenvectors:
  assignment of, 384, 392
  definition, 20
  of lateral dynamics, 211, 392
  modal decomposition using-, 153
  short period, 210, 391, 399
Elevator:
  control effectiveness, 95
  -pulse simulation, 198
Envelope, *see* Aircraft, envelope, altitude-Mach
Equilibrium point, 2, 116
Equilibrium, stable/unstable, 62
Error:
  coefficients, 224
  covariance, 551
Estimate of state vector, 539
Euler:
  angles, 25
  -angles from rotation matrix, 28–29
  equations of motion, 45
  integration, 2, 172
  kinematical equations, 27–28
  rotation sequence, 25–27
  theorem, 24
Exogenous input, 456
Expected value, 547

Feedback:
  of airspeed, 340
  of altitude, 335
  of altitude-rate, 363
  of angle of attack, 293–298, 310–313
  linearization, 484, 487
  of normal acceleration, 313–320
  output, 387
  of pitch attitude, 327–334
  of pitch rate, 293–298, 309–313
  of roll rate, 299–308
  state-, 387
  of yaw rate, 299–308
Feedback control:
  closed-loop equations, 219
  compensation, 228
  configurations, 218
  definitions, 216
  design, *see* Root-locus; Frequency domain
  stability, 225
  steady-state error, 223
  -system type, 223
  with unity feedback, 220
Flaps, 71, 75, 83, 94, 95, 105
Flare:
  controller, 362
  LQR design of, 470
Flat-Earth equations, *see* State equations
Flexibility, structural, 1, 293, 316, 535, 577
Flight-path angle, 101, 188–189
Flying qualities, *see* Handling qualities
Force:
  apparent, 12
  centrifugal, 12
  Coriolis, 12
  definitions of symbols, 74
  mass attraction, 14, 40
  specific, 14, 40, 314
Frames:
  inertial, 3, 37
  moving, 10
  reference, 3, 37
Free fall, 15
Frequency domain design, 237
Frequency response:
  definition of, 159
  elevator to pitch-rate, 213
  examples, 164–169
  from poles and zeros, 159
  short-period approximation, 213–214
Friction drag, 79–80
F-16 aircraft model, 184

Gain:
  margin, 237–238, 243, 283, 332–333, 336–337, 342–343
  scheduling, 259, 351, 419, 506, 509, 533, 591
Gaussian pdf, 548
g-command, *see* Control augmentation systems
Geodesy, 34
Geodetic coordinates:
  from cartesian, 38–39
  definition of, 37
Glidepath:
  controller, 359–362
  coupler, LQR design, 452
Goldstein, H., 7, 24, 42
Gravitation, 40–41
Gravity vector, 15, 41
Ground effect, 83, 85, 95
Gyros:
  attitude reference-, 255, 327, 345
  pitch rate-, 293
  roll rate-, 300, 345
  yaw rate-, 300
Gyroscopic coupling, *see* Inertia coupling

INDEX **659**

Hamiltonian, 406, 427, 439, 554
Handling qualities:
　control feel, 287
　Cooper-Harper rating of, 279–280
　definitions, 280, 289
　equivalent low-order system, 282–283
　factors influencing-, 279
　frequency response specifications for, 282
　human operator-model based, 285
　MIL-F-1797, 288
　MIL-F-8785C, 288
　military specifications for-, 288–291
　Neal-Smith criterion, 286–287
　pole-zero specifications for-, 281
　stick-force per g, 287
　time-response specifications for-, 284
Harmonic oscillator, 507
Heading-hold autopilot, 348
Helix angle, 77
H-infinity norm, 520
Homogeneous solution, 152
Horizontal tail:
　downwash, 78, 84, 95, 102, 104–105
　efficiency, 95, 103, 104–105
　volume ratio, 103, 134
Human-operator models, *see* Handling qualities
　optimal control model, 286
　transfer function model, 285
Hydraulic actuator, *see* Actuator
Hypersonic aircraft, 54

Implicit model-following, *see* Model following
Implicit state equations, *see* State equations
Inertia coupling, 45, 255, 320
Inertial navigation, 11, 28, 55
Inertial position vector, 48, 51
Inertial reference frame, 3, 11, 12, 37, 42
Initial conditions for LQR problem, 407, 480
Initial stabilizing gain, 410, 436
Inertia matrix:
　definition, 43
　inverse of, 44
　in principal axes, 45
　in wind/stability axes, 113–114
Integration:
　Adams-Bashforth, 175
　Euler, 2, 172
　linear-multistep methods, 174
　Runge Kutta, 172
　trapezoidal, 173

Jacobian matrix:
　calculation of, 202
　defined, 119
　numerical examples, 205–208
Jordan-form matrix, *see* Matrix

Kalman,
　filter, 386, 546
　gain, 479, 555

Kinematic:
　coupling, 320
　linearized equations, 122

Lagrange multiplier, 406
Lanchester's phugoid approximation, 267
Landing, automatic-, 359, 362
Laplace transform:
　initial and final value theorems, 160
　solution of state equations, 154–156
　time-response from-, 169–172
Lateral:
　acceleration, *see* Acceleration, lateral
　-dynamics Bode plot, 519
　eigenvectors, 211, 391
　regulator, LQR design, 413, 441
Lateral-directional dimensionless derivatives, 136–138
Lateral-directional SAS:
　description, 299–303
　example, 303–308
Latitude, 36, 37, 39, 41
Latitude rate, 38, 48
Lift:
　coefficient of aircraft, 83–85
　coefficient of airfoil, 66
　-curve slope, 66, 68, 132
　definition, 61
　dynamic, 76
　-over-drag, 68–71
Limit-cycle oscillation, 181, 237
Limiters, command-, 321, 373
Linearization:
　Algebraic *versus* numerical, 205
　algebraic-, of aircraft state equations, 118–128
　numerical, 201–205
　numerical examples, 205–208
Linear quadratic regulator (LQR):
　loop transfer recovery (LTR), 563
　with output feedback, 408
　with state feedback, 478
Load factor, 281, 314
Longitude, 36, 40
Longitude rate, 38, 48
Longitudinal dimensionless derivatives, 129–136
Loop gain, multivariable, 514
Lyapunov equation, 406, 438

Mach cone, 65
Mach number:
　critical-, 65
　defined, 65
　drag-divergence-, 65
　effect of, 67
　-hold autopilot, 334
Maneuverability, 309
Margins, gain and phase-, 237–238
Matched pole-zero (MPZ) discretization, 601

Matrix:
    companion form, 150
    cross-product, 22
    exponential function, 151
    Jordan-form, 21, 153
    modal, 21
    transition, 152, 153, 155
MIL-F-1797, *see* Handling qualities
MIL-F-8785C, *see* Handling qualities
MIMO:
    systems, 144
    transfer functions, 156
Minimality, 390
Minimization:
    gradient method, 409, 427
    of performance index, 404
    simplex-, 191–192, 409, 427
Modal coordinates, 21
Modal decomposition, 153–154, 209–210, 389
Model following:
    control, 385
    explicit-, 463, 473
    implicit-, 413, 469
Model reduction, 527
Mode, *see*
    Dutch-roll
    Natural modes
    Phugoid
    Rigid-body, modes
    Roll-subsidence
    Short-period
    Spiral
    Third oscillatory mode
Moments of inertia *see* Inertia matrix
Multiplicative uncertainty, 526

Natural frequency, 164, 167, 171
Natural modes:
    algebraic derivation of-, 262, 272–275
    of business jet, 275–276
    definition, 22
    effect of flight conditions, 265, 267–268, 273–275
    of F-16 model, 209–213, 277–278
    in multivariable design, 384
    numerical evaluation of-, 277–279
    stick-fixed, 263
    third oscillatory mode, 269, 294
    of transport aircraft, 198–200, 214–216
Neal-Smith criterion, *see* Handling qualities
Neutral point, 106
Newton's system:
    LQR design, 482
    observer, 543
NLSIM program, 179
Noise:
    measurement, 512, 547
    nonwhite, 550
    nonzero mean, 592

    process, 547
    white, 549
Nonlinear simulation, *see* Simulation
Non-minimum-phase (NMP), *see* Zeros
    LTR design, 572
Normal acceleration, *see* Acceleration
Normal force coefficient, 73, 84
Normal pdf, *see* Gaussian pdf
Numerical integration *see* State equations, numerical
Nyquist:
    criterion, 226
    frequency, 599
    plot example, 241

Observability:
    in PI weighting matrices, 412, 441
    properties, 389, 411, 481, 542
Observer, 386, 539
Operator gain, L2, 520
Optimal return difference relation, LQR, 564
Orthogonal:
    matrix, 20, 24
    random processes, 549
Output:
    injection, 541
    measured, 421
    performance, 421
Output feedback:
    design, 388
    eigenvector assignment, 396
    LQR, 408
    robust, 534
Overshoot, in step response, 161, 171, 228, 229, 243–245

Padé approximant, 615
Parameter variation modeling error, 533
Partial fraction expansion, 157, 390, 528
Performance analysis, 101
Performance index, LQR:
    constrained feedback, 435
    derivative weighting, 437, 447
    introduced, 384, 404
    time weighting, 438, 443, 458
    tracker, 426
    weighting matrices, 411, 415, 431, 441, 484
Performance specifications, frequency-domain, 520
Phase lag:
    compensation, 230, 237–238, 239, 285, 298, 336
    network, 162, 169
Phase lead:
    compensation, 229, 232, 237–238, 241, 244, 285, 332, 336, 342
    network 162, 168,
    simple-, 162, 166

Phase margin, 237, 239–243, 244, 283, 332–333, 336–337, 342–343
Phugoid mode:
  approximation, 265
    accuracy of, 268
  damping, 267
  dependence on flight conditions, 267–268
  eigenvectors, 210
  -mode described, 199–200
  natural frequency, 267
PID controller:
  antiwindup protection, 624
  digital, 602
Pitch-attitude hold, 327
Pitch damping derivative, 133
Pitching moment coefficient:
  of aircraft, 92–95
  of airfoil section, 66
Pitch pointing, 398
Pitch-rate CAS:
  classical design, 309–313
  LQR design, 444, 535, 604, 617
  simulation, 350
Pitch-rate gyro:
  feedback, 293–298
  location, 293
Pitch SAS:
  description, 292–293
  examples, 293–299
Pitch stability:
  dynamic, 269–270
  static, 103 ff
Pitch stiffness derivative, 133
Poisson's kinematical equations, 28, 46
Polar plot, lead compensator-, 239
Pole-placement design, 384
Polynomial, characteristic, 20
Power-boosted controls, 255, 256, 287
Power curve:
  back side of, 339
  for F-16 model, 196–197
Power spectral density, 222–223
Prandtl-Glauert correction, 67
Prandtl's lifting line theory, 68
Principal axes, 45
Probability density function (PDF), 547
Probability theory, review, 547

Quaternion:
  coordinate rotation, 29
  definition, 15
  direction cosine matrix from-, 31
  from direction cosine matrix, 32
  from Euler angles, 31
  kinematical equations, 33, 46–47, 50
  properties, 16
  rotation of a vector, 17

Radius of curvature:
  meridian, 37
  prime-vertical, 38
Random process, 548
Rate of climb, 189
Reachability, *see* Controllability
Regulator:
  dynamic, 386, 397, 559
  problem, 217, 384
  reduced-order, 572
Relaxed static stability, 104, 106, 183, 257
Remnant (human operator model), 285
Residues, 157
Resonance, 160
Resonant frequency, 166–167
Return difference, multivariable, 514
Reynolds number:
  definition, 64
  dependence of friction drag on-, 80
  effect of, 67
Riccati equation, 480, 555
Rigid-body:
  angular motion, 42
  definition, 1
  modes, 261
  translation, 47
Robust design:
  concepts, 386
  with output feedback, 534
Robustness:
  of LQR, 564
  performance, 509, 522
  to plant parameter variations, 533
  stability, 509, 526
Roll:
  angle, 26
  -angle hold autopilot, 345, 448
  -damping derivative, 91, 137
  stability-axis-, 320
  stiffness, *see* Stiffness
Rolling moment coefficient, 87–92
Roll-rate feedback, 299–308
Roll-subsidence mode:
  approximation, 273–275
  accuracy of-, 275
  dependence on flight conditions, 274–275, 300–301
  eigenvector, 211
  -mode described, 273
  time constant, 274, 300–301
Roll-yaw coupling, 271
Root locus:
  design examples, 232–237
  rules, 230–232
Rotation matrix:
  body-fixed to wind axes, 73
  definition and properties, 23–25
  ECEF to NED, 40
  ECI to ECEF, 39

Rotation matrix (*continued*)
  from Euler angles, 26
  from quaternion, 31
  NED to body-fixed, 26
Rudder:
  deflection, 75
  feedback to-, 303–308
  rolling moment due to-, 91
  sideforce due to-, 85–87
  yawing moment of, 98
Runge-Kutta integration, 172–174, 176–177, 179, 595

Sampling:
  frequency, 598
  period, 177, 181, 610
  theorem, 610
Scalar product, 22
Sensitivity, multivariable, 514
Separation principle, 561
Service ceiling, 258
Servo design, 385
Shaping filter, 550, 557
Shock wave, 65
Short-period mode:
  approximation, 263–265
  damping, 264
  dependence on flight conditions, 265
  eigenvectors, 210, 390, 398
  flying-qualities requirements, 282, 289
  -mode described (aircraft motion), 199
  natural frequency, 264
Sideforce coefficient, 85–87, 99
Sideslip:
  adverse, 320
  -angle definition, 72
  -angle feedback, 323
Similarity:
  geometric-, 64
  transformation, 20–21, 153
Simplex algorithm, 191, 192, 645
Simpson's rule, 152, 175
Simulation:
  diagram, 149
  digital *see* State equations, numerical solution of-
  of limiting behavior, 370
  nonlinear, 349
  numerical examples, 178–181, 198, 199, 200, 307–308, 312–313, 350, 359, 363, 366, 370
  time-history, program for, 179
Singular points, 116
Singular value decomposition:
  balancing, 524
  introduced, 515
  LQR constraint, 565
SISO:
  systems, 144
  transfer functions, 157

Spectral density matrix, 549
Speed:
  -damping derivative, 135
  of sound, 65
  stability, 288
Spinning rotors:
  angular momentum of, 46
Spiral mode:
  approximation, 273–275
  accuracy of-, 275
  dependence on flight conditions, 274–275
  described, 273–274
  eigenvector, 211
  stability of, 275
  time constant, 274
Spoilers, 91–92
Stability:
  axes, 72
  -axis roll, 320
  derivatives, *see* Derivatives
  directional, 95–96, 98, 136–137
  Nyquist test for-, 226
  pitch-, 62, 103, 132, 269–270
  roll-, 91, 136
  speed, 288
  static, 99, 103–107
  static *versus* dynamic, 63
  using LQR design, 481
Stability augmentation systems (SAS) 261, 291.
    See also
  Lateral-directional SAS, 261, 299–308, 413, 441
  Pitch SAS, 261, 292–299
Stability derivatives:
  dimensional, 120–128
  dimensionless, 128–137
Stabilizability, 481
Stagnation point, 60
State equations:
  aircraft, linear-, 119–128
  aircraft, nonlinear:
    in body axes, 107–114
    in wind or stability axes, 109–114
  decoupled, linear, 126
  decoupled, nonlinear, 114–116
  definition, 2
  of electrical networks, 147–148, 162
  explicit time-domain solution, 151
  with feedback, 219, *see also*
    Output feedback
    State feedback
  flat-Earth, nonlinear, 52–53
  general forms, 2, 143
  implicit, 116
  Laplace transform solution, 154
  linearization of-, 118
  of mechanical systems, 145–146
  numerical solution of-, 172
  from ODE's, 148–151
  partitioned, 263
  recursive solution, 152–153

for rigid-body angular motion, 42–47
for round-the-Earth translational motion, 47–51
from transfer functions, 163
State feedback:
  eigenvector assignment, 392
  introduced, 387
  LQR, 478
Static analysis, 99–107
Static equilibrium, 100
Static loop sensitivity, 158
Static margin, 106
Stationary random process, 548
Steady-state control error, 223, 424, 522
Steady-state flight:
  definition of-, 117, 119
  determination of conditions for-, 187
  numerical examples of-, 191–197
  rate-of-climb constraint, 189
  trim algorithm for-, 191–192, 642
  turn coordination constraint, 190, 347
Step:
  -response of quadratic lag, 171
  Unit-, 160
Stick force per g, 287
Stiff equations, 175
Stiffness:
  pitch, 63, 103, 132
  roll, 91, 136
  yaw, 98, 136–137
Strakes, 71,
Strapdown equation, *see* Poisson's kinematic equations
Successive loop closure, 220, 341, 383

Target feedback loop, LTR, 578
Taylor series, 118, 172, 202
Third oscillatory mode, 269, 294
Throttle:
  automatic control of, 341
  -pulse response simulation, 199
Thrust:
  coefficient, 76, 84, 93–95, 97–98
  components, 74–75
  derivatives, 120, 124, 129
  effect on downwash, 103
  effect of tail efficiency, 105
  installed, 109
  moment, 94–95, 101, 127
  variation with altitude and speed, 183, 186, 258, 267–268
  vector, 101, 111–112
Time-domain solution of state equations, 151
Time-history, *see* Simulation
Time-response from transfer function, 169
Time-scaling of aircraft equations, 210
Time weighting in LQ performance index, 438, 443
Tracker, LQ:
  introduced, 385
  with output feedback, 419, 428

Transfer function:
  aileron to roll rate-, 271, 272, 302
  aileron to yaw rate-, 302
  algebraic derivation, 262, 270
  coupling numerators, 220
  elevator to alpha-, 264
  elevator to normal acceleration-, 315–316
  elevator to pitch attitude-, 329
  elevator to pitch rate-, 212–214, 262, 264
  frequency response 159, 164
  geometrical interpretation of-, 167
  interpretation of, 158, 211
  low-pass, 169
  matrix, 156
  MIMO, 156
  models, 156
  of networks, 162
  poles, 157
  quadratic lag, 162, 164, 167–168
  relative degree, 157
  rudder to roll rate-, 302
  rudder to yaw rate-, 302
  SISO, 157
  s-plane vector representation, 159
  standard forms, 161
  throttle to speed-, 214–216, 342
  zeros, 158
Transformation:
  congruence, 20
  coordinate, 23–27, 29–30
  linear, 19
  similarity, 20
Transition matrix:
  continuous-time, 152
  discrete-time, 153
Transport aircraft model, 182
Trapezoidal integration, 173
Trimmed/untrimmed coefficients, 78
Tuck:
  derivative, 134
  effect, 134–135, 270
Turbulence:
  in flow, 61
Turn:
  coordination/turn compensation, 190, 323, 347, 574
  simulation of, 200
  teady-state, 117, 188–191
Tustin's approximation, 598
Type, feedback control system-, 223

Unbiased estimate, 552
Unmodeled dynamics, 509, 526
Unsteady aerodynamic effects, 76

Van der Pol equation, 178–181
Vector:
  angular velocity, 9
  components, 4
  cross-product, 6

definition, 3
derivative, 4, 9
dot product of, 6
pseudo-, 5
rotation of a-, 7
unit-, 5
Volume ratio, *see* Horizontal tail
VOR-hold autopilot, 348
Vortex:
  lift, 69, 71
  sheet, 68

Washout filter, 300, 303
Weathercock stability, *see* Stability, directional
Wind:
  axes, 72
  gusts, 53, 535, 556, 578
  included in state equations, 51, 52
Windup, integrator-, 225, 622
Wing:
  aspect ratio, 68–70
  bending mode filter, 300
  delta-, 69
  finite-, 67
  -leveler autopilot, 345, 448
  planform parameters, 68

rock, 77
sweep, 68–70
World geodetic system-84 (WGS-84), 34

X-series aircraft, 255

Yaw:
  adverse, 96
  angle, 26
  damper control system, *see* Lateral-directionl SAS
  damping derivative, 98, 137
  stability, *see* Stability, directional
  stiffness derivative, 98, 136
Yawing moment coefficient, 95–98

Zero-angle root locus, 231
Zero dynamics, 487, 496
Zero-order hold, 177, 595, 611
Zeros:
  effect of accelerometer position on-, 315–316
  in LQR design, 424, 429, 455
  non-minimum-phase (NMP), 161, 302–303, 315, 324–325, 572
  of MIMO systems, 220

T_2 zero, 281–282, 297, 298–299